Table entry for z is the area to the left of z.

Areas of a Standard Normal Distribution

(a) Table of Areas to the Left of z

z	.00	.01	.02	.03	.04	.05	.06	.07	.08	.09
−3.4	.0003	.0003	.0003	.0003	.0003	.0003	.0003	.0003	.0003	.0002
−3.3	.0005	.0005	.0005	.0004	.0004	.0004	.0004	.0004	.0004	.0003
−3.2	.0007	.0007	.0006	.0006	.0006	.0006	.0006	.0005	.0005	.0005
−3.1	.0010	.0009	.0009	.0009	.0008	.0008	.0008	.0008	.0007	.0007
−3.0	.0013	.0013	.0013	.0012	.0012	.0011	.0011	.0011	.0010	.0010
−2.9	.0019	.0018	.0018	.0017	.0016	.0016	.0015	.0015	.0014	.0014
−2.8	.0026	.0025	.0024	.0023	.0023	.0022	.0021	.0021	.0020	.0019
−2.7	.0035	.0034	.0033	.0032	.0031	.0030	.0029	.0028	.0027	.0026
−2.6	.0047	.0045	.0044	.0043	.0041	.0040	.0039	.0038	.0037	.0036
−2.5	.0062	.0060	.0059	.0057	.0055	.0054	.0052	.0051	.0049	.0048
−2.4	.0082	.0080	.0078	.0075	.0073	.0071	.0069	.0068	.0066	.0064
−2.3	.0107	.0104	.0102	.0099	.0096	.0094	.0091	.0089	.0087	.0084
−2.2	.0139	.0136	.0132	.0129	.0125	.0122	.0119	.0116	.0113	.0110
−2.1	.0179	.0174	.0170	.0166	.0162	.0158	.0154	.0150	.0146	.0143
−2.0	.0228	.0222	.0217	.0212	.0207	.0202	.0197	.0192	.0188	.0183
−1.9	.0287	.0281	.0274	.0268	.0262	.0256	.0250	.0244	.0239	.0233
−1.8	.0359	.0351	.0344	.0336	.0329	.0322	.0314	.0307	.0301	.0294
−1.7	.0446	.0436	.0427	.0418	.0409	.0401	.0392	.0384	.0375	.0367
−1.6	.0548	.0537	.0526	.0516	.0505	.0495	.0485	.0475	.0465	.0455
−1.5	.0668	.0655	.0643	.0630	.0618	.0606	.0594	.0582	.0571	.0559
−1.4	.0808	.0793	.0778	.0764	.0749	.0735	.0721	.0708	.0694	.0681
−1.3	.0968	.0951	.0934	.0918	.0901	.0885	.0869	.0853	.0838	.0823
−1.2	.1151	.1131	.1112	.1093	.1075	.1056	.1038	.1020	.1003	.0985
−1.1	.1357	.1335	.1314	.1292	.1271	.1251	.1230	.1210	.1190	.1170
−1.0	.1587	.1562	.1539	.1515	.1492	.1469	.1446	.1423	.1401	.1379
−0.9	.1841	.1814	.1788	.1762	.1736	.1711	.1685	.1660	.1635	.1611
−0.8	.2119	.2090	.2061	.2033	.2005	.1977	.1949	.1922	.1894	.1867
−0.7	.2420	.2389	.2358	.2327	.2296	.2266	.2236	.2206	.2177	.2148
−0.6	.2743	.2709	.2676	.2643	.2611	.2578	.2546	.2514	.2483	.2451
−0.5	.3085	.3050	.3015	.2981	.2946	.2912	.2877	.2843	.2810	.2776
−0.4	.3446	.3409	.3372	.3336	.3300	.3264	.3228	.3192	.3156	.3121
−0.3	.3821	.3783	.3745	.3707	.3669	.3632	.3594	.3557	.3520	.3483
−0.2	.4207	.4168	.4129	.4090	.4052	.4013	.3974	.3936	.3897	.3859
−0.1	.4602	.4562	.4522	.4483	.4443	.4404	.4364	.4325	.4286	.4247
−0.0	.5000	.4960	.4920	.4880	.4840	.4801	.4761	.4721	.4681	.4641

For values of z less than −3.49, use 0.000 to approximate the area.

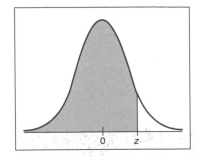

Table entry for z is the area to the left of z.

Areas of a Standard Normal Distribution *continued*

z	.00	.01	.02	.03	.04	.05	.06	.07	.08	.09
0.0	.5000	.5040	.5080	.5120	.5160	.5199	.5239	.5279	.5319	.5359
0.1	.5398	.5438	.5478	.5517	.5557	.5596	.5636	.5675	.5714	.5753
0.2	.5793	.5832	.5871	.5910	.5948	.5987	.6026	.6064	.6103	.6141
0.3	.6179	.6217	.6255	.6293	.6331	.6368	.6406	.6443	.6480	.6517
0.4	.6554	.6591	.6628	.6664	.6700	.6736	.6772	.6808	.6844	.6879
0.5	.6915	.6950	.6985	.7019	.7054	.7088	.7123	.7157	.7190	.7224
0.6	.7257	.7291	.7324	.7357	.7389	.7422	.7454	.7486	.7517	.7549
0.7	.7580	.7611	.7642	.7673	.7704	.7734	.7764	.7794	.7823	.7852
0.8	.7881	.7910	.7939	.7967	.7995	.8023	.8051	.8078	.8106	.8133
0.9	.8159	.8186	.8212	.8238	.8264	.8289	.8315	.8340	.8365	.8389
1.0	.8413	.8438	.8461	.8485	.8508	.8531	.8554	.8577	.8599	.8621
1.1	.8643	.8665	.8686	.8708	.8729	.8749	.8770	.8790	.8810	.8830
1.2	.8849	.8869	.8888	.8907	.8925	.8944	.8962	.8980	.8997	.9015
1.3	.9032	.9049	.9066	.9082	.9099	.9115	.9131	.9147	.9162	.9177
1.4	.9192	.9207	.9222	.9236	.9251	.9265	.9279	.9292	.9306	.9319
1.5	.9332	.9345	.9357	.9370	.9382	.9394	.9406	.9418	.9429	.9441
1.6	.9452	.9463	.9474	.9484	.9495	.9505	.9515	.9525	.9535	.9545
1.7	.9554	.9564	.9573	.9582	.9591	.9599	.9608	.9616	.9625	.9633
1.8	.9641	.9649	.9656	.9664	.9671	.9678	.9686	.9693	.9699	.9706
1.9	.9713	.9719	.9726	.9732	.9738	.9744	.9750	.9756	.9761	.9767
2.0	.9772	.9778	.9783	.9788	.9793	.9798	.9803	.9808	.9812	.9817
2.1	.9821	.9826	.9830	.9834	.9838	.9842	.9846	.9850	.9854	.9857
2.2	.9861	.9864	.9868	.9871	.9875	.9878	.9881	.9884	.9887	.9890
2.3	.9893	.9896	.9898	.9901	.9904	.9906	.9909	.9911	.9913	.9916
2.4	.9918	.9920	.9922	.9925	.9927	.9929	.9931	.9932	.9934	.9936
2.5	.9938	.9940	.9941	.9943	.9945	.9946	.9948	.9949	.9951	.9952
2.6	.9953	.9955	.9956	.9957	.9959	.9960	.9961	.9962	.9963	.9964
2.7	.9965	.9966	.9967	.9968	.9969	.9970	.9971	.9972	.9973	.9974
2.8	.9974	.9975	.9976	.9977	.9977	.9978	.9979	.9979	.9980	.9981
2.9	.9981	.9982	.9982	.9983	.9984	.9984	.9985	.9985	.9986	.9986
3.0	.9987	.9987	.9987	.9988	.9988	.9989	.9989	.9989	.9990	.9990
3.1	.9990	.9991	.9991	.9991	.9992	.9992	.9992	.9992	.9993	.9993
3.2	.9993	.9993	.9994	.9994	.9994	.9994	.9994	.9995	.9995	.9995
3.3	.9995	.9995	.9995	.9996	.9996	.9996	.9996	.9996	.9996	.9997
3.4	.9997	.9997	.9997	.9997	.9997	.9997	.9997	.9997	.9997	.9998

For z values greater than 3.49, use 1.000 to approximate the area.

Areas of a Standard Normal Distribution *continued*

(b) Confidence Interval Critical Values z_c

Level of Confidence c	Critical Value z_c
0.75, or 75%	1.15
0.80, or 80%	1.28
0.85, or 85%	1.44
0.90, or 90%	1.645
0.95, or 95%	1.96
0.98, or 98%	2.33
0.99, or 99%	2.58

Areas of a Standard Normal Distribution *continued*

(c) Hypothesis Testing, Critical Values z_0

Level of Significance	$\alpha = 0.05$	$\alpha = 0.01$
Critical value z_0 for a left-tailed test	−1.645	−2.33
Critical value z_0 for a right-tailed test	1.645	2.33
Critical values $\pm z_0$ for a two-tailed test	±1.96	±2.58

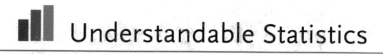

Understandable Statistics

SEVENTH EDITION

Understandable Statistics

Concepts and Methods

Advanced Placement High School Edition

Charles Henry Brase
Regis University

Corrinne Pellillo Brase
Arapahoe Community College

HOUGHTON MIFFLIN COMPANY
Boston   New York

Publisher: Jack Shira
Sponsoring Editor: Lauren Schultz
Editorial Associate: Marika Hoe
Assistant Editor: Jennifer King
Senior Project Editor: Nancy Blodget
Editorial Assistant: Kristin Penta
Production/Design Coordinator: Lisa Jelly Smith
Senior Manufacturing Coordinator: Priscilla Bailey
Marketing Manager: Ben Rivera

Cover photographer: Harold Burch, NYC
Viewpoint artist: Lauren Arnest

A complete list of photo credits appears in the back of the book, immediately preceding the appendices.

ComputerStat displays for Windows reprinted with permission of Houghton Mifflin Company.
TI-83 is a registered trademark of Texas Instruments, Inc.
Minitab is a registered trademark of Minitab, Inc.
Microsoft Excel screen shots reprinted by permission from Microsoft Corporation. Excel, Microsoft, and Windows are either registered trademarks or trademarks of Microsoft Corporation in the United States and/or other countries.

Printed in the U.S.A.

Library of Congress Control Number: 2001133233

ISBN
Student Text: 0-618-20554-3
Instructor's Annotated Edition: 0-618-20555-1
Advanced Placement High School Edition: 0-618-26509-0

3 4 5 6 7 8 9 – DOW – 06 05 04

This book is dedicated to the memory of
a great teacher, mathematician, and friend

Burton W. Jones

Professor Emeritus, University of Colorado

Contents

7 *Introduction to Sampling Distributions* 334

8 *Estimation* 372

9 *Hypothesis Testing* 452

Appendix I: *Additional Topics*   **A1**

Appendix II: *Tables*   **A13**

Welcome to the exciting world of statistics! We have written this text to make statistics accessible to everyone, including those with a limited mathematics background. Statistics affects all aspects of our lives. Whether we are testing new medical devices or determining what will entertain us, applications of statistics are so numerous that, in a sense, we are limited only by our own imagination.

Overview

The seventh edition of *Understandable Statistics: Concepts and Methods* continues to emphasize concepts of statistics. Statistical methods are carefully presented with a focus on understanding both the *suitability of the method* and the *meaning of the result*. Statistical methods and measurements are developed in the context of applications.

We have retained and expanded features that made the first six editions of the text readable. Examples, exercises, and problems touch on applications appropriate to a broad range of interests.

Content Changes in the Seventh Edition

With each new edition we reevaluate the scope, appropriateness, and effectiveness of the text's presentation and reflect on extensive user feedback. Revisions have been made throughout the text to clarify explanations of important concepts and to update problems.

Sections and Organization

- Chapter 1, Getting Started, has been organized to include Section 1.2, Random Samples (formerly Section 2.1 in the sixth edition), and Section 1.3, Introduction to Experimental Design. This new section discusses data collection, experiments, observations, and randomized two-treatment experiments.

- Section 7.3, Sampling Distributions for Proportions, is a new section in Chapter 7, Introduction to Sampling Distributions. The class project illustrating the central limit theorem is now in the Using Technology section.

- Section 10.4, Inferences Concerning Regression Parameters, replaces Section 10.4 of the sixth edition. This section includes inferences for the correlation coefficient as well as for the slope of the least-squares line.

Table for Standard Normal Distribution (cumulative area to the left of *z*)

New with the seventh edition is a two-page table that gives the cumulative area to the left of a specified *z* value. This type of table is easy to use and is also compatible with cumulative distribution results given on computers and calculators. The *Standard Normal Distribution Table* giving the area between 0 and *z* that was used in previous editions is included in Appendix I with instructions for its use.

New Topics

Brief treatments and discussions of dotplots, back-to-back stem plots, moving averages, odds, and linear combinations of two independent random variables are now included.

Features in the Seventh Edition

Chapter and Section Lead-ins

- NEW! *Preview Questions* at the beginning of each chapter are keyed to the sections.

- NEW! *Focus Points* at the beginning of each section describe the primary learning objectives of the section.

Carefully Developed Pedagogy

- *Examples* show students how to select and use appropriate procedures.

- *Guided exercises* within the sections give students an opportunity to work with a new concept. Completely worked-out solutions appear beside each exercise to give immediate reinforcement.

- NEW! *Labels* for each example or guided exercise highlight the technique, concept, or process illustrated by the example or guided exercise. In addition, new labels for section and chapter problems describe the field of application and show the wide variety of subjects in which statistics are used.

- *Section and chapter problems* require the student to use all the new concepts mastered in the section or chapter. The problem sets include a variety of real-world applications with data or setting from identifiable sources. Key steps and solutions to odd-numbered problems appear at the end of the book.

- *Data Highlights* and *Linking Concepts* provide group projects and writing projects.

- *Viewpoints* are brief essays presenting diverse situations in which statistics are used.

- *Enhanced design and photos* provide improved readability.

Technology within the Text

- NEW! *Tech Notes* within sections provide brief point-of-use instructions for the TI-83Plus calculator, Excel, and Minitab. These Notes replace the Calculator Notes of the sixth edition.

- *Using Technology* sections have been revised to show the use of Excel as well as the TI-83Plus calculator, Minitab, and ComputerStat.

Alternate Routes Through the Text

Understandable Statistics: Concepts and Methods, Seventh Edition, is designed to be flexible. It offers the professor a choice of teaching possibilities. In most one-semester courses, it is not practical to cover all the material in depth. However, depending on the emphasis of the course, the professor may choose to cover various topics. For help in topic selection, refer to the *Table of Prerequisite Material* on page 1.

- *Introducing linear regression early.* For courses requiring an early presentation of linear regression, the descriptive components of linear regression (Sections 10.1, 10.2, and 10.3) can be presented any time after Chapter 3. However, inference topics including the confidence bounds in Section 10.2 require an introduction to confidence intervals (Section 8.1) and to hypothesis testing (Sections 9.1 and 9.2).

- *Probability.* For courses requiring minimal probability, Section 4.1, What Is Probability? and the first part of Section 4.2, Some Probability Rules—Compound Events, will be sufficient.

Acknowledgments

It is our pleasure to acknowledge the prepublication reviewers of this text. All of their insights and comments have been very valuable to us. Reviewers of this text include:

Delores Anderson, Truett-McConnell College
Lynda L. Ballou, Kansas State University
John Bray, Broward Community College
Jennifer M. Dollar, Grand Rapids Community College
Andrew Ellett, Indiana University
Mary Fine, Moberly Area Community College
Rene Garcia, Miami-Dade Community College
Larry Green, Lake Tahoe Community College
Jane Keller, Metropolitan Community College
Raja Khoury, Collin County Community College
Michael R. Lloyd, Henderson State University
Beth Long, Pellissippi State Technical and Community College
Darcy P. Mays, Virginia Commonwealth University
Charles C. Okeke, College of Southern Nevada, Las Vegas
Peg Pankowski, Community College of Allegheny County
Michael L. Russo, Suffolk County Community College
Janel Schultz, Saint Mary's University of Minnesota
Winson Taam, Oakland University
Jennifer L. Taggart, Rockford College
William Truman, University of North Carolina at Pembroke
Jim Wienckowski, State University of New York at Buffalo
Stephen M. Wilkerson, Susquehanna University
Hongkai Zhang, East Central University
Shunpu Zhang, University of Alaska Fairbanks

We would especially like to thank Helen Medley for her careful accuracy review of this text, and Lauren Arnest for her creative illustrations that accompany the Viewpoint essays. Finally, we acknowledge the cooperation of Minitab, Inc., Texas Instruments, and Microsoft Excel.

Charles Henry Brase
Corrinne Pellillo Brase

A User's Guide to Features

The seventh edition of *Understandable Statistics* includes a variety of features designed to enhance a student's understanding by providing overviews and summaries of concepts and methods, interesting real-world problems using real data sets, and information about using technology. The newly enhanced design highlights important features and provides visual interest.

Key features of the text are described on the following pages. New features have been added to the seventh edition, while other hallmark features have been retained.

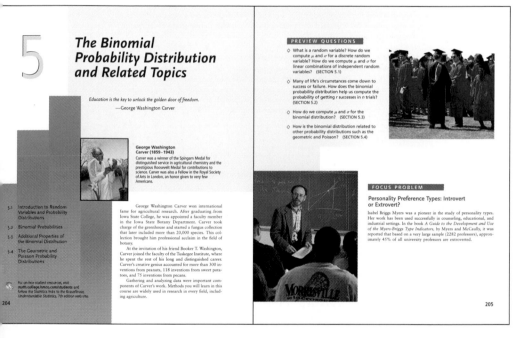

◀ Preview Questions

Newly designed chapter openers now contain a helpful list of Preview Questions that focus on the main objectives of the chapter. Section references appear next to each question to show students where they will learn the answers to develop an understanding of these topics.

◀ Focus Problems

"What kinds of problems will this chapter help me solve?" The Focus Problems motivate students by showing examples of work they can do once they have mastered the skills in the chapter. A Focus Problem appears in each chapter opener.

Focus Points ▶

Located at the beginning of each section, Focus Points briefly list the main objectives of the section for easy review and reference.

GUIDED EXERCISE 2

Using a sample space

Professor Gutierrez is making up a final exam for a course in literature of the Southwest. He wants the last three questions to be of the true–false type. To guarantee that the answers do not follow his favorite pattern, he lists all possible true–false combinations for three questions on slips of paper and then picks one at random from a hat.

(a) Finish listing the outcomes in the given sample space.

$\implies$ The missing outcomes are FFT and FFF.

TTT	FTT	TFT	____
TTF	FTF	TFF	____

(b) What is the probability that all three items will be false? Use the formula

$$P(\text{all F}) = \frac{\text{No. of favorable outcomes}}{\text{Total no. of outcomes}}$$

$\implies$ There is only one outcome, FFF, favorable to all false, so

$$P(\text{all F}) = \frac{1}{8}$$

(c) What is the probability that exactly two items will be true?

$\implies$ There are three outcomes that have exactly two true items: TTF, TFT, and FTT. Thus,

$$P(\text{two T}) = \frac{\text{No. of favorable outcomes}}{\text{Total no. of outcomes}} = \frac{3}{8}$$

◀ **Guided Exercises**

These unique exercises appear after selected examples in the text. Each Guided Exercise encourages students to examine and analyze a problem similar to the preceding example. Completely worked-out solutions appear beside each exercise to give immediate reinforcement in the learning process. In this way, students have a chance to work through and learn a new concept before being presented with additional concepts.

where $SS_x = \Sigma(x - \mu)^2 = \Sigma x^2 - \frac{(\Sigma x)^2}{N}$ ◇

at the standard deviation (sample or population) is a measure of data spread. We will use the standard deviation extensively in later chapters. In Chapter 6 we will use it to study standard z values and areas under normal curves. In Chapters 8 and 9 we will use it to study the inferential statistics topics of estimation and testing. The standard deviation will appear again in our study of regression and correlation.

 TECH NOTE Most scientific or business calculators have a statistics mode, and provide the mean and sample standard deviation directly. The TI-83Plus, Excel, and Minitab provide the median and several other measures as well.

Many technologies display only the sample standard deviation s. You can quickly compute σ if you know s by using the formula

$$\sigma = s\sqrt{\frac{n-1}{n}}$$

The mean given in displays can be interpreted as the sample mean $\bar{x}$ or the population mean μ as appropriate.

The following three displays show output for the hybrid rose data of Guided Exercise 3.

TI-83Plus Display
Press **STAT** ➤ **CALC** ➤ **1:1-Var Stats**. S_x is the sample standard deviation. σ_x is the population standard deviation.

```
1-Var Stats
x̄=6
Σx=48
Σx²=296
Sx=1.069044968
σx=1
↓n=8
```

Tech Notes ▶

Tech Notes provide optional information about using the TI-83Plus graphing calculator, Excel, Minitab, and the text-specific program ComputerStat.

SECTION 3.1 PROBLEMS

1. *Agriculture: Growing Season* The average length of the growing season is often measured in average number of frost-free days. The front range of Colorado (Fort Collins, Boulder, Denver, Colorado Springs, Pueblo) was studied by J. F. Benci and T. B. McKee, from the Department of Atmospheric Science at Colorado State University. Based on data from their Climatology Report No. 77-3, different locations in the Colorado front range had the following average number of frost-free days per year:

156	161	152	162	144	153
148	157	168	157	161	157

Compute the mean, median, and mode. Write a brief description of the meaning of these numbers from the point of view of a gardener.

2. *Baseball: Home Runs* Babe Ruth was the American League Home Run Champion 12 times (during the period from 1918 to 1931). The number of home runs he hit to earn the 12 titles were

11	29	54	59	41	46
47	60	54	46	49	46

Find the mean, median, and mode of the number of home runs.

3. *Environmental Studies: Death Valley* How hot does it get in Death Valley? The following data are taken from a study conducted by the National Park System, of which Death Valley is a unit. The ground temperatures (°F) were taken from May to November in the vicinity of Furnace Creek.

146	152	168	174	180	178	179
180	178	168	165	152	144	

Compute the mean, median, and mode for these ground temperatures.

4. *Ecology: Wolf Packs* How large is a wolf pack? The following information is from a random sample of winter wolf packs in regions of Alaska, Minnesota, Michigan, Wisconsin, Canada, and Finland. (Source: *The Wolf*, by L. D. Mech, University of Minnesota Press.) Winter pack size:

13	10	7	5	7	7	2	4	3
2	3	15	4	4	2	8	7	8

Compute the mean, median, and mode for the size of winter wolf packs.

5. *Medical: Injuries* The Grand Canyon and the Colorado River are beautiful, rugged, and sometimes dangerous. Thomas Myers is a physician at the park clinic in Grand Canyon Village. Dr. Myers has recorded (for a 5-year period) the number of visitor injuries at different landing points for commercial boat trips down the Colorado River in both the upper and lower Grand Canyon (Source: *Fateful Journey* by Myers, Becker, Stevens).

Upper Canyon: Number of Injuries per Landing Point Between North Canyon and Phantom Ranch

2	3	1	1	3	4	6	9	3	1	3

Lower Canyon: Number of Injuries per Landing Point Between Bright Angel and Lava Falls

8	1	1	0	6	7	2	14	3	0	1	13	2	1

◀ **Real-World Exercises**

Numerous application problems utilizing real data and real-world situations are included in the text. These exercises use identifiable sources, including some web sites, and also cover a wide range of fields, such as natural science, business, economics, medicine, social science, archaeology, and consumer interest.

Viewpoints ▶

These brief illustrated essays show the broad scope of statistical applications to a variety of human experiences and endeavors. In many cases, Internet web site URLs are provided for students interested in a further explanation of topics. The Viewpoint feature is located immediately before most section problem sets and chapter problem sets.

VIEWPOINT

The First Measured Century

The 20th century saw measurements of aspects of American life that had never been systematically studied before. Social conditions involving crime, sex, food, fun, religion, and work have been numerically investigated. The measurements and survey responses taken over the entire century reveal unsuspected statistical trends. *The First Measured Century* is a book by Caplow, Hicks, and Wattenberg. It is also a PBS documentary available on video. For more information, visit the Brase/Brase statistics site at http://math.college.hmco.com/students and find the link to the PBS first measured century documentary.

▼ End-of-Chapter Material

The following features are included at the end of each chapter: a brief chapter **Summary**, a list of **Important Words & Symbols** grouped by section for easy review, and **Chapter Review Problems**.

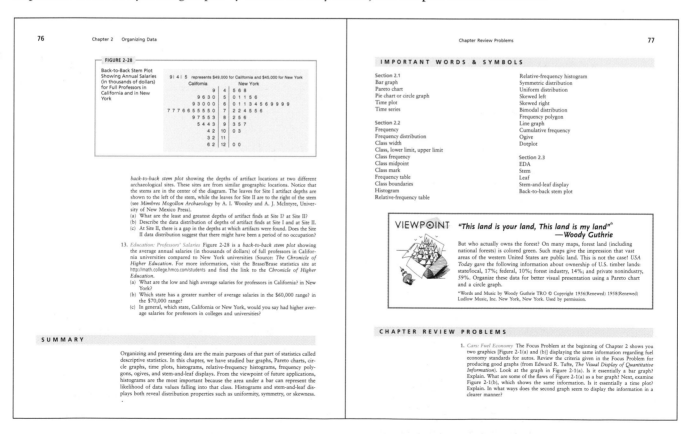

Data Highlights: Group Projects ▶

These in-depth projects give students an additional opportunity to practice their skills by asking them to solve problems using appropriate methods from the chapter. Newspapers, magazines, and journals are the sources for these projects.

Linking Concepts: Writing Projects ▶

These questions help students extend and integrate their thinking to develop a broader conceptual understanding of statistics. Students are asked to discuss and write about key concepts from the chapter and related topics from prior chapters.

◀ **Using Technology**

These features have been revised for this edition to include information on using the TI-83Plus graphing calculator, Excel, Minitab, and ComputerStat to solve statistical problems.

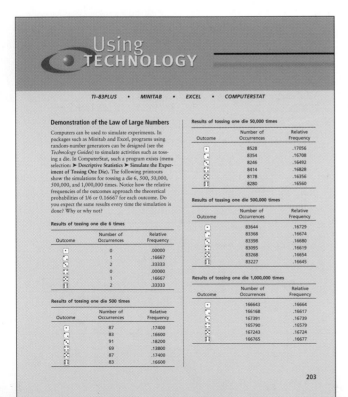

Table of Prerequisite Material

Chapter	Prerequisite Sections
1 Getting Started	None
2 Organizing Data	1.1, 1.2
3 Averages and Variation	1.1, 1.2, 2.2
4 Elementary Probability Theory	1.1, 1.2, 2.2, 3.1, 3.2
5 The Binomial Probability Distribution and Related Topics	1.1, 1.2, 2.2, 3.1, 3.2, 4.1, 4.2 4.3 useful but not essential
6 Normal Distributions (omit 6.4) (include 6.4)	1.1, 1.2, 2.2, 3.1, 3.2, 4.1, 4.2, 5.1 also 5.2, 5.3
7 Introduction to Sampling Distributions (omit 7.3) (include 7.3)	1.1, 1.2, 2.2, 3.1, 3.2, 4.1, 4.2, 5.1, 6.1, 6.2, 6.3 also 6.4
8 Estimation (omit 8.3 and parts of 8.4 and 8.5) (include 8.3 and parts of 8.4 and 8.5)	1.1, 1.2, 2.2, 3.1, 3.2, 4.1, 4.2, 5.1, 6.1, 6.2, 6.3, 7.1, 7.2 also 5.2, 5.3, 6.4
9 Hypothesis Testing (omit 9.5 and part of 9.7) (include 9.5 and all of 9.7)	1.1, 1.2, 2.2, 3.1, 3.2, 4.1, 4.2, 5.1, 6.1, 6.2, 6.3, 7.1, 7.2 also 5.2, 5.3, 6.4
10 Regression and Correlation (omit part of 10.2, 10.4, and 10.5) (include all of 10.2, 10.4, and 10.5)	1.1, 1.2, 3.1, 3.2 also 4.1, 4.2, 5.1, 6.1, 6.2, 6.3, 7.1, 7.2, 8.1, 9.1, 9.2
11 Chi-Square and *F* Distributions (omit 11.3) (include 11.3)	1.1, 1.2, 2.2, 3.1, 3.2, 4.1, 4.2, 5.1, 6.1, 6.2, 6.3, 7.1, 7.2, 9.1 also 8.1
12 Nonparametric Statistics	1.1, 1.2, 2.2, 3.1, 3.2, 4.1, 4.2, 5.1, 6.1, 6.2, 6.3, 7.1, 7.2, 9.1, 9.5

1

Getting Started

To guess is cheap,
To guess wrongly is expensive.

Tell me, I'll forget.
Show me, I may remember.
But involve me and I'll understand.

—Old Chinese Proverbs

**Dragon Gate,
Chinatown
San Francisco**

Dragon Gate is one of the many landmarks in San Francisco that exemplifies the city's ethnic diversity.

Most of the important decisions in life involve incomplete information. Such decisions often involve so many complicated factors that a complete analysis is not practical or even possible. We are often forced into the position of making a guess based on limited information. However, as the first proverb implies, a blind guess is not the best solution. Statistical methods, such as you will learn in this book, can help you make the best "educated guess."

The authors of this book want you to understand and enjoy statistics. The reading material will *tell you* about the subject. The examples will *show you* how it works. To understand, however, the second proverb says you must *get involved*. Guided exercises, calculator and computer applications, section and chapter problems, and writing exercises are all designed to get you involved in the subject. As you grow in your understanding of statistics, we believe you will enjoy learning a subject that has a world full of good applications.

PREVIEW QUESTIONS

◇ Why is statistics important? (SECTION 1.1)

◇ What is the nature of data? (SECTION 1.1)

◇ How can we draw a random sample? (SECTION 1.2)

◇ What are other sampling techniques? (SECTION 1.2)

◇ How can we design ways to collect data? (SECTION 1.3)

Adapted from Ohio State University Firefly Files logo

FOCUS PROBLEM

Where Have All the Fireflies Gone?

A feature article in the *Wall Street Journal* discusses the alarming disappearance of fireflies. In the article, Professor Sara Lewis (of Tufts University) and other scholars express concern about the decline in the worldwide population of fireflies. So far the evidence is anecdotal: stories from around the world about the alarming decline of the fireflies. Japan has a tradition that nighttime fireflies are a beautiful reminder of the souls of friends and family who have passed away. For reasons such as this, and possibly for the simple joy of watching fireflies, Japan has a number of special sanctuaries to protect the declining population of fireflies.

There are a number of possible explanations for the world decline in the firefly population. Artificial nighttime lighting might be interfering with the Morse code-like mating ritual of the fireflies. Woodlands, wetlands, and open fields have given way to cities and subdivisions. Chemicals and pesticides may be killing the fireflies.

Fireflies are commercially valuable because they contain two rare chemicals used in research on cancer, multiple sclerosis, cystic fibrosis, and heart disease. The chemicals are also used on spacecraft in electronic detectors designed to look for earth-life forms in outer space. Near St. Louis there is a bounty on fireflies: A chemical

3

company pays 1 cent for each firefly. One family was so successful in collecting fireflies that they could send three children to college from the proceeds!

What does any of this have to do with statistics?

The truth, at this time, is that no one really knows (a) how much the world firefly population has declined or (b) how to explain the decline. The population of all fireflies is simply too large to study in its entirety. In any study of fireflies, we must rely on incomplete information from samples. Furthermore, from these samples we must draw realistic conclusions that have statistical integrity.

You may be sure that a number of very fine scholars in universities across the world are beginning to examine the possible world decline in fireflies. You also may be sure that these scholars will use statistical methods when they collect, analyze, and investigate their data about fireflies.

Ohio State University Department of Entomology does research on anthropods and maintains extensive information about fireflies. For more information about fireflies, visit the Brase/Brase statistics site at http://math.college.hmco.com/students and find the link to the Ohio State University Department of Entomology. Then search for firefly in the index of the Ohio State site.

1.1
What Is Statistics?

FOCUS POINTS
✓ Identify variables in a statistical study.
✓ Distinguish between quantitative and qualitative variables.
✓ Identify populations and samples.
✓ Determine the level of measurement.
✓ Compare descriptive and inferential statistics.

Introduction

Decision making is an important aspect of our lives. We make decisions based on the information we have, our attitudes, and our values. Statistical methods help examine information. Moreover, statistics can be used for making decisions when we are faced with uncertainties. For instance, if we wish to estimate the proportion of people who will have a severe reaction to a flu shot without giving the shot to everyone who wants it, statistics provides appropriate methods. Statistical methods enable us to look at information from a small collection of people or items and make inferences about a larger collection of people or items.

Procedures for analyzing data, together with rules of inference, are central topics in the study of statistics.

Statistics

> **Statistics** is the study of how to collect, organize, analyze, and interpret numerical information from data.

The statistical procedures you will learn in this book should supplement your built-in system of inference—that is, the results from statistical procedures and good sense should dovetail. Of course, statistical methods themselves have no power to work miracles. These methods can help us make some decisions, but not all conceivable decisions. Remember, a properly applied statistical procedure is no more accurate than the data, or facts, on which it is based. Finally, statistical results should be interpreted by one who understands not only the methods, but also the subject matter to which they have been applied.

The general prerequisite for statistical decision making is the gathering of data. First, we need to identify the individuals or objects to be included in the study and the characteristics or features of the individuals that are of interest.

Individual
Variable

> **Individuals** are the people or objects included in the study.
> A **variable** is the characteristic of the individual to be measured or observed.

For instance, if we want to do a study about the people who have climbed Mt. Everest, then the individuals in the study are all people who have actually made it to the summit. One variable might be the height of such individuals. Other variables might be age, weight, gender, nationality, income, and so on. Regardless of the variables we use, we would not include measurements or observations from people who have not climbed the mountain.

The variables in a study may be *quantitative* or *qualitative* in nature.

Quantitative variable
Qualitative variable

> A **quantitative variable** has a value or numerical measurement for which operations such as addition or averaging make sense. A **qualitative variable** describes an individual by placing the individual into a category or group such as male or female.

For the Mt. Everest climbers, variables such as heights, weights, age, or income are *quantitative* variables. *Qualitative variables* involve nonnumerical observations such as gender or nationality. Sometimes qualitative variables are referred to as *categorical*.

Another important issue regarding data is their source. Do the data comprise information from *all* individuals of interest, or from just *some* of the individuals?

Population data
Sample data

> In **population data,** the variable is from *every* individual of interest.
> In **sample data,** the variable is from *only some* of the individuals of interest.

For instance, if we have data from *all* the individuals who have climbed Mt. Everest, then we have population data. On the other hand, if our data come from just some of the climbers, we have sample data.

It is interesting to note that the population is defined in terms of our *desire for knowledge.* The population can be thought of as measurements or observations for the entire group of objects or individuals about which information is desired. In this sense, a population can be an existing set of data, or it can be a set of data that is clear in our understanding but is not yet complete. The ages of all U.S. presidents at the time of inauguration—from George Washington to George W. Bush—can be thought of as an existing set of data. However, the set of ages of all presidents at the time of inauguration from George Washington into the future is not complete. Nevertheless, we have a clear understanding of how this population is to be constructed. In a way, we can think of such an incomplete population as being open-ended.

EXAMPLE 1
Using basic terminology

The Hawaii Department of Tropical Agriculture is doing a study of ready-to-harvest pineapples in an experimental field.

(a) The pineapples are the *objects* (individuals) of the study. If the researchers are interested in the individual weights of pineapples in the field, then the *variable* consists of weights. At this point, it is important to specify units of measurement

and degree of accuracy of measurement. The weights could be to the nearest ounce or gram. Weight is a *quantitative* variable since it is a numerical measure. If weights of *all* the ready-to-harvest pineapples in the field are included in the data, then we have a *population*.

(b) Suppose the researchers also want data on taste. A panel of tasters rates the pineapples according to the categories poor, acceptable, good. Only some of the pineapples are included in the taste test. In this case, the *variable* is taste. This is a *qualitative* or *categorical* variable. Because only some of the pineapples in the field are included in the study, we have a *sample*.   ◇

Throughout this text, you will encounter *guided exercises* embedded in the reading material. These exercises are included to give you an opportunity to work immediately with new ideas. The questions guide you through appropriate analysis. Cover the answers on the right side (an index card will fit this purpose). After you have thought about or written down *your own response*, check the answers. If there are several parts to an exercise, check each part before you continue. You should be able to answer most of these exercise questions, but don't skip them—they are important.

GUIDED EXERCISE 1

Using basic terminology

Television station QUE wants to know the proportion of TV owners in Virginia who watch the station's new program at least once a week. The station asked a group of 1000 TV owners in Virginia if they watch the program at least once a week.

(a) Identify the individuals of the study and the variable. ⟹ The individuals are the 1000 TV owners surveyed. The variable is the response does, or does not, watch the new program at least once a week.

(b) Do the data comprise a sample? If so, what is the underlying population? ⟹ The data comprise a sample of the population of responses from all TV owners in Virginia.

(c) Is the variable qualitative or quantitative? ⟹ Qualitative—the categories are the two possible responses, does or does not watch the program.

(d) Identify a quantitative variable that might be of interest. ⟹ Age or income might be of interest.

Levels of Measurement

We have categorized data as either qualitative or quantitative. Another way to classify data is according to one of the following four *levels of measurement:*

| Nominal | Ordinal | Interval | Ratio |

Nominal level

Since the nominal level is the lowest level, let's examine it first. A dictionary meaning of the word *nominal* is "in name only." This is an easy way to remember the meaning of the nominal level of measurement. Data at this level of measurement consist of "names only," or qualities, with no implied criteria by which the data can be identified as greater than or less than other data items.

EXAMPLE 2

Nominal level

The following are examples of data at the nominal level of measurement.

(a) Aspen, Vail, and Breckenridge are names of three ski resorts from the population of names of all ski resorts in Colorado.

(b) Taos, Acoma, Zuni, and Cochiti are names of four Native American pueblos from the population of all names of Native American pueblos in Arizona and New Mexico.   ◇

It is clear that the nominal data in Example 2 are not intended for numerical calculation. The specific names or qualities do not contain any implied ordering or numerical significance.

Ordinal level

The next level of measurement is the *ordinal level*. Data at the ordinal level may be arranged in some order, but actual differences between data values either cannot be determined or are meaningless.

EXAMPLE 3

Ordinal level

The following are examples of data at the ordinal level of measurement.

(a) In a fishing tackle catalogue, 17 fishing reels are advertised. Of these reels, 6 were rated as good quality, 4 were rated as better quality, and 7 were rated as best quality.

(b) In a high school graduating class of 319 students, Jim ranked 25th, June ranked 19th, Walter ranked 10th, and Julia ranked 4th.   ◇

In Example 3, we should not try to determine specific quantitative differences between "good," "better," and "best." The difference between June's and Jim's rank was 6, and this is the same difference that exists between Walter's and Julia's ranks. However, this difference doesn't really mean anything significant. For instance, if you look at grade point average, Walter and Julia may have had a big gap between them, whereas June and Jim may have been closer together. In any ranking system, it is only the relative standing that matters. Differences in ranks can be meaningless.

In general, the ordinal level of measurement provides information about relative comparisons, but exact differences are not computed.

Interval level

The *interval level of measurement* is like the ordinal level, but it has the additional property that meaningful differences between data values can be computed. However, interval-level data may not have an intrinsic zero or starting point. Consequently, *differences* are meaningful, but *ratios* of data values are not.

EXAMPLE 4

Interval level

The following are examples of data at the interval level of measurement.

(a) Years in which Democrats won presidential elections.

(b) Body temperatures (in degrees Celsius) of trout in the Yellowstone River.   ◇

Temperature readings in Celsius (or Fahrenheit) are examples of data at the interval level of measurement. Such values are certainly ordered, and we can compute meaningful differences. However, for Celsius-scale temperatures, there is not an inherent starting point. The value 0°C may seem to be a starting point, but the value of 0°C does not indicate the state of "no heat." Furthermore, it is not correct to say 20°C is twice as hot as 10°C. Calendar times are also interval measurements, since the date 0 A.D. does not signify "no time." However, a time lapse is at a higher level of measurement and is, in fact, an example of our top level of measurement, the *ratio level*.

Ratio level

The *ratio level of measurement* is the highest level. The ratio level is similar to the interval level, but it includes an inherent zero as a starting point for all measurements. Consequently, at this level, both differences *and* ratios are meaningful.

EXAMPLE 5

Ratio level

The following are examples of data at the ratio level of measurement.

(a) The core temperatures of stars in the Milky Way when measured in degrees Kelvin. Notice that in the Kelvin scale of measurement 0°K means "no heat." This is a special temperature scale used primarily by scientists.

(b) Length of trout swimming in the Yellowstone River. A trout 18 inches long is three times as long as a 6-inch trout. Observe that we can divide 6 into 18 to determine the *ratio* of the trout lengths.   ◇

In summary, there are four levels of measurement. The nominal is considered the lowest, and in ascending order we have the ordinal, interval, and ratio levels. In general, calculations based on one level of measurement may not be appropriate for a lower level.

Level of Measurement	Suitable Calculation
Nominal	We can put the data into categories.
Ordinal	We can order the data from smallest to largest or "worst" to "best." Each data value can be *compared* with another data value.
Interval	We can order the data and also take the differences between data values. At this level, it makes sense to compare the differences between data values. For instance, we can say that one data value is 5 more than another or 12 less than another data value.
Ratio	We can order the data, take differences, and also find the ratio between data values. For instance, it makes sense to say that one data value is twice as large as another.

GUIDED EXERCISE 2

Levels of measurement

The following describe different data associated with a state senator. For each data entry, indicate the corresponding *level of measurement*.

(a) The senator's name is Sam Wilson. ⟹ Nominal level

(b) The senator is 58 years old. ⟹ Ratio level

(c) The years in which the senator was elected to the senate are 1980, 1986, 1992, and 1998. ⟹ Interval level

(d) His total taxable income last year was $878,314.19. ⟹ Ratio level

(e) The senator sponsored a bill to protect water rights. Out of 1100 voters in his district, 400 said they strongly favored the bill, 300 said they favored the bill, 200 said they were neutral, 150 said they did not favor the bill, and 50 said they strongly did not favor the bill. ⟹ The opinions about the bill are at the ordinal level.

(f) The senator is married now. ⟹ Nominal level

(g) A leading news magazine claims the senator is ranked seventh for his voting record on bills regarding public education. ⟹ Ordinal level

Looking Ahead

The purpose of collecting and analyzing data is to obtain information. Statistical methods provide us tools to obtain information from data. These methods break into two branches.

Descriptive statistics

Inferential statistics

Descriptive statistics involves methods of organizing, picturing, and summarizing information from samples or populations.

Inferential statistics involves methods of using information from a sample to draw conclusions regarding the population.

We will look at methods of descriptive statistics in the next two chapters and in Chapter 10. These methods may be applied to data from samples or populations.

Sometimes we do not have access to the entire population. At other times, the difficulties or expense of working with the entire population are prohibitive. In such cases, we will use inferential statistics together with probability. These are the topics of Chapters 4 through 12.

The First Measured Century

The 20th century saw measurements of aspects of American life that had never been systematically studied before. Social conditions involving crime, sex, food, fun, religion, and work have been numerically investigated. The measurements and survey responses taken over the entire century reveal unsuspected statistical trends. *The First Measured Century* is a book by Caplow, Hicks, and Wattenberg. It is also a PBS documentary available on video. For more information, visit the Brase/Brase statistics site at http://math.college.hmco.com/students and find the link to the PBS first measured century documentary.

SECTION 1.1   PROBLEMS

1. *Marketing: Fast Food* USA Today reported that 44.9% of those surveyed (1261 adults) ate in fast-food restaurants from one to three times each week.
   (a) Identify the variable.
   (b) Is the variable quantitative or qualitative?
   (c) What is the implied population?

2. *Advertising: Auto Mileage* What is the average miles per gallon (mpg) for all new 2001 cars? Using *Consumer Reports,* a random sample of 35 new 2001 cars gave an average of 21.1 mpg.
   (a) Identify the variable.
   (b) Is the variable quantitative or qualitative?
   (c) What is the implied population?

3. *Academic: Student Fees* The students at Eastmore College are concerned about the level of student fees. They took a random sample of 30 colleges and universities throughout the nation and obtained information about the student fees at these institutions. From this information they concluded that their student fees are higher than those of most colleges in the nation.
   (a) Identify the variable.
   (b) Is the variable quantitative or qualitative?
   (c) What is the implied population?

4. *Quality Control: Shelf Life* The quality-control department at Healthy Crunch, Inc., wants to estimate the shelf life of all Healthy Crunch granola bars. A random sample of 10 of these bars was tested, and the shelf life was determined. From the sample results, a shelf life for all Healthy Crunch granola bars was estimated.
   (a) Identify the variable.
   (b) Is the variable quantitative or qualitative?
   (c) What is the implied population?

5. *Insurance: Claim Payment* An insurance company wants to determine the time interval between the arrival of an insurance payment check and the time that the check clears. A central payment office processes the payments for a five-state region. A random sample of 32 payment checks from this five-state region was received and processed. The time interval between receipt and check clearance was determined for each check. From this information the company estimated the time interval necessary for all checks sent to this office to clear.

(a) Identify the variable.
(b) Is the variable quantitative or qualitative?
(c) What is the implied population?

6. *Education: Teacher Evaluation* If you were going to apply *statistical methods* to analyze teacher evaluations, which question form, A or B, would be better?
   Form A: In your own words, tell how this teacher compares with other teachers you have had.
   Form B: Use the following scale to rank your teacher as compared with other teachers you have had.

1	2	3	4	5
worst	below average	average	above average	best

7. *Student Life: Levels of Measurement* Categorize these measurements associated with student life according to level: nominal, ordinal, interval, or ratio.
   (a) Length of time to complete an exam
   (b) Time of first class
   (c) Major field of study
   (d) Course evaluation scale: poor, acceptable, good
   (e) Score on last exam (based on 100 possible points)
   (f) Age of student

8. *Business: Levels of Measurement* Categorize these measurements associated with a robotics company according to level: nominal, ordinal, interval, or ratio.
   (a) Salesperson's performance: below average, average, above average
   (b) Price of company's stock
   (c) Names of new products
   (d) Room temperature (°F) in CEO's private office
   (e) Gross income for each of past 5 years
   (f) Color of packaging

9. *Fishing: Levels of Measurement* Categorize these measurements associated with fishing according to level: nominal, ordinal, interval, or ratio.
   (a) Species of fish caught: perch, bass, pike, trout
   (b) Cost of rod and reel
   (c) Time of return home
   (d) Guidebook rating of fishing area: poor, fair, good
   (e) Number of fish caught
   (f) Temperature of water

1.2
Random Samples

FOCUS POINTS

✓ Explain the importance of random samples.
✓ Construct a simple random sample using random numbers.
✓ Simulate a random process.
✓ Describe stratified sampling, cluster sampling, systematic sampling, and convenience sampling.

Simple Random Samples

Eat lamb—20,000 coyotes can't be wrong!

This slogan is sometimes found on bumper stickers in the western United States. The slogan indicates the trouble that ranchers have experienced in protecting their flocks from predators. Based on their experience with this sample of the coyote population, the ranchers concluded that *all* coyotes are dangerous to their flocks and should be eliminated! The ranchers used a special poison bait to get rid of the coyotes. Not only was this poison distributed on ranch land, but with government cooperation it also was distributed widely on public lands.

The ranchers found that the results of the widespread poisoning were not very effective. The sheep-eating coyotes continued to thrive while the general population of coyotes and other predators declined. What was the problem? The sheep-eating coyotes the ranchers observed were not a representative sample of all coyotes. Modern methods of predator control target the sheep-eating coyotes. To a certain extent the new methods have come about through a closer examination of the sampling techniques used.

In this section, we will examine several widely used sampling techniques. One of the most important sampling techniques is a *simple random sample*.

Simple random sample

> A **simple random sample** of n measurements from a population is a subset of the population selected in a manner such that
>
> (a) every sample of size n from the population has an equal chance of being selected and
>
> (b) every member of the population has an equal chance of being included in the sample.

Notice that two criteria are necessary for a simple random sample of a specified size. Not only must every member of the population have an equal chance of being included, but every sample of the specified size must have an equal chance of being included as well.

GUIDED EXERCISE 3

Simple random sample

Is open space around metropolitan areas important? Players of the Colorado Lottery might think so because some of the proceeds of the game go to fund open space and outdoor recreational space. To play the game, you pay one dollar and choose any 6 different numbers from the group of numbers 1 through 42. If your group of 6 numbers matches the winning group of 6 numbers selected by simple random sampling, then you are a winner of a grand prize of at least 1.5 million dollars.

(a) Is the number 25 as likely to be selected in the winning group of six numbers as the number 5?

⟹ Yes, since the winning numbers constitute a simple random sample, each number from 1 through 42 has an equal chance of being selected.

(b) Could all the winning numbers be even?

⟹ Yes, since 6 even numbers is one of the possible groups of six numbers.

(c) Your friend always plays these numbers

   1 2 3 4 5 6

   Could she ever win?

⟹ Yes, in a simple random sample the listed group of six numbers is *as likely as any* of the 5,245,786 groups of six numbers to be selected as the winner. (See Section 4.3 to learn how to compute the number of groups of 6 numbers selected from 42 numbers.)

How do we get random samples? Suppose you need to know if the emission system of the latest shipment of Toyotas satisfies pollution-control standards. You want to pick a random sample of 30 cars from this shipment of 500 cars and test them. One way to pick a random sample is to number the cars 1 through 500. Write these numbers on cards, mix up the cards, and then draw 30 numbers. The sample will consist of the cars with the chosen numbers. If you mix the cards sufficiently, this procedure produces a random sample.

Random-number table

An easier way to select the numbers is to use a *random-number table*. You can make one yourself by writing the digits 0 through 9 on separate cards and mixing up these cards in a hat. Then draw a card, record the digit, return the card, and mix up the cards again. Draw another card, record the digit, and so on. However, Table 1 in Appendix II is a ready-made random-number table (adapted from Rand Corporation, *A Million Random Digits with 100,000 Normal Deviates*). Let's see how to pick our random sample of 30 Toyotas by using this random-number table.

EXAMPLE 6

Random-number table

Use a random-number table to pick a random sample of 30 cars from a population of 500 cars.

SOLUTION: Again, we assign each car a different number between 1 and 500, inclusive. Then we use the random-number table to choose the sample. Table 1 in Appendix II has 50 rows and 10 blocks of five digits each; it can be thought of as a solid mass of digits that has been broken up into rows and blocks for user convenience.

You read the digits by beginning anywhere in the table. We dropped a pin on the table, and the head of the pin landed in row 15, block 5. We'll begin there and list all the digits in that row. If we need more digits, we'll move on to row 16, and so on. The digits we begin with are

    99281   59640   15221   96079   09961   05371

Since the highest number assigned to a car is 500, and this number has three digits, we regroup our digits into blocks of 3:

   992   815   964   015   221   960   790   996   105   371

To construct our random sample, we use the first 30 car numbers we encounter in the random-number table when we start at row 15, block 5. We skip the first three groups—992, 815, and 964—because these numbers are all too large. The next group of three digits is 015, which corresponds to 15. Car number 15 is the first car included in our sample, and the next is car number 221. We skip the next three groups and then include car numbers 105 and 371. To get the rest of the cars in the sample, we continue to the next line and use the random-number table in this fashion. If we encounter a number we've used before, we'll skip it.   ◇

◇ **COMMENT** When we use the term *(simple) random sample*, we have very specific criteria in mind for selecting the sample. One proper method for selecting a simple random sample is to use a computer-based or calculator-based random-number generator or to use a table of random numbers as we have done in the example. The term *random* should not be confused with *haphazard!*   ◇

Simulation

Another important use of random-number tables is in *simulation*. We use the word *simulation* to refer to the process of providing arithmetic imitations of "real" phenomena. Simulation methods have been productive in studying a diverse array of subjects such as nuclear reactors, cloud formation, cardiology (and medical science in general), highway design, production control, shipbuilding, airplane design, war games, economics, and electronics. A complete list would probably include something from every aspect of modern life. In Example 7 and Guided Exercise 4 we'll perform a brief simulation.

EXAMPLE 7
Simulation

A well-known theory in stock market analysis is the "random walk" theory (see *A Random Walk Down Wall Street,* 6th Edition, Burton Malkiel, W. W. Norton & Co.). The term *random walk,* as applied to stock prices, means that short-term changes in stock prices cannot be predicted but rather are random. In particular, according to the random walk theory, the next move in the price of a stock (up or down) is completely unpredictable on the basis of what price changes happened before.

Let's use a very simplified model to *simulate* the stock price changes of a hypothetical company, Fun Boards (maker of skate boards, surf boards, snow boards, and in-line skates). Suppose the initial price of Fun Boards stock is $50 per share. Use the random-number table to simulate daily price changes for the next 15 trading days in the following way. Notice we are interested in price *changes*. Days during which the stock does not change price will be ignored.

SOLUTION: The daily stock price change will be dictated by a number from the random-number table. When you encounter an even digit (0, 2, 4, 6, 8), increase the stock price by $1. When you encounter an odd digit (1, 3, 5, 7, 9), decrease the stock price by $1. Do this for a sequence of 15 trading days during which the price changed.

Beginning with line 6, block 2 in Table 1 of Appendix II, we see the 15 random digits

51709   94456   48396

Since the first random digit is odd, we will decrease the price by $1 on the first day. In fact, the next two digits are also odd, so we will decrease the price again by $1 on day 2 and then again on day 3. The next digit is even, so we will increase the price by $1 on day 4. Table 1-1 shows the simulated price changes for the 15-trading-day period. ◇

Trading floor, New York Stock Exchange

◇ **COMMENT** Recall that the random samples we have been constructing so far are called *simple random samples.* Throughout this text we will use the term *random sample* to mean simple random sample. All the statistical methods in this text assume that a simple random sampling has been used to collect the data. ◇

TABLE 1-1 Simulated Price Moves of Fun Boards Stock

Price	50	49	48	47	48	47	46	47	48	47	48	49	50	49	48	49
Day	Initial	1	2	3	4	5	6	7	8	9	10	11	12	13	14	15
Digit		5	1	7	0	9	9	4	4	5	6	4	8	3	9	6

GUIDED EXERCISE 4

Simulation

Use a random-number table to simulate the outcomes of tossing a balanced (that is, fair) penny 10 times.

(a) How many outcomes are possible when you toss a coin once?

⟹ Two; heads or tails

(b) There are several ways to assign numbers to the two outcomes. Because we assume a fair coin, assign an even digit to the outcome heads and an odd digit to the outcome tails. Then, starting at block 3 of row 2 of Table 1 in Appendix II, list the first 10 single digits.

⟹ 7 1 5 4 9 4 4 8 4 3

(c) What are the outcomes associated with the 10 digits?

⟹ T T T H T H H H H T

(d) If you start in a different block and row of Table 1 in Appendix II, will you get the same sequence of outcomes?

⟹ It is possible, but not very likely. (In Section 4.3 you will learn how to determine that there are 1024 possible sequences of outcomes for 10 tosses of a coin.)

Sampling with replacement

TECH NOTE Most statistical software packages, spreadsheet programs, and statistical calculators generate random numbers. In general, these devices sample *with replacement*. Sampling with replacement means that although a number is selected for the sample, it is *not removed* from the population. Therefore, the same number may be selected for the sample more than once. If you need to sample without replacement, generate more items than you need for the sample. Then sort the sample and remove duplicate values. Specific procedures for generating random samples on the TI-83Plus calculator, Excel, Minitab, and ComputerStat are shown in Using Technology at the end of this chapter. More details are given in the separate Technology Guides for each of these technologies.

Other Sampling Techniques

Although we will always assume that (simple) random samples are used throughout this text, other methods of sampling are also widely used. Appropriate statistical techniques exist for these sampling methods, but they are beyond the scope of this text.

Stratified sampling

One of these sampling methods is called *stratified sampling*. Groups or classes inside a population that share a common characteristic are called *strata* (plural of *stratum*). For example, in the population of all undergraduate college students, some strata might be freshmen, sophomores, juniors, or seniors. Other strata might be men or women or in-state students or out-of-state students, and so on. In the method of stratified sampling, the population is divided into at least two distinct strata. Then a (simple) random sample of a certain size is drawn from each stratum, and the information obtained is carefully adjusted or weighted in all resulting calculations.

The groups or strata are often sampled in proportion to their actual percentages of occurrence in the overall population. However, other (more sophisticated) ways to determine the optimal sample size in each stratum may give the best results. In general, statistical analysis and tests based on data obtained from stratified samples are somewhat different from techniques discussed in an introductory course in statistics. Such methods for stratified sampling will not be discussed in this text.

Systematic sampling

Another popular method of sampling is called *systematic sampling*. In this method, it is assumed that the elements of the population are arranged in some natural sequential order. Then we select a (random) starting point and select every kth element for our sample. For example, people lining up to buy rock concert tickets are "in order." To generate a systematic sample of these people (and ask questions regarding topics such as age, smoking habits, income level, etc.), we could include every 5th person in line. The "starting" person could be selected at random from the first five.

The advantage of a systematic sample is that it is easy to get. However, there are dangers in using systematic sampling. When the population is repetitive or cyclic in nature, systematic sampling should not be used. For example, consider a fabric mill that produces dress material. Suppose the loom that produces the material makes a mistake every 17th yard, but we check only every 16th yard with an automated electronic scanner. In this case, a random starting point may or may not result in detection of fabric flaws before a large amount of fabric is produced.

Cluster sampling

Cluster sampling is a method used extensively by government agencies and certain private research organizations. In cluster sampling we begin by dividing the demographic area into sections. Then we randomly select sections or clusters. Every member of the cluster is included in the sample. For example, in conducting a survey of school children in a large city, we could first randomly select 5 schools and then include all the children from each selected school.

Convenience sampling

Convenience sampling simply uses results or data that are conveniently and readily obtained. In some cases, this may be all that is available, and in many cases, it is better than no information at all. However, convenience sampling does run the risk of being severely biased. For instance, consider a newsperson who wishes to get the "opinions of the people" about a proposed seat tax to be imposed on tickets to all sporting events. The revenues from the seat tax will then be used to support the local symphony. The newsperson stands in front of a classical music store at noon and surveys the first five people coming out of the store who will cooperate. This method of choosing a sample will produce some opinions, and perhaps some human interest stories, but it certainly has bias. It is hoped that the city council will not use these opinions as the sole basis for a decision about the proposed tax. It is good advice to be very cautious indeed when the data come from the method of convenience sampling.

Our discussion of sampling methods is intended to be brief and general. This is not the place for an extensive treatment of theory and practice of sampling. However, the interested reader is referred to the book *A Sampler on Sampling,* by Bill Williams, John Wiley and Sons, Inc.

VIEWP⊙INT *Extraterrestrial Life?*

Do you believe intelligent life exists on other planets? Using methods of random sampling, a Fox News opinion poll found that about 54% of all U.S. men do believe in intelligent life on other planets, whereas only 47% of women believe there is such life. How could you conduct a random survey of students on your campus regarding belief in extraterrestrial life?

SECTION 1.2   PROBLEMS

1. *Sampling: Random*
   (a) In your own words, explain the meaning of the terms *random numbers* and *random samples*.
   (b) Why are random samples so important in statistics?

2. *Sampling: Random* Use a random-number table to get a list of eight random numbers from 1 to 976. Explain your work.

3. *Sampling: Random* Use a random-number table to get a list of six random numbers from 1 to 8615. Explain your work.

4. *Sampling: Random* Use a random-number table to get a list of five random numbers from 43 to 719. Explain your work.

5. *Psychology: Random Selection* How do colds affect analytical thinking performance? Results of a study conducted by McGraw and Schleser were reported in *Psychology Today*. The study showed that under certain conditions, persons with colds do better than their healthy colleagues. The study considered 62 subjects: 40 healthy men and women and 22 suffering from colds or flu. A key component in this study was the formation of two groups of equal size from the 62 participants, with each group containing both healthy and sick participants.
   (a) Describe how you could take the 40 healthy subjects and randomly divide them into two groups of equal size using the random-number table.
   (b) Repeat part (a) for the 22 sick subjects.
   (c) How would you combine the groups of healthy and sick subjects found in parts (a) and (b) to form two groups of equal size so that each group contained both healthy and sick subjects?

6. *Simulation: Coin Toss* Use a random-number table to simulate the outcomes of tossing a quarter 25 times. Assume that the quarter is balanced (i.e., fair).

7. *Computer Simulation: Roll of a Die* A die is a cube with dots on each face. The faces have 1, 2, 3, 4, 5, or 6 dots. The table below is a computer simulation (from the software package Minitab) of the results of rolling a fair die 20 times.

DATA DISPLAY

ROW	C1	C2	C3	C4	C5	C6	C7	C8	C9	C10
1	5	2	2	2	5	3	2	3	1	4
2	3	2	4	5	4	5	3	5	3	4

(a) Assume that each number corresponds to the number of dots on the face of the die. Is it appropriate that the same number appear more than once? Why? What is the outcome of the 4th roll?

(b) If we simulate more "rolls of the die," do you expect to get the same sequence of outcomes? Why or why not?

8. *Simulation: Birthday Problem* Suppose there are 30 people at a party. Do you think any two share the same birthday? Let's use the random-number table to simulate the birthdays of the 30 people at the party. Ignoring leap year, let's assume that the year has 365 days. Number the days, with 1 representing January 1, 2 representing January 2, and so forth, with 365 representing December 31. Draw a random sample of 30 days (with replacement). These days represent the birthdays of the people at the party. Were any two of the birthdays the same? Compare your results with those obtained by other students in the class. Would you expect the results to be the same or different?

9. *Sampling: Students in Class* Suppose you are given the number 1, and each of the other students in your statistics class calls out consecutive numbers until each person in class has his or her own number. Explain how you could get a random sample of four students from your statistics class.

(a) Explain why the first four students walking into the classroom would not necessarily form a random sample.

(b) Explain why four students coming in late would not necessarily form a random sample.

(c) Explain why four students sitting in the back row would not necessarily form a random sample.

(d) Explain why the four tallest students may not necessarily form a random sample.

10. *Sampling: Quality Control* Products are inspected during production, and equipment is adjusted to correct defects. It is usually not possible to examine every product, so a random sample is examined. For each of the following, give a detailed explanation of how you could get the requested random sample. Be sure to include the random numbers you use to get the random sample.

(a) How would you draw a random sample of 6 of the next 500 stereo headsets coming off an assembly line?

(b) How would you obtain a random sample of 10 men's dress shirts coming off an assembly line from 8 A.M. to 12 noon?

(c) Serial numbers are placed on radios as they come off an assembly line. How could you get a random sample of nine radios with serial numbers from 21942 to 98756?

(d) A truck has just delivered 800 cartons of eggs to a supermarket. How would you get a random sample of 12 cartons to check for broken eggs?

11. *Sampling: General* For each of the following, give a detailed explanation of how you would get the requested random sample. Be sure to include the random numbers you use to get the random sample.

(a) You are a veterinarian. How would you get a random sample of 15 sheep to check for ticks on a farm that has 250 sheep?

(b) You are a medical records technician. Patients from 1992 to the present were given file numbers starting at 1024 and ending with 8342. How would you get a random sample of 10 of these patients?

(c) You are a security agent for an airline. How could you get a random sample of five pieces of luggage that are moving on a conveyor belt in the next 25 minutes?

(d) You are conducting a survey for your sociology class project. How can you get a random sample of 12 adults walking past the information booth at a shopping center between 6 and 7 P.M.?

12. *Education: Test Construction* Professor Gill uses true–false questions. She wishes to place 20 such questions on the next test. To decide whether to place a true statement or false statement in each of the 20 questions, she uses a random-number table. She selects 20 digits from the table. An even digit tells her to use a true statement. An odd digit tells her to use a false statement. Use a random-number table to pick a sequence of 20 digits, and describe the corresponding sequence of 20 true–false questions. What would the test key for your sequence look like?

13. *Education: Test Construction* Professor Gill is designing a multiple-choice test. There are to be 10 questions. Each question is to have five choices for answers. The choices are to be designated by the letters *a, b, c, d,* and *e.* Professor Gill wishes to use a random-number table to determine which letter choice should contain the correct answer for a question. Using the number correspondence 1 for *a,* 2 for *b,* 3 for *c,* 4 for *d,* and 5 for *e,* use a random-number table to determine the letter choice for the correct answer in each of the 10 questions.

14. *Sampling Methods: Health Care* Modern Managed Hospitals (MMH) is a national for-profit chain of hospitals. Management wants to survey patients discharged this past year to obtain patient satisfaction profiles. They wish to use a sample of such patients. Several sampling techniques are described below. Categorize each technique as *simple random sample, stratified sample, systematic sample, cluster sample,* or *convenience sample.*
    (a) Obtain a list of patients discharged from all MMH facilities. Divide the patients according to length of hospital stay (2 days or less, 3–7 days, 8–14 days, more than 14 days). Draw simple random samples from each group.
    (b) Obtain lists of patients discharged from all MMH facilities. Number these patients, and then use a random-number table to obtain the sample.
    (c) Randomly select some MMH facilities from each of five geographic regions, and then include all the patients on the discharge lists of the selected hospitals.
    (d) At the beginning of the year, instruct each MMH facility to survey every 500th patient discharged.
    (e) Instruct each MMH facility to survey 10 discharged patients this week and send in the results.

15. *Sampling Methods: Benefits Package* An important part of employee compensation is a benefits package that might include health insurance, life insurance, child care, vacation days, retirement plan, parental leave, bonuses, etc. Suppose you want to conduct a survey of benefit packages available in private businesses in Hawaii. You want a sample size of 100. Some sampling techniques are described below. Categorize each technique as *simple random sample, stratified sample, systematic sample, cluster sample,* or *convenience sample.*
    (a) Assign each business in the Island Business Directory a number, and then use a random-number table to select the businesses to be included in the sample.
    (b) Use the postal ZIP Codes to divide the state into regions. Pick a random sample of 10 ZIP Code areas and then include all the businesses in each selected ZIP Code area.
    (c) Send a team of five research assistants to Bishop Street in downtown Honolulu. Let each assistant select a block or building and interview an employee from each business found. Each researcher can have the rest of the day off after getting responses from 20 different businesses.
    (d) Use the Island Business Directory. Number all the businesses. Select a starting place at random, and then use every 50th business listed until you have 100 businesses.
    (e) Group the businesses according to type: medical, shipping, retail, manufacturing, financial, construction, restaurant, hotel, tourism, other. Then select a random sample of 10 businesses from each business type.

1.3
Introduction to Experimental Design

FOCUS POINTS

✓ Discuss what it means to take a census.

✓ Describe simulations, observational studies, and experiments.

✓ Identify control groups, placebo effects, and randomized two-treatment design.

✓ Discuss potential pitfalls that might make your data unreliable.

Planning a Statistical Study

Planning a statistical study and gathering data are essential components for obtaining reliable information. Depending on the nature of the statistical study, a great deal of expertise and resources may be required during the planning stage. In this section, we look at some of the basics for planning a statistical study.

> **Basic guidelines for planning a statistical study**
>
> 1. First, identify the individuals or objects of interest.
>
> 2. Specify the variables as well as protocols for taking measurements or making observations.
>
> 3. Determine if you will use an entire population or a representative sample. If using a sample, decide on a viable sampling method.
>
> 4. Collect the data.
>
> 5. Use appropriate descriptive statistics methods (Chapters 2, 3, 10) and make decisions using appropriate inferential statistics methods (Chapters 8–12).
>
> 6. Finally, note any concerns you might have about your data collection methods and list any recommendations for future studies.

One issue to consider is whether to use the entire population in a study or a representative sample. If we use data from the entire population, we have a *census*.

Census

> In a **census**, measurements or observations from the *entire* population are used.

When the population is small and easily accessible, a census is very useful because it gives complete information about the population. However, obtaining a census can be both expensive and difficult. Every 10 years, the U.S. Department of Commerce, Bureau of Census is required to conduct a census of the United States. However, contacting some members of the population—such as the homeless—is almost impossible. Sometimes members of the population will not respond. In such cases, statistical estimates for the missing responses are often supplied.

If we use data from only part of the population of interest, we have a *sample*.

Sample

> In a **sample**, measurements or observations from a *representative part* of the population should be used.

In the previous section, we examined several sampling strategies: simple random sampling, stratified sampling, cluster sampling, systematic sampling, and convenience sampling. In this text, we will study methods of inferential statistics based on simple random samples.

Simulation

As discussed in Section 1.2., *simulation* is a numerical facsimile of real-world phenomena. Sometimes simulation is called a "dry lab" approach, in the sense that it is an arithmetic imitation of a real situation. Advantages of simulation are that arithmetic and statistical simulations can fit real-world problems extremely well. The researcher can explore procedures in simulation that might be very dangerous in real life. In the real world you might not want to introduce a high level of a drug into a diabetic person's bloodstream. However, you might want to simulate the injection statistically and study the results. You will harm no one, and the information gained may be of real medical value. Similarly, you might test the effect of wind sheer on an airplane wing in a simulated environment rather than with an actual airplane in flight.

Experiments and Observation

When gathering data for a statistical study, we want to distinguish between observational studies and experiments.

> In an **observational study,** observations and measurements of individuals are conducted in a way that doesn't change the response or the variable being measured.
>
> In an **experiment,** a *treatment* is deliberately imposed on the individuals in order to observe a possible change in the response or variable being measured.

EXAMPLE 8
Experiment

In 1778, Captain James Cook landed in what we now call the Hawaiian Islands. He gave the islanders a present of several goats, and over the years these animals multiplied into wild herds totaling several thousand. They eat almost anything, including the famous silver sword plant, which was once unique to Hawaii. At one time, the silver sword grew abundantly on the island of Maui (in Haleakala, a national park on that island, the silver sword can still be found), but each year there seemed to be fewer and fewer plants. Biologists suspected that the goats were partially responsible for the decline in the plants and conducted a statistical study that verified their theory.

(a) To test the theory, park biologists set up stations in remote areas of Haleakala. At each station two plots of land similar in soil conditions, climate, and plant count were selected. One plot was fenced to keep out the goats, while the other was not. At regular intervals a plant count was made in each plot. This study involves an *experiment* because a *treatment* (the fence) was imposed on one plot.

(b) The experiment involved two plots at each station. The plot that was not fenced represents the *control* plot. This is the plot where a treatment was specifically not imposed, but it was similar to the fenced plot in every other way.   ◆

Silver sword plant, Haleakala National Park

Placebo

Statistical experiments are commonly used to determine the effect of a treatment. However, the design of the experiment needs to *control* for other possible causes of the effect. For instance, in medical experiments the *placebo* effect is the improvement or change that is the result of patients just believing in the treatment whether the treatment itself is effective or not. To account for the placebo effect, patients are divided into two groups. One group receives the prescribed treatment. The other group, called the *control group,* receives a dummy or placebo treatment that is disguised to look like the real treatment. Finally, after the treatment cycle, the medical condition of the patients in the *treatment group* is compared to that of the patients in the control group.

It is difficult to account for all variables that might influence a patient's response to a treatment. Sometimes patients assigned to a treatment group and a control group are carefully matched by age, gender, level of medical condition, smoker, etc. However, a more common way to assign patients to the treatment and control groups is by using a random process. This is the essence of a *randomized two-treatment experiment.*

Randomized two-treatment experiment

EXAMPLE 9

Randomized two-treatment experiment

Can chest pain be relieved by drilling holes in the heart? Since 1980, surgeons have been using a laser procedure to drill holes in the heart. Many patients report a lasting and dramatic decrease in angina (chest pain) symptoms. Is the relief due to the procedure, or is it a placebo effect? A recent research project at Lenox Hill Hospital in New York City provided some information about this issue by using a randomized two-treatment experiment. The laser treatment was applied through a less invasive (catheter laser) process. A group of 298 volunteers with severe, untreatable chest pain were randomly assigned to get the laser or not. The patients were sedated but awake. They could hear the doctors discuss the laser process. Each patient thought he or she was receiving the treatment.

The experimental design can be pictured as

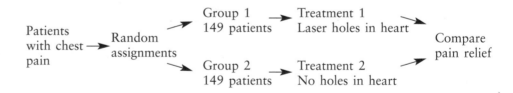

The laser patients did well. But shockingly, the placebo group showed more improvement in pain relief. The medical impacts of this study are still being investigated. ◊

Double-blind experiment

Many experiments are also *double-blind.* This means that neither the individuals in the study nor the observers know which subjects are receiving the treatment. Double-blind experiments help control for subtle biases that a doctor might pass on to a patient.

We have explored only a few features of experimental design. When the effects of several treatments are all being examined at the same time, more complicated designs including *blocking* are required. Section 11.6, Introduction to Two-Way ANOVA, has further examples for such situations.

The study cited in Example 9 has many features of good experimental design.

There is a **control group.** This group received a dummy treatment enabling the researchers to control for the placebo effect. In general, a control group is used to account for the influence of other known or unknown variables that might be an underlying cause of a change in response in the experimental group. Such variables are called **lurking** or **confounding variables.**

Randomization is used to assign individuals to the two treatment groups. This helps prevent bias in selecting members for each group.

Replication of the experiment on many patients reduces the possibility that the differences in pain relief for the two groups occurred by chance alone.

GUIDED EXERCISE 5

Collecting data

Which technique (sampling, experiment, simulation, or census) for gathering data do you think might be the most appropriate for the following studies?

(a) Study of the effect of stopping the cooling process of a nuclear reactor.

⟹ Probably simulation, since you may not want to risk a nuclear meltdown.

(b) Study of the amount of time college students taking a full course load spend watching television.

⟹ Sampling and using an observational study would work well. Notice that obtaining the information from a student will probably not change the amount of time a student spends watching television.

(c) Study of the effect of a calcium supplement given to young girls on bone mass.

⟹ Experimentation. A study by Tom Lloyd reported in the *Journal of the American Medical Association* utilized 94 young girls. Half were randomly selected and given a placebo. The other half were given calcium supplements to bring their daily calcium intake up to about 1400 milligrams per day. The group getting the experimental treatment of calcium gained 1.3% more bone mass in a year than the girls getting the placebo.

(d) Study of the credit hours load of *each* student enrolled at your college at the end of the drop/add period this semester.

⟹ Census. The registrar can obtain records for *every* student.

Potential pitfalls

Nonresponse

Voluntary response

Hidden bias

Other variables

Surveys

Once you decide you are going to use sampling, census, observation, or experiments, a common means to gather data about people is to ask them questions. This process is the essence of *surveying*. Sometimes the possible responses are simply yes or no. Other times the respondents choose a number on a scale that represents their feelings from, say, strongly disagree to strongly agree. In the case of an open-ended, discussion-type response, the researcher must determine a way to convert the response to a category or number.

A number of issues can arise when using a survey. Are the questions asked in a neutral way, or is conscious or unconscious bias built into the wording? In the case of an interviewer, is he or she giving the respondents subtle feedback that might influence the responses? How can you be sure the respondents are answering truthfully? Is your sample representative of the population? For instance, when conducting election polls, some studies use only registered voters because these are the only people eligible to vote. Other polls use only "likely" voters. They first inquire if the respondent is planning to vote.

Other problems arise if the selected respondent cannot be contacted or refuses to respond. This is known as *nonresponse*. If the nonresponse is sufficiently high, the study may be biased and some sort of adjustment may need to be made.

Voluntary response samples often overrepresent people with strong opinions. A Colorado newspaper with statewide distribution asked the question, "Should grazing fees for use of public lands be increased?" Of the people who called in, about 85% said, "No!" However, the results were misleading as an indicator of the opinions held by the population of *all* people in Colorado. The sample who responded were self-selected people. The sample consisted mainly of ranchers who felt strongly enough to call the newspaper. These people did not want their grazing fees on public land increased.

Data from voluntary responses can be useful and interesting. However, the information given is anecdotal in nature. It would be questionable to generalize the results to the entire population of interest. Surveys must be designed and administered carefully for the results to generalize.

Whenever you gather data, whether by sampling a population, by results of experiment, or by simulation, you should view the data with a critical eye. The way the data are gathered may produce a *hidden bias*. This means that in reality you are not actually measuring what you hoped to measure. For instance, if a uniformed police officer conducts a survey regarding opinions about and use of illegal drugs, the responses might be different than if a casually dressed person asks the same questions. Asking students if parking lots are large enough and safe before asking if they would approve an increase in parking fees might bias responses. Care must be taken to deal with all units or subjects in the same way so that no (conscious or unconscious) preferential treatment or selection can occur.

Sometimes our goal is to understand the cause-and-effect relationships between two variables (as might occur in regression and correlation of two variables presented in Chapter 10). However, the effect of one variable on another can be hidden by other variables for which no data have been obtained. These variables are called *lurking* or *confounding* variables. For instance, a study of ticket price and attendance at a sporting event might show that higher ticket prices and higher attendance are related, since events with higher ticket prices seem to have greater attendance. One might be led to conclude that if you want to increase attendance at an event, you should raise the price of the tickets. What is missing from this analysis is the lurking variable of *event popularity*. A Super Bowl football game is so special that people are willing to pay higher prices just to be there. A preseason football game, however, does not have the drawing power and may need lower ticket prices to ensure a reasonable crowd at the game. Other variables—such as the location of the game, weather, and records of the teams involved—also might influence attendance.

The problem of other variables that influence one or both of the original variables often can be overcome if the researcher not only is familiar with statistics but also is well versed in the field of investigation.

Generalizing results

Some researchers want to generalize their findings to a situation wider than that of the actual data setting. The true scope of a new discovery must be determined by repeated studies in various real-world settings. Statistical experiments showing that a drug had a certain effect on a collection of laboratory rats do not guarantee that the drug will have a similar effect on a herd of wild horses in Montana.

GUIDED EXERCISE 6

Cautions about data

Comment on the usefulness of the data collected as described.

(a) A uniformed police officer interviews a group of 20 college freshmen. She asks each one his or her name and then if he or she has used an illegal drug in the last month.

➡ Respondents may not answer truthfully. Some may refuse to participate.

(b) Jessica saw some data that show that cities with more low-income housing have more homeless people. Does building low-income housing cause homelessness?

➡ There may be some other confounding or lurking variables such as the size of the city. Larger cities may have more low-income housing and more homeless.

(c) A survey about food in the student cafeteria was conducted by having forms available for customers to pick up at the cash register. A drop box for completed forms was available outside the cafeteria.

➡ The voluntary response will likely produce more negative comments.

(d) Extensive studies on coronary problems were conducted using men over age 50 as the subjects.

➡ Conclusions for men over age 50 may or may not generalize to other age and gender groups. These results may be useful for women or younger people, but studies specifically involving these groups may need to be performed.

VIEWPOINT   *Is the Placebo Effect a Myth?*

Henry Beecher, Chief of Anesthesiology at Massachusetts General Hospital, published a paper in the *Journal of the American Medical Association* (1955) in which he claimed that the placebo effect is so powerful that about 35% of patients would improve simply if they believed a dummy treatment (placebo) was real. However, two Danish medical researchers refute this widely accepted claim in the *New England Journal of Medicine* (May 2001). They say the placebo effect is nothing more than a "regression effect," referring to a well-known statistical observation that patients who feel especially bad one day will almost always feel better the next day no matter what is done for them. However, other respected statisticians question the findings of the Danish researchers. Regardless of the new controversy surrounding the placebo effect, medical researchers agree that placebos are still needed in clinical research. Double-blind research using placebos prevents the researchers from inadvertently biasing results.

SECTION 1.3   PROBLEMS

1. *Ecology: Gathering Data* Which technique (observational study or experiment) for gathering data do you think was used in the following studies?
   (a) The Colorado Division of Wildlife netted and released 774 fish at Quincy Reservoir. There were 219 perch, 315 blue gill, 83 pike, and 157 rainbow trout.
   (b) The Colorado Division of Wildlife caught 41 bighorn sheep on Mt. Evans and gave each one an injection to prevent heart worm. A year later, 38 of these sheep did not have heart worm, while the other three did.
   (c) The Colorado Division of Wildlife imposed special fishing regulations on the Deckers section of the South Platt River. All trout under 15 inches must be released. A study of trout before and after the regulation went into effect showed the average length increased by 4.2 inches after the new regulation.
   (d) An ecology class used binoculars to watch 23 turtles at Lowell Ponds. It was found that 18 were box turtles, and five were snapping turtles.

2. *General: Gathering Data* Which technique (sampling, experiment, simulation, or census) for gathering data do you think was used in the following studies?
   (a) One way to find information on the Super Bowl football game is to look at the NFL web site. Visit the Brase/Brase statistics site at http://math.college.hmco.com/students and find the link to the NFL site. Explore the Super Bowl results. There the winning scores for all the Super Bowl games played to date are given. Using the data for all the games, we find that as of Super Bowl XXXIV, the average score for the winning teams was 31.
   (b) A sample of 82 healthy female and male subjects was recruited to participate in a study on pain (*Physical Therapy*, Vol. 70, No. 1). The subjects were divided into two groups. The experimental group received laser stimulation, and the control group received sham stimulation. Tests of pain tolerance were then conducted on each group.
   (c) Computer imaging of runners shows the effect of stride length on running efficiency.

(d) Do the Chinese like chocolate? Gallup Chinese is conducting surveys in China to answer the question for the U.S. Chocolate Manufacturers Association. Gallup Chinese is surveying a portion of the Chinese population to determine whether there is a market for chocolate in China (*Wall Street Journal*).

3. *General: Gathering Data* Which technique (sampling, experiment, simulation, or census) for gathering data do you think was used in the following studies?
   (a) An analysis of a sample of 31,000 patients from New York hospitals suggests that the poor and the elderly sue for malpractice at one-fifth the rate of wealthier patients (*Journal of the American Medical Association*).
   (b) The effects of wind sheer on airplanes during both landing and takeoff are studied by using complex computer programs that mimic actual flight.
   (c) A study of football scores attained through touchdowns and field goals was conducted by the National Football League to determine whether field goals account for more scoring events than touchdowns (*USA Today*).
   (d) An Australian study included 588 men and women who already had some precancerous skin lesions. Half got a skin cream containing a sunscreen with a sun protection factor of 17; half got an inactive cream. After 7 months, those using the sunscreen with the sun protection had fewer new precancerous skin lesions (*New England Journal of Medicine*).

4. *Surveys: Manipulation* The *New York Times* did a special report on polling that was carried in papers across the nation. The article points out how readily the results of a survey can be manipulated. Some features that can influence the results of a poll include the following: the number of possible responses, the phrasing of the question, the sampling techniques used (voluntary response or sample designed to be representative), the fact that words may mean different things to different people, the questions that precede the question of interest, and finally, the fact that respondents can offer opinions on issues that they know nothing about.
   (a) Consider the expression "over the last few years." Do you think that this expression means the same time span to everyone? What would be a more precise phrase?
   (b) Consider this question: "Do you think fines for running stop signs should be doubled?" Do you think the response would be different if the question "Have you ever run a stop sign?" preceded the question about fines?
   (c) Consider this question: "Do you watch too much television?" What do you think the responses would be if the only responses possible were yes or no? What do you think the responses would be if the possible responses were rarely, sometimes, or frequently?

5. *General: Randomized Two-Treatment Experiment* How would you use a randomized two-treatment experiment in each of the following settings? Is a placebo being used or not? Be specific and give details.
   (a) A veterinarian wants to test a strain of antibiotic on calves to determine their resistance to common infection. In a pasture are 22 newborn calves. There is enough vaccine for 10 calves. However, blood tests to determine resistance to infection can be done on all calves.
   (b) The Denver Police Department wants to improve its image with teenagers. A uniformed officer is sent to a school one day a week for ten weeks. Each day the officer visits with students, eats lunch with students, attends pep rallies, and so on. There are 18 schools, but the police department can visit only half of these schools this semester. A survey regarding how teenagers view police is sent to all 18 schools at the end of the semester.
   (c) A skin patch contains a new drug to help people quit smoking. A group of 75 cigarette smokers have volunteered as subjects to test the new skin patch. For

one month 40 of the volunteers receive skin patches with the new drug. The other volunteers receive skin patches with no drugs. At the end of two months, each subject is surveyed regarding his or her current smoking habits.

6. *General: Randomized Two-Treatment Experiment* How would you use a randomized two-treatment experiment in each of the following settings? Is this a double-blind experiment or not? Be specific and give details.

   (a) A new high-temperature bonding process is used to manufacture automobile tires. A group of 43 people volunteered to test the new tires on their cars. Out of this group, 25 cars are given the new high-temperature bond tires. The other cars are given new tires without the high-temperature bond. Only the serial numbers on the tires tell which type is being used, and this information is held confidential. After 35,000 miles, all tires are thoroughly examined by the manufacturer.

   (b) The FBI wants to test the Miami Airport security system. The FBI has inserted weapons into 10 out of 250 carry-on bags that agents take through security checks. The agent carrying the bag knows whether or not the bag contains a weapon. Those bags that get through the security check and those that do not are observed by a remote security camera.

   (c) The medical school is investigating new eye drops as a treatment for glaucoma. Out of 63 volunteers, 35 will get the new eye drops. The others will get the currently used (not new) eye drops. The new eye drops come in a grey plastic bottle and the old drops come in a red plastic bottle. Neither the patient nor the doctor knows which color contains which eye drops. After six months, eye pressure on each patient is measured and a sealed report revealing medication is opened.

SUMMARY

In this chapter, we've seen that statistics is the study of how to collect, organize, analyze, and interpret numerical information from populations or samples. First, we looked at classifying data as quantitative or qualitative. Next, we examined levels of measurement. The technique of drawing a simple random sample was discussed along with other sampling methods. Finally, we looked at methods of collecting data and some of the issues involved in experimental design and surveys.

IMPORTANT WORDS AND SYMBOLS

Section 1.1*
Statistics
Individual
Variable
Quantitative variable
Qualitative variable
Population data
Sample data
Levels of measurement
   Nominal

Ordinal
Interval
Ratio
Descriptive statistics
Inferential statistics

Section 1.2
Simple random sample
Random-number table
Simulation

*Indicates section of first appearance.

Stratified sample
Systematic sample
Cluster sample
Convenience sample

Section 1.3
Census
Observational study
Experiment
Placebo
Randomized two-treatment experiment

Control group
Treatment group
Replication
Confounding variable
Lurking variable
Double-blind
Survey
Nonresponse
Voluntary response
Hidden bias

VIEWPOINT

Is Chocolate Good for Your Heart?

A study of 7,841 Harvard alumni showed the death rate was 30% lower in those who ate candy compared with those who abstained. It turns out that candy, especially chocolate, contains antioxidants that help slow the aging process. Also, chocolate, like aspirin, reduces the activity of blood platelets that contribute to plaque and blood clotting. Furthermore, chocolate seems to raise high-density lipoprotein (HDL), the good cholesterol. However, these results are all preliminary. The investigation is far from complete. A wealth of information on this topic was published in the August 2000 issue of the *Journal of Nutrition*. Statistical studies and reliable experimental design are indispensable in this type of research.

CHAPTER REVIEW PROBLEMS

1. *General: Samples and Variables* Find a newspaper or web site article that uses statistics. Are the data from the entire population or just from a sample? What are the variables?

2. *Radio Talk Show: Sample Bias* A radio talk show asked listeners to respond either yes or no to the question, Is the candidate who spends the most on a campaign the most likely to win? Fifteen people called in and nine said yes. What is the implied population? What is the variable? Can you detect any bias in the selection of the sample?

3. *Essay: Levels of Measurement* In your own words, give a complete and careful description of the four levels of measurement. Which level is the highest? Which is the lowest? What are the different suitable uses for each level of measurement?

4. *Personal Data: Levels of Measurement* Write a brief description of yourself in which you list your name, age, year of birth, height, Social Security number, color of your hair and eyes, address, phone number, place of birth, letter grade on test, intended college major (if decided), distance you live from college, and so forth. In one column, list each item in the descriptions of yourself. In a second column to the right of the first, list the level of measurement corresponding to each item in the description.

5. *Colorado Lotto: Simulation* Lotto is the name of the Colorado lottery. The Lotto boards consist of 42 numbers (from 1 to 42). To play, you select six distinct

numbers. Every week a drawing machine randomly selects six numbered Ping-Pong balls. If one of your boards contains all six winning numbers, in any order, you've hit the jackpot! You can pick your numbers any way you wish. However, suppose you want to use a random-number table to pick your six numbers. Describe how you would do so, and list your selected numbers. (To play, you must pay $1 to have your selections entered into a computer for a specified week's drawing.)

6. *General: Type of Sampling* Categorize the type (simple random, stratified, systematic, cluster, or convenience) of sampling used in each of the following situations.
   (a) To conduct a preelection opinion poll on a proposed amendment to the state constitution, a random sample of 10 telephone prefixes (first three digits of the phone number) was selected, and all households from the phone prefixes selected were called.
   (b) To conduct a study on depression among the elderly, a sample of 30 patients in one nursing home was used.
   (c) To maintain quality control in a brewery, every 20th bottle of beer coming off the production line was opened and tested.
   (d) Subscribers to the magazine *Sound Alive* were assigned numbers. Then a sample of 30 subscribers was selected by using a random-number table. The subscribers in the sample were invited to rate new compact disc players for a "What the Subscribers Think" column.
   (e) To judge the appeal of a proposed television sitcom, a random sample of 10 people from each of three different age categories was selected and those chosen were asked to rate a pilot show.

7. *General: Gathering Data* Which technique (observational study or experiment) for gathering data do you think was used in the following studies? Explain your answer.
   (a) The U.S. Census Bureau tracks population age. In 1900, the percentage of the population that was nineteen years old or younger was 44.4%. In 1930, the percentage was 38.8%; in 1970, the percentage was 37.9%; and in 2000, the percentage in the age group was down to 28.5% (*The First Measured Century*, T. Caplow, L. Hicks, B. J. Wattenberg).
   (b) After receiving the same lessons, a class of 100 students was randomly divided into two groups of 50 each. One group was given a multiple-choice exam over the material in the lessons. The other group was given an essay exam. The average test scores for the two groups were then compared.

8. *General: Randomized Two-Treatment Experiment* How would you use a randomized two-treatment experiment in each of the following settings? Is a placebo being used or not? Be specific and give details.
   (a) A charitable nonprofit organization wants to test two methods of fund raising. From a list of 1,000 past donors, half will be sent literature about the successful activities of the charity and asked to make another donation. The other 500 donors will be contacted by phone and asked to make another donation. The percentage of people from each group who make a new donation will be compared.
   (b) A tooth-whitening gel is to be tested for effectiveness. A group of 85 adults have volunteered to participate in the study. Of these, 43 are to be given a gel that contains the tooth-whitening chemicals. The remaining 42 are to be given a similar-looking package of gel that does not contain the tooth-whitening chemicals. A standard method will be used to evaluated the whiteness of teeth for all participants. Then the results for the two groups will be compared. How could this experiment be designed to be double-blind?

9. *Student Life: Data Collection Project* Make a statistical profile of your own statistics class. Items of interest might be
   (a) Height, age, gender, pulse, number of siblings, marital status

(b) Number of college credit hours completed (as of beginning of term); grade point average
(c) Major; number of credit hours enrolled in this term
(d) Number of scheduled hours working per week
(e) Distance from residence to first class; time it takes to travel from residence to first class
(f) Year, make, and color of car usually driven

What directions would you give to people answering these questions? For instance, how accurate should the measurements be? Should age be recorded as of last birthday?

10. *Census: Web Site Census and You*, a publication of the Bureau of the Census, indicates that "Wherever your Web journey ends up, it should start at the Census Bureau's site." Visit the Brase/Brase statistics site at http://math.college.hmco.com/students and find a link to the Census Bureau's site. The site touts itself as the source of "official statistics." But it is willing to share the spotlight. The web site now has links to other "official" sources: other federal agencies, foreign statistical agencies, and state data centers. If you have access to the Internet, try the Census Bureau's site.

DATA HIGHLIGHTS: GROUP PROJECTS

1. Use a random-number table or random-number generator to simulate tossing a fair coin 10 times. Generate 20 such simulations of 10 coin tosses. Compare the simulations. Are there any strings of 10 heads? of 4 heads? Does it seem that in most of the simulations half the outcomes are heads? half are tails? In Chapter 5, we will study the probability of getting from 0 to 10 heads in such a simulation.

2. Use a random-number table or random-number generator to get a random sample of 30 distinct values from the set of integers from 1 to 100. Instructions for doing this using the TI-83, Excel, Minitab, or ComputerStat are in Using Technology at the end of this chapter. Generate five such samples. How many of the samples include the number 1? the number 100? Comment about the differences among the samples. How well do the samples seem to represent the numbers between 1 and 100?

LINKING CONCEPTS: WRITING PROJECTS

Discuss each of the following topics in class or review the topics on your own. Then write a brief but complete essay in which you summarize the main points. Please include formulas and graphs as appropriate.

1. What does it mean to say we are going to use a sample to draw an inference about the population? Why is a random sample so important for this process? If we wanted a random sample of students in the cafeteria, why couldn't we just take the students who order Diet Pepsi with their lunch? Comment on the statement, "A random sample is kind of a miniature population, whereas samples that are not random are likely to be biased." Why would the students who order Diet Pepsi with lunch not be a random sample of students in the cafeteria?

2. In your own words, explain the differences among the following sampling techniques: simple random sample, stratified sample, systematic sample, cluster sample, and convenience sample. Describe situations in which each type might be useful.

Using TECHNOLOGY

TI-83PLUS • EXCEL • MINITAB • COMPUTERSTAT

General spreadsheet programs such as Microsoft's Excel, specific statistical software packages such as Minitab, and graphing calculators such as the TI-83Plus all offer computing support for statistical methods. Applications in this section may be completed using software or calculators with statistical functions. Select keystroke or menu choices are shown for the TI-83Plus, Minitab, Excel, and ComputerStat in the Technology Hints portion of this section. More details can be found in the software-specific Technology Guides that accompany this text.

APPLICATIONS

Most software packages sample *with replacement*. That is, the same number may be used more than once in the sample. If your applications require sampling without replacement, draw more items than you need. Then use sort commands in the software to put the data in order, and delete repeated data.

1. Simulate the results of tossing a fair die 18 times. Repeat the simulation. Are the results the same? Do you expect them to be the same? Why or why not? Do there appear to be equal numbers of outcomes 1 through 6 in each simulation? In Chapter 4, we will see the law of large numbers, which tells us that we would expect equal numbers of outcomes only when the simulation is very large.

2. A college has 5,000 students, and the registrar wishes to use a random sample of 50 students to examine credit hour enrollment for this semester. Write a brief description of how a random sample can be drawn. Draw a random sample of 50 students. Are you sampling with or without replacement?

Technology Hints: Random Numbers

TI-83Plus

To select a random set of integers between two specified values, press the **MATH** key and highlight **PRB** with **5:randInt** (low value, high value, sample size). Press Enter and fill in the low value, high value, and sample size. To store the sample in list L1, press the **STO➡** key and then L1. The screen display shows two random samples of size 5 drawn from the integers between 1 and 100.

```
randInt(1,100,5)
{63 89 13 46 47}
randInt(1,100,5)
{29 82 99 50 41}
```

Excel

To select a random number between two specified values, type the command **=Randbetween**(low value, high value) in the formula bar. Alternatively, access a dialogue box for the command by clicking on the **paste function** key on the menu bar. Then choose **All** in the left drop-down menu and **RANDBETWEEN** in the right menu. Fill in the dialogue box.

A5	=	=RANDBETWEEN(1,100)			
	A	B	C	D	E
1	99				
2	58				
3	69				
4	86				
5	13				

Minitab

To generate random integers between specified values, use the menu selection **Calc ➤ Random Data ➤ Integer.** Fill in the dialogue box to get 5 random numbers between 1 and 100.

Worksheet 2 ***	C1
↓	
1	8
2	35
3	33
4	9
5	15

ComputerStat

In ComputerStat, select the program Random Samples found in the Descriptive Statistics menu. This program lets you select the range of integers from which you wish to draw random samples. It also gives you the option of sampling with or without replacement.

2

Organizing Data

In dwelling upon the vital importance of sound observation, it must never be lost sight of what observation is for. It is not for the sake of piling up miscellaneous information or curious facts, but for the sake of saving life and increasing health and comfort.

—Florence Nightingale, *Notes on Nursing*

**Florence Nightingale
(1820–1910)**

2.1 Bar Graphs, Circle Graphs, and Time Plots

2.2 Frequency Distributions and Histograms

2.3 Stem-and-Leaf Displays

Florence Nightingale has been described as a "passionate statistician" and a "relevant statistician." She viewed statistics as a science that allows one to transcend his or her narrow individual experience and aspire to the broader service of humanity. She was one of the first nurses to use graphic representation of statistics, illustrating with charts and diagrams how improved sanitation decreased the rate of mortality. Her statistical reports about the appalling sanitary conditions at Scutari (the main British hospital during the Crimean War) were taken very seriously by the English Secretary at War, Sidney Herbert. When sanitary reforms recommended by Nightingale were instituted in military hospitals, the mortality rate dropped from an incredible 42.7% to only 2.2%.

For on-line student resources, visit **math.college.hmco.com/students** and follow the Statistics links to the Brase/Brase, *Understandable Statistics*, 7th edition web site.

◇ How can we select graphs appropriate for given data sets? (SECTION 2.1)

◇ What are histograms and ogives? Where are they used? (SECTION 2.2)

◇ What are common distribution shapes? (SECTION 2.2)

◇ How can we quickly order data and, at the same time, reveal the distribution shape? (SECTION 2.3)

FOCUS PROBLEM

Say It with Pictures

Edward R. Tufte, in his book *The Visual Display of Quantitative Information*, has a number of guidelines for producing good graphics. According to the criteria, a graphical display should

- show the data;

- induce the viewer to think about the substance of the graphic rather than about the methodology, the design, the technology, or other production devices;

- avoid distorting what the data have to say.

As an example of a graph that violates some of the criteria, Tufte includes a graphic that appeared in a well-known newspaper. Figure 2-1(a) shows a facsimile of the problem graphic, whereas part (b) of the figure shows a better rendition of the data display. Look at both graphics, and identify some of the problems in part (a). Examine part (b). How is this graphic different from part (a), and what features better highlight the information displayed?

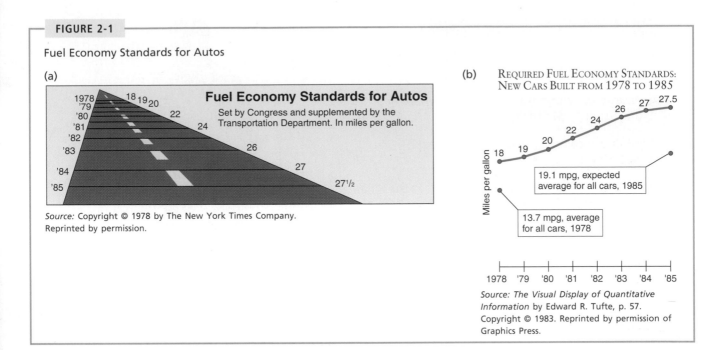

FIGURE 2-1

Fuel Economy Standards for Autos

(a)

Fuel Economy Standards for Autos

Set by Congress and supplemented by the Transportation Department. In miles per gallon.

1978 '79 18 19 20 22 24 26 27 27½
'80 '81 '82 '83 '84 '85

Source: Copyright © 1978 by The New York Times Company. Reprinted by permission.

(b)   REQUIRED FUEL ECONOMY STANDARDS: NEW CARS BUILT FROM 1978 TO 1985

18 19 20 22 24 26 27 27.5

Miles per gallon

19.1 mpg, expected average for all cars, 1985

13.7 mpg, average for all cars, 1978

1978 '79 '80 '81 '82 '83 '84 '85

Source: The Visual Display of Quantitative Information by Edward R. Tufte, p. 57. Copyright © 1983. Reprinted by permission of Graphics Press.

2.1
Bar Graphs, Circle Graphs, and Time Plots

FOCUS POINTS

✓ Determine types of graphs appropriate for specific data.

✓ Construct bar graphs, Pareto charts, circle graphs, and time plots.

✓ Interpret information displayed in graphs.

Open almost any magazine, newspaper, or large web site, and you will find graphs. In fact, because graphs display information so effectively, most word processors and spreadsheet programs include extensive chart- and graph-making tools. In this section, we identify several common types of graphs, show how to construct them, and show how to interpret information presented in graphical form.

Let's start with *bar graphs*. These are graphs that can be used to display quantitative or qualitative data.

Bar graphs

Features of a bar graph

1. Bars can be vertical or horizontal.

2. Bars are of uniform width and uniformly spaced.

3. The lengths of the bars represent values of the variable being displayed, the frequency of occurrence, or the percentage of occurrence. The same measurement scale is used for the length of each bar.

4. The graph is well annotated with title, labels for each bar, and vertical scale or actual value for the length of each bar.

EXAMPLE 1
Bar graph

What's a square foot worth? In retail, annual sales per square foot is an important industry measurement. A recent article in *The Wall Street Journal* gave average sales per square foot for nationwide department stores: $147 for Dillard's Inc., $279 for

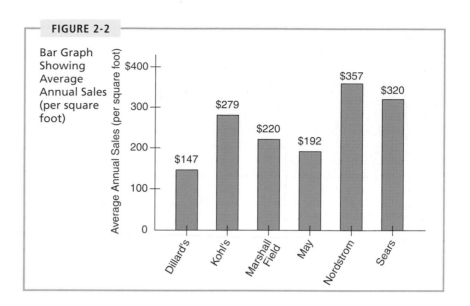

FIGURE 2-2

Bar Graph Showing Average Annual Sales (per square foot)

Changing scales

Kohl's Corp., $220 for Marshall Field, $192 for May Department Stores, $357 for Nordstrom, Inc., and $320 for Sears. To make a bar graph of this information, we let the length of the bars in Figure 2-2 represent the sales (in dollars) per square foot. Each store has its own bar.

Usually, it is not necessary to display both the vertical scale and the individual height of each bar. However, the height label gives the reader exact information while the scale display gives the viewer a sense of the range of data values. In addition, the graph is titled and the bars are labeled by store.

From the graph, it is readily apparent that Nordstrom and Sears have the highest sales per square foot. *The Wall Street Journal* reports that Nordstrom's upscale merchandise and Sears' appliance sales push their sales figures up. Among the other four stores that carry similar merchandise, we see that Kohl's has the highest average sales per square foot.   ◇

There are several variations of bar graphs. Figure 2-3 on the next page shows life expectancy for men and women born in specific years. Notice the bars for men and women are clustered over the year of birth, and a legend identifies which bars are for men and which are for women.

An important feature illustrated in Figure 2-3(b) is that of a *changing scale.* Notice that the scale between 0 and 65 is compressed. The changing scale amplifies the apparent difference between life spans for men and women, as well as the increase in life spans from those born in 1970 to the projected span of those born in 2010. Whenever you use a change in scale, warn the viewer by using a squiggle ∿ on the changed axis. Sometimes, if a single bar is unusually long, the bar length is compressed with a squiggle in the bar itself.

Quality control is an important aspect of today's production and service industries. Dr. W. Edwards Deming was one of the developers of total quality management (TQM). In his book *Out of Crisis* he outlines many strategies for monitoring and improving service and production industries. In particular, Dr. Deming recommends the use of some statistical methods to organize and analyze data from

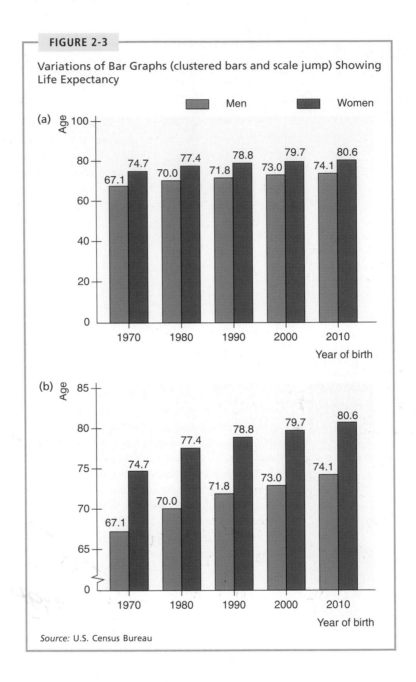

FIGURE 2-3

Variations of Bar Graphs (clustered bars and scale jump) Showing Life Expectancy

Source: U.S. Census Bureau

Pareto charts

industries so that sources of problems can be identified and then corrected. *Pareto* (pronounced "Pah-rāy-tō) *charts* are among the many techniques used in quality-control programs.

A **Pareto chart** is a bar graph in which the bar height represents frequency of an event. In addition, the bars are arranged from left to right according to decreasing height.

GUIDED EXERCISE 1

Pareto charts

This exercise is adapted from *The Deming Management Method* by Mary Walton. Suppose you want to arrive at the college 15 minutes before your first class so that you can feel relaxed when you walk into class. An early arrival time also allows room for unexpected delays. However, you always find yourself arriving "just in time" or slightly late. What causes you to be late? Charlotte made a list of possible causes and then kept a checklist for 2 months (Table 2-1) On some days more than one item was checked because several events occurred that caused her to be late.

TABLE 2-1   Causes for Lateness
(September–October)

Cause	Frequency
Snoozing after alarm goes off	15
Car trouble	5
Too long over breakfast	13
Last-minute studying	20
Finding something to wear	8
Talking too long with roommate	9
Other	3

(a) Make a Pareto chart showing the causes for lateness. Be sure to label the causes, and draw the bars using the same vertical scale.

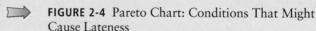

FIGURE 2-4 Pareto Chart: Conditions That Might Cause Lateness

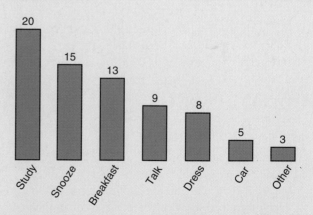

(b) Looking at the Pareto chart, what recommendations do you have for Charlotte?

According to the chart, rearranging study time, or getting up earlier to allow for studying, would cure her most frequent cause for lateness. Repairing the car might be important, but for getting to campus early, it would not be as effective as adjusting study time.

Circle graphs or pie charts

Another popular pictorial representation of data is the *circle graph*, or *pie chart*. It is relatively safe from misinterpretation and is especially useful for showing the division of a total quantity into its component parts. The total quantity, or 100%, is represented by the entire circle. Each wedge of the circle represents a component part of the total. These proportional segments are usually labeled with corresponding percentages of the total. Guided Exercise 2 shows how to make a circle graph.

GUIDED EXERCISE 2

Circle graph

How long do we spend talking on the telephone after hours (at home after 5 P.M.)? The results from a recent survey of 500 people (as reported in *USA Today*) are shown in Table 2-2. We'll make a circle graph to display these data.

TABLE 2-2   Time Spent on Home Telephone After 5 P.M.

Time	Number	Fractional Part	Percentage	Number of Degrees
Less than ½ hour	296	296/500	59.2	59.2% × 360° ≈ 213°
½ hour to 1 hour	83	83/500	16.6	16.6% × 360° ≈ 60°
More than 1 hour	121	_____	_____	_____
Total	_____		_____	_____

(a) Fill in the missing parts in Table 2-2 for "More than 1 hour." Remember that the central angle of a circle is 360°. Round to the nearest degree.

⇨ For "More than 1 hour," Fractional Part = 121/500; Percentage = 24.2%; Number of Degrees = 24.2% × 360° ≈ 87°. The symbol ≈ means approximately equal.

(b) Fill in the totals. What is the total number of responses? Do the percentages total 100% (within rounding error)? Do the numbers of degrees total 360° (within rounding error)?

⇨ The total number of responses is 500. The percentages total 100%. You must have such a total in order to create a circle graph. The numbers of degrees total 360°.

(c) Draw a circle graph. Divide the circle into pieces with the designated numbers of degrees. Label each piece, and show the percentage corresponding to each piece. The numbers of degrees are usually omitted from pie charts shown in newspapers, magazines, journals, and reports.

⇨ **FIGURE 2-5** Hours on Home Telephone After 5 P.M.

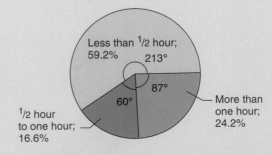

TABLE 2-3   Distance in Miles Walked/Jogged in 30 Minutes

Week	1	2	3	4	5	6	7	8	9	10
Distance	1.5	1.4	1.7	1.6	1.9	2.0	1.8	2.0	1.9	2.0
Week	11	12	13	14	15	16	17	18	19	20
Distance	2.1	2.1	2.3	2.3	2.2	2.4	2.5	2.6	2.4	2.7

Suppose you begin an exercise program that involves walking or jogging for 30 minutes. You exercise several times a week but monitor yourself by logging the distance you cover in 30 minutes each Saturday. How do you display these data in a meaningful way? Making a bar chart showing the frequency of distances you cover might be interesting, but it does not really show how the distance you cover in 30 minutes has changed over time. A graph showing the distance covered on each date will let you track your performance over time.

Time plot

We will use a *time plot*. A time plot is a graph showing data measurements in chronological order. To make a time plot, we put time on the horizontal scale and the variable being measured on the vertical scale. In a basic time plot, we connect the data points by lines.

EXAMPLE 2
Time plot

Suppose you have been in the walking/jogging exercise program for 20 weeks, and for each week you have recorded the distance you covered in 30 minutes. Your data log is shown in Table 2-3.

(a) Make a time plot.

SOLUTION: The data are appropriate for a time plot because they represent the same measurement (distance covered in a 30-minute period) taken at different times. The measurements are also recorded at equal time intervals (every week). To make our time plot, we list the weeks in order on the horizontal scale. Above each week, plot the distance covered that week on the vertical scale. Then connect the dots. Figure 2-6 shows the time plot. Be sure the scales are labeled.

FIGURE 2-6

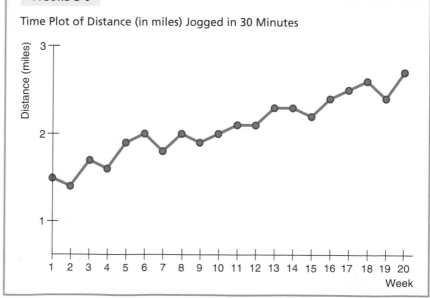

Time Plot of Distance (in miles) Jogged in 30 Minutes

(b) From looking at Figure 2-6, can you detect any patterns?

> **SOLUTION:** There seems to be an upward trend in distance covered. The distances covered in the last few weeks are about a mile farther than those for the first few weeks. However, we cannot conclude that this trend will continue. Perhaps you have reached your goal for this training activity and now wish to maintain a distance of about 2.5 miles in 30 minutes.   ◇

Time series

Data sets composed of similar measurements taken at regular intervals over time are called *time series*. Time series are often used in economics, finance, sociology, medicine, and any situation where we want to study or monitor a similar measure over a period of time. A time plot can reveal some of the main features of time series.

We've seen several styles of graphs. Which kinds are suitable for a specific data collection?

Bar graphs are useful for quantitative or qualitative data. With qualitative data, the frequency or percentage of occurrence can be displayed. With quantitative data, the measurement itself can be displayed as was done in the bar graph showing life expectancy. Watch that the measurement scale is consistent or that a jump scale squiggle is used.

Pareto charts identify the frequency of events or categories in decreasing order of frequency of occurrence.

Circle graphs display how a *total* is dispersed into several categories. The circle graph is very appropriate for qualitative data, or any data where percentage of occurrence makes sense. Circle graphs are most effective when the number of segments are ten or fewer.

Time plots display how data change over time. It is best if the units of time are consistent in a given plot. For instance, measurements taken every day should not be mixed on the same plot with data taken every week.

For any graph: Provide a title, label axes, identify units of measure. As Edward Tufte suggests in his book *The Visual Display of Quantitative Information*, don't let artwork or skewed perspective cloud the clarity of the information displayed.

TECH NOTES   *Bar graphs, circle graphs, time plot*

TI-83Plus Only time plot. Place consecutive values 1 through the number of time segments in list L1 and data in L2. Press **Stat Plot** and highlight an *xy* line plot.

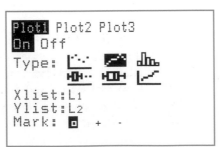

Excel Use the chart wizard on the toolbar [icon]. Select the desired option and follow the instructions in the dialogue boxes.

Minitab Use the menu selection **Graph.** Select the desired option and follow the instructions in the dialogue boxes.

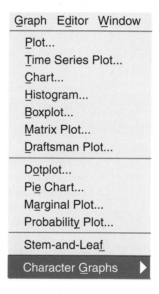

VIEWPOINT

Do Ethical Standards Vary by the Situation?

Lutheran Brotherhood did a national survey and found that nearly 60% of all U.S. adults claim that ethics vary by the situation; 33% claim that there is only one ethical standard; and 7% were not sure. How could you draw a circle graph to make a visual impression of Americans' views on ethical standards?

SECTION 2.1   PROBLEMS

1. *Education: Does College Pay Off?* It is costly in both time and money to go to college. Does it pay off? According to the Bureau of Census, the answer is yes. The average annual income (in thousands of dollars) of a *household* headed by a person with the stated education level is as follows: 16.1 if 9th grade is the highest level achieved, 34.3 for high school graduates, 48.6 for those holding associate degrees, 62.1 for those with bachelor's degrees, 71.0 for those with master's degrees, and 84.1 for those with doctoral degrees. Make a bar graph showing household income for each education level.

2. *Accidents: Child Deaths* How safe is the world for kids? Unfortunately, some children between the ages of 1 and 14 die of injuries every year. United Nations data show that by nation, the annual number of deaths from injuries per 100,000 children are as follows: Australia, 9.5; Canada, 9.7; Denmark, 8.1; France, 9.1; Germany, 8.3; Hungary, 10.8; Ireland, 8.3; Italy, 6.1; Japan, 8.4; Korea, 25.6; Mexico, 19.8; New Zealand, 13.7; Netherlands, 6.6; Poland, 13.4; Portugal, 17.8; Spain, 8.1; Sweden, 5.2; Switzerland, 9.6; U.K., 6.1; United States, 14.1. Display these data in a Pareto chart.

3. *Airlines: Fear of Flying* In an article entitled "How to Cure the Fear of Flying," *Fortune* magazine gave the following information: The number of people who died in a calendar year while on a commercial flight, 33; in the bathtub, 318; by poisoning, 5900; as pedestrians, 6500; murdered by spouses, 7000; from falls, 12,400; in motor vehicle accidents, 33,800. Make a Pareto chart showing activities causing death.

4. *Business: Company Failures* Suppose you are interested in investing in a business corporation. What are some pitfalls that might cause a business corporation to fail? To answer this question, we can look at history. Buccino & Associates surveyed more than 1300 business professionals and asked them about leading causes of business failures. According to the report in *USA Today*, 88% of those surveyed cited internal problems, not external factors, as the leading cause of business failure. Of the internal problems listed, 13% of those interviewed cited insufficient management experience as the leading cause of business failure, 13% cited poor business planning, 18% cited inadequate leadership, and 29% cited excessive debt.
   (a) Make a Pareto chart for the causes of business failure.
   (b) Does your chart show *all* the causes for business failure cited by business professionals interviewed? Which internal cause was cited most frequently?

5. *Lifestyle: Hide the Mess!* A survey of 1000 adults (reported in *USA Today*) uncovered some interesting housekeeping secrets. When unexpected company comes, where do we hide the mess? The survey showed that 68% of the respondents toss their mess in the closet, 23% shove things under the bed, 6% put things in the bathtub, and 3% put the mess in the freezer. Make a circle graph to display this information.

6. *Lifestyle: Fast Food* What meal are we most likely to eat in a fast-food restaurant? A survey of 1261 adults (reported in *USA Today*) revealed that 48.9% of the respondents are most likely to eat lunch at a fast-food restaurant; 7.7%, breakfast; 31.6%, dinner; 10%, a snack; and 1.8% answered "don't know." Display this information in a circle graph.

7. *Education: College Professors' Time* How do college professors spend their time? *The National Education Association Almanac of Higher Education* gives the following average distribution of professional time allocation: teaching, 51%; research,

16%; professional growth, 5%; community service, 11%; service to the college, 11%; and consulting outside the college, 6%. Make a pie chart showing the allocation of professional time for college professors.

8. *Education: College Professors' Age* How old are college professors? The NEA reference in Problem 7 gave the following age distribution in years: under 35 years, 8%; 35–44 years, 29%; 45–54 years, 37%; 55–59 years, 13%; 60–64 years, 9%; and 65 years and over, 4%. Make a pie chart showing the age distribution of college professors.

9. *Merchandise: Consumer Electronics* Telephone gadgets fill our home. Consumer Electronics Manufacturers Association reported that 94.1% of U.S. households have corded phones, 78% have cordless phones, 74% have a telephone answering device, 51% have a cellular phone, 28% have pagers, and 51% have a modem or fax machine. Display this information in a bar chart showing percentage of households owning each gadget. Could the information as reported be put into a circle graph? Explain.

10. *Driving: Bad Habits* Driving would be more pleasant if we didn't have to put up with bad habits of other drivers. *USA Today* reported the results of a Valvoline Oil Company survey of 500 drivers in which the drivers marked their complaints about other drivers. The top complaints turned out to be tailgating, marked by 22% of the respondents; not using turn signals, marked by 19%; 16% marked being cut off; 11% complained about other drivers driving too slowly; and 8% complained about other drivers being inconsiderate. Make a Pareto chart showing percentage of drivers listing each stated complaint. Could this information as reported be put in a circle graph? Why or why not?

11. *Ecology: Lakes* Pyramid Lake, Nevada, is described as the pride of the Paiute Indian Nation. It is a beautiful desert lake famous for very large trout. The elevation of the lake surface (feet above sea level) varies according to the annual flow of the Truckee River from Lake Tahoe. The U.S. Geological Survey provided the following data:

Year	Elevation	Year	Elevation	Year	Elevation
1986	3817	1992	3798	1998	3811
1987	3815	1993	3797	1999	3816
1988	3810	1994	3795	2000	3817
1989	3812	1995	3797		
1990	3808	1996	3802		
1991	3803	1997	3807		

Make a time plot displaying the data. For more information, visit the Brase/Brase statistics site at http://math.college.hmco.com/students and find the link to the Pyramid Lake Fisheries.

12. *Vital Statistics: Height* How does average height for boys change as the boy gets older? According to *Physician's Handbook*, the heights at different ages are as follows:

Age (years)	0.5	1	2	3	4	5	6	7
Height (inches)	26	29	33	36	39	42	45	47

Age (years)	8	9	10	11	12	13	14
Height (inches)	50	52	54	56	58	60	62

Make a time plot for average height for age 0.5 through 14 years.

13. *Financial: Two Common Stocks* Thirsty or hungry? Many people think of having a Coke or going to McDonald's when thirst or hunger strikes. However, other people consider the value of the company stock. Which company, Coca-Cola or McDonald's, would be a better investment? To determine the answer, some investors look at historical performance. You can find this kind of information on the web by visiting the Brase/Brase statistics site at http://math.college.hmco.com/students and finding the link to Money Net. Figure 2-7 shows the weekly close of each company for a 1-year period. The top of the vertical line represents the highest price of the week, whereas the bottom represents the lowest price of the week. In

FIGURE 2-7

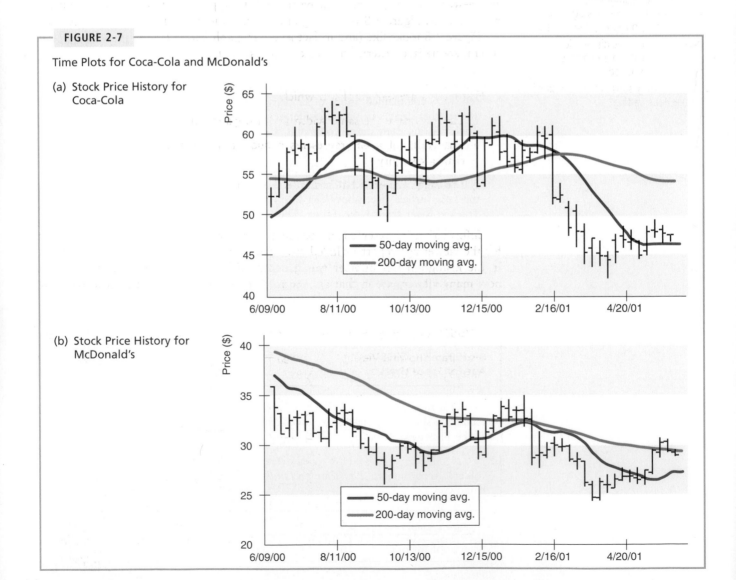

Time Plots for Coca-Cola and McDonald's

(a) Stock Price History for Coca-Cola

(b) Stock Price History for McDonald's

addition, two moving averages are shown. Look at the two parts of the figure, and comment about the general price trend over the 1-year period shown. Note that during that period, the Dow Jones Industrial Average increased by about 2.3%. However, both of these stocks declined. Comment on the volatility and performance difference of the stocks. Which declined by the smallest percentage during the period shown?

2.2
Frequency Distributions and Histograms

Histograms and Relative-Frequency Histograms

FOCUS POINTS

✓ Organize raw data using a frequency table.

✓ Construct histograms, relative-frequency histograms, frequency polygons, and ogives.

✓ Recognize basic distribution shapes: uniform, symmetric, bimodal, skewed.

✓ Interpret graphs in the context of the data setting.

Healthy Crunch cereal is about to take over sponsorship of the TV program *Space Voyage*. The advertising manager has requested a report on the age distribution of the viewers so the spot ads can be tailored to appeal to the age groups with the most viewers. Figure 2-8 is a histogram showing the frequency distribution of ages.

Figure 2-8 looks like (and in fact is) a bar graph. However, the graph also meets several specialized criteria that make it a histogram.

Histograms are bar graphs in which

(a) The bars have the same width and always touch.

(b) The width of a bar represents a quantitative value, such as age, rather than a category.

(c) The height of each bar indicates frequency.

Information is presented in condensed form in a histogram. The original data for the viewer age report included the *exact* number of 21-year-olds in the sample. In the histogram this number was grouped with the ages 14.5–24.5. We can tell how many viewers are in that age group, but we cannot tell exactly how many are

FIGURE 2-8

Histogram Showing Viewer Age for *Space Voyage*

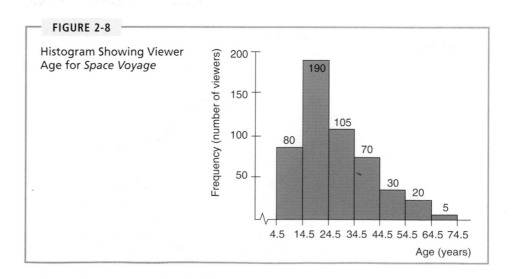

TABLE 2-4   One-Way Commuting Distances in Miles for 60 Workers in Downtown Dallas

13	47	10	3	16	20	17	40	4	2
7	25	8	21	19	15	3	17	14	6
12	45	1	8	4	16	11	18	23	12
6	2	14	13	7	15	46	12	9	18
34	13	41	28	36	17	24	27	29	9
14	26	10	24	37	31	8	16	12	16

Procedure to make histogram for integer data

21 years old. However, the condensed information in the histogram can be assimilated more quickly than the same information in more detailed form.

If you are given many pieces of data, how do you condense the information to make a histogram? The task force to encourage car pools did a study of one-way commuting distances for workers in the downtown Dallas area. A random sample of 60 of these workers was taken. The commuting distances of the workers in the sample are given in Table 2-4.

The first thing to do is to decide *how many bars or classes* you want in the histogram. Five to 15 classes are usually used. If you use fewer than 5 classes, you risk losing too much information, but if you use more than 15 classes, the clarity of the diagram might be sacrificed for detail. Let the spread of the data and the purpose of the histogram be your guide when selecting the number of classes.

Next, find a convenient *class width*. To do this, find the difference between the largest and smallest data values and divide by the number of classes. If you want the class width to be a whole number, always *increase* the result to the next whole number so that the classes cover the data.

Class width

1. Compute $\dfrac{\text{Largest data value} - \text{Smallest data value}}{\text{Desired number of classes}}$

2. Increase the value computed to the next highest whole number.

Note: To ensure that the classes cover the data, we need to increase the result of step 1 to the *next* whole number, even if step 1 produced a whole number. For instance, if the calculation in step 1 produces the value 5, we make the class width 6.

In this case, let's use 6 classes. The largest distance commuted is 47 miles and the smallest is 1 mile.

$$\text{Class width} = \frac{47 - 1}{6} \approx 7.7 \quad \text{increase to} \quad 8$$

The lowest and highest values that can fit into a class are called the *lower class limit* and *upper class limit*, respectively. The *class width* is the difference between the lower class limit of one class and the lower class limit of the next class. Each class should have the same width, although it is not uncommon to see either the first or the last class width a little longer or shorter than the others. The center of

TABLE 2-5   Frequency Table of One-Way Commuting Distances for 60 Downtown Dallas Workers (data in miles)

Class Limits Lower–Upper	Class Boundaries Lower–Upper	Tally*	Frequency	Class Midpoint
1–8	0.5–8.5	ⵀⵀ ⵀⵀ IIII	14	4.5
9–16	8.5–16.5	ⵀⵀ ⵀⵀ ⵀⵀ ⵀⵀ I	21	12.5
17–24	16.5–24.5	ⵀⵀ ⵀⵀ I	11	20.5
25–32	24.5–32.5	ⵀⵀ I	6	28.5
33–40	32.5–40.5	IIII	4	36.5
41–48	40.5–48.5	IIII	4	44.5

*The tally column is included simply as an aid for determining the frequencies. It is not a necessary part of a frequency table.

the class is called the *midpoint* (or *class mark*). This is found by adding the lower and upper class limits of one class and dividing by 2.

$$\text{Midpoint} = \frac{\text{Lower class limit } + \text{ Upper class limit}}{2}$$

The midpoint is often used as a representative value of the entire class.

Now we can organize the commuting distance data into a *frequency table* (Table 2-5). Such a table shows the limits of each class, the frequency with which the data fall into a class, and the class midpoint. A tally will help us find the frequencies.

We're almost ready to make a histogram. But in a histogram we want the bars to touch. There is a space between the upper limit of one class and the lower limit of the next class. The halfway points of these intervals are called *class boundaries*. We use the class boundaries as the endpoints of the bars in the histogram. Then there is no space between the bars. Figure 2-9 is the histogram of commuter

FIGURE 2-9

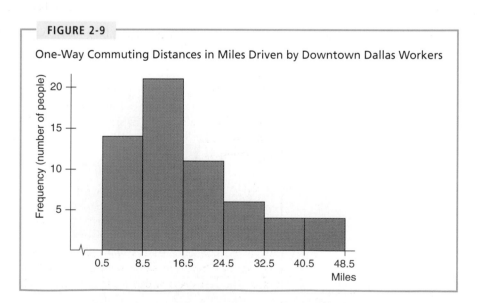

One-Way Commuting Distances in Miles Driven by Downtown Dallas Workers

TABLE 2-6   Relative Frequencies of One-Way Commuting Distances

Class	Frequency f	Relative Frequency f/n
1–8	14	14/60 ≈ 0.23
9–16	21	21/60 ≈ 0.35
17–24	11	11/60 ≈ 0.18
25–32	6	6/60 ≈ 0.10
33–40	4	4/60 ≈ 0.07
41–48	4	4/60 ≈ 0.07

distances. The class boundaries are shown. (Sometimes only the class midpoints are labeled.)

Relative-frequency tables and histograms

Other useful tools for organizing data are *relative-frequency tables* and *relative-frequency histograms*. Once we have made a frequency table, it is easy to construct a relative-frequency table. The relative frequency for a particular class is found by dividing the class frequency by the total of all frequencies (sample size).

$$\text{Relative frequency} = \frac{f}{n} = \frac{\text{Class frequency}}{\text{Total of all frequencies}}$$

Table 2-6 shows the relative frequencies for the commuter data of Table 2-4. Since we already have the frequency table (Table 2-5), the relative-frequency table is obtained easily. The sample size is $n = 60$. Notice that the sample size is the total

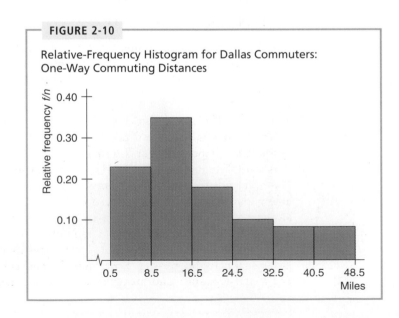

FIGURE 2-10

Relative-Frequency Histogram for Dallas Commuters: One-Way Commuting Distances of all the frequencies. Therefore, the relative frequency for the first class (the class from 1 to 8) is

$$\text{Relative frequency} = \frac{f}{n} = \frac{14}{60} \approx 0.23$$

The symbol $\approx$ means "approximately equal to." We use the symbol because we rounded the relative frequency. Relative frequencies for the other classes are computed in a similar way.

The total of the relative frequencies should be 1. However, rounded results may make the total slightly higher or lower than 1.

Using Table 2-6 and Figure 2-9, we can quickly make a relative-frequency histogram (Figure 2-10). The horizontal scale will be the same, but the vertical scale will be marked with *relative frequencies f/n* instead of the actual frequencies *f*. The basic shape of the two graphs otherwise will be the same.

In Guided Exercise 3 we will ask you to make a frequency table, histogram, relative-frequency table, and relative-frequency histogram. There are a number of steps to follow. The individual steps are not difficult, but you need to keep them all in mind. The steps are listed for your convenience.

To make a frequency table

1. Determine the class width.

2. Create the distinct classes. We use the convention that the *lower class limit* of the first class is the smallest data value. Add the class width to *this* number to get the *lower class limit* of the next class.

3. Tally the data into classes. Each data value should fall into exactly one class. Total the tallies to obtain each *class frequency*.

4. Compute the *midpoint* (class mark) for each class.

5. Determine the *class boundaries*.

To make a relative-frequency table

6. For each class, compute the *relative frequency f/n,* where *f* is the class frequency and *n* is the total sample size.

◊ **COMMENT** The use of class boundaries in histograms assures us that the bars of the histogram touch and that no data fall on the boundaries. Both these features are important. But a histogram displaying class boundaries may look awkward. For instance, the age range of 14.5 to 24.5 years shown in Figure 2-8 isn't as natural a choice as an age range of 15 to 25 years. For this reason, many magazines and newspapers do not use class boundaries as labels on a histogram. Instead, some use lower class limits as labels, with the convention that *a data value falling on the class limit is included in the next higher class (class to the right of the limit).* Another convention is to label midpoints instead of class boundaries. Determine the convention being used before creating frequency tables and histograms on a computer.   ◊

GUIDED EXERCISE 3

Histogram and relative-frequency histogram

One irate customer called Dollar Day Mail Order Company 40 times during the last two weeks to see why his order had not arrived. Each time he called, he recorded the length of time he was put "on hold" before being allowed to talk to a customer representative.

TABLE 2-7 Length of Time on Hold, in Minutes

1	5	5	6	7	4	8	7	6	5
5	6	7	6	6	5	8	9	9	10
7	8	11	2	4	6	5	12	13	6
3	7	8	8	9	9	10	9	8	9

(a) What are the largest and smallest values in Table 2-7? If we want five classes, what should the class width be?

The largest value is 13; the smallest value is 1. The class width is

$$\frac{13 - 1}{5} = 2.4 \approx 3$$

Note that we *increase* the value to 3.

(b) Complete the following frequency table.

TABLE 2-8 Time on Hold

Class Limits				
Lower–Upper		Tally	Frequency	Midpoint
1	– 3	_____	_____	_____
4	– ___	_____	_____	_____
___	– 9	_____	_____	_____
___	– ___	_____	_____	_____
___	– ___	_____	_____	_____

TABLE 2-9 Completion of Table 2-8

Class Limits			
Lower–Upper	Tally	Frequency	Midpoint
1–3	＼ 卅 ＼	3	2
4–6	卅卅卅卅	15	5
7–9	IIII IIII IIII II	17	8
10–12	IIII	4	11
13–15	I	1	14

(c) Recall that the class boundary is halfway between the upper limit of one class and the lower limit of the next. Use this fact to find the class boundaries and to complete the partial histogram in Figure 2-11.

TABLE 2-10 Class Boundaries

Class Limits	Class Boundaries
1–3	0.5–3.5
4–6	3.5–6.5
7–9	6.5–____
10–12	____–____
13–15	____–____

TABLE 2-11 Completion of Table 2-10

Class Limits	Class Boundaries
1–3	0.5–3.5
4–6	3.5–6.5
7–9	6.5–9.5
10–12	9.5–12.5

Continued

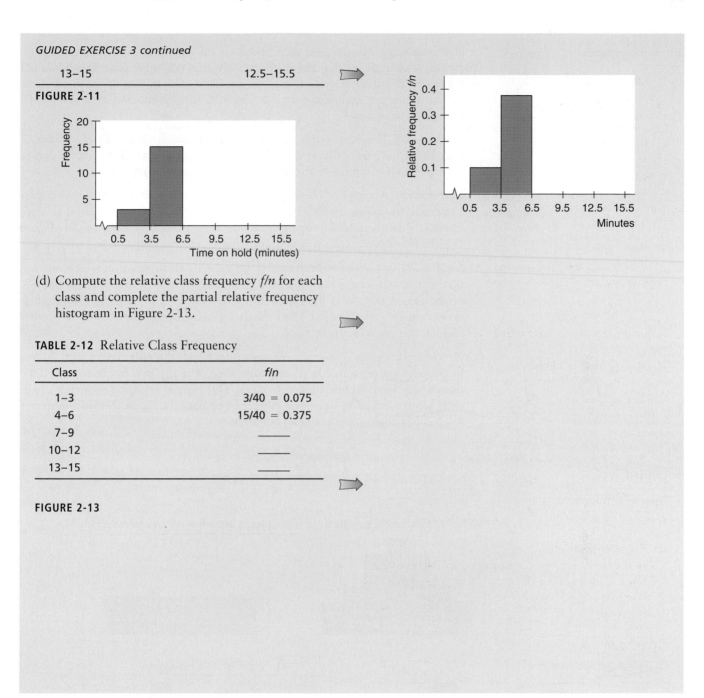

GUIDED EXERCISE 3 continued

13–15	12.5–15.5

FIGURE 2-11

(d) Compute the relative class frequency *f/n* for each class and complete the partial relative frequency histogram in Figure 2-13.

TABLE 2-12 Relative Class Frequency

Class	f/n
1–3	3/40 = 0.075
4–6	15/40 = 0.375
7–9	_____
10–12	_____
13–15	_____

FIGURE 2-13

We will see relative-frequency distributions again when we study probability in Chapter 4. There we will see that if a random sample is large enough, then we can estimate the probability of an event by the relative frequency of the event. The relative-frequency distribution then can be interpreted as a *probability distribution*. Such distributions will form the basis of our work in inferential statistics.

Distribution shapes

Histograms are valuable and useful tools. If the raw data came from a random sample of population values, the histogram constructed from your sample values should have a distribution shape that is reasonably similar to that of the population.

Several terms are commonly used to describe histograms and their associated population distribution.

(a) *Symmetrical:* This term refers to a histogram in which both sides are (more or less) the same when the graph is folded vertically down the middle. Figure 2-15(a) shows a typical histogram with a symmetrical shape.

(b) *Uniform* or *rectangular:* These terms refer to a histogram in which every class has equal frequency. From one point of view, a uniform distribution is symmetrical with the added property that the bars are of the same height. Figure 2-15(b) illustrates a typical histogram with a uniform shape.

(c) *Skewed left* or *skewed right:* These terms refer to a histogram in which one tail is stretched out longer than the other. The direction of skewness is on the side of the *longer* tail. So if the longer tail is on the left, we say the histogram is skewed to the left. Figure 2-15(c) shows a typical histogram skewed to the left and another skewed to the right.

(d) *Bimodal:* This term refers to a histogram in which the two classes with largest frequencies are separated by at least one class. The top two frequencies of these classes may have slightly different values. This type of situation sometimes indicates we are sampling from two different populations. Figure 2-15(d) illustrates a typical histogram with a bimodal shape.

FIGURE 2-15

Types of Histograms

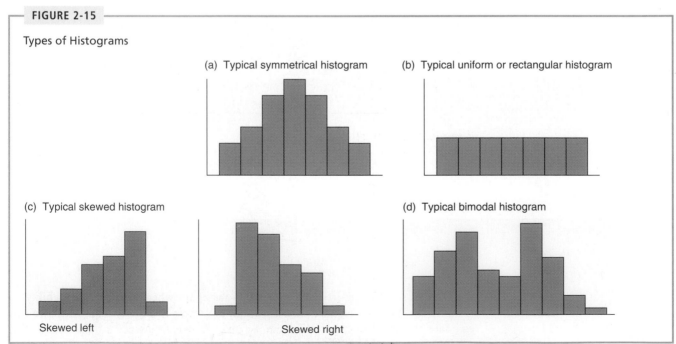

(a) Typical symmetrical histogram

(b) Typical uniform or rectangular histogram

(c) Typical skewed histogram

Skewed left

Skewed right

(d) Typical bimodal histogram

TECH NOTE The TI-83Plus, Excel, and Minitab all create histograms. However, each technology automatically selects the number of classes to use. In Using Technology at the end of this chapter, you will see instructions for specifying the number of classes yourself and for generating histograms such as we create "by hand."

The displays show the default histograms for the commuting data of Table 2-4. For each technology, enter the data and then follow the designated menu/key choices.

TI-83Plus **Stat Plot ➤ On;Histogram Type ➤ Zoom ➤ 9:ZoomStat.** Use **Trace** to see the number in a class.

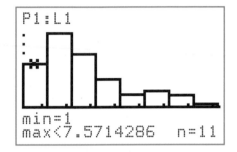

Excel **Tools ➤ Data Analysis ➤ Histogram** Select data range for input and check Chart. In Excel, classes are called bins.

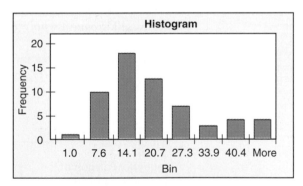

Minitab **Graph ➤ Histogram**

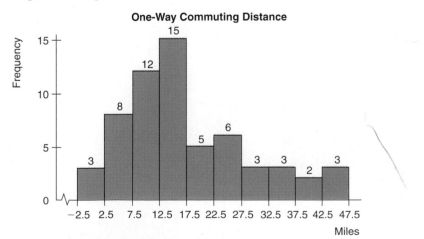

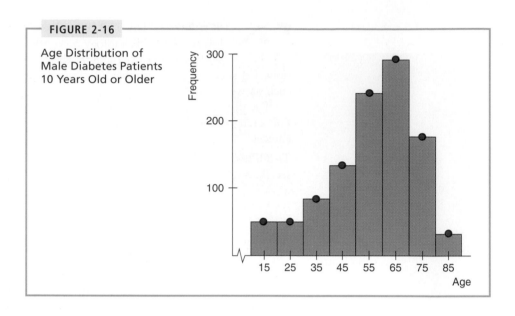

FIGURE 2-16

Age Distribution of
Male Diabetes Patients
10 Years Old or Older

Related Graphs: Frequency Polygons and Ogives

A histogram gives the impression that frequencies jump suddenly from one class to the next. If you want to emphasize the *continuous* rise or fall of the frequencies, you can use a *frequency polygon* or *line graph*.

Frequency polygons

EXAMPLE 3

Frequency polygon

Figure 2-16 shows a histogram of the age distribution of male diabetes patients 10 years and older. Figure 2-17 shows the corresponding frequency polygon.

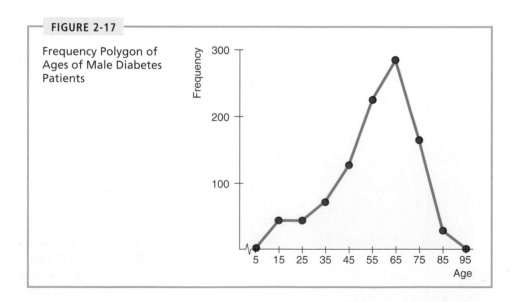

FIGURE 2-17

Frequency Polygon of
Ages of Male Diabetes
Patients

(a) Notice the heavy dots in Figure 2-17 occur at points with coordinates (class midpoint, class frequency). These dots are connected.

(b) Consecutive class midpoints differ by the class width. From the histogram, we see that the class width is 10 years. Subtracting 10 from the first class midpoint gives the starting point at 5. Likewise, adding 10 to the last class midpoint gives the ending point at 95.   ◇

To make a frequency polygon

(a) Make a frequency table that shows class midpoints and class frequencies.

(b) For each class, make a dot over the class midpoint at the height of the class frequency. The coordinates of these dots are (class midpoint, class frequency). Connect these dots with line segments.

(c) By convention, frequency polygons begin and end with a frequency of zero. On the left end, draw a line segment that meets the horizontal axis *one class width* to the left of the first midpoint. On the right end, draw a line segment meeting the horizontal axis *one class width* to the right of the last midpoint.

Sometimes we want to study *cumulative totals* instead of frequencies or relative frequencies. Cumulative totals can be used to determine how many scores are *above* or *below* a set level.

Cumulative-frequency table

Once we have made a frequency table, it is not hard to make a cumulative-frequency table. For example, let's consider the climatologic data published by the U.S. government that gives the daily high temperatures (°F) during the ski season in Aspen, Colorado. The *cumulative frequency* for a class is the sum of the frequencies for *that class and all previous classes*. In Table 2-14 we see that the cumulative frequency of the second class is 66. This is the sum of the frequencies 23 and 43 from the first and second classes. Likewise, the cumulative frequency for class three is the sum of the frequencies in the first three classes.

Ogives

An *ogive* (pronounced "oh-jive") is a graph that displays cumulative frequencies. It is especially useful for quickly determining the number of data values above or below a specified level.

TABLE 2-14   High Temperatures During the Aspen Ski Season (°F)

Class Boundaries		Frequency	Cumulative Frequency
Lower	Upper		
10.5	20.5	23	23
20.5	30.5	43	66 (sum 23 + 43)
30.5	40.5	51	117 (sum 66 + 51)
40.5	50.5	27	144 (sum 117 + 27)
50.5	60.5	7	151 (sum 144 + 7)

To make an ogive

(a) Make a frequency table showing class boundaries and cumulative frequencies.

(b) For each class, make a dot over the *upper class boundary* at the height of the cumulative class frequency. The coordinates of the dots are (upper class boundary, cumulative class frequency). Connect these dots with line segments.

(c) By convention, an ogive begins on the horizontal axis at the lower class boundary of the first class.

Table 2-14 shows the cumulative frequencies for high temperatures during the Aspen ski season. The ogive based on this frequency table is shown in Figure 2-18 of Guided Exercise 4. Notice how the points shown in the ogive reflect the information in Table 2-14.

GUIDED EXERCISE 4

Ogive

Ogives are helpful when we want to know how many of our scores are *above* or *below* some level. For instance, the Aspen–Ashcroft area is a world famous center for cross-country skiing on wilderness trails. If the daily high temperature is above 40°F, the surface snow tends to melt. Then at night it freezes again. This can result in a snow crust that makes cross-country skiing faster and more challenging. It also can increase avalanche danger. Let's use the ogive in Figure 2-18 to estimate the percentage of a typical ski season during which the daily high temperatures go above 40°F.

(a) Use Figure 2-18 and find 40.5°F on the horizontal scale. What is the cumulative frequency of number of days with high temperatures less than or equal to 40°F?

The upper class boundary for temperatures below or at 40°F falls at 40.5°F (Figure 2-19). The cumulative frequency corresponding to 40.5 is 117 (see Figure 2-18).

FIGURE 2-18 Ogive for Daily High Temperatures (°F) During Aspen Ski Season

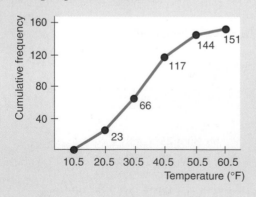

FIGURE 2-19 Segment of Figure 2-18

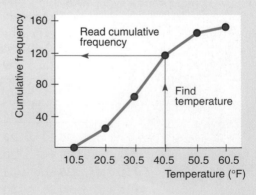

Continued

GUIDED EXERCISE 4 continued

(b) What does the cumulative frequency of 117 at the 40.5°F mark tell us?

⇒ The cumulative frequency 117 tells us that there are 117 days in the season during which the high temperature is 40°F or lower.

(c) How many days are represented in the data set?

⇒ The highest cumulative frequency is 151 (see dot over 60.5).

(d) Use the information in parts (b) and (c) to determine how many days the daily high temperature equals or exceeds 40°F. What percentage of the season has temperatures exceeding 40°F?

⇒ Because 151 days are represented in the sample and 117 have daily high temperatures at or below 40°F, there are $151 - 117 = 34$ days during which the daily high exceeds 40°F. This means that about 23% of the days have temperatures above 40°F.

◇ **COMMENT** Frequency polygons and ogives show different aspects of the data distribution. Among the technical differences, recall that

(a) for a frequency polygon, the points are graphed above the class midpoints, whereas the vertical coordinates are the class frequency.

(b) for an ogive, the points are graphed above the upper class boundaries, whereas the vertical coordinates are the cumulative frequency. ◇

VIEWPOINT   *Mush, You Huskies!*

In 1925, the village of Nome, Alaska, had a terrible diphtheria epidemic. Serum was available in Anchorage but had to be brought to Nome by dogsled over the 1161-mile Iditarod Trail. Since 1973, the Iditarod Dog Sled Race from Anchorage to Nome has been an annual sporting event with a current winner's purse of more than $350,000. Winning times range from more than 20 days to a little over 9 days.

To collect data on winning times, visit the Brase/Brase statistics site at http://math.college.hmco.com/students and find the link to the Iditarod. Make a frequency distribution for these times.

SECTION 2.2   PROBLEMS

For Problems 1–6, use the specified number of classes to do the following:
(a) Find the class width.
(b) Make a frequency table showing class limits, class boundaries, midpoints, frequencies, relative frequencies, cumulative frequencies.

(c) Draw a histogram.
(d) Draw a frequency polygon.
(e) Draw a relative-frequency histogram.
(f) Draw an ogive.

1. *Sports: Dog Sled Racing* How long does it take to finish the 1161-mile Iditarod Dog Sled Race from Anchorage to Nome, Alaska (*see* Viewpoint)? Finish times (to the nearest hour) for 57 dog-sled teams are shown below.

261	271	236	244	279	296	284	299	288	288	247	256
338	360	341	333	261	266	287	296	313	311	307	307
299	303	277	283	304	305	288	290	288	289	297	299
332	330	309	328	307	328	285	291	295	298	306	315
310	318	318	320	333	321	323	324	327			

Use five classes.

2. *Sports: Skiing* Recreational skiing is big business in some of the western and New England states. Many recreational skiers are beyond the beginning level and want intermediate or "more difficult" terrain (but not the most difficult). *Snow Country* described the 35 top-rated ski areas and gave the percentage of the skiing terrain that was at the more difficult level. The percentages follow:

36	54	51	49	30	43	40	46	35	40	52	28
57	51	40	40	45	40	60	20	25	50	40	65
58	43	59	49	55	30	33	60	30	46	65	

Use five classes.

3. *Sociology: Denver Neighborhoods* A random sample of 50 Denver neighborhoods gave the following information regarding children as a percent of total population (Source: The Piton Foundation). For more information, visit the Brase/Brase statistics site at http://math.college.hmco.com/students and find the link to the Piton Foundation.

30	39	32	35	31	24	19	28	5	10	8	30
12	14	18	21	19	33	38	37	22	20	27	29
39	33	27	42	16	39	20	19	20	30	24	36
24	12	37	15	40	32	29	7	35	14	33	27
53	21										

Use seven classes.

4. *Business: Fast-Food* Franchise and Business Opportunities *Annual Report* contains information about franchise opportunities in the United States and Canada. A franchise fee is one of the expenses associated with owning a franchise. Other expenses include startup, advertising, royalties, and so on. A large category of franchises is the fast-food business, which includes franchises such as baked goods, donuts, hamburgers, chicken, and hot dogs. Franchise fees (in thousands of dollars) for the fast-food category are as follows:

21	25	25	18	44	20	25	15	19	24	10
28	30	25	25	25	25	10	25	25	20	20

20	15	10	20	25	25	13	30	15	28	15
15	35	24	40	15	20	35	5	50	30	15
25	40	15	25	15	75	33	23	30	10	15
8	25	10	8	20	25	20	30			

Use five classes.

5. *Nursing: Workload* Nurses on the eighth floor of Community Hospital believe they need extra staffing at night. To estimate the night workload, a random sample of 35 nights was used. For each night, the total number of room calls to the nurses' station on the eighth floor was recorded as follows:

68	60	69	70	83	58	90	86	71	71	92
95	70	74	46	18	84	82	75	63	101	77
102	80	86	85	73	86	62	100	90	37	88
70	87									

Use five classes.

6. *Advertising: Readability* "Readability Levels of Magazine Ads," by F. K. Shuptrine and D. D. McVicker, is an article in the *Journal of Advertising Research*. (For more information, visit the Brase/Brase statistics site at http://math.college.hmco.com/students and find the link to DASL, the Carnegie Mellon University Data and Story Library. Look in Data Subjects under Consumer and then Magazine Ads Readability file.) The following is a list of the number of three-syllable (or longer) words in advertising copy of randomly selected magazine advertisements:

34	21	37	31	10	24	39	10	17	18	32
17	3	10	6	5	6	6	13	22	25	3
5	2	9	3	0	4	29	26	5	5	24
15	3	8	16	9	10	3	12	10	10	10
11	12	13	1	9	43	13	14	32	24	15

Use eight classes.

7. *U.S. Senators: Age* Overlaid frequency polygons can be useful for visually comparing two distributions. Let's use frequency polygons to compare the age distribution of U.S. senators in different congresses. *The Macmillan Visual Almanac* gives the following age distributions for the senators in the 95th Congress and in the 103rd Congress:

Age (years)	95th		Age (years)	103rd
30–39	6		30–39	1
40–49	26		40–49	16
50–59	35		50–59	49
60–69	21		60–69	22
70–79	10		70–79	11
≥80	2		≥80	1

(a) Find the class midpoints for the ages, using 84.5 for the class midpoint of the last class.

(b) Make a frequency polygon showing the ages of the senators of the 95th Congress. On the same diagram, superimpose the frequency polygon showing the ages of the senators of the 103rd Congress.

(c) Comment on the age differences of the senators in the two congresses.

8. *U.S. Representatives: Age* How do ages of U.S. representatives in the house vary from congress to congress? *The Macmillan Visual Almanac* gave the following age distributions for the U.S. representatives in the 95th and 103rd Congresses:

Age (years)	95th	Age (years)	103rd
30–39	81	30–39	47
40–49	121	40–49	153
50–59	147	50–59	131
60–69	71	60–69	89
70–79	15	70–79	12
≥80	0	≥80	3

(a) Find the class midpoints for the ages, using 84.5 for the class midpoint of the last class.

(b) Make a frequency polygon showing the ages of the representatives of the 95th Congress. On the same diagram, superimpose the frequency polygon showing the ages of the representatives of the 103rd Congress.

(c) Comment on the age differences of the representatives in the two congresses.

9. *Business: Profits and Sales Fortune* published statistics on its Fortune 500 companies. Forty-eight of the companies were categorized as food companies (including companies such as RJR Nabisco, Pillsbury, General Mills, Quaker Oats, Tyson Foods, Hershey Foods). For 39 of these companies, profits were given as a percentage of sales (with a negative number indicating a loss). The data follow:

2	8	3	2	7	4	5	1	6	5	7	6	5
6	11	6	2	0	8	3	1	4	2	3	0	2
4	2	−1	2	2	10	−3	6	4	1	0	0	1

Also listed in the Fortune 500 companies were 45 electronic companies (including companies such as General Electric, Westinghouse Electronic, Texas Instruments). For 44 of these companies, profit as a percentage of sales was also given. The data follow:

7	7	5	6	−6	4	8	6	1	2	9	5	10	16	12
3	5	1	8	1	7	2	5	5	6	5	2	2	11	−3
2	3	5	6	4	−2	2	1	12	5	3	11	6	12

(a) Make a frequency table and histogram for the food companies, and then make another frequency table and histogram for the electronic companies. Use five classes for each.

(b) By looking at the two histograms, can we determine which category (food or electronics) has the greatest profits as a percentage of sales? Discuss some of the problems involved in comparing these two distributions.

10. *Football: Weights of Professional Players* How do football teams stack up weight-wise? Let's look at two teams in the National Football League. *Lindy's Pro Football*

gave the preseason player statistics for all the NFL teams. For the 70 players with the Miami Dolphins, the weights (in pounds) came in as follows:

242	220	224	200	220	222	222	185	223	192
225	196	208	193	202	232	200	225	226	220
198	185	193	190	190	210	192	240	238	227
235	228	237	264	228	235	221	275	275	289
280	275	280	275	295	290	282	295	290	282
295	277	248	182	175	184	234	190	180	238
240	260	270	275	255	280	265	273	280	249

For the 72 players with the San Diego Chargers, the weights were as follows:

213	196	185	119	208	205	203	222	220	214
198	195	180	192	170	200	185	202	198	184
207	223	212	188	206	240	255	242	242	236.
246	230	230	230	275	278	280	310	260	280
293	310	295	275	270	275	230	265	305	248
248	165	210	195	188	277	184	300	259	291
237	245	267	271	209	170	285	279	233	195
200	205								

(a) Make frequency tables and histograms for each of the two teams using six classes.

(b) From the histograms, can you tell if either team seems to have the heavier distribution of players' weights? Discuss some of the problems involved in comparing these distributions.

11. *Horse Racing Winning Times:* The Kentucky Derby has been run annually since 1900 at Churchill Downs, Louisville, Kentucky. The distance is 1¼ miles. Since 1900, all the winning times have been over 2 minutes, except for the record time of 1 minute and 59.2 seconds run by Secretariat in 1973 and the 2001 time of 1 minute 59.97 seconds run by Monarchos. (For more information, visit the Brase/Brase statistics site at http://math.college.hmco.com/students and find the link to the Kentucky Derby.) The ogive in Figure 2-20 on the next page shows the *seconds over 2 minutes* for the winning times. A winning time of 2 minutes and 8.4 seconds has a data entry of 8.4 seconds. Secretariat's record time of 1 minute and 59.2 seconds is a data entry of −0.8 second, since the time is *below* 2 minutes.

(a) Use Figure 2-20 to estimate the number of winning times less than 7.15 seconds (over 2 minutes). There are 101 data values represented. What percentage of the winning times are under 2 minutes and 7.15 seconds?

(b) Use Figure 2-20 to estimate the number of winning times *between* 5.15 and 11.15 seconds (over 2 minutes). What percentage of the winning times are between 2 minutes and 5.15 seconds and 2 minutes and 11.15 seconds?

12. *Insurance: Hospital Costs* The Health Insurance Association of America *Source Book of Health Insurance Data* gave information about the average cost per day per patient in hospitals by state, including the District of Columbia. Figure 2-21 on the next page shows a histogram of these data.

(a) Use the information given in the histogram to construct an ogive.

(b) How many states (including the District of Columbia) have average costs per day less than $690.50?

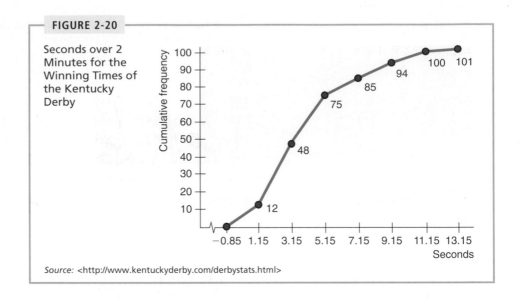

FIGURE 2-20

Seconds over 2 Minutes for the Winning Times of the Kentucky Derby

Source: <http://www.kentuckyderby.com/derbystats.html>

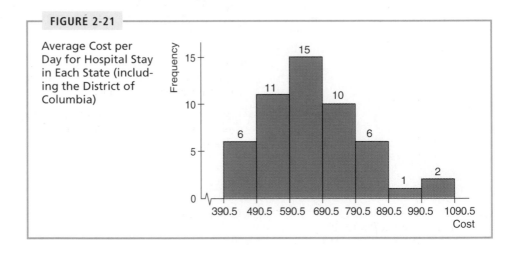

FIGURE 2-21

Average Cost per Day for Hospital Stay in Each State (including the District of Columbia)

13. *Education: Testing* Professor Silva teaches anatomy and physiology. He has developed five different versions of a test on the same material. On giving each version to a different sample of 60 students, he discovered that the test score distributions looked like those shown in Figure 2-22.
   (a) Categorize the distribution shapes as uniform, symmetric, bimodal, skewed left, or skewed right.
   (b) Comment on some advantages or problems with each test version. As a student, which version might you prefer? Which version would you like the least?

14. *Consumer: Warranty Cards* Many products come with owner registration or warranty cards. Usually, the consumer is asked a few questions about his or her family and household income. Random samples of warranty or registration cards for the indicated product revealed the household income distribution shown in Figure 2-23.
   (a) Categorize the distribution shapes as uniform, symmetric, bimodal, skewed left, or skewed right.

FIGURE 2-22

Test Score Distributions

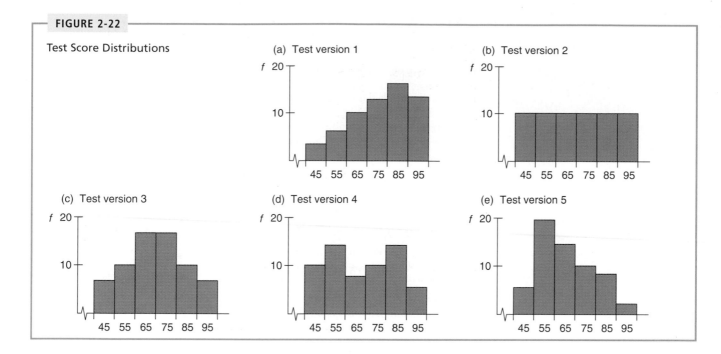

(b) If you were in charge of advertising, how would you use income-distribution information of present customers to target ads for the indicated product?

(c) How valid do you think income information is on warranty cards?

15. *Agriculture: Wheat Harvest* The following data represent tonnes of wheat harvested each year (1894–1925) from Plot 19 at the Rothamsted Agricultural Experiment Stations, England.

FIGURE 2-23

Annual Household Income Shown on Product Warranty Card (in thousands of dollars)

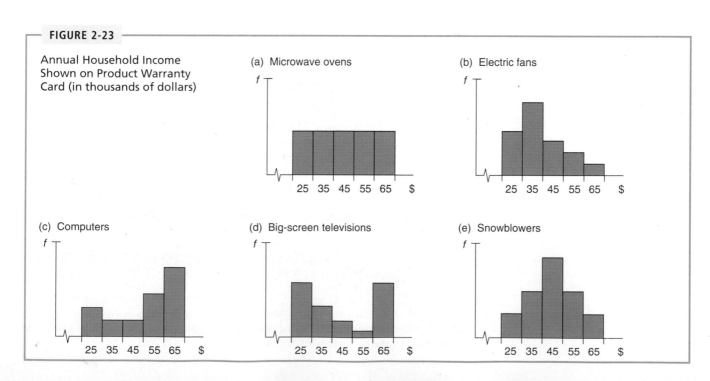

2.71	1.62	2.60	1.64	2.20	2.02	1.67	1.99	2.34	1.26	1.31
1.80	2.82	2.15	2.07	1.62	1.47	2.19	0.59	1.48	0.77	2.04
1.32	0.89	1.35	0.95	0.94	1.39	1.19	1.18	0.46	0.70	

(a) Multiply each data value by 100 to "clear" the decimals.

(b) Use the standard procedures of this section to make a frequency table and histogram with your whole-number data. Use six classes.

(c) Divide class limits, class boundaries, and class midpoints by 100 to get back to your original data values.

16. *Baseball: Batting Averages* The following data represent baseball batting averages for a random sample of National League players near the end of the baseball season. The data are from the baseball statistics section of *The Denver Post*.

0.194	0.258	0.190	0.291	0.158	0.295	0.261	0.250	0.181
0.125	0.107	0.260	0.309	0.309	0.276	0.287	0.317	0.252
0.215	0.250	0.246	0.260	0.265	0.182	0.113	0.200	

(a) Multiply each data value by 1000 to "clear" the decimals.

(b) Use the standard procedures of this section to make a frequency table and histogram with your whole-number data. Use five classes.

(c) Divide class limits, class boundaries, and class midpoints by 1000 to get back to your original data.

17. *Dotplots: Minitab* Another display technique that is somewhat similar to a histogram is a *dotplot*. In a dotplot, the data values are displayed along the horizontal axis. A dot is then plotted over each data value in the data set. The next display shows a dotplot generated by Minitab (➤**Graph** ➤**Dotplot**) for the number of licensed drivers per 1000 residents by state, including the District of Columbia (Source: U.S. Department of Transportation).

Dotplot for Licensed Drivers per 1000 Residents

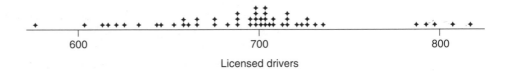

Licensed drivers

(a) From the dotplot, how many states have 600 or fewer licensed drivers per 1000 residents?

(b) About what percentage of the states (out of 51) seem to have close to 800 licensed drivers per 1000 residents?

(c) Consider the intervals 550 to 650, 650 to 750, and 750 to 850 licensed drivers per 1000 residents. In which interval do most of the states fall?

18. *Sports: Dog Sled Racing* Make a dotplot for the data in Problem 1 regarding the finish time (number of hours) for the Iditarod Dog Sled Race. Compare the dotplot to the histogram of Problem 1.

19. *Sociology: Denver Neighborhoods* Make a dotplot for the data in Problem 3 regarding the percentage of children in the population of Denver neighborhoods. Compare the dotplot to the histogram of Problem 3.

2.3
Stem-and-Leaf Displays

Exploratory Data Analysis

Together with histograms and other graphics techniques, the stem-and-leaf display is one of many useful ways of studying data in a field called *exploratory data analysis* (often abbreviated as *EDA*). John W. Tukey wrote one of the definitive books on the subject, *Exploratory Data Analysis* (Addison-Wesley). Another very useful reference for EDA techniques is the book *Applications, Basics, and Computing of Exploratory Data Analysis,* by Paul F. Velleman and David C. Hoaglin (Duxbury Press). Exploratory data analysis techniques are particularly useful for detecting patterns and extreme data values. They are designed to help us explore a data set, to ask questions we had not thought of before, or to pursue leads in many directions.

EDA techniques are similar to those of an explorer. An explorer has a general idea of destination but is always alert to the unexpected. An explorer needs to assess situations quickly and often simplify and clarify them. An explorer makes pictures—that is, maps showing the relationships of landscape features. The aspects of rapid implementation, visual displays such as graphs and charts, data simplification, and robustness (that is, analysis that is not influenced much by extreme data values) are key ingredients of EDA techniques. In addition, these techniques are good for exploration because they require very few prior assumptions about the data.

EDA methods are especially useful when our data have been gathered for general interest and observation of subjects. For example, we may have data regarding the age of applicants to graduate programs. We don't have a specific question in mind. We want to see what the data reveal. Are the ages fairly uniform or spread out? Are there exceptionally young or old applicants? If there are, we might look at other characteristics of these applicants, such as field of study. EDA methods help us quickly absorb some aspects of the data and then may lead us to ask specific questions for which we might apply methods of traditional statistics.

In contrast, when we design an experiment to produce data to answer a specific question, we focus on particular aspects of the data that are useful to us. If we want to determine the average highway gas mileage of a specific sports car, we use that model car in well-designed tests. We don't need to worry about unexpected road conditions, poorly trained drivers, different fuel grades, sudden stops and starts, etc. Our experiment is designed to control outside factors. Consequently, we do not need to "explore" our data as much. We can often make valid assumptions about the data. Methods of traditional statistics will be very useful to analyze such data and answer our specific questions.

Stem-and-Leaf Display

In this text, we will introduce two EDA techniques: stem-and-leaf displays and, in Section 3.4, box-and-whisker plots. Let's first look at a stem-and-leaf display. We know that frequency distributions and histograms provide a useful organization and summary of data. However, in a histogram, we lose most of the specific data values. A stem-and-leaf display is a device that organizes and groups data but allows us to recover the original data if desired. In the next example, we will make a stem-and-leaf display.

TABLE 2-15 Weights of Carry-On Luggage in Pounds

30	27	12	42	35	47	38	36	27	35
22	17	29	3	21	0	38	32	41	33
26	45	18	43	18	32	31	32	19	21
33	31	28	29	51	12	32	18	21	26

EXAMPLE 4

Stem-and-leaf display

Many airline passengers seem weighted down with their carry-on luggage. Just how much weight are they carrying? The carry-on luggage weights for a random sample of 40 passengers returning from a vacation to Hawaii were recorded in pounds (see Table 2-15).

To make a stem-and-leaf display, we break the digits of each data value into *two* parts. The left group of digits is called a *stem,* and the remaining group of digits on the right is called a *leaf.* We are free to choose the number of digits to be included in the stem.

The weights in our example consist of two-digit numbers. For a two-digit number, the stem selection is obviously the left digit. In our case, the tens digits will form the stems, and the units digits will form the leaves. For example, for the weight 12, the stem is 1, and the leaf is 2. For the weight 18, the stem is again 1, but the leaf is 8. In the stem-and-leaf display, we list each possible stem once on the left and all its leaves in the same row on the right as in Figure 2-24(a). Finally, we order the leaves as shown in Figure 2-24(b).

Figure 2-24 shows a stem-and-leaf display for the weights of carry-on luggage. From the stem-and-leaf display in Figure 2-24, we see that two bags weighed 27 lb, one weighed 3 lb, one weighed 51 lb, and so on. We see that most of the weights were in the 30-lb range, only two were less than 10 lb, and six were over 40 lb. Note that the length of line containing the leaves gives the visual impression that a sideways histogram would present.

As a final step, we need to indicate the scale. This is usually done by indicating the value represented by the stem and one leaf.   ◇

FIGURE 2-24

Stem-and-Leaf Displays of Airline Carry-On Luggage Weights

(a) Leaves Not Ordered

```
   3 | 2   represents 32 lb
Stem | Leaves
   0 | 3 0
   1 | 2 7 8 8 9 2 8
   2 | 7 7 2 9 1 6 1 8 9 1 6
   3 | 0 5 8 6 5 8 2 3 2 1 2 3 1 2
   4 | 2 7 1 5 3
   5 | 1
```

(b) Final Display with Leaves Ordered

```
   3 | 2   represents 32 lb
Stem | Leaves
   0 | 0 3
   1 | 2 2 7 8 8 8 9
   2 | 1 1 1 2 6 6 7 7 8 9 9
   3 | 0 1 1 2 2 2 2 3 3 5 5 6 8 8
   4 | 1 2 3 5 7
   5 | 1
```

There are no firm rules for selecting the group of digits for the stem. But whichever group you select, you must list all the possible stems from smallest to largest in the data collection.

To make a stem-and-leaf display

1. Divide the digits of each data value into two parts. The leftmost part is called the *stem* while the rightmost part is called the *leaf*.

2. Align all the stems in a vertical column from smallest to largest. Draw a vertical line to the right of all the stems.

3. Place all the leaves with the same stem on the same row as the stem, and arrange the leaves in increasing order.

4. Use a label to indicate the magnitude of the numbers in the display. We include the decimal position in the label rather than with the stems or leaves.

In the example that follows, we show a stem-and-leaf display for data with three digits.

EXAMPLE 5

Selecting the stem

What does it take to win at sports? If you're talking about basketball, one sports writer gave the answer. He listed the winning scores of the conference championship games over the last 35 years. The scores for those games follow below.

132	118	124	109	104	101	125	83	99
131	98	125	97	106	112	92	120	103
111	117	135	143	112	112	116	106	117
119	110	105	128	112	126	105	102	

To make a stem-and-leaf display, we'll use the first *two* digits as the stems (see Figure 2-25). Notice that the distribution of scores is fairly symmetrical.   ◊

FIGURE 2-25

Winning Scores of
Conference Basketball
Championship Games

```
08 | 3  represents 083 or 83 points

08 | 3
09 | 2 7 8 9
10 | 1 2 3 4 5 5 6 6 9
11 | 0 1 2 2 2 6 7 7 8 9
12 | 0 4 5 5 6 8
13 | 1 2 5
14 | 3
```

GUIDED EXERCISE 5

Stem-and-leaf display

Tel-a-Message is experimenting with computer-delivered telephone advertisements. Of primary concern is how much of the 4-minute advertisement is heard. A study was done to see how long the advertisement ran before the listeners hung up. A random sample of 30 calls gave the information in Table 2-16.

TABLE 2-16   Time Spent Listening to Advertisement (in minutes)

1.3	0.7	2.1	0.5	0.2	0.9	1.1	3.2	4.0	3.8
1.4	3.1	2.5	0.6	0.5	2.1	4.0	4.0	0.3	1.2
1.0	1.5	0.4	4.0	2.3	2.7	4.0	0.7	0.5	4.0

(a) We'll make a stem-and-leaf display using the first digit as the stem and the second as a leaf. What is the leaf unit?

⟹ The trailing digit is in the tenths position so

1 leaf unit = 0.1 min

(b) List all the stem values.

⟹ Stem: 0
1
2
3
4

(c) Complete the stem-and-leaf display including the unit designation. Order the leaves.

⟹ **FIGURE 2-26** Time Before Hang-Up

```
1 | 3  represents 1.3 min
0 | 2 3 4 5 5 5 6 7 7 9
1 | 0 1 2 3 4 5
2 | 1 1 3 5 7
3 | 1 2 8
4 | 0 0 0 0 0 0
```

(d) Looking at the stem-and-leaf display, what could you say about the time intervals before people hung up?

⟹ Most people hung up before the end of the advertisement. Of those people, more hung up within the first minute than within any other 1-minute interval. Six people listened to the entire advertisement.

◊ **COMMENT** Stem-and-leaf displays organize the data, let the data analyst spot extreme values, and are easy to create. In fact, they can be used to organize data so that frequency tables are easier to make. However, at this time, histograms are used more often in formal data presentations, whereas stem-and-leaf displays are used by data analysts to gain initial insights about the data. ◊

TECH NOTE *Stem-and-Leaf Display*

TI-83Plus Does not support stem-and-leaf displays. You can sort the data by using keys **Stat ➤ Edit ➤ 2:SortA.**

Excel Does not support stem-and-leaf displays. You can sort the data by using menu choices **Data ➤ Sort.**

Minitab Use the menu selections **Graph ➤ Stem-and-Leaf** and fill in the dialogue box.

Minitab Release 12 Stem-and-Leaf Display (for Data in Table 2-16)

```
Stem-and-Leaf   of   C1        N=30
LeafUnit=0.10

    10                  0    2345556779
   (6)                  1    012345
    14                  2    11357
     9                  3    128
     6                  4    000000
```

The values shown in the left column represent depth. Numbers above the value in parentheses show the cumulative number of values from the top to the middle value. Numbers below the value in parentheses show the cumulative number of values from the bottom to the middle value. The number in parentheses shows how many values are on the same line as the middle value.

VIEWPOINT

What Does It Take to Win?

Scores for NFL Super Bowl games can be found at the NFL web site. Visit the Brase/Brase statistics site at http://math.college.hmco.com/students and find the link to the NFL. Once at the NFL web site, follow links to the Super Bowl. Of special interest in football statistics is the spread, or difference, between scores of the winning and losing teams. If the spread is too large, the game can appear to be lopsided, and TV viewers become less interested in the game (and accompanying commercial ads). Make a stem-and-leaf display of the spread for the NFL Super Bowl games and analyze the results.

SECTION 2.3   PROBLEMS

1. *Cowboys: Longevity* How long did *real* cowboys live? One answer may be found in the book *The Last Cowboys* by Connie Brooks (University of New Mexico Press). This delightful book presents a thoughtful sociological study of cowboys in West

Texas and Southeastern New Mexico around the year 1890. A sample of 32 cowboys gave the following years of longevity:

58	52	68	86	72	66	97	89	84	91	91
92	66	68	87	86	73	61	70	75	72	73
85	84	90	57	77	76	84	93	58	47	

(a) Make a stem-and-leaf display for these data.

(b) Consider the following quote from Baron von Richthofen in his *Cattle Raising on the Plains of North America:* "Cowboys are to be found among the sons of the best families. The truth is probably that most were not a drunken, gambling lot, quick to draw and fire their pistols." Does the data distribution of longevity lend credence to this quote?

2. *Ecology: Habitat* Wetlands offer a diversity of benefits. They provide habitat for wildlife, spawning grounds for U.S. commercial fish, and renewable timber resources. In the last 200 years the United States has lost more than half its wetlands. *Environmental Almanac* gives the percentage of wetlands lost in each state in the last 200 years. For the lower 48 states, the percentage loss of wetlands per state is as follows:

46	37	36	42	81	20	73	59	35	50
87	52	24	27	38	56	39	74	56	31
27	91	46	9	54	52	30	33	28	35
35	23	90	72	85	42	59	50	49	
48	38	60	46	87	50	89	49	67	

Make a stem-and-leaf display of these data. Be sure to indicate the scale. How are the percentages distributed? Is the distribution skewed? Are there any gaps?

State	No. of Hospitals	Average Length of Stay	State	No. of Hospitals	Average Length of Stay	State	No. of Hospitals	Average Length of Stay
Alabama	119	7.0	Kentucky	107	6.9	N. Dakota	47	11.1
Alaska	16	5.7	Louisiana	136	6.7	Ohio	193	6.6
Arizona	61	5.5	Maine	38	7.2	Oklahoma	113	6.7
Arkansas	88	7.0	Maryland	51	6.8	Oregon	66	5.3
California	440	6.0	Massachusetts	101	7.0	Pennsylvania	236	7.5
Colorado	71	6.8	Michigan	175	7.3	Rhode Island	12	6.9
Connecticut	35	7.4	Minnesota	148	8.7	S. Carolina	68	7.1
Delaware	8	6.8	Mississippi	102	7.2	S. Dakota	52	10.3
Dist. of Columbia	11	7.5	Missouri	133	7.4	Tennessee	122	6.8
Florida	227	7.0	Montana	53	10.0	Texas	421	6.2
Georgia	162	7.2	Nebraska	90	9.6	Utah	42	5.2
Hawaii	19	9.4	Nevada	21	6.4	Vermont	15	7.6
Idaho	41	7.1	New Hampshire	27	7.0	Virginia	98	7.0
Illinois	209	7.3	New Jersey	96	7.6	Washington	92	5.6
Indiana	113	6.6	New Mexico	37	5.5	W. Virginia	59	7.1
Iowa	123	8.4	New York	231	9.9	Wisconsin	129	7.3
Kansas	133	7.8	N. Carolina	117	7.3	Wyoming	27	8.5

3. *Health Care: Hospitals* The American Medical Association Center for Health Policy Research included data, by state, on the number of community hospitals and the average patient stay (in days) in its publication *State Health Care Data: Utilization, Spending, and Characteristics*. The data (by state) are shown in the table. Make a stem-and-leaf display of the data for the average length of stay in days. Comment about the general shape of the distribution.

4. *Health Care: Hospitals* Using the number of hospitals per state listed in the table, make a stem-and-leaf display for the number of community hospitals per state. Which states have an unusually high number of hospitals?

5. *Boston Marathon: Winning Times* The Boston Marathon is the oldest and best known U.S. marathon. It covers a route from Hopkinton, Massachusetts, to downtown Boston. The distance is approximately 26 miles. Visit the Brase/Brase statistics site at http://math.college.hmco.com/students and find the link to Boston Marathon. Search the site for the marathon to find a wealth of information about the history of the race. In particular, it gives the winning times for the Boston Marathon. They are all over 2 hours. The following data are the minutes over 2 hours for the winning male runner:

   1961–1980

23	23	18	19	16	17	15	22	13	10
18	15	16	13	9	20	14	10	9	12

   1981–2000

9	8	9	10	14	7	11	8	9	8
11	8	9	7	9	9	10	7	9	9

   (a) Make a stem-and-leaf diagram for the minutes over 2 hours of the winning times for the years 1961 to 1980. Use two lines per stem. For instance, to use two lines for the stem value 0, we designate the first stem as 0* and put leaf digits 0 through 4 on this line. The second line for stem 0 is designated with stem 0˙, and we put leaf digits 5 through 9 on this line.

   (b) Make a stem-and-leaf diagram for the minutes over 2 hours of the winning times for the years 1981 to 2000. Use two lines per stem.

   (c) Compare the two distributions. How many times under 15 minutes are in each distribution?

6. *Golf: Scores* The U.S. Open Golf Tournament was played at Congressional Country Club, Bethesda, Maryland, with prizes ranging from $465,000 for first place to $5000. Par for the course is 70. The tournament consists of four rounds played on different days. The scores for each round of the 32 players who placed in the money (more than $17,000) were given on a web site. For more information, visit the Brase/Brase statistics site at http://math.college.hmco.com/students and find the link to golf. The scores for the first round were as follows:

71	65	67	73	74	73	71	71	74	73	71
70	75	71	72	71	75	75	71	71	74	75
66	75	75	75	71	72	72	73	71	67	

   The scores for the fourth round for these players were as follows:

69	69	73	74	72	72	70	71	71	70	72
73	73	72	71	71	71	69	70	71	72	73
74	72	71	68	69	70	69	71	73	74	

(a) Make a stem-and-leaf diagram for the first-round scores. Use two lines per stem. (See Problem 5.)

(b) Make a stem-and-leaf diagram for the fourth-round scores. Use two lines per stem.

(c) Compare the two distributions. How do the highest scores compare? How do the lowest scores compare?

7. *Astronomy: Motions of Stars* Proper motions are the angular motions of stars across the sky due to effects other than the motion of the earth. These motions primarily reflect the stars' own velocities in space. *Centennial proper motion* indicates that the units are arc seconds per century (per information in an article by K. M. Cudworth, *Astronomical Journal,* Vol. 81, pp. 975–982). The following data represent centennial proper motion for a sample of stars from the global star cluster M92, some of the oldest stars in our galaxy.

1.260	0.251	0.788	0.038	0.050	0.057	0.008
0.219	0.667	0.014	0.759	1.438	0.623	0.173
1.169	1.660	1.808	1.057	1.024	0.216	0.430
0.351	1.616	1.169	0.888	1.260	0.042	0.369

Make a stem-and-leaf diagram describing the centennial proper motion for these stars using the first two digits as the stem and the last two digits as the leaf.* Do there appear to be any gaps in centennial proper motion for these stars? Would such a discovery be of interest to an astronomer?

8. *Astronomy: Cosmic Radio Signals* Is there an association between the location of cosmic radio sources and particular galaxies? A partial answer might be found in *Astrophysics Journal,* Vol. 148, pp. 321–365. Galaxies and radio sources were randomly studied in the celestial equator between longitude 10h and 16h and latitude −16° and 16°. The radio brightness (units 10^{-26} watts per square meter per hertz) of the closest radio sources are

9.0	9.5	67.0	11.5	44.0	13.6	9.0	12.5	10.5	11.0
16.5	11.5	9.8	44.0	16.5	16.5	13.5	20.0	9.5	
44.0	28.0	20.0	9.4	11.5	12.5	44.0	13.7	9.5	

After gathering appropriate data, the first activity in many research projects is to perform an initial organization of the data. Make a stem-and-leaf plot using one digit as the stem and two digits as the leaf.* *Hint:* 9.0 would be coded as 09.0, so the first digit is considered to be 0. Be sure to label the unit measure. Are there any unusually high or low data values recorded?

Are cigarettes bad for people? Cigarette smoking involves tar, carbon monoxide, and nicotine. The first two are definitely not good for a person's health, and the last ingredient can cause addiction. Problems 9, 10, and 11 refer to Table 2-17, which was taken from the web site maintained by the *Journal of Statistics Education.* For more information, visit the

*Technically, if there is more than one digit remaining in the leaf portion, only the leftmost of the leaf digits is considered to be a leaf, and the other digits are truncated. The appropriate scale is listed in the label. For instance, if numbers such as 12,472 are in the data set, and the first two digits comprise the stem, the leaf is the next digit, or 4. Then a label indicates that 12 | 4 represents 12,400. Computer packages such as Minitab follow the convention of truncating numbers so that only one digit is in the leaf portion. In Problems 7 and 8 you are asked to use two digits for each leaf. Place the two digits together and then leave a larger space before listing the next leaf "group" of two digits. In this way you can retrieve actual data values from the diagram rather than truncated values.

TABLE 2-17 Milligrams of Tar, Nicotine, and Carbon Monoxide (CO) Content for One Cigarette

Brand	Tar	Nicotine	CO	Brand	Tar	Nicotine	CO
Alpine	14.1	0.86	13.6	Merit	7.8	0.57	10.0
Benson & Hedges	16.0	1.06	16.6	MultiFilter	11.4	0.78	10.2
Bull Durham	29.8	2.03	23.5	Newport Lights	9.0	0.74	9.5
Camel Lights	8.0	0.67	10.2	Now	1.0	0.13	1.5
Carlton	4.1	0.40	5.4	Old Gold	17.0	1.26	18.5
Chesterfield	15.0	1.04	15.0	Pall Mall Lights	12.8	1.08	12.6
Golden Lights	8.8	0.76	9.0	Raleigh	15.8	0.96	17.5
Kent	12.4	0.95	12.3	Salem Ultra	4.5	0.42	4.9
Kool	16.6	1.12	16.3	Tareyton	14.5	1.01	15.9
L&M	14.9	1.02	15.4	True	7.3	0.61	8.5
Lark Lights	13.7	1.01	13.0	Viceroy Rich Light	8.6	0.69	10.6
Marlboro	15.1	0.90	14.4	Virginia Slim	15.2	1.02	13.9
				Winston Lights	12.0	0.82	14.9

Source: Journal of Statistics Education web site http://www.stat.nesu.edu/info/jse/homepage.html. Reprinted with permission.

Brase/Brase statistics site at http://math.college.hmco.com/students and find the link to the *Journal of Statistics Education*. Follow the links to the cigarette data.

9. *Health: Cigarette Smoke* Use the data in Table 2-17 to make a stem-and-leaf display for milligrams of tar per cigarette smoked.

10. *Health: Cigarette Smoke* Use the data in Table 2-17 to make a stem-and-leaf display for milligrams of carbon monoxide per cigarette smoked.

11. *Health: Cigarette Smoke* Use the data in Table 2-17 to make a stem-and-leaf display for milligrams of nicotine per cigarette smoked. In this case, truncate the measurements at the tenth position and use two lines per stem (see Problem 5, part a).

12. *Archaeology: Depth of Artifacts* In archaeology, the depth (below surface grade) at which artifacts are found is very important. Greater depths sometimes indicate older artifacts, perhaps from a different archaeological period. Figure 2-27 is a

---

FIGURE 2-27

Depth (in cm) of Artifact Location

```
                        5 | 2 | 0    = 25 cm at Site I and 20 cm at Site II
                Site I                Site II

                    5    2    0 5 5
                  5 0    3    0 0 0 0 5 5
                5 5 5    4    0 0 0 5
                  5 0    5    0 0 5 5
            5 5 5 5 5 0   6    0 0 0 5 5 5
        5 5 5 5 5 5 0 0   7
            5 5 0 0 0     8
            5 5 0 0       9
                5 5      10
                  0      11    0 0 5 5 5
                         12    0 0 0 0 5
```

FIGURE 2-28

Back-to-Back Stem Plot
Showing Annual Salaries
(in thousands of dollars)
for Full Professors in
California and in New
York

9 | 4 | 5   represents $49,000 for California and $45,000 for New York

California		New York
9	4	5 6 8
9 6 3 0	5	0 1 1 5 6
9 3 0 0 0	6	0 1 1 3 4 5 6 9 9 9 9
7 7 7 6 6 5 5 5 5 0	7	2 2 4 5 5 6
9 7 5 5 3	8	2 5 6
5 4 4 3	9	3 5 7
4 2	10	0 3
3 2	11	
6 2	12	0 0

---

back-to-back stem plot showing the depths of artifact locations at two different archaeological sites. These sites are from similar geographic locations. Notice that the stems are in the center of the diagram. The leaves for Site I artifact depths are shown to the left of the stem, while the leaves for Site II are to the right of the stem (see *Mimbres Mogollon Archaeology* by A. I. Woosley and A. J. McIntyre, University of New Mexico Press).

(a) What are the least and greatest depths of artifact finds at Site I? at Site II?

(b) Describe the data distribution of depths of artifact finds at Site I and at Site II.

(c) At Site II, there is a gap in the depths at which artifacts were found. Does the Site II data distribution suggest that there might have been a period of no occupation?

13. *Education: Professors' Salaries* Figure 2-28 is a *back-to-back stem plot* showing the average annual salaries (in thousands of dollars) of full professors in California universities compared to New York universities (Source: *The Chronicle of Higher Education*. For more information, visit the Brase/Brase statistics site at http://math.college.hmco.com/students and find the link to the *Chronicle of Higher Education*.

(a) What are the low and high average salaries for professors in California? in New York?

(b) Which state has a greater number of average salaries in the $60,000 range? in the $70,000 range?

(c) In general, which state, California or New York, would you say had higher average salaries for professors in colleges and universities?

SUMMARY

Organizing and presenting data are the main purposes of that part of statistics called descriptive statistics. In this chapter, we have studied bar graphs, Pareto charts, circle graphs, time plots, histograms, relative-frequency histograms, frequency polygons, ogives, and stem-and-leaf displays. From the viewpoint of future applications, histograms are the most important because the area under a bar can represent the likelihood of data values falling into that class. Histograms and stem-and-leaf displays both reveal distribution properties such as uniformity, symmetry, or skewness.

IMPORTANT WORDS & SYMBOLS

Section 2.1
Bar graph
Pareto chart
Pie chart or circle graph
Time plot
Time series

Section 2.2
Frequency
Frequency distribution
Class width
Class, lower limit, upper limit
Class frequency
Class midpoint
Class mark
Frequency table
Class boundaries
Histogram
Relative-frequency table

Relative-frequency histogram
Symmetric distribution
Uniform distribution
Skewed left
Skewed right
Bimodal distribution
Frequency polygon
Line graph
Cumulative frequency
Ogive
Dotplot

Section 2.3
EDA
Stem
Leaf
Stem-and-leaf display
Back-to-back stem plot

VIEWPOINT   *"This land is your land, This land is my land"*[*]
                   —Woody Guthrie

But who actually owns the forest? On many maps, forest land (including national forests) is colored green. Such maps give the impression that vast areas of the western United States are public land. This is not the case! *USA Today* gave the following information about ownership of U.S. timber lands: state/local, 17%; federal, 10%; forest industry, 14%; and private nonindustry, 59%. Organize these data for better visual presentation using a Pareto chart and a circle graph.

*Words and Music by Woody Guthrie TRO © Copyright 1956(Renewed) 1958(Renewed) Ludlow Music, Inc. New York, New York. Used by permission.

CHAPTER REVIEW PROBLEMS

1. *Cars: Fuel Economy* The Focus Problem at the beginning of Chapter 2 shows you two graphics [Figure 2-1(a) and (b)] displaying the same information regarding fuel economy standards for autos. Review the criteria given in the Focus Problem for producing good graphs (from Edward R. Tufte, *The Visual Display of Quantitative Information*). Look at the graph in Figure 2-1(a). Is it essentially a bar graph? Explain. What are some of the flaws of Figure 2-1(a) as a bar graph? Next, examine Figure 2-1(b), which shows the same information. Is it essentially a time plot? Explain. In what ways does the second graph seem to display the information in a clearer manner?

FIGURE 2-29

Number of State and
Federal Prisoners per
100,000 Population

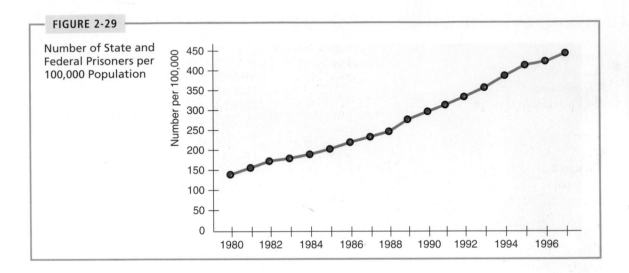

2. *Criminal Justice: Prisoners* The time plot in Figure 2-29 gives the number of state and federal prisoners per 100,000 population (Source: *Statistical Abstract of the United States,* 120th Edition).
   (a) Estimate the number of prisoners per 100,000 people for 1980 and for 1997.
   (b) During the time period shown there was increased prosecution of drug offenses, longer sentences for common crimes, and reduced access to parole. What does the time plot say about the prison population change per 100,000 people?
   (c) In 1997, the U.S. population was approximately 266,574,000 people. At the rate of 444 prisoners per 100,000 population, about how many prisoners were in the system? The projected U.S. population for the year 2020 is 323,724,000. If the rate of prisoners per 100,000 stays the same as in 1997, about how many prisoners do we expect to have in the system? To obtain the most recent information, visit the Brase/Brase statistics site at http://math.college.hmco.com/students and find the link to the Census Bureau.

3. *IRS: Tax Returns* Almost everyone files (or will sometime file) a federal income tax return. A research poll for Turbo Tax (a computer software package to aid in tax-return preparation) asked what aspect of filing a return people thought to be the most difficult. The results showed that 43% of the people said understanding the IRS jargon, 28% said knowing deductions, 10% said getting the right form, 8% said calculating the numbers, and 10% didn't know. Make a circle graph to display this information. *Note:* Percentages will not total 100% because of rounding.

4. *Law Enforcement: DUI* Driving under the influence of alcohol (DUI) is a serious offense. The following data give the ages of a random sample of 50 drivers arrested while driving under the influence of alcohol. This distribution is based on the age distribution of DUI arrests given in the *Statistical Abstract of the United States* (112th Edition).

46	16	41	26	22	33	30	22	36	34
63	21	26	18	27	24	31	38	26	55
31	47	27	43	35	22	64	40	58	20
49	37	53	25	29	32	23	49	39	40
24	56	30	51	21	45	27	34	47	35

(a) Make a stem-and-leaf display of the age distribution.
(b) Make a frequency table using seven classes.
(c) Make a histogram showing class boundaries.
(d) Make an ogive. From the ogive, estimate the percentage of drivers who are aged 29 or under and arrested while DUI.

5. *Medical: Glucose Blood Test* A fasting glucose blood test was given to 53 (non-pregnant) women at Boston City Hospital. The results of the blood tests (milligrams of glucose per 100 milliliters of blood) are shown below. (Source: *American Journal of Clinical Nutrition,* Vol. 19, pp. 345–351.)

66	59	69	84	85	96	85	82	80	90
75	80	73	69	85	82	91	84	77	86
75	83	87	74	80	83	77	69	78	93
93	87	70	85	76	87	87	79	91	86
82	82	84	85	90	79	83	65	87	70
89	81	74							

(a) Make a frequency table with seven classes showing class limits, class boundaries, midpoints, frequencies, relative frequencies, and cumulative frequencies.
(b) Draw a histogram.
(c) Draw a frequency polygon.
(d) Draw a relative-frequency histogram.
(e) Draw an ogive.

6. *Law: Corporation Lawsuits* Many people say the civil justice system is overburdened. Many cases center on suits involving businesses. The following data are based on a *Wall Street Journal* report. Researchers conducted a study of lawsuits involving 1908 businesses ranked in the Fortune 1000 over a 20-year period. They found the following distribution of civil justice caseloads brought before the federal courts involving the businesses:

Case Type	Number of Filings (in thousands)
Contracts	107
General torts (personal injury)	191
Asbestos liability	49
Other product liability	38
All other	21

Note: Contracts cases involve disputes over contracts between businesses.
(a) Make a Pareto chart of the caseloads. Which type of cases occur most frequently?
(b) Make a circle chart showing the percentage of cases in each type.

7. *Archaeology: Tree-Ring Data The Sand Canyon Archaeological Project,* edited by W. D. Lipe and published by Crow Canyon Archaeological Center, contains the stem-and-leaf diagram in Figure 2-30 on the next page. The study uses tree rings to accurately determine the year in which a tree was cut. The figure gives the tree-ring-cutting dates for samples of timbers found in the architectural units at Sand Canyon Pueblo. The text referring to the figure says, "The three-digit numbers in

---

FIGURE 2-30

Tree-Ring-Cutting Dates from Architectural Units at Sand Canyon Pueblo: *The Sand Canyon Archaeological Project*

```
119 | 5 6
120 | 0 0 1 2 3 3 3 3 3 3 3 3 3 3 3 3 3 3 3 3 3 3 3 3 3 3 3 3 3 3
120 |
121 | 2
121 | 5 5
122 | 0 0 1 1 1 1 2 2 3 4 4 4 4 4 4 4
122 | 5 8 9
123 | 0 1 2 3 3 4
123 | 5 5 5 5 5 5 5 5 5 5 5 5 5 5 5 6 8 8 9
124 | 1 2 2 2 2 2 2 2 2 2 2 2 2 2 2 2 2 2 2 2 2 2 3 4 4
124 | 5 6 8 9 9 9 9 9 9 9 9 9
125 | 0 0 0 0 0 0 0 0 0 0 0 0 0 0 0 1 1 1 1 1 1 1 2 2 2
125 |
126 | 0 0 0 1 2 2 2 2 2 2 2 2 2 2 2 2 2 2 2 4 4 4 4 4 4 4
126 | 5 5 5 6 6 7
127 | 0 1 1 1 1 4 4
```

---

the left column represent centuries and decades A.D. The numbers to the right represent individual years, with each number derived from an individual sample. Thus, **124 2 2 2** represents three samples dated to A.D. 1242." Use Figure 2-30 and the verbal description to answer the following questions.

(a) Which decade contained the most samples?

(b) How many samples had a tree-ring-cutting date between (and including) 1200 A.D. and 1239 A.D.?

(c) What are the dates of the longest interval in which no tree-cutting samples occurred? What might this indicate about new construction or renovation of the pueblo structures during this period?

8. *Education: Grades* The following relative-frequency histogram is based on information from the *Statistical Abstract of the United States* (112th Edition). In Figure 2-31, x = high-school grade point average for a person who has received a bachelor's degree. We use a 4-point scale for grading, so A = 4, B+ = 3.5, B = 3, C+ = 2.5, C = 2, D+ = 1.5, and D = 1. The vertical axis gives the relative frequency for the high-school grade point average of students who graduated from college.

(a) Comment on the shape of the distribution. Is it skewed? Is it symmetrical? Is it uniform?

(b) The class boundaries are halfway between the class midpoints. Find the class boundaries. *Hint:* Use the fact that the classes are all the same width to find the lower boundary of the first class and the upper boundary of the last class.

(c) Convert the relative frequencies into percentages. Then find the percentage of college graduates who had high-school grade point averages of less than 3.25 (low B). What percentage had high-school grade point averages of less than 3.75 (high B)?

9. *Health Care: Age of Patients* The following data are based on information from the *Statistical Abstract of the United States* (112th Edition). In Figure 2-32, the horizontal axis represents the age of hospital patients (from 5 years of age on). The vertical axis gives the relative frequencies.

(a) Which age group is the most frequent?

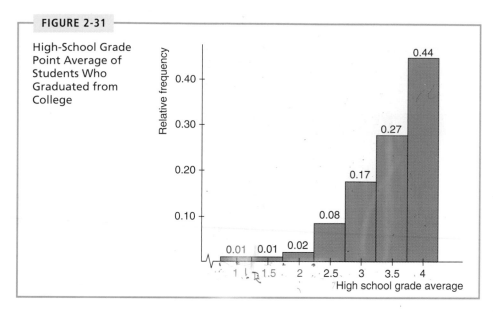

FIGURE 2-31

High-School Grade
Point Average of
Students Who
Graduated from
College

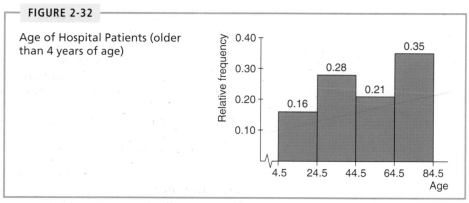

FIGURE 2-32

Age of Hospital Patients (older
than 4 years of age)

(b) Convert the relative frequencies into percentages. Then find the percentage of patients older than age 44.

(c) What percentage of patients is 44 years old or younger?

10. *World's Wealthiest People: Age Forbes Richest People* gives the profile of the world's wealthiest men and women. For more information, visit the Brase/Brase statistics site at http://math.college.hmco.com/students and find the link to Forbes. Do you have to be old to be worth at least $2 billion? You can answer this question yourself by studying the following data—ages of men and women worth at least $2 billion:

40	66	43	82	52	58	77	52	50	48	47
68	66	73	76	53	67	88	40	79	73	66
65	70	72	77	48	75	82	54	76	41	93
65	60	57	74	70	83	67	68	77	66	34
66	59	48	56	71	40	53	63	52	57	83
52	60	56	71	64	61	53	53	73	70	

(a) Make a stem-and-leaf diagram.

(b) Make a histogram using seven classes. Describe the shape of the distribution (that is, indicate if it is symmetrical, skewed, or bimodal).

(c) Make an ogive. Estimate the percentage of these very rich people aged 51 or under.

DATA HIGHLIGHTS: GROUP PROJECTS

Break into small groups and discuss the following topics. Organize a brief outline in which you summarize the main points of your group discussion.

1. Examine Figure 2-33, "Slobs make worst roommates." This is a double bar graph because two percentages are given for each response category: responses from men and responses from women. Comment about how the artistic rendition has slightly changed the format of a bar graph. Do the bars seem to have lengths that accurately reflect the relative percentages of the responses? In your own opinion, does the artistic rendition enhance or confuse the information? Explain. Which characteristic of "worst roommates" does the graphic seem to illustrate? Can this graph be considered a Pareto chart for men? For women? Why or why not? From the information given in the figure, do you think the survey just listed the four given annoying characteristics? Do you think a respondent could choose more than one characteristic? Explain your answer in terms of the percentages given and in terms of the explanation given in the graphic. Could this information also be displayed in one circle graph for men and another for women? Explain.

2. Examine Figure 2-34, "Global teen worries." How many countries were contained in the sample? The graph contains bars and a circle. Which bar is the longest? Which bar represents the greatest percentage? Is this a bar graph or not? If not, what changes would need to be made to put the information into a bar graph? Could it be made into a Pareto chart? Could it be made into a circle graph? Explain.

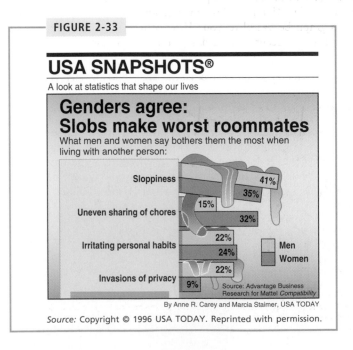

FIGURE 2-33

Source: Copyright © 1996 USA TODAY. Reprinted with permission.

FIGURE 2-34

Source: Copyright © 1996 USA TODAY. Reprinted with permission.

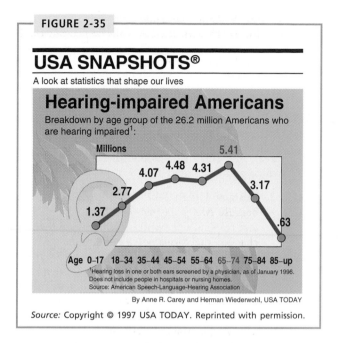

FIGURE 2-35

USA SNAPSHOTS®

A look at statistics that shape our lives

Hearing-impaired Americans

Breakdown by age group of the 26.2 million Americans who are hearing impaired[1]:

Millions

5.41
4.48 4.31
4.07
2.77
3.17
1.37
.63

Age 0–17 18–34 35–44 45–54 55–64 65–74 75–84 85–up

[1]Hearing loss in one or both ears screened by a physician, as of January 1996. Does not include people in hospitals or nursing homes.
Source: American Speech-Language-Hearing Association

By Anne R. Carey and Herman Wiederwohl, USA TODAY

Source: Copyright © 1997 USA TODAY. Reprinted with permission.

3. Examine Figure 2-35, "Hearing-impaired Americans." Is this figure best categorized as a time plot, an ogive, or a frequency polygon? What is the midpoint for each class? What would be a reasonable midpoint for the last class? Look at the age classes. For which classes are the class widths equal? Redraw the polygon so that the horizontal scale more accurately represents all the listed intervals. Compare your graph with the given figure. Discuss some reasons you think the designers of the original graphic made the first two classes longer.

4. Examine Figure 2-36, "Dow Jones Industrial Averages" (*The Wall Street Journal*). Explain why this could be considered a time plot. If you were a trader on the floor of the exchange, a broker in Miami, or just an individual who bought and sold stocks, how could this graph be useful to you? How could the table insert, which is a breakdown of the 5 most recent trading days, be useful? Explain.

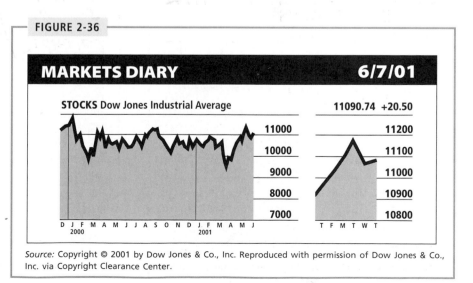

FIGURE 2-36

MARKETS DIARY   6/7/01

STOCKS Dow Jones Industrial Average

11090.74 +20.50

11000
10000
9000
8000
7000

11200
11100
11000
10900
10800

D J F M A M J J A S O N D J F M A M J
2000   2001

T F M T W T

Source: Copyright © 2001 by Dow Jones & Co., Inc. Reproduced with permission of Dow Jones & Co., Inc. via Copyright Clearance Center.

LINKING CONCEPTS: WRITING PROJECTS

Discuss each of the following topics in class or review the topics on your own. Then write a brief but complete essay in which you summarize the main points. Please include formulas and graphs as appropriate.

1. In your own words, explain the differences among bar graphs, circle graphs, time plots, Pareto charts, ogives, histograms, relative-frequency histograms, and stem-and-leaf plots. If you have nominal data, which graphic displays might be useful? What if you have ordinal, interval, or ratio data?

2. What do we mean when we say a histogram is skewed to the left? to the right? What is a bimodal histogram? Discuss the following statement: "A bimodal histogram usually results if we draw a sample from two populations at once." Suppose you took a sample of weights of college football players and with this sample you included weights of cheerleaders. Do you think a histogram made from the combined weights would be bimodal? Explain.

3. Discuss the statement that stem-and-leaf displays are quick and easy to construct. How can we use a stem-and-leaf display to make the construction of a frequency table easier? How does a stem-and-leaf display help you spot extreme values fast?

4. Go to the library and pick up a current issue of *The Wall Street Journal, Newsweek, Time, USA Today,* or other news media. Examine each newspaper or magazine for graphs of the type discussed in this chapter. List the variables used, method of data collection, and general type of conclusion drawn from the graphs. Another source for information is the Internet. Explore several web sites, and categorize the graphs you find as you did for the print media. For interesting web sites, visit the Brase/Brase statistics site at http://math.college.hmco.com/students and find links to the Social Statistics Briefing Room, to law enforcement, and to golf.

Using TECHNOLOGY

TI-83PLUS • *MINITAB* • *EXCEL* • *COMPUTERSTAT*

APPLICATIONS

The following tables show the first-round winning scores of the NCAA men's and women's basketball teams (March 2001).

TABLE 2-18 Men's Winning First-Round NCAA Tournament Scores

95	70	79	99	83	72	79	101
69	82	86	70	79	69	69	70
95	70	77	61	69	68	69	72
89	66	84	77	50	83	63	58

TABLE 2-19 Women's Winning First-Round NCAA Tournament Scores

80	68	51	80	83	75	77	100
96	68	89	80	67	84	76	70
98	81	79	89	98	83	72	100
101	83	66	76	77	84	71	77

1. Use the software or method of your choice to construct separate histograms for the men's and women's winning scores. Try 5, 7, and 10 classes for each. Which number of classes seems to be the best choice? Why?
2. Use the same class boundaries for histograms of men's and of women's scores. How do the scores for the two groups compare? What general shape do the histograms follow?
3. Use the software or method of your choice to make stem-and-leaf diagrams for each set of scores. If your software does not make stem-and-leaf displays, sort the data first and then make a back-to-back display by hand. Do there seem to be any extreme values in either set? How do the data sets compare?

Technology Hints: Creating Histograms

The default histograms produced by the TI-83Plus, Minitab, and Excel all determine the number of classes to use automatically. To control the number of classes the technology uses, follow the key steps as indicated. The display screens are generated for data found in Table 2-4—Commuting Distances of Dallas Workers—using five classes.

TI-83Plus

Determine the class width for the number of classes you want and the lower class boundary for the first class. Enter the data in list L1.

Press **STATPLOT** and highlight On and the histogram plot.

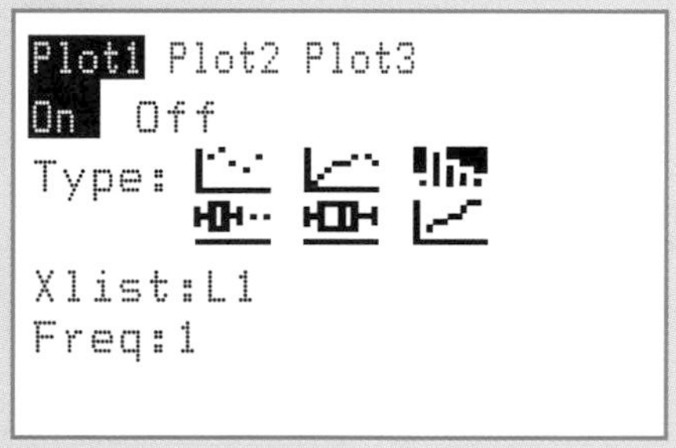

Plot1 Plot2 Plot3
On Off
Type:
Xlist:L1
Freq:1

Press **WINDOW** and set Xmin = lowest class boundary, Xscl = class width. Use appropriate values for the other settings.

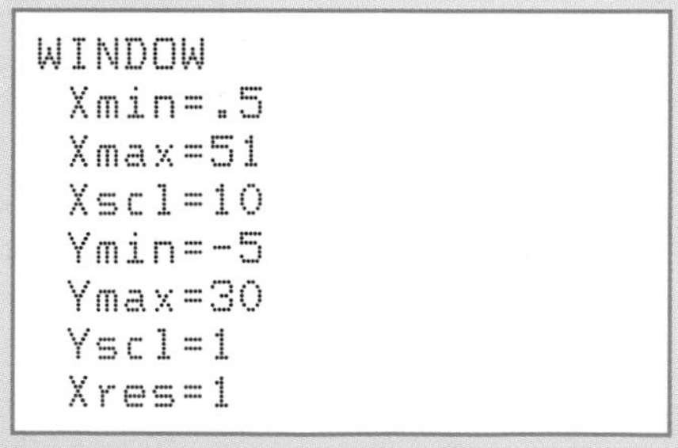

WINDOW
Xmin=.5
Xmax=51
Xscl=10
Ymin=-5
Ymax=30
Yscl=1
Xres=1

Press **GRAPH. TRACE** gives boundaries and frequency.

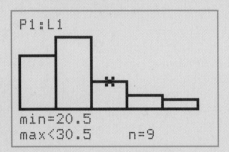

```
P1:L1



min=20.5
max<30.5      n=9
```

Excel

Determine the upper class boundaries for the five classes. Enter the data. In a separate column, enter the upper class boundaries. Use the menu selection **Tools ➤ Data Analysis ➤ Histogram**.

Put the data range in the Input Range. Put the upper class boundaries range for the Bin Range.

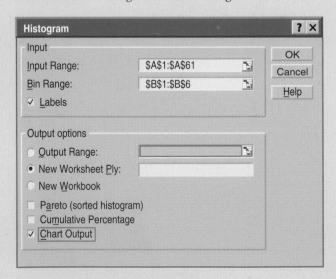

To make bars touch, right click on a bar, select **Format Data Series ➤ Options tab**. Set **gap width** to 0.

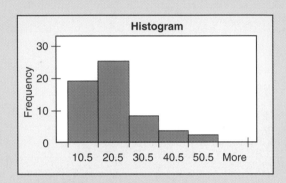

Minitab

Determine the class boundaries. Enter the data. Use the menu selection **Graph ➤ Histogram**.

In the Dialogue Box, press **Options**. Then select cutpoints, and enter the class boundaries as cutpoint positions.

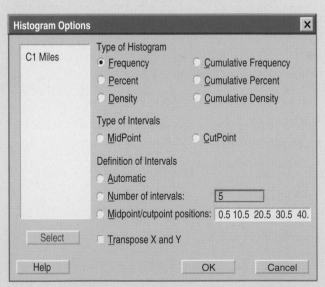

Select **Frame** to adjust scales. Otherwise, press **OK**.

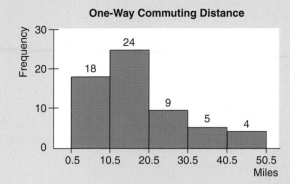

One-Way Commuting Distance

ComputerStat

Use menu selections **Descriptive Statistics ➤ Frequency Distributions and Grouped Data.** Enter the data, and follow the instructions on the screen. You may select the number of classes to use up to 10.

3 Averages and Variation

While the individual man is an insolvable puzzle,
in the aggregate he becomes
a mathematical certainty. You can,
for example, never foretell what any one man
will do, but you can say
with precision what an average number will be up to.

— Arthur Conan Doyle, *The Sign of Four*

Sherlock Holmes

For on-line student resources, visit **math.college.hmco.com/students** and follow the Statistics links to the Brase/Brase, *Understandable Statistics,* 7th edition web site.

Sherlock Holmes spoke these words to his colleague Dr. Watson as the two were unraveling a mystery. The detective was commenting that if a single member is drawn at random from a population, we cannot predict *exactly* what that member will look like. However, there are some "average" features of the entire population that an individual is likely to possess. The degree of certainty with which we would expect to observe such average features in any individual depends on our knowledge of the variation among individuals in the population. Sherlock Holmes has led us to two of the most important statistical concepts: average and variation.

◇ What are commonly used measures of central tendency? What do they tell us? (SECTION 3.1)

◇ How do variance and standard deviation measure data spread? Why is this important? (SECTION 3.2)

◇ When data have already been grouped in a frequency table or histogram, how can we estimate the mean and standard deviation? (SECTION 3.3)

◇ How do we make a box-and-whisker plot, and what does it tell us about the spread of the data? (SECTION 3.4)

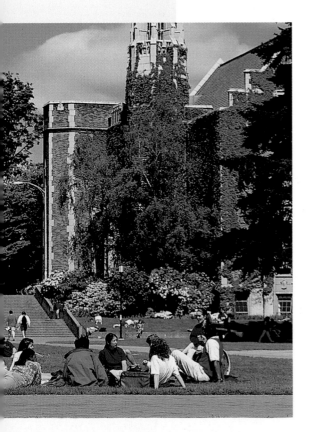

FOCUS PROBLEM

Why Bother!

Is it really worth all the effort to get a college degree? From a philosophical point of view, the love of learning is sufficient reason to get a college degree. However, the U.S. Census Bureau also makes another relevant point. Annually college graduates (bachelor's) earn on average $17,583 more than high school graduates. This means college graduates earn about 77% more than high school graduates, and according to the USA SNAPSHOT on the next page, the gap in earnings is increasing. Furthermore, as the College Board indicates, for most Americans, college remains relatively affordable.

How would you answer the following questions?

(a) Does a college degree *guarantee* someone a 77% increase in earnings over a high school degree? Remember, we are using only *averages* from census data.

(b) Using census data (not shown in the USA SNAPSHOT), it is estimated that the standard deviation of college-graduate earnings is about $8,500. Compute a 75% Chebyshev confidence interval centered on the mean ($40,478) for bachelor's degree earnings.

(c) How much does college tuition cost? That depends, of course, on where you go to college. Let's construct a weighted average. Using

89

the data from USA SNAPSHOT, "College affordable for most," we estimate mid-points for the cost intervals. Say 51% are about $3,500; 21% are about $6,000; 6% are about $10,000; 8% are about $14,000; and 7% are about $18,000. Compute the weighted average of college tuition charged at all colleges.

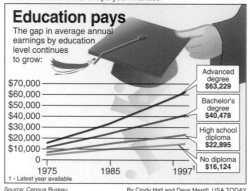

USA SNAPSHOTS®

A look at statistics that shape your finances

Education pays

The gap in average annual earnings by education level continues to grow:

$70,000		Advanced degree **$63,229**
$60,000		
$50,000		Bachelor's degree **$40,478**
$40,000		
$30,000		
$20,000		High school diploma **$22,895**
$10,000		
0		No diploma **$16,124**

1975   1985   1997[1]

1 - Latest year available

Source: Census Bureau   By Cindy Hall and Dave Merrill, USA TODAY

Source: Copyright © 1999 USA TODAY. Reprinted with permission.

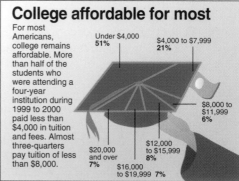

USA SNAPSHOTS®

A look at statistics that shape the nation

College affordable for most

For most Americans, college remains affordable. More than half of the students who were attending a four-year institution during 1999 to 2000 paid less than $4,000 in tuition and fees. Almost three-quarters pay tuition of less than $8,000.

Under $4,000 **51%**
$4,000 to $7,999 **21%**
$8,000 to $11,999 **6%**
$20,000 and over **7%**
$12,000 to $15,999 **8%**
$16,000 to $19,999 **7%**

Source: The College Board   By James Abundis and Genevieve Lynn, USA TODAY

Source: Copyright © 1999 USA TODAY. Reprinted with permission.

3.1 Measures of Central Tendency: Mode, Median, and Mean

FOCUS POINTS

✓ Compute mean, median, and mode from raw data.

✓ Interpret what mean, median, and mode tell us.

✓ Explain how mean, median, and mode can be affected by extreme data values.

✓ Why do we trim data? Compute a trimmed mean.

The average price of an ounce of gold is $295. The Zippy car averages 39 miles per gallon on the highway. A survey showed the average shoe size for women is size 8.

In each of the preceding statements, *one* number is used to describe the entire sample or population. Such a number is called an *average*. There are many ways to compute averages, but we will study only three of the major ones.

The easiest average to compute is the *mode*.

> The **mode** of a data set is the value that occurs most frequently.

EXAMPLE 1

Mode

Count the letters in each word of this sentence and give the mode. The number of letters in the words of the sentence are

5   3   7   2   4   4   2   4   8   3   4   3   4

Scanning the data, we see that 4 is the mode because more words have 4 letters than any other number. For larger data sets, it is useful to order—or sort—the data before scanning them for the mode.   ◊

Not every data set has a mode. For example, if Professor Fair gives equal numbers of As, Bs, Cs, Ds, and Fs, then there is no modal grade. In addition, the mode is not very stable. Changing just one number in a data set can change the mode dramatically. However, the mode is a useful average when we want to know the most frequently occurring data value, such as the most frequently requested shoe size.

Median

Another average that is useful is the *median,* or central value, of an ordered distribution. When you are given the median, you know there are an equal number of data values in the ordered distribution that are above it and below it.

> The **median** is the central value of an ordered distribution. To find it,
>
> (a) Order the data from smallest to largest.
>
> (b) For an *odd* number of data values in the distribution,
>
> $$\text{Median} = \text{Middle data value}$$
>
> (c) For an *even* number of data values in the distribution,
>
> $$\text{Median} = \frac{\text{Sum of middle two values}}{2}$$

EXAMPLE 2

Median

What do barbecue-flavored potato chips cost? According to *Consumer Reports,* Volume 66, No. 5, the price per ounce in cents of the rated chips are

19   19   27   28   18   35

(a) To find the median, we first order the data, and then note that there are an even number of entries, so the median is constructed using the two middle values

18   19   19   27   28   35

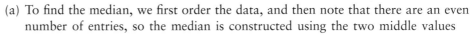

middle values

$$\text{Median} = \frac{19 + 27}{2} = 23 \text{ cents}$$

(b) According to *Consumer Reports,* the brand with the lowest overall taste rating costs 35 cents per ounce. Eliminate that brand, and find the median price per ounce for the remaining barbecue-flavored chips. Again order the data. Note that there are an odd number of entries, so the median is simply the middle value.

18   19   19   27   28

↑

middle value

Median = Middle value = 19 cents

(c) One ounce of potato chips is considered a small serving. Is it reasonable to budget about $10.45 to serve the barbecue-flavored chips to 55 people?

  Yes, since the median price of the chips is 19 cents per small serving. This budget for chips assumes that there is plenty of other food!   ◇

The median uses the *position* rather than the specific value of each data entry. If the extreme values of a data set change, the median usually does not change. This is why the median is often used as the average for house prices. If one mansion costing several million dollars sells in a community of much lower-priced homes, the median selling price for houses in the community would be affected very little, if at all.

GUIDED EXERCISE 1

Median and mode

Belleview College must make a report to the budget committee about the average credit hour load a full-time student takes. (A 12-credit-hour load is the minimum requirement for full-time status. For the same tuition, students may take up to 20 credit hours.) A random sample of 40 students yielded the following information (in credit hours):

17	12	14	17	13	16	18	20	13	12
12	17	16	15	14	12	12	13	17	14
15	12	15	16	12	18	20	19	12	15
18	14	16	17	15	19	12	13	12	15

(a) Organize the data from smallest to largest number of credit hours.

➡
12 12 12 12 12 12 12 12 12 12
13 13 13 13 14 14 14 14 15 ⑮
⑮ 15 15 15 16 16 16 16 17 17
17 17 17 18 18 18 19 19 20 20

(b) Since there are an _____ (odd, even) number of values, we add the two middle values and divide by 2 to get the median. What is the median credit hour load?

➡ There are an even number of entries. The two middle values are circled in part (a).

$$\text{Median} = \frac{15 + 15}{2} = 15$$

(c) What is the mode of this distribution? Is it different from the median? If the budget committee is going to fund the school according to the average student credit hour load (more money for higher loads), which of these two averages do you think the college will use?

➡ The mode is 12. It is different from the median. Since the median is higher, the school will probably use it and indicate that the average being used is the median.

Mean

An average that uses the exact value of each entry is the *mean* (sometimes called the *arithmetic mean*). To compute the mean, we add the values of all the entries and then divide by the number of entries.

$$\text{Mean} = \frac{\text{Sum of all the entries}}{\text{Number of entries}}$$

The mean is the average usually used to compute a test average.

EXAMPLE 3

Mean

To graduate, Linda needs at least a B in biology. She did not do very well on her first three tests; however, she did well on the last four. Here are her scores:

58   67   60   84   93   98   100

Compute the mean and determine if Linda's grade will be a B (80 to 89 average) or a C (70 to 79 average).

SOLUTION:

$$\text{Mean} = \frac{\text{Sum of scores}}{\text{Number of scores}} = \frac{58 + 67 + 60 + 84 + 93 + 98 + 100}{7}$$

$$= \frac{560}{7} = 80$$

Since the average is 80, Linda will get the needed B.   ◇

◇ **COMMENT** When we compute the mean, we sum the given data. There is a convenient notation to indicate the sum. Let x represent any value in the data set. Then the notation

Σx (read "the sum of all given x values")

means that we are to sum all the data values. In other words, we are to sum all the entries in the distribution. The *summation symbol* Σ means *sum the following* and is capital sigma, the *S* of the Greek alphabet.   ◇

Formulas for the mean

The symbol for the mean of a *sample* distribution of x values is denoted by $\bar{x}$ (read "x bar"). If your data comprises the entire *population*, we use the symbol μ (lowercase Greek letter mu, pronounced "mew") to represent the mean. The procedure to compute the mean is the same regardless of whether we have population or sample data. If we let n represent the number of entries in a *sample* data set and N represent the number of entries in a *population* data set, the formulas are

$$\text{Sample mean} = \bar{x} = \frac{\Sigma x}{n} \qquad\qquad \text{Population mean} = \mu = \frac{\Sigma x}{N} \qquad (1)$$

● **CALCULATOR NOTE** It is very easy to compute the mean on *any* calculator: Simply add the data values and divide the total by the number of data. However, on calculators with a statistics mode, you place the calculator in that mode, *enter* the data, and then press the key for the mean. The key is usually designated $\bar{x}$. Because the formula for the population mean is the same as that of the sample mean, the same key gives the value for μ.

We have seen three averages: the mode, the median, and the mean. For later work, the mean is the most important. A disadvantage of the mean, however, is that it can be affected by exceptional values.

Resistant measures

A *resistant measure* is one that is not influenced by extremely high or low data values. The mean is not a resistant measure of center because we can make the mean as large as we want by increasing the size of only one data value. The median, on the other hand, is more resistant. However, a disadvantage of the median is that it is not sensitive to the specific size of a data value.

Trimmed mean

A measure of center that is more resistant than the mean but still sensitive to specific data values is the *trimmed mean*. A trimmed mean is the mean of the data values left after "trimming" a specified percentage of the smallest and largest data values from the data set. Usually a 5% trimmed mean is used. This implies that we trim the lowest 5% of the data as well as the highest 5% of the data. A similar procedure is used for a 10% trimmed mean.

To compute a 5% trimmed mean

1. Order the data from smallest to largest.

2. Delete the bottom 5% of the data and the top 5% of the data. *Note:* If the calculation of 5% of the number of data values does not produce a whole number, *round* to the nearest integer.

3. Compute the mean of the remaining 90% of the data.

GUIDED EXERCISE 2

Mean and trimmed mean

Barron's Profiles of American Colleges, 19th Edition, lists average class size for introductory lecture courses at each of the profiled institutions. A sample of 20 colleges and universities in California showed class size for introductory lecture courses to be

⑭	20	20	20	20	23	25	30	30	30
35	35	35	40	40	42	50	50	80	⑧⓪

(a) Compute the mean for the entire sample.

⟹ Add all the values and divide by 20:

$$\bar{x} = \frac{\Sigma x}{n} = \frac{719}{20} \approx 36.0$$

(b) Compute a 5% trimmed mean for the sample.

⟹ The data are already ordered. Since 5% of 20 is 1, we eliminate one data value from the bottom of the list and one from the top. These values are circled in the data set. Then take the mean of the remaining 18 entries.

$$5\% \text{ trimmed mean} = \frac{\Sigma x}{n} = \frac{625}{18} \approx 34.7$$

(c) Find the median for the original data set.

⟹ Note the data are already ordered.

$$\text{Median} = \frac{30 + 35}{2} = 32.5$$

(d) Find the median of the 5% trimmed data set. Does the median change because of trimming the data?

⟹ The median is still 32.5. Notice that trimming the same number of entries from both ends leaves the middle position of the data set unchanged.

(e) Is the trimmed mean or the original mean closer to the median?

⟹ The trimmed mean is closer to the median.

TECH NOTE   Minitab, Excel, and the TI-83Plus all provide the mean and median of a data set. Minitab and Excel also provide the mode. The TI-83Plus sorts data, so you can easily scan the sorted data for the mode. Minitab provides the 5% trimmed mean, as does Excel.

All this technology is a wonderful aid for analyzing data. However, *a measurement has no meaning if you do not know what it represents or how a change in data values might affect the measurement.* The defining formulas and procedures for computing the measures tell you a great deal about the measures. Even if you use a calculator to evaluate all the statistical measures, pay attention to the information the formulas and procedures give you about the components or features of the measurement.

◊ **COMMENT**   In Chapter 1, we examined four levels of data: nominal, ordinal, interval, and ratio. The mode (if it exists) can be used with all four levels, including nominal. For instance, the modal color of all passenger cars sold last year might be blue. The median may be used with data at the ordinal level or above. If we ranked the passenger cars in order of customer satisfaction level, we could identify the median satisfaction level. For the mean, our data need to be at the interval or ratio level (although there are exceptions in which the mean of ordinal-level data is computed). We can certainly find the mean model year of used passenger cars sold or the mean price of new passenger cars.   ◊

VIEWP⊙INT

What's Wrong with Pitching Today?

One way to answer this question is to look at *averages.* Batting averages and average hits per game are shown for select years 1901 to 2000 (Source: *The Wall Street Journal*).

Year	1901	1920	1930	1941	1951	1961	1968	1976	1986	2000
B.A.	0.277	0.284	0.288	0.267	0.263	0.256	0.231	0.256	0.262	0.276
Hits	19.2	19.2	20.0	18.4	17.9	17.3	15.2	17.3	17.8	19.1

A quick scan of the averages shows that batting averages and average hits per game are virtually the same as almost 100 years ago. It seems there is *nothing* wrong with today's pitching! So what's changed? For one thing, the rules have changed! The strike zone is considerably smaller than it once was, and the pitching mound is lower. Both give the hitter an advantage over the pitcher. Even so, pitchers don't give up hits with any greater frequency than they did a century ago (look at the averages). However, modern hits go much farther, which is something a pitcher can't control.

SECTION 3.1   PROBLEMS

1. *Agriculture: Growing Season* The average length of the growing season is often measured in average number of frost-free days. The front range of Colorado (Fort Collins, Boulder, Denver, Colorado Springs, Pueblo) was studied by J. F. Benci and T. B. McKee, from the Department of Atmospheric Science at Colorado State University. Based on data from their Climatology Report No. 77-3, different locations in the Colorado front range had the following average number of frost-free days per year:

156	161	152	162	144	153
148	157	168	157	161	157

   Compute the mean, median, and mode. Write a brief description of the meaning of these numbers from the point of view of a gardener.

2. *Baseball: Home Runs* Babe Ruth was the American League Home Run Champion 12 times (during the period from 1918 to 1931). The number of home runs he hit to earn the 12 titles were

11	29	54	59	41	46
47	60	54	46	49	46

   Find the mean, median, and mode of the number of home runs.

3. *Environmental Studies: Death Valley* How hot does it get in Death Valley? The following data are taken from a study conducted by the National Park System, of which Death Valley is a unit. The ground temperatures (°F) were taken from May to November in the vicinity of Furnace Creek.

146	152	168	174	180	178	179
180	178	178	168	165	152	144

   Compute the mean, median, and mode for these ground temperatures.

4. *Ecology: Wolf Packs* How large is a wolf pack? The following information is from a random sample of winter wolf packs in regions of Alaska, Minnesota, Michigan, Wisconsin, Canada, and Finland. (Source: *The Wolf,* by L. D. Mech, University of Minnesota Press.) Winter pack size:

13	10	7	5	7	7	2	4	3
2	3	15	4	4	2	8	7	8

   Compute the mean, median, and mode for the size of winter wolf packs.

5. *Medical: Injuries* The Grand Canyon and the Colorado River are beautiful, rugged, and sometimes dangerous. Thomas Myers is a physician at the park clinic in Grand Canyon Village. Dr. Myers has recorded (for a 5-year period) the number of visitor injuries at different landing points for commercial boat trips down the Colorado River in both the upper and lower Grand Canyon (Source: *Fateful Journey* by Myers, Becker, Stevens).

   Upper Canyon: Number of Injuries per Landing Point Between North Canyon and Phantom Ranch

2	3	1	1	3	4	6	9	3	1	3

   Lower Canyon: Number of Injuries per Landing Point Between Bright Angel and Lava Falls

8	1	1	0	6	7	2	14	3	0	1	13	2	1

(a) Compute the mean, median, and mode for injuries per landing point in the Upper Canyon.

(b) Compute the mean, median, and mode for injuries per landing point in the Lower Canyon.

(c) Compare the results of parts (a) and (b).

(d) The Lower Canyon stretch had some extreme data values. Compute a 5% trimmed mean for this region, and compare this result to the mean for the Upper Canyon computed in part (a).

6. *Football: Age of Professional Players* How old are professional football players? The 11th Edition of *The Pro Football Encyclopedia* gave the following information. Random sample of pro football player ages in years:

24	23	25	23	30	29	28	26	33	29
24	37	25	23	22	27	28	25	31	29
25	22	31	29	22	28	27	26	23	21
25	21	25	24	22	26	25	32	26	29

(a) Compute the mean, median, and mode of the ages.

(b) Compare the averages. Does one seem to represent the age of the pro football players most accurately? Explain.

7. *Salaries: Bank Employees* Brookridge National Bank is a small bank in a rural Iowa town. The 12 people who work at the bank are the president, the vice president (his son-in-law), eight tellers, and two secretaries. The annual salaries for these people in thousands of dollars are as follows:

President: 93

Vice President: 80

Tellers: 15, 25, 14, 18, 21, 16, 19, 20

Secretaries: 12, 13

(a) Compute the mean of all 12 salaries.

(b) Compute the median of all 12 salaries, and compare your answers with the mean of part (a). Which average best describes the salaries of the *majority* of employees?

(c) Omit the salaries of the president and vice president. Calculate the mean and median for the remaining 10 people.

(d) Compare your answers from part (c) with those of parts (a) and (b). Comment on the effect of extreme values on the mean and median.

8. *Car Theft: Honolulu* A reporter for *Honolulu Star Bulletin* was doing a news article about car theft in Honolulu. For a given 10-day period, the police reported the following number of car thefts:

9	6	10	8	10	8	4	8	3	8

Then, for the next 3 days, for an unexplained reason, the number of car thefts jumped to 36, 51, and 30.

(a) Compute the mean, median, and mode for the first 10-day period.

(b) Compute the mean, median, and mode for the entire 13-day period.

(c) Comment on the effect of extreme values on the mean, median, and mode in this problem.

9. *Astronomy: Meteoroids* The College Astronomy Club has been counting meteoroids each night for the past week. Between the hours of 10:00 P.M. and 1:00 A.M., the meteoroid count for each night was

| 15 | 12 | 15 | 10 | 17 | 18 | 15 |

Then, for the next two nights there was a meteoroid shower, and the club counted 57 and 62 meteoroids.

(a) Compute the mean, median, and mode for the meteoroid counts on the first seven nights.

(b) Compute the mean, median, and mode for all nine nights.

(c) Comment on the effect of extreme values on the mean, median, and mode in this problem.

10. *Psychology: Performance/Reward Theory* In a course entitled "Experimental Psychology" at Regis University, students train white rats to do various tasks based on a performance/reward theory (for rats).

(a) One event is the hurdles. Times in seconds for seven rats to run the hurdles were

| 5.2 | 3.3 | 3.3 | 2.9 | 2.8 | 2.3 | 1.8 |

Compute the mean, median, and mode.

(b) Another event is the ladder climb. Times in seconds for eight rats to do the ladder climb were

| 41.9 | 7.7 | 7.9 | 6.9 | 6.6 | 6.6 | 5.5 | 5.1 |

Compute the mean, median, and mode for the ladder climb. The first time (41.9 seconds) was for a rat who got distracted in the middle of the performance. Omit the time for this rat, and recalculate the mean, median, and mode. Comment on the effect on the mean, median, and mode when we leave out the extreme value 41.9.

11. *Leisure: Maui Vacation* How expensive is Maui? If you want a vacation rental condominium (up to four people), visit the Brase/Brase statistics site at http://math.college.hmco.com/students, find the link to Maui, and then search for accommodations. *The Maui News* gave the following cost in dollars per day for a random sample of condominiums located throughout the island of Maui.

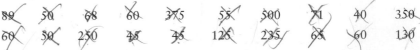

| 89 | 50 | 68 | 60 | 375 | 55 | 500 | 71 | 40 | 350 |
| 60 | 50 | 250 | 45 | 45 | 125 | 235 | 65 | 60 | 130 |

(a) Compute the mean, median, and mode for the data.

(b) Compute a 5% trimmed mean for the data, and compare it with the mean computed in part (a). Does the trimmed mean more accurately reflect the general level of the daily rental costs?

(c) If you were a travel agent, and a client asked about the daily cost of renting a condominium on Maui, what average would you use? Explain. Is there any other information about the costs that you think might be useful, such as the spread of the costs?

12. *Financial: Credit Cards* Consider the following types of data that were obtained from a random sample of 49 credit card accounts. Identify all the averages (mean, median, or mode) that can be used to summarize the data.

(a) Outstanding balance on each account.

(b) Name of credit card (e.g., MasterCard, Visa, American Express, etc.).

(c) Dollar amount due on next payment.

13. *Student Life: Schedules* Consider the following types of data about courses obtained from a random sample of 35 student schedules. Identify all the averages (mean, median, or mode) that can be used to summarize the data.

(a) Name of department in which each course is offered.
(b) Number of credit hours each student is taking.
(c) Starting time of earliest class on each schedule.

14. *Customer Service: Satisfaction Thriving on Chaos,* by Tom Peters, has some excellent cautions about utilizing measurements and averages. In discussing ways to provide superior service, he says, "The use of averages is downright dangerous." He illustrates with examples. Suppose a manufacturing company claims, "on average, we ship parts within 37 hours of order entry." But a careful look at data shows that for the "worst-off 10 percent of customers," the shipping time was within 89 hours of order entry. Peters' advice is to "focus attention on the worst-off 1, 5, 10, or 25 percent of customers" instead of on the average. Comment on this advice.

15. *General: Mean and Median* Consider a data set of 15 distinct measurements with mean A and median B.
    (a) If the highest number were increased, what would be the effect on the median and mean? Explain.
    (b) If the highest number were decreased to a value that is still larger than B, what would be the effect on the median and mean?
    (c) If the highest number were decreased to a value smaller than B, what would be the effect on the median and mean?

16. *General: Mean, Median, Mode* Create a data set with five numbers in which
    (a) the mean, median, and mode are all equal.
    (b) the mean is greater than the median.
    (c) the mean is less than the median.
    (d) the mode is higher than the median or mean.
    (e) all the numbers are distinct and the mean and median are both zero.

17. *Lifestyle: Gather Your Own Data* For the next week, record the length of time you spend on each telephone call you make or receive. Round the time to the nearest minute. Compute the mean, median, and mode. What accounts for some of your shorter phone calls (leaving messages, responding to marketing calls, etc.)?

3.2
Measures of Variation

FOCUS POINTS
✓ Find the range, variance, and standard deviation.
✓ Compute the coefficient of variation from raw data. Why is the coefficient of variation important?
✓ Apply Chebyshev's theorem to raw data. What does a Chebyshev interval tell us?

An average is an attempt to summarize a set of data in just one number. As some of our examples have shown, an average taken by itself may not always be very meaningful. We need a statistical cross-reference that measures the spread of the data.

The range is one such measure of variation.

> The **range** is the difference between the largest and smallest values of a data distribution.

EXAMPLE 4

Range

A large bakery regularly orders cartons of Maine blueberries. The average weight of the cartons is supposed to be 22 ounces. Random samples of cartons from two suppliers were weighed. The weights in ounces of the cartons were

| **Supplier I:** | 17 | 22 | 22 | 22 | 27 |
| **Supplier II:** | 17 | 19 | 20 | 27 | 27 |

Blueberry patch

Sample standard deviation

(a) Compute the range of carton weights from each supplier.

$$\text{Range} = \text{Largest value} - \text{Smallest value}$$
$$\text{Supplier I range} = 27 - 17 = 10 \text{ ounces}$$
$$\text{Supplier II range} = 27 - 17 = 10 \text{ ounces}$$

(b) Compute the mean weight of cartons from each supplier. In both cases the mean is 22 ounces.

(c) Look at the two samples again. The samples have the same range and mean. How do they differ? The bakery uses one carton of blueberries in each blueberry muffin recipe. It is important that the cartons be of consistent weight so that the muffins turn out right.

Supplier I provides more cartons that have weights closer to the mean. Or put another way, the weights of cartons from Supplier I are more clustered around the mean. The bakery might find Supplier I more satisfactory.   ◊

As we see in Example 4, although the range tells the difference between the largest and smallest values in a distribution, it does not tell us how much other values vary from one another or from the mean.

A measurement that will give you a better idea of how the data entries differ from the mean is the *standard deviation*. The formula for the standard deviation differs slightly depending on whether you are using an entire population or just a sample. At the moment, we will compute the standard deviation for sample data only. When we have sample data, we use the letter s to denote the standard deviation. The formula for the sample standard deviation is

$$\text{Sample standard deviation} = s = \sqrt{\frac{\Sigma(x - \bar{x})^2}{n - 1}} \qquad (2)$$

where x is any entry in the distribution, $\bar{x}$ is the mean, and n is the number of entries. The sum is taken over all data values.

Notice that the standard deviation uses the difference between each entry x and the mean $\bar{x}$. This quantity $(x - \bar{x})$ will be negative if the mean $\bar{x}$ is greater than the entry x. If you take the sum

$$\Sigma(x - \bar{x})$$

then the negative values will cancel the positive values, leaving you with a variation measure of 0 even if some entries vary greatly from the mean.

In the formula for the standard deviation, the quantities $(x - \bar{x})$ are *squared* before they are summed. This device eliminates the possibility of having some negative values in the sum. So, in the formula, we have the quantity

$$\Sigma(x - \bar{x})^2$$

Then we divide this sum by $n - 1$ to get the quantity under the square root sign:

$$\frac{\Sigma(x - \bar{x})^2}{n - 1}$$

These three steps have given us a quantity called the *variance* of a sample, denoted by s^2:

$$\text{Sample variance} = s^2 = \frac{\Sigma(x - \bar{x})^2}{n - 1} \tag{3}$$

The blueberry data in Example 4 initially described the weights of the cartons in ounces, but the variance s^2 of these data would be in *square ounces*. Square ounces—what's that? We need to take the square root of the variance to return to ounces. This brings us to the standard deviation of a sample.

$$\text{Sample standard deviation} = s = \sqrt{\frac{\Sigma(x - \bar{x})^2}{n - 1}}$$

The next example shows how to use this formula.

EXAMPLE 5

Sample standard deviation

Big Blossom Greenhouse was commissioned to develop an extra large rose for the Rose Bowl Parade. A random sample of blossoms from Hybrid A bushes yielded these diameters (in inches) for mature peak blossoms:

2   3   4   5   6   8   10   10

Find the standard deviation.

SOLUTION: Several steps are involved in computing the standard deviation, and a table will be helpful (see Table 3-1). Since $n = 8$, we take the sum of the entries in the first column of Table 3-1 and divide by 8 to find the mean $\bar{x}$.

$$\bar{x} = \frac{\Sigma x}{n} = \frac{48}{8} = 6.0$$

Using this value for $\bar{x}$, we obtain Column II. Square each value in the second column to obtain Column III, and then add the values in Column III. To get the variance, divide the sum of Column III by $n - 1$. Since $n = 8$, $n - 1 = 7$.

$$s^2 = \frac{\Sigma(x - \bar{x})^2}{n - 1} = \frac{66}{7} \approx 9.43$$

TABLE 3-1   Diameter of Rose Blossoms (in inches)

Column I x	Column II $x - \bar{x}$	Column III $(x - \bar{x})^2$
2	$2 - 6 = -4$	$(-4)^2 = 16$
3	$3 - 6 = -3$	$(-3)^2 = 9$
4	$4 - 6 = -2$	$(-2)^2 = 4$
5	$5 - 6 = -1$	$(-1)^2 = 1$
6	$6 - 6 = 0$	$(0)^2 = 0$
8	$8 - 6 = 2$	$(2)^2 = 4$
10	$10 - 6 = 4$	$(4)^2 = 16$
10	$10 - 6 = 4$	$(4)^2 = 16$
$\Sigma x = 48$		$\Sigma(x - \bar{x})^2 = 66$

Now obtain the standard deviation by taking the square root of the variance.

$$s = \sqrt{s^2} = \sqrt{9.43} \approx 3.07$$

(Use a calculator to compute a square root. Because of rounding, we use the approximately equal symbol, $\approx$.)   ◊

GUIDED EXERCISE 3

Sample standard deviation

Big Blossom Greenhouse gathered another random sample of mature peak blooms from Hybrid B. The eight blossoms had these widths (in inches):

5   5   5   6   6   6   7   8

(a) Again, we will construct a table so we can find the mean, variance, and standard deviation more easily. In this case, what is the value of n? Find the sum of Column I in Table 3-2, and compute the mean.

➡ $n = 8$. The sum of Column I is $\Sigma x = 48$, so the mean is

$$\bar{x} = \frac{48}{8} = 6 \text{ in.}$$

TABLE 3-2   Complete Columns II and III

I x	II $x - \bar{x}$	III $(x - \bar{x})^2$
5	_____	_____
5	_____	_____
5	_____	_____
6	_____	_____
6	_____	_____
6	_____	_____
7	_____	_____
8	_____	_____
$\Sigma x =$ _____		$\Sigma(x - \bar{x})^2 =$ _____

TABLE 3-3   Completion of Table 3-2

I x	II $x - \bar{x}$	III $(x - \bar{x})^2$
5	-1	1
5	-1	1
5	-1	1
6	0	0
6	0	0
6	0	0
7	1	1
8	2	4
$\Sigma x = 48$		$\Sigma(x - \bar{x})^2 = 8$

(b) What is the value of $n - 1$? Divide the total sum of Column III by $n - 1$ to find the variance.

➡ $n - 1 = 7$

$$\text{Variance} = s^2 = \frac{\Sigma(x - \bar{x})^2}{n - 1} = \frac{8}{7} \approx 1.14$$

(c) Use a calculator to find the square root of the variance. Is this the standard deviation?

➡ $\sqrt{\text{variance}} = \sqrt{s^2} = \sqrt{1.14} \approx 1.07 \text{ in.}$ The square root of the variance is the standard deviation.

Let's summarize and compare the results of Guided Exercise 3 and Example 5. The greenhouse found the following blossom diameters for Hybrid A and Hybrid B:

Hybrid A: mean, 6.0 in.; standard deviation, 3.07 in.

Hybrid B: mean, 6.0 in.; standard deviation, 1.07 in.

In both cases, the means are the same: 6 in. But the first hybrid has a larger standard deviation. This means that the blossoms of Hybrid A are less consistent than those of Hybrid B. If you want a rosebush that occasionally has 10-in. blooms and 2-in. blooms, use the first hybrid. But if you want a bush that consistently produces roses close to 6 in. across, use Hybrid B.

◇ **COMMENT** The value $\Sigma(x - \overline{x})^2$ is used so commonly in other measures in statistics that it has its own name, the *sum of squares,* and symbol SS_x. A computationally simpler (requires fewer subtractions) version of the expression that is often used is

$$\text{Sum of squares} = SS_x = \Sigma(x - \overline{x})^2 = \Sigma x^2 - \frac{(\Sigma x)^2}{n} \tag{4}$$

Then $s = \sqrt{\dfrac{SS_x}{n - 1}}$. We will see the sum of squares SS_x when we study linear regression and analysis of variance.   ◇

● **ROUNDING NOTE**   Rounding errors cannot be completely eliminated, even if a computer or calculator does all the computations. However, software and calculator routines are designed to minimize the error. If the mean is rounded, the value of the standard deviation will change slightly depending on how much the mean is rounded. If you do your calculations "by hand" or reenter intermediate values into a calculator, try to carry one or two more digits than occur in the original data. If your resulting answers vary slightly from those in this text, do not be overly concerned. The text answers are computer- or calculator-generated.

Population mean and standard deviation

In almost all applications of statistics, we work with a random sample of data rather than the entire population of *all* possible data values. However, if we do in fact have data for the entire population, we can compute the *population mean* μ (lowercase Greek letter mu, pronounced "mew") *and population standard deviation* σ (lowercase Greek letter sigma) using the following formulas:

$$\mu = \frac{\Sigma x}{N} \quad \text{population mean}$$

$$\sigma = \sqrt{\frac{\Sigma(x - \mu)^2}{N}} \quad \text{population standard deviation}$$

where N is the number of data values in the population, and x represents the individual data values of the population. We note that the formula for μ is the same as the formula for $\overline{x}$ (the sample mean) and the formula for σ is the same as the formula for s (the sample standard deviation), except that the population size N is used instead of $n - 1$. Also, μ is used instead of $\overline{x}$ in the formula for σ.

In the formulas for s and σ we use $n - 1$ to compute s, and N to compute σ. Why? The reason is that N (capital letter) represents the *population size*, while n (lowercase letter) represents the sample size. Since a random sample usually will not contain extreme data values (large or small), we divide by $n - 1$ in the formula for s to make s a little larger than it would have been had we divided by n. Courses in advanced theoretical statistics show that this procedure will give us the best possible estimate for the standard deviation σ. In fact, s is called the *unbiased estimate* for σ. If we have the population of all data values, then extreme data values are, of course, present, so we divide by N instead of $N - 1$.

◇ **COMMENT** The computation formula for the population standard deviation is

$$\sigma = \sqrt{\frac{SS_x}{N}} \quad \text{where} \quad SS_x = \Sigma(x - \mu)^2 = \Sigma x^2 - \frac{(\Sigma x)^2}{N} \quad ◇$$

We've seen that the standard deviation (sample or population) is a measure of data spread. We will use the standard deviation extensively in later chapters. In Chapter 6 we will use it to study standard z values and areas under normal curves. In Chapters 8 and 9 we will use it to study the inferential statistics topics of estimation and testing. The standard deviation will appear again in our study of regression and correlation.

 TECH NOTE Most scientific or business calculators have a statistics mode, and provide the mean and sample standard deviation directly. The TI-83Plus, Excel, and Minitab provide the median and several other measures as well.

Many technologies display only the sample standard deviation s. You can quickly compute σ if you know s by using the formula

$$\sigma = s\sqrt{\frac{n - 1}{n}}$$

The mean given in displays can be interpreted as the sample mean $\bar{x}$ or the population mean μ as appropriate.

The following three displays show output for the hybrid rose data of Guided Exercise 3.

TI-83Plus Display
Press **STAT** ➤ **CALC** ➤ **1:1-Var Stats.** S_x is the sample standard deviation. σ_x is the population standard deviation.

```
1-Var Stats
 x̄=6
 Σx=48
 Σx²=296
 Sx=1.069044968
 σx=1
↓n=8
```

Excel Display

Menu choices: **Tools ➤ Data Analysis ➤ Descriptive Statistics.** The standard deviation is the sample standard deviation.

Mean	6
Standard Error	0.377964
Median	6
Mode	5
Standard Deviation	1.069045
Sample Variance	1.142857
Kurtosis	0.35
Skewness	0.935414
Range	3
Minimum	5
Maximum	8
Sum	48
Count	8

Minitab Display

Menu choices: **Stat ➤ Basic Statistics ➤ Display Descriptive Statistics.** StDev is the sample standard deviation. TrMean is a 5% trimmed mean.

```
Descriptive Statistics
    N          Mean        Median        TrMean         StDev         SE Mean
    8          6.000        6.000         6.000         1.069         0.378
 Minimum        Maximum          Q1            Q3
  5.000          8.000        5.000         6.750
```

Now let's look at two immediate applications of the standard deviation. The first is the coefficient of variation, and the second is Chebyshev's theorem.

Coefficient of Variation

A disadvantage of the standard deviation as a comparative measure of variation is that it depends on the units of measurement. This means that it is difficult to use the standard deviation to compare measurements from different populations. For this reason, statisticians have defined the *coefficient of variation,* which expresses the standard deviation as a percentage of the sample or population mean.

If $\bar{x}$ and s represent the sample mean and sample standard deviation, then the sample **coefficient of variation** CV is defined to be

$$CV = \frac{s}{\bar{x}} \cdot 100$$

If μ and σ represent the population mean and standard deviation, then the population coefficient of variation CV is defined to be

$$CV = \frac{\sigma}{\mu} \cdot 100$$

Notice that the numerator and denominator in the definition of *CV* have the same units, so *CV* itself has no units of measurement. This gives us the advantage of being able to directly compare the variability of two different populations using the coefficient of variation.

In the next example and guided exercise, we will compute the *CV* of a population and of a sample and then compare the results.

EXAMPLE 6

Coefficient of variation

The Trading Post on Grand Mesa is a small, family-run store in a remote part of Colorado. The Grand Mesa region contains many good fishing lakes, so the Trading Post sells spinners (a type of fishing lure). The store has a very limited selection of spinners, however. In fact, the Trading Post has only eight different types of spinners for sale. The prices (in dollars) are

2.10   1.95   2.60   2.00   1.85   2.25   2.15   2.25

Since the Trading Post has only eight different kinds of spinners for sale, we consider the eight data values to be the *population*.

(a) Use a calculator with appropriate statistics keys to verify that for the Trading Post data, $\mu = \$2.14$ and $\sigma = \$0.22$.

SOLUTION: Since the computation formula for $\bar{x}$ and μ are identical, most calculators provide the value of $\bar{x}$ only. Use the output of this key for μ. The computation formulas for the sample standard deviation s and the population standard deviation σ are slightly different. Be sure that you use the key for σ (sometimes designated as σ_n or σ_x).

(b) Compute the *CV* of prices for the Trading Post and comment on the meaning of the result.

SOLUTION:

$$CV = \frac{\sigma}{\mu} \times 100 = \frac{0.22}{2.14} \times 100 = 10.28\%$$

The coefficient of variation can be thought of as a measure of the spread of the data relative to the average of the data. Since the Trading Post is very small, it carries a small selection of spinners that are all priced similarly. The CV tells us that the standard deviation of the spinner prices is only 10.28% of the mean.   ◊

GUIDED EXERCISE 4

Coefficient of variation

Cabela's in Sidney, Nebraska, is a very large outfitter that carries a broad selection of fishing tackle. It markets its products nationwide through a catalog service. A random sample of 10 spinners from Cabela's extensive spring catalog gave the following prices (in dollars):

1.69   1.49   3.09   1.79   1.39
2.89   1.49   1.39   1.49   1.99

GUIDED EXERCISE 4 continued

(a) Use a calculator with sample mean and sample standard deviation keys to compute $\bar{x}$ and s.

⟹ $\bar{x} = \$1.87$ and $s = \$0.62$.

(b) Compute the CV for the spinner prices at Cabela's.

⟹ $CV = \dfrac{s}{\bar{x}} \times 100 = \dfrac{0.62}{1.87} \times 100 = 33.16\%.$

(c) Compare the mean, standard deviation, and CV for the spinner prices at the Grand Mesa Trading Post (Example 6) and Cabela's. Comment on the differences.

⟹   The CV's for Cabela's and the Trading Post are pure numbers (no units), so a direct comparison is possible. The CV for Cabela's is more than three times the CV for the Trading Post. Why? First, because of the remote location, the Trading Post tends to have somewhat higher prices (larger μ). Second, the Trading Post is very small, so it has a rather limited selection of spinners with a smaller variation in price. For Cabela's, however, the average price $\bar{x}$ is lower and the variety larger (larger s). It makes sense that the CV for Cabela's is larger than the CV for the Trading Post.

Chebyshev's Theorem

From our earlier discussion about standard deviation, we recall that the spread or dispersion of a set of data about the mean will be small if the standard deviation is small, and it will be large if the standard deviation is large. If we are dealing with a symmetrical bell-shaped distribution, then we can make very definite statements about the proportion of the data that must lie within a certain number of standard deviations on either side of the mean. This will be discussed in detail in Chapter 6 when we talk about normal distributions.

However, the concept of data spread about the mean can be expressed quite generally for *all data distributions* (skewed, symmetric, or other shape) by using the remarkable theorem of Chebyshev:

Chebyshev's theorem

For *any* set of data (either population or sample) and for any constant k greater than 1, the proportion of the data that must lie within k standard deviations on either side of the mean is *at least*

$$1 - \frac{1}{k^2}$$

Results of Chebyshev's theorem

For *any* set of data:

- at *least* 75% of the data fall in the interval from $\mu - 2\sigma$ to $\mu + 2\sigma$
- at *least* 88.9% of the data fall in the interval from $\mu - 3\sigma$ to $\mu + 3\sigma$
- at *least* 93.8% of the data fall in the interval from $\mu - 4\sigma$ to $\mu + 4\sigma$

The results of Chebyshev's theorem can be derived by using the theorem and a little arithmetic. For instance, if we create an interval $k = 2$ standard deviations on either side of the mean, Chebyshev's theorem tells us that

$$1 - \frac{1}{2^2} = 1 - \frac{1}{4} = \frac{3}{4} \text{ or } 75\%$$

is the minimum percentage of data in the $\mu - 2\sigma$ to $\mu + 2\sigma$ interval.

Notice that Chebyshev's theorem refers to the *minimum* percentage of data that must fall within the specified number of standard deviations of the mean. If the distribution is mound-shaped, an even *greater* percentage of data will fall into the specified intervals (see the Empirical Rule in Section 6.1).

EXAMPLE 7

Chebyshev's theorem

Students Who Care is a student volunteer program in which college students donate work time to various community projects such as planting trees. Professor Gill is the faculty sponsor for this student volunteer program. For several years, Dr. Gill has kept a careful record of x = total number of work hours volunteered by a student in the program each semester. For a random sample of students in the program, the mean number of hours was $\bar{x} = 29.1$ hours each semester, with a standard deviation $s = 1.7$ hours each semester. Find an interval A to B for the number of hours volunteered into which at least 75% of the students in this program would fit.

SOLUTION: According to results of Chebyshev's theorem, at least 75% of the data must fall within 2 standard deviations of the mean. Because the mean $\bar{x} = 29.1$ and the standard deviation $s = 1.7$, the interval is

$$\bar{x} - 2s \text{ to } \bar{x} + 2s$$
$$29.1 - 2(1.7) \text{ to } 29.1 + 2(1.7)$$
$$25.7 \text{ to } 32.5$$

At least 75% of the students would fit into the group that volunteered from 25.7 to 32.5 hours each semester.   ◊

GUIDED EXERCISE 5

Chebyshev interval

The *East Coast Independent News* periodically runs ads in its own classified section offering a month's free subscription to those who respond. In this way, management can get a sense about the number of subscribers who read the classified section each day. Over a period of 2 years, careful records have been kept. The mean number of responses per ad is $\bar{x} = 525$ with standard deviation $s = 30$.

Determine a Chebyshev interval about the mean in which at least 88.9% of the data fall.	By Chebyshev's theorem, at least 88.9% of the data fall into the interval $\bar{x} - 3s$ to $\bar{x} + 3s$ Because $\bar{x} = 525$ and $s = 30$, the interval is $525 - 3(30)$ to $525 + 3(30)$ or from 435 to 615 responses per ad.

VIEWP◉INT   Socially Responsible Investing

Make a difference *and* make money! Socially responsible mutual funds tend to screen out corporations that sell tobacco, weapons, and alcohol, as well as companies that are environmentally unfriendly. In addition, these funds screen out companies that use child labor in sweatshops. There are 68 socially responsible funds tracked by the Social Investment Forum. For more information, visit the Brase/Brase statistics site at http://math.college.hmco.com/students and find the link to social investing.

How do these funds rate compared to other funds? One way to answer this question is to study the annual percent returns of the funds using both the *mean and standard deviation*. (See Problems 5 and 6 of this section.)

SECTION 3.2   PROBLEMS

1. *Ecology: Deer* By sampling different landscapes in Mesa Verde National Park over a 2-year period, the number of deer per square kilometer was determined (*The Mule Deer of Mesa Verde National Park,* by G. W. Mieran and J. L. Schmidt, published by Mesa Verde Museum Association, 1981). The results were (deer per square kilometer)

30	20	5	29	58	7
20	18	4	29	22	9

   (a) Compute the range, sample mean, sample variance, and sample standard deviation.

(b) Compute the coefficient of variation. Is the coefficient of variation a fairly large number? Would you say there was a considerable variation in the distribution of deer from one section of the park to another? Explain.

2. *Wildlife Management: Canada Geese* Hatching success of game birds is a topic discussed in the book *Wildlife Management Techniques,* edited by R. H. Giles (The Wildlife Society Press, Washington, D.C.). What percentage of Canada goose nests are successful (at least one gosling survives)? Studies in regions of Montana, Illinois, Wyoming, Utah, and California gave the following percentage of successful nests:

23.9   52.5   60.0   68.5   78.6   71.0   17.8   57.5   59.0   52.0

(a) Compute the range and mean.
(b) Compute the sample variance and standard deviation.
(c) Compute the coefficient of variation. Write a brief explanation of the meaning of this number.

3. *Wildlife Management: Mallard Ducks* Hatching success for mallard duck nests is also given in the same reference as in Problem 2. A successful nest means at least one duckling survives. The following are percentages of successful mallard duck nests:

55.7   85.2   51.5   12.7   38.5   90.3   70.0   39.8   65.0   48.0

(a) Compute the range and mean.
(b) Compute the sample variance and standard deviation.
(c) Compute the coefficient of variation. Write a brief explanation of the meaning of this number. Compare this value with the coefficient of variation for Canada geese in Problem 2. What does this mean about relative variability of nesting success rates for Canada geese compared with mallard ducks?

4. *Business Administration: Profits* Jobs and productivity! How productive are American workers? One way to answer this question is to examine annual profits per employee. The results may change, however, from one industry group to another. This problem is based on information taken from *Forbes Top Companies* (edited by J. T. Davis, John Wiley Publisher).
(a) A random sample of seven automotive parts companies gave the following information about *profits* per employee (in thousands of dollars per employee):

14.1   12.4   7.7   6.9   9.0   6.8   6.8

Compute the range, mean, sample variance, sample standard deviation, and coefficient of variation.
(b) For a random sample of seven computer companies, the profits per employee in the same units were

29.0   24.5   31.0   29.8   21.8   27.7   19.1

Compute the range, mean, sample variance, sample standard deviation, and coefficient of variation.
(c) Examine your answers for parts (a) and (b). Although the standard deviation for part (a) is the smaller of the two, the mean for part (b) is the larger of the two. The *CV* for part (b) is almost half that of part (a) In this case, why would a small *CV* and high mean indicate more reliable productivity from the point of view of annual profits per employee? Explain.

5. *Investing: Socially Responsible Mutual Funds* Pax World Balanced is a highly respected, socially responsible mutual fund of stocks and bonds (see Viewpoint). Vanguard Balanced Index is another highly regarded fund that represents the entire

U.S. stock and bond market (an index fund). The mean and standard deviation of annualized percent returns are shown below. The annualized mean and standard deviation are based on 36 months of data prior to May 2001 (Source: Morningstar).

Pax World Balanced: $\bar{x} = 11.69\%$; $s = 11.56\%$

Vanguard Balanced Index: $\bar{x} = 5.61\%$; $s = 12.50\%$

(a) Compute the coefficient of variation for each fund. If $\bar{x}$ represents return and s represents risk, then explain why the coefficient of variation can be taken to represent risk per unit of return. From this point of view, which fund appears to be better? Explain.

(b) Compute a 75% Chebyshev interval around the mean for each fund. Use the intervals to compare the two funds. As usual, past performance does not guarantee future performance.

6. *Investing: Moving Averages* You do not need a lot of money to invest in a mutual fund. However, if you decide to put some money into an investment, you are usually advised to leave it in for (at least) several years. Why? Because good years tend to cancel out the bad years, giving you a better overall return with less risk. To see what we mean, let's use a 3-year *moving average* on the Calvert Social Balanced Fund (a socially responsible fund).

Year:	1990	1991	1992	1993	1994	1995	1996	1997	1998	1999	2000
% Return:	1.78	17.79	7.46	5.95	−4.74	25.85	9.03	18.92	17.49	6.80	−2.38

Source: Morningstar

(a) Use a calculator with mean and standard deviation keys to verify that the mean annual return for all 11 years is approximately 9.45%, with standard deviation 9.57%.

(b) To compute a 3-year moving average for 1992, we take the data value for 1992 and the data for the prior two years and average them. To compute a 3-year moving average for 1993, we take the data values for 1993 and the prior two years and average them. Verify that the following 3-year moving averages are correct.

Year:	1992	1993	1994	1995	1996	1997	1998	1999	2000
3-year moving average:	9.01	10.40	2.89	9.02	10.05	17.93	15.15	14.40	7.30

(c) Use a calculator with mean and standard deviation keys to verify that for the 3-year moving average, the mean is 10.68% with sample standard deviation 4.53%.

(d) Compare the results of parts (a) and (c). Suppose we take the point of view that risk is measured by standard deviation. Is the risk (standard deviation) of the 3-year moving average considerably smaller? This is an example of a general phenomenon that will be studied in more detail in Chapter 7.

7. *Astronomy: Sun Spot Cycles* The National Aeronautics and Space Administration (NASA) has studied data on sun spot cycles collected for the years 1745 to the present. During this time, the mean length of a cycle (maximum to maximum) was 11.01 years, with a standard deviation 2.17 years.

(a) Use Chebyshev's theorem to find an interval centered about the mean for the cycle length in which you would expect at least 75% of the cycles to fall.

(b) Use Chebyshev's theorem to find an interval centered about the mean for the cycle length in which you would expect at least 93.8% of the cycles to fall.

8. *Meteorology: Tornados* The U.S. Weather Bureau has provided the following information about the total annual number of reported tornados in the United States for the years 1956 to 1975:

504	856	564	604	616	697	657	464	704	906
585	926	660	608	653	888	741	1102	947	918

(a) Use a calculator with mean and standard deviation keys to verify that the mean number of tornados per year is 730, with a sample standard deviation of 172 tornados.

(b) Use Chebyshev's theorem to find an interval centered about the mean for the annual number of tornados in which you would expect at least 75% of the years to fall.

(c) Use Chebyshev's theorem to find an interval centered about the mean in which you would expect at least 88.9% of the years to fall.

9. *History: Billy the Kid* In 1881, Billy the Kid killed two deputies and escaped from a jail cell in the Lincoln County Courthouse, New Mexico Territory. Famous frontier personalities such as Kit Carson, Jesse Chisum, and Sheriff Pat Garrett (who eventually shot Billy the Kid) also were involved in the notorious Lincoln County Cattle Baron Wars. Before the 1985 renovation of the now famous courthouse, the Museum of New Mexico commissioned anthropologist/historian Yvonne Oakes to do a thorough analysis of the area. The distribution of artifacts (of the 1880s' period) for seven excavation sites were

851	596	444	956	576	219	326

(Source: *Museum of New Mexico: Laboratory of Anthropology Notes No. 357.*)

(a) Compute the range and mean.
(b) Compute the sample variance and standard deviation.
(c) Compute the coefficient of variation. Write a brief explanation of the meaning of this number in the context of this problem.
(d) Use Chebyshev's theorem to find an interval centered about the mean in which at least 75% of the artifact counts for all such excavation sites would fall.

History: Lincoln Country Courthouse Excavation artifacts at the Lincoln County Courthouse (see Problem 9) were examined and labeled according to their functional typology. Table 3-4 gives the mean and coefficient of variation for percentage of artifacts found in excavation sites at the Lincoln County Courthouse. This table will be used in Problems 10 through 14.

10. Suppose you were looking for antique liquor bottles, spittoons, and so forth (i.e., indulgence typology). With museum permission, you start a new excavation site at the Lincoln County Courthouse. Let P represent the percentage of total artifacts that are of the indulgence typology. Can we estimate a range of values for P before we start the excavation? Use the information of Table 3-4 with Chebyshev's theorem to find an interval centered about the mean in which you expect at least 75% of the Ps (percentage of indulgence artifacts at different sites) to fall.

11. Repeat Problem 10 assuming that you were interested in domestic routine typology.

12. Repeat Problem 10 assuming that you were interested in arms typology.

13. Examine Table 3-4. Would you agree with the statement that construction artifacts were one of the more likely and relatively dependable artifacts to be found? Use the coefficient of variation to explain your answer. (*Hint:* Which typology has the smallest *CV* and largest mean?)

TABLE 3-4   Artifacts Found at Lincoln County Courthouse*

Artifact Typology	Mean Percentage	Standard Deviation	CV
Foodstuffs (food storage, processing)	12.5	4.5	36.0
Indulgences (liquor, smoking items)	4.0	1.2	30.0
Domestic routine (tableware, furniture, lamps)	9.3	2.6	28.0
Construction (hardware, tools)	54.1	6.2	11.5
Personal effects (clothing, jewelry)	3.1	1.2	38.7
Entertainment (toys, games, musical instruments)	1.6	0.4	25.0
Arms (ammunition, guns)	0.4	0.1	25.0
Stable	0.2	0.1	50.0
Unknown	27.2	8.2	30.2

*The means were averaged over numerous excavation sites, and they need not total to 100%.

14. Examine Table 3-4. Arms and stable artifacts have the same standard deviation, but one CV was twice the other. How can this be explained using the mean percentage of artifacts for each?

15. *Financial: Wal-Mart and Disney Stock* The majority of U.S. households now own stocks. Two widely held stocks are Wal-Mart and Disney. Let's look at daily closing price per share (in dollars) for the two stocks during the period May 11 to June 9, 2001.

    Wal-Mart, $: 54.10 54.35 52.00 51.65 51.76 52.04 53.20 53.52 52.52 52.89
    51.20 51.21 51.47 51.75 51.72 50.99 51.30 50.75 51.10 51.02

    Disney, $: 31.27 30.95 31.10 32.06 32.51 32.60 34.28 34.50 33.11 33.33
    32.64 32.62 31.78 31.62 31.72 31.62 31.35 31.60 32.15 31.85

    (a) Use a calculator to verify the following means and standard deviations: Wal-Mart: $\bar{x} \approx \$52.03$, $s \approx \$1.06$; Disney: $\bar{x} \approx \$32.23$, $s \approx \$0.98$.

    (b) Compute the coefficient of variation for Wal-Mart and for Disney. If $\bar{x}$ represents return and s represents risk, then explain why the coefficient of variation can be taken to represent risk per unit of return. From this point of view, do the stocks appear to be about equally attractive (in this period)? Explain.

    (c) Compute an 88.9% Chebyshev interval for Wal-Mart and for Disney. Many stock brokers refer to *support* and *resistance* as the trading range of a stock. The *support* is a value the broker is "confident" the stock will not sink below. The *resistance* is a value the broker is "confident" the stock will not rise above. There are many different ways to compute supports and resistances. An 88.9% Chebyshev interval is one way, with the support being the lower end of the interval, and the resistance being the upper end. Compute the support and resistance for Wal-Mart and for Disney using this method.

16. *Medical: Physician Visits* In some reports, the mean and coefficient of variation are given. For instance, in *Statistical Abstract of the United States*, 116th Edition, one report gives the average number of physician visits by males per year. The average reported is 2.2, and the reported coefficient of variation is 1.5%. Use this

information to determine the standard deviation of the number of visits to physicians made by males.

17. *Lifestyle: Gathering Data* For the next 2 weeks, record how long it takes you to go from your residence (or job) to your first class. Record the values to the nearest minute. Compute the range, mean, sample standard deviation, and coefficient of variation. Use Chebyshev's theorem to compute an interval in which at least 88.9% of the travel times fall. Were there any unusual data values in your data set? How would you explain them?

3.3
Mean and Standard Deviation of Grouped Data

FOCUS POINTS

✓ Estimate the mean, variance, and standard deviation from grouped data.

✓ Compute a weighted average.

✓ Discuss applications of weighted averages.

If you have a great many data values, it can be quite tedious to compute the mean and standard deviation. Even if you have a calculator, you must punch in the data. In many cases, a close approximation to the mean and standard deviation is all that is needed, and it is not difficult to approximate these two values from a frequency distribution.

The basic plan is as follows:

1. Make a frequency table corresponding to the histogram.

2. Compute the midpoint for each class and call it x.

3. Count the number of entries in each class and denote the number by f.

4. Add the number of entries from each class together to find the total number of entries n in the sample distribution.

Treat each entry of a class as though it falls on the midpoint (x) of that class. Then the midpoint times the number of entries in a class (xf) represents the sum of the observations in the class. The formulas for the *mean of grouped data* and *standard deviation of grouped data* are as follows:

Approximating $\bar{x}$ and s from grouped data

Sample mean for a frequency distribution

$$\bar{x} = \frac{\Sigma xf}{n} \tag{5}$$

Sample standard deviation for a frequency distribution

$$s = \sqrt{\frac{\Sigma(x - \bar{x})^2 f}{n - 1}} \tag{6}$$

Computation formula for the standard deviation

$$s = \sqrt{\frac{SS_x}{n - 1}} \tag{7}$$

where $SS_x = \Sigma x^2 f - \frac{(\Sigma xf)^2}{n}$

x is the midpoint of a class,
f is the number of entries in that class,
n is the total number of entries in the distribution, and $n = \Sigma f$
The summation Σ is over all classes in the distribution.

EXAMPLE 8

Grouped data

FIGURE 3-1

Time in Minutes Before Checkout Begins

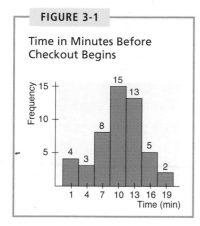

The manager of Pantry Queen Supermarket wants to hire one more checkout clerk. To justify his request to the regional manager, the manager chose a random sample of 50 customers and timed how long each stood in line before a clerk could begin checking the customer out. The written request contained the histogram in Figure 3-1. Approximate the mean and standard deviation of the distribution.

SOLUTION: First, make a table with all the columns necessary to compute the mean and standard deviation (see Table 3-5). (Columns V, VI, and VII are filled in after the mean is computed.) Use the first four columns to find the mean. The value of n is found by summing Column II.

$$n = \Sigma f = 50$$

The mean is

$$\bar{x} = \frac{\Sigma xf}{n} = \frac{509}{50} \text{ (from the sum of Column IV)}$$

$$= 10.2 \text{ min}$$

Once you have the value of the mean, you can complete Columns V, VI, and VII. (They have already been completed for our convenience.)

$$s = \sqrt{\frac{\Sigma(x - \bar{x})^2 f}{n - 1}}$$

$$= \sqrt{\frac{\text{Sum of Column VII}}{(\text{Sum of Column II}) - 1}}$$

$$= \sqrt{\frac{961.40}{50 - 1}}$$

$$= \sqrt{19.62}$$

$$= 4.43$$

Finding $\bar{x}$ and s with repeated data

In the case where our data are not grouped but there are several repeated data values, we can use the techniques of grouped data to find the mean and standard deviation fairly quickly. In such cases, we do not need to find a midpoint of a class interval, since each class consists of a single data value.

Checkout lines

TABLE 3-5   Time in Minutes Before Checkout Begins

I	II	III	IV	V	VI	VII
	Frequency	Midpoint				
Class	f	x	xf	$x - \bar{x}$	$(x - \bar{x})^2$	$(x - \bar{x})^2 f$
0–2	4	1	4	−9.2	84.64	338.56
3–5	3	4	12	−6.2	38.44	115.32
6–8	8	7	56	−3.2	10.24	81.92
9–11	15	10	150	−0.2	0.04	0.60
12–14	13	13	169	2.8	7.84	101.92
15–17	5	16	80	5.8	33.64	168.20
18–20	2	19	38	8.8	77.44	154.88
	$\Sigma f = 50$		$\Sigma xf = 509$			$\Sigma(x - \bar{x})^2 f = 961.40$

GUIDED EXERCISE 6

Mean and standard deviation for repeated data (computation formula)

A random sample of 60 college football players gave the following information (Table 3-6) about recovery time from shoulder injuries, where

x = number of weeks for recovery; f = number of injured players

TABLE 3-6   Recovery Times from Shoulder Injuries

x	1	2	3	4	5	6	7	8
f	5	8	12	19	7	4	3	2

Use the computation formula for grouped data to find the mean and standard deviation of the recovery times.

(a) Looking at the formula for $\bar{x}$ and s,

we see $\bar{x} = \dfrac{\Sigma xf}{n}$

and $s = \sqrt{\dfrac{SS_x}{n-1}}$ where

$SS_x = \Sigma x^2 f - \dfrac{(\Sigma xf)^2}{n}$ and $n = \Sigma f$

Notice that we sum the values f, xf, and $x^2 f$. Make a computation table with x, f, xf, and $x^2 f$ as headers and fill in the values and required totals.

➡️ **TABLE 3-7   Recovery Times** (column headers for computation formula for s)

x	f	xf	$x^2 f$
1	5	5	5
2	8	16	32
3	12	36	108
4	19	76	304
5	7	35	175
6	4	24	144
7	3	21	147
8	2	16	128
	$\Sigma f = 60$	$\Sigma xf = 229$	$\Sigma x^2 f = 1043$

(b) Use the formulas to find s and $\bar{x}$.

➡️ $n = \Sigma f = 60$

$\bar{x} = \dfrac{\Sigma xf}{n} = \dfrac{229}{60} \approx 3.82$ weeks

$SS_x = \Sigma x^2 f - \dfrac{(\Sigma xf)^2}{n} = 1043 - \dfrac{229^2}{60} \approx 168.98$

$s = \sqrt{\dfrac{SS_x}{n-1}} = \sqrt{\dfrac{168.98}{59}} \approx 1.69$

Continued

GUIDED EXERCISE 6 continued

(c) To find the mean and standard deviation of these data, do you think it would be easier to enter all 60 data values directly into your calculator and have the calculator do the work, or use the shortcut method for grouped data?

 When you gain skill in using the grouped-data formula, you will find that it is very efficient. Some calculators, including the TI-83Plus, allow you to enter data values and their frequencies as in Table 3-6 instead of entering all 60 data values. Then the calculator will quickly give you values for $\bar{x}$ and s.

Weighted average

Sometimes we wish to average numbers, but we want to assign more importance or weight to some of the numbers. For instance, suppose your professor tells you that your grade will be based on a midterm and a final exam, each of which has 100 possible points. However, the final exam will be worth 60% of the grade and the midterm only 40%. How could you determine your average score to reflect these different weights? The average you need is the *weighted average*.

If we view the weight of a measurement as a "frequency," then we discover that the formula for the mean of a frequency distribution gives us the weighted average.

$$\text{Weighted average} = \frac{\Sigma xw}{\Sigma w}$$

where w is the weight of the data value x.

EXAMPLE 9

Weighted average

Suppose your midterm test score is 83 and your final exam score is 95. Using the weights of 40% for the midterm and 60% for the final exam, compute the weighted average of your scores. If the minimum average for an A is 90, will you earn an A?

SOLUTION: By the formula, we multiply each score by its weight and add the results together. Then we divide by the sum of all the weights. Converting the percentages to decimal notation, we get

$$\text{Weighted average} = \frac{83(0.40) + 95(0.60)}{0.40 + 0.60}$$
$$= \frac{33.2 + 57}{1} = 90.2$$

Your average is high enough to earn an A.   ◊

GUIDED EXERCISE 7

Weighted average

In an investment portfolio, stocks are rated on a scale of 1 to 10 for dividend earning, security, and capital growth potential. On the scale, 1 equals very poor and 10 equals excellent. In one investment strategy favoring security, the dividend rating is given a weight of 2, the security a weight of 5, and the capital growth potential a weight of 3.

(a) Stock A has the ratings shown in Table 3-8. Complete the table and find the weighted average rating of the stock.

TABLE 3-8   Stock A Rating

	Rating x	Weight w	xw
Dividend	7	2	14
Security	8	5	40
Growth	4	3	12
		$\Sigma w = 10$	$\Sigma xw = 66$

  $\Sigma w = 10$

The last column has entries 14, 40, and 12, and the sum $\Sigma xw = 66$

$$\text{Weighted average} = \frac{\Sigma xw}{\Sigma w}$$

$$= \frac{66}{10}$$

$$= 6.6$$

(b) Suppose the weight given for dividend increases to 4 while the weight for growth decreases to 1. Would the new weighted average be lower, higher, or the same? Why?

Higher, since the new weights give more weight to dividend, which has a higher rating and less weight to growth, which has a lower rating. The different ratings will increase the weighted average. The new average is 7.2.

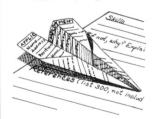

VIEWP●INT   Are Students Ready for Work?

Of all high school seniors, 62% say they have "very good" mathematical skills. Only 8% of all employers agree with this claim (*USA Today*). Employers appear to be telling us they need people with *more training* in mathematics, statistics, and general quantitative reasoning skills.

SECTION 3.3   PROBLEMS

1. *Crime: Shoplifting* What is the age distribution of adult shoplifters (21 years of age or older) in supermarkets? The following is based on information taken from the National Retail Federation. A random sample of 895 incidents of shoplifting gave the following age distribution:

Age range (years)	21–30	31–40	41 and over
Number of shoplifters	260	348	287

Estimate the mean age, sample variance, and sample standard deviation for the shoplifters. For the class 41 and over, use 45.5 as the class midpoint.

2. *Anthropology: Navajo Reservation* What was the age distribution of prehistoric Native Americans? Extensive anthropologic studies in the southwestern United States gave the following information about a prehistoric extended family group of 80 members on what is now the Navajo Reservation in northwestern New Mexico. (Source: Based on information taken from *Prehistory in the Navajo Reservation District,* by F. W. Eddy, Museum of New Mexico Press.)

Age range (years)	1–10*	11–20	21–30	31 and over
Number of individuals	34	18	17	11

*Includes infants.

Estimate the mean age expressed in years for this community, the sample variance, and the sample standard deviation. For the class 31 and over, use 35.5 as the class midpoint.

3. *Business Administration: Profits/Assets* What are the big corporations doing with their wealth? One way to answer this question is to examine profits as percentage of assets. A random sample of 50 *Fortune 500* companies gave the following information. (Source: Based on information from *Fortune 500,* Vol. 135, No. 8.)

Profit as percentage of assets	8.6–12.5	12.6–16.5	16.6–20.5	20.6–24.5	24.6–28.5
Number of companies	15	20	5	7	3

Estimate the sample mean, sample variance, and sample standard deviation for profit as percentage of assets.

4. *Agriculture: Water Table* The Bureau of Land Management (BLM) did a study of the water table near Custer, Wyoming, in the month of June. Based on data from the BLM, a random sample of 20 water wells showed that the distance from the ground to the water level (in feet) is

Distance from ground to water level (ft), *x*	12–14	15–17	18–20	21–23	24–26
Number of wells, *f*	1	3	8	2	6

Using the midpoints of the depth intervals, estimate the mean depth, the standard deviation, and the coefficient of variation.

5. *Sales: Catalog Shoppers* Based on data from *USA Today,* the ages of a random sample of 300 adults who shop by catalog are

Age	18–24	25–34	35–44	45–54	55–64	65–80
Number	78	75	48	33	33	33

Estimate the mean age of the adults who shop by catalog. Estimate the standard deviation of the age of the shoppers and the coefficient of variation.

6. *Income: Men versus Women* Based on data from the *Statistical Abstract of the United States,* 116th Edition, a histogram of annual earnings in thousands of dollars for a random sample of 1000 men at least 15 years old is shown in Figure 3-2. The histogram in Figure 3-3 shows the earnings in thousands of dollars for a random sample of 1000 women who are at least 15 years old.
(a) Estimate the mean earnings, standard deviation, and coefficient of variation for men.
(b) Estimate the mean earnings, standard deviation, and coefficient of variation for women.

7. *Medical: Hours of Sleep per Day* Alexander Borbely is a professor at the University of Zurich Medical School, where he is director of the sleep laboratory. The histogram in Figure 3-4 is based on information in his book *Secrets of Sleep.* The histogram displays hours of sleep per day for a random sample of 200 subjects. Estimate the mean hours of sleep, standard deviation of hours of sleep, and coefficient of variation.

8. *Ecology: Winter Wolf Packs* How big are winter wolf packs in Minnesota? The following table is based on information taken from *The Wolf,* by L. D. Mech, published by the University of Minnesota Press.

Pack size	1	2	3	4	5	6	7	8	9
Number of packs	15	15	7	9	13	13	9	8	9

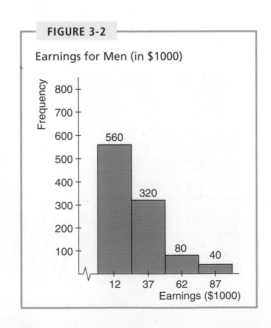

FIGURE 3-2

Earnings for Men (in $1000)

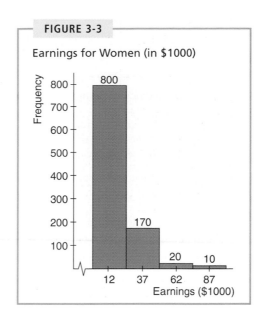

FIGURE 3-3

Earnings for Women (in $1000)

Find the sample mean size of the wolf packs, the sample variance, and the sample standard deviation.

9. *Ecology: Life Span of Deer* How long do deer live? In Mesa Verde National Park, deer are not hunted, so the life span of deer in a national park might give a reasonable estimate for the natural life span of deer. The data on the next page are from *The Mule Deer of Mesa Verde National Park,* by G. W. Mieran and J. L. Schmidt, published by the Mesa Verde Museum Association.

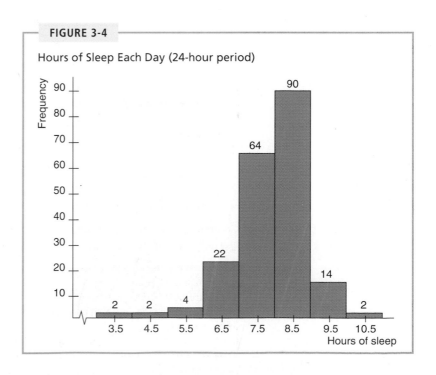

FIGURE 3-4

Hours of Sleep Each Day (24-hour period)

Age in years	1	2	3	4	5	6	7	8	9	10
Number of deer	3	7	6	5	4	2	0	1	2	1

(a) Use the methods of grouped data to compute the mean and sample standard deviation of the ages.

(b) Enter the 31 ages individually in a calculator and compute the mean and sample standard deviation of the ages. Are your results the same as in part (a)? Explain why the grouped-data method should give the same results.

10. *Fishing: Bait* What is the mortality rate (percent of fish that die) for trout caught and released using bait with a barbed hook? The following table is based on information taken from regions in British Columbia, Colorado, Michigan, Wisconsin, and Wyoming. (Source: *Proceedings of National Symposium on Catch and Release Fishing,* sponsored by Humboldt State University.)

Percent mortality	35.4	23.0	29.4	38.3	61.5
Number of fish	79	565	136	103	400

Compute the sample mean, sample variance, and sample standard deviation for the mortality rate of trout caught and released (bait on a barbed hook).

11. *Fishing: Lures* What is the mortality rate (percent of fish that die) for trout caught and released using artificial lures with barbed treble hooks? The following table is based on information taken from regions in British Columbia, Colorado, Michigan, Wisconsin, and Wyoming (see source in Problem 10).

Percent mortality	2.8	6.3	1.8	4.8	3.0
Number of fish	145	270	224	271	67

Compute the sample mean, sample variance, and sample standard deviation for the mortality rate of trout caught and released (artificial lure with barbed treble hook).

12. *Fishing: Flies* What is the mortality rate (percent of fish that die) for trout caught and released using artificial flies on barbed hooks? The following table is based on information taken from regions in British Columbia, Colorado, Michigan, Wisconsin, and Wyoming (see source in Problem 10).

Percent mortality	1.3	8.7	11.3	5.9	3.3
Number of fish	75	190	80	51	181

Compute the sample mean, sample variance, and sample standard deviation for the mortality rate of trout caught and released (artificial flies with barbed hook).

13. *Grades: Weighted Average* In your biology class, your final grade is based on several things: a lab score, scores on two major tests, and your score on the final exam. There are 100 points available for each score. However, the lab score is worth 25% of your total grade, each major test is worth 22.5%, and the final exam is worth 30%. Compute the weighted average for the following scores: 92 on the lab, 81 on the first major test, 93 on the second major test, and 85 on the final exam.

14. *Grades: Weighted Average* Suppose the weighting for the activities in your biology class changed so that the lab score was still worth 25%, but each of the two major tests also were worth 25%, and the final was worth only 25%. Compute the weighted average for the same set of scores as those in Problem 13: Use 92 on lab, 81 on the first major test, 93 on the second major test, and 85 on the final exam. Is the weighted average different from that in Problem 13? Since the weights are all the same, did you really need to use a weighted average, or could you simply have taken the mean of the four scores?

15. *Merit Pay Scale: Weighted Average* At General Hospital, nurses are given performance evaluations to determine eligibility for merit pay raises. The supervisor rates them on a scale of 1 to 10 (10 being the highest rating) for several activities: promptness, record keeping, appearance, and bedside manner with patients. Then an average is determined by giving a weight of 2 for promptness, 3 for record keeping, 1 for appearance, and 4 for bedside manner with patients. What is the average rating for a nurse with ratings of 9 for promptness, 7 for record keeping, 6 for appearance, and 10 for bedside manner?

16. *Athletic Awards: Weighted Average* The alumni club of Jefferson College gives a $10,000 award to the most outstanding athlete each year. Since competitive sports include football, basketball, baseball, swimming, and tennis, the rating scale is a way of comparing athletes who participate in any of the sports. Each candidate is rated on a scale of 1 to 10 (with 10 being the best) for individual performance, win–loss record of team, grade point average, and sportsmanship. The ratings are then averaged using a weight of 5 for individual performance, 2 for win–loss record of team, 1 for grade point average, and 3 for sportsmanship.
    (a) One athlete had the following ratings: 9 for individual performance, 7 for team record, 6 for grade point average, 8 for sportsmanship. Compute the weighted average of the ratings.
    (b) Another athlete had ratings of 8 for individual performance, 9 for team record, 5 for grade point average, and 9 for sportsmanship. Compute the weighted average of these ratings. Which athlete had the higher average rating?

17. *Bowling Balls: Weighted Average* Brunswick Corporation gave the following information about weights and sales of bowling balls.

Weight (lb)	6–8	10–12	13–15	16–17
Sales	15%	39%	27%	19%

Estimate the weighted average of the weight (in pounds) of the bowling balls sold. *Hint:* The weights of the bowling balls are the data values, while the percentages represent the weights in the weighted average formula.

18. *Study Time: Weighted Average* An Attitude Research Specialist poll (*USA Today*) of more than 2000 college students reported the study time outside the classroom per week to be

Study time (hours)	2–5	6–10	11 or more
Students	26%	40%	34%

To compute a weighted average for study time, we need an upper limit on the 11 or more class.

(a) Use 15 hours as the upper limit for the 11 or more class to make the last class from 11 to 15 hours. Compute the weighted mean of the study times.
(b) Use 20 hours as the upper limit for the 11 or more class to make the last class from 11 to 20 hours. Compute the weighted mean of the study times.
(c) Compare the results of parts (a) and (b) and comment on the difference you find.
(d) Keep track of your own outside-of-class study time for a week. How does your study time compare with the averages of parts (a) and (b)?

3.4
Percentiles and Box-and-Whisker Plots

FOCUS POINTS

✓ Interpret the meaning of percentile scores.

✓ Compute the median, quartiles, and five-number summary from raw data.

✓ Make a box-and-whisker plot. Interpret the results.

✓ Describe how a box-and-whisker plot indicates spread of data around the median.

We've seen measures of central tendency and spread for a set of data. The arithmetic mean $\bar{x}$ and the standard deviation s will be very useful in later work. However, because they each utilize every data value, they can be heavily influenced by one or two extreme data values. In cases where our data distributions are heavily skewed or even bimodal, we often get a better summary of the distribution by utilizing relative position of data rather than exact values.

Recall that the median is an average computed by using relative position of the data. If we are told that 81 is the median score on a biology test, we know that after the data have been ordered, 50% of the data fall at or below the median value of 81. The median is an example of a *percentile*; in fact, it is the 50th percentile. The general definition of the Pth percentile follows.

For whole numbers P (where $1 \leq P \leq 99$), the Pth **percentile** of a distribution is a value such that $P\%$ of the data fall at or below it and $(100 - P)\%$ of the data fall at or above it.

In Figure 3-5, we see the 60th percentile marked on a histogram. We see that 60% of the data lie below the mark and 40% lie above it.

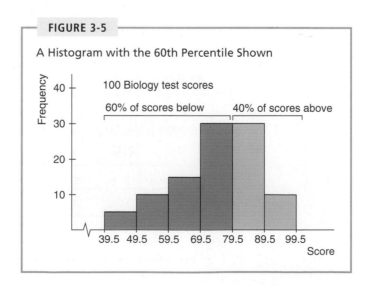

FIGURE 3-5

A Histogram with the 60th Percentile Shown

GUIDED EXERCISE 8

Percentiles

You took the English achievement test to obtain college credit in freshman English by examination.

(a) If your score was at the 89th percentile, what percentage of scores are at or below yours?

⟹ The percentile means that 89% of the scores were at or below yours.

(b) If the scores ranged from 1 to 100 and your raw score is 95, does this necessarily mean that your score is at the 95th percentile?

⟹ No, the percentile gives an indication of relative position of the scores. The determination of your percentile has to do with the number of scores at or below yours. If everyone did very well and only 80% of the scores fell at or below yours, you would be at the 80th percentile even though you got 95 out of 100 points on the exam.

There are 99 percentiles, and in an ideal situation, the 99 percentiles divide the data set into 100 equal parts. (See Figure 3-6.)

However, if the number of data elements is not exactly divisible by 100, the percentiles will not divide the data into equal parts.

There are several widely used conventions for finding percentiles. They lead to slightly different values for different situations, but these values are close together. For all conventions, the data are first *ranked* or ordered from smallest to largest. A natural way to find the Pth percentile is to then find a value so that P% of the data fall at or below it. This will not always be possible, so we take the nearest value satisfying the criterion. It is at this point that there are a variety of processes to determine the exact value of the percentile.

We will not be very concerned about exact procedures for evaluating percentiles in general. However, *quartiles* are special percentiles used so frequently that we want to adopt a specific procedure for their computation.

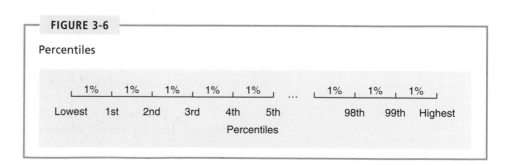

FIGURE 3-6

Percentiles

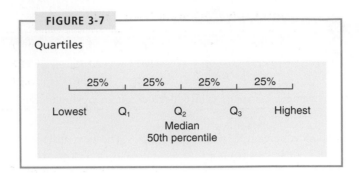

FIGURE 3-7

Quartiles

25%   25%   25%   25%

Lowest Q_1 Q_2 Q_3 Highest
Median
50th percentile

Quartiles

Quartiles are those percentiles that divide the data into fourths. The *first quartile* Q_1 is the 25th percentile, the *second quartile* Q_2 is the median, and the *third quartile* Q_3 is the 75th percentile. (See Figure 3-7.)

Again, several conventions are used for computing quartiles, but the following one utilizes the median and is widely adopted.

Procedure to compute quartiles

1. Order the data from smallest to largest,

2. Find the median. This is the 2nd quartile.

3. The first quartile Q_1 is then the median of the lower half of the data; that is, it is the median of the data falling *below* the Q_2 position (and not including Q_2).

4. The third quartile Q_3 is the median of the upper half of the data; that is, it is the median of the data falling *above* the Q_2 position (and not including Q_2).

In short, all we do to find the quartiles is to find three medians.

Interquartile range

The median, or second quartile, is a popular measure of the center utilizing relative position. A useful measure of data spread utilizing relative position is the *interquartile range (IQR)*. It is simply the difference between the third and first quartiles.

$$\text{Interquartile range} = Q_3 - Q_1$$

The interquartile range tells us the spread of the middle half of the data. Now let's look at an example to see how to compute all these quantities.

EXAMPLE 10

Quartiles

In a hurry? On the run? Hungry as well? How about an ice cream bar as a snack? Ice cream bars are popular among all age groups. *Consumer Reports* did a study

TABLE 3-9   Cost of Ice Cream Bars (in dollars)

0.99	1.07	1.00	0.50	0.37	1.03	1.07	1.07
0.97	0.63	0.33	0.50	0.97	1.08	0.47	0.84
1.23	0.25	0.50	0.40	0.33	0.35	0.17	0.38
0.20	0.18	0.16					

of ice cream bars. Twenty-seven bars with taste ratings of at least "fair" were listed, and cost per bar was included in the report. Just how much will an ice cream bar cost? The data, expressed in dollars, appear in Table 3-9. As you can see, the cost varies quite a bit, partly because the bars are not of uniform size.

(a) Find the quartiles.

   SOLUTION: We first order the data from smallest to largest.
   Next, we find the median. Since the number of data values is 27, there are an odd number of data, and the median is simply the center or 14th value. The value is shown boxed in Table 3-10.

   $$\text{Median} = Q_2 = 0.50$$

There are 13 values below the median position, and Q_1 is the median of these values. It is the middle or 7th value and is shaded in Table 3-10.

   $$\text{First quartile} = Q_1 = 0.33$$

There are also 13 values above the median position. The median of these is the 7th value from the right end. This value is also shaded in Table 3-10.

   $$\text{Third quartile} = Q_3 = 1.00$$

(b) Find the interquartile range.

   SOLUTION:

   $$\begin{aligned} IQR &= Q_3 - Q_1 \\ &= 1.00 - 0.33 \\ &= 0.67 \end{aligned}$$

This means that the middle half of the data has a cost spread of 67¢.   ◇

TABLE 3-10   Ordered Cost of Ice Cream Bars (in dollars)

0.16	0.17	0.18	0.20	0.25	0.33	0.33	0.35
0.37	0.38	0.40	0.47	0.50	0.50	0.50	0.63
0.84	0.97	0.97	0.99	1.00	1.03	1.07	1.07
1.07	1.08	1.23					

GUIDED EXERCISE 9

Quartiles

Many people consider the number of calories in an ice cream bar as important as, if not more important than, the cost. The *Consumer Reports* article also included the calorie count of the rated ice cream bars (Table 3-11). There were 22 vanilla-flavored bars rated. Again, the bars varied in size, and some of the smaller bars had fewer calories. The calorie counts for the vanilla bars follow.

TABLE 3-11   Calories in Vanilla-Flavored Ice Cream Bars

342	377	319	353	295
234	294	286	377	182
310	439	111	201	182
197	209	147	190	151
131	151			

(a) Our first step is to order the data. Do so.

TABLE 3-12   Ordered Data

111	131	147	151	151	182
182	190	197	201	209	234
286	294	295	310	319	342
353	377	377	439		

(b) There are 22 data values. Find the median.

Average the 11th and 12th data values boxed together in Table 3-12.

$$\text{Median} = \frac{209 + 234}{2}$$
$$= 221.5$$

(c) How many values are below the median position? Find Q_1.

Since the median lies halfway between the 11th and 12th values, there are 11 values below the median position. Q_1 is the median of these values.

$Q_1 = 182$

(d) There are the same number of data above as below the median. Use this fact to find Q_3.

Q_3 is the median of the upper half of the data. There are 11 values in the upper portion.

$Q_3 = 319$

(e) Find the interquartile range and comment on its meaning.

$IQR = Q_3 - Q_1$
$= 319 - 182$
$= 137$

The middle portion of the data has a spread of 137 calories.

Box-and-Whisker Plots

The quartiles together with the low and high data values give us a very useful *five-number summary* of the data and their spread.

Five-number summary

Lowest value, Q_1, median, Q_3, highest value

We will use these five numbers to create a graphic sketch of the data called a *box-and-whisker plot*. Box-and-whisker plots provide another useful technique from exploratory data analysis (EDA) for describing data.

To make a box-and-whisker plot

1. Draw a vertical scale to include the lowest and highest data values.

2. To the right of the scale draw a box from Q_1 to Q_3.

3. Include a solid line through the box at the median level.

4. Draw solid lines, called *whiskers*, from Q_1 to the lowest value and from Q_3 to the highest value.

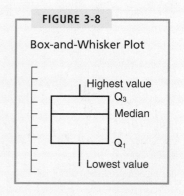

FIGURE 3-8

Box-and-Whisker Plot

Highest value
Q_3
Median
Q_1
Lowest value

The next example demonstrates the process of making a box-and-whisker plot.

EXAMPLE 11

Box-and-whisker plot

Using the data shown in Guided Exercise 9, make a box-and-whisker plot showing the calories in vanilla-flavored ice cream bars. Use the plot to make observations about the distribution of calories.

(a) In Guided Exercise 9, We ordered the data (see Table 3-12) and found the values of the median, Q_1 and Q_3. From this previous work we have the five-number summary:

low value = 111; Q_1 = 182; median = 221.5; Q_3 = 319; high value = 439

(b) We select an appropriate vertical scale and make the plot (Figure 3-9 on the next page).

(c) A quick glance at the box-and-whisker plot reveals the following:

   (i) The box tells us where the middle half of the data lies, so we see that half of the ice cream bars have between 182 and 319 calories, with an interquartile range of 137 calories.

   (ii) The median is slightly closer to the lower part of the box. This means that the lower calorie counts are more concentrated. The calorie counts above the median are more spread out, indicating that the distribution is slightly skewed toward the higher values.

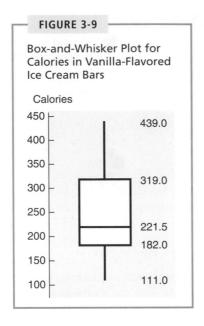

FIGURE 3-9

Box-and-Whisker Plot for
Calories in Vanilla-Flavored
Ice Cream Bars

Calories

439.0

319.0

221.5
182.0

111.0

(iii) The upper whisker is also longer than the lower, which again emphasizes
skewness toward the higher values.   ◇

◇ **COMMENT** In exploratory data analysis, *hinges* rather than quartiles are used to
create the box. Hinges are computed in a manner similar to the way we compute quartiles. However, in the case of an odd number of data values, include
the median itself in both the lower and upper halves of the data (see *Applications, Basics, and Computing of Exploratory Data Analysis,* by Paul Velleman
and David Hoaglin, Duxbury Press). This has the effect of shrinking the box
and moving the ends of the box slightly toward the median. For an even number of data, the quartiles as we computed them equal the hinges.   ◇

GUIDED
EXERCISE **10**

Box-and-whisker plot

The Renata College Development Office sent salary surveys to alumni who graduated two and
five years ago. The voluntary responses received are summarized in the box-and-whisker plots
shown in Figure 3-10.

(a) From Figure 3-10, estimate the median and
extreme values of salaries of alumni graduating
two years ago. In what range are the middle half
of the salaries?

    The median seems to be about $44,000. The
extremes are about $33,000 and $54,000. The
middle half of the salaries fall between $40,000
and $47,000.

Continued

GUIDED EXERCISE 10 continued

FIGURE 3-10 Box-and-Whisker Plots for Alumni Salaries (in thousands of dollars)

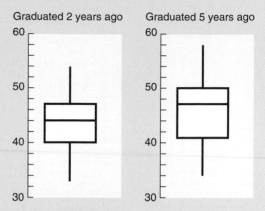

(b) From Figure 3-10, estimate the median and the extreme values of salaries of alumni graduating five years ago. What is the location of the middle half of the salaries?

The median seems to be $47,000. The extremes are $34,000 and $58,000. The middle half of the data is enclosed by the box with low side at $41,000 and high side at $50,000.

(c) Compare the two box plots and make comments about the salaries of alumni graduating two and five years ago.

The salaries of the alumni graduating five years ago have a larger spread but begin slightly higher and extend to levels about $4,000 above those graduating two years ago. The middle half of the data is also more spread out with higher boundaries and a higher median.

We have developed the skeletal box-and-whisker plot. Other variations may include *fences,* which are marks placed on either side of the box to represent various portions of the data. Values that lie outside the fences are called *outliers* (see Problem 14 of this section). These values seem to stand by themselves, away from most of the data. They might be exceptional values and deserve closer study. Or they may be the result of data entry error. For a more complete discussion of outliers and variations of box plots, see *Applications, Basics, and Computing of Exploratory Data Analysis,* by Velleman and Hoaglin.

TECH NOTE *Box-and-Whisker Plot* Both Minitab and the TI-83Plus support box-and-whisker plots. On the TI-83Plus, the quartiles Q_1 and Q_3 are calculated as we calculate them in this text. In Minitab and Excel, they are calculated using a slightly different process.

TI-83Plus

STATPLOT ➤ On. Highlight box plot. Use **Trace** and arrow keys to display values of five-number summary. The display shows plot for calories in ice cream bars.

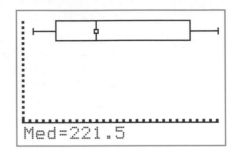

Med=221.5

Excel

Does not produce plot. **Paste Function** [f_x] ➤ **Statistics** ➤ **Quartiles** gives five-number summary.

Minitab

Graph ➤ Boxplot. In dialogue box set Display to IQRange Box.

VIEWPOINT

Is Shorter Higher?

Can you estimate a person's *height* from the *pitch* of his or her voice? Is a soprano shorter than an alto? Is a bass taller than a tenor? A statistical study of singers in the New York Choral Society provided information. For more information, visit the Brase/Brase statistics site at http://math.college.hmco.com/students and find the link to DASL, the Carnegie Mellon University Data and Story Library. From the Data Subjects, select music and then singers. Methods of this chapter can be used with new methods we will learn in Chapters 8 and 9 to examine such questions from a statistical point of view.

SECTION 3.4   PROBLEMS

1. *Education: Aptitude Test* Angela took a general aptitude test and scored in the 82nd percentile for aptitude in accounting. What percentage of the scores were at or below her score? What percentage were above?

2. *Education: College Admission* One standard for admission to Redfield College is that the student must rank in the upper quartile of his or her graduating high school class. What is the minimal percentile rank of a successful applicant?

3. *Education: Competency Exam* The town of Butler, Nebraska, decided to give a teacher-competency exam and defined the passing scores to be those in the 70th percentile or higher. The raw test scores ranged from 0 to 100. Was a raw score of 82 necessarily a passing score? Explain.

4. *Education: Test Scores* Clayton and Timothy took different sections of Introduction to Economics. Each section had a different final exam. Timothy scored 83 out of 100 and had a percentile rank in his class of 72. Clayton scored 85 out of 100 but his percentile rank in his class was 70. Who performed better with respect to the rest of the students in the class: Clayton or Timothy? Explain your answer.

5. *Consumer: Pizza Cost* Want a takeout dinner or a dinner that is easy to prepare? Pizza comes to mind for many people. *Consumer Reports* (Vol. 62, No. 1) rated a variety of pizzas from supermarkets. For a 5-ounce serving of supermarket cheese pizza, the costs are (in dollars)

0.90	0.85	0.80	1.00	0.69	0.75	0.81	0.90
1.92	1.64	1.54	0.72	1.15	0.52	0.72	1.50

Compute the five-number summary and interquartile range. Then make a box-and-whisker plot.

6. *Consumer: Pizza Calories* How many calories are in a serving of cheese pizza? The article in *Consumer Reports* referred to in Problem 5 also gave the calories in a 5-ounce serving of supermarket cheese pizza. The calories are

332	364	393	347	350	353	357	296
358	322	337	323	333	299	316	275

Compute the five-number summary and interquartile range. Then make a box-and-whisker plot.

7. *Health Care: Nurses* At Center Hospital there is some concern about the high turnover of nurses. A survey was done to determine how long (in months) nurses had been in their current positions. The responses of 20 nurses were (in months):

23	2	5	14	25	36	27	42	12	8
7	23	29	26	28	11	20	31	8	36

Make a box-and-whisker plot of the data. Find the interquartile range.

8. *Health Care: Staff* Another survey was done at Center Hospital to determine how long (in months) clerical staff had been in their current positions. The responses of 20 clerical staff members were (in months)

25	22	7	24	26	31	18	14	17	20
31	42	6	25	22	3	29	32	15	72

(a) Make a box-and-whisker plot. Find the interquartile range.
(b) Compare this plot with the one in Problem 7. Discuss the location of the medians, the location of the middle half of the data banks, and the distance from Q_1 and Q_3 to the extreme values.

9. *Consumer: Auto Insurance* Is there a difference in auto insurance cost if you reside in a suburban area or an urban area? *Consumer Reports* (Vol. 62, No. 1) gave suburban and urban costs of auto insurance for several states. In Illinois, the least expensive costs for similar policies were (in dollars)

Suburban   **Urban**

1170	1216	1211	1282	1292	2356	2584	2674	2840	2910
808	874	972	986	992	1768	1968	1968	2083	2107

For each group, compute the five-number summary and interquartile range. Then make a box-and-whisker plot for each group and comment on the differences between urban and suburban auto insurance rates.

10. *Sociology: High-school Dropouts* What percentage of the general U.S. population are high-school dropouts? The *Statistical Abstract of the United States,* 120th Edition, gives the percentage of high-school dropouts by state.

13	11	14	14	10	9	10	14	14	14	8	10
11	11	7	9	13	13	8	11	9	10	6	12
11	8	7	15	9	10	12	10	13	5	9	10
12	9	11	12	8	13	13	9	8	10	11	11
7	7										

(a) Make a box-and-whisker plot and find the interquartile range.
(b) Wyoming has a dropout rate of about 7%. Into what quartile does this rate fall?

11. *Sociology: College Graduates* What percentage of the general U.S. population have bachelor's degrees? The *Statistical Abstract of the United States,* 120th Edition, gives the percentage of bachelor's degrees by state.

22	26	24	17	27	38	34	24	22	22	26	21	26
18	22	26	20	21	23	35	31	21	32	19	23	24
20	20	27	31	25	27	24	22	26	24	27	24	27
21	26	18	24	28	28	32	29	18	24	22		

(a) Make a box-and-whisker plot and find the interquartile range.
(b) Illinois has a bachelor's degree percentage rate of about 26%. Into what quartile does this rate fall?

12. *Financial: Interpret Graphs* Box-and-whisker plots for the weekly percentage change in the closing prices of Coca-Cola stock, McDonald's stock, and Disney stock (*Historical Quotes, Dow Jones News Retrieval Service*) are shown in Figure 3-11. Compare the plots. Answer each of the following questions, and explain each answer.
(a) For all the stocks, does the percentage change distribution appear to be skewed right? Explain.
(b) Which stock has a percentage change distribution that is most spread?
(c) Which stock was most volatile during the period shown?
(d) Which stock had a median percentage change that was negative? Which stock had more weekly declines than weekly increases?
(e) If you were a conservative investor interested in purchasing one of these stocks but you wanted the one that had percentage declines that were smallest, which would you select? How do the sizes of the percentage increases in this stock compare with those of the other two?
(f) If you were an investor willing to take risks and you wanted a stock with high growth, which of these stocks would you select? Which one had the highest percentage increases? How do the percentage decreases of this stock compare with those of the other two?

13. *Auto Insurance: Interpret Graphs* *Consumer Reports* rated automobile insurance companies and gave annual premiums for top-rated companies in several states. Figure 3-12 shows box plots for annual premiums for urban customers (married couple with one 17-year-old son) in three states. The box plots in Figure 3-12 are all drawn on the same scale on a TI-83Plus calculator.

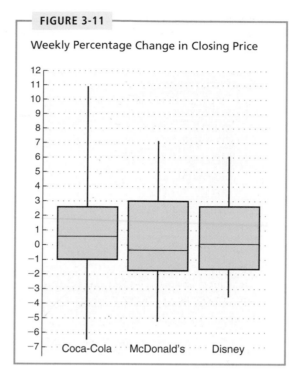

FIGURE 3-11

Weekly Percentage Change in Closing Price

(a) Which state has the lowest premium? the highest?

(b) Which state has the highest median premium?

(c) Which state has the smallest range of premiums? the smallest interquartile range?

(d) Figure 3-13 on page 136 gives the five-number summaries generated on the TI-83 for the box plots of Figure 3-12. Match the five-number summaries to the appropriate box plot.

14. *General: Outliers* Some data sets include values so high or so low that they seem to stand apart from the rest of the data. These data are called *outliers*. Outliers may be from data collection errors, data entry errors, or simply valid but unusual data values. Regardless of the reason, it is important to identify outliers in the data set and examine the outliers carefully to determine if they are in error. One way to detect outliers is to use a box-and-whisker plot. Data values that fall beyond the limits,

Lower Limit: $Q_1 - 1.5 \times (IQR)$

Upper Limit: $Q_3 + 1.5 \times (IQR)$

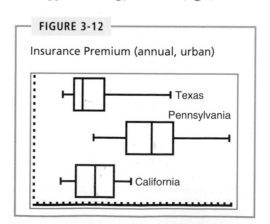

FIGURE 3-12

Insurance Premium (annual, urban)

FIGURE 3-13

Five-Number Summaries for Insurance Premiums

(a)
```
1-Var Stats
↑n=10
  minX=2382
  Q₁=2758
  Med=2991
  Q₃=3652
  maxX=5715
```

(b)
```
1-Var Stats
↑n=10
  minX=3314
  Q₁=4326
  Med=5116.5
  Q₃=5801
  maxX=7527
```

(c)
```
1-Var Stats
↑n=10
  minX=2323
  Q₁=2801
  Med=3377.5
  Q₃=3966
  maxX=4482
```

---

where IQR is the interquartile range, are suspected outliers. In the computer software package Minitab, values beyond these limits are plotted with asterisks (*).

Students from a statistics class were asked to record their heights in inches. The heights (as recorded) were

65	72	68	64	60	55	73	71	52	63	61	74
69	67	74	50	4	75	67	62	66	80	64	65

(a) Make a box-and-whisker plot of the data.
(b) Find the value of the interquartile range (IQR).
(c) Multiply the IQR by 1.5 and find the lower and upper limits.
(d) Are there any data values below the lower limit? above the upper limit? List any suspected outliers. What might be some explanations for the outliers?

15. *Salary Increases: Interpret Graphs* The following Minitab display shows box-and-whisker plots for the percentage increases in annual salaries for the university and college faculty in Tennessee (by rank). Source: *Academe: Bulletin of the American Association of University Professors,* Volume 85, Number 2.

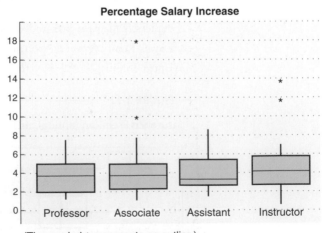

Percentage Salary Increase

(The symbol * represents an outlier.)

Minitab menu selections: ➤ **Graph** ➤ **Boxplot.**

Note: Minitab uses a procedure to find Q_1 and Q_3 that is slightly different from the method shown in this text. As a result, Minitab values for Q_1 and Q_3 will differ slightly from text or TI-83 values.

(a) Which faculty rank had the smallest median percentage salary increase? Which faculty rank had the single highest percentage salary increase?

(b) Which faculty rank had the largest spread between the first and third quartiles?

(c) Which faculty rank had the smallest spread for the lower 50% of the percentage salary increases?

(d) Which faculty rank had the most symmetric percentage salary increases? If the outliers for the associate professors were omitted, would that distribution appear to be symmetric?

(e) Look at the following descriptive statistics from Minitab.

```
Minitab Display: Descriptive Statistics
VARIABLE       N        MEAN      MEDIAN    TRMEAN    STDEV    SE MEAN
Professo      20        3.685     3.700     3.611     1.766    0.395
Associat      24        4.362     3.750     3.927     3.592    0.733
Assistan      23        4.330     3.500     4.262     2.007    0.419
Instruct      21        4.781     4.100     4.568     3.132    0.684
VARIABLE    MINIMUM    MAXIMUM      Q₁        Q₃
Prefesso      1.200     7.500     2.100     4.975
Associat      0.600    17.700     2.350     5.075
Assistan      1.500     8.600     2.900     5.500
Instruct      0.200    13.400     2.850     5.800
```

For the ranks of associate professor and instructor, compute the upper limit for outliers (see Problem 14). Do these ranks show outliers?

SUMMARY

To characterize numerical data, we use both measures of center and of variation. An average is an attempt to summarize the data into just one number. We studied several important averages: the mode, the median, the mean, and weighted averages. We also looked at trimmed means, which are more resistant to the effects of extreme values. However, an average alone can be misleading; we really need another statistical cross-reference. The measure of data spread, or variation, satisfies the purpose. The variations that we looked at most carefully were the range, the variance, the standard deviation, and the interquartile range. Chebyshev's theorem enables us to estimate the data spread. The coefficient of variation lets us compare relative spread from different data sets. A box-and-whisker plot gives a good visual impression of the range of the data and the location of the middle half of the data.

In later work the average we will use most is the mean; the measure of variation we will use most is the standard deviation. Because the mean and standard deviation are so important, we learned how to estimate these values from data already organized in a frequency distribution.

IMPORTANT WORDS AND SYMBOLS

Section 3.1
Average
Mode
Median
Sample mean, $\bar{x}$
Population mean, μ
Summation symbol, Σ
Resistant measure
Trimmed mean

Section 3.2
Range
Sample standard deviation, s
Sample variance, s^2
Sum of squares, SS_x
Population standard deviation, σ
Population size, N

Coefficient of variation, CV
Chebyshev's theorem

Section 3.3
Mean of grouped data
Standard deviation of grouped data
Weighted average

Section 3.4
Percentile
Quartile
Five-number summary
Interquartile range, IQR
Box-and-whisker plot
Whisker
Outlier

VIEWPOINT

The Fujita Scale

How do you measure a tornado? Professor Fujita and Allen Pearson (Director of the National Severe Storm Forecast Center) developed a measure based on wind speed and type of damage done by a tornado. The result is an excellent example of both descriptive and inferential statistical methods. For more information, visit the Brase/Brase statistics site at http://math.college.hmco.com/students and find the link to the tornado project. Then look up Fujita scale. If we group the data a little, the scale becomes

FS	WS	%
F0 & F1	40–112	67
F2 & F3	113–206	29
F4 & F5	207–318	4

where *FS* represents Fujita scale; *WS*, wind speed in miles per hour; and %, percentage of all tornados. Out of 100 tornados, what would you estimate for the mean and standard deviation of wind speed?

CHAPTER REVIEW PROBLEMS

1. *Consumer: Power Surge Protector* June purchased a new home computer and has been having trouble with voltage spikes on the power line. Such voltage jumps can be caused by the operation of appliances, such as clothes dryers and electric irons,

or just by a power surge on the outside power line. Her friend Jim is an electronics technician and has obtained the following data about voltages when certain electric appliances are turned on and off. Remember, the normal line voltage is 110 volts. All measurements are taken from the line and measured in volts.

73   140   78   142   80   140   90   133

(a) Compute the sample mean, sample standard deviation, coefficient of variation, and range.

Jim advised June to buy a device called a *power surge protector* that protects the computer from strong voltage spikes. Using the power surge protector, Jim again measured voltages to the computer when the same appliances were turned on and off. The results in voltage were

100   120   108   114   105   117   103   114

(b) Compute the sample mean, sample standard deviation, coefficient of variation, and range of the voltages using the power surge protector.
(c) Compare your answers for parts (a) and (b). Were the means about the same? Were the voltage distributions different with and without the power surge protector? How did the standard deviation, coefficient of variation, and range reflect this when the mean did not? Explain your answer.

2. *Consumer: Radon Gas* "Radon: The Problem No One Wants to Face" is the title of an article appearing in *Consumer Reports*. Radon is a gas emitted from the ground that can collect in houses and buildings. At certain levels it can cause lung cancer. Radon concentrations are measured in picocuries per liter (pCi/L). A radon level of 4 pCi/L is considered "acceptable." Radon levels in a house vary from week to week. In one house, a sample of 8 weeks had the following readings for radon level (in pCi/L):

1.9   2.8   5.7   4.2   1.9   8.6   3.9   7.2

(a) Find the mean, median, and mode.
(b) Find the sample standard deviation, coefficient of variation, and the range.

3. *Political Science: Georgia Democrats* How Democratic is Georgia? County-by-county results are shown for a recent election. (Source: *County and City Data Book*, 12th edition, U.S. Census Bureau.)

Percentage of Democratic Vote by Counties in Georgia

42 53 45 45 46 38 49 43 34 40 35 53 41 41 40 46
51 46 57 41 59 49 31 46 44 47 41 35 53 49 66 68
38 53 53 56 41 41 35 40 48 38 55 62 41 42 36 39
44 57 52 40 66 52 34 33 50 49 56 44

(a) Make a box-and-whisker plot of the data. Find the interquartile range.
(b) Make a frequency table using 5 classes. Then estimate the mean and sample standard deviation using the frequency table. Compute a 75% Chebyshev interval centered about the mean.
(c) If you have a statistical calculator or computer, use it to find the actual sample mean and sample standard deviation.

4. *Grades: Weighted Average* Professor Cramer determines a final grade based on attendance, two papers, three major tests, and a final exam. Each of these activities has a total of 100 possible points. However, the activities carry different weights.

Attendance is worth 5%, each paper is worth 8%, each test is worth 15%, and the final is worth 34%.

(a) What is the average for a student with 92 on attendance, 73 on the first paper, 81 on the second paper, 85 on test 1, 87 on test 2, 83 on test 3, and 90 on the final exam?

(b) Compute the average for a student with the above scores on the papers, tests, and final exam, but with a score of only 20 on attendance.

5. *General: Average Weight* An elevator is loaded with 16 people and is at its load limit of 2500 lb. What is the mean weight of these people?

6. *Ecology: Blue Spruce Trees* The circumferences (distance around) of 94 blue spruce trees selected at random in Roosevelt National Forest were measured. The results to the nearest inch were grouped in Table 3-13. Using the midpoints of the tree circumference classes, find the mean and standard deviation.

TABLE 3-13 Circumference of Blue Spruce Trees

Circumference (in.)	Number of Trees
10–24 17	6
25–39 31.5	20
40–54 47	52
55–69 62	16

7. *Fishing: Line Strength* A certain brand of nylon monofilament fishing line is known to deteriorate in very cold temperatures. A spool of 10-pound test monofilament line was left out overnight at Fairbanks, Alaska, when temperatures dropped to −35°F. A random sample of six pieces of line gave the following breaking strengths (in pounds):

10.1   6.2   9.8   5.3   9.9   5.7

(a) Compute the sample mean, sample standard deviation, coefficient of variation, and range.

A second spool of this line that had not been subjected to extreme cold temperatures gave the following breaking strengths (in pounds) for a random sample of six pieces of line:

10.2   9.7   9.8   10.3   9.6   10.1

(b) Compute the sample mean, sample standard deviation, coefficient of variation, and range for these values.

(c) Compare your answers for parts (a) and (b) and comment on the observed differences. Which line had the more consistent performance? How was this reflected in the sample standard deviations and in the coefficients of variation? In the ranges?

8. *Medical: Glucose Blood Level* The data in Table 3-14 represents glucose blood level (mg/100 ml) after a 12-hour fast for a random sample of 70 women (based on information from the *American Journal of Clinical Nutrition*, Vol. 19, 345–351). Compute the five-number summary and the interquartile range, and make a box-and-whisker plot.

TABLE 3-14 Glucose Blood Level (mg/100 ml)

45	66	83	71	76	64	59	59	76	82	80
81	85	77	82	90	87	72	79	69	83	71
87	69	81	76	96	83	67	94	101	94	89
94	73	99	93	85	83	80	78	80	85	83
84	74	81	70	65	89	70	80	84	77	65
46	80	70	75	45	101	71	109	73	73	80
72	81	63	74							

9. *Medical: Glucose Blood Level* The glucose blood level data of Table 3-14 can be organized into a frequency table as follows, where x = milligrams (mg) of glucose per 100 ml of blood.

x	45–55	56–66	67–77	78–88	89–99	100–110
Frequency	3	7	22	26	9	3

(a) Use the method of grouped data to estimate the sample mean, sample variance, sample standard deviation, and coefficient of variation.

(b) Use Chebyshev's theorem to estimate an interval centered at the mean in which at least 75% of the glucose blood level measurements will fall.

10. *Fishing: Atlantic Salmon* What is the mortality rate (percent of fish that die) for Atlantic salmon caught and released using artificial flies with barbed hooks? The following information is based on studies done in Maine. (Source: *Proceedings of National Symposium on Catch and Release Fishing*, sponsored by Humboldt State University.)

Percent mortality	4.6	12.0	2.6	4.1	3.9
Number of fish	304	52	39	319	77

Compute the sample mean, sample variance, and sample standard deviation for the mortality rate of salmon caught and released (artificial flies with barbed hooks).

11. *Performance Rating: Weighted Average* A performance evaluation for new sales representatives at Office Automation Incorporated involves several ratings done on a scale of 1 to 10, with 10 the highest rating. The activities rated include new contacts, successful contacts, total contacts, dollar volume of sales, and reports. Then an overall rating is determined by using a weighted average. The weights are 2 for new contacts, 3 for successful contacts, 3 for total contacts, 5 for dollar value of sales, and 3 for reports. What would the overall rating be for a sales representative with ratings of 5 for new contacts, 8 for successful contacts, 7 for total contacts, 9 for dollar volume of sales, and 7 for reports?

12. *Agriculture: Bell Peppers* The pathogen *Phytophthora capsici* causes bell pepper plants to wilt and die. A research project was designed to study the effect of soil water content and the spread of the disease in fields of bell peppers (Source: *Journal of Agricultural, Biological, and Environmental Statistics*, Vol. 2, No. 2). It is thought that too much water helps spread the disease. The fields were divided into rows and quadrants. The soil water content (percent of water by volume of soil) was determined for each plot. An important first step in such a research project is to give a statistical description of the data.

Soil Water Content for Bell Pepper Study

```
15  14  14  14  13  12  11  11  11  11  10  11  13  16  10
 9  15  12   9  10   7  14  13  14   8   9   8  11  13  13
15  12   9  10   9   9  16  16  12  10  11  11  12  15   6
10  10  10  11   9
```

(a) Make a box-and-whisker plot of the data. Find the interquartile range.

(b) Make a frequency table using 4 classes. Then estimate the mean and sample standard deviation using the frequency table. Compute a 75% Chebyshev interval centered about the mean.

(c) If you have a statistical calculator or computer, use it to find the actual sample mean and sample standard deviation.

13. *General: Create Examples*
   (a) Is it possible that the range and standard deviation can be equal? If your answer is yes, give an example of a data set where they are equal.
   (b) Is it possible that the mean, median, and mode can all be equal? Is it possible that they can all be different? If your answer is yes to either question, give an example to illustrate your answer.

DATA HIGHLIGHTS: GROUP PROJECTS

Old Faithful Geyser, Yellowstone National Park

Break into small groups and discuss the following topics. Organize a brief outline in which you summarize the main points of your group discussion.

1. *The Story of Old Faithful* is a short book written by George Marler and published by the Yellowstone Association. Chapter 7 of this interesting book talks about the effect of the 1959 earthquake on eruption intervals for Old Faithful Geyser. Dr. John Rinehart (a senior research scientist with the National Oceanic and Atmospheric Administration) has done extensive studies of the eruption intervals before and after the 1959 earthquake. Examine Figure 3-14. Notice the general shape. Is the graph more or less symmetrical? Does it have a single mode frequency? The mean interval between eruptions has remained steady at about 65 minutes for the past 100 years. Therefore, the 1959 earthquake did not significantly change the mean, but it did change the distribution of eruption intervals. Examine Figure 3-15. Would you say there are really two frequency modes? One shorter and the other longer? Explain. The overall mean is about the same for both graphs, but one graph has a much larger standard deviation (for eruption intervals) than the other. Do no calculations, just look at both graphs, and then explain which graph has the smaller and which has the larger standard deviation. Which distribution will have the larger coefficient of variation? In everyday terms, what would this mean if you were

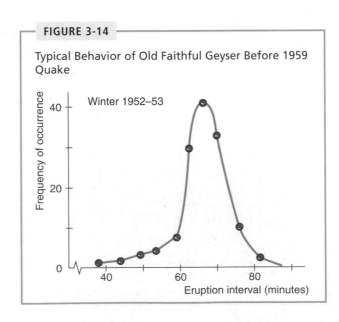

FIGURE 3-14

Typical Behavior of Old Faithful Geyser Before 1959 Quake

FIGURE 3-15

Typical Behavior of Old Faithful Geyser After 1959 Quake

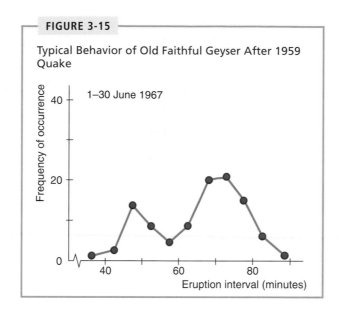

actually at Yellowstone waiting to see the next eruption of Old Faithful? Explain your answer.

2. Most academic advisors tell students to major in a field the student really loves. After all, it is true that money cannot buy happiness! Nevertheless, it is interesting to at least look at some of the higher-paying fields of study. After all, a field like mathematics can be a lot of fun, once you get into it. We see that women's salaries tend to be less than men's salaries. However, women's salaries are rapidly catching up, and this benefits the entire work force in different ways. Figure 3-16 shows the

FIGURE 3-16

Lucrative College Majors

USA SNAPSHOTS®

A look at statistics that shape the nation

Lucrative majors for women

The U.S. median income for women age 30 or older with bachelor's degrees and earning wages or salaries was $31,848[1]. Fields of study with highest median pay:

Major	Median income
Pharmacy	$47,567
Engineering	$46,389
Computer/ information sciences	$41,559
Physical therapy/ related services	$40,491
Nursing	$40,096

1–Excludes self-employed, through 1993 (latest year available)

Source: Bureau of Labor Statistics *Occupational Outlook Quarterly, Summer 1998*

By Anne R. Carey and Grant Jerding, USA TODAY

Source: Copyright © 1997 USA TODAY. Reprinted with permission.

USA SNAPSHOTS®

A look at statistics that shape the nation

Bachelor's degree pay for men

The U.S. median income for men age 30 or older with bachelor's degrees and earning wages or salaries was $43,856[1]. Fields of study with highest median pay:

Major	Median income
Engineering	$52,998
Mathematics	$52,316
Physics	$51,819
Pharmacy	$50,805
Economics	$50,360

1–Excludes self-employed through 1993 (latest year available)

Source: Bureau of Labor Statistics *Occupational Outlook Quarterly, Summer 1998*

By Anne R. Carey and Grant Jerding, USA TODAY

Source: Copyright © 1997 USA TODAY. Reprinted with permission.

median incomes for college graduates with different majors. The employees in the sample are all at least 30 years old. Does it seem reasonable to assume that many of the employees are in jobs beyond the entry level? Explain. Compare the median incomes shown for all women aged 30 or older holding bachelor's degrees with the median income for men of similar age holding bachelor's degrees. Look at the particular majors listed. What percentage of men holding bachelor's degrees in mathematics make $52,316 or more? What percentage of women holding computer/information science degrees make $41,559 or more? How do median incomes for men and women holding engineering degrees compare? What about pharmacy degrees?

LINKING CONCEPTS: WRITING PROJECTS

Discuss each of the following topics in class or review the topics on your own. Then write a brief but complete essay in which you summarize the main points. Please include formulas and graphs as appropriate.

1. An average is an attempt to summarize a collection of data into just *one* number. Discuss how the mean, median, and mode all represent averages in this context. Also discuss the differences among these averages. Why is the mean a balance point? Why is the median a midway point? Why is the mode the most common data point? List three areas of daily life where you think either the mean, median, or mode would be the best choice to describe an "average."

2. Why do we need to study the variation of a collection of data? Why isn't the average by itself adequate? We have studied three ways to measure variation. The range, standard deviation, and to a large extent, a box-and-whisker plot all indicate the variation within a data collection. Discuss similarities and differences among these ways to measure data variation. Why would it seem reasonable to pair the median with a box-and-whisker plot and to pair the mean with the standard deviation? What are the advantages and disadvantages of each method of describing data spread? Comment on statements such as the following: (a) The range is easy to compute, but it doesn't give much information; (b) although the standard deviation is more complicated to compute, it has some significant applications; (c) the box-and-whisker plot is fairly easy to construct, and it gives a lot of information at a glance.

3. Why is the coefficient of variation important? What do we mean when we say the coefficient of variation has no units? What advantage can there be in having no units? Why is *relative size* important?

   Consider robin eggs; the mean weight of a collection of robin eggs is 0.72 ounce and the standard deviation is 0.12 ounce. Now consider elephants; the mean weight of elephants in the zoo is 6.42 tons, with a standard deviation 1.07 tons. The units of measurement are different and there is a great deal of difference between the size of an elephant and a robin's egg. Yet the coefficient of variation is about the same for both. Comment on this from the viewpoint of the size of the standard deviation relative to the mean.

4. What is Chebyshev's theorem? Suppose you have a friend who knows very little about statistics. Write a paragraph or two in which you describe Chebyshev's theorem for your friend. Keep the discussion as simple as possible, but be sure to get the main ideas across to your friend. Suppose he or she asks, "What is this stuff good for?" and suppose you respond (a little sarcastically) that Chebyshev's theorem applies to everything from butterflies to the orbits of the planets! Would you be correct? Explain.

5. Have each student count the amount of loose change (not paper money) brought to class today. Use a calculator with mean and standard deviation buttons to compute the mean and standard deviation for the class. (Perhaps the professor can enter the data into the calculator while the students tell him or her the numbers.)

   (a) Compute the endpoints of the interval $\bar{x} - 2s$ to $\bar{x} + 2s$. Have each student raise a hand if the amount of loose change that was counted falls into this interval. Explain why you expect about 75% or more of the class to raise their hands.

   (b) Compute the endpoints of the interval $\bar{x} - \sqrt{2}s$ to $\bar{x} + \sqrt{2}s$. Repeat part (a) for this interval. Explain why you expect about 50% or more of the class to raise their hands.

   (c) Suppose we take the point of view that our statistics class is a sample of the entire student body. What percentage of the entire student body do you expect will have from $\bar{x} - 3s$ to $\bar{x} + 3s$ cents in loose change? Is your answer a lower estimate for the actual percentage? Explain.

   (d) Write a paragraph or two in which you discuss how in part (c) we have used inferential statistics to estimate the population (i.e., student body) distribution of loose change using a sample (i.e., our statistics class) from that population.

Using TECHNOLOGY

Raw Data

APPLICATION 1

Using the software or calculator available to you, do the following.

1. Trade winds are one of the beautiful features of island life in Hawaii. The following data represent total air movement in miles each day over a weather station in Hawaii as determined by a continuous anemometer recorder. The period of observation is January 1 to February 15, 1971.

26	14	18	14	113	50	13	22
27	57	28	50	72	52	105	138
16	33	18	16	32	26	11	16
17	14	57	100	35	20	21	34
18	13	18	28	21	13	25	19
11	19	22	19	15	20		

Source: United States Department of Commerce, National Oceanic and Atmospheric Administration, Environmental Data Service. *Climatological Data, Annual Summary, Hawaii,* Vol. 67, No. 13. Asheville: National Climatic Center, 1971, pp. 11, 24.

(a) Enter the data.
(b) Use the computer to find the sample mean, median, and (if it exists) mode.
(c) Use the computer to find the range, sample variance, and sample standard deviation.
(d) We studied the box-and-whisker plot as a topic in exploratory data analysis (EDA). Use the five-number summary provided by the computer to make a box-and-whisker plot of total air movement over the weather station.
(e) Four data values are exceptionally high: 113, 105, 138, and 100. The strong winds of January 5 (113 reading) brought in a cold front that dropped snow on Haleakala National Park (at the 8000 ft elevation). The residents were so excited that they drove up to see the snow and caused such a massive traffic jam that the Park Service had to close the road. The winds of January 15 and 16 (readings 105 and 138) brought in a storm that created more damaging funnel clouds than any other storm to that date in the recorded history of Hawaii. The strong winds of January 28 (reading 100) accompanied a storm with funnel clouds that did much damage. Eliminate these values (i.e., 100, 105, 113, 138) from the data bank and redo parts (a) through (d). Compare your results with those previously obtained. Which average is most affected? What happens to the standard deviation? How do the two box-and-whisker plots compare?

(f) What is "normal" weather in a region? This is a relevant question since climate is an important factor in agriculture, commerce, transportation, and tourism. The National Climate Data Center keeps temperature, moisture, wind, and other climate records. It establishes "climatic normals." A "climatic normal" is an arithmetic average of a meteorological element over 30 years. For more information, visit the Brase/Brase statistics site at http://math.college.hmco.com/students and find the links to NOAA (National Oceanic and Atmospheric Administration) and *USA Today*. Once at the NOAA site, use the index to find "climatic normals." At the *USA Today* site, select weather.

Technology Hints: Raw Data

The Tech Note of Section 3.2 gives brief instructions for finding summary statistics for raw data using the TI-83, Excel, and Minitab. The Tech Note of Section 3.4 gives brief instructions for constructing box plots on the TI-83 and in Minitab. In ComputerStat, select Descriptive Statistics, then Frequency Distribution.

Grouped Data

Using your software or calculator, do the following.

1. The summit of Longs Peak in Colorado (elevation 14,256 ft) is very beautiful, and in winter very austere. In the following data, x = hourly peak gusts in miles per hour of winter wind on the summit of Longs Peak and f = frequency of occurrence for a sample period of 1518 observation hours.
   (a) Using the class limits, compute each class midpoint. (*Hint:* Compute the first class midpoint. To get the others, just keep adding 10 to the previous class midpoint.)
   (b) Find the approximate sample mean, sample variance, and sample standard deviation of hourly peak gusts on the summit of Longs Peak.

Hourly Peak Gust, mph	Total Hours of Occurrence
0–9	141
10–19	217
20–29	376
30–39	290
40–49	195
50–59	143
60–69	61
70–79	37
80–89	21
90–99	11
100–109	12
110–119	3
120–129	5
130–139	3
140–149	3

(In addition to the values shown in the table, there were three incidences where the wind gusts exceeded 150 mph.)

Source: D. E. Glidden, *Winter Wind Studies in Rocky Mountain National Park.* Estes Park, Colo.: Rocky Mountain Nature Association, 1982, p. 23.

Technology Hints: Grouped Data

TI-83Plus

Enter midpoints in list L1 and frequencies in list L2. Then select 1-Var Stats for L1,L2. The final screen will give you the mean and standard deviation for the grouped data.

Excel and Minitab

Neither Excel nor Minitab gives results for grouped data directly. Those familiar with column operations and formula procedures can generate programs to find the mean and standard deviation for grouped data.

```
Grouped-Data Program for Minitab
MTB > # enter midpoints in C1
MTB > # enter frequencies in C2
MTB > SUM C2 put in K1
MTB > Let C3 = C1*C2
MTB > SUM C3 put in K2
MTB > NAME K3 = 'MEAN'
MTB > LET K3 = K2/K1
MTB > LET C4 = (C1 − K3)**2*C2
MTB > SUM C4 put in K4
MTB > LET K5 = SQRT(K4/(K1 − 1))
MTB > NAME K5 = 'STDEV'
MTB > PRINT K3, K5
```

ComputerStat

Select Descriptive Statistics and then Averages and Variation for Grouped Data.

4

Elementary Probability Theory

We see that the theory of probabilities is at bottom only common sense reduced to calculation; it makes us appreciate with exactitude what reasonable minds feel by a sort of instinct, often without being able to account for it.

—Pierre-Simon Laplace

**Pierre-Simon Laplace
(1749–1827)**

This renowned French astronomer and mathematician wrote *The Analytical Theory of Probability.*

4.1   What is Probability?

4.2   Some Probability Rules—Compound Events

4.3   Trees and Counting Techniques

This is how the great mathematician Laplace described the theory of mathematical probability. The discovery of the mathematical theory of probability was shared by two Frenchmen: Blaise Pascal and Pierre Fermat. These seventeenth-century scholars were attracted to the subject by the inquiries of the Chevalier de Méré, a gentleman gambler.

Although the first applications of probability were to games of chance and gambling, today the subject seems to pervade almost every aspect of modern life. Everything from the orbits of spacecraft to the social behavior of woodchucks is described in terms of probabilities.

For on-line student resources, visit **math.college.hmco.com/students** and follow the Statistics links to the Brase/Brase, *Understandable Statistics,* 7th edition web site.

◇ Why do we need probability? *Hint:* Many important issues in life involve uncertain answers.   (SECTION 4.1)

◇ What are the basic definitions and rules of probability?   (SECTION 4.2)

◇ What are counting techniques, trees, permutations, and combinations? (SECTION 4.3)

FOCUS PROBLEM

How Often Do Lie Detectors Lie?

James Burke is an educator who is known for his interesting science-related radio and television shows aired by the British Broadcasting Corporation. His book *Chances: Risk and Odds in Everyday Life* (Virginia Books, London) contains a great wealth of fascinating information about probabilities. The following quote is from Professor Burke's book:

> *If I take a polygraph test and lie, what is the risk I will be detected?* According to some studies there's about a 72 percent chance you will be caught by the machine.

> *What is the risk that if I take a polygraph test it will incorrectly say that I lied?* At least 1 in 15 will be thus falsely accused.

Both of these statements contain conditional probabilities, which we will study in Section 4.2. Information from that section will allow us to answer the following:

Suppose a person answers 10% of a long battery of questions with lies. Assume that the remaining 90% of the questions are answered truthfully.

1. Estimate the percentage of answers the polygraph will *wrongly* indicate as lies.

2. Estimate the percentage of answers the polygraph will *correctly* indicate as lies.

If the polygraph indicated that 30% of the questions were answered as lies, what would you estimate for the *actual* percentage of questions the person answered as lies?

4.1
What Is Probability?

FOCUS POINTS

✓ Assign probabilities to events.

✓ Explain how the law of large numbers relates to relative frequencies.

✓ Apply basic rules of probability in everyday life.

✓ Explain the relationship between statistics and probability.

Basic concepts

We encounter statements in terms of probability all the time. An excited sports announcer claims that Sheila has a 90% chance of breaking the world record in the upcoming 100-yard dash. Henry figures that if he guesses on a true–false question, the probability of getting it right is 1/2. The Right to Health Lobby claims the probability is 0.40 of getting an erroneous report from a medical laboratory in one low-cost health center. It is consequently lobbying for a federal agency to license and monitor all medical laboratories.

When we use probability in a statement, we're using a *number between 0 and 1* to indicate the likelihood of an event. We'll use the notation $P(A)$ (read, "P of A") to denote the *probability of event A*. The closer to 1 the probability assignment is, the more likely the event is to occur. If the event A is certain to occur, then $P(A)$ should be 1.

It is important to know what probability statements mean and how to compute or assign probabilities to events because probability is the language of inferential statistics. For instance, suppose a college counselor claims that 70% of first-year students receive counseling to help plan their schedules. Because of the high percentage of students needing help, he is requesting that an additional counselor be hired. You want to test the counselor's claim. In doing so, you pick a random sample of 30 first-year students and find that only 3 of them got help from a counselor. Can you challenge the counselor's claim on the basis of this random sample in which only 10% of the students got counselor help with their schedules? To answer this question, we need to find the *probability* of picking a random sample of first-year students in which only 10% got counselor help under the assumption that the counselor's claim is correct. This is the kind of question we will consider in hypothesis testing (Chapter 9).

In the meantime, we need to learn how to find probabilities or assign them to events. There are three major methods. One is *intuition*. The sports announcer probably used Sheila's performances in past track events and his own confidence in her running ability as a basis for his prediction that she has a 90% chance of breaking the world record. In other words, the announcer feels that the probability is 0.90 that Sheila will break the world record.

Relative frequency

The Right to Health Lobby used another method to arrive at its probability statement. It took the *relative frequency* with which erroneous laboratory reports occurred. From a random sample of $n = 100$, it found $f = 40$ erroneous laboratory reports. From this it computed the relative frequency of erroneous laboratory reports via Formula (1).

Probability formula for relative frequency

$$\text{Probability of an event} = \text{Relative frequency} = \frac{f}{n} \qquad (1)$$

where f is the frequency of an event,
  n is the sample size.

In the case of the laboratory reports, we have

$$\text{Relative frequency} = \frac{f}{n} = \frac{40}{100} = 0.40$$

The relative frequency of erroneous laboratory reports was used as the *probability* of erroneous reports.

The technique of using the relative frequency of an event as the probability of that event is a common way of assigning probabilities and will be used a great deal in later chapters. The underlying assumption we make is that if events occurred a certain percentage of times in the past, they will occur about the same percentage of times in the future. In fact, this can be strengthened to a very general statement called the *law of large numbers*.

Law of large numbers

Law of large numbers

In the long run, as the sample size increases and increases, the relative frequencies of outcomes get closer and closer to the theoretical (or actual) probability value.

The law of large numbers is the reason such businesses as health insurance, automobile insurance, and gambling casinos can exist and make a profit. In Central City, Colorado, there are many casinos with many slot machines. The winnings of a gambler on a single play or even a few plays are uncertain (small sample size). This is one of the reasons gambling is exciting. However, on tens of thousands of plays, the theoretical or actual probability of winning favors the casino. That's why the casino and its owners regard gambling as a business. The house is guaranteed a profit in the long run.

Henry used the third method of assigning probabilities when he determined the probability of correctly guessing the answer to a true–false question. Essentially, he used the probability formula for *equally likely outcomes*.

Equally likely outcomes

Probability formula when outcomes are equally likely

$$\text{Probability of an event} = \frac{\text{Number of outcomes favorable to event}}{\text{Total number of outcomes}} \qquad (2)$$

In Henry's case, there are two possible outcomes. A test answer will be either correct or incorrect. Since he is guessing, we assume that the outcomes are equally likely, and only one is "favorable" to being correct. So, by Formula (2),

$$P(\text{correct answer}) = \frac{\text{Number of favorable outcomes}}{\text{Total number of outcomes}}$$

$$= \frac{1}{2}$$

We've seen three ways to assign probabilities: intuition, relative frequency, and—when outcomes are equally likely—a formula. Which do we use? Most of the time it depends on the information that is at hand or that can be feasibly obtained. Our choice of methods also depends on the particular problem. In Guided Exercise 1, you will see three different situations, and you will decide which way to assign the probabilities. *Remember, probabilities are numbers between 0 and 1, so don't assign probabilities outside this range.*

GUIDED EXERCISE 1

Determine a probability

Assign a probability to the indicated event on the basis of the information provided. Indicate the technique you use: intuition, relative frequency, or the formula for equally likely outcomes.

(a) The director of the Readlot College Health Center wishes to open an eye clinic. To justify the expense of such a clinic, the director reports the probability that a student selected at random from the college roster needs corrective lenses. She took a random sample of 500 students to compute this probability and found that 375 of them needed corrective lenses. What is the probability that a Readlot College student selected at random needs corrective lenses?

In this case we are given a sample size of 500, and we are told that 375 of these students need glasses. It is appropriate to use a relative frequency for the desired probability:

$$P(\text{student needs glasses}) = \frac{f}{n}$$

$$= \frac{375}{500}$$

$$= 0.75$$

(b) The Friends of the Library host a fund-raising barbecue. George is on the cleanup committee. There are four members on this committee, and they draw lots to see who will clean the grills. Assuming that each member is equally likely to be drawn, what is the probability that George will be assigned the grill-cleaning job?

There are four people on the committee, and each is equally likely to be drawn. It is appropriate to use the formula for equally likely events. George can be drawn in only one way, so there is only one outcome favorable to that event.

$$P(\text{George}) = \frac{\text{No. of favorable outcomes}}{\text{Total no. of outcomes}}$$

$$= \frac{1}{4}$$

$$= 0.25$$

Continued

GUIDED EXERCISE 1 continued

(c) Joanna photographs whales for Sea Life Adventure Films. On her next expedition, she is to film blue whales feeding. Her boss asks her what she thinks the probability of success will be for this particular assignment. She gives an answer based on her knowledge of the habits of blue whales and the region she is to visit. She is almost certain she will be successful. What specific number do you suppose she gave for the probability of success, and how do you suppose she arrived at it?

 Since Joanna is almost certain of success, she should make the probability close to 1. We would say *P*(success) is above 0.90 but less than 1. We think the probability assignment was based on intuition.

No matter how we compute probabilities, it is useful to know what outcomes are possible in a given setting. For instance, if you are going to decide the probability that Hardscrabble will win the Kentucky Derby, you need to know which other horses will be running.

Sample space

A *statistical experiment* or *statistical observation* can be thought of as any random activity that results in a definite outcome. Usually, the outcome is in the form of a description, count, or measurement. For example, tossing a coin can be thought of as an experiment. There are only two possible outcomes: heads or tails. The set of all possible distinct outcomes is called the *sample space*. If you toss a coin, the sample space for that experiment consists of the two outcomes (heads or tails).

It is especially convenient to know the sample space in the case where all outcomes are equally likely because then we can compute probabilities of various events by using Formula (2).

$$P(\text{event } A) = \frac{\text{Number of outcomes favorable to } A}{\text{Total number of outcomes}} \qquad (2)$$

To use this formula, we need to know the sample space so that we can determine which outcomes are favorable to the event in question as well as the total number of outcomes.

EXAMPLE 1

Using a sample space

Human eye color is controlled by a single pair of genes (one from the father and one from the mother) called a *genotype*. Brown eye color, B, is dominant over blue eye color, ℓ. Therefore, in the genotype Bℓ, consisting of one brown gene B and one blue gene ℓ, the brown gene dominates. A person with a Bℓ genotype has brown eyes.

If both parents have brown eyes and have genotype Bℓ, what is the probability that their child will have blue eyes? What is the probability the child will have brown eyes?

SOLUTION: To answer these questions, we need to look at the sample space of all possible eye-color genotypes for the child. They are given in Table 4-1.

According to genetics theory, the four possible genotypes for the child are equally likely. Therefore, we can use Formula (2) to compute probabilities. Blue eyes

can occur only with the $\ell\ell$ genotype, so there is only one outcome favorable to blue eyes. By Formula (2),

$$P(\text{blue eyes}) = \frac{\text{Number of favorable outcomes}}{\text{Total number of outcomes}}$$

$$= \frac{1}{4}$$

TABLE 4-1   Eye Color Genotypes for Child

	Mother	
Father	B	ℓ
B	BB	Bℓ
ℓ	ℓB	$\ell\ell$

Brown eyes occur with the three remaining genotypes: BB, Bℓ, and ℓB. By Formula (2),

$$P(\text{brown eyes}) = \frac{\text{Number of favorable outcomes}}{\text{Total number of outcomes}} = \frac{3}{4}$$ ◇

GUIDED EXERCISE 2

Using a sample space

Professor Gutierrez is making up a final exam for a course in literature of the Southwest. He wants the last three questions to be of the true–false type. To guarantee that the answers do not follow his favorite pattern, he lists all possible true–false combinations for three questions on slips of paper and then picks one at random from a hat.

(a) Finish listing the outcomes in the given sample space.

TTT	FTT	TFT	_____
TTF	FTF	TFF	_____

⟹ The missing outcomes are FFT and FFF.

(b) What is the probability that all three items will be false? Use the formula

$$P(\text{all F}) = \frac{\text{No. of favorable outcomes}}{\text{Total no. of outcomes}}$$

⟹ There is only one outcome, FFF, favorable to all false, so

$$P(\text{all F}) = \frac{1}{8}$$

(c) What is the probability that exactly two items will be true?

⟹ There are three outcomes that have exactly two true items: TTF, TFT, and FTT. Thus,

$$P(\text{two T}) = \frac{\text{No. of favorable outcomes}}{\text{Total no. of outcomes}} = \frac{3}{8}$$

Complement of an event

There is one more important point about probability assignments. The sum of all the probabilities assigned to outcomes in a sample space must be 1. This makes sense. If you think the probability is 0.65 that you will win a tennis match, you assume the probability is 0.35 that your opponent will win. This fact is particularly useful, for if the probability that an event occurs is denoted by p and the probability that it *does not* occur is denoted by q, we have

$p + q = 1$ because the sum of the probabilities of the outcomes must be 1

or

$$q = 1 - p \qquad (3)$$

For an event A, the event *not A* is called the *complement of A.** To compute the probability of the complement of A, we use Formula (3) and find

$$P(not\ A) = 1 - P(A)$$

EXAMPLE 2

Complement of an event

The probability that a college student without a flu shot will get the flu is 0.45. What is the probability that a college student will *not* get the flu if the student has not had the flu shot?

SOLUTION: In this case, we have

$P(\text{will get flu}) = p = 0.45$

$P(\text{will } not \text{ get flu}) = q = 1 - p = 1 - 0.45 = 0.55$ ◇

GUIDED EXERCISE 3

Complement of an event

A veterinarian tells you that if you breed two cream-colored guinea pigs, the probability that an offspring will be pure white is 0.25. What is the probability that it will not be pure white?

(a) $P(\text{pure white}) + P(not \text{ pure white}) = $ _____   ⟹   1

(b) $P(not \text{ pure white}) = $ _____   ⟹ $1 - 0.25$, or 0.75

The important facts about probabilities we have seen in this section are

1. The probability of an event A is denoted by $P(A)$.

2. The probability of any event is a number between 0 and 1. The closer to 1 the probability is, the more likely the event is.

3. The sum of the probabilities of outcomes in a sample space is 1.

4. Probabilities can be assigned by using three methods: intuition, relative frequencies, or the formula for equally likely outcomes.

5. The probability event A does not occur is denoted by $P(not\ A)$.

6. $P(A) + P(not\ A) = 1$.

*The complement of event A is often symbolized by the compact notation A^c or $\overline{A}$. However, in this text, we will continue to use the more expanded description *not A* to refer to the complement of A. That is, *not A* refers to all the outcomes of the sample space that are not favorable to event A. (For Bayes's theorem in Appendix I we use A^c for the complement of A.)

Probability Related to Statistics

We conclude this section with a few comments on the nature of statistics versus probability. Although statistics and probability are closely related fields of mathematics, they are nevertheless separate fields. It can be said that probability is the medium through which statistical work is done. In fact, if it were not for probability theory, inferential statistics would not be possible.

Put very briefly, probability is the field of study that makes statements about what will occur when samples are drawn from a *known population*. Statistics is the field of study that describes how samples are to be obtained and how inferences are to be made about *unknown populations*.

A simple but effective illustration of the difference between these two subjects can be made by considering how we treat the following examples:

Example of a probability application

Condition: We *know* the exact makeup of the *entire* population.
Example: Given 3 green marbles, 5 red marbles, and 4 white marbles in a bag, draw 6 marbles at random from the bag. What is the probability none of the marbles is red?

Example of a statistical application

Condition: We have only *samples* from an otherwise *unknown* population.
Example: Draw a random sample of 6 marbles from (unknown) population of all marbles in a bag and observe the colors. Based on the sample results, make a conjecture about the colors and numbers of marbles in the entire population of all marbles in the bag.

In another sense, probability and statistics are like flip sides of the same coin. On the probability side, you know the overall description of the population. The central problem is to compute the likelihood that a specific outcome will happen. On the statistics side, you know only the results of a sample drawn from the population. The central problem is to describe the sample (descriptive statistic) and to draw conclusions about the population based on the sample results (inferential statistics).

In statistical work, the inferences we draw about an unknown population are not claimed to be absolutely correct. Since the population remains unknown (in a theoretical sense), we must accept a "best guess" for our conclusions and act using the most probable answer rather than absolute certainty.

Probability is the topic of this chapter. However, we will not study probability just for its own sake. Probability is a wonderful field of mathematics, but we will study mainly the ideas from probability that are needed for a proper understanding of statistics.

Even though we have only a brief introduction to probability, we will be able to answer questions raised in situations such as the following:

Salmon moving upstream

- Commercial salmon fishing is very important in Alaska. Since the salmon are sold by weight, it is useful to know the average weight of a freshly caught salmon. How large a sample of freshly caught salmon is needed if we want to be 95% sure the mean weight of the sample is within plus or minus 1 ounce of the mean weight of all catchable salmon in Alaskan waters? For the answer to this type question, see Example 6 in Section 8.4, in which we discuss probability as applied to sample size.

- Retail sales account for about two-thirds of all the money flow in the U.S. economy. For their first job after graduation, many college graduates are either in sales or in charge of a sales team. Whether you are a salesperson or in charge of a sales team, *sales quotas* are important. Suppose you are a bond broker and your job is to sell a particular bond issue by phone. History shows that the probability is about 20% that any one phone call will result in a bond sale. Your sales supervisor claims that you should make at least nine sales each day. How many phone calls should you make if you want to be 99% sure of getting your nine sales? For the answer to this type of question, see Section 5.3, in which we discuss probabilities and quota problems.

- Many practical situations call on us to choose between two competing statements. *Consumer Reports* did a study on the maintenance records of two competing brands of cellular phones. Each brand claims to have the lowest maintenance costs. Is one brand actually better than the other? If there is an apparent difference, is the difference statistically significant at, say, a 1% level (of risk)? For answers to these kinds of questions, see Section 9.7, in which we discuss tests for differences.

All the listed examples require some use of probability. This is why we encourage you to study this chapter carefully. Your time in doing so will be well invested.

VIEWP◉INT   *What Makes a Good Teacher?*

A survey of 735 students at nine colleges in the United States was taken to determine instructor behaviors that help students succeed. Data for this survey can be found by visiting the Brase/Brase statistics site at http://math.college.hmco.com/students and finding the link to DASL, the Carnegie Mellon University Data and Story Library. Once at the DASL site, select Data Subjects, then Psychology, and then Instructor Behavior. You can estimate the probability of how a student would respond (very positive, neutral, very negative) to different instructor behaviors. For example, more than 90% of the students responded "very positive" to the instructor's use of real-world examples in the classroom.

SECTION 4.1   PROBLEMS

1. *General: Concepts* In your own words, carefully answer the question: What is probability? List three methods of assigning probabilities.

2. *General: Provide Examples* List examples of where probability might be applied in business, medicine, social science, and natural science. Identify some ways that probability will be useful in the study of statistics.

3. *General: Valid Probabilities* Which of the following numbers cannot be the probability of some event?

   (a) 0.71   (b) 4.1   (c) $\frac{1}{8}$ (d) -0.5

   (e) 0.5   (f) 0   (g) 1   (h) 150%

4. *General: Valid Probabilities*
   (a) Explain why -0.41 cannot be the probability of some event.
   (b) Explain why 1.21 cannot be the probability of some event.
   (c) Explain why 120% cannot be the probability of some event.
   (d) Can the number 0.56 be the probability of an event? Explain.

5. *Probability Estimate: Wiggle Your Ears* Can you wiggle your ears? Use the students in your statistics class (or a group of friends) to estimate the percentage of people who can wiggle their ears. How can your result be thought of as an estimate for the probability that a person chosen at random can wiggle his or her ears? *Comment:* National statistics indicate that about 13% of Americans can wiggle their ears (Source: Bernice Kanner, *Are You Normal?* St. Martin's Press, New York).

6. *Probability Estimate: Raise One Eyebrow* Can you raise one eyebrow at a time? Use the students in your statistics class (or a group of friends) to estimate the percentage of people who can raise one eyebrow at a time. How can your result be thought of as an estimate for the probability that a person chosen at random can raise one eyebrow at a time? *Comment:* National statistics indicate that about 30% of Americans can raise one eyebrow at a time (see source in Problem 5).

7. *Myers–Briggs: Personality Types* Isabel Briggs Myers was a pioneer in the study of personality types. The personality types are broadly defined according to four main preferences. Do married couples choose similar or different personality types in their mates? The following data give an indication (Source: I. B. Myers and M. H. McCaulley, *A Guide to the Development and Use of the Myers–Briggs Type Indicators*).

   Similarities and Differences in a Random Sample of 375 Married Couples

Number of Similar Preferences	Number of Married Couples
All four	34
Three	131
Two	124
One	71
None	15

   Suppose that a married couple is selected at random.
   (a) Use the data to estimate the probability that they will have 0, 1, 2, 3, or 4 personality preferences in common.
   (b) Do the probabilities add up to 1? Why should they? What is the sample space in this problem?

8. *Sociology: Dating Couples* Do couples get engaged or not? If they are engaged, how long did they date before becoming engaged? A poll of 1000 couples conducted by Bruskin and Goldring Research for Korbel Champagne Cellars gave the following information (*USA Today*):

Length of Dating Time Before Engagement

Time	Number of Couples
Never engaged	200
Less than 1 year	240
1 to 2 years	210
More than 2 years	350

(a) Use the data to estimate the probability of each event for a dating couple chosen at random: The couple is not engaged, dated less than 1 year before getting engaged, dated 1 to 2 years before getting engaged, or dated more than 2 years before getting engaged.
(b) Do the probabilities of part (a) add up to 1? Why should they? What is the sample space in this problem?

9. *Psychology: Creativity* When do creative people get their *best* ideas? *USA Today* did a survey of 966 inventors (who hold U.S. patents) and obtained the following information:

Time of Day When Best Ideas Occur

Time	Number of Inventors
6 A.M.–12 noon	290
12 noon–6 P.M.	135
6 P.M.–12 midnight	319
12 midnight–6 A.M.	222

(a) Assuming that the time interval includes the left limit and all the times up to but not including the right limit, estimate the probability that an inventor has a best idea during each time interval: from 6 A.M. to 12 noon, from 12 noon to 6 P.M., from 6 P.M. to 12 midnight, from 12 midnight to 6 A.M.
(b) Do the probabilities of part (a) add up to 1? Why should they? What is the sample space in this problem?

10. *General: Roll a Die*
(a) If you roll a single die and count the number of dots on top, what is the sample space of all possible outcomes? Are the outcomes equally likely?
(b) Assign probabilities to the outcomes of the sample space of part (a). Do the probabilities add up to 1? Should they add up to 1? Explain.
(c) What is the probability of getting a number less than 5 on a single throw?
(d) What is the probability of getting 5 or 6 on a single throw?

11. *Health Care: Flu* "Oh, leave me alone!" How do people want to be treated when they have the flu? A Sterling Health poll of 1000 people gave the information shown at the top of the next page (*USA Today*).

Want to Be . . .	Number of People
Left alone	770
Waited on hand and foot	160
Treated differently	70

(a) Consider the events: want to be left alone, want to be waited on hand and foot, want to be treated differently. Do these events form a sample space for the way people who have the flu wish to be treated? Use relative frequencies to assign probabilities to these events. Do the probabilities add up to 1? Why should they?

(b) Find the probability of the event do *not* want to be left alone. Find the probability of the event do *not* want to be waited on hand and foot.

12. *Agriculture: Cotton* A botanist has developed a new hybrid cotton plant that can withstand insects better than other cotton plants. However, there is some concern about the germination of seeds from the new plant. To estimate the probability that a seed from the new plant will germinate, a random sample of 3000 seeds was planted in warm, moist soil. Of these seeds, 2430 germinated.

(a) Use relative frequencies to estimate the probability that a seed will germinate. What is your estimate?

(b) Use relative frequencies to estimate the probability that a seed will *not* germinate. What is your estimate?

(c) Either a seed germinates, or it does not. What is the sample space in this problem? Do the probabilities assigned to the sample space add up to 1? Should they add up to 1? Explain.

(d) Are the outcomes in the sample space of part (c) equally likely?

13. *General: Odds in Favor* Sometimes probability statements are expressed in terms of *odds*.

The *odds in favor* of an event A is the ratio $\dfrac{P(A)}{P(not\ A)}$.

For instance, if $P(A) = 0.60$, then $P(not\ A) = 0.40$ and the odds in favor of A are

$\dfrac{0.60}{0.40} = \dfrac{6}{4} = \dfrac{3}{2}$, written as 3 to 2 or 3:2

(a) Show that if we are given the *odds in favor* of event A as $n{:}m$, the probability of event A is given by $P(A) = \dfrac{n}{n + m}$. *Hint:* Solve the equation $\dfrac{n}{m} = \dfrac{P(A)}{1 - P(A)}$ for $P(A)$.

(b) A telemarketing supervisor tells a new worker that the odds of making a sale on a single call are 2 to 15. What is the probability of a successful call?

(c) A sports announcer says that the odds a basketball player will make a free throw shot are 3 to 5. What is the probability the player will make the shot?

14. *Horse Racing: Betting Odds Against* Betting odds are usually stated *against* the event happening (against winning).

The odds *against* event W is the ratio $\dfrac{P(not\ W)}{P(W)}$.

In horse racing, the betting odds are based on the probability that the horse does *not* win.

(a) Show that if we are given the odds *against* an event W as $a{:}b$, the probability of *not* W is $P(not\ W) = \dfrac{a}{a+b}$. *Hint:* Solve the equation $\dfrac{a}{b} = \dfrac{P(not\ W)}{1 - P(not\ W)}$ for $P(not\ W)$.

(b) In a recent Kentucky Derby, the betting odds for the favorite horse, Point Given, were 9 to 5. Use these odds to compute the probability that Point Given would lose the race. What is the probability that Point Given would win the race?

(c) In the same race, the betting odds for the horse Monarchos were 6 to 1. Use these odds to estimate the probability that Monarchos would lose the race. What is the probability that Monarchos would win the race?

(d) Invisible Ink was a long shot, with betting odds of 30 to 1. Use these odds to estimate the probability that Invisible Ink would lose the race. What is the probability the horse would win the race? For further information on the Kentucky Derby, visit the Brase/Brase statistics site at http://math.college.hmco.com/students and find the link to the Kentucky Derby.

15. *Business: Customers* John runs a computer software store. Yesterday he counted 127 people who walked by his store, 58 of whom came into the store. Of the 58, only 25 bought something in the store.

(a) Estimate the probability that a person who walks by the store will enter the store.

(b) Estimate the probability that a person who walks into the store will buy something.

(c) Estimate the probability that a person who walks by the store will come in *and* buy something.

(d) Estimate the probability that a person who comes into the store will buy nothing.

4.2
Some Probability Rules—Compound Events

You roll two dice. What is the probability that you will get a 5 on each die? You draw two cards from a well-shuffled, standard deck without replacing the first card before drawing the second. What is the probability that they will both be aces?

It seems that these two problems are nearly alike. They are alike in the sense that in each case you are to find the probability of two events occurring *together*. In the first problem, you are to find

$P(5$ on 1st die *and* 5 on 2nd die$)$

In the second, you want

$P($ace on 1st card *and* ace on 2nd card$)$

Independent events

The two problems differ in one important aspect, however. In the dice problem, the outcome of a 5 on the first die does not have any effect on the probability of getting a 5 on the second die. Because of this, the events are *independent*.

> Two events are **independent** if the occurrence or nonoccurrence of one does *not* change the probability that the other will occur.

In the card problem, the probability of an ace on the first card is 4/52, since there are 52 cards in the deck and 4 of them are aces. If you get an ace on the first card, then the probability of an ace on the second is changed to 3/51, because one ace has already been drawn and only 51 cards remain in the deck. Therefore, the two events in the card-draw problem are *not* independent. They are, in fact, *dependent,* since the outcome of the first draw changes the probability of getting an ace on the second card.

Probability of *A* and *B*

Why does the *independence* or *dependence* of two events matter? The type of events determines the way we compute the probability of the two events happening together. If two events *A* and *B* are *independent,* then we use Formula (4) to compute the probability of the event *A* and *B*:

For independent events,

$$P(A \ and \ B) = P(A) \cdot P(B) \tag{4}$$

If the events are *dependent,* then we must take into account the changes in the probability of one event caused by the occurrence of the other event. The notation *P(A, given B)* denotes the probability that event *A* will occur, *given* that event *B* has occurred. This is called a *conditional probability.* We read *P(A, given B)* as "probability of *A* given *B*." If *A* and *B* are dependent events, then $P(A) \neq P(A, given \ B)$ because the occurrence of event *B* has changed the probability that event *A* will occur. A standard notation for *P(A, given B)* is $P(A \mid B)$. However, we will use the more expanded notation *P(A, given B)* to remind you that we assume that event *B* has already occurred. In Appendix I, where Bayes's theorem is discussed, we will revert to the more standard notation $P(A \mid B)$. We use either Formula (5) or Formula (6) to compute the probability of *A* and *B* when the events *A* and *B* are dependent.

For dependent events,

$$P(A \ and \ B) = P(A) \cdot P(B, given \ that \ A \ has \ occurred) \tag{5}$$

$$P(A \ and \ B) = P(B) \cdot P(A, given \ that \ B \ has \ occurred) \tag{6}$$

We will use either Formula (5) or Formula (6) according to the information available.

Formulas (4), (5), and (6) constitute the *multiplication rules* of probability. They help us compute the probability of events happening together when the sample space is too large for convenient reference or when it is not completely known.

Let's use the multiplication rules to complete the dice and card problems. We'll compare the results with those obtained by using the sample space directly.

EXAMPLE 3

Multiplication rule

Suppose you are going to throw two fair dice. What is the probability of getting a 5 on each die?

SOLUTION USING THE MULTIPLICATION RULE: The two events are independent, so we should use Formula (4). *P*(5 on 1st die *and* 5 on 2nd die) = *P*(5 on 1st) · *P*(5 on 2nd). To finish the problem, we need only compute the probability of getting a 5 when we throw one die.

There are six faces on a die, and on a fair die each is equally likely to come up when you throw the die. Only one face has five dots, so by Formula (2) for equally likely outcomes,

$$P(5 \text{ on die}) = \frac{1}{6}$$

Now we can complete the calculation.

$$P(5 \text{ on 1st die } and \text{ 5 on 2nd die}) = P(5 \text{ on 1st}) \cdot P(5 \text{ on 2nd})$$
$$= \frac{1}{6} \cdot \frac{1}{6}$$
$$= \frac{1}{36}$$

SOLUTION USING SAMPLE SPACE: The first task is to write down the sample space. Each die has six equally likely outcomes, and each outcome of the second die can be paired with each of the first. The sample space is shown in Figure 4-1. The total number of outcomes is 36, and only one is favorable to a 5 on the first die *and* a 5 on the second. The 36 outcomes are equally likely, so by Formula (2) for equally likely outcomes,

$$P(5 \text{ on 1st } and \text{ on 2nd}) = \frac{1}{36}$$

The two methods yield the same result. The multiplication rule was easier to use because we did not need to look at all 36 outcomes in the sample space for tossing two dice.   ◊

EXAMPLE 4

Dependent events

Compute the probability of drawing two aces from a well-shuffled deck of 52 cards if the first card is not replaced before the second card is drawn.

FIGURE 4-1

Sample Space for Two Dice

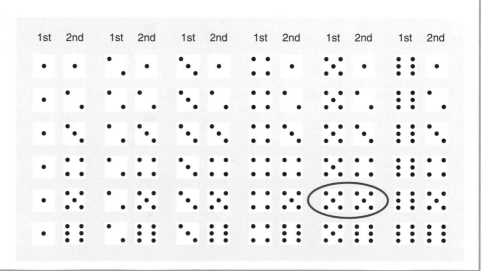

MULTIPLICATION RULE METHOD: These events are *dependent*. The probability of an ace on the first card is 4/52, but on the second card the probability of an ace is only 3/51 if an ace was drawn for the first card. An ace on the first draw changes the probability for an ace on the second draw. By the multiplication rule for dependent events,

$$P(\text{ace on 1st } and \text{ ace on 2nd}) = P(\text{ace on 1st}) \cdot P(\text{ace on 2nd, } given \text{ ace on 1st})$$

$$= \frac{4}{52} \cdot \frac{3}{51} = \frac{12}{2652} = 0.0045$$

SAMPLE SPACE METHOD: We won't actually look at the sample space because each of the 51 possible outcomes for the second card must be paired with each of the 52 possible outcomes for the first card. This gives us a total of 2652 outcomes in the sample space! We'll just think about the sample space and try to list all the outcomes favorable to the event of aces on both cards. The 12 favorable outcomes are shown in Figure 4-2. By the formula for equally likely outcomes,

$$P(\text{ace on 1st card } and \text{ ace on 2nd card}) = \frac{12}{2652} = 0.0045$$

Again, the two methods agree.   ◊

The multiplication rules apply whenever we wish to determine the probability of two events happening *together*. To indicate together, we use *and* between the

---

FIGURE 4-2

Outcomes Favorable to
Drawing Two Aces

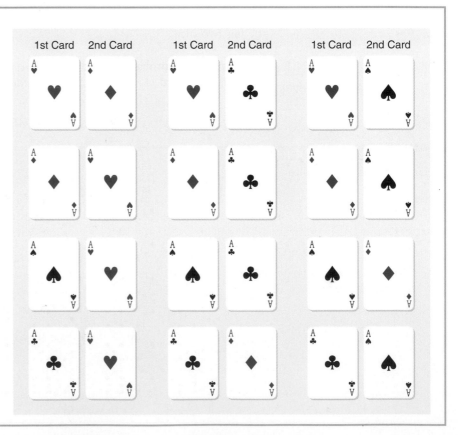

events. But before you use a multiplication rule to compute the probability of *A and B,* you must determine if *A* and *B* are independent or dependent events. Let's practice using the multiplication rule.

GUIDED EXERCISE 4

Multiplication rule

Andrew is 55, and the probability that he will be alive in 10 years is 0.72. Ellen is 35, and the probability that she will be alive in 10 years is 0.92. Assuming that the life span of one will have no effect on the life span of the other, what is the probability they will both be alive in 10 years?

(a) Are these events dependent or independent?

⟹ Since the life span of one does not affect the life span of the other, the events are independent.

(b) Use the appropriate multiplication rule to find

P(Andrew alive in 10 years *and* Ellen alive in 10 years)

⟹ We use the rule for independent events:

$P(A \text{ and } B) = P(A) \cdot P(B)$

$P(\text{Andrew alive } and \text{ Ellen alive})$

$= P(\text{Andrew alive}) \cdot P(\text{Ellen alive})$

$= (0.72)(0.92) = 0.66$

GUIDED EXERCISE 5

Dependent events

A quality-control procedure for testing Ready-Flash disposable cameras consists of drawing two cameras at random from each lot of 100 without replacing the first camera before drawing the second. If both are defective, the entire lot is rejected. Find the probability that both cameras are defective if the lot contains 10 defective cameras. Since we are drawing the cameras at random, assume that each camera in the lot has an equal chance of being drawn.

(a) What is the probability of getting a defective camera on the first draw?

⟹ The sample space consists of all 100 cameras. Since each is equally likely to be drawn and there are 10 defective ones,

$$P(\text{defective camera}) = \frac{10}{100} = \frac{1}{10}$$

Continued

GUIDED EXERCISE 5 continued

(b) The first camera drawn is not replaced, so there are only 99 cameras for the second draw. What is the probability of getting a defective camera on the second draw if the first camera was defective?

➡ If the first camera is defective, then there are only 9 defective cameras left among the 99 remaining cameras in the lot.

P(defective camera on 2nd draw, *given*

defective camera on 1st) $= \dfrac{9}{99} = \dfrac{1}{11}$

(c) Are the probabilities computed in parts (a) and (b) different? Does drawing a defective camera on the first draw change the probability of getting a defective camera on the second draw? Are the events dependent?

➡ The answer to all these questions is yes.

(d) Use the formula for dependent events,

$P(A \text{ and } B) = P(A) \cdot P(B, \text{ given } A \text{ has occurred})$

to compute P(1st camera defective *and* 2nd camera defective).

➡ P(1st defective *and* 2nd defective) $= \dfrac{1}{10} \cdot \dfrac{1}{11}$

$= \dfrac{1}{110}$

$= 0.009$

One of the multiplication rules can be used any time we are trying to find the probability of two events happening *together*. Pictorially, we are looking for the probability of the shaded region in Figure 4-3.

Probability of *A* or *B*

Another way to combine events is to consider the possibility of one event *or* another occurring. For instance, if a sports car saleswoman gets an extra bonus if she sells a convertible or a car with leather upholstery, she is interested in the probability that you will buy a car that is a convertible *or* has leather upholstery. Of course, if you bought a convertible with leather upholstery, that would be fine, too. Pictorially, the shaded portion of Figure 4-4 represents the outcomes satisfying the

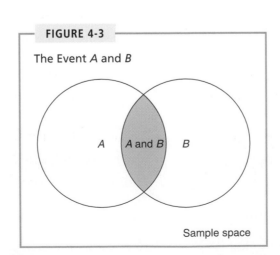

FIGURE 4-3

The Event *A* and *B*

Sample space

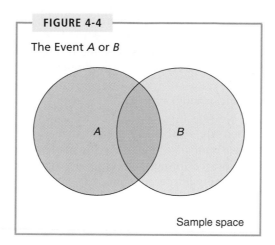

FIGURE 4-4

The Event *A or B*

Sample space

or condition. Notice that the condition *A or B* is satisfied by any one of the following conditions:

1. Any outcome in *A* occurs.

2. Any outcome in *B* occurs.

3. Any outcome in both *A* and *B* occurs.

It is important to distinguish between the *or* combinations and the *and* combinations because we apply different rules to compute their probabilities.

GUIDED EXERCISE **6**

Combining events

Indicate how each of the following pairs of events are combined. Use either the *and* combination or the *or* combination.

(a) Satisfying the humanities requirement by taking a course in the history of Japan or by taking a course in classical literature

    ⟹   Use the *or* combination.

(b) Buying new tires and aligning the tires

    ⟹   Use the *and* combination.

(c) Getting an A not only in psychology but also in biology

    ⟹   Use the *and* combination.

(d) Having at least one of these pets: cat, dog, bird, rabbit

    ⟹   Use the *or* combination.

Once you decide that you are to find the probability of an *or* combination rather than an *and* combination, what formula do you use? Again, it depends on the situation. If you want to compute the probability of drawing either a jack or a king on a single draw from a well-shuffled deck of cards, the formula is simple:

$$P(\text{jack } or \text{ king}) = P(\text{jack}) + P(\text{king}) = \frac{4}{52} + \frac{4}{52} = \frac{8}{52} = \frac{2}{13}$$

since there are 4 jacks and 4 kings in a deck of 52 cards.

If you want to compute the probability of drawing a king or a diamond on a single draw, the formula is a bit more complicated. We have to take the overlap of the two events into account so that we do not count the outcomes twice. We can see the overlap of the two events in Figure 4-5.

$$P(\text{king}) = \frac{4}{52} \quad P(\text{diamond}) = \frac{13}{52} \quad P(\text{king } and \text{ diamond}) = \frac{1}{52}$$

If we simply add $P(\text{king})$ and $P(\text{diamond})$, we're including $P(\text{king } and \text{ diamond})$ twice in the sum. To compensate for this double summing, we simply subtract $P(\text{king } and \text{ diamond})$ from the sum. Therefore,

$$P(\text{king } or \text{ diamond}) = P(\text{king}) + P(\text{diamond}) - P(\text{king } and \text{ diamond})$$

$$= \frac{4}{52} + \frac{13}{52} - \frac{1}{52}$$

$$= \frac{16}{52} = \frac{4}{13}$$

FIGURE 4-5

Drawing a King or a Diamond from a Standard Deck

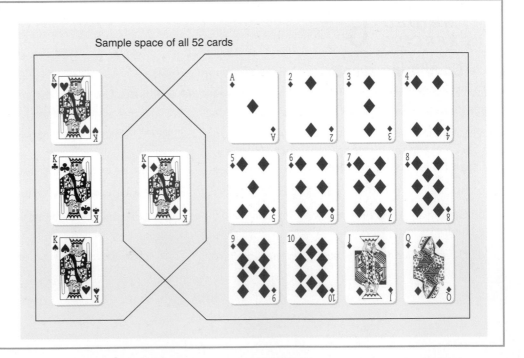

Sample space of all 52 cards

Mutually exclusive events

We say the events A and B are *mutually exclusive* or *disjoint* if they cannot occur together. This means that A and B have no outcomes in common or, put another way, that $P(A \text{ and } B) = 0$. Formula (7) is the *addition rule for mutually exclusive events* A and B.

For *mutually exclusive* events A and B

$$P(A \text{ or } B) = P(A) + P(B) \qquad\qquad (7)$$

If the events are not mutually exclusive, we must use the more general Formula (8), which is the *general addition rule for any events* A and B.

For any events A and B,

$$P(A \text{ or } B) = P(A) + P(B) - P(A \text{ and } B) \qquad\qquad (8)$$

You may ask: Which formula should we use? The answer is: Use Formula (7) only if you know that A and B are mutually exclusive (i.e., cannot occur together); if you do not know whether A and B are mutually exclusive, then use Formula (8). Formula (8) is valid either way. Notice that when A and B are mutually exclusive, then $P(A \text{ and } B) = 0$, so Formula (8) reduces to Formula (7).

GUIDED EXERCISE 7

Mutually exclusive events

The Cost Less Clothing Store carries seconds in slacks. If you buy a pair of slacks in your regular waist size without trying them on, the probability that the waist will be too tight is 0.30 and the probability that it will be too loose is 0.10.

(a) Are the events too tight or too loose mutually exclusive?

⟹ The waist cannot be both too tight and too loose at the same time, so the events are mutually exclusive.

(b) If you choose a pair of slacks at random in your regular waist size, what is the probability that the waist will be too tight or too loose?

⟹ Since the events are mutually exclusive,

$P(\text{too tight } or \text{ too loose})$

$= P(\text{too tight}) + P(\text{too loose})$

$= 0.30 + 0.10$

$= 0.40$

GUIDED EXERCISE 8

General addition rule

Professor Jackson is in charge of a program to prepare people for a high school equivalency exam. Records show that 80% of the students need work in math, 70% need work in English, and 55% need work in both areas.

(a) Are the events *need math* and *need English* mutually exclusive?

⇒ These events are not mutually exclusive, since some students need both. In fact,

$$P(\text{need math } and \text{ need English}) = 0.55$$

(b) Use the appropriate formula to compute the probability that a student selected at random needs math *or* needs English.

⇒ Since the events are not mutually exclusive, we use Formula (8):

$$P(\text{need math } or \text{ need English})$$
$$= P(\text{need math}) + P(\text{need English})$$
$$\quad - P(\text{need math } and \text{ English})$$
$$= 0.80 + 0.70 - 0.55$$
$$= 0.95$$

Combination of several events

The addition rule for mutually exclusive events can be extended so that it applies to the situation in which we have more than two events that are each mutually exclusive to all the other events.

EXAMPLE 5

Mutually exclusive events

Laura is playing Monopoly. On her next move she needs to throw a sum bigger than 8 on the two dice in order to land on her own property and pass Go. What is the probability that Laura will roll a sum bigger than 8?

SOLUTION: When two dice are thrown, the largest sum that can come up is 12. Consequently, the only sums larger than 8 are 9, 10, 11, and 12. These outcomes are mutually exclusive, since only one of these sums can possibly occur on one throw of the dice. The probability of throwing more than 8 is the same as

$$P(9 \text{ or } 10 \text{ or } 11 \text{ or } 12)$$

Since the events are mutually exclusive,

$$P(9 \text{ or } 10 \text{ or } 11 \text{ or } 12) = P(9) + P(10) + P(11) + P(12)$$
$$= \frac{4}{36} + \frac{3}{36} + \frac{2}{36} + \frac{1}{36}$$
$$= \frac{10}{36} = \frac{5}{18}$$

To get the specific values of $P(9)$, $P(10)$, $P(11)$, and $P(12)$, we used the sample space for throwing two dice (see Figure 4-1 on page 163). There are 36 equally likely outcomes—for example, those favorable to 9 are 6, 3; 3, 6; 5, 4; and 4, 5. So $P(9) = 4/36$. The other values can be computed in a similar way. ◊

The multiplication rule for independent events also extends to more than two independent events. If you toss a fair coin, then roll a fair die, and finally draw a card from a standard deck of bridge cards, the three events are independent. To compute the probability of the outcome heads on the coin *and* five on the die *and* an ace for the card, we use the extended multiplication rule for independent events together with the facts

$$P(\text{head}) = \frac{1}{2} \quad P(5) = \frac{1}{6} \quad P(\text{ace}) = \frac{4}{52} = \frac{1}{13}$$

Then

$$P(\text{head } and \text{ five } and \text{ ace}) = \frac{1}{2} \cdot \frac{1}{6} \cdot \frac{1}{13}$$

$$= \frac{1}{156}$$

Further Examples

Most of us have been asked to participate in a survey. Schools, retail stores, news media, and government offices all conduct surveys. There are many types of surveys, and it is not our intention to give a general discussion of this topic. Let us study a very popular method called the *simple tally survey*. Such a survey consists of questions for which the responses can be recorded in rows and columns of a table. These questions are appropriate to the information you want and are designed to cover the *entire* population of interest. In addition, the questions should be designed so that we can partition the sample space of responses into distinct (that is, mutually exclusive) sectors.

If the survey includes responses from a reasonably large random sample, then the results should be representative of your population. In this case, we can estimate simple probabilities, conditional probabilities, and the probabilities of some combinations of events directly from the results of the survey.

EXAMPLE 6
Survey

At Hopewell Electronics, all 140 employees were asked about their political affiliation. The employees were grouped by type of work, as executives or production workers. The results with row and column totals are shown in Table 4-2.

TABLE 4-2   Employee Type and Political Affiliation

Employee Type	Political Affiliation				
	Democrat (D)	Republican (R)	Independent (I)	Row Total	
Executive (*E*)	5	34	9	48	
Production worker (*PW*)	63	21	8	92	
Column Total	68	55	17	140	Grand Total

Suppose an employee is selected at random from the 140 Hopewell employees. Let us use the following notation to represent different events of choosing: E = executive; PW = production worker; D = Democrat; R = Republican; I = Independent.

(a) Compute $P(D)$ and $P(E)$.

SOLUTION: To find these probabilities, we look at the *entire* sample space.

$$P(D) = \frac{\text{Number of Democrats}}{\text{Number of employees}} = \frac{68}{140} = 0.486$$

$$P(E) = \frac{\text{Number of executives}}{\text{Number of employees}} = \frac{48}{140} = 0.343$$

(b) Compute $P(D, given\ E)$.

SOLUTION: For the conditional probability, we *restrict* our attention to the portion of the sample space satisfying the condition of being an executive.

$$P(D, given\ E) = \frac{\text{Number of executives who are Democrats}}{\text{Number of executives}} = \frac{5}{48} = 0.104$$

(c) Are the events D and E independent?

SOLUTION: One way to determine if the events D and E are independent is to see if $P(D) = P(D, given\ E)$ [or equivalently, if $P(E) = P(E, given\ D)$]. Since $P(D) = 0.486$ and $P(D, given\ E) = 0.104$, we see that $P(D) \neq P(D, given\ E)$. This means that the events D and E are *not* independent. The probability of event D "depends on" whether or not event E has occurred.

(d) Compute $P(D\ and\ E)$.

SOLUTION: This probability is not conditional, so we must look at the entire sample space.

$$P(D\ and\ E) = \frac{\text{Number of executives who are Democrats}}{\text{Total number of employees}} = \frac{5}{140} = 0.036$$

Let's recompute this probability using the rules of probability for dependent events.

$$P(D\ and\ E) = P(E) \cdot P(D, given\ E) = \frac{48}{140} \cdot \frac{5}{48} = \frac{5}{140} = 0.036$$

The results using the rules are consistent with those using the sample space.

(e) Compute $P(D\ or\ E)$.

SOLUTION: From part (d) we know that the events Democrat or executive are not mutually exclusive, because $P(D\ and\ E) \neq 0$. Therefore,

$$P(D\ or\ E) = P(D) + P(E) - P(D\ and\ E)$$

$$= \frac{68}{140} + \frac{48}{140} - \frac{5}{140} = \frac{111}{140} = 0.793$$

◊

GUIDED EXERCISE 9

Survey

Using Table 4-2 on page 171, let's consider other probabilities regarding the type of employees at Hopewell and their political affiliation. This time let's consider the production worker and the affiliation of Independent. Suppose an employee is selected at random from the group of 140.

(a) Compute $P(I)$ and $P(PW)$.

$$P(I) = \frac{\text{No. of independents}}{\text{Total no. of employees}}$$

$$= \frac{17}{140} = 0.121$$

$$P(PW) = \frac{\text{No. of production workers}}{\text{Total no. of employees}}$$

$$= \frac{92}{140} = 0.657$$

(b) Compute $P(I, given\ PW)$. This is a conditional probability. Be sure to restrict your attention to production workers since that is the condition given.

$$P(I, given\ PW) = \frac{\text{No. of independent production workers}}{\text{No. of production workers}}$$

$$= \frac{8}{92} = 0.087$$

(c) Compute $P(I\ and\ PW)$. In this case, look at the entire sample space and at the number of employees who are both Independent and in production.

$$P(I\ and\ PW) = \frac{\text{No. of independent production workers}}{\text{Total no. employees}}$$

$$= \frac{8}{140} = 0.057$$

(d) Use the multiplication rule for dependent events to calculate $P(I\ and\ PW)$. Is the result the same as that of part (c)?

By the multiplication rule,

$$P(I\ and\ PW) = P(PW) \cdot P(I, given\ PW)$$

$$= \frac{92}{140} \cdot \frac{8}{92} = \frac{8}{140} = 0.057$$

The results are the same.

(e) Compute $P(I\ or\ PW)$. Are the events mutually exclusive?

Since the events are not mutully exclusive,

$$P(I\ or\ PW) = P(I) + P(PW) - P(I\ and\ PW)$$

$$= \frac{17}{140} + \frac{92}{140} - \frac{8}{140}$$

$$= \frac{101}{140} = 0.721$$

Basic probability rules

As you apply probability to various settings, keep the following rules in mind.

Summary of basic probability rules for events *A* and *B*

A statistical experiment or statistical observation is any random activity that results in a recordable outcome. The sample space is the set of all possible distinct outcomes. Events are subsets of the sample space.

1. $P(\text{entire sample space}) = 1$

2. For any event A: $0 \leq P(A) \leq 1$

3. Complement of A: $P(not\ A) = 1 - P(A)$

4. Events A and B are independent events if $P(A) = P(A,\ given\ B)$

5. Multiplication Rules
   If A and B are independent events: $P(A\ and\ B) = P(A) \cdot P(B)$
   If A and B are dependent events: $P(A\ and\ B) = P(A) \cdot P(B,\ given\ A)$
   or equivalently; $P(B,\ given\ A) = P(A\ and\ B)/P(A)$

6. Events A and B are mutually exclusive if $P(A\ and\ B) = 0$

7. Addition Rules
   If A and B are mutually exclusive events: $P(A\ or\ B) = P(A) + P(B)$
   If A and B are not mutually exclusive:

   $$P(A\ or\ B) = P(A) + P(B) - P(A\ and\ B)$$

In this section, we have studied some important rules that are valid in all probability spaces. The rules and definitions of probability not only are interesting, but they also have extensive *applications* in our everyday lives. If you are inclined to continue your study of probability a little farther, we recommend *Bayes's theorem* in Appendix I. The Reverend Thomas Bayes (1702–1761) was an English mathematician who discovered an important relation for conditional probabilities.

VIEWPOINT   *The Psychology of Odors*

The Smell and Taste Treatment Research Foundation of Chicago collected data on the time required to complete a maze while subjects were smelling different scents. Data for this survey can be found by visiting the Brase/Brase statistics site at http://math.college.hmco.com/students and finding the link to DASL, the Carnegie Mellon University Data and Story Library. Once at the DASL site, select Data Subjects, then Psychology, and then Scents. You can estimate conditional probabilities regarding response times for smokers, nonsmokers, and types of scents.

SECTION 4.2   PROBLEMS

In Problems 1–12 use the appropriate addition or multiplication rules. When possible, verify results by considering the sample space.

1. *General: Candy Colors* M&M plain candies come in various colors. According to the M&M/Mars Department of Consumer Affairs (link to the Mars company web site from the Brase/Brase statistics site at http://math.college.hmco.com/students), the distribution of colors for plain M&M candies is

Color	Brown	Yellow	Red	Orange	Green	Blue
Percentage	30%	20%	20%	10%	10%	10%

   Suppose you have a large bag of plain M&M candies and you take one candy at random. Find
   (a) *P*(orange *or* blue candy). Are these outcomes mutually exclusive? Why?
   (b) *P*(yellow candy *or* red candy). Are these outcomes mutually exclusive? Why?
   (c) *P*(*not* brown candy).

2. *General: Candy Colors* According to the Department of Consumer Affairs of M&M/Mars, the color distribution of peanut M&M candies is

Color	Brown	Yellow	Red	Orange	Green	Blue
Percentage	20%	20%	20%	10%	10%	20%

   Suppose you have a large bag of peanut M&M candies and you take one candy at random. Compute the probabilities in parts (a) through (c) of Problem 1 for peanut M&M candies. Compare the results with those of plain M&M candies. Do you expect any differences? Why or why not?

3. *General: Candy Colors* Almond M&M candies have another color distribution, utilizing only five colors. The color distribution for almond M&M candies is uniform. There are only five colors: brown, red, yellow, green, and blue. Each color comprises 20% of the almond M&M mix. Compute the probabilities in parts (a) through (c) of Problem 1 for almond M&M candies. Compare the results with those of plain M&M candies. Do you expect any differences? Why or why not?

4. *Environmental: Land Formations* Arches National Park is located in southern Utah. The park is famous for its beautiful desert landscape and its many natural sandstone arches. Park Ranger Edward McCarrick started an inventory (not yet complete) of natural arches within the park that have an opening of at least 3 feet. The following table is based on information taken from the book *Canyon Country Arches and Bridges*, by F. A. Barnes. The height of the arch opening is rounded to the nearest foot.

Height of arch, feet	3–9	10–29	30–49	50–74	75 and higher
Number of arches in park	111	96	30	33	18

   For an arch chosen at random in Arches National Park, use the preceding information to estimate the probability that the height of the arch opening is

(a) 3 to 9 feet tall   (d) 10 to 74 feet tall
(b) 30 feet or taller   (e) 75 feet or taller
(c) 3 to 49 feet tall

5. *General: Roll Two Dice* You roll two fair dice, a green one and a red one.
   (a) Are the outcomes on the dice independent?
   (b) Find P(5 on green die *and* 3 on red die).
   (c) Find P(3 on green die *and* 5 on red die).
   (d) Find P((5 on green die *and* 3 on red die) *or* (3 on green die *and* 5 on red die)).

6. *General: Roll Two Dice* You roll two fair dice, a green one and a red one.
   (a) Are the outcomes on the dice independent?
   (b) Find P(1 on green die *and* 2 on red die).
   (c) Find P(2 on green die *and* 1 on red die).
   (d) Find P((1 on green die *and* 2 on red die) *or* (2 on green die *and* 1 on red die)).

7. *General: Roll Two Dice* You roll two fair dice, a green one and a red one.
   (a) What is the probability of getting a sum of 6?
   (b) What is the probability of getting a sum of 4?
   (c) What is the probability of getting a sum of 6 *or* 4? Are these outcomes mutually exclusive?

8. *General: Roll Two Dice* You roll two fair dice, a green one and a red one.
   (a) What is the probability of getting a sum of 7?
   (b) What is the probability of getting a sum of 11?
   (c) What is the probability of getting a sum of 7 *or* 11? Are these outcomes mutually exclusive?

9. *General: Deck of Cards* You draw two cards from a standard deck of 52 cards without replacing the first one before drawing the second.
   (a) Are the outcomes on the two cards independent? Why?
   (b) Find P(ace on 1st card *and* king on 2nd).
   (c) Find P(king on 1st card *and* ace on 2nd).
   (d) Find the probability of drawing an ace *and* a king in either order.

10. *General: Deck of Cards* You draw two cards from a standard deck of 52 cards without replacing the first one before drawing the second.
    (a) Are the outcomes on the two cards independent? Why?
    (b) Find P(3 on 1st card *and* 10 on 2nd).
    (c) Find P(10 on 1st card *and* 3 on 2nd).
    (d) Find the probability of drawing a 10 *and* a 3 in either order.

11. *General: Deck of Cards* You draw two cards from a standard deck of 52 cards, but before you draw the second card, you put the first one back and reshuffle the deck.
    (a) Are the outcomes on the two cards independent? Why?
    (b) Find P(ace on 1st card *and* king on 2nd).
    (c) Find P(king on 1st card *and* ace on 2nd).
    (d) Find the probability of drawing an ace *and* a king in either order.

12. *General: Deck of Cards* You draw two cards from a standard deck of 52 cards, but before you draw the second card, you put the first one back and reshuffle the deck.
    (a) Are the outcomes on the two cards independent? Why?
    (b) Find P(3 on 1st card *and* 10 on 2nd).
    (c) Find P(10 on 1st card *and* 3 on 2nd).
    (d) Find the probability of drawing a 10 *and* a 3 in either order.

13. *Marketing: Toys USA Today* gave the information shown at the top of the next page about ages of children receiving toys. The percentages represent all toys sold.

Age (years)	Percentage of Toys
2 and under	15%
3–5	22%
6–9	27%
10–12	14%
13 and over	22%

What is the probability that a toy is purchased for someone
(a) 6 years or older?
(b) 12 years or younger?
(c) between 6 and 12 years old?
(d) between 3 and 9 years old?

A child between 10 and 12 years old looks at this probability distribution and asks, "Why are people more likely to buy toys for kids older than I am (13 and over) than for kids in my age group (10–12)?" How would you respond?

14. *Health Care: Flu* Based on data from the *Statistical Abstract of the United States,* 112th Edition, only about 14% of senior citizens (65 years or older) get the flu each year. However, about 24% of the people under 65 years old get the flu each year. In the general population, there are 12.5% senior citizens (65 years or older).
    (a) What is the probability that a person selected at random from the general population is a senior citizen who will get the flu this year?
    (b) What is the probability that a person selected at random from the general population is a person under age 65 who will get the flu this year?
    (c) Answer parts (a) and (b) for a community that has 95% senior citizens.
    (d) Answer parts (a) and (b) for a community that has 50% senior citizens.

15. *Psychology: Lie Detector Test* In this problem, you are asked to solve part of the Focus Problem at the beginning of this chapter. In his book *Chances: Risk and Odds in Everyday Life*, James Burke says that there is a 72% chance a polygraph test (lie detector test) will catch a person who is in fact lying. Furthermore, there is approximately a 7% chance that the polygraph will falsely accuse someone of lying.
    (a) Suppose that a person answers 90% of a long battery of questions truthfully. What percentage of the story will the polygraph *wrongly* indicate is a lie?
    (b) Suppose that a person answers 10% of a long battery of questions with lies. What percentage of the story will the polygraph *correctly* indicate is a lie?
    (c) Repeat parts (a) and (b) if 50% of the questions are answered truthfully and 50% are answered with lies.
    (d) Repeat parts (a) and (b) if 15% of the questions are answered truthfully and the rest are answered with lies.

16. *Psychology: Lie Detector Test* This problem continues the Focus Problem. The solution involves applying several basic probability rules and a little algebra to solve an equation.
    (a) If the polygraph (of Problem 15) said 30% of the questions were answered with lies, what would you estimate for the actual percentage of lies in the story? *Hint:* Let B = event detector indicates a lie. We are given $P(B) = 0.30$. Let A = event person is lying, so *not A* = event person is not lying. Then

$$P(B) = P(A \text{ and } B) + P(not\ A \text{ and } B)$$

$$P(B) = P(A)P(B, given\ A) + P(not\ A)P(B, given\ not\ A)$$

Replacing $P(not\ A)$ by $1 - P(A)$ gives

$$P(B) = P(A)P(B, given\ A) + [1 - P(A)]P(B, given\ not\ A)$$

Substitute known values for $P(B)$, $P(B, given\ A)$, and $P(B, given\ not\ A)$ into the last equation and solve for $P(A)$.
    (b) If the polygraph (of Problem 15) said 70% of questions were answered with lies, what would you estimate for the actual percentage of lies in the story?

17. *Life Style: Glasses or Contacts* In the book *Chances: Risk and Odds in Every-day Life,* James Burke says that 56% of the general population wears eyeglasses, while only 3.6% wears contacts. He also says that of those who do wear glasses, 55.4% are women and 44.6% are men. Of those who wear contacts, 63.1% are women and 36.9% are men. Assume that no one wears both glasses and contacts. For the next person you encounter at random, what is the probability that this person is
    (a) a woman wearing glasses?
    (b) a man wearing glasses?
    (c) a woman wearing contacts?
    (d) a man wearing contacts?
    (e) none of the above?

18. *Environment: Alternative Fuels* Gasoline-powered automobiles are a major source of pollution. Alternative fuels such as ethanol or methanol lower some of the hydro-carbon emissions. However, building cars that run on alternative fuels will cost more. A survey in *USA Today* reported how much a person selected at random would be willing to pay for the option of using alternative fuels. As reported, 27.5% of the respondents would not be willing to pay anything, 9.6% would be willing to pay less than $200, 13.5% would be willing to pay from $200 to $599, 2.1% would be willing to pay from $600 to $999, 26.2% would be willing to pay $1000 or more, and 21.1% don't know. Based on this survey, if we picked an adult at random, find the probability that this person would be willing to pay $600 or more for the option of using alternative fuels. What is the probability the person would be willing to pay no more than $199?

19. *Survey: Sales Approach* In a sales effectiveness seminar, a group of sales representatives tried two approaches to selling a customer a new automobile: the aggressive approach and the passive approach. From 1160 customers, the following record was kept:

	Sale	No Sale	Row Total
Aggressive	270	310	580
Passive	416	164	580
Column Total	686	474	1160

Suppose that a customer is selected at random from the 1160 participating customers. Let us use the following notation for events: A = aggressive approach, Pa = passive approach, S = sale, N = no sale. So P(A) is the probability that an aggressive approach was used, and so on.
(a) Compute P(S), P(S, given A), and P(S, given Pa).
(b) Are the events S = sale and Pa = passive approach independent? Explain.
(c) Compute P(A and S) and P(Pa and S).
(d) Compute P(N) and P(N, given A).
(e) Are the events N = no sale and A = aggressive approach independent? Explain.
(f) Compute P(A or S).

20. *Survey: Medical Tests* Diagnostic tests of medical conditions have several results. The test result can be positive or negative, whether or not a patient has the condition (+ indicates a patient has the condition). Consider a random sample of 200 patients, some of whom have a medical condition and some of whom do not. Results of a new diagnostic test for the condition are shown.

	Condition Present	Condition Absent	Row Total
Test Result +	110	20	130
Test Result −	20	50	70
Column Total	130	70	200

Assume the sample is representative of the entire population. For a person selected at random, compute the following probabilities:

(a) $P(+,$ *given* condition present); this is known as the *sensitivity* of a test.

(b) $P(-,$ *given* condition present); this is known as the false-negative rate.

(c) $P(-,$ *given* condition absent); this is known as the *specificity* of a test.

(d) $P(+,$ *given* condition absent); this is known as the false-positive rate.

(e) $P($condition present *and* $+)$; this is the predictive value of the test.

(f) $P($condition present *and* $-)$.

21. *Survey: Lung/Heart* In an article entitled "Diagnostic accuracy of fever as a measure of postoperative pulmonary complications" (*Heart Lung* 10, No. 1:61), J. Roberts and colleagues discuss using a fever of 38°C or higher as a diagnostic indicator of postoperative atelectasis (collapse of the lung) as evidenced by x-ray observation. For fever ≥ 38°C as the diagnostic test, the results for postoperative patients are

	Condition Present	Condition Absent	Row Total
Test Result +	72	37	109
Test Result −	82	79	161
Column Total	154	116	270

Complete parts (a) through (f) from Problem 20.

22. *Survey: Customer Loyalty* Are customers more loyal in the East or in the West? The following table is based on information from *Trends in the United States*, published by the Food Marketing Institute, Washington, D.C. The columns represent length of customer loyalty (years) at a primary supermarket. The rows represent regions of the United States.

	Less Than 1 Year	1–2 Years	3–4 Years	5–9 Years	10–14 Years	15 or More Years	Row Total
East	32	54	59	112	77	118	452
Midwest	31	68	68	120	63	173	523
South	53	92	93	158	106	158	660
West	41	56	67	78	45	86	373
Column Total	157	270	287	468	291	535	2008

What is the probability that a customer chosen at random

(a) has been loyal 10 to 14 years?

(b) has been loyal 10 to 14 years, given that he or she is from the East?

(c) has been loyal *at least* 10 years?

(d) has been loyal *at least* 10 years, given that he or she is from the West?

(e) is from the West, given that he or she has been loyal less than 1 year?

(f) is from the South, given that he or she has been loyal less than 1 year?

(g) has been loyal *1 or more years*, given that he or she is from the East?

(h) has been loyal *1 or more years*, given that he or she is from the West?

(i) Are the events from the East and loyal 15 or more years independent? Explain.

23. *Survey: Trips to Supermarket* How many times do shoppers go to the supermarket each week? The age group of customers with frequent visits can influence store inventory and marketing methods. The following table is based on information from *Trends in the United States* (see reference in Problem 22). The columns represent number of visits to the primary supermarket in an average week. The rows represent age distribution of customers in years.

Age (years)	One Visit	Two Visits	Three Visits	Four Visits	Five Visits	Six or More Visits	Row Total
18–24	65	58	12	5	4	4	148
25–39	386	230	69	22	17	15	739
40–49	210	161	36	13	9	5	434
50–64	186	102	35	14	7	5	349
65 and over	115	69	18	12	7	3	224
Column Total	962	620	170	66	44	32	1894

What is the probability that a customer chosen at random

(a) has been to the supermarket *at least* 2 times this past week?

(b) has been to the supermarket *at least* 2 times this past week, given that he or she is 25 to 39 years old?

(c) has been to the supermarket *more than* 3 times this past week?

(d) has been to the supermarket *more than* 3 times this past week, given that he or she is *65 or older*?

(e) is *40 or older*?

(f) is *40 or older*, given that he or she has visited the supermarket 4 times this past week?

(g) Are the events age 25–39 years and visits *more than* once a week independent? Explain.

24. *Education: College of Nursing* At Litchfield College of Nursing, 85% of incoming freshmen nursing students are female and 15% are male. Recent records indicate that 70% of the entering female students will graduate with a BSN degree, while 90% of the male students will obtain a BSN degree. If an incoming freshman nursing student is selected at random, find

(a) *P*(student will graduate, *given* student is female).

(b) *P*(student will graduate *and* student is female).

(c) *P*(student will graduate, *given* student is male).

(d) *P*(student will graduate *and* student is male).

(e) *P*(student will graduate). Note that those who will graduate are either males who will graduate or females who will graduate.

(f) The events described by the phrases "will graduate *and* is female" and "will graduate, *given* female" seem to be describing the same students. Why are the probabilities *P*(will graduate *and* is female) and *P*(will graduate, *given* female) different?

25. *Franchise Stores: Profits* Wing Foot is a shoe franchise commonly found in shopping centers across the United States. Wing Foot knows that its stores will not show a profit unless they gross over $940,000 per year. Let *A* be the event that a new Wing Foot store grosses over $940,000 its first year. Let *B* be the event

that a store grosses over $940,000 its second year. Wing Foot has an administrative policy of closing a new store if it does not show a profit in *either* of the first 2 years. The accounting office at Wing Foot provided the following information: 65% of *all* Wing Foot stores show a profit the first year; 71% of *all* Wing Foot stores show a profit the second year (this includes stores that did not show a profit in the first year); however, 87% of Wing Foot stores that showed a profit the first year also showed a profit the second year. Compute the following:

(a) $P(A)$

(b) $P(B)$

(c) $P(B,$ *given* $A)$

(d) $P(A$ *and* $B)$

(e) $P(A$ *or* $B)$

(f) What is the probability that a new Wing Foot store will not be closed after 2 years? What is the probability that a new Wing Foot store will be closed after 2 years?

26. *Therapy: Alcohol Recovery* The Eastmore Program is a special program to help alcoholics. In the Eastmore Program, an alcoholic lives at home but undergoes a two-phase treatment plan. Phase I is an intensive group-therapy program lasting 10 weeks. Phase II is a long-term counseling program lasting 1 year. Eastmore Programs are located in most major cities, and past data gave the following information, based on percentages of success and failure collected over a long period of time: The probability that a client will have a relapse in phase I is 0.27; the probability that a client will have a relapse in phase II is 0.23. However, if a client did not have a relapse in phase I, then the probability that this client will not have a relapse in phase II is 0.95. If a client did have a relapse in phase I, then the probability that this client will have a relapse in phase II is 0.70. Let A be the event that a client has a relapse in phase I and B be the event that a client has a relapse in phase II. Let C be the event that a client has no relapse in phase I and D be the event that a client has no relapse in phase II. Compute the following:

(a) $P(A)$, $P(B)$, $P(C)$, and $P(D)$

(b) $P(B,$ *given* $A)$ and $P(D,$ *given* $C)$

(c) $P(A$ *and* $B)$ and $P(C$ *and* $D)$

(d) $P(A$ *or* $B)$

(e) What is the probability that a client will go through both phase I and phase II without a relapse?

(f) What is the probability that a client will have a relapse in both phase I and phase II?

(g) What is the probability that a client will have a relapse in either phase I or phase II?

27. *Medical: Tuberculosis* The state medical school has discovered a new test for tuberculosis. (If the test indicates a person has tuberculosis, the test is positive.) Experimentation has shown that the probability of a positive test is 0.82, given that a person has tuberculosis. The probability is 0.09 that the test registers positive, given that the person does not have tuberculosis. Assume that in the general population the probability that a person has tuberculosis is 0.04. What is the probability that a person chosen at random will

(a) have tuberculosis and a positive test?

(b) not have tuberculosis?

(c) not have tuberculosis and have a positive test?

4.3
Trees and Counting Techniques

Tree diagrams

When outcomes are equally likely, we compute the probability of an event by using the formula

$$P(A) = \frac{\text{Number of outcomes favorable to the event } A}{\text{Number of outcomes in the sample space}}$$

The probability formula requires that we be able to determine the number of outcomes in the sample space. In the problems we have done in previous sections, this task has not been difficult because the number of outcomes was small or the sample space consisted of fairly straightforward events. The tools we present in this section will help you count the number of possible outcomes in larger sample spaces or those formed by more complicated events.

A *tree diagram* helps us display the outcomes of an experiment consisting of a series of activities. The total number of outcomes corresponds to the total number of final branches in the tree. Perhaps the best way to learn to make a tree diagram is to see one. In the next example we will see a tree diagram and analyze its parts.

EXAMPLE 7

Tree diagram

Jacqueline is in the nursing program and is required to take a course in psychology and one in anatomy and physiology (A and P) next semester. She also wants to take Spanish II. If there are four sections of psychology, two of A and P, and three of

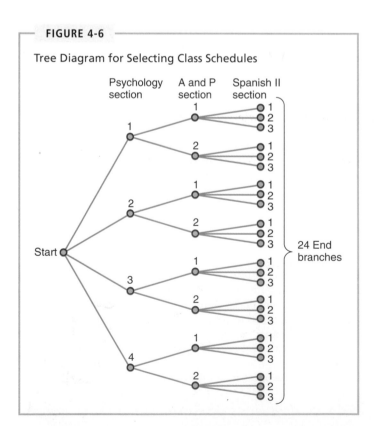

FIGURE 4-6

Tree Diagram for Selecting Class Schedules

24 End branches

TABLE 4-3 Schedules Utilizing Section 1 of Psychology

Psychology Section	A and P Section	Spanish II Section
1	1	1
1	1	2
1	1	3
1	2	1
1	2	2
1	2	3

TABLE 4-4 Schedules Utilizing Section 2 of Psychology

Psychology Section	A and P Section	Spanish II Section
2	1	1
2	1	2
2	1	3
2	2	1
2	2	2
2	2	3

Spanish, how many different class schedules can Jacqueline choose from? (Assume that the times of the sections do not conflict with each other.) Figure 4-6 shows a tree diagram for Jacqueline's possible schedules.

SOLUTION: Let's study the tree diagram and see how it shows Jacqueline's schedule choices. There are four branches from Start. These branches indicate the four possible choices for psychology sections. No matter which section of psychology Jacqueline chooses, she can choose from the two available A and P sections. Therefore, we have two branches leading from *each* psychology branch. Finally, after the psychology and A and P sections are selected, there are three choices for Spanish II. That is why there are three branches from *each* A and P section.

The tree ends with a total of 24 branches. This number of end branches tells us the number of possible schedules. The outcomes themselves can be listed from the tree by following each series of branches from Start to End. For instance, the top branch from Start generates the schedules shown in Table 4-3.

Following the second branch from Start, we see all the possible schedules utilizing Section 2 of psychology (see Table 4-4). The other 12 schedules can be listed in a similar manner.   ◊

We draw a tree diagram in stages, indicating the possible outcomes for the first event, second event, and so forth. Guided Exercise 10 will lead you through the process.

GUIDED
EXERCISE **10**

Tree diagram

Louis plays three tennis matches. Use a tree diagram to list the possible win and loss sequences Louis can experience for the set of three matches.

(a) On the first match Louis can win or lose. From Start, indicate these two branches.

⇨ **FIGURE 4-7** W = Win, L = Lose

1st Match
W

Start

L

(b) Regardless of whether Louis wins or loses the first match, he plays the second and can again win or lose. Attach branches representing these two outcomes to *each* of the first match results.

⇨ **FIGURE 4-8**

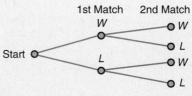

1st Match   2nd Match
W   W
             L
Start
L   W
             L

Continued

GUIDED EXERCISE 10 continued

(c) Louis may win or lose the third match. Attach branches representing these two outcomes to *each* of the second match results.

 **FIGURE 4-9**

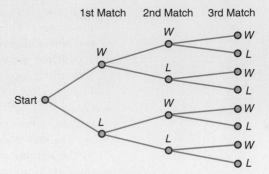

(d) How many possible win–lose sequences are there for the three matches?

Since there are eight branches at the end, there are eight sequences.

(e) Complete this list of win–lose sequences.

The last four sequences all involve a loss on Match 1.

1st	2nd	3rd
W	W	W
W	W	L
W	L	W
W	L	L

1st	2nd	3rd
L	W	W
L	W	L
L	L	W
L	L	L

Tree diagrams help us display the outcomes of an experiment involving several stages. If we label each branch of the tree with an appropriate probability, we can use the tree diagram to help us compute probabilities of an outcome displayed on the tree. One of the easiest ways to illustrate this feature of tree diagrams is to use an experiment of drawing balls out of an urn. We do this in the next example.

EXAMPLE 8
Probability

Suppose there are five balls in an urn. They are identical except in color. Three of the balls are red and two are blue. You are instructed to draw out one ball, note its color, and set it aside. Then you are to draw out another ball and note its color. What are the outcomes of the experiment? What is the probability of each outcome?

SOLUTION: The tree diagram in Figure 4-10 will help us answer these questions. Notice that since you did not replace the first ball before drawing the second one, the two stages of the experiment are dependent. The probability associated with the color of the second ball depends on the color of the first ball. For instance, on the top branches, the color of the first ball drawn is red, so we compute the probabilities of the colors on the second ball accordingly. The tree diagram helps us organize the probabilities.

From the diagram, we see that there are four possible outcomes to the experiment. They are

RR = red on 1st *and* red on 2nd

RB = red on 1st *and* blue on 2nd

BR = blue on 1st *and* red on 2nd

BB = blue on 1st *and* blue on 2nd

To compute the probability of each outcome, we will use the multiplication rule for dependent events. As we follow the branches for each outcome, we will find the necessary probabilities.

$$P(R \text{ on 1st } and \text{ } R \text{ on 2nd}) = P(R) \cdot P(R, given \text{ } R)$$
$$= \frac{3}{5} \cdot \frac{2}{4} = \frac{3}{10}$$
$$P(R \text{ on 1st } and \text{ } B \text{ on 2nd}) = P(R) \cdot P(B, given \text{ } R)$$
$$= \frac{3}{5} \cdot \frac{2}{4} = \frac{3}{10}$$
$$P(B \text{ on 1st } and \text{ } R \text{ on 2nd}) = P(B) \cdot P(R, given \text{ } B)$$
$$= \frac{2}{5} \cdot \frac{3}{4} = \frac{3}{10}$$
$$P(B \text{ on 1st } and \text{ } B \text{ on 2nd}) = P(B) \cdot P(B, given \text{ } B)$$
$$= \frac{2}{5} \cdot \frac{1}{4} = \frac{1}{10}$$

Notice that the probabilities of the outcomes in the sample space add to 1, as they should. ◇

FIGURE 4-10

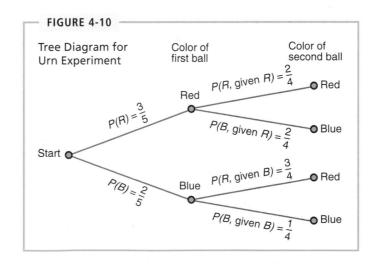

Tree Diagram for Urn Experiment

Color of first ball

Color of second ball

Start

$P(R) = \frac{3}{5}$

$P(B) = \frac{2}{5}$

Red

Blue

$P(R, given \text{ } R) = \frac{2}{4}$ Red

$P(B, given \text{ } R) = \frac{2}{4}$ Blue

$P(R, given \text{ } B) = \frac{3}{4}$ Red

$P(B, given \text{ } B) = \frac{1}{4}$ Blue

GUIDED EXERCISE 11

Probability

Repeat the urn experiment with the five balls, three of which are red and two of which are blue. This time replace the first ball before drawing the second.

(a) Draw a tree diagram for the outcomes of this experiment. Show the probabilities of each stage on the appropriate branch. (*Hint:* Are the stages dependent or independent?)

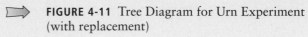

FIGURE 4-11 Tree Diagram for Urn Experiment (with replacement)

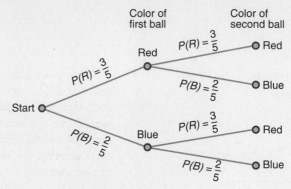

(b) List the four possible outcomes of the experiment (i.e., list the sample space).

> Red on 1st *and* red on 2nd
> Red on 1st *and* blue on 2nd
> Blue on 1st *and* red on 2nd
> Blue on 1st *and* blue on 2nd

(c) Use the multiplication rule for independent events and the probabilities shown on your tree to compute the probability of each outcome.

> $P(\text{1st } R \text{ and 2nd } R) = P(R) \cdot P(R)$
> $$= \frac{3}{5} \cdot \frac{3}{5} = \frac{9}{25}$$
>
> $P(\text{1st } R \text{ and 2nd } B) = P(R) \cdot P(B)$
> $$= \frac{3}{5} \cdot \frac{2}{5} = \frac{6}{25}$$
>
> $P(\text{1st } B \text{ and 2nd } R) = P(B) \cdot P(R)$
> $$= \frac{2}{5} \cdot \frac{3}{5} = \frac{6}{25}$$
>
> $P(\text{1st } B \text{ and 2nd } B) = P(B) \cdot P(B)$
> $$= \frac{2}{5} \cdot \frac{2}{5} = \frac{4}{25}$$

(d) Do the probabilities of the outcomes in the sample space add up to 1?

> Yes, as they should.

Continued

GUIDED EXERCISE 11 continued

(e) Compare the tree diagram of this exercise with that of the previous example, in which the first ball was not replaced before the second was drawn. Are the outcomes the same? Are the probabilities of the corresponding outcomes the same?

 The outcomes are the same: *RR, RB, BR,* and *BB.* However, because one experiment did not permit replacement of the first ball before the second was drawn and the other experiment required replacement, the corresponding probabilities of the second stages of the experiment are different. The corresponding probabilities of the final outcomes are also different.

Multiplication rule of counting

When an outcome is composed of a series of events, tree diagrams tell us how many possible outcomes there are. They also help us list the individual outcomes and organize the probabilities associated with each stage of the outcomes. However, if we are interested only in the number of different outcomes created by a series of events, the multiplication rule will give us the total number of outcomes more directly. We state the multiplication rule for an outcome composed of a series of two events.

Multiplication rule of counting

If there are n possible outcomes for event E_1 and m possible outcomes for event E_2, then there are a total of $n \times m$ or nm possible outcomes for the series of events E_1 followed by E_2.

The rule extends to outcomes created by a series of three, four, or more events. We simply multiply the number of outcomes possible for each step in the series of events to get the total number of outcomes for the series.

EXAMPLE 9

Multiplication rule

The Night Hawk is the new car model produced by Limited Motors, Inc. It comes with a choice of two body styles, three interior package options, and four different colors, as well as the choice of automatic or standard transmission. Select-an-Auto Car Dealership wants to carry one of each of the different types of Night Hawks. How many cars are required?

SOLUTION: There are four items to select. We take the product of the number of choices for each item.

$$\binom{\text{No. of body}}{\text{styles}} \binom{\text{No. of}}{\text{interiors}} \binom{\text{No. of}}{\text{colors}} \binom{\text{No. of transmission}}{\text{types}}$$
$$(2)(3)(4)(2) = 48$$

Select-an-Auto must stock 48 cars to have one of each possible type.   ◇

GUIDED EXERCISE 12

Multiplication rule of counting

The Old Sage Inn offers a special dinner menu each night. There are two appetizers to choose from, three main courses, and four desserts. A customer can select one item from each category. How many different meals can be ordered from the special dinner menu?

(a) Each special dinner consists of three items. List the item and the number of choices per item.

⟹   Appetizer—2; main course—3; dessert—4

(b) To find the number of different dinners composed of the three items, multiply the number of choices per item together.

⟹   (2)(3)(4) = 24

There are 24 different dinners that can be ordered from the special dinner menu.

Sometimes when we consider *n* items, we need to know the number of different *ordered arrangements* of the *n* items that are possible. The multiplication rules can help us find the number of possible ordered arrangements. Let's consider the classic example of determining the number of different ways in which eight people can be seated at a dinner table. For the first chair at the head of the table, there are eight choices. For the second chair, there are seven choices, since one person is already seated. For the third chair, there are six choices, since two people are already seated. By the time we get to the last chair, there is only one person left for that seat. We can view each arrangement as an outcome of a series of eight events. Event 1 is *fill the first chair,* event 2 is *fill the second chair,* and so forth. The multiplication rule will tell us the number of different outcomes.

Choices for	1st	2nd	3rd	4th	5th	6th	7th	8th	Chair position
	↓	↓	↓	↓	↓	↓	↓	↓	
	(8)	(7)	(6)	(5)	(4)	(3)	(2)	(1)	= 40,320

In all, there are 40,320 different seating arrangements for eight people. It is no wonder that it takes a little time to seat guests at a dinner table!

Factorial notation

The multiplication pattern shown above is not unusual. In fact, it is an example of the multiplication indicated by the factorial notation 8!.

! is read "factorial"

8! is read "8 factorial"

$8! = 8 \cdot 7 \cdot 6 \cdot 5 \cdot 4 \cdot 3 \cdot 2 \cdot 1$

In general, *n*! indicates the product of *n* with each of the positive counting numbers less than *n*. *By special definition 0! = 1.*

Factorial notation

For a counting number n,

$$n! = n(n - 1)(n - 2) \cdots 1$$
$$0! = 1$$
$$1! = 1$$

GUIDED
EXERCISE **13**

Factorial

(a) Evaluate 3!.

$\Rightarrow$ $3! = 3 \cdot 2 \cdot 1 = 6$

(b) How many different ways can three objects be arranged in order? How many choices do you have for the first position? for the second position? for the third position?

$\Rightarrow$ We have three choices for the first position, two for the second position, and one for the third position. By the multiplication rule, we have

$$(3)(2)(1) = 3! = 6 \text{ arrangements}$$

(c) Verify step (b) with a three-stage tree diagram.

$\Rightarrow$ **FIGURE 4-12** Three Choices

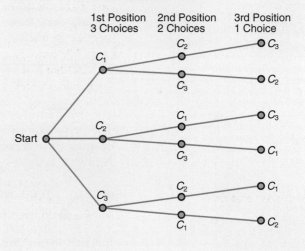

Permutations

We have considered the number of ordered arrangements of n objects taken as an entire group. Specifically, we considered a dinner party for eight and found the number of ordered seating arrangements for all eight people. However, suppose you have an open house and have only five chairs. How many ways can five of the eight people seat themselves in the chairs? The formula we use to compute this number is called the *permutation formula*. As we see in the next example, the *permutations rule* is really another version of the multiplication rule.

Counting rule for permutations

The number of ways to *arrange in order n* distinct objects, taking them *r* at a time, is

$$P_{n,r} = \frac{n!}{(n-r)!} \tag{9}$$

where *n* and *r* are whole numbers and $n \geq r$. Another commonly used notation for permutations is nPr.

EXAMPLE 10

Permutations rule

Let's compute the number of ordered seating arrangements we have for eight people in five chairs.

SOLUTION: In this case, we are considering a total of $n = 8$ different people, and we wish to arrange $r = 5$ of these people. Substituting into Formula (9), we have

$$P_{n,r} = \frac{n!}{(n-r)!}$$

$$P_{8,5} = \frac{8!}{(8-5)!}$$

$$= \frac{8!}{3!}$$

$$= \frac{40,320}{6}$$

$$= 6720$$

Using the multiplication rule, we get the same results

Chair	1		2		3		4		5	
Choices for	8	×	7	×	6	×	5	×	4	= 6720

The permutations rule has the advantage of using factorials. Most scientific calculators have a factorial key ! as well as a permutations key nPr (see the Tech Note). ◊

TECH NOTE   Most scientific calculators have a factorial key, often designated x! or n!. Many of these same calculators have the permutation function built in, often labeled nPr. They also have the combination function that is discussed after the next exercise. The combination function is often labeled nCr.

TI-83Plus The factorial, permutation, and combination functions are all under **MATH**, then **PRB**.

Excel Use the **paste function** $\boxed{f_x}$ then **all**. **Fact** gives factorials, **Permut** gives permutations, **Combin** gives combinations.

GUIDED EXERCISE 14

Permutation

The board of directors of Belford Community Hospital has 12 members. Three officers—president, vice president, and treasurer—must be elected from the members. How many different possible slates of officers are there? We will view a slate of officers as a list of three people with one person for president listed first, one person for vice president listed second, and one person for treasurer listed third. For instance, if Mr. Acosta, Ms. Hill, and Mr. Smith wish to be on a slate together, there are several different slates possible, depending on which one will run for president, which for vice president, and which for treasurer. Not only are we asking for the number of different groups of three names for a slate, we are also concerned about order, since it makes a difference which name is listed in which position.

(a) What is the size of the group from which the slates of officers will be selected? This is the value of n.

$\Longrightarrow$ $n = 12$

(b) How many people will be selected for each slate of officers? This is the value of r.

$\Longrightarrow$ $r = 3$

(c) Each slate of officers is composed of three candidates. Different slates occur as we arrange the three candidates in the positions of president, vice president, and treasurer. For this reason, we need to consider the number of *permutations* of 12 items arranged in groups of 3. Compute $P_{n,r}$.

$\Longrightarrow$

$$P_{n,r} = \frac{n!}{(n-r)!}$$

$$P_{12,3} = \frac{12!}{(12-3)!}$$

$$= \frac{12!}{9!}$$

$$= \frac{479,001,600}{362,880}$$

$$= 1320$$

There are 1320 different possible slates of officers. An alternative is to use your calculator to compute $P_{12,3}$ directly.

Combinations

In each of our previous counting formulas, we have taken the *order* of the objects or people into account. But what if order is not important? For instance, suppose we need to choose 3 members from the 12-member board of directors of Belford Community Hospital to go to a convention. We are interested in *different groupings* of 12 people so that each group contains 3 people. The order is of no concern, since all 3 will go to the convention. In other words, we need to consider the number of different *combinations* of 12 people taken 3 at a time. Our next formula will help us compute this number of different combinations.

Counting rule for combinations

The number of *combinations* of n objects taken r at a time is

$$C_{n,r} = \frac{n!}{r!(n-r)!} \qquad (10)$$

where n and r are whole numbers and $n \geq r$. Other commonly used notations for combinations include $_nC_r$ and $\binom{n}{r}$.

Notice the difference between the concepts of permutations and combinations. When we consider permutations, we are considering groupings *and order*. When we consider combinations, we are considering only the number of different groupings. For combinations, order within the groupings is not considered. As a result, the number of combinations of n objects taken r at a time is generally smaller than the number of permutations of the same n objects taken r at a time. In fact, the combinations formula is simply the permutations formula with the number of permutations of each distinct group divided out. In the formula for combinations, notice the factor of $r!$ in the denominator.

Now let's look at an example in which we use the *combinations rule* to compute the number of *combinations* of 12 people taken 3 at a time.

EXAMPLE 11

Combinations

Three members from the group of 12 on the board of directors at Belford Community Hospital will be selected to go to a convention with all expenses paid. How many different groups of 3 are there?

SOLUTION: In this case, we are interested in *combinations* rather than permutations of 12 people taken 3 at a time. Using Formula (10), we get

$$C_{n,r} = \frac{n!}{r!(n-r)!} \quad \text{or}$$

$$C_{12,3} = \frac{12!}{3!(12-3)!} = \frac{12!}{3!9!} = \frac{479,001,600}{(6)(362,880)} = 220$$

There are 220 different possible groups of 3 to go to the convention.

Another way to get the solution is to use your calculator to evaluate $C_{12,3}$ directly. Since order is not considered, this number is much smaller than the number of different slates of 3 officers we computed in Guided Exercise 14.   ◊

We have different formulas for permutations and combinations of n objects taken r at a time. How do you decide which one to use? Always ask yourself if order in the groups of r objects is relevant. If it is, use $P_{n,r}$. If order is not relevant, use $C_{n,r}$.

We have introduced you to three counting formulas: the multiplication rule, the permutations rule, and the combinations rule. Other rules apply when the objects are not distinct. Many counting problems are easy to state and fairly difficult to solve. Some have you combine several counting rules. However, the problems for this section are all straightforward. Some ask you to use your counting abilities to compute probabilities.

GUIDED EXERCISE 15

Combinations

In your political science class, you are given a list of 10 books. You are to select 4 to read during the semester. How many different *combinations* of 4 books are available from the list of 10?

(a) Is the order in which you read the books relevant to the task of selecting the books?

➡ No.

(b) Do we use the number of permutations or combinations of 10 books taken 4 at a time?

➡ Since consideration of order in which the books are selected is not relevant, we compute the number of *combinations* of 10 books taken 4 at a time.

(c) How many books are available from which to select? How many must you read? What is the value of n? of r?

➡ There are 10 books among which you must select 4 to read. $n = 10$ and $r = 4$.

(d) Compute $C_{10,4}$ to determine the number of different groups of 4 books from the list of 10.

➡
$$C_{n,r} = \frac{n!}{r!(n-r)!}$$

$$C_{10,4} = \frac{10!}{4!(10-4)!}$$

$$= \frac{10!}{4!6!}$$

$$= \frac{3,628,800}{(24)(720)}$$

$$= 210$$

There are 210 different groups of 4 books to select from the list of 10. An alternate method of solution is to use the $_nC_r$ key on your calculator.

VIEWP⊙INT   *Powerball*

Powerball is a multistate lottery game that consists of drawing five distinct whole numbers from numbers 1 through 53. Then one more number from numbers 1 through 42 is selected as the Powerball number (this number could be one of the original five). Powerball numbers are drawn every Wednesday and Saturday. If you match all six numbers, you win the jackpot, which is worth at least 10 million dollars. Use methods of this section to show that there are 120,526,770 possible Powerball plays. This means that the odds of winning the jackpot are 1 to 120,526,770. For more information about the game of Powerball and the probability of winning different prizes, visit the Brase/Brase statistics site at http://math.college.hmco.com/students and find the link to the Multi-State Lottery Association. Then select Powerball.

SECTION 4.3   PROBLEMS

1. *Tree Diagram*
   (a) Draw a tree diagram to display all the possible head–tail sequences that can occur when you flip a coin three times.
   (b) How many sequences contain exactly two heads?
   (c) *Probability extension:* Assuming the sequences are all equally likely, what is the probability that you will get exactly two heads when you toss a coin three times?

2. *Tree Diagram*
   (a) Draw a tree diagram to display all the possible outcomes that can occur when you flip a coin and then toss a die.
   (b) How many outcomes contain a head and a number greater than four?
   (c) *Probability extension:* Assuming the outcomes displayed in the tree diagram are all equally likely, what is the probability that you will get a head *and* a number greater than four when you flip a coin and toss a die?

3. *Tree Diagram* There are six balls in an urn. They are identical except for color. Two are red, three are blue, and one is yellow. You are to draw a ball from the urn, note its color, and set it aside. Then you are to draw another ball from the urn and note its color.
   (a) Make a tree diagram to show all possible outcomes of the experiment. Label the probability associated with each stage of the experiment on the appropriate branch.
   (b) *Probability extension:* Compute the probability for each outcome of the experiment.

4. *Tree Diagram* Repeat the experiment described in Problem 3. However, replace the first ball before you draw the second one.
   (a) Make a tree diagram to show all possible outcomes of the experiment. Label the probability associated with each stage of the experiment on the appropriate branch.
   (b) *Probability extension:* Compute the probability for each outcome of the experiment.

5. *Tree Diagram* Consider three true–false questions. There are two possible outcomes for each question: true or false.
   (a) Draw a tree diagram showing all possible sequences of responses for the three questions. Does your tree diagram look similar to the one in Problem 1? Why would you expect this result?
   (b) *Probability extension:* Only one sequence will contain all three correct answers. Assuming you are guessing and all of the sequences are equally likely to occur when you guess, what is the probability of getting all three questions correct?

6. *Tree Diagram*
   (a) Make a tree diagram to show all the possible sequences of answers for three multiple-choice questions, each with four possible responses.
   (b) *Probability extension:* Assuming that you are guessing the answers so that all outcomes listed in the tree are equally likely, what is the probability that you will guess the one sequence that contains all three correct answers?

7. *Multiplication Rule* Four wires (red, green, blue, and yellow) need to be attached to a circuit board. A robotic device will attach the wires. The wires can be attached in any order, and the production manager wishes to determine which order would be fastest for the robot to use. Use the multiplication rule of counting to determine the number of all the possible sequences of assembly that must be tested. (*Hint:*

There are four choices for the first wire, three for the second, two for the third, and only one for the fourth.)

8. *Multiplication Rule* A sales representative must visit four cities: Omaha, Dallas, Wichita, and Oklahoma City. There are direct air connections between each of the cities. Use the multiplication rule of counting to determine the number of different choices the sales representative has for the order in which to visit the cities. How is this problem similar to Problem 7?

9. *Multiplication Rule* You have two decks of cards (52 cards per deck), and you draw one card from each deck.
   (a) Use the multiplication rule of counting to determine the number of pairs of cards possible.
   (b) There are four kings in each deck. How many pairs of kings are possible?
   (c) *Probability extension:* Assuming all pairs are equally likely to be drawn, what is the probability of drawing two kings?

10. *Multiplication Rule* You toss a pair of dice.
    (a) Use the multiplication rule of counting to determine the number of possible pairs of outcomes. (Recall that there are six possible outcomes for each die.)
    (b) There are three even numbers on each die. How many outcomes are possible with even numbers appearing on each die?
    (c) *Probability extension:* What is the probability that both dice will show an even number?

11. *Multiplication Rule* Barbara is a research biologist for Green Carpet Lawns. She is studying the effects of fertilizer type, temperature at time of application, and water treatment after application. She has four fertilizer types, three temperature zones, and three water treatments to test. Use the multiplication rule of counting to determine the number of different lawn plots she needs in order to test each fertilizer type, temperature range, and water treatment configuration.

12. *Multiplication Rule: Menu Choices* The Deli Special lunch offers a choice of three different sandwiches, four kinds of salads, and five different desserts. Use the multiplication rule of counting to determine the number of different lunches that can be ordered using the Deli Special lunch option, if each lunch consists of one sandwich, one salad, and one dessert.

13. Compute $P_{5,2}$.   17. Compute $C_{5,2}$.

14. Compute $P_{8,3}$.   18. Compute $C_{8,3}$.

15. Compute $P_{7,7}$.   19. Compute $C_{7,7}$.

16. Compute $P_{9,9}$.   20. Compute $C_{8,8}$.

21. *Permutation: Hiring* There are three nursing positions to be filled at Lilly Hospital. Position one is the day nursing supervisor; position two is the night nursing supervisor; and position three is the nursing coordinator position. There are 15 candidates qualified for all three of the positions. Use the permutations rule to determine the number of different ways the positions can be filled by these applicants.

22. *Permutation: Lottery* In the Cash Now lottery game there are 10 finalists who submitted entry tickets on time. From these 10 tickets, three grand prize winners will be drawn. The first prize is one million dollars, the second prize is one hundred thousand dollars, and the third prize is ten thousand dollars. Use the permutations rule to determine the total number of different ways the winners can be drawn. (Assume that the tickets are not replaced after they are drawn.)

23. *Permutation: Objective Exam* Matching questions are sometimes used on objective tests.
    (a) If eight words are to be matched with eight definitions (numbered 1 through 8), use the permutations rule to determine the number of possible word-to-definition matches. (Assume each word corresponds to exactly one definition.)
    (b) If there are eight words but only five definitions, use the permutations rule to determine the number of possible word-to-definition matches.

24. *Counting: Shelving* To emphasize the importance of correct shelving of books, a librarian tells a group of students the number of possible orders in which just six books may be placed on a shelf. What is that number?

25. *Counting: Sports* The University of Montana ski team has five entrants in a men's downhill ski event. The coach would like the first, second, and third places to go to the team members. In how many ways can the five team entrants achieve first, second, and third places?

26. *Counting: Sales* During the Computer Daze special promotion, a customer purchasing a computer and printer is given a choice of three free software packages. There are 10 different software packages from which to select. How many different groups of software packages can be selected?

27. *Counting: Hiring* There are 15 qualified applicants for five trainee positions in a fast-food management program. How many different groups of trainees can be selected? (*Hint:* Is order important? If not, use the formula for combinations.)

28. *Counting: Grading* One professor grades homework by randomly choosing 5 out of 12 homework problems to grade.
    (a) How many different groups of 5 problems are there from the 12 problems?
    (b) *Probability extension:* Jerry did only 5 problems of one assignment. What is the probability that the problems he did comprised the group that was selected to be graded?
    (c) Silvia did 7 problems. How many different groups of 5 did she complete? What is the probability that one of the groups of 5 she completed comprised the group selected to be graded?

29. *Counting: Hiring* The qualified applicant pool for six management trainee positions consists of seven women and five men.
    (a) How many different groups of applicants can be selected for the positions?
    (b) How many different groups of trainees would consist entirely of women?
    (c) *Probability extension:* If the applicants are equally qualified and the trainee positions are selected by drawing the names at random so that all groups of six are equally likely, what is the probability that the trainee class will consist entirely of women?

30. *Counting: Lottery* In the Colorado State Lotto game, there are 42 numbers. Players choose any 6. Then the state selects 6 of the numbers at random. The winning tickets (for the grand prize) are those on which the player's 6 numbers match the state's 6 numbers.
    (a) From 42 numbers, how many groups of 6 are possible?
    (b) *Probability extension:* If you buy one lottery ticket, what is the probability of winning the grand prize?
    (c) *Probability extension:* If you buy 10 lottery tickets, what is the probability of winning the grand prize?

SUMMARY

In this chapter we first examined the question: What is probability? We found that probabilities can be assigned to events by intuition, by the method of relative frequency, or by the method of equally likely outcomes. Next, we studied some probability rules. The most important rules are the multiplication rules for independent and dependent events and the addition rules for mutually exclusive and general events. We also looked at some counting techniques useful in computing probabilities. These techniques included tree diagrams, the multiplication rule for counting, combinations, and permutations.

IMPORTANT WORDS & SYMBOLS

Section 4.1
Probability of an event A, $P(A)$
Relative frequency
Law of large numbers
Equally likely outcomes
Sample space
Statistical experiment
Complement of A

Section 4.2
Independent events
Dependent events
A, *given B*
Conditional probability

Multiplication rules of probability (for independent and dependent events)
A and B
Mutually exclusive events
A or B
Addition rules (for mutually exclusive and general events)

Section 4.3
Tree diagram
Multiplication rule of counting
Permutations rule
Combinations rule

VIEWPOINT *Deathday and Birthday*

Can people really postpone death? If so, how much can the timing of death be influenced by psychological, social, or other influential factors? One special event is a birthday. Will famous people try to postpone their death until an important birthday? Both Thomas Jefferson and John Adams died on July 4, 1826, when the United States was celebrating its 50th birthday. Is this only a strange coincidence, or is there an unexpected connection between birthdays and deathdays? The probability associated with a death rate decline of famous people just before important birthdays has been studied by Professor D. P. Phillips, of the State University of New York, and is presented in the book *Statistics, A Guide to the Unknown,* edited by J. M. Tanur.

CHAPTER REVIEW PROBLEMS

1. *Salary Raise: Women* Does it pay to ask for a raise? A national survey of heads of households showed the percentage of those who asked for a raise and the percentage who got one (*USA Today*). According to the survey, of the women interviewed, 24% had asked for a raise, and of those women who had asked for a raise, 45% received the raise. If a woman is selected at random from the survey population of women, find the following probabilities: *P*(woman asked for a raise); *P*(woman received raise, *given* she asked for one); *P*(woman asked for raise *and* received raise).

2. *Salary Raise: Men* According to the same survey quoted in Problem 1, of the men interviewed, 20% had asked for a raise and 59% of the men who had asked for a raise *received* the raise. If a man is selected at random from the survey population of men, find the following probabilities: *P*(man asked for a raise); *P*(man received raise, *given* he asked for one); *P*(man asked for raise *and* received raise).

3. *General: Deck of Cards* Two cards are drawn at random from a standard deck. (A standard deck has 52 cards: 13 hearts, 13 diamonds, 13 clubs, and 13 spades.)
   (a) Are the outcomes of the two cards independent? Why?
   (b) If the first card is replaced before the second is drawn, what is the probability that both cards will be hearts?
   (c) If the first card is not replaced before the second is drawn, what is the probability that both cards will be hearts?

4. *General: Die and Coin* Suppose you throw a fair die and flip a fair coin. Let's represent the outcomes of 3 on the die and heads on the coin by 3H.
   (a) One outcome is 3H. What are the other outcomes? What is the sample space?
   (b) Are all outcomes in the sample space equally likely? Explain.
   (c) What is the probability of getting heads and a number less than 3?

5. *General: Thumbtack*
   (a) Describe how you could use a relative frequency to estimate the probability that a thumbtack will land with its flat side down.
   (b) What is the sample space of outcomes for the thumbtack?
   (c) How would you make a probability assignment to this sample space if, when you drop 500 tacks, 340 land flat side down?

6. *Survey: Reaction to Poison Ivy* Allergic reactions to poison ivy can be miserable. Plant oils cause the reaction. Researchers at Allergy Institute did a study to see the effects of washing the oil off within 5 minutes of exposure. A random sample of 1000 people with known allergies to poison ivy participated in the study. Oil from the poison ivy plant was rubbed on a patch of skin. For 500 of the subjects, it was washed off *within* 5 minutes. For the other 500 subjects, the oil was washed off *after* 5 minutes. The results are summarized in Table 4-5.

TABLE 4-5   Time in Which Oil Was Washed Off

Reaction	Within 5 Minutes	After 5 Minutes	Row Total
None	420	50	470
Mild	60	330	390
Strong	20	120	140
Column Total	500	500	1000

Let's use the following notation for the various events: W = washing oil off within 5 minutes, A = washing oil off after 5 minutes, N = no reaction, M = mild reaction, S = strong reaction. Find the following probabilities for a person selected at random from this sample of 1000 subjects.

(a) $P(N)$, $P(M)$, $P(S)$

(b) $P(N$, given $W)$, $P(S$, given $W)$

(c) $P(N$, given $A)$, $P(S$, given $A)$

(d) $P(N$ and $W)$, $P(M$ and $W)$

(e) $P(N$ or $M)$ Are the events N = no reaction and M = mild reaction mutually exclusive? Explain.

(f) Are the events N = no reaction and W = washing oil off within 5 minutes independent? Explain.

7. *General: Two Dice* In a game of craps you roll two fair dice. Whether you win or lose depends on the sum of the numbers occurring on the tops of the dice. Let x be the random variable that is the sum of the numbers on the tops of the dice.

(a) What values can x take on?

(b) What is the probability distribution of these x values (that is, what is the probability that $x = 2, 3$, etc.)?

8. *Academic: Passing French* Class records at Rockwood College indicate a student selected at random has probability 0.77 of passing French 101. For the student who passes French 101, the probability is 0.90 that he or she will pass French 102. What is the probability that a student selected at random will pass both French 101 and 102?

9. *Combination: City Council* There is money to send two of eight city council members to a conference in Honolulu. All want to go, so they decide to choose the members to go to the conference by a random process. How many different combinations of two council members can be selected from the eight who want to go to the conference?

10. Compute

(a) $P_{7,2}$ (b) $C_{7,2}$ (c) $P_{3,3}$ (d) $C_{4,4}$

11. *Inspection: Food Processing* Freeze Dry Food, Inc., packages all its foods in clear plastic that is sealed. The quality control for the packaging process checks for three items: (1) that the weight shown is correct, (2) that the label is correct, and (3) that the package is properly sealed. These three processes can be done in any order. A computer-operated device directs the packages to the three inspection stations according to backlog in that area. If there is a larger backlog in one area, products are sent to one of the other two areas first. In how many different ways can a package be cycled through the three inspection stations?

12. *Scheduling: College Courses* A student must satisfy the literature, social science, and philosophy requirements this semester. There are four literature courses to select from, three social science courses, and two philosophy courses. Make a tree diagram showing all the possible sequences of literature, social science, and philosophy courses.

13. *General: Multiplication Rule* There are five multiple-choice questions on an exam, each with four possible answers. Use the multiplication rule of counting to determine the number of possible answer sequences for the five questions. Only one of the sets can contain all five correct answers. If you are guessing, so that you are as likely to choose one sequence of answers as another, what is the probability of getting all five answers correct?

14. *General: Multiplication Rule* A coin is tossed six times. Use the multiplication rule of counting to determine the number of possible head–tail sequences that can occur.

15. *General: Combination Lock* To open a combination lock, you turn the dial to the right and stop at a number; then you turn it to the left and stop at a second number. Finally, you turn it back to the right and stop at a third number. If you used the correct sequence of numbers, the lock opens. If the dial of the lock contains 10 numbers, 0 through 9, use the multiplication rule to determine the number of different combinations possible for the lock. (*Note:* The same number can be reused.)

16. *General: Combination Lock* You have a combination lock. Again, to open it you turn the knob to the right and stop at a first number; then you turn it to the left and stop at a second number. Finally, you turn it to the right and stop at a third number. Suppose you remember that the three numbers for your lock are 2, 9, and 5, but you don't remember the order in which the numbers occur. How many permutations of these three numbers are possible?

DATA HIGHLIGHTS: GROUP PROJECTS

Break into small groups and discuss the following topics. Organize a brief outline in which you summarize the main points of your group discussion.

1. Look at Figure 4-13, "Peeking at workers' e-mail." What group of people was surveyed? Estimate the probability that an executive selected at random from the survey population works at a company that uses e-mail. Estimate the probability that an executive peeks at employee e-mail, given that the company uses e-mail. Compute the probability that an executive from the survey population works at a company that uses e-mail *and* peeks at the employees' e-mail.

FIGURE 4-13

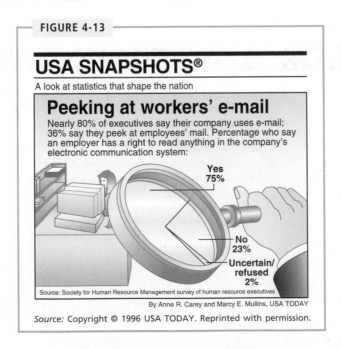

USA SNAPSHOTS®

A look at statistics that shape the nation

Peeking at workers' e-mail

Nearly 80% of executives say their company uses e-mail; 36% say they peek at employees' mail. Percentage who say an employer has a right to read anything in the company's electronic communication system:

Yes 75%

No 23%

Uncertain/refused 2%

Source: Society for Human Resource Management survey of human resource executives

By Anne R. Carey and Marcy E. Mullins, USA TODAY

Source: Copyright © 1996 USA TODAY. Reprinted with permission.

FIGURE 4-14

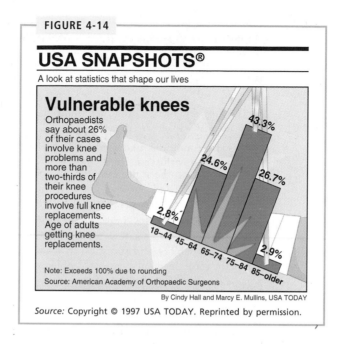

USA SNAPSHOTS®

A look at statistics that shape our lives

Vulnerable knees

Orthopaedists say about 26% of their cases involve knee problems and more than two-thirds of their knee procedures involve full knee replacements. Age of adults getting knee replacements.

2.8% 18–44
24.6% 45–64
43.3% 65–74
26.7% 75–84
2.9% 85–older

Note: Exceeds 100% due to rounding
Source: American Academy of Orthopaedic Surgeons

By Cindy Hall and Marcy E. Mullins, USA TODAY

Source: Copyright © 1997 USA TODAY. Reprinted by permission.

2. Consider the information given in Figure 4-14, "Vulnerable knees." What is the probability an orthopaedic case selected at random involves knee problems? Of those cases, estimate the probability that the case requires full knee replacement. Compute the probability that an orthopaedic case selected at random involves a knee problem *and* requires a full knee replacement. Next, look at the probability distribution for ages of patients requiring full knee replacement. Medicare insurance coverage begins when a person reaches age 65. What is the probability that the age of a person receiving a knee replacement is 65 or older?

LINKING CONCEPTS: WRITING PROJECTS

Discuss each of the following topics in class or review the topics on your own. Then write a brief but complete essay in which you summarize the main points. Please include formulas as appropriate.

1. Discuss the following concepts and give examples from everyday life where you encounter each concept. *Hint:* For instance, consider the "experiment" of arriving for class. Some possible outcomes are not arriving (that is, missing class), arriving on time, and arriving late.

   (a) Sample space.

   (b) Probability assignment to a sample space. In your discussion, be sure to include answers to the following questions.

      (i) Is there more than one valid way to assign probabilities to a sample space? Explain and give an example.

      (ii) How can probabilities be estimated by relative frequencies? How could probabilities be computed if events are equally likely?

2. Discuss the concepts of mutually exclusive events and independent events. List several examples of each type of event from everyday life.

   (a) If A and B are mutually exclusive events, does it follow that A and B *cannot* be independent events? Give an example to demonstrate your answer. *Hint:* Discuss an election where only one person can win the election. Let A be the event that party A's candidate wins, let B be the event that party B's candidate wins. Does the outcome of one event determine the outcome of the other event? Are A and B mutually exclusive events?

   (b) Discuss conditions under which $P(A \text{ and } B) = P(A) \cdot P(B)$ is true. Under what conditions is this not true?

   (c) Discuss conditions under which $P(A \text{ or } B) = P(A) + P(B)$ is true. Under what conditions is this not true?

3. Although we learn a good deal about probability in this course, the main emphasis is on statistics. Write a few paragraphs in which you talk about the distinction between probability and statistics. In what types of problems would probability be the main tool? In what types of problems would statistics be the main tool? Give some examples of both types of problems. What kind of outcome or conclusions do we expect from each type of problem?

Using TECHNOLOGY

Demonstration of the Law of Large Numbers

Computers can be used to simulate experiments. In packages such as Minitab and Excel, programs using random-number generators can be designed (see the *Technology Guides*) to simulate activities such as tossing a die. In ComputerStat, such a program exists (menu selection: ➤ **Descriptive Statistics** ➤ **Simulate the Experiment of Tossing One Die**). The following printouts show the simulations for tossing a die 6, 500, 50,000, 500,000, and 1,000,000 times. Notice how the relative frequencies of the outcomes approach the theoretical probabilities of 1/6 or 0.16667 for each outcome. Do you expect the same results every time the simulation is done? Why or why not?

Results of tossing one die 6 times

Outcome	Number of Occurrences	Relative Frequency
⚀	0	.00000
⚁	1	.16667
⚂	2	.33333
⚃	0	.00000
⚄	1	.16667
⚅	2	.33333

Results of tossing one die 500 times

Outcome	Number of Occurrences	Relative Frequency
⚀	87	.17400
⚁	83	.16600
⚂	91	.18200
⚃	69	.13800
⚄	87	.17400
⚅	83	.16600

Results of tossing one die 50,000 times

Outcome	Number of Occurrences	Relative Frequency
⚀	8528	.17056
⚁	8354	.16708
⚂	8246	.16492
⚃	8414	.16828
⚄	8178	.16356
⚅	8280	.16560

Results of tossing one die 500,000 times

Outcome	Number of Occurrences	Relative Frequency
⚀	83644	.16729
⚁	83368	.16674
⚂	83398	.16680
⚃	83095	.16619
⚄	83268	.16654
⚅	83227	.16645

Results of tossing one die 1,000,000 times

Outcome	Number of Occurrences	Relative Frequency
⚀	166643	.16664
⚁	166168	.16617
⚂	167391	.16739
⚃	165790	.16579
⚄	167243	.16724
⚅	166765	.16677

5

The Binomial Probability Distribution and Related Topics

Education is the key to unlock the golden door of freedom.

—George Washington Carver

George Washington Carver (1859–1943)

Carver was a winner of the Spingarn Medal for distinguished service in agricultural chemistry and the prestigious Roosevelt Medal for contributions to science. Carver was also a Fellow in the Royal Society of Arts in London, an honor given to very few Americans.

For on-line student resources, visit **math.college.hmco.com/students** and follow the Statistics links to the Brase/Brase, *Understandable Statistics,* 7th edition web site.

George Washington Carver won international fame for agricultural research. After graduating from Iowa State College, he was appointed a faculty member in the Iowa State Botany Department. Carver took charge of the greenhouse and started a fungus collection that later included more than 20,000 species. This collection brought him professional acclaim in the field of botany.

At the invitation of his friend Booker T. Washington, Carver joined the faculty of the Tuskegee Institute, where he spent the rest of his long and distinguished career. Carver's creative genius accounted for more than 300 inventions from peanuts, 118 inventions from sweet potatoes, and 75 inventions from pecans.

Gathering and analyzing data were important components of Carver's work. Methods you will learn in this course are widely used in research in every field, including agriculture.

◇ What is a random variable? How do we compute μ and σ for a discrete random variable? How do we compute μ and σ for linear combinations of independent random variables? (SECTION 5.1)

◇ Many of life's circumstances come down to success or failure. How does the binomial probability distribution help us compute the probability of getting r successes in n trials? (SECTION 5.2)

◇ How do we compute μ and σ for the binomial distribution? (SECTION 5.3)

◇ How is the binomial distribution related to other probability distributions such as the geometric and Poisson? (SECTION 5.4)

Personality Preference Types: Introvert or Extrovert?

Isabel Briggs Myers was a pioneer in the study of personality types. Her work has been used successfully in counseling, educational, and industrial settings. In the book *A Guide to the Development and Use of the Myers-Briggs Type Indicators,* by Myers and McCaully, it was reported that based on a very large sample (2282 professors), approximately 45% of all university professors are extroverted.

Suppose you have classes with six different professors.

1. What is the probability that all six are extroverts?

2. What is the probability that none of your professors is an extrovert?

3. What is the probability that at least two of your professors are extroverts?

4. In a group of six professors selected at random, what is the *expected number* of extroverts? What is the *standard deviation* of the distribution?

5. Suppose you were assigned to write an article for the student newspaper and you were given a quota (by the editor) of interviewing at least three extroverted professors. How many professors selected at random would you need to interview to be at least 90% sure of filling the quota?

◇ **COMMENT** Both extroverted and introverted professors can be excellent teachers.   ◇

5.1   Introduction to Random Variables and Probability Distributions

FOCUS POINTS

✓ Distinguish between discrete and continuous random variables.

✓ Graph discrete probability distributions.

✓ Compute μ and σ for a discrete probability distribution.

✓ Compute μ and σ for a linear function of a random variable x.

✓ Compute μ and σ for a linear combination of two independent random variables.

For our purposes, we will say that a *statistical experiment* or *observation* is any process by which measurements are obtained. Examples are

1. Counting the number of eggs in a robin's nest

2. Measuring daily rainfall in inches

3. Counting the number of defective light bulbs in a case of bulbs

4. Measuring the weight in kilograms of a polar bear cub

Let x represent a quantitative variable that is measured or observed in an experiment. We are interested in the numerical values that x can take on. So $x =$ number of eggs in a robin's nest and $x =$ weight in kilograms of a polar bear cub would be examples of such quantitative variables. Furthermore, we say that the quantitative variable x is a *random variable* because the value that x takes on in a given experiment is a chance or random outcome. We will study two types of random variables: *discrete random variables* and *continuous random variables*.

Discrete random variable

> **Definition**
>
> When the observations of a quantitative random variable can take on only a finite number of values or a countable number of values, we say that the variable is a *discrete random variable*.

We know what a *finite* number of values is, but what is a *countable* number of values? As an example, let the random variable x be the number of wells an oil prospector drills until the first productive well is found. Then x could be any of the values 1, 2, 3. . . . In theory, we have an infinite number of possibilities for the values of x. The set of values of x corresponds to the set of counting numbers. Therefore, this type of infinity is called *countable,* and we say x has a countable number of values. This is an intuitive approach to the concept of a countable set. The reader interested in a more rigorous discussion is referred to the advanced text *Introduction to Mathematical Statistics* by Hogg and Craig (Macmillan Publishing).

In most of the cases we will consider, a *discrete random variable* will be the result of a count (the terms *countable* and *discrete,* however, have different mathematical meanings). For instance, the number of students in a certain section of a statistics course this term is a discrete random variable. The value must be a counting number such as 25, or 57, or 135, and so forth. The values 25.34 or $25\frac{1}{2}$ are not possible. The cost of tuition to the nearest dollar or nearest cent is another example of a discrete random variable. In this case, we are counting dollars or cents.

Continuous random variable

> **Definition**
>
> When the observations of a quantitative random variable can take on any of the countless number of values in a line interval, we say that the variable is a *continuous random variable.*

For our purposes, we will see most *continuous random variables* occurring as the result of a measurement. For example, the air pressure in an automobile tire represents a continuous random variable. The air pressure could in theory take on any value from 0 lb/in^2 (psi) to the bursting pressure of the tire. Values such as 20.126 psi, 20.12678 psi, and so forth are possible. Another example is the height of students in your statistics class. The height could in theory take on any value from a low of, say, 3 feet to a high of, say, 7.25 feet.

The distinction between discrete and continuous random variables is important because of the different mathematical techniques associated with the two kinds of random variables. Although we will not discuss these techniques at great length in this book, the distinction is very important in the study of advanced mathematical statistics.

In general, measurements of quantities such as length, weight, volume, temperature, or time yield continuous random variables. If the temperature changes from 12°C to 13°C, for example, it must take on all the temperature values between 12 and 13. Temperatures cannot just jump from one reading to the next. Discrete random variables often come from counts, such as the number of passing scores on an exam or the number of weeds in a garden.

GUIDED EXERCISE 1

Discrete or continuous random variables

Which of the following random variables are discrete and which are continuous?

(a) *Measure* the time it takes a student selected at random to register for the fall term.

⟹ Time can take on any value, so this is a continuous random variable.

(b) *Count* the number of bad checks drawn on Upright Bank on a day selected at random.

⟹ The number of bad checks can be only a whole number such as 0, 1, 2, 3, etc. This is a discrete variable.

Continued

GUIDED EXERCISE 1 continued

(c) *Measure* the amount of gasoline needed to drive your car 200 miles.

⟹ We are measuring volume, which can assume any value, so this is a continuous random variable.

(d) Pick a random sample of 50 registered voters in a district and find the number who voted in the last county election.

⟹ This is a count, so the variable is discrete.

Probability distribution

A random variable has a probability distribution whether it is discrete or continuous. The *probability distribution* is simply an assignment of probabilities to the specific values of the random variable or to a range of values of the random variable.

1. The probability distribution of a *discrete* random variable has a probability assigned to *each* value of the random variable.

2. The sum of these probabilities must be 1.

Let's look at a discrete probability distribution and its graph.

EXAMPLE 1

Discrete probability distribution

Dr. Fidgit developed a test to measure boredom tolerance. He administered it to a group of 20,000 adults between the ages of 25 and 35. The possible scores were 0, 1, 2, 3, 4, 5, and 6, with 6 indicating the highest tolerance for boredom. The test results for this group are shown in Table 5-1.

(a) If a subject is chosen at random from this group, the probability that he or she will have a score of 3 is 6000/20,000, or 0.30. In a similar way, we can use the relative frequency to compute the probabilities for the other scores (Table 5-2). These probability assignments make up the probability distribution. Notice that the scores are mutually exclusive: No one subject has two scores. The sum of the probabilities of all the scores is 1.

(b) The graph of this distribution is simply a relative-frequency histogram (see Figure 5-1) in which the height of the bar over a score represents the probability

TABLE 5-1   Boredom Tolerance Test Scores for 20,000 Subjects

Score	Number of Subjects
0	1400
1	2600
2	3600
3	6000
4	4400
5	1600
6	400

TABLE 5-2   Probability Distribution of Scores on Boredom Tolerance Test

Score x	Probability $P(x)$
0	0.07
1	0.13
2	0.18
3	0.30
4	0.22
5	0.08
6	0.02
	$\Sigma P(x) = 1$

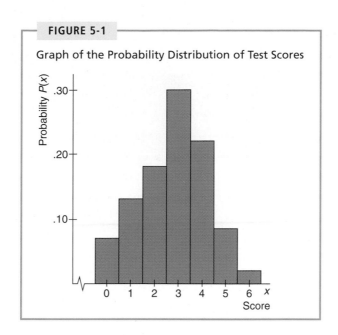

FIGURE 5-1

Graph of the Probability Distribution of Test Scores

of that score. Since each bar is one unit wide, the area of the bar over a score equals the height and thus represents the probability of that score. Since the sum of the probabilities is 1, the area under the graph is also 1.

(c) The Topnotch Clothing Company needs to hire someone with a score on the boredom tolerance test of 5 or 6 to operate the fabric press machine. Since the scores 5 and 6 are mutually exclusive, the probability that someone in the group who took the boredom tolerance test made either a 5 or a 6 is the sum

$$P(5 \text{ or } 6) = P(5) + P(6)$$
$$= 0.08 + 0.02 = 0.10$$

Notice that to find $P(5 \text{ or } 6)$, we could have simply added the *areas* of the bars over 5 and over 6. One out of 10 of the group who took the boredom tolerance test would qualify for the position at Topnotch Clothing. ◊

GUIDED
EXERCISE **2**

Discrete probability distribution

One of the elementary tools of cryptanalysis (the science of code breaking) is to use relative frequencies of occurrence of different letters in the alphabet to break standard English alphabet codes. Large samples of plain text such as newspaper stories generally yield about the same relative frequencies for letters. A sample 1000 letters long yielded the information in Table 5-3.

Continued

GUIDED EXERCISE 2 continued

(a) Use the relative frequencies to compute the omitted probabilities in Table 5-3.

 Table 5-4 shows the completion of Table 5-3.

TABLE 5-3   Frequency of Letters in a 1000-Letter Sample

Letter	Freq.	Prob.	Letter	Freq.	Prob.
A	73	——	N	78	0.078
B	9	0.009	O	74	——
C	30	0.030	P	27	0.027
D	44	0.044	Q	3	0.003
E	130	——	R	77	0.077
F	28	0.028	S	63	0.063
G	16	0.016	T	93	0.093
H	35	0.035	U	27	——
I	74	——	V	13	0.013
J	2	0.002	W	16	0.016
K	3	0.003	X	5	0.005
L	35	0.035	Y	19	0.019
M	25	0.025	Z	1	0.001

Source: From *Elementary Cryptanalysis. A Mathematical Approach,* by Abraham Sinkov. Copyright © 1968 by Yale University. Reprinted by permission of Random House, Inc.

TABLE 5-4   Completion of Table 5-3

Letter	Relative Frequency	Probability
A	$\dfrac{73}{1{,}000}$	0.073
E	$\dfrac{130}{1{,}000}$	0.130
I	$\dfrac{74}{1{,}000}$	0.074
O	$\dfrac{74}{1{,}000}$	0.074
U	$\dfrac{27}{1{,}000}$	0.027

(b) Do the probabilities of all the individual letters add up to 1?

Yes.

(c) If a letter is selected at random from a newspaper story, what is the probability that the letter will be a vowel?

If a letter is selected at random,

$$P(a,\ e,\ i,\ o,\ or\ u) = P(a) + P(e) + P(i) + P(o) + P(u)$$
$$= 0.073 + 0.130 + 0.074 + 0.074 + 0.027$$
$$= 0.378$$

Mean and standard deviation of a discrete probability distribution

A probability distribution can be thought of as a relative-frequency distribution based on a very large *n*. As such, it has a mean and standard deviation. If we are referring to the probability distribution of a *population,* then we use the Greek letters μ for the mean and σ for the standard deviation. When we see the Greek letters used, we know the information given is from the *entire population* rather than just a sample. If we have a sample probability distribution, we use $\bar{x}$ (*x* bar) and *s*, respectively, for the mean and standard deviation. For a given population, μ and σ are fixed numbers and are sometimes called the *parameters* of the population.

Definition

The *mean* and the *standard deviation of a discrete population probability distribution* are found by using these formulas:

$\mu = \Sigma x P(x)$ is the expected value of x

$\sigma = \sqrt{\Sigma (x - \mu)^2 P(x)}$ is the standard deviation of x

where x is the value of a random variable,
$P(x)$ is the probability of that variable,
the sum Σ is taken for all the values of the random variable.

Note: μ is the *population mean* and σ is the underlying *population standard deviation* because the sum Σ is taken over *all* values of the random variable (i.e., the entire sample space).

Expected value

The mean of a probability distribution is often called the *expected value* of the distribution. This terminology reflects the idea that the mean represents a "central point" or "cluster point" for the entire distribution. Of course, the mean or expected value is an average value, and as such, it *need not be a point of the sample space.*

EXAMPLE 2

Expected value, standard deviation

Are we influenced to buy a product by an ad we saw on TV? National Infomercial Marketing Association determined the number of times *buyers* of a product watched a TV infomercial *before* purchasing the product. The results are shown here:

Number of Times Buyers Saw Infomercial	1	2	3	4	5*
Percentage of Buyers	27%	31%	18%	9%	15%

* This category was 5 or more, but will be treated as 5 in this example.

We can treat the information shown as an estimate of the probability distribution because the events are mutually exclusive and the sum of the percentages is 100%. Compute the mean and standard deviation of the distribution.

SOLUTION: We put the data in the first two columns of a computation table and then fill in the other entries (see Table 5-5).

TABLE 5-5   Number of Times Buyers View Infomercial Before Making Purchase

x (number of viewings)	$P(x)$	$xP(x)$	$x - \mu$	$(x - \mu)^2$	$(x - \mu)^2 P(x)$
1	0.27	0.27	−1.54	2.372	0.640
2	0.31	0.62	−0.54	0.292	0.091
3	0.18	0.54	0.46	0.212	0.038
4	0.09	0.36	1.46	2.132	0.192
5	0.15	0.75	2.46	6.052	0.908
		$\mu = \Sigma x P(x) = 2.54$			$\Sigma (x - \mu)^2 P(x) = 1.869$

The average number of times a buyer views the infomercial before purchase is

$$\mu = \Sigma x P(x) = 2.54 \text{ (sum of column 3)}$$

To find the standard deviation, we take the square root of the sum of column 6:

$$\sigma = \sqrt{\Sigma(x - \mu)^2 P(x)} \approx \sqrt{1.869} \approx 1.37 \qquad \Diamond$$

● **CALCULATOR NOTE**   Some calculators, including the TI-83, accept fractional frequencies. If yours does, you can get μ and σ directly by using techniques for grouped data and the calculator's STAT mode.

GUIDED EXERCISE 3

Expected value

At a carnival, you pay $2.00 to play a coin-flipping game with three fair coins. On each coin one side has the number 0 and the other side has the number 1. You flip the three coins at one time and you win $1.00 for every 1 that appears on top. Are your expected earnings equal to the cost to play? We'll answer this question in several steps.

(a) In this game, the random variable of interest counts the number of 1s that show. What is the sample space for the values of this random variable?

→ The sample space is {0, 1, 2, 3}, since any of these numbers of 1s can appear.

(b) There are eight equally likely outcomes for throwing three coins. They are 000, 001, 010, 011, 100, 101, _____, and _____.

→ 110 and 111.

(c) Complete Table 5-6.

TABLE 5-6

Number of 1s, x	Frequency	P(x)	xP(x)
0	1	0.125	0
1	3	0.375	___
2	3	___	___
3	___	___	___

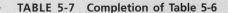

TABLE 5-7   Completion of Table 5-6

x	Frequency	P(x)	xP(x)
0	1	0.125	0
1	3	0.375	0.375
2	3	0.375	0.750
3	1	0.125	0.375

(d) The expected value is the sum

$$\mu = \Sigma x P(x)$$

Sum the appropriate column of Table 5-6 to find this value. Are your expected earnings less than, equal to, or more than the cost of the game?

The expected value can be found by summing the last column of Table 5-7. The expected value is $1.50. It cost $2.00 to play the game; the expected value is less than the cost. The carnival is making money. In the long run, the carnival can expect to make an average of about 50 cents per player.

We have seen probability distributions of discrete variables and the formulas to compute the mean and standard deviation of discrete population probability distributions. Probability distributions of continuous random variables are similar except that the probability assignments are made to intervals of values rather than to specific values of the random variable. We will see an important example of a discrete probability distribution, the binomial distribution, in the next section and one of a continuous probability distribution in Chapter 6 when we study the normal distribution.

We conclude this section with some results that will be important for later work.

Linear Functions of a Random Variable

Let a and b be any constants, and let x be a random variable. Then the new random variable $L = a + bx$ is called a *linear function of x*. Using some more advanced mathematics, the following can be proved.

> Let x be a random variable with mean μ and standard deviation σ. Then the linear function $L = a + bx$ has mean, variance, and standard deviation as follows:
>
> $$\mu_L = a + b\mu$$
> $$\sigma_L^2 = b^2\sigma^2$$
> $$\sigma_L = \sqrt{b^2\sigma^2} = |b|\sigma$$

Linear Combinations of Independent Random Variables

Independent random variables

Suppose we have two random variables x_1 and x_2. These variables are *independent* if any event involving x_1 by itself is *independent* of any event involving x_2 by itself. Sometimes, we want to combine independent random variables and examine the mean and standard deviation of the resulting combination.

Let x_1 and x_2 be independent random variables. Let a and b be any constants. Then the new random variable $W = ax_1 + bx_2$ is called a *linear combination of x_1 and x_2*. Using some more advanced mathematics, the following can be proved.

> Let x_1 and x_2 be independent random variables with respective means μ_1, μ_2, and variances σ_1^2 and σ_2^2. For the linear combination $W = ax_1 + bx_2$, the mean, variance, and standard deviation are as follows:
>
> $$\mu_W = a\mu_1 + b\mu_2$$
> $$\sigma_W^2 = a^2\sigma_1^2 + b^2\sigma_2^2$$
> $$\sigma_W = \sqrt{a^2\sigma_1^2 + b^2\sigma_2^2}$$

Note: The formula for the mean of a linear combination of random variables is valid regardless of whether the variables are independent or not. However, *the formulas for the variance and standard deviation are valid* only if x_1 and x_2 are *independent* random variables. In later work (Chapter 7 on), we will use independent random samples to ensure that the resulting variables (usually means, proportions, etc.) are statistically independent.

EXAMPLE 3

Linear combinations of independent random variables

Let x_1 and x_2 be independent random variables with respective means $\mu_1 = 75$, $\mu_2 = 50$, and standard deviations $\sigma_1 = 16$, $\sigma_2 = 9$.

(a) Let $L = 3 + 2x_1$. Compute the mean, variance, and standard deviation of L.

SOLUTION: L is a linear function of the random variable x_1. Using the formulas with $a = 3$ and $b = 2$, we have

$$\mu_L = 3 + 2\mu_1 = 3 + 2(75) = 153$$
$$\sigma_L^2 = 2^2\sigma_1^2 = 4(16)^2 = 1024$$
$$\sigma_L = |2|\sigma_1 = 2(16) = 32$$

Notice that the variance and standard deviation of the linear function are influenced only by the coefficient of x_1 in the linear function.

(b) Let $W = x_1 + x_2$. Find the mean, variance, and standard deviation of W.

SOLUTION: W is a linear combination of the independent random variables x_1 and x_2. Using the formulas with both a and b equal to 1, we have

$$\mu_W = \mu_1 + \mu_2 = 75 + 50 = 125$$
$$\sigma_W^2 = \sigma_1^2 + \sigma_2^2 = 16^2 + 9^2 = 337$$
$$\sigma_W = \sqrt{\sigma_1^2 + \sigma_2^2} = \sqrt{337} \approx 18.36$$

(c) Let $W = x_1 - x_2$. Find the mean, variance, and standard deviation of W.

SOLUTION: W is a linear combination of the independent random variables x_1 and x_2. Using the formulas with $a = 1$ and $b = -1$, we have

$$\mu_W = \mu_1 - \mu_2 = 75 - 50 = 25$$
$$\sigma_W^2 = 1^2\sigma_1^2 + (-1)^2\sigma_2^2 = 16^2 + 9^2 = 337$$
$$\sigma_W = \sqrt{\sigma_1^2 + \sigma_2^2} = \sqrt{337} \approx 18.36$$

(d) Let $W = 3x_1 - 2x_2$. Find the mean, variance, and standard deviation of W.

SOLUTION: W is a linear combination of the independent random variables x_1 and x_2. Using the formulas with $a = 3$ and $b = -2$

$$\mu_W = 3\mu_1 - 2\mu_2 = 3(75) - 2(50) = 125$$
$$\sigma_W^2 = 3^2\sigma_1^2 + (-2)^2\sigma_2^2 = 9(16^2) + 4(9^2) = 2628$$
$$\sigma_W = \sqrt{2628} \approx 51.26$$

◊ COMMENT: Problem 16 of Section 10.3 shows how to find the mean, variance, and standard deviation of a linear combination of two *linearly dependent* random variables. ◊

SECTION 5.1   PROBLEMS

1. *Random Variables: Classification* Which of the following are continuous variables, and which are discrete?
   (a) Number of traffic fatalities per year in the state of Florida
   (b) Distance a golf ball travels after being hit with a driver
   (c) Time required to drive from home to college on any given day
   (d) Number of ships in Pearl Harbor on any given day
   (e) Your weight before breakfast each morning

VIEW⊙POINT   *The Rosetta Project*

Around 196 B.C. Egyptian priests inscribed a decree on a granite slab affirming the rule of 13-year-old Ptolemy V. The proclamation was in Egyptian hieroglyphics with another translation in a form of ancient Greek. By 1799, the meaning of Egyptian hieroglyphics had been lost for many centuries. However, Napoleon's troops discovered the granite slab (Rosetta Stone). Linguists used the Rosetta Stone and their knowledge of ancient Greek to unlock the meaning of the Egyptian hieroglyphics.

Linguistic experts say that because of industrialization and globalization, by the year 2100 as many as 90% of the world's languages may be extinct. To help preserve some of these languages for future generations, 1000 translations of the first three chapters of Genesis have been inscribed in tiny text onto 3-inch nickel disks and encased in hardened glass balls that are expected to last at least 1000 years. Why Genesis? Because it is the most translated text in the world. The Rosetta Project is sending the disks to libraries and universities all over the world. It is very difficult to send information into the future. However, if in the year 2500 linguists are using the "Rosetta Disks" to unlock the meaning of a lost language, you may be sure they will use statistical methods of cryptanalysis (*see* Guided Exercise 2). To find out more about the Rosetta Project, visit the Brase/Brase statistics site at http://math.college.hmco.com/students and find the link to the Rosetta Project site.

2. *Random Variables: Classification* Which of the following are continuous variables, and which are discrete?
   (a) Speed of an airplane
   (b) Age of a college professor chosen at random
   (c) Number of books in the college bookstore
   (d) Weight of a football player chosen at random
   (e) Number of lightning strikes in Rocky Mountain National Park on a given day

3. *Probability Distribution* Consider each distribution. Determine if it is a valid probability distribution or not, and explain your answer.

(a)

x	0	1	2
$P(x)$	0.25	0.60	0.15

(b)

x	0	1	2
$P(x)$	0.25	0.60	0.20

4. *Marketing: Age* What is the age distribution of promotion-sensitive shoppers? A *supermarket super shopper* is defined as a shopper for whom at least 70% of the items purchased were on sale or purchased with a coupon. The following table is based on information taken from *Trends in the United States* (Food Marketing Institute, Washington, D.C.).

Age range, years	18–28	29–39	40–50	51–61	62 and over
Midpoint x	23	34	45	56	67
Percent of super shoppers	7%	44%	24%	14%	11%

For the 62 and over group, use the midpoint 67 years.
(a) Using the age midpoints x and the percentage of super shoppers, do we have a valid probability distribution? Explain.
(b) Use a histogram to graph the probability distribution of part (a).
(c) Compute the expected age μ of a super shopper.
(d) Compute the standard deviation σ for ages of super shoppers.

5. *Marketing: Income* What is the income distribution of super shoppers (*see* Problem 4). In the following table, income units are in thousands of dollars, and each interval goes up to but does not include the given high value. The midpoints are given to the nearest thousand dollars.

Income range	5–15	15–25	25–35	35–45	45–55	55 or more
Midpoint x	10	20	30	40	50	60
Percent of super shoppers	21%	14%	22%	15%	20%	8%

(a) Using the income midpoints x and the percent of super shoppers, do we have a valid probability distribution? Explain.
(b) Use a histogram to graph the probability distribution of part (a).
(c) Compute the expected income μ of a super shopper.
(d) Compute the standard deviation σ for the income of super shoppers.

6. *Sociology: Family Size* The following data are based on information taken from the *Statistical Abstract of the United States* (112th Edition). In this table, x = size of family. The percentage data are the percentages of U.S. families of this size.

x	2	3	4	5	6	7 or more
%	42%	23%	21%	10%	3%	1%

(a) Convert the percentage data to probabilities and make a histogram of the probability distribution for family size.
(b) What is the probability that a family selected at random will have only two members?
(c) What is the probability that a family selected at random will have more than three members?
(d) Compute μ, the expected family size (round families of size 7 or more to size 7).
(e) Compute σ, the standard deviation (round families of size 7 or more to size 7).

7. *Working Conditions: Nursing* The head nurse on the third floor of a community hospital is interested in the number of nighttime room calls requiring a nurse. For a random sample of 208 nights (9:00 P.M. to 6:00 A.M.), the following information was obtained, where x = number of room calls requiring a nurse and f = frequency with which this many calls occurred (i.e., number of nights).

x	36	37	38	39	40	41	42	43	44	45
f	6	10	11	20	26	32	34	28	25	16

(a) If a night is chosen at random from these 208 nights, use relative frequencies to find $P(x)$ when x = 36, 37, 38, 39, 40, 41, 42, 43, 44, and 45.
(b) Use a histogram to graph the probability distribution of part (a).
(c) Assuming that these 208 nights represent the population of all nights at the community hospital, what do you estimate the probability is that, on a randomly selected night, there will be from 39 to 43 (including 39 and 43) room calls requiring a nurse?

(d) What do you estimate the probability is that there will be from 36 to 40 (including 36 and 40) room calls requiring a nurse?

(e) Find the expected number of room calls requiring a nurse.

(f) Find the standard deviation of the x distribution.

8. *History: Florence Nightingale* What was the age distribution of nurses in Great Britain at the time of Florence Nightingale? Thanks to Florence Nightingale and the British census of 1851, we have the following information (based on data from the classic text *Notes on Nursing*, by Florence Nightingale). *Note:* In 1851 there were 25,466 nurses in Great Britain. Furthermore, Nightingale made a strict distinction between nurses and domestic servants.

Age range (yr)	20–29	30–39	40–49	50–59	60–69	70–79	80+
Midpoint x	24.5	34.5	44.5	54.5	64.5	74.5	84.5
Percent of nurses	5.7%	9.7%	19.5%	29.2%	25.0%	9.1%	1.8%

(a) Using the age midpoints x and the percent of nurses, do we have a valid probability distribution? Explain.

(b) Use a histogram to graph the probability distribution of part a.

(c) Find the probability that a British nurse selected at random in 1851 would be 60 years of age or older.

(d) Compute the expected age μ of a British nurse contemporary to Florence Nightingale.

(e) Compute the standard deviation σ for ages of nurses shown in the distribution.

9. *Fishing: Trout* The following data are based on information taken from *Daily Creel Summary*, published by the Paiute Indian Nation, Pyramid Lake, Nevada. Movie stars and U.S. presidents have fished Pyramid Lake. It is one of the best places in the lower 48 states to catch trophy cutthroat trout. In this table, x = number of fish caught in a 6-hour period. The percentage data are the percentages of fishermen who caught x fish in a 6-hour period while fishing from shore.

x	0	1	2	3	4 or more
%	44%	36%	15%	4%	1%

(a) Convert the percentages to probabilities and make a histogram of the probability distribution.

(b) Find the probability that a fisherman selected at random fishing from shore catches one or more fish in a 6-hour period.

(c) Find the probability that a fisherman selected at random fishing from shore catches two or more fish in a 6-hour period.

(d) Compute μ, the expected value of the number of fish caught per fisherman in a 6-hour period (round 4 or more to 4).

(e) Compute σ, the standard deviation of the number of fish caught per fisherman in a 6-hour period (round 4 or more to 4).

10. *Criminal Justice: Parole* USA Today reported that approximately 25% of all state prison inmates released on parole become repeat offenders while on parole. Suppose the parole board is examining five prisoners up for parole. Let x = number of prisoners out of five on parole who become repeat offenders. The methods of Section 5.2 can be used to compute the probability assignments for the x distribution.

x	0	1	2	3	4	5
P(x)	0.237	0.396	0.264	0.088	0.015	0.001

(a) Find the probability that one or more of the five parolees will be repeat offenders. How does this number relate to the probability that none of the parolees will be repeat offenders?
(b) Find the probability that two or more of the five parolees will be repeat offenders.
(c) Find the probability that four or more of the five parolees will be repeat offenders.
(d) Compute μ, the expected number of repeat offenders out of five.
(e) Compute σ, the standard deviation of the number of repeat offenders out of five.

11. *Fund Raiser: Hiking Club* The college hiking club is having a fund raiser to buy new equipment for fall and winter outings. The club is selling Chinese fortune cookies at a price of $1 per cookie. Each cookie contains a piece of paper with a different number written on it. A random drawing will determine which number is the winner of a dinner for two at a local Chinese restaurant. The dinner is valued at $35. Since the fortune cookies were donated to the club, we can ignore the cost of the cookies. The club sold 719 cookies before the drawing.
(a) Lisa bought 15 cookies. What is the probability she will win the dinner for two? What is the probability she will not win?
(b) Lisa's expected earnings can be found by multiplying the value of the dinner by the probability that she will win. What are Lisa's expected earnings? How much did she effectively contribute to the hiking club?

12. *Spring Break: Caribbean Cruise* The college student senate is sponsoring a spring break Caribbean cruise raffle. The proceeds are to be donated to the Samaritan Center for the Homeless. A local travel agency donated the cruise, valued at $2000. The students sold 2852 raffle tickets at $5 per ticket.
(a) Kevin bought six tickets. What is the probability that Kevin will win the spring break cruise to the Caribbean? What is the probability that Kevin will not win the cruise?
(b) Expected earnings can be found by multiplying the value of the cruise by the probability that Kevin will win. What are Kevin's expected earnings? Is this more or less than the amount Kevin paid for the six tickets? How much did Kevin effectively contribute to the Samaritan Center for the Homeless?

13. *Expected Value: Life Insurance* Jim is a 60-year-old Anglo male in reasonably good health. He wants to take out a $50,000 term (that is, straight death benefit) life insurance policy until he is 65. The policy will expire on his 65th birthday. The probability of death in a given year is provided by the Vital Statistics Section of the *Statistical Abstract of the United States* (116th Edition).

x = age	60	61	62	63	64
P(death at this age)	0.01191	0.01292	0.01396	0.01503	0.01613

Jim is applying to Big Rock Insurance Company for his term insurance policy.
(a) What is the probability that Jim will die in his 60th year? Using this probability and the $50,000 death benefit, what is the expected loss to Big Rock Insurance?
(b) Repeat part (a) for years 61, 62, 63, and 64. What would be the total expected loss to Big Rock Insurance over the years 60 through 64?
(c) If Big Rock Insurance wants to make a profit of $700 above the expected total loss paid out for Jim's death, how much should it charge for the policy?
(d) If Big Rock Insurance Company charges $5000 for the policy, how much profit does the company expect to make?

14. *Expected Value: Life Insurance* Sara is a 60-year-old Anglo female in reasonably good health. She wants to take out a $50,000 term (that is, straight death benefit) life insurance policy until she is 65. The policy will expire on her 65th birthday. The probability of death in a given year is provided by the Vital Statistics Section of the *Statistical Abstract of the United States* (116th Edition).

x = age	60	61	62	63	64
P(death at this age)	0.00756	0.00825	0.00896	0.00965	0.01035

Sara is applying to Big Rock Insurance Company for her term insurance policy.
(a) What is the probability that Sara will die in her 60th year? Using this probability and the $50,000 death benefit, what is the expected loss to Big Rock Insurance?
(b) Repeat part (a) for years 61, 62, 63, and 64. What would be the total expected loss to Big Rock Insurance over the years 60 through 64?
(c) If Big Rock Insurance wants to make a profit of $700 above the expected total loss paid out for Sara's death, how much should it charge for the policy?
(d) If Big Rock Insurance Company charges $5000 for the policy, how much profit does the company expect to make?

15. *Combination of Random Variables: Golf* Norb and Gary are entered in a local golf tournament. Both have played the local course many times. Their scores are random variables with the following means and standard deviations.

$$\text{Norb, } x_1: \mu_1 = 115; \ \sigma_1 = 12 \qquad \text{Gary, } x_2: \mu_2 = 100; \ \sigma_2 = 8$$

In the tournament, Norb and Gary are not playing together, and we will assume their scores vary independently of each other.
(a) The difference between their scores is $W = x_1 - x_2$. Compute the mean, variance, and standard deviation for the random variable W.
(b) The average of their scores is $W = 0.5x_1 + 0.5x_2$. Compute the mean, variance, and standard deviation for the random variable W.
(c) The tournament rules have a special handicap system for each player. For Norb, the handicap formula is $L = 0.8x_1 - 2$. Compute the mean, variance, and standard deviation for the random variable L.
(d) For Gary, the handicap formula is $L = 0.95x_2 - 5$. Compute the mean, variance, and standard deviation for the random variable L.

16. *Combination of Random Variables: Repair Service* A computer repair shop has two work centers. The first center examines the computer to see what is wrong, and the second center repairs the computer. Let x_1 and x_2 be random variables representing the length of time in minutes to examine a computer (x_1) and to repair a computer (x_2). Assume x_1 and x_2 are independent random variables. Long-term history has shown the following times:

Examine computer, $x_1: \mu_1 = 28.1$ minutes; $\sigma_1 = 8.2$ minutes

Repair computer, $x_2: \mu_2 = 90.5$ minutes; $\sigma_2 = 15.2$ minutes

(a) Let $W = x_1 + x_2$ be a random variable representing the total time to examine and repair the computer. Compute the mean, variance, and standard deviation of W.
(b) Suppose it costs $1.50 per minute to examine the computer and $2.75 per minute to repair the computer. Then $W = 1.50x_1 + 2.75x_2$ is a random variable representing the service charges (without parts). Compute the mean, variance, and standard deviation of W.
(c) There is a flat rate of $1.50 per minute to examine the computer, and if no repairs are ordered, there is also an additional $50 service charge. Let $L = 1.5x_1 + 50$. Compute the mean, variance, and standard deviation of L.

17. *Combination of Random Variables: Insurance Risk* Insurance companies know the *risk* of insurance is greatly reduced if we insure not just one person, but many people. How does this work? Let x be a random variable representing the expectation of life in years for a 25-year-old male (i.e., number of years until death). Then the mean and standard deviation of x are $\mu = 50.2$ years, $\sigma = 11.5$ years (Vital Statistics Section of the *Statistical Abstract of the United States*, 116th Edition).

Suppose Big Rock Insurance Company has sold life insurance policies to Joel and David. Both are 25 years old, unrelated, live in different states, and have about the same health record. Let x_1 and x_2 be random variables representing Joel and David's life expectancy. It is reasonable to assume x_1 and x_2 are independent.

$$\text{Joel, } x_1: \mu_1 = 50.2 \qquad \sigma_1 = 11.5$$
$$\text{David, } x_2: \mu_2 = 50.2 \qquad \sigma_2 = 11.5$$

If life expectancy can be predicted with more accuracy, Big Rock will have less risk in its insurance business. Risk in this case is measured by σ (larger σ means more risk).

(a) The average life expectancy for Joel and David is $W = 0.5x_1 + 0.5x_2$. Compute the mean, variance, and standard deviation of W.

(b) Compare the mean life expectancy for a single policy (x_1) with that of two policies (W).

(c) Compare the standard deviation of the life expectancy for a single policy (x_1) with that of two policies (W).

(d) The mean life expectancy is the same for a single policy (x_1) as for two policies (W), but the standard deviation is smaller for two policies. What happens to the mean life expectancy and the standard deviation when we include more policies issued to people whose life expectancies have the same mean and standard deviation (i.e., 25-year-old males)? For instance, for 3 policies, $W = (\mu + \mu + \mu)/3 = \mu$ and $\sigma_W^2 = (1/3)^2\sigma^2 + (1/3)^2\sigma^2 + (1/3)^2\sigma^2 = (1/3)^2(3\sigma^2) = (1/3)\sigma^2$ and $\sigma_W = \dfrac{1}{\sqrt{3}}\sigma$. Likewise, for n such policies, $W = \mu$ and $\sigma_W^2 = (1/n)\sigma^2$ and $\sigma_W = \dfrac{1}{\sqrt{n}}\sigma$. Looking at the general result, is it appropriate to say that when we increase the number of policies to n, the risk decreases by a factor of $\dfrac{1}{\sqrt{n}}$?

5.2
Binomial Probabilities

Binomial Experiment

On a TV quiz show each contestant has a try at the wheel of fortune. The wheel of fortune is a roulette wheel with 36 slots, one of which is gold. If the ball lands in the gold slot, the contestant wins $50,000. No other slot pays. What is the probability that the quiz show will have to pay the fortune to three contestants out of 100?

In this problem, the contestant and the quiz show sponsors are concerned about only two outcomes from the wheel of fortune: The ball lands on the gold, or the ball does not land on the gold. This problem is typical of an entire class of problems that are characterized by the feature that there are exactly two possible outcomes (for each trial) of interest. These problems are called *binomial experiments*, or *Bernoulli experiments*, after the Swiss mathematician Jacob Bernoulli, who studied them extensively in the late 1600s.

Features of a binomial
experiment

> **Features of a binomial experiment**
>
> 1. There are a fixed number of trials. We denote this number by the letter n.
>
> 2. The n trials are independent and repeated under identical conditions.
>
> 3. Each trial has only two outcomes: success, denoted by S, and failure, denoted by F.
>
> 4. For each individual trial, the probability of success is the same. We denote the probability of success by p and that of failure by q. Since each trial results in either success or failure, $p + q = 1$ and $q = 1 - p$.
>
> 5. The central problem of a binomial experiment is to find the probability of r successes out of n trials.

EXAMPLE 4

Binomial experiment

Let's see how the wheel of fortune problem meets the criteria of a binomial experiment. We'll take the criteria one at a time.

1. Each of the 100 contestants has a trial at the wheel, so there are $n = 100$ trials in this problem.

2. Assuming that the wheel is fair, the *trials are independent,* since the result of one spin of the wheel has no effect on the results of other spins.

3. We are interested in only two outcomes on each spin of the wheel: The ball either lands on the gold, or it does not. Let's call landing on the gold *success* (*S*) and not landing on the gold *failure* (*F*). In general, the assignment of the terms *success* and *failure* to outcomes does not imply good or bad results. These terms are assigned simply for the user's convenience.

4. On each trial the probability p of success (landing on the gold) is $1/36$, since there are 36 slots and only one of them is gold. Consequently, the probability of failure is

$$q = 1 - p = 1 - \frac{1}{36} = \frac{35}{36}$$

on each trial.

5. We want to know the probability of 3 successes out of 100 trials, so $r = 3$ in this example. It turns out that the probability the quiz show will have to pay the fortune to 3 contestants out of 100 is about 0.23. Later in this section we'll see how this probability was computed.   ◇

Anytime we make selections from a population *without replacement, we do not have independent trials.* However, replacement is often not practical. If the number of trials is quite small with respect to the population, we *almost* have independent trials, and we can say the situation is *closely approximated* by a binomial experiment. For instance, suppose we select 20 tuition bills at random from a collection of 10,000 bills issued at one college and observe if the bill is in error or not. If 600 of the 10,000 bills are in error, then the probability that the first one selected is in error is 600/10,000, or 0.0600. If the first is in error, then the probability that the second is in error is 599/9999, or 0.0599. Even if the first 19 bills selected are in error, the probability that the 20th is also in error is 581/9981, or 0.0582. All these probabilities round to 0.06, and we can say that the independence condition is approximately satisfied.

GUIDED EXERCISE 4

Binomial experiment

Let's analyze the following binomial experiment to determine p, q, n, and r:

According to the *Textbook of Medical Physiology*, 5th Edition, by Arthur Guyton, 9% of the population has blood type B. Suppose we choose 18 people at random from the population and test the blood type of each. What is the probability that 3 of these people have blood type B? (*Note:* Independence is approximated because 18 people is an extremely small sample with respect to the entire population.)

(a) In this experiment, we are observing whether or not a person has type B blood. We will say we have a success if the person has type B blood. What is failure?

➡ Failure occurs if a person does not have type B blood.

(b) The probability of success is 0.09, since 9% of the population has type B blood. What is the probability of failure, q?

➡ The probability of failure is
$$q = 1 - p$$
$$= 1 - 0.09 = 0.91$$

(c) In this experiment, there are $n = \rule{1cm}{0.4pt}$ trials.

➡ In this experiment, $n = 18$.

(d) We wish to compute the probability of 3 successes out of 18 trials. In this case, $r = \rule{1cm}{0.4pt}$.

➡ In this case, $r = 3$.

Next, we will see how to compute the probability of r successes out of n trials when we have a binomial experiment.

Computing Probabilities for a Binomial Experiment

The central problem of a binomial experiment is to find the probability of r successes out of n trials. Now we'll see how to find these probabilities.

A model with three trials

Suppose you are taking a timed final exam. You have three multiple-choice questions left to do. Each question has four suggested answers, and only one of the answers is correct. You have only 5 seconds left to do these three questions, so you decide to mark answers on the answer sheet without even reading the questions. Assuming that your answers are randomly selected, what is the probability that you get zero, one, two, or all three questions correct?

This is a binomial experiment. Each question can be thought of as a trial, so there are $n = 3$ trials. The possible outcomes on each trial are success S,

TABLE 5-8   Outcomes for a Binomial Experiment with $n = 3$ Trials

Outcome	Probability of Outcome	r (number of successes)
SSS	$P(SSS) = P(S)P(S)P(S) = p^3 = (0.25)^3 \approx 0.016$	3
SSF	$P(SSF) = P(S)P(S)P(F) = p^2q = (0.25)^2(0.75) \approx 0.047$	2
SFS	$P(SFS) = P(S)P(F)P(S) = p^2q = (0.25)^2(0.75) \approx 0.047$	2
FSS	$P(FSS) = P(F)P(S)P(S) = p^2q = (0.25)^2(0.75) \approx 0.047$	2
SFF	$P(SFF) = P(S)P(F)P(F) = pq^2 = (0.25)(0.75)^2 \approx 0.141$	1
FSF	$P(FSF) = P(F)P(S)P(F) = pq^2 = (0.25)(0.75)^2 \approx 0.141$	1
FFS	$P(FFS) = P(F)P(F)P(S) = pq^2 = (0.25)(0.75)^2 \approx 0.141$	1
FFF	$P(FFF) = P(F)P(F)P(F) = q^3 = (0.75)^3 \approx 0.422$	0

indicating a correct response, or failure F, meaning a wrong answer. The trials are independent—the outcome of any one trial does not affect the outcome of the others.

What is the probability of success on any question? Since you are guessing and there are four answers from which to select, the probability of a correct answer is 0.25. The probability q of a wrong answer is then 0.75. In short, we have a binomial experiment with $n = 3$, $p = 0.25$, and $q = 0.75$.

Now what are the possible outcomes in terms of success or failure for these three trials? Let's use the notation SSF to mean success on the first question, success on the second, and failure on the third. There are eight possible combinations of S's and F's. They are

$$SSS \quad SSF \quad SFS \quad FSS \quad SFF \quad FSF \quad FFS \quad FFF$$

To compute the probability of each outcome, we can use the multiplication law because the trials are independent. For instance, the probability of success on the first two questions and failure on the last is

$$P(SSF) = P(S) \cdot P(S) \cdot P(F) = p \cdot p \cdot q = p^2q = (0.25)^2(0.75) = 0.047$$

In a similar fashion, we can compute the probability of each of the eight outcomes. These are shown in Table 5-8, along with the number of successes r associated with each trial.

Now we can compute the probability of r successes out of three trials for $r = 0, 1, 2,$ or 3. Let's compute $P(1)$. The notation $P(1)$ stands for the probability of one success. For three trials, there are three different outcomes that show exactly one success. They are the outcomes SFF, FSF, and FFS. Since the outcomes are mutually exclusive, we can add the probabilities. So

$$P(1) = P(SFF \text{ or } FSF \text{ or } FFS) = P(SFF) + P(FSF) + P(FFS)$$
$$= pq^2 + pq^2 + pq^2$$
$$= 3pq^2$$
$$= 3(0.25)(0.75)^2$$
$$= 0.422$$

In the same way, we can find $P(0)$, $P(2)$, and $P(3)$. These values are shown in Table 5-9.

TABLE 5-9 $P(r)$ for $n = 3$ Trials, $p = 0.25$

r (number of successes)	P(r) (probability of r successes in 3 trials)		P(r) for p = 0.25
0	$P(0) = P(FFF)$	$= q^3$	0.422
1	$P(1) = P(SFF) + P(FSF) + P(FFS)$	$= 3pq^2$	0.422
2	$P(2) = P(SSF) + P(SFS) + P(FSS)$	$= 3p^2q$	0.141
3	$P(3) = P(SSS)$	$= p^3$	0.016

We have done quite a bit of work to determine your chances of $r = 0, 1, 2,$ or 3 successes on three multiple-choice questions if you are just guessing. Now we see that there is only a small chance (about 0.016) that you will get them all correct.

The model we constructed in Table 5-9 to compute the probability of r successes out of three trials can be used for any binomial experiment with $n = 3$ trials. Simply change the values of p and q to fit the experiment. In Guided Exercise 5 we use this model again.

GUIDED EXERCISE 5

Compute P(r)

Maria is doing a study on the issue of the quarter system versus the semester system. To obtain faculty input, she mails out questionnaires to the faculty. The probability that a faculty member returns the completed questionnaire is 0.65. Three faculty members chosen at random from the foreign language department are sent questionnaires. Compute the probability that *exactly two* completed questionnaires are returned and the probability that *all three* are returned. We'll do these computations in steps.

(a) In this problem, what are the values of n, p, q?

$\implies$ $n = 3$, $p = 0.65$, $q = 1 - 0.65 = 0.35$

(b) The probability that exactly two questionnaires will be returned is $P(\underline{})$. In this case, $r = \underline{}$. By Table 5-9,

$P(2) = 3p^2q$ for $n = 3$ trials

Use this formula to compute $P(2)$.

$\implies$ We want $P(2)$, so $r = 2$:

$P(2) = 3(0.65)^2(0.35) = 0.444$

(c) Use the appropriate formula from Table 5-9 to compute the probability that all three questionnaires will be returned.

$\implies$ $P(3) = p^3 = (0.65)^3 = 0.275$

Table 5-9 can be used as a model for computing the probability of r successes out of only *three* trials. How can we compute the probability of 7 successes out of 10 trials? We can develop a table for $n = 10$, but this would be a tremendous task because there are 1024 possible combinations of successes and failures on 10 trials.

Fortunately, mathematicians have given us a direct formula to compute the probability of r successes for any number of trials.

General formula for binomial probability distribution

Formula for the binomial probability distribution

$$P(r) = C_{n,r}p^r q^{n-r}$$

where $C_{n,r} = \dfrac{n!}{r!(n-r)!}$ is the *binomial coefficient*. Values of $C_{n,r}$ for various values of n and $r = 0, 1, 2, \ldots, n$ can be found in Table 2 of Appendix II.

Those of you who studied the section on counting techniques (Section 4.3) will recognize the symbol $C_{n,r}$ as the symbol used for the number of combinations of n objects taken r at a time.

Table for $C_{n,r}$

Table 2 of Appendix II gives the values of the binomial coefficient $C_{n,r}$ for selected values of n and r. However, you can compute $C_{n,r}$ directly from a formula or on your calculator. (See Section 4.3 for the formula for $C_{n,r}$ and examples showing how to use the formula.)

In the meantime, let's look more carefully at the formula itself. There are two main parts. The expression $p^r q^{n-r}$ is the probability of getting one outcome with r successes and $n - r$ failures. The binomial coefficient $C_{n,r}$ counts the number of outcomes that have r successes and $n - r$ failures. For instance, in the case of $n = 3$ trials, we saw in Table 5-8 that the probability of getting an outcome with one success and two failures was pq^2. This is the value of $p^r q^{n-r}$ when $r = 1$ and $n = 3$. We also observed that there were three outcomes with one success and two failures, so $C_{3,1}$ is 3.

Now let's take a look at an application of the binomial distribution formula in Example 5.

EXAMPLE 5

Compute P(r)

Privacy is a concern for many users of the Internet. One survey showed that 59% of Internet users are somewhat concerned about the confidentiality of their e-mail. Based on this information, what is the probability that for a random sample of 10 Internet users, 6 are concerned about the privacy of their e-mail?

SOLUTION:

(a) This is a binomial experiment with 10 trials. If we assign success to an Internet user being concerned about the privacy of e-mail, the probability of success is 59%. We are interested in the probability of 6 successes. We have

$$n = 10 \qquad p = 0.59 \qquad q = 0.41 \qquad r = 6$$

By the formula,

$$P(6) = C_{10,6}(0.59)^6(0.41)^{10-6}$$

$$= 210(0.59)^6(0.41)^4 \qquad \text{Use Table 2 of Appendix II or a calculator.}$$

$$= 210(0.0422)(0.0283) \qquad \text{Use a calculator.}$$

$$= 0.25$$

There is a 25% chance that *exactly* 6 of the 10 Internet users are concerned about the privacy of e-mail.

FIGURE 5-2

T1-83Plus Display

```
10 nCr 6*.59^6*.
41^(10-6)
        .250303245
```

(b) Many calculators have a built-in combinations function. On the TI-83Plus you can find the combinations function under the math menu. It is designated nCr. Fig. 5-2 displays the process for computing $P(6)$ directly on the TI-83Plus.   ◊

In many cases, we will be interested in the probability of a range of successes rather than in the probability of an exact number of successes. For instance, we might wish to compute the probability that *at least* 6 of the 10 Internet users have concerns about the privacy of e-mail. In such a case, we need to use the addition rule for mutually exclusive events.

$$P(at\ least\ 6\ successes) = P(r \geq 6)$$
$$= P(r = 6\ or\ 7\ or\ 8\ or\ 9\ or\ 10)$$
$$= P(6) + P(7) + P(8) + P(9) + P(10)$$

Table for $P(r)$

It would be quite a task to compute all the required probabilities by using the formula. Table 3 of Appendix II gives values for $P(r)$ for selected p and values of n through 20. To use the table, find the section labeled with your value of n. Then find the entry in the column headed by your value of p and the row labeled by the r value of interest.

Table 3 of Appendix II has only a limited selection of values for p. In fact, for Example 5 the probability of success, $p = 0.59$, is not available in the table. However, for other problems in this text, you will find the specified value of p. Example 6 demonstrates the use of Table 3 (Appendix II) to find binomial probabilities.

EXAMPLE 6

Using the binomial distribution table to find $P(r)$

A biologist is studying a new hybrid tomato. It is known that the seeds of this hybrid tomato have probability 0.70 of germinating. The biologist plants 10 seeds.

(a) What is the probability that *exactly* 8 seeds will germinate?

SOLUTION: This is a binomial experiment with $n = 10$ trials. Each seed planted represents an independent trial. We'll say germination is success, so the probability for success on each trial is 0.70.

$$n = 10 \qquad p = 0.70 \qquad q = 0.30 \qquad r = 8$$

We wish to find $P(8)$, the probability of exactly eight successes.

In Table 3, Appendix II, find the section with $n = 10$. Then find the entry in the column headed by $p = 0.70$ and the row headed by the r value 8. This entry is 0.233.

$$P(8) = 0.233$$

(b) What is the probability that *at least* 8 seeds will germinate?

SOLUTION: In this case, we are interested in the probability of 8 or more seeds germinating. This means we are to compute $P(r \geq 8)$. Since the events are mutually exclusive, we can use the addition rule

$$P(r \geq 8) = P(r = 8 \quad or \quad r = 9 \quad or \quad r = 10) = P(8) + P(9) + P(10)$$

We already know the value of $P(8)$. We need to find $P(9)$ and $P(10)$.

Use the same part of the table but find the entries in the row headed by the r value 9 and then the r value 10. Be sure to use the column headed by the value of p, 0.70.

$$P(9) = 0.121 \quad and \quad P(10) = 0.028$$

Now we have all the parts necessary to compute $P(r \geq 8)$.

$$
\begin{aligned}
P(r \geq 8) &= P(8) + P(9) + P(10) \\
&= 0.233 + 0.121 + 0.028 \\
&= 0.382
\end{aligned}
$$

In Guided Exercise 6 you'll practice using the formula for $P(r)$ in one part and then use Table 3 (Appendix II) for $P(r)$ values in the second part.

GUIDED EXERCISE 6

Find P(r)

A rarely performed and somewhat risky eye operation is known to be successful in restoring the eyesight of 30% of the patients who undergo the operation. A team of surgeons has developed a new technique for this operation that has been successful for four of six operations. Does it seem likely that the new technique is much better than the old? We'll use the binomial probability distribution to answer this question. We'll compute the probability of at least four successes in six trials for the old technique.

(a) Each operation is a binomial trial. In this case,
$n = \underline{\hspace{1cm}}, p = \underline{\hspace{1cm}}, q = \underline{\hspace{1cm}}, r = \underline{\hspace{1cm}}.$

➡ $n = 6, p = 0.30, q = 1 - 0.30 = 0.70, r = 4$

(b) Use your values of n, p, and q, as well as Table 2 of Appendix II (or your calculator) to compute $P(4)$ from the formula:

$$P(r) = C_{n,r}p^r q^{n-r}$$

➡ $$
\begin{aligned}
P(4) &= C_{6,4}(0.30)^4(0.70)^2 \\
&= 15(0.0081)(0.490) \\
&= 0.060
\end{aligned}
$$

(c) Compute the probability of *at least* four successes out of the six trials.

$$
\begin{aligned}
P(r \geq 4) &= P(r = 4 \; or \; r = 5 \; or \; r = 6) \\
&= P(4) + P(5) + P(6)
\end{aligned}
$$

Use Table 3 of Appendix II to find values of $P(4)$, $P(5)$, and $P(6)$. Then use these values to compute $P(r \geq 4)$.

➡ To find $P(4)$, $P(5)$, and $P(6)$ in Table 3, we look in the section labeled $n = 6$. Then we find the column headed by $p = 0.30$. To find $P(4)$, we use the row labeled $r = 4$. For the values of $P(5)$ and $P(6)$, use the same column but change the row headers to $r = 5$ and $r = 6$, respectively.

$$
\begin{aligned}
P(r \geq 4) &= P(4) + P(5) + P(6) \\
&= 0.060 + 0.010 + 0.001 = 0.071
\end{aligned}
$$

Continued

GUIDED EXERCISE 6 continued

(d) Under the older operation technique, the probability that at least four patients out of six regain their eyesight is _____. Does it seem that the new technique is better than the old? Would you encourage the surgeon team to do more work on the new technique?

 It seems the new technique is better than the old, since, by pure chance, the probability of four or more successes out of six trials is only 0.071 for the old technique. This means one of the following two things may be happening:

(i) The new method is no better than the old method, and our surgeons have encountered a rare event (probability 0.071), or

(ii) The new method is in fact better. We think it is worth encouraging the surgeons to do more work on the new technique.

Common expressions and corresponding inequalities

Many times we are asked to compute the probability of a range of successes. For instance, in a binomial experiment with n trials, we may be asked to compute the probability of four or more successes. Table 5-10 shows how common English expressions such as "four or more successes" translate to inequalities involving r.

TABLE 5-10   Common English Expressions and Corresponding Inequalities (consider a binomial experiment with n trials and r successes)

Expression	Inequality
Four or more successes At least four successes No fewer than four successes Not less than four successes	$r \geq 4$ That is, $r = 4, 5, 6, \ldots, n$
Four or fewer successes At most four successes No more than four successes The number of successes does not exceed four	$r \leq 4$ That is, $r = 0, 1, 2, 3,$ or 4
More than four successes The number of successes exceeds four	$r > 4$ That is, $r = 5, 6, 7, \ldots, n$
Fewer than four successes The number of successes is not as large as four	$r < 4$ That is, $r = 0, 1, 2, 3$

TECH NOTE The software packages Minitab and Excel as well as the TI-83Plus calculator include built-in binomial probability distribution options. These options give the probability $P(r)$ of a specific number of successes r, as well as the cumulative total probability for r or fewer successes.

TI-83Plus Press **DIST** key, scroll to **binompdf**(n, p, r). Enter the number of trials n, the probability of success on a single trial p, and the number of successes r. This gives $P(r)$. For the cumulative probability that there are r or fewer successes, use **binomcdf**(n, p, r).

$P(r) = 4$

$P(r \leq 4)$

```
binompdf(6,.3,4)

              .059535
binomcdf(6,.3,4)

              .989065
```

Excel Menu Choice: **Paste Function** ☐ f_x ➤ **Statistical** ➤ **Binomdist** In the dialogue box, fill in the values r, n, and p. For $P(r)$, use false; and for P(at least r successes), use true.

Minitab First, enter the r values 0, 1, 2, . . . , n in a column. Then use menu choice: **Calc** ➤ **Probability Distribution** ➤ **Binomial** In the dialogue box, select Probability for $P(r)$ or Cumulative for P(at least r successes). Enter the number of trials n, the probability of success p, and the column containing the r values. A sample print-out is shown in Problem 17 at the end of this section.

VIEWP●INT

Lies! Lies!! Lies!!! The Psychology of Deceit

This is the title of an intriguing book by C. V. Ford, professor of psychiatry. The book recounts the true story of Floyd "Buzz" Fay, who was falsely convicted of murder on the basis of a failed polygraph examination. During his $2\frac{1}{2}$ years of wrongful imprisonment, Buzz became a polygraph expert. He taught inmates, who freely confessed guilt, how to pass a polygraph examination. (For more information on this topic, *see* Problem 7.)

SECTION 5.2   PROBLEMS

In each of the following problems, the binomial distribution will be used. Please answer the following questions and then complete the problem.
(a) What makes up a trial? What is a success? What is a failure?
(b) What are the values of n, p, and q?

1. *Binomial Probabilities: Coin Flip* A fair quarter is flipped three times. For each of the following probabilities, use the formula for the binomial distribution and a

calculator to compute the requested probability. Next, look up the probability in Table 3 of Appendix II and compare the table result with the computed result.
(a) Find the probability of getting exactly three heads.
(b) Find the probability of getting exactly two heads.
(c) Find the probability of getting two or more heads.
(d) Find the probability of getting exactly three tails.

2. *Binomial Probabilities: Multiple-Choice Quiz* Richard has just been given a 10-question multiple-choice quiz in his history class. Each question has 5 answers, of which only one is correct. Since Richard has not attended class recently, he doesn't know any of the answers. Assuming that Richard guesses on all 10 questions, find the indicated probabilities.
   (a) What is the probability that he will answer all questions correctly?
   (b) What is the probability that he will answer all questions incorrectly?
   (c) What is the probability that he will answer at least one of the questions correctly? Compute this probability two ways. First, use the rule for mutually exclusive events and the probabilities shown in Table 3 of Appendix II. Then use the fact that $P(r \geq 1) = 1 - P(r = 0)$. Compare the two results. Should they be equal? Are they equal? If not, how do you account for the difference?
   (d) What is the probability that Richard will answer at least half the questions correctly?

3. *Sociology: Marriage* The percentage of American men who say they would marry the same woman if they had it to do all over again is 80%. The percentage of American women who say they would marry the same man again is 50% (Source: *Harper's Index*).
   (a) What is the probability that in a group of 10 married men, at least 7 will claim that they would marry the same woman again? What is the probability that less than half will say this?
   (b) What is the probability that in a group of 10 married women, at least 7 will claim they would marry the same man again? What is the probability that less than half will say this?

4. *Sociology: Ethics* The one-time fling! Have you ever purchased an article of clothing (dress, sports jacket, etc.), worn the item *once* to a party, and then returned the purchase? This is called a *one-time fling*. About 10% of all adults deliberately do a one-time fling and feel no guilt about it! (Source: *Are You Normal?*, by Bernice Kanner, St. Martin's Press.) In a group of seven adult friends, what is the probability that
   (a) no one has done a one-time fling?
   (b) at least one person has done a one-time fling?
   (c) no more than two people have done a one-time fling?

5. *Sociology: Mother-in-Law* The ★#@&#★ mother-in-law! Sociologists say that 90% of married women claim that their husband's mother is the biggest bone of contention in their marriages (sex and money are lower-rated areas of contention). (See the source in Problem 4.) Suppose that six married women are having coffee together one morning. What is the probability that
   (a) all of them dislike their mother-in-law?
   (b) none of them dislike their mother-in-law?
   (c) at least four of them dislike their mother-in-law?
   (d) no more than three of them dislike their mother-in-law?

6. *Sociology: Dress Habits* A research team at Cornell University conducted a study showing that approximately 10% of all businessmen who wear ties wear them so tight that they actually reduce blood flow to the brain, diminishing cerebral functions (Source: *Chances: Risk and Odds in Everyday Life*, by James Burke). At a board meeting of 20 businessmen, all of whom wear ties, what is the probability that

(a) at least one tie is too tight?
(b) more than two ties are too tight?
(c) none are too tight?
(d) at least 18 are *not* too tight?

7. *Psychology: Deceit* Aldrich Ames is a convicted traitor who leaked American secrets to a foreign power. Yet Ames took routine lie detector tests and each time passed them. How can this be done? Recognizing control questions, employing unusual breathing patterns, biting one's tongue at the right time, pressing one's toes hard to the floor, and counting backwards by 7 are countermeasures that are difficult to detect but can change the results of a polygraph examination (Source: *Lies! Lies!! Lies!!! The Psychology of Deceit,* by C. V. Ford, professor of psychiatry, University of Alabama.) In fact, it is reported in Professor Ford's book that after only 20 minutes of instruction by "Buzz" Fay (a prison inmate), 85% of those trained were able to pass the polygraph examination even when guilty of a crime. Suppose that a random sample of nine students (in a psychology laboratory) are told a "secret" and then given instructions on how to pass the polygraph examination without revealing their knowledge of the secret. What is the probability that
(a) all the students are able to pass the polygraph examination?
(b) more than half the students are able to pass the polygraph examination?
(c) no more than four of the students are able to pass the polygraph examination?
(d) all the students fail the polygraph examination?

8. *Survey: Cellular Phones* According to an article appearing in *The Wall Street Journal,* about 35% of all U.S. households have a cellular phone. Suppose that you are conducting a survey of customer satisfaction regarding their cellular phone. If you called 11 households selected at random, what is the probability that
(a) every household has a cellular phone?
(b) more than four households have a cellular phone?
(c) fewer than 5 do not have a cellular phone?
(d) more than 7 do not have a cellular phone?

9. *Restaurants: Income* After examining daily receipts over the past year, it was found that the Green Parrot Italian Restaurant has been grossing over $2200 a day for about 85% of its business days. Using this as a reasonably accurate measure, find the probability that the Green Parrot will gross over $2200
(a) at least 5 days in the next 7 business days.
(b) at least 5 days in the next 10 business days.
(c) fewer than 3 days in the next 5 business days.
(d) fewer than 7 days in the next 10 business days.
(e) fewer than 3 days in the next 7 business days. If this actually happened, might it shake your confidence in the statement $p = 0.85$? Might you suspect that p is less than 0.85? Explain.

10. *Hardware Store: Income* Trevor is interested in purchasing the local hardware/sporting goods store in the small town of Dove Creek, Montana. After examining accounting records for the past several years, he found that the store has been grossing over $850 per day about 60% of the business days it is open. Estimate the probability the store will gross over $850
(a) at least 3 out of 5 business days.
(b) at least 6 out of 10 business days.
(c) fewer than 5 out of 10 business days.
(d) fewer than 6 out of the next 20 business days. If this actually happened, might it shake a person's confidence in the statement $p = 0.60$? Might it make a person suspect that p is less than 0.60? Explain.

(e) more than 17 out of the next 20 business days. If this actually happened, might a person suspect that p is greater than 0.60? Explain.

11. *Fishing: Northern Pike* Manitoba northern pike are hardy, tough fish! Using artificial lures with barbed treble hooks, it was found that the hooking mortality rate was only about 5%. This means that only 5% of pike that were caught and released died. (Source: *Proceedings of National Symposium on Catch and Release Fishing,* sponsored by Humboldt State University.) Suppose that a group of anglers caught and released 16 northern pike in Manitoba. What is the probability that
(a) none of the fish died?
(b) less than 3 of the fish died?
(c) all the fish lived?
(d) more than 14 fish lived?

12. *Marketing: Coffee Tasting* The Tasty Bean Coffee Company claims that its coffee is so good that you can distinguish it from any other coffee. Five different brands of coffee (one of them Tasty Bean) are set before tasters who are to pick the one that tastes the best. Suppose that there is really no difference in the way any of these coffees taste; however, each of four tasters picks one coffee anyway (not knowing which is which, because the coffee is in identical cups). What is the probability that
(a) all four tasters choose Tasty Bean?
(b) none of them chooses Tasty Bean?
(c) at least three choose Tasty Bean?

13. *Psychology: Myers-Briggs* Approximately 75% of all marketing personnel are extroverts, whereas about 60% of all computer programmers are introverts (Source: *A Guide to the Development and Use of the Myers-Briggs Type Indicator,* by Myers and McCaulley).
(a) At a meeting of 15 marketing personnel, what is the probability that 10 or more are extroverts? What is the probability that 5 or more are extroverts? What is the probability that all are extroverts?
(b) In a group of 5 computer programmers, what is the probability that none are introverts? What is the probability that 3 or more are introverts? What is the probability that all are introverts?

14. *Sociology: Dating* In *Chances: Risks and Odds in Everyday Life,* James Burke claims that about 70% of all single men would welcome a woman taking the initiative in asking for a date. A random sample of 20 single men was asked if they would welcome a woman taking the initiative in asking for a date. What is the probability that
(a) at least 18 of the men will say yes?
(b) fewer than 3 of the men will say yes?
(c) none of the men will say yes?
(d) at least 5 of the men will say no?

15. *Health Care: Diabetes* People with diabetes may develop other health complications associated with the disease. The following information is based on a feature in *USA Today* entitled "A Look at Statistics That Shape Our Lives." About 40% of all people with diabetes will also develop hypertension (blood pressure problems), and about 30% of people with diabetes will develop an eye disease. Suppose that you are the director of a health care center that has 10 people with diabetes and no other related health problems. Part of your duties is to monitor these patients for symptoms of new illnesses related to diabetes so that corrective measures can be started. What is the probability that
(a) none of the diabetes patients will ever develop related hypertension?
(b) fewer than 5 of the diabetes patients will ever develop related hypertension?

(c) no more than 2 of the diabetes patients will ever develop a related eye disease?

(d) at least 6 of the diabetes patients will never develop a related eye disease?

16. *Business Ethics: Privacy* Are your finances, buying habits, medical records, and phone calls really private? A real concern for many adults is that computers and the Internet are reducing privacy. A survey conducted by Peter D. Hart Research Associates for the Shell Poll was reported in *USA Today*. According to the survey, 37% of the adults are concerned that employers are monitoring phone calls. Use the binomial distribution formula to calculate the probability that

(a) out of 5 adults, none is concerned that employers are monitoring phone calls.

(b) out of 5 adults, all are concerned that employers are monitoring phone calls.

(c) out of 5 adults, exactly 3 are concerned that employers are monitoring phone calls.

17. *Business Ethics: Privacy* According to the same poll quoted in Problem 16, 53% of the adults are concerned that Social Security numbers are used for general identification. For a group of 8 adults selected at random, we used Minitab to generate the binomial probability distribution and the cumulative binomial probability distribution (menu selections ➤ **Calc** ➤ **Probability Distributions** ➤ **Binomial**).

Number	r	P(r)	P(<=r)
	0	0.002381	0.00238
	1	0.021481	0.02386
	2	0.084781	0.10864
	3	0.191208	0.29985
	4	0.269521	0.56937
	5	0.243143	0.81251
	6	0.137091	0.94960
	7	0.044169	0.99377
	8	0.006226	1.00000

Find the probability that out of 8 adults selected at random,

(a) at most 5 are concerned about Social Security numbers being used for identification. Do the problem by adding the probabilities $P(r = 0)$ through $P(r = 5)$. Is this the same as the cumulative probability $P(r <= 5)$?

(b) more than 5 are concerned about Social Security numbers being used for identification. First, do the problem by adding the probabilities $P(r = 6)$ through $P(r = 8)$. Then do the problem by subtracting the cumulative probability $P(r <= 5)$ from 1. Do you get the same results?

18. *Ecology: Wolves* The following is based on information taken from *The Wolf in the Southwest: The Making of an Endangered Species,* edited by David Brown (University of Arizona Press). Before 1918, approximately 55% of the wolves in the New Mexico and Arizona region were male, and 45% were female. However, cattle ranchers in this area have made a determined effort to exterminate wolves. From 1918 to the present, approximately 70% of wolves in the region are male, and 30% are female. Biologists suspect that male wolves are more likely than females to return to an area where the population has been greatly reduced.

(a) Before 1918, in a random sample of 12 wolves spotted in the region, what is the probability that 6 or more were male? What is the probability that 6 or more were female? What is the probability that fewer than 4 were female?

(b) Answer part a for the period from 1918 to the present.

19. *Archaeology: Stone Tools* The following is based on information from *Bandelier Archaeological Excavation Project: Summer 1989 Excavations at Burnt Mesa Pueblo,* edited by Kohler (Washington State University). At an archaeological site in Bandelier National Monument, approximately 15% of the chipped stone tools are made from Jemez obsidian, and 55% are made from basalt.
   (a) If 11 chipped stone tools are discovered, what is the probability that at least 3 will be made from Jemez obsidian?
   (b) If 5 chipped stone tools are discovered, what is the probability that at least 2 will be made from basalt?
   (c) If 10 chipped stone tools are discovered, what is the probability that at least 4 are neither Jemez obsidian nor basalt?

20. *Health Care: Office Visits* What is the age distribution of patients who make office visits to a doctor or nurse? The following table is based on information taken from the Medical Practice Characteristics section of the *Statistical Abstract of the United States* (116th Edition).

Age group, years	Under 15	15–24	25–44	45–64	65 and older
Percent of office visitors	20%	10%	25%	20%	25%

Suppose that you are a district manager of a health management organization (HMO) who is monitoring the office of a local doctor or nurse in general family practice. This morning the office you are monitoring has eight office visits on the schedule. What is the probability that
   (a) at least half the patients are under 15 years old? First, explain how this can be modeled as a binomial distribution with 8 trials, where success is visitor age is under 15 years old and the probability of success is 20%.
   (b) from 2 to 5 patients are 65 or older (include 2 and 5)?
   (c) from 2 to 5 patients are 45 or older (include 2 and 5)? (*Hint:* Success is 45 or older. Use the table to compute the probability of success on a single trial.)
   (d) all the patients are under 25 years of age?
   (e) all the patients are 15 or older?

21. *Binomial Distribution Table: Symmetry* Study the binomial distribution table (Table 3, Appendix II). Notice that the probability of success on a single trial p ranges from 0.01 to 0.95. Some binomial distribution tables stop at 0.50 because of the symmetry in the table. Let's look for that symmetry. Consider the section of the table with $n = 5$. Look at the numbers in the columns headed by $p = 0.30$ and $p = 0.70$. Do you detect any similarities? Consider the following probabilities for a binomial experiment with five trials.
   (a) Compare $P(3$ successes$)$ where $p = 0.30$ with $P(2$ successes$)$ where $p = 0.70$.
   (b) Compare $P(3$ or more successes$)$ where $p = 0.30$ with $P(2$ or fewer successes$)$ with $p = 0.70$.
   (c) Find the value of $P(4$ successes$)$ with $p = 0.30$. For what value of r is $P(r$ successes$)$ the same using $p = 0.70$?
   (d) What column is symmetrical with the one headed by $p = 0.20$?

22. *Binomial Distribution: Control Charts* This problem will be referred to in the study of control charts (Section 6.1). In the binomial probability distribution, let the number of trials be $n = 3$, and let the probability of success be $p = 0.0228$. Use a calculator to compute
   (a) the probability of two successes.
   (b) the probability of three successes.
   (c) the probability of two or three successes.

5.3
Additional Properties of the Binomial Distribution

FOCUS POINTS

✓ Make histograms for binomial distributions.

✓ Compute μ and σ for a binomial distribution.

✓ Compute minimal number of trials n to achieve a given probability of success $P(r)$.

Any probability distribution may be represented in graphic form. How should we graph the binomial distribution? Remember, the binomial distribution tells us the probability of r successes out of n trials. Therefore, we'll place values of r along the horizontal axis and values of $P(r)$ on the vertical axis. The binomial distribution is a *discrete* probability distribution because r can assume only whole-number values such as 0, 1, 2, 3, ... Therefore, a histogram is an appropriate graph of a binomial distribution. Let's look at an example to see exactly how we'll make these histograms.

EXAMPLE 7

Graph of a binomial distribution

A waiter at the Green Spot Restaurant has learned from long experience that the probability that a lone diner will leave a tip is only 0.7. During one lunch hour he serves six people who are dining by themselves. Make a graph of the binomial probability distribution that shows the probabilities that 0, 1, 2, 3, 4, 5, or all 6 lone diners leave tips.

SOLUTION: This is a binomial experiment with $n = 6$ trials. Success is achieved when the lone diner leaves a tip, so the probability of success is 0.7 and that of failure is 0.3:

$$n = 6 \qquad p = 0.7 \qquad q = 0.3$$

We want to make a histogram showing the probability of r successes when $r = 0, 1, 2, 3, 4, 5$, or 6. It is easier to make the histogram if we first make a table of r values and the corresponding $P(r)$ values (Table 5-11). We'll use Table 3 of Appendix II to find the $P(r)$ values for $n = 6$ and $p = 0.70$.

To construct the histogram, we'll put r values on the horizontal axis and $P(r)$ values on the vertical axis. Our bars will be one unit wide and will be centered over the appropriate r value. The height of the bar over a particular r value tells the probability of that r (see Figure 5-3).

FIGURE 5-3

Graph of the Binomial Distribution for $n = 6$ and $p = 0.7$

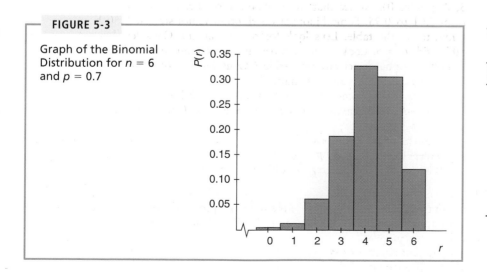

TABLE 5-11   Binomial Distribution for $n = 6$ and $p = 0.70$

r	$P(r)$
0	0.001
1	0.010
2	0.060
3	0.185
4	0.324
5	0.303
6	0.118

The probability of a particular value of r is given not only by the height of the bar over that r value but also by the *area* of the bar. Each bar is only one unit wide, so the area (area = height times width) equals its height. Since the area of each bar represents the probability of the r value under it, the sum of the areas of the bars must be 1. In this example, the sum turns out to be 1.001. It is not exactly equal to 1 because of rounding error.   ◇

Guided Exercise 7 illustrates another binomial distribution with $n = 6$ trials. The graph will be different from that of Figure 5-3 because the probability of success p is different.

In Example 7 and Guided Exercise 7, we see the graphs of two binomial distributions associated with $n = 6$ trials. The two graphs are different because the

GUIDED EXERCISE 7

Graph of a binomial distribution

Jim enjoys playing basketball. He figures that he makes about 50% of the field goals he attempts during a game. Make a histogram showing the probability that Jim will make 0, 1, 2, 3, 4, 5, or 6 shots out of six attempted field goals.

(a) This is a binomial experiment with $n =$ _____ trials. In this situation, we'll say success occurs when Jim makes an attempted field goal. What is the value of p?

⟹ In this example, $n = 6$ and $p = 0.5$.

(b) Use Table 3 of Appendix II to complete Table 5-12 of $P(r)$ values for $n = 6$ and $p = 0.5$.

TABLE 5-12

r	$P(r)$
0	0.016
1	0.094
2	0.234
3	_____
4	_____
5	_____
6	_____

⟹ TABLE 5-13   Completion of Table 5-12

r	$P(r)$
·	·
·	·
·	·
3	0.312
4	0.234
5	0.094
6	0.016

(c) Use the values of $P(r)$ given in Table 5-13 to complete the histogram in Figure 5-4.

Continued

GUIDED EXERCISE 7 continued

FIGURE 5-4 Beginning of Graph of Binomial Distribution for $n = 6$ and $p = 0.5$

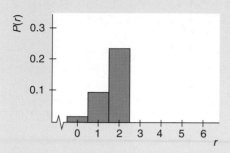

FIGURE 5-5 Completion of Figure 5-4

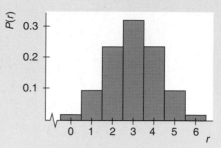

(d) The area of the bar over $r = 2$ is 0.234. What is the area of the bar over $r = 4$? How does the probability that Jim makes exactly two field goals out of six compare with the probability that he makes exactly four field goals out of six?

The area of the bar over $r = 4$ is also 0.234. Jim is as likely to make two out of six field goals attempted as he is to make four out of six.

probability of success p is different in the two cases. In Example 7, $p = 0.7$ and the graph is skewed to the left—that is, the left tail is longer. In Guided Exercise 7, p is equal to 0.5 and the graph is symmetrical—that is, if we fold it in half, the two halves coincide exactly. Whenever p equals 0.5, *the graph of the binomial distribution will be symmetrical no matter how many trials we have.* In Chapter 6, we will see that if the number of trials n is quite large, the binomial distribution is almost symmetrical even when p is not close to 0.5.

Mean and standard deviation of binomial probability distributions

Two other features that help describe the graph of any distribution are the balance point of the distribution and the spread of the distribution about that balance point. The *balance point* is the mean μ of the distribution, and the *measure of spread* that is most commonly used is the standard deviation σ. The mean μ is the *expected value* of the number of successes.

For the binomial distribution, we can use two special formulas to compute the mean μ and the standard deviation σ. These are easier to use than the general formulas in Section 5.1 for μ and σ of any discrete probability distribution.

For the binomial distribution

$\mu = np$ is the expected number of successes r

$\sigma = \sqrt{npq}$ is the standard deviation for the number of successes r

where n is the number of trials,
    p is the probability of success,
    q is the probability of failure ($q = 1 - p$).

EXAMPLE 8

Compute μ and σ

Let's compute the mean and standard deviation for the distribution of Example 7 that describes that probabilities of lone diners leaving tips at the Green Spot Restaurant.

SOLUTION: In that example,

$$n = 6 \qquad p = 0.7 \qquad q = 0.3$$

For the binomial distribution,

$$\mu = np$$
$$= 6(0.7) = 4.2$$

The balance point of the distribution is at $\mu = 4.2$. The standard deviation is given by

$$\sigma = \sqrt{npq}$$
$$= \sqrt{6(0.7)(0.3)}$$
$$= \sqrt{1.26}$$
$$\approx 1.12$$

The mean μ is not only the balance point of the distribution; it is also the *expected value* of r. Specifically, in Example 7, the waiter can expect 4.2 lone diners out of 6 to leave a tip. (The waiter would probably round the expected value to 4 tippers out of 6.)

GUIDED EXERCISE 8

Expected value and standard deviation

When Jim (of Guided Exercise 7) shoots field goals in basketball games, the probability that he makes a shot is only 0.5.

(a) The mean of the binomial distribution is the expected value of r successes out of n trials. Out of six throws, what is the expected number of goals Jim will make?

➡ The expected value is the mean μ:

$$\mu = np = 6(0.5) = 3$$

Jim can expect to make three goals out of six tries.

(b) For six trials, what is the standard deviation of the binomial distribution of the number of successful field goals Jim makes?

➡ $\sigma = \sqrt{npq} = \sqrt{6(0.5)(0.5)} = \sqrt{1.5} = 1.22$

Finding minimal value of n for given $P(r)$

In applications, you do not want to confuse the expected value of r with certain probabilities associated with r. Guided Exercise 9 illustrates this point.

Find the minimum value of n for given P(r)

A satellite requires three solar cells for its power. The probability that any one of these cells will fail is 0.15, and the cells operate or fail independently.

Part I: In this part, we want to find the least number of cells the satellite should have so that the *expected value* of the number of working cells is no smaller than three. In this situation, n represents the number of cells, r is the number of successful or working cells, p is the probability that a cell will work, q is the probability that a cell will fail, and μ is the expected value that should be no smaller than three.

(a) What is the value of q? of p?

⟹ $q = 0.15$ as given in the problem. p must be 0.85, since $p = 1 - q$.

(b) The expected value μ for the number of working cells is given by $\mu = np$. The expected value of the number of working cells should be no smaller than three, so

$$3 \leq \mu = np$$

From part (a), we know the value of p. Solve the inequality $3 \leq np$ for n.

⟹ $3 \leq np$

$3 \leq n(0.85)$

$\dfrac{3}{0.85} \leq n$ Divide both sides by 0.85.

$3.53 \leq n$

(c) Since n is between 3 and 4, would you round it to 3 or 4 to make sure that μ is at least 3?

⟹ n should be at least 3.53. Since we can't have a fraction of a cell, we had best make $n = 4$. For $n = 4$, $\mu = 4(0.85) = 3.4$. This value satisfies the condition that μ be at least 3.

Part II: In this part, we want to find the smallest number of cells the satellite should have to be 97% sure that there will be adequate power—that is, that at least three cells work.

(a) The letter r has been used to denote the number of successes. In this case, r represents the number of working cells. We are trying to find the number n of cells necessary to ensure that (choose the correct statement)

(i) $P(r \geq 3) = 0.97$ or

(ii) $P(r \leq 3) = 0.97$

⟹ $P(r \geq 3) = 0.97$

(b) We need to find a value for n so that

$$P(r \geq 3) = 0.97$$

Try $n = 4$. Then $r \geq 3$ means $r = 3$ or 4 so

$$P(r \geq 3) = P(3) + P(4)$$

Use Table 3 (Appendix II) with $n = 4$ and $p = 0.85$ to find values of $P(3)$ and $P(4)$. Then compute $P(r \geq 3)$ for $n = 4$. Will $n = 4$ guarantee that $P(r \geq 3)$ is at least 0.97?

⟹ $P(3) = 0.368$

$P(4) = 0.522$

$P(r \geq 3) = 0.368 + 0.522 = 0.890$

Thus $n = 4$ is *not* sufficient to be 97% sure that at least three cells will work. For $n = 4$, the probability that at least three will work is only 0.890.

Continued

GUIDED EXERCISE 9 continued

(c) Now try $n = 5$ cells. For $n = 5$,

$$P(r \geq 3) = P(3) + P(4) + P(5)$$

since r can be 3, 4, or 5. Are $n = 5$ cells adequate? [Be sure to find new values of $P(3)$ and $P(4)$ since we now have $n = 5$.]

$$P(r \geq 3) = P(3) + P(4) + P(5)$$
$$= 0.138 + 0.392 + 0.444$$
$$= 0.974$$

Thus $n = 5$ cells are required if we want to be 97% sure that there will be at least three working cells.

In part I and part II, we got different values for n. Why? In part I, we had $n = 4$ and $\mu = 3.4$. This means that if we put up lots of satellites with four cells, we can expect that an *average* of 3.4 cells will be working per satellite. But for $n = 4$ cells there is a probability of only 0.89 that at least three cells will work in any one satellite. In part II, we are trying to find the number of cells necessary so that the probability is 0.97 that at least three will work in any *one* satellite. If we use $n = 5$ cells, then we can satisfy this requirement.

Quota problems

Quotas occur in many aspects of everyday life. The manager of a sales team gives every member of the team a weekly sales quota. In some districts, police have a monthly quota for the number of traffic tickets issued. Nonprofit organizations have recruitment quotas for donations or new volunteers. The basic ideas used to compute quotas also can be used in medical science (how frequently checkups should occur), quality control (how many production flaws should be expected), or risk management (how many bad loans a bank should expect in a certain investment group). In fact, Part II of Guided Exercise 9 is a *quota problem*. To have adequate power, a satellite must have a quota of 3 working solar cells. Such problems come from many different sources, but they all have one thing in common: They are solved using the binomial probability distribution.

Example 9 is a quota problem. Junk bonds are sometimes controversial. In some cases, junk bonds have been the salvation of a basically good company that has had a run of bad luck. From another point of view, junk bonds are not much more than a gambler's effort to make money by shady ethics.

The book *Liar's Poker,* by Michael Lewis, is an exciting and sometimes humorous description of his career as a Wall Street bond broker. Most bond brokers, including Mr. Lewis, are ethical people. However, the book does contain an interesting discussion of Michael Milken and shady ethics. In the book, Mr. Lewis says, "If it was a good deal, the brokers kept it for themselves; if it was a bad deal, they'd try to sell it to their customers. In Example 9 we use some binomial probabilities for a brief explanation of what Mr. Lewis's book is talking about.

EXAMPLE 9
Quota

Junk bonds can be profitable as well as risky. Why are investors willing to consider junk bonds? Suppose you can buy junk bonds at a tremendous discount. You try to choose "good" companies with a "good" product. The company should have done well but for some reason did not. Suppose you consider only companies with a 35% estimated risk of default, and your financial investment goal requires four bonds to be "good" bonds in the sense that they will not default before a certain date. Remember, junk bonds that do not default are usually very profitable because they carry a very high rate of return. The other bonds in your investment group can default (or not) without harming your investment plan. Suppose you want to be 95% certain of meeting your goal (quota) of at least four good bonds. How many junk bond issues should you buy to meet this goal?

SOLUTION: Since the probability of default is 35%, the probability of a "good" bond is 65%. Let success S represent a good bond. Let n be the number of bonds purchased, and let r be the number of good bonds in this group. We want

$$P(r \geq 4) \geq 0.95$$

This is equivalent to

$$1 - P(0) - P(1) - P(2) - P(3) \geq 0.95$$

Since the probability of success is $p = P(S) = 0.65$, we need to look in the binomial table under $p = 0.65$ and different values of n to find the *smallest value* of n that will satisfy the preceding relation. Table 3 in Appendix II shows that if $n = 10$ when $p = 0.65$, then

$$1 - P(0) - P(1) - P(2) - P(3) = 1 - 0 - 0 - 0.004 - 0.021 = 0.975$$

The probability 0.975 satisfies the condition of being greater than or equal to 0.95. We see that 10 is the smallest value of n for which the condition

$$P(r \geq 4) \geq 0.95$$

is satisfied. Under the given conditions (a good discount on price, no more than 35% chance of default, and a fundamentally good company), you can be 95% sure of meeting your investment goal with $n = 10$ (carefully selected) junk bond issues.

In this example, we see that by carefully selecting junk bonds, there is a high probability of getting some good bonds that will produce a real profit. What do you do with the other bonds that aren't so good? Perhaps the quote from *Liar's Poker* will suggest what is sometimes attempted.   ◇

VIEWP⊙INT

Kodiak Island, Alaska

Kodiak Island is famous for its giant brown bears. The sea surrounding the island is also famous for its king crab. The state of Alaska, Department of Fish and Game, has collected a huge amount of data regarding ocean latitude, longitude, and size of king crab. Of special interest to commercial fishing skippers is the size of crab. Those too small must be returned to the sea. To find locations and sizes of king crab catches near Kodiak Island, visit the Brase/Brase statistics site at http://math.college.hmco.com/students and find the link to the StatLib site hosted by the Department of Statistics at Carnegie Mellon University. Once at StatLib, go to crab data. From this information, it is possible to use methods of this chapter and Chapter 8 to estimate the proportion of legal crab in a sea skipper's catch.

SECTION 5.3   PROBLEMS

1. *Binomial Distribution: Histograms* Consider a binomial distribution with $n = 5$ trials. Use the probabilities given in Table 3 in Appendix II to make histograms showing the probability of $r = 0, 1, 2, 3, 4, 5$ successes for each of the following. Comment on the skewness of each distribution.
   (a) The probability of success is $p = 0.50$.
   (b) The probability of success is $p = 0.25$.
   (c) The probability of success is $p = 0.75$.
   (d) What is the relationship between the distributions shown in parts (b) and (c)?
   (e) If the probability of success is $p = 0.73$, do you expect the distribution to be skewed to the right or to the left? Why?

2. *Binomial Distributions: Histograms* Figure 5-6 shows histograms of several binomial distributions with $n = 6$ trials. Match the given probability of success with the best graph.
   (a) $p = 0.30$ goes with graph _____.
   (b) $p = 0.50$ goes with graph _____.
   (c) $p = 0.65$ goes with graph _____.
   (d) $p = 0.90$ goes with graph _____.
   (e) In general, when the probability of success p is close to 0.5, would you say that the graph is more symmetrical or more skewed? In general, when the probability of success p is close to 1, would you say that the graph is skewed to the right or to the left? What about when p is close to 0?

FIGURE 5-6

Binomial Probability Distributions with $n = 6$ (generated on the TI-83Plus calculator)

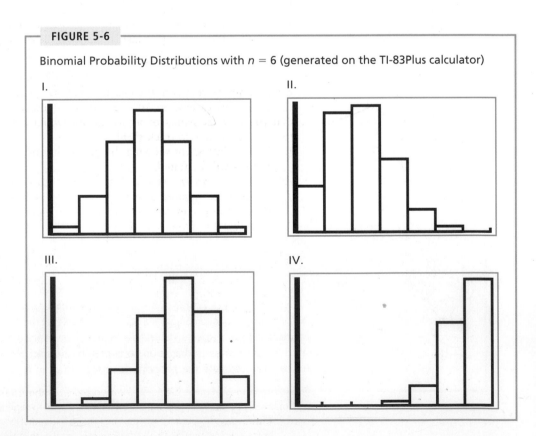

I.

II.

III.

IV.

3. *Marketing: Photography* Does the *kid factor* make a difference? If you are talking photography, the answer may be yes! The following table is based on information from *American Demographics* (Vol. 19, No. 7).

Age of children in household, years	Under 2	None under 21
Percent of U.S. households that buy film	80%	50%

Let us say you are a market research person who interviews a random sample of 10 households.

(a) Suppose that the 10 households you interview are chosen to have children under the age of 2 years. Let r represent the number of such households that buy film. Make a histogram showing the probability distribution of r for $r = 0$ through $r = 10$. Find the mean and standard deviation of this probability distribution.

(b) Suppose that the 10 households are chosen to have no children under 21 years old. Let r represent the number of such households that buy film. Make a histogram showing the probability distribution of r for $r = 0$ through $r = 10$. Find the mean and standard deviation of this probability distribution.

(c) Compare the distributions shown in parts (a) and (b). You are designing TV ads to sell film. Could you justify featuring ads of parents taking pictures of toddlers? Explain your answer.

4. *Quality Control: Syringes* The quality-control inspector of a production plant will reject a batch of syringes if two or more defectives are found in a random sample of eight syringes taken from the batch. Suppose the batch contains 1% defective syringes.

(a) Make a histogram showing the probability of $r = 0, 1, 2, 3, 4, 5, 6, 7, 8$ defective syringes in a random sample of eight syringes.

(b) Find μ. What is the expected number of defective syringes the inspector will find?

(c) What is the probability that the batch will be accepted?

(d) Find σ.

5. *Private Investigation: Locating People* Old Friends Information Service is a California company that is in the business of finding addresses of long-lost friends. Old Friends claims to have a 70% success rate (Source: *Wall Street Journal*). Suppose that you have the names of six friends for whom you have no addresses and decide to use Old Friends to track them.

(a) Make a histogram showing the probability of $r = 0$ to 6 friends for whom an address will be found.

(b) Find the mean and standard deviation of this probability distribution. What is the expected number of friends for whom addresses will be found?

(c) How many names would you have to submit to be 97% sure that at least two addresses will be found?

6. *Insurance: Auto* The Mountain States Office of State Farm Insurance Company reports that approximately 85% of all automobile damage liability claims were made by people under 25 years of age. A random sample of five automobile insurance liability claims is under study.

(a) Make a histogram showing the probability that $r = 0$ to 5 claims are made by people under 25 years of age.

(b) Find the mean and standard deviation of this probability distribution. For samples of size 5, what is the expected number of claims made by people under 25 years of age?

7. *Education: Illiteracy* USA Today reported that about 20% of all people in the United States are illiterate. Suppose that you take seven people at random off a city street.

(a) Make a histogram showing the probability distribution of the number of illiterate people out of the seven people in the sample.

(b) Find the mean and standard deviation of this probability distribution. Find the expected number of people in this sample who are illiterate.

(c) How many people would you need to interview to be 98% sure that at least seven of these people can read and write (are not illiterate)?

8. *Rude Drivers: Tailgating* Do you tailgate the car in front of you? About 35% of all drivers will tailgate before passing, thinking they can make the car in front of them go faster (Source: Bernice Kanner, *Are You Normal?* St. Martin's Press). Suppose that you are driving a considerable distance on a two-lane highway and are passed by 12 vehicles.

(a) Let r be the number of vehicles that tailgate before passing. Make a histogram showing the probability distribution of r for $r = 0$ through $r = 12$.

(b) Compute the expected number of vehicles out of 12 that will tailgate.

(c) Compute the standard deviation of this distribution.

9. *Hype: Improved Products* The *Wall Street Journal* reported that approximately 25% of the people who are told a product is *improved* will believe that it is in fact improved. The remaining 75% believe that this is just hype (the same old thing with no real improvement). Suppose a marketing study consisted of a random sample of eight people who are given a sales talk about a new, *improved* product.

(a) Make a histogram showing the probability that $r = 0$ to 8 people believe the product is in fact improved.

(b) Compute the mean and standard deviation of this probability distribution.

(c) How many people are needed in the marketing study to be 99% sure that at least one person believes the product to be improved?

10. *Forest Fires: Satellite Detection* The National Forest Service uses satellites with infrared sensors to detect forest fires in remote wilderness areas. The satellites are very sensitive and do find all the hot spots. However, The satellites register geothermal hot spots, cloud cover, manmade bonfires, etc., as well as hot spots caused by forest fires. Over a period of years, the Forest Service has decided that only 85% of the hot spots reported by the satellite are forest fires. The usual procedure is to send out a reconnaissance plane to check on what the satellite has detected. Let us say that the satellite was successful if it reports a hot spot that is a forest fire.

Recently, the satellite reported nine hot spots in the remote Seward Peninsula of Alaska.

(a) Find the probability $P(r)$ of r successes for r ranging from 0 to 9.

(b) Make a histogram for the probability distribution of part (a).

(c) Based on the satellite information, what is the expected number μ of real forest fires on the Seward Peninsula?

(d) What is the standard deviation σ?

(e) How many hot spots must the satellite report to be 99.9% sure of at least one real forest fire?

11. *Quota Problem: Sales* Vince is a computer software salesman who has a history of making a successful sales call 40% of the time. If Vince has a sales quota of at least five sales this week, how many sales calls must he make to be 95% sure of meeting the quota?

12. *Quota Problem: Sales* June solicits contributions for charity by phone. She has a history of being successful at getting a donor on 55% of her calls. If June has a quota to get at least four donors each day, how many calls must she make to be 96.4% sure of meeting the quota?

13. *Archaeology: Arrow Points* An archaeological excavation at Burnt Mesa Pueblo showed that about 10% of the flaked stone objects were finished arrow points (Source: *Bandelier Archaeological Excavation Project: Summer 1990 Excavations at*

Burnt Mesa Pueblo, edited by Kohler, Washington State University). How many flaked stone objects need to be found to be 90% sure that at least one is a finished arrow point? (*Hint:* Use a calculator.)

14. *Law Enforcement: Speed Traps* Do you warn oncoming traffic of a police speed trap? About 40% of all drivers will flash their lights to warn oncoming traffic of a speed trap ahead. (*See* reference in Problem 8.) Suppose seven cars have just driven past a police speed trap.
    (a) What is the probability at least one of the drivers will warn oncoming traffic?
    (b) What is the expected number of these seven drivers who will warn oncoming traffic of a police speed trap ahead? What is the standard deviation?
    (c) How many cars *n* would need to go by the speed trap to be 99.8% sure that at least one driver will warn oncoming traffic of the speed trap?

15. *Criminal Justice: Parole USA Today* reports that about 25% of all prison parolees become repeat offenders. Alice is a social worker whose job is to counsel people on parole. Let us say success means a person does not become a repeat offender. Alice has been given a group of four parolees.
    (a) Find the probability *P(r)* for *r* ranging from 0 to 4.
    (b) Make a histogram for the probability distribution of part (a).
    (c) What is the expected number of parolees in Alice's group who will not be repeat offenders? What is the standard deviation?
    (d) How large a group should Alice have to be about 98% sure three or more will not become repeat offenders?

16. *Defense: Radar Stations* The probability that a single radar station will detect an enemy plane is 0.65.
    (a) How many such stations are required to be 98% certain that an enemy plane flying over will be detected by at least one station?
    (b) If four stations are in use, what is the expected number of stations that will detect an enemy plane?

17. *Criminal Justice: Jury Duty* Have you ever tried to get out of jury duty? About 25% of those called will find an excuse (work, poor health, travel out of town, etc.) to avoid jury duty (Source: Bernice Kanner, *Are You Normal?* St. Martin's Press, New York). If 12 people are called for jury duty,
    (a) what is the probability that all 12 will be available to serve on the jury?
    (b) what is the probability that 6 or more will *not* be available to serve on the jury?
    (c) find the expected number of those available to serve on the jury. What is the standard deviation?
    (d) how many people *n* must the jury commissioner contact to be 95.9% sure of finding at least 12 people who are available to serve?

18. *Public Safety: 911 Calls The Denver Post* reported that a recent audit of the Los Angeles 911 calls showed that 85% were not emergencies. Suppose that the 911 operators in Los Angeles have just received four calls.
    (a) What is the probability all four calls are in fact emergencies?
    (b) What is the probability that three or more calls are not emergencies?
    (c) How many calls *n* would the 911 operators need to answer to be 96% (or more) sure that at least one call was in fact an emergency?

19. *Law Enforcement: Property Crime* Does crime pay? The *FBI Standard Survey of Crimes* showed that for about 80% of all property crimes (burglary, larceny, car theft, etc.), the criminals are never found and the case is never solved (Source: *True Odds,* by James Walsh, Merrit Publishing). Suppose that a neighborhood district in a large city has repeated property crimes, not always by the same criminals. The police are investigating six property crime cases in this district.

(a) What is the probability that none of the crimes will ever be solved?

(b) What is the probability that at least one crime will be solved?

(c) What is the expected number of crimes that will be solved? What is the standard deviation?

(d) How many property crimes n must the police investigate before they can be at least 90% sure of solving one or more cases?

20. *Security: Burglar Alarms* A large bank vault has several automatic burglar alarms. The probability is 0.55 that a single alarm will detect a burglar.

(a) How many such alarms should be used to be 99% certain that a burglar trying to enter is detected by at least one alarm?

(b) Suppose the bank installs nine alarms. What is the expected number of alarms that would detect a burglar?

21. *Criminal Justice: Convictions* Innocent until proven guilty? In Japanese criminal trials, about 95% of the defendants are found guilty. In the United States, about 60% of the defendants are found guilty in criminal trials (Source: *The Book of Risks*, by Larry Laudan, John Wiley and Sons). Suppose you are a news reporter following seven criminal trials.

(a) If the trials were in Japan, what is the probability that all the defendants would be found guilty? What is this probability if the trials were in the United States?

(b) Of the seven trials, what is the expected number of guilty verdicts in Japan? What is the expected number in the United States? What is the standard deviation in each case?

(c) As a U.S. news reporter, how many trails n would you need to cover to be at least 99% sure of two or more convictions? How many trials n would you need if you covered trials in Japan?

22. *Focus Problem: Personality Types* Solve the Focus Problem at the beginning of this chapter.

23. *Quota Problem: Motel Rooms* The owners of a motel in Florida have noticed that in the long run about 40% of the people who stop and inquire about a room for the night actually rent a room.

(a) How many inquiries must the owner answer to be 99% sure of renting at least one room?

(b) If 25 separate inquiries are made about rooms, what is the expected number that will result in room rentals?

5.4
The Geometric and Poisson Probability Distributions

FOCUS POINTS

✓ In many activities, the *first* to succeed wins everything! Use the geometric distribution to compute the probability that the *n*th trial is our first success.

✓ Use the Poisson distribution to compute the probability of the occurrence of events spread out over time or space.

✓ Use the Poisson distribution to approximate the binomial distribution when the number of trials is large and the probability of success is small.

In this chapter, we have studied binomial probabilities for the discrete random variable r, the number of successes in n binomial trials. Before we continue in Chapter 6 with continuous random variables, let us examine two other discrete probability distributions, the *geometric* and the *Poisson* probability distributions. These are both related to the binomial distribution. (The *hypergeometric distribution* is another discrete probability distribution related to the binomial distribution. A discussion of the hypergeometric distribution can be found in Appendix I.)

Geometric Distribution

Suppose that we have an experiment where we repeat binomial trials until we get our *first success*, and then we stop. Let n be the number of the trial on which we get our *first success*. In this context, n is not a fixed number. In fact, n could be any of the numbers 1, 2, 3, . . . and so on. What is the probability our first success comes on the *n*th trial? The answer is given by the *geometric probability distribution.*

Geometric probability distribution

$$P(n) = p(1 - p)^{n-1}$$

where *n* is the number of the trial on which the *first success* occurs (*n* = 1, 2, 3, . . .) and *p* is the probability of success on each trial. *Note: p* must be the same for each trial.

Using some mathematics (involving infinite series), it can be shown that the population mean and standard deviation of the geometric distribution are

$$\mu = \frac{1}{p} \quad \text{and} \quad \sigma = \frac{\sqrt{1-p}}{p}$$

In many real-life situations, we keep on trying until we achieve success. This is true in areas as diverse as diplomacy, military science, real estate sales or general marketing strategies, medical science, engineering, and technology.

To Engineer Is Human: The Role of Failure in Successful Design is a fascinating book by Henry Petroski (a professor of engineering at Duke University). Reviewers for the *Los Angeles Times* describe this book as serious, amusing, probing, and sometimes frightening. The book examines topics such as the collapse of the Tacoma Narrows suspension bridge, the collapse of the Kansas City Hyatt Regency walk-way, and the explosion of the space shuttle *Challenger*. Professor Petroski discusses such topics as "success in foreseeing failure" and the "limits of design." What is meant by expressions such as "foreseeing failure" and the "limits of design"? In the next example, we will see how the geometric probability distribution might help us "forecast" failure.

EXAMPLE 10

First success

An automobile assembly plant produces sheet metal door panels. Each panel moves on an assembly line. As the panel passes a robot, a mechanical arm will perform spot welding at different locations. Each location has a magnetic dot painted where the weld is to be made. The robot is programmed to locate the magnetic dot and perform the weld. However, experience shows that on each trial the robot is only 85% successful at locating the dot. If it cannot locate the magnetic dot, it is programmed to *try again*. The robot will keep trying until it finds the dot (and does the weld) or the door panel passes out of the robot's reach.

(a) What is the probability that the robot's first success will be on attempts *n* = 1, 2, or 3?

> **SOLUTION:** Since the robot will keep trying until it is successful, the geometric distribution is appropriate. In this case, success S means that the robot finds the correct location. The probability of success is $p = P(S) = 0.85$. The probabilities are

n	$P(n) = p(1 - p)^{n-1} = 0.85(0.15)^{n-1}$
1	$0.85(0.15)^0 = 0.85$
2	$0.85(0.15)^1 = 0.1275$
3	$0.85(0.15)^2 = 0.0191$

(b) The assembly line moves so fast that the robot has a maximum of only three chances before the door panel is out of reach. What is the probability that the robot will be successful before the door panel is out of reach?

SOLUTION: Since $n = 1$ or 2 or 3 are mutually exclusive, then

$$
\begin{aligned}
P(n = 1 \text{ or } 2 \text{ or } 3) &= P(1) + P(2) + P(3) \\
&= 0.85 + 0.1275 + 0.0191 \\
&= 0.9966
\end{aligned}
$$

This means that the weld should be correctly located about 99.7% of the time.

(c) What is the probability that the robot will not be able to locate the correct spot within three tries? If 10,000 panels are made, what is the expected number of defectives? Comment on the meaning of this answer in the context of "forecasting failure" and the "limits of design."

SOLUTION: The probability that the robot will correctly locate the weld is 0.9966 from part (b). Therefore, the probability that it cannot do so (after three unsuccessful tries) is $1 - 0.9966 = 0.0034$. If we made 10,000 panels, we would expect (forecast) $(10,000)(0.0034) = 34$ defectives. We could reduce this by inspecting every door, but such a solution is most likely too costly. If a defective weld of this type is not considered too dangerous, we can accept an expected 34 failures out of 10,000 panels due to the limits of our production design— that is, the speed of the assembly line and the accuracy of the robot. If this is not acceptable, a new (perhaps more costly) design is needed.   ◇

Poisson Probability Distribution

If we examine the binomial distribution as the number of trials n gets larger and larger while the probability of success p gets smaller and smaller, we obtain the *Poisson distribution*. Siméon Denis Poisson (1781–1840) was a French mathematician who studied probabilities of rare events that occur infrequently in space, time, volume, and so forth. The Poisson distribution applies to accident rates, arrival times, defect rates, the occurrence of bacteria in the air, and many other areas of everyday life.

As with the binomial distribution, we assume only two outcomes: A particular event occurs (success) or does not occur (failure) during the specified time period or space. The events need to be independent so that one success does not change the probability of another success during the specified interval. We are interested in computing the probability of r occurrences in the given time period, space, volume, or specified interval.

Poisson distribution

Let λ (Greek letter lambda) be the mean number of successes over time, volume, area, and so forth. Let r be the number of successes ($r = 0, 1, 2, 3, \ldots$) in a corresponding interval of time, volume, area, and so forth. Then the probability of r successes in the interval is

$$
P(r) = \frac{e^{-\lambda}\lambda^r}{r!}
$$

where e is approximately equal to 2.7183.

Using some mathematics (involving infinite series), it can be shown that the population mean and standard deviation of the Poisson distribution are

$$
\mu = \lambda \qquad \text{and} \qquad \sigma = \sqrt{\lambda}
$$

Note: e^x is a key commonly found on most calculators. Simply use 1 as the exponent, and the calculator will display a decimal approximation for *e*.

There are many applications of the Poisson distribution. For example, if we take the point of view that waiting time can be subdivided into many small intervals, then the actual arrival (of whatever we are waiting for) during any one of the very short intervals could be thought of as an infrequent (or rare) event. This means that the Poisson distribution can be used as a mathematical model to describe the probability of arrivals such as cars to a gas station, planes at an airport, calls to a fire station, births of babies, and even fish arriving on a fisherman's line.

EXAMPLE 11

Poisson distribution

Pyramid Lake, Nevada

Pyramid Lake is located in Nevada on the Paiute Indian Reservation. The lake is described as a lovely jewel in a beautiful desert setting. In addition to its natural beauty, the lake contains some of the world's largest cutthroat trout. Eight- to ten-pound trout are not uncommon and 12- to 15-pound trophies are taken each season. The Paiute Nation uses highly trained fish biologists to study and maintain this famous fishery. In one of their publications, *Creel Chronicle* (Vol. 3, No. 2), the following information was given about the November catch for boat fishermen.

Total fish per hour = 0.667

Suppose that you decide to fish Pyramid Lake for 7 hours during the month of November.

(a) Use the information provided by the fishery biologist in *Creel Chronicle* to find a probability distribution for *r* the number of fish (of all sizes) you catch in a period of 7 hours.

SOLUTION: For fish of all sizes, the mean success rate per hour is 0.667.

$$\lambda = 0.667/1 \text{ hour}$$

Since we want to study a *7-hour interval*, we use a little arithmetic to adjust λ to 7 hours. That is, we adjust λ so that it represents the average number of fish expected in a 7-hour period.

$$\lambda = \frac{0.667}{1 \text{ hour}} \cdot \left(\frac{7}{7}\right) = \frac{4.669}{7 \text{ hour}}$$

For convenience, let us use the rounded value $\lambda = 4.7$ for a 7-hour period. Since *r* is the number of successes (fish caught) in the corresponding 7-hour period and $\lambda = 4.7$ for this period, we use the Poisson distribution to get

$$P(r) = \frac{e^{-\lambda}\lambda^r}{r!} = \frac{e^{-4.7}(4.7)^r}{r!}$$

Recall that $e = 2.7183 \ldots$. Most calculators have e^x, y^x, and **n!** keys (see your calculator manual), so the Poisson distribution is not hard to compute.

(b) What is the probability that in 7 hours you will get 0, 1, 2, or 3 fish of any size?

SOLUTION: Using the result of part (a), we get

$$P(0) = \frac{e^{-4.7}(4.7)^0}{0!} = 0.0091 \approx 0.01$$

$$P(1) = \frac{e^{-4.7}(4.7)^1}{1!} = 0.0427 \approx 0.04$$

$$P(2) = \frac{e^{-4.7}(4.7)^2}{2!} = 0.1005 \approx 0.10$$

$$P(3) = \frac{e^{-4.7}(4.7)^3}{3!} = 0.1574 \approx 0.16$$

The probabilities of getting 0, 1, 2, or 3 fish are about 1%, 4%, 10%, and 16%, respectively.

(c) What is the probability that you will get 4 or more fish in the 7-hour fishing period?

SOLUTION: The sample space of all r values is $r = 0, 1, 2, 3, 4, 5, \ldots$ and so on. The probability in the entire sample space is 1, and these events are mutually exclusive. Therefore,

$$1 = P(0) + P(1) + P(2) + P(3) + P(4) + P(5) + \cdots$$

So

$$P(r \geq 4) = P(4) + P(5) + \cdots = 1 - P(0) - P(1) - P(2) - P(3)$$
$$= 1 - 0.01 - 0.04 - 0.10 - 0.16$$
$$= 0.69$$

There is about a 69% chance that you will catch 4 or more fish in a 7-hour period.   ◇

Use of tables

Table 4 in Appendix II is a table of the Poisson probability distribution for selected values of λ and the number of successes r. Table 5-14 is an excerpt from that table.

To find the value of $P(r = 2)$ when $\lambda = 0.3$, look in the column headed by 0.3 and the row headed by 2.

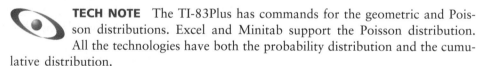 **TECH NOTE**   The TI-83Plus has commands for the geometric and Poisson distributions. Excel and Minitab support the Poisson distribution. All the technologies have both the probability distribution and the cumulative distribution.

TI-83Plus Use **DISTR** key and scroll to **geometpdf**(p,n) for probability of first success on trial number n; scroll to **geomecdf**(p,n) for probability of first success on

TABLE 5-14   Excerpt from Appendix II, Table 4, "Poisson Probability Distribution"

	λ				
r	.1	.2	.3	.4	.5 ....
0	.9048	.8187	.7408	.6703	.6065
1	.0905	.1637	.2222	.2681	.3033
2	.0045	.0164	.0333		
3	.0002	.0011	.0033	.0072	.0126
4	.0000	.0001	.0003	.0007	.0016

trial number $\leq n$. Use **poissonpdf**(λ,r) for probability of r successes. Use **poissoncdf**(λ,r) for probability of at least r successes. For example, when $\lambda = 0.25$ and $r = 2$, we get the results

$P(r = 2)$

$P(r \leq 2)$

```
poissonpdf(.25,2)
        .0243375245
poissoncdf(.25,2)
        .9978385033
```

Excel Menu choice: **Paste Function** $\boxed{f_x}$ ➤ **Statistical** ➤ **Poisson.** Enter trial number *r*, and λ for mean. False gives the probability $P(r)$ and True gives the cumulative probability P(at least r).

Minitab Put *r* values in a column. Then use menu choice **Calc** ➤ **Probability Distribution** ➤ **Poisson.** In the dialogue box, select probability or cumulative, enter the column number containing the *r* values, and use λ for the mean.

Poisson Approximation to the Binomial Probability Distribution

In the preceding examples, we have seen how the Poisson distribution can be used over intervals of time, space, area, and so on. However, the Poisson distribution also can be used as a probability distribution for "rare" events. In the binomial distribution, if the number of trials *n* is large while the probability *p* of success is quite small, we call the event (success) a "rare" event. Put another way, it can be shown that for most practical purposes, the *Poisson* distribution will be a very good *approximation to the binomial* distribution provided that the number of trials *n* is larger than or equal to 100 and $\lambda = np$ is less than 10. As *n* gets larger and *p* gets smaller, the approximation becomes better and better.

Consider the binomial distribution with

n = number of trials

r = number of successes

p = probability of success on each trial

If $n \geq 100$ and $np < 10$, then *r* has a binomial distribution that is approximated by a Poisson distribution with $\lambda = np$.

EXAMPLE 12

Poisson approximation to the binomial

Isabel Briggs Myers was a pioneer in the study of personality types. Today the Myers-Briggs Type Indicator is used in many career counseling programs as well as in many industrial settings where people must work closely together as a team. The 16 personality types are discussed in detail in the book *A Guide to the Development and Use of the Myers-Briggs Type Indicators,* by Myers and McCaulley. Each personality type has its own special contribution in any group activity. One of the more "rare" types is INFJ (introverted, intuitive, feeling, judgmental), which occurs in only about 2.1% of the population. Suppose that a high-school graduating class has 167 students, and suppose that we call success the event that a student is of personality type INFJ.

(a) Let r be the number of successes (INFJ students) in the $n = 167$ trials (graduating class). If $p = P(S) = 0.021$, will the Poisson distribution be a good approximation to the binomial?

SOLUTION: Since $n = 167$ is greater than 100 and $\lambda = np = 167(0.021) \approx 3.5$ is less than 10, the Poisson distribution should be a good approximation to the binomial.

(b) Estimate the probability that this graduating class has 0, 1, 2, 3, or 4 people who have the INFJ personality type.

SOLUTION: Since Table 4 (Appendix II) for the Poisson distribution includes the values $\lambda = 3.5$ and $r = 0, 1, 2, 3,$ or 4, we may simply look up the values for $P(r)$, $r = 0, 1, 2, 3, 4$:

$$P(r = 0) = 0.0302 \qquad P(r = 1) = 0.1057$$
$$P(r = 2) = 0.1850 \qquad P(r = 3) = 0.2158$$
$$P(r = 4) = 0.1888$$

Since the outcomes $r = 0, 1, 2, 3,$ or 4 successes are mutually exclusive, we can compute the probability of 4 or fewer INFJ types by the addition rule for mutually exclusive events:

$$\begin{aligned} P(r \leq 4) &= P(0) + P(1) + P(2) + P(3) + P(4) \\ &= 0.0302 + 0.1057 + 0.1850 + 0.2158 + 0.1888 \\ &= 0.7255 \end{aligned}$$

The probability that the graduating class will have 4 or fewer INFJ personality types is about 0.73.

(c) Estimate the probability that this class has 5 or more INFJ personality types.

SOLUTION: Because the outcomes of a binomial experiment are all mutually exclusive, we have

$$P(r \leq 4) + P(r \geq 5) = 1$$
$$\text{or } P(r \geq 5) = 1 - P(r \leq 4) = 1 - 0.7255 = 0.2745$$

The probability is approximately 0.27 that there will be 5 or more INFJ personality types in the graduating class.   ◊

Summary

In this section, we have studied two discrete probability distributions. The Poisson distribution gives us the probability of r successes in an interval of time or space. The Poisson distribution also can be used to approximate the binomial distribution when $n \geq 100$ and $np < 10$. The geometric distribution gives us the probability that our first success will occur on the nth trial. In the next guided exercise, we will see situations in which each of these distributions applies.

GUIDED EXERCISE 10

Select appropriate distribution

For each problem, first identify the type of probability distribution needed to solve the problem: geometric or Poisson. Then solve the problem.

(i) Denver, Colorado, is prone to severe hail storms. Insurance agents claim that a home owner in Denver can expect to replace his or her roof (due to hail damage) once every 10 years. What is the probability that in 12 years a homeowner in Denver will need to replace the roof twice because of hail?

(ii) A telephone network substation will keep trying to connect a long-distance call to a trunk line until the 4th attempt has been made. After the 4th unsuccessful attempt, the call number goes into a buffer memory bank, and the caller gets a recorded message to be patient. During peak calling periods, the probability of a call connecting into a trunk line is 65% on each try. What percentage of all calls during peak time will wind up in the buffer memory bank?

(a) Consider the problem stated in part (i). What is success in this case? We are interested in the probability of two successes over a specified time interval of 12 years. Which distribution should we use?

⟹ Here we can say success is needing to replace a roof because of hail. Because we are interested in the probability of two successes over a time interval, we use the Poisson distribution.

(b) In part (i), we are told that the average roof replacement is once every 10 years. What is the average number of times the roof needs to be replaced in 12 years? What is λ for the 12-year period?

⟹ We are given a value of $\lambda = 1$ for 10 years. To compute λ for 12 years, we convert the denominator to 12 years.

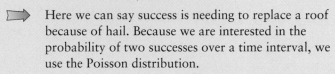

$$\lambda = \frac{0.1}{1 \text{ year}} \cdot \frac{12}{12} = \frac{1.2}{12 \text{ years}}$$

For 12 years, we have $\lambda = 1.2$.

(c) To finish part (i), use the Poisson distribution to find the probability of two successes in the 12-year period.

⟹ We may use Table 4 in Appendix II because $\lambda = 1.2$ and $r = 2$ are values in the table. The table gives $P(r = 2) = 0.2169$. Using the formula and a calculator gives the same result.

$$P(r) = \frac{e^{-\lambda}\lambda^r}{r!}$$

$$P(r = 2) = \frac{e^{-1.2}(1.2)^2}{2!} = 0.2169$$

There is about a 21.7% chance the roof will be damaged twice by hail during a 12-year period.

Continued

GUIDED EXERCISE 10 continued

(d) Consider part (ii). What is success? We are interested in the probability that a call will wind up in the buffer memory. This will occur if success does not occur before which trial? Since we are looking at the probability of a first success by a specified trial number, which probability distribution do we use?

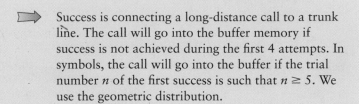 Success is connecting a long-distance call to a trunk line. The call will go into the buffer memory if success is not achieved during the first 4 attempts. In symbols, the call will go into the buffer if the trial number n of the first success is such that $n \geq 5$. We use the geometric distribution.

(e) What is the probability of success on a single trial? Use this information and the formula for the geometric distribution to compute the probability that the first success occurs on trial 1, 2, 3, or 4. Then use this information to compute $P(n \geq 5)$, where n is the trial number of the first success. What percentage of the calls go to the buffer?

Success means the call connects to the trunk line. According to the description in the problem,

$$P(S) = 0.65 = p$$

By the formula for the geometric probability distribution, when n represents the trial number of the first success,

$$P(n) = p(1 - p)^{n-1}$$

Therefore,

$$P(1) = (0.65)(0.35)^0 = 0.65$$
$$P(2) = (0.65)(0.35)^1 = 0.2275$$
$$P(3) = (0.65)(0.35)^2 = 0.0796$$
$$P(4) = (0.65)(0.35)^3 = 0.0279$$
$$P(n \geq 5) = 1 - P(1) - P(2) - P(3) - P(4)$$
$$= 1 - 0.65 - 0.2275 - 0.0796$$
$$- 0.0279$$
$$= 0.015$$

About 1.5% of the calls go to the buffer.

VIEWP⊙INT   *When Do Cracks Become Breakthroughs?*

No one *wants* to learn by mistakes! However, learning by our successes will not take us beyond the state of the art! Each new idea, technology, social plan, or engineering structure can be considered a new trial. In the meantime, we the laypeople whose spokesperson is often a poet or writer will be threatened by both *failures* and *successes*. This is the nature not only of science, technology, and engineering but also of all human endeavors. [For more discussion on this topic, *see* Problem 12 as well as *To Engineer Is Human: The Role of Failure in Successful Design*, by Professor Petroski (Duke University Press).]

SECTION 5.4   PROBLEMS

1. *College: Core Requirement* Susan is taking Western Civilization this semester on a pass/fail basis. The department teaching the course has a history of passing 77% of the students in Western Civilization each term. Let $n = 1, 2, 3, \ldots$ represent the number of times a student takes Western Civilization until the *first* passing grade is received. (Assume the trials are independent.)
   (a) Write out a formula for the probability distribution of the random variable n.
   (b) What is the probability that Susan passes on the first try ($n = 1$)?
   (c) What is the probability that Susan first passes on the second try ($n = 2$)?
   (d) What is the probability that Susan needs three or more tries to pass Western Civilization?
   (e) What is the expected number of attempts at Western Civilization Susan must make to have her (first) pass? *Hint:* Use μ for the geometric distribution and round.

2. *Law: Bar Exam* Bob is a recent law school graduate who intends to take the state bar exam. According to the National Conference on Bar Examiners, about 57% of all people who take the state bar exam pass (Source: *The Book of Odds,* by Shook and Shook, Signet). Let $n = 1, 2, 3, \ldots$ represent the number of times a person takes the bar exam until the *first* pass.
   (a) Write out a formula for the probability distribution of the random variable n.
   (b) What is the probability that Bob first passes the bar exam on the second try ($n = 2$)?
   (c) What is the probability that Bob needs three attempts to pass the bar exam?
   (d) What is the probability that Bob needs more than three attempts to pass the bar exam?
   (e) What is the expected number of attempts at the state bar exam Bob must make for his (first) pass? *Hint:* Use μ for the geometric distribution and round.

3. *Anthropology: Pot Shards* Anthropological studies at Casa del Rito indicate that approximately 5% of all pot shards at the excavation site are from the traditional type of pot known as *Socorro black on white* (Source: *Bandelier Archaeological Excavation Project: Summer 1990 Excavations at Burnt Mesa Pueblo and Casa del Rito,* edited by Kohler, Washington State University Department of Anthropology). Let $n = 1, 2, 3, \ldots$ represent the number of pot shards that must be discovered (and examined) until the *first* Socorro black on white is found.
   (a) Write out a formula for the probability distribution of the random variable n.
   (b) What is the probability that the first Socorro black on white pot shard is found on trial $n = 5$?
   (c) What is the probability that the first Socorro black on white pot shard is found on trial $n = 10$?
   (d) What is the probability that more than three pot shards have to be examined before finding the first Socorro black on white?
   (e) What is the expected number of pot shards that must be examined to find the first Socorro black on white? *Hint:* Use μ for the geometric distribution and round.

4. *Sociology: Hawaiians* On the leeward side of the island of Oahu, in the small village of Nanakuli, about 80% of the residents are of Hawaiian ancestry (Source: *The Honolulu Advertiser*). Let $n = 1, 2, 3, \ldots$ represent the number of people you must meet until you encounter the *first* person of Hawaiian ancestry in the village of Nanakuli.
   (a) Write out a formula for the probability distribution of the random variable n.
   (b) Compute the probability that $n = 1$, $n = 2$, $n = 3$.

(c) Compute the probability that $n \geq 4$.

(d) In Waikiki it is estimated that about 4% of the residents are of Hawaiian ancestry. Repeat parts (a), (b), and (c) for Waikiki.

5. *Sociology: Values* Approximately 71% of all modern college students claim that the main reason they are attending college is to make more money after graduation. In 1967, approximately 83% of all college students claimed that they were attending college primarily to develop a meaningful philosophy of life (Source: *Chances: Risk and Odds in Everyday Life,* by James Burke). Let $n = 1, 2, 3, \ldots$ represent the *first* person who is a college student selected at random whom you encounter who says he or she is in college primarily to make more money after graduation.

(a) Write out a formula for the probability distribution of the random variable n.

(b) Find the probability that $n = 1$, $n = 2$, $n \geq 3$.

(c) Repeat parts a and b when n represents the *first* person who was a college student in 1967 selected at random who claimed the primary reason for attending college was to develop a meaningful philosophy of life.

6. *Agriculture: Apples* Approximately 3.6% of all (untreated) Jonathan apples had bitter pit in a study conducted by the botanists Ratkowsky and Martin (Source: *Australian Journal of Agricultural Research,* Vol. 25, pp. 783–790). (Bitter pit is a disease of apples resulting in a soggy core, which can be caused either by overwatering the apple tree or by a calcium deficiency in the soil.) Let n be a random variable that represents the first Jonathan apple chosen at random that has bitter pit.

(a) Write out a formula for the probability distribution of the random variable n.

(b) Find the probability that $n = 3$, $n = 5$, $n = 12$.

(c) Find the probability that $n \geq 5$.

(d) What is the expected number of apples that must be examined to find the first one with bitter pit? *Hint:* Use μ for the geometric distribution and round.

7. *Fishing: Lake Trout* At Fontaine Lake Camp on Lake Athabasca in northern Canada, history shows that about 30% of the guests catch lake trout over 20 pounds on a 4-day fishing trip (Source: Athabasca Fishing Lodges, Saskatoon, Canada). Let n be a random variable that represents the *first* trip to Fontaine Lake Camp on which a guest catches a lake trout over 20 pounds.

(a) Write out a formula for the probability distribution of the random variable n.

(b) Find the probability that a guest catches a lake trout weighing at least 20 pounds for the *first* time on trip number 3.

(c) Find the probability that it takes more than three trips for a guest to catch a lake trout weighing at least 20 pounds.

(d) What is the expected number of fishing trips that must be taken to catch the first lake trout over 20 pounds? *Hint:* Use μ for the geometric distribution and round.

8. *Archaeology: Artifacts* At Burnt Mesa Pueblo in one of the archaeological excavation sites, the artifact density (number of prehistoric artifacts per 10 liters of sediment) was 1.5 (*see* source in Problem 3). Suppose you are going to dig up and examine 50 liters of sediment at this site. Let $r = 0, 1, 2, 3, \ldots$ be a random variable that represents the number of prehistoric artifacts found in your 50 liters of sediment.

(a) Explain why the Poisson distribution would be a good choice for the probability distribution of r. What is λ? Write out the formula for the probability distribution of the random variable r.

(b) Compute the probability that in your 50 liters of sediment you will find two prehistoric artifacts; that you will find three artifacts; that you will find four artifacts.

(c) Find the probability that you will find three or more artifacts in the 50 liters of sediment.

(d) Find the probability that you will find fewer than three prehistoric artifacts in the 50 liters of sediment.

9. *Ecology: River Otters* In his doctoral thesis, L. A. Beckel (University of Minnesota, 1982) studied the social behavior of river otters during the mating season. An important role in the bonding process of river otters is social grooming. After extensive observations, Dr. Beckel found that one group of river otters under study had a frequency of grooming of approximately 1.7 for each 10 minutes. Suppose that you are observing river otters for 30 minutes. Let $r = 0, 1, 2, \ldots$ be a random variable that represents the number of times (in a 30-minute interval) that one otter grooms another.
   (a) Explain why the Poisson distribution would be a good choice for the probability distribution of r. What is λ? Write out the formula for the probability distribution of the random variable r.
   (b) Find the probability that in your 30 minutes of observation, one otter will groom another four times, five times, six times.
   (c) Find the probability that one otter will groom another four or more times during the 30-minute observation period.
   (d) Find the probability that one otter will groom another less than four times during the 30-minute observation period.

10. *Law Enforcement: Shoplifting* The *Denver Post* reported that a large shopping center had an incident of shoplifting (that was caught by security) on the average of once every three hours. The shopping center is open from 10 A.M. to 9 P.M. (11 hours). Let r be the number of shoplifting incidents caught by security in an 11-hour period during which the center is open.
   (a) Explain why the Poisson probability distribution would be a good choice for the random variable r. What is λ?
   (b) What is the probability that from 10 A.M. to 9 P.M. there will be at least one shoplifting incident caught by security?
   (c) What is the probability that from 10 A.M. to 9 P.M. there will be at least three shoplifting incidents caught by security?
   (d) What is the probability that from 10 A.M. to 9 P.M. there will be no shoplifting incidents caught by security?

11. *Vital Statistics: Birth Rate* USA Today reported that the U.S. (annual) birth rate is about 16 per 1000 people and the death rate is about 8 per 1000 people.
   (a) Explain why the Poisson probability distribution would be a good choice for the random variable $r =$ number of births (or deaths) for a community of a given population size.
   (b) In a community of 1000 people, what is the probability of 10 births? What is the probability of 10 deaths? What is the probability of 16 births? 16 deaths?
   (c) Repeat part (b) for a community of 1500 people. You will need to use a calculator to compute $P(10 \text{ births})$ and $P(16 \text{ births})$.
   (d) Repeat part (b) for a community of 750 people.

12. *Engineering: Cracks* Henry Petroski is a professor of civil engineering at Duke University. In his book *To Engineer Is Human: The Role of Failure in Successful Design*, Professor Petroski says that up to 95% of all structural failures, including those of bridges, airplanes, and other commonplace products of technology, are believed to be the result of crack growth. In most cases, the cracks grow slowly. It is only when the cracks reach intolerable proportions and still go undetected that catastrophe can occur. In a cement retaining wall, occasional hairline cracks are normal and nothing to worry about. If these cracks are spread out and not too close together, the wall is considered safe. However, if a number of cracks group together in a small region, there may be real trouble. Suppose that a given cement retaining wall is

considered safe if hairline cracks are evenly spread out and occur on the average of 4.2 cracks per 30-foot section of the wall.

(a) Explain why a Poisson probability distribution would be a good choice for the random variable r = number of hairline cracks for a given length of retaining wall.

(b) In a 50-foot section of safe wall, what is the probability of three (evenly spread out) hairline cracks? What is the probability of three or *more* (evenly spread out) hairline cracks?

(c) Answer part (b) for a 20-foot section of wall.

(d) Answer part (b) for a 2-foot section of wall. Round λ to the nearest tenth.

(e) Consider your answers for parts (b), (c), and (d). If you had three hairline cracks evenly spread out over a 50-foot section of wall, should this be cause for concern? The probability is low. Could this mean that you are lucky to have so few cracks? On a 20-foot section of wall [part (c)] the probability of three cracks is higher. Does this mean that this distribution of cracks is closer to what we should expect? For part (d), the probability is very small. Could this mean you are not so lucky and have something to worry about? Explain your answers.

13. *Meteorology: Winter Conditions* Much of Trail Ridge Road in Rocky Mountain National Park is over 12,000 feet high. Although it is a beautiful drive in summer months, in winter the road is closed because of severe weather conditions. *Winter Wind Studies in Rocky Mountain National Park,* by Glidden (published by Rocky Mountain Nature Association), states that sustained gale-force winds (over 32 miles per hour and often over 90 miles per hour) occur on the average of once every 60 hours at a Trail Ridge Road weather station.

(a) Let r = frequency with which gale-force winds occur in a given time interval. Explain why the Poisson probability distribution would be a good choice for the random variable r.

(b) For an interval of 108 hours, what is the probability that r = 2, 3, or 4? What is the probability that $r < 2$?

(c) For an interval of 180 hours, what is the probability that r = 3, 4, or 5? What is the probability that $r < 3$?

14. *Earthquakes: San Andreas Fault* USA Today reported that Parkfield, California, is dubbed the world's earthquake capital because it sits on top of the notorious San Andreas fault. Since 1857, Parkfield has had a major earthquake on the average of once every 22 years.

(a) Explain why a Poisson probability distribution would be a good choice for r = number of earthquakes in a given time interval.

(b) Compute the probability of at least one major earthquake in the next 22 years. Round λ to the nearest hundredth, and use a calculator.

(c) Compute the probability there will be no major earthquake in the next 22 years. Round λ to the nearest hundredth, and use a calculator.

(d) Compute the probability of at least one major earthquake in the next 50 years. Round λ to the nearest hundredth, and use a calculator.

(e) Compute the probability of no major earthquakes in the next 50 years. Round λ to the nearest hundredth, and use a calculator.

15. *Real Estate: Sales* Jim is a real estate agent who sells large commercial buildings. Because his commission is so large on a single sale, he does not need to sell many buildings to make a good living. History shows that Jim has a record of selling an average of eight large commercial buildings in 275 days.

(a) Explain why a Poisson probability distribution would be a good choice for r = number of buildings sold in a given time interval.

(b) In a 60-day period, what is the probability that Jim will make no sales? one sale? two or more sales?

(c) In a 90-day period, what is the probability that Jim will make no sales? two sales? three or more sales?

$\dfrac{661}{100,000}$ $\overset{\sim}{316}$

$\lambda = 2.09$

16. *Law Enforcement: Burglaries* The Honolulu Advertiser stated that in Honolulu there was an average of 661 burglaries per 100,000 households in a given year. In the Kohola Drive neighborhood there are 316 homes. Let r = number of these homes that will be burglarized in a year.
    (a) Explain why the Poisson approximation to the binomial would be a good choice for the random variable r. What is n? What is p? What is λ to the nearest tenth?
    (b) What is the probability that there will be no burglaries this year in the Kohola Drive neighborhood?
    (c) What is the probability that there will be no more than one burglary in the Kohola Drive neighborhood?
    (d) What is the probability there will be two or more burglaries in the Kohola Drive neighborhood?

17. *Criminal Justice: Drunk Drivers* Harper's Index reported that the number of (Orange County, California) convicted drunk drivers whose sentence included a tour of the morgue was 569, out of which only 1 became a repeat offender.
    (a) Suppose that out of 1000 newly convicted drunk drivers, all were required to take a tour of the morgue. Let us assume that the probability of a repeat offender is still $p = 1/569$. Explain why the Poisson approximation to the binomial would be a good choice for r = number of repeat offenders out of 1000 convicted drunk drivers who toured the morgue. What is λ to the nearest tenth?
    (b) What is the probability that $r = 0$?
    (c) What is the probability that $r > 1$?
    (d) What is the probability that $r > 2$?
    (e) What is the probability that $r > 3$?

18. *Airlines: Lost Bags* USA Today reported that for all airlines, the number of lost bags was

    May: 6.02 per 1000 passengers   December: 12.78 per 1000 passengers

    Note: A passenger could lose more than one bag.
    (a) Let r = number of bags lost per 1000 passengers in May. Explain why the Poisson distribution would be a good choice for the random variable r. What is λ to the nearest tenth?
    (b) In the month of May, what is the probability that out of 1000 passengers, no bags are lost? 3 or more bags are lost? 6 or more bags are lost?
    (c) In the month of December, what is the probability that out of 1000 passengers, no bags are lost? 6 or more bags are lost? 12 or more bags are lost? (Round λ to the nearest whole number.)

19. *Law Enforcement: Officers Killed* Chances: Risk and Odds in Everyday Life, by James Burke, reports that the probability that a police officer will be killed in the line of duty is 0.5% (or less).
    (a) In a police precinct with 175 officers, let r = number of police officers killed in the line of duty. Explain why the Poisson approximation to the binomial would be a good choice for the random variable r. What is n? What is p? What is λ to the nearest tenth?
    (b) What is the probability that no officer in this precinct will be killed in the line of duty?
    (c) What is the probability one or more officers in this precinct will be killed in the line of duty?

(d) What is the probability that two or more officers in this precinct will be killed in the line of duty?

20. *Business Franchise: Shopping Center Chances: Risk and Odds in Everyday Life,* by James Burke, reports that only 2% of all local franchises are business failures. A Colorado Springs shopping complex has 137 franchises (restaurants, print shops, convenience stores, hair salons, etc.).
   (a) Let r be the number of these franchises that are business failures. Explain why a Poisson approximation to the binomial would be appropriate for the random variable r. What is n? What is p? What is λ (round to the nearest tenth)?
   (b) What is the probability that none of the franchises will be a business failure?
   (c) What is the probability that two or more franchises will be business failures?
   (d) What is the probability that four or more franchises will be business failures?

21. *Poisson Approximation to Binomial: Comparisons*
   (a) For $n = 100$, $p = 0.02$, and $r = 2$, compute $P(r)$ using the formula for the binomial distribution and your calculator:

   $$P(r) = C_{n,r}p^r(1 - p)^{n-r}$$

   (b) For $n = 100$, $p = 0.02$, and $r = 2$, estimate $P(r)$ using the Poisson approximation to the binomial.
   (c) Compare the results of parts (a) and (b). Does it appear that the Poisson distribution with $\lambda = np$ provides a good approximation for $P(r = 2)$?
   (d) Repeat parts (a) to (c) for $r = 3$.

SUMMARY

The concepts of discrete and continuous random variables were introduced in this chapter. We looked at the general probability distribution for a discrete random variable and saw how to compute the expected value μ and the standard deviation σ of a discrete probability distribution.

Then we turned our attention to a special discrete probability distribution, the binomial probability distribution. This distribution is used to determine the probability of outcomes from a binomial experiment.

A binomial experiment must meet the following criteria:

1. There are a fixed number of trials denoted by n.

2. The trials are independent and repeated under identical conditions.

3. Each trial has only two outcomes: success S and failure F.

4. For each trial the probability p of success remains the same. The probability of failure is $q = 1 - p$.

5. The central problem is to find the probability of r successes out of n trials.

   The formula for the binomial probability distribution is

   $$P(r) = C_{n,r}p^rq^{n-r}$$

where $C_{n,r}$ is the binomial coefficient as found in Table 2 of Appendix II. For certain values of p and n, $P(r)$ can be found directly in Table 3 of Appendix II.

The mean or expected value and standard deviation of the binomial distribution are given by the formulas

$$\mu = np$$
$$\sigma = \sqrt{npq} \qquad \text{where } q = 1 - p$$

The binomial probability distribution is a distribution of a discrete random variable. Other distributions that relate to the binomial distribution are the geometric distribution and the Poisson distribution. These are also probability distributions of discrete random variables. The geometric distribution is used to find the probability that the first success of a binomial experiment occurs on trial number n. The Poisson distribution is used to compute the probability of r successes in an interval of time, volume, area, and so forth. We also use the Poisson distribution as an approximation to the binomial when n is large and the probability of success p on a single trial is small. Appendix I contains a discussion of another discrete probability distribution, the hypergeometric distribution.

IMPORTANT WORDS & SYMBOLS

Section 5.1
Random variable
  Discrete
  Continuous
Population parameters
Mean μ of a probability distribution
Standard deviation σ of a probability distribution
Expected value μ
  Linear function of a random variable
  Linear combination of two independent random
    variables

Section 5.2
Binomial experiment
Independent trials
Successes and failures in a binomial experiment
Probability of success $P(S) = p$
Probability of failure $P(F) = q = 1 - p$
Binomial coefficient $C_{n,r}$
Binomial probability distribution $P(r) = C_{n,r}p^r q^{n-r}$

Section 5.3
Mean for the binomial distribution $\mu = np$
Standard deviation for binomial distribution
  $\sigma = \sqrt{npq}$
Quota problem

Section 5.4
Geometric probability distribution
Poisson probability distribution
Poisson approximation to the binomial

VIEWPOINT

What's Your Type?

Are students and professors *really* compatible? One way of answering this question is to look at Myers-Briggs Type Indicators for personality preferences. What is the probability that your professor is introverted and judgmental? What is the probability that you are extroverted and perceptive? Are most of the leaders in student government extroverted and judgmental? Is it true that members of Phi Beta Kappa have personality types more like the professors'? We will consider questions such as these in more detail in Chapter 8 (estimation) and Chapter 9 (hypothesis testing), where we will continue our work with binomial probabilities. In the meantime, you can find many answers regarding careers, probability, and personality types in *Applications of the Myers-Briggs Type Indicator in Higher Education,* edited by J. Provost and S. Anchors.

CHAPTER REVIEW PROBLEMS

1. *Probability Distribution: Auto Leases* Consumer Banker Association released a report showing the length of automobile leases for new automobiles. The results are

Lease Length in Months	Percent of Leases
13–24	12.7%
25–36	37.1%
37–48	28.5%
49–60	21.5%
More than 60	0.2%

(a) Use the midpoint of each class, and call the midpoint of the last class 66.5 months for purposes of computing the expected lease term. Also find the standard deviation of the distribution.

(b) Sketch a graph of the probability for the duration of new auto leases.

2. *Ecology: Predator and Prey* Isle Royale, an island in Lake Superior, has provided an important study site of wolves and prey. In the National Park Service Scientific Monograph Series 11, *Wolf Ecology and Prey Relationships on Isle Royale,* Peterson gives results of many wolf–moose studies. Of special interest is the study of the number of moose killed by wolves. In the period from 1958 to 1974, there were 296 moose deaths identified as wolf kills. The age distribution of the kills is

Age of Moose in Years	Number Killed by Wolves
Calf (0.5 yr)	112
1–5	53
6–10	73
11–15	56
16–20	2

(a) For each age group, compute the probability that a moose in that age group is killed by a wolf.

(b) Consider all ages in a class equal to the class midpoint. Find the expected age of a moose killed by a wolf and the standard deviation of the ages.

3. *Insurance: Auto* State Farm Insurance studies show that in Colorado, 55% of the auto insurance claims submitted for property damage were submitted by males under 25 years of age. Suppose 10 property damage claims involving automobiles are selected at random.

(a) Let r be the number of claims that are made by males under age 25. Make a histogram for the r-distribution probabilities.

(b) What is the probability that six or more claims are made by males under age 25?

(c) What is the expected number of claims made by males under age 25? What is the standard deviation of the r-probability distribution?

4. *Quality Control: Pens* A stationery store has decided to accept a large shipment of ball-point pens if an inspection of 20 randomly selected pens yields no more than two defective pens.

(a) Find the probability that this shipment is accepted if 5% of the total shipment is defective.

(b) Find the probability that this shipment is not accepted if 15% of the total shipment is defective.

5. *Criminal Justice: Inmates* According to *Harper's Index*, 50% of all federal inmates are serving time for drug dealing. A random sample of 16 federal inmates is selected.

(a) What is the probability that 12 or more are serving time for drug dealing?

(b) What is the probability that 7 or fewer are serving time for drug dealing?

(c) What is the expected number of inmates serving time for drug dealing?

6. *Airlines: On-Time Arrivals* *Consumer Reports* rated airlines and found that 80% of the flights involved in the study arrived on time (that is, within 15 minutes of scheduled arrival time). Assuming that the on-time arrival rate is representative of the entire commercial airline industry, consider a random sample of 200 flights. What is the expected number that will arrive on time? What is the standard deviation of this distribution?

7. *Agriculture: Grapefruit* It is estimated that 75% of a grapefruit crop is good; the other 25% have rotten centers that cannot be detected unless the grapefruit are cut open. The grapefruit are sold in sacks of 10. Let r be the number of good grapefruit in a sack.

(a) Make a histogram of the probability distribution of r.

(b) What is the probability of getting no more than one bad grapefruit in a sack? What is the probability of getting at least one good grapefruit in a sack?

(c) What is the expected number of good grapefruit in a sack?

(d) What is the standard deviation of the r probability distribution?

8. *Restaurants: Reservations* The Orchard Café has found that about 5% of the parties who make reservations don't show up. If 82 party reservations have been made, how many can be expected to show up? Find the standard deviation of this distribution.

9. *College Life: Student Government* The student government claims that 85% of all students favor an increase in student fees to buy indoor potted plants for the classrooms. A random sample of 12 students produced 2 in favor of the project. What is the probability that 2 or fewer in the sample will favor the project, assuming that

the student government is correct? Do the data support the student government claim, or does it seem that the percentage favoring the increase in fees is less than 85%?

10. *Financial: Bonds* Suppose you are a (junk) bond broker who buys only bonds that have a 50% chance of default. You want a portfolio with at least five bonds that do not default. You can dispose of the other bonds in the portfolio with no great loss. How many such bonds should you buy if you want to be 94.1% sure that five or more will not default?

11. *Theater: Coughs* A person with a cough is a *persona non grata* on airplanes, elevators, or at the theater. In theaters especially, the irritation level rises with each muffled explosion. According to Dr. Brian Carlin, a Pittsburgh pulmonologist, in any large audience you'll hear about 11 coughs per minute (Source: *USA Today*).
    (a) Let r = number of coughs in a given time interval. Explain why the Poisson distribution would be a good choice for the probability distribution of r.
    (b) Find the probability of 3 or fewer coughs (in a large auditorium) in a 1-minute period.
    (c) Find the probability of at least 3 coughs (in a large auditorium) in a 30-second period.

12. *Accident Rate: Small Planes* Flying over the western states with mountainous terrain in a small aircraft is 40% riskier than flying over similar distances in flatter portions of the nation according to a General Accounting Office study completed in response to a congressional request. The accident rate for small aircraft in the 11 mountainous western states is 2.4 per 100,000 flight operations (Source: *Denver Post*).
    (a) Let r = number of accidents for a given number of operations. Explain why the Poisson distribution would be a good choice for the probability distribution of r.
    (b) Find the probability of no accidents in 100,000 flight operations.
    (c) Find the probability of at least 4 accidents in 200,000 flight operations.

13. *Banking: Loan Defaults* Records over the past year show that 1 out of 350 loans made by Mammon Bank have defaulted. Find the probability that 2 or more out of 300 loans will default. *Hint:* Is it appropriate to use the Poisson approximation to the binomial distribution?

14. *Colorado Lottery* In the Colorado Lottery, the probability of winning the grand prize with one ticket is about 1/5,000,000 (1 in 5 million). Suppose you join a pool that buys 500,000 tickets (at a cost of $1 per ticket). What is the probability that *none* of the tickets purchased by the pool will be a grand prize winner? What is the probability that one will be? *Hint:* Is it appropriate to use the Poisson approximation to the binomial?

15. *General: Coin Flip* An experiment consists of tossing a coin a specified number of times and recording the outcomes.
    (a) What is the probability that the *first* head will occur on the second trial? Does this probability change if we toss the coin three times? What if we toss the coin four times? What probability distribution model do we use to compute these probabilities?
    (b) What is the probability that the *first* head will occur on the 4th trial? after the 4th trial?

16. *Testing: CPA Exam* Cathy is planning to take the Certified Public Accountant Examination (CPA exam). Records kept by the college of business from which she graduated indicate that 83% of the students who have graduated pass the CPA exam. Assume that the exam is changed each time it is given. Let n = 1, 2, 3, . . . represent

the number of times a person takes the CPA exam until the *first* pass. (Assume the trials are independent.)

(a) What is the probability that Cathy passes the CPA exam on the first try?

(b) What is the probability that Cathy passes the CPA exam on the second or third try?

DATA HIGHLIGHTS: GROUP PROJECTS

Break into small groups and discuss the following topics. Organize a brief outline in which you summarize the main points of your group discussion.

1. Powerball! Imagine, you could win a jackpot worth at least $10 million. Some jackpots have been more than $250 million! Powerball is a multistate lottery. To play Powerball, you purchase a $1 ticket. On the ticket you select five distinct white balls (numbered 1 through 53) and then one red Powerball (numbered 1 through 42). The red Powerball number may be any of the numbers 1 through 42, including any of the numbers you selected for the white balls. Every Wednesday and Saturday there is a drawing. If your chosen numbers match those drawn, you win! Figure 5-7 shows all the prizes and the probability of winning each prize and specifies how many numbers on your ticket must match those drawn to win the prize. The Multi-State Lottery Association maintains a web site that displays the results of each drawing, as well as a history of the results of previous drawings. To find out more about Powerball, visit the Brase/Brase statistics site at http://math.college.hmco.com/students and find the link to the Multi-State Lottery Association.

    (a) Assume that the jackpot is $10 million and that there will be only one jackpot winner. Figure 5-7 lists the prizes and the probability of winning each prize. What is the probability of *not winning* any prize? Consider all the prizes and their respective probabilities and the prize of $0 (no win) and its probability. Use all these values to estimate your expected winnings μ if you play one ticket. How much do you effectively contribute to the state where you purchased the ticket (ignoring the overhead cost of operating Powerball)?

    (b) Suppose that the jackpot increased to $25 million (and that there were to be only one winner). Compute your expected winnings if you buy one ticket. Does the probability of winning the jackpot change because the jackpot is higher?

    (c) Pretend that you are going to buy 10 Powerball tickets when the jackpot is $10 million. Use the random-number table to select your numbers. Check the Multi-State Lottery Association web site (or any other Powerball site) for the most recent drawing results to see if you would have won a prize.

    (d) The probability of winning any prize is about 0.0277. Suppose that you decide to buy five tickets. Use the binomial distribution to compute the probability of winning (any prize) at least once. *Note:* You will need to use the binomial formula. Carry at least three digits after the decimal.

    (e) The probability of winning *any* prize is about 0.0286. Suppose that you play Powerball 100 times. Explain why it is appropriate to use the Poisson approximation for the binomial to compute the probability of winning at least one prize. Compute $\lambda = np$. Use the Poisson table to estimate the probability of winning at least one prize.

FIGURE 5-7

Ways to Win Powerball

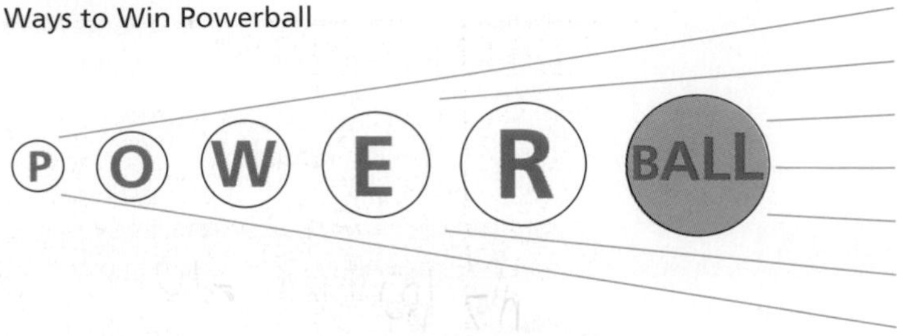

Match	Approximate Probability	Prize
5 white balls + **Powerball**	0.0000000083	Jackpot*
5 white balls	0.000000340	$100,000
4 white balls + **Powerball**	0.00000199	$5,000
4 white balls	0.0000816	$100
3 white balls + **Powerball**	0.0000936	$100
3 white balls	0.0038	$7
2 white balls + **Powerball**	0.0014	$7
1 white ball + **Powerball**	0.0888	$4
0 white balls + **Powerball**	0.0142	$3
Overall chance of winning	0.0277 (one play)	

*The Jackpot will be divided equally (if necessary) among multiple winners and is paid in 30 equal annual installments or in a reduced lump sum.

2. Would you like to travel in space, if given a chance? According to Opinion Research for Space Day Partners, you are not alone. Forty-four percent of adults surveyed agreed that they would travel in space if given a chance. Look at Figure 5-8, and use the information presented to answer the following questions.

(a) According to Figure 5-8, the probability that an adult selected at random agrees with the statement that humanity should explore planets is 64%. Round this probability to 65%, and use this estimate with the binomial distribution table to determine the probability that of 10 adults selected at random, at least half agree that humanity should explore planets.

(b) Does space exploration have an impact on daily life? Find the probability that of 10 adults selected at random, at least 9 agree that space exploration does have an impact on daily life. *Hint:* Use the formula for the binomial distribution.

(c) In a room of 35 adults, what is the expected number who would travel in space, given a chance? What is the standard deviation?

(d) What is the probability that the first adult (selected at random) you asked would agree with the statement that space will be colonized in the person's lifetime? *Hint:* Use the geometric distribution.

FIGURE 5-8

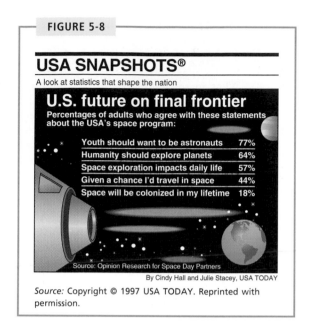

USA SNAPSHOTS®
A look at statistics that shape the nation

U.S. future on final frontier
Percentages of adults who agree with these statements about the USA's space program:

Youth should want to be astronauts	77%
Humanity should explore planets	64%
Space exploration impacts daily life	57%
Given a chance I'd travel in space	44%
Space will be colonized in my lifetime	18%

Source: Opinion Research for Space Day Partners

By Cindy Hall and Julie Stacey, USA TODAY

Source: Copyright © 1997 USA TODAY. Reprinted with permission.

LINKING CONCEPTS: WRITING PROJECTS

Discuss each of the following topics in class or review the topics on your own. Then write a brief but complete essay in which you summarize the main points. Please include formulas and graphs as appropriate.

1. Discuss what we mean by a binomial experiment. As you can see, a binomial process or binomial experiment involves a lot of assumptions! For example, all the trials are supposed to be independent and repeated under identical conditions. Is this always true? Can we always be completely certain the probability of success does not change from one trial to the next? In the real world there is almost nothing we can be absolutely sure about, so the *theoretical* assumptions of the binomial probability distribution often will not be completely satisfied. Does that mean we cannot use the binomial distribution to solve practical problems? Looking at this chapter, the answer that appears is that we can indeed use the binomial distribution even if not all the assumptions are *exactly* met. We find in practice that the conclusions are sufficiently accurate for our intended application. List three applications of the binomial distribution for which you think, although some of the assumptions are not exactly met, there is adequate reason to apply the binomial distribution anyhow.

2. Why do we need to learn the formula for the binomial probability distribution? Using the formula repeatedly can be very tedious. To cut down on tedious calculations, most people will use a binomial table such as found in Appendix II of this book.
   (a) However, there are many applications where a table in the back of *any* book is not adequate. For instance, compute

   $$P(r = 3) \quad \text{where } n = 5 \text{ and } p = 0.735$$

   Do you find the result in the table? Do the calculation by using the formula. List some other situations in which a table might not be adequate to solve a particular binomial distribution problem.

(b) The formula itself also has limitations. For instance, consider the difficulty of computing

$$P(r \geq 285) \quad \text{where } n = 500 \text{ and } p = 0.6$$

What are some of the difficulties you run into? Consider the calculation of $P(r = 285)$. You will be raising 0.6 and 0.4 to very high powers; this will give you very, very small numbers. Then you need to compute $C_{500,285}$, which is a very, very large number. When combining extremely large and extremely small numbers in the same calculation, most accuracy is lost unless you carry a huge number of significant digits. If this isn't tedious enough, then consider the steps that you need to compute

$$P(r \geq 285) = P(r = 285) + P(r = 286) + \cdots + P(r = 500)$$

Does it seem clear that we need a better way to compute $P(r \geq 285)$? In Chapter 6, you will find a much better way to compute binomial probabilities when the number of trials is large.

3. In Chapter 3, we learned about means and standard deviations. In Section 5.1, we learned that probability distributions also can have a mean and standard deviation. Discuss what is meant by the expected value and standard deviation of a binomial distribution. How does this relate back to the material we learned in Chapter 3 and Section 5.1?

4. In Chapter 2, we looked at the shapes of distributions. Review the concepts of skewness and symmetry; then categorize the following distributions as to skewness or symmetry:
   (a) A binomial distribution with $n = 11$ trials and $p = 0.50$
   (b) A binomial distribution with $n = 11$ trials and $p = 0.10$
   (c) A binomial distribution with $n = 11$ trials and $p = 0.90$

In general, does it seem true that binomial probability distributions in which the probability of success is close to 0 are skewed right, whereas those with probability of success close to 1 are skewed left?

Using TECHNOLOGY

Binomial Distributions

Although tables of binomial probabilities can be found in most libraries, such tables are often inadequate. Either the value of p (the probability of success on a trial) you are looking for is not in the table, or the value of n (the number of trials) you are looking for is too large for the table. In Chapter 6, we will study the normal approximation to the binomial. This approximation is a great help in many practical applications. Even so, we sometimes use the formula for the binomial probability distribution on a computer or graphing calculator to compute the probability we want.

APPLICATIONS

The following percentages were obtained over many years of observation by the U.S. Weather Bureau. All data listed are for the month of December.

Location	Long-Term Mean % of Clear Days in Dec.
Juneau, Alaska	18%
Seattle, Washington	24%
Hilo, Hawaii	36%
Honolulu, Hawaii	60%
Las Vegas, Nevada	75%
Phoenix, Arizona	77%

Adapted from *Local Climatological Data*, U.S. Weather Bureau publication, "Normals, Means, and Extremes" Table.

In the locations listed, the month of December is a relatively stable month with respect to weather. Since weather patterns from one day to the next are more or less the same, it is reasonable to use a binomial probability model.

1. Let r be the number of clear days in December. Since December has 31 days, $0 \leq r \leq 31$. Using appropriate computer software or calculators available to you, find the probability $P(r)$ for each of the listed locations when $r = 0, 1, 2, \ldots, 31$.

2. For each location, what is the expected value of the probability distribution? What is the standard deviation?

You may find that using cumulative probabilities and appropriate subtraction of probabilities will make the solution of Applications 3 to 7 easier than adding probabilities.

3. Estimate the probability that Juneau will have at most 7 clear days in December.

4. Estimate the probability that Seattle will have from 5 to 10 (including 5 and 10) clear days in December.

5. Estimate the probability that Hilo will have at least 12 clear days in December.

6. Estimate the probability that Phoenix will have 20 or more clear days in December.

7. Estimate the probability that Las Vegas will have from 20 to 25 (including 20 and 25) clear days in December.

Technology Hints

The Tech Note in Section 5.2 gives specific instructions for binomial distribution functions on the TI-83Plus, Excel, and Minitab. For ComputerStat, select Binomial Coefficients and Probability Distributions under the Probability Distribution menu.

6

Normal Distributions

One cannot escape the feeling that these mathematical formulas have an independent existence and an intelligence of their own, that they are wiser than we are, wiser even than their discoverers, that we get more out of them than was originally put into them.

—Heinrich Hertz

How can it be that mathematics, a product of human thought independent of experience, is so admirably adapted to the objects of reality?

—Albert Einstein

Heinrich Rudolf Hertz (1857–1894)

This German physicist was largely responsible for doing pioneering work in electromagnetic theory.

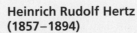

For on-line student resources, visit **math.college.hmco.com/students** and follow the Statistics links to the Brase/Brase, *Understandable Statistics,* 7th edition web site.

Heinrich Hertz was a pioneer in the study of radio waves. His work and the later work of Maxwell and Marconi led the way to modern radio, television, and radar. Albert Einstein is world renowned for his great discoveries in relativity and quantum mechanics. Everyone who has worked in both mathematics and real-world applications cannot help but marvel at how the "pure thought" of the mathematical sciences can predict and explain events in other realms. In this chapter, we will study the most important type of probability distribution in all of mathematical statistics: the normal distribution. Why is the normal distribution so important? Two of the reasons are that it applies to a wide variety of situations, and other distributions tend to become normal under certain conditions.

◇ What are some characteristics of a normal distribution? What does the empirical rule tell us about the data spread around the mean? How can this information be used in quality control?   (SECTION 6.1)

◇ Can we compare apples and oranges, or maybe elephants and butterflies? In most cases, the answer is no, . . . unless we first *standardize* our measurements. What are a *standard normal distribution* and a *standard z score*?   (SECTION 6.2)

◇ How do we convert *any* normal distribution to a *standard normal* distribution? How do we find probabilities of "standardized events"?   (SECTION 6.3)

◇ The binomial and normal are two of the most important probability distributions in statistics. Under certain limiting conditions, the binomial can be thought to evolve (or envelope) into the normal distribution. How can we apply this in the real world?   (SECTION 6.4)

Large Auditorium Shows: How Many Will Attend?

For many years, Denver, as well as most other cities, has hosted large exhibition shows in big auditoriums. These shows include house and gardening shows, fishing and hunting shows, car shows, boat shows, Native American powwows, and so on. Information provided by Denver exposition sponsors indicates that most shows have an average attendance of about 8000 people per day with an estimated standard deviation of about 500 people. Suppose that the daily attendance figures follow a normal distribution.

1. What is the probability that the daily attendance will be fewer than 7200 people?

2. What is the probability that the daily attendance will be more than 8900 people?

3. What is the probability that the daily attendance will be between 7200 and 8900 people?

Most exhibition shows open in the morning and close in the late evening. A study of Saturday arrival times showed that the average arrival time was 3 hours and 48 minutes after the doors open, and the standard deviation was estimated at about 52 minutes. Suppose that the arrival times follow a normal distribution.

4. At what time after the doors open will 90% of the people who are coming to the Saturday show have arrived?

5. At what time after the doors open will only 15% of the people who are coming to the Saturday show have arrived?

6. Do you think the probability distribution of arrival times for Friday might be different from the distribution of arrival times for Saturday? Explain.

6.1
Graphs of Normal Probability Distributions

FOCUS POINTS

✓ Graph a normal curve and summarize its important properties.

✓ Apply the empirical rule to solve real-world problems.

✓ Use control limits to construct control charts. Examine the chart for three possible out-of-control signals.

One of the most important examples of a continuous probability distribution is the *normal distribution*. This distribution was studied by the French mathematician Abraham de Moivre (1667–1754) and later by the German mathematician Carl Friedrich Gauss (1777–1855), whose work is so important that the normal distribution is sometimes called *gaussian*. The work of these mathematicians provided a foundation on which much of the theory of statistical inference is based.

Applications of a normal probability distribution are so numerous that some mathematicians refer to it as "a veritable Boy Scout knife of statistics." However, before we can apply it, we must examine some of the properties of a normal distribution.

A rather complicated formula, presented later in this section, defines a normal distribution in terms of μ and σ, the mean and standard deviation of the population distribution. It is only through this formula that we can verify if a distribution is normal. However, we can look at the graph of a normal distribution and get a good pictorial idea of some of the essential features of any normal distribution.

Normal curve

The graph of a normal distribution is called a *normal curve*. It possesses a shape very much like the cross section of a pile of dry sand. Because of its shape, blacksmiths would sometimes use a pile of dry sand in the construction of a mold for a bell. Thus the normal curve is also called a *bell-shaped curve* (see Figure 6-1).

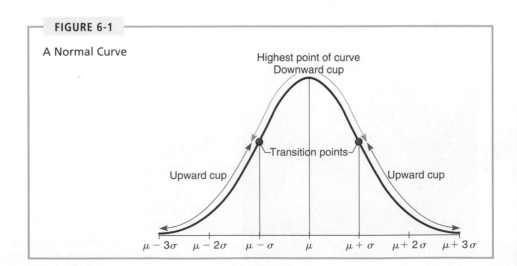

FIGURE 6-1

A Normal Curve

Highest point of curve
Downward cup

Transition points

Upward cup   Upward cup

$\mu - 3\sigma$ $\mu - 2\sigma$ $\mu - \sigma$ μ $\mu + \sigma$ $\mu + 2\sigma$ $\mu + 3\sigma$

We see that a general normal curve is smooth and symmetrical about the vertical line over the mean μ. Notice that the highest point of the curve occurs over μ. If the distribution were graphed on a piece of sheet metal, cut out, and placed on a knife edge, the balance point would be at μ. We also see that the curve tends to level out and approach the horizontal (x axis) like a glider making a landing. However, in mathematical theory, such a glider would never quite finish its landing because a normal curve never touches the horizontal axis.

The parameter σ controls the spread of the curve. The curve is quite close to the horizontal axis at $\mu + 3\sigma$ and $\mu - 3\sigma$. Thus, if the standard deviation σ is large, the curve will be more spread out; if it is small, the curve will be more peaked. Figure 6-1 shows the normal curve cupped downward for an interval on either side of the mean μ. Then it begins to cup upward as we go to the lower part of the bell. The exact places where the transition between the upward and downward cupping occur are above the points $\mu + \sigma$ and $\mu - \sigma$.

Important properties of a normal curve

1. The curve is bell-shaped with the highest point over the mean μ.

2. It is symmetrical about a vertical line through μ.

3. The curve approaches the horizontal axis but never touches or crosses it.

4. The transition points between cupping upward and downward occur above $\mu + \sigma$ and $\mu - \sigma$.

GUIDED
EXERCISE **1**

Curves that are not normal

Each of the curves in Figure 6-2 fails to be a normal curve. Give reasons why these curves are not normal curves.

FIGURE 6-2

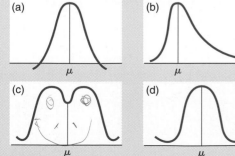

(a) A normal curve gets closer and closer to the horizontal axis, but it never touches it or crosses it.

(b) A normal curve must be symmetrical. This curve is not.

(c) A normal curve is bell-shaped with one peak. Because this curve has two peaks, it is not normal.

(d) The tails of a normal curve must get closer and closer to the x axis. In this curve the tails are going away from the x axis.

GUIDED EXERCISE 2

Identify μ and σ on a normal curve

The points *A*, *B*, and *C* are indicated on the normal curve in Figure 6-3. One of these points is μ, one is $\mu + \sigma$, and one is $\mu - 2\sigma$.

FIGURE 6-3 A Normal Curve

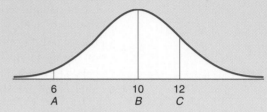

(a) Which point corresponds to the mean? What is the value of μ?

⇒ The mean μ is under the peak of the normal curve. The point *B* corresponds to the mean, so $\mu = 10$.

(b) Which point corresponds to $\mu + \sigma$? Use the values of $\mu + \sigma$ and μ to compute σ.

⇒ The point *C* where the curve changes from cupped down to cupped up is one standard deviation σ from the mean. The point *C* is $\mu + \sigma$. Since $\mu + \sigma = 12$ and $\mu = 10$, $\sigma = 2$.

(c) Which point corresponds to $\mu - 2\sigma$?

⇒ Since $\mu = 10$ and $\sigma = 2$, we see that

$$\mu - 2\sigma = 10 - 2(2) = 6$$

Point *A* corresponds to $\mu - 2\sigma$.

The parameters that control the shape of a normal curve are the mean μ and the standard deviation σ. When both μ and σ are specified, a specific normal curve is determined. In brief, μ locates the balance point, and σ determines the extent of the spread.

GUIDED EXERCISE 3

Identify μ and σ on a normal curve

Look at the normal curves in Figure 6-4.

Continued

GUIDED EXERCISE 3 continued

FIGURE 6-4

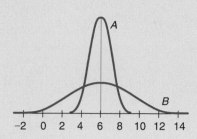

(a) Do these distributions have the same mean? If so, what is it?

⇨ The means are the same, since both graphs have the high point over 6. $\mu = 6$.

(b) One of the curves corresponds to a normal distribution with $\sigma = 3$ and the other to one with $\sigma = 1$. Which curve has which σ?

⇨ Curve *A* has $\sigma = 1$, and curve *B* has $\sigma = 3$. (Since curve *B* is more spread out, it has the larger σ value.)

◇ **COMMENT** The normal distribution curve is always above the horizontal axis. The area beneath the curve and above the axis is exactly one. As such, the normal distribution curve is an example of a *density curve*. The formula used to generate the shape of the normal distribution curve is called the *normal density function*. If *x* is a normal random variable with mean μ and standard deviation σ, the formula for the normal density function is

$$f(x) = \frac{e^{-\frac{1}{2}\left(\frac{x-\mu}{\sigma}\right)^2}}{\sigma\sqrt{2\pi}}$$

In this text, we will not use this formula explicitly. However, we will use tables of areas based on the normal density function.   ◇

Empirical rule

The total area under any normal curve studied in this book will *always* be 1. The graph of the normal distribution is important because the portion of the *area* under the curve above a given interval represents the *probability* that a measurement will lie in that interval.

In Section 3.2, we studied Chebyshev's theorem. This theorem gives us information about the *smallest* proportion of data that lies within 2, 3, or *k* standard deviations of the mean. This result applies to *any* distribution. However, for normal distributions, we can get a much more precise result, which is given by the *empirical rule*.

Empirical rule

For a distribution that is symmetrical and bell-shaped (in particular, for a normal distribution):

Approximately 68% of the data values will lie within one standard deviation on each side of the mean.

Approximately 95% of the data values will lie within two standard deviations on each side of the mean.

Approximately 99.7% (or almost all) of the data values will be within three standard deviations on each side of the mean.

The preceding statement is called the *empirical rule* because, for symmetrical, bell-shaped distributions, the given percentages are observed in practice. Furthermore, for the normal distribution, the empirical rule is a direct consequence of the very nature of the distribution (see Figure 6-5). Notice that the empirical rule is a stronger statement than Chebyshev's theorem in that it gives *definite percentages*, not just lower limits. Of course, the empirical rule applies only to normal or symmetrical, bell-shaped distributions, whereas Chebyshev's theorem applies to all distributions.

EXAMPLE 1

Empirical rule

The playing life of a Sunshine radio is normally distributed with mean $\mu = 600$ hours and standard deviation $\sigma = 100$ hours. What is the probability that a radio selected at random will last from 600 to 700 hours?

SOLUTION: The probability that the playing time will be between 600 and 700 hours is equal to the percentage of the total area under the curve that is shaded in Figure 6-6. Since $\mu = 600$ and $\mu + \sigma = 600 + 100 = 700$, we see that the shaded area is simply the area between μ and $\mu + \sigma$. The area from μ to $\mu + \sigma$ is 34% of the total area. This tells us that the probability a Sunshine radio will last between 600 and 700 playing hours is about 0.34.   ◊

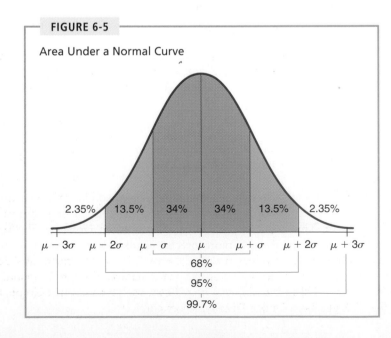

FIGURE 6-5

Area Under a Normal Curve

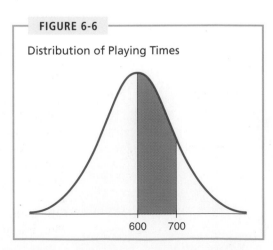

FIGURE 6-6

Distribution of Playing Times

GUIDED EXERCISE 4

Empirical rule

The yearly wheat yield per acre on a particular farm is normally distributed with mean $\mu = 35$ bushels and standard deviation $\sigma = 8$ bushels.

(a) Shade the area under the curve in Figure 6-7 that represents the probability that an acre will yield between 19 and 35 bushels.

See Figure 6-8.

(b) Is the area the same as the area between $\mu - 2\sigma$ and μ?

Yes, since $\mu = 35$ and $\mu - 2\sigma = 35 - 2(8) = 19$.

FIGURE 6-7

FIGURE 6-8 Completion of Figure 6-7

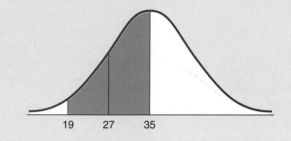

(c) Use Figure 6-5 to find the percentage of area over the interval between 19 and 35.

The area between the values $\mu - 2\sigma$ and μ is 47.5% of the total area.

(d) What is the probability that the yield will be between 19 and 35 bushels per acre?

It is 47.5% of the total area, which is 1. Therefore, the probability is 0.475 that the yield will be between 19 and 35 bushels.

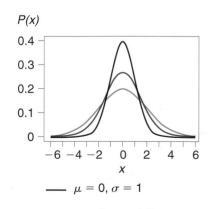

$\mu = 0, \sigma = 1$

$\mu = 0, \sigma = 1.5$

$\mu = 0, \sigma = 2$

TECH NOTE   We can graph normal distributions on the TI-83Plus, Excel, and Minitab. In each technology, set the range of x values between -3.5σ and 3.5σ. Then use the built-in normal density functions to generate the corresponding y values.

TI-83Plus Press the **Y =** key. Then, under **DISTR**, select **1:normalpdf(x,μ,σ)** and fill in desired μ and σ values. Press the **WINDOW** key. Set **Xmin** to $\mu - 3\sigma$ and **Xmax** to $\mu + 3\sigma$. Finally, press the **ZOOM** key and select option **0:ZoomFit.**

Excel In one column, enter x values from -3.5σ to 3.5σ in increments of 0.2σ. In the next column, enter y values by using the menu choices **Paste function** [f_x] ➤ **Statistical** ➤ **NORMDIST(x, μ, σ, false)**. Next, use the chart wizard, and select **XY(scatter).** Choose the first picture with the dots connected and fill in the dialogue boxes.

Minitab In one column, enter x values from -3.5σ to 3.5σ in increments of 0.2σ. In the next column, enter y values by using the menu choices **Calc** ➤ **Probability Distribution** ➤ **Normal.** Fill in the dialogue box. Next, use menu choices **Graph** ➤ **Plot.** Fill in the dialogue box. Under Display, select connect.

Control Charts

If we are examining data over a period of equally spaced time intervals or in some sequential order, then *control charts* are especially useful. Business managers and people in charge of production processes are aware that there exists an inherent amount of variability in any sequential set of data. For example, the sugar content of bottled drinks taken sequentially off a production line, the extent of clerical errors in a bank from day to day, advertising expenses from month to month, or even the number of new customers from year to year are examples of sequential data. There is a certain amount of variability in each.

A random variable x is said to be in *statistical control* if it can be described by the *same* probability distribution when it is observed at successive points in time. Control charts combine graphic and numerical descriptions of data with probability distributions.

Control charts were invented in the 1920s by Walter Shewhart at Bell Telephone Laboratories. Since a control chart is a *warning device,* it is not absolutely necessary that our assumptions and probability calculations be precisely correct. For example, the x distributions need not follow a normal distribution exactly. Any mound-shaped and more or less symmetrical distribution will be good enough.

How do we make a control chart? A control chart for a variable x is a plot of the observed x values (on the vertical scale) in time sequence order (the horizontal scale represents time). There is a center line at height μ and dashed control limits at $\mu \pm 2\sigma$ and at $\mu \pm 3\sigma$. How do we pick values for μ and σ? In most practical cases, values for μ (population mean) and σ (population standard deviation) are computed from past data for which the process we are studying was known to be *in control.* Methods for choosing the sample size to fit given error tolerances can be found in Chapter 8.

Sometimes values for μ and σ are chosen as *target values.* That is, μ and σ values are chosen as set goals or targets that reflect the production level or service level at which a company hopes to perform. To be realistic, such target assignments for μ and σ should be reasonably close to actual data taken when the process was operating at a satisfactory production level. In Example 2, we will make a control chart; then we discuss ways to analyze it to see if a process or service is "in control."

EXAMPLE 2
Control chart

Susan Tamara is director of personnel at the Antlers Lodge in Denali National Park, Alaska. Every summer Ms Tamara hires many part-time employees from all over the United States. Most are college students seeking summer employment. Perhaps the biggest activity for Ms Tamara's staff is that of "making up" the rooms each day. The lodge has 385 rooms, and from long experience Ms Tamara has determined that by 3:30 P.M. each day the average number of rooms not made up is $\mu = 19.3$ with standard deviation $\sigma = 4.7$. Although the lodge has a policy that guest rooms be made up by 3:30 P.M., Ms Tamara knows that there will always be a few rooms not made up by this time because there is a high personnel turnover and corresponding reassignment of staff to new jobs for which they are in a training period. Every 15 days Ms Tamara has a general staff meeting. Before the meeting, she examines several control charts for the restaurant, the gift shop, and, of course, housekeeping. For the past 15 days, the housekeeping unit has reported the following number of rooms not ready by 3:30 P.M. (Table 6-1). Make a control chart for these data.

TABLE 6-1   Number of Rooms not Made Up by 3:30 P.M.

Day	1	2	3	4	5	6	7	8
x = number of rooms	11	20	25	23	16	19	8	25

Day	9	10	11	12	13	14	15	
x = number of rooms	17	20	23	29	18	14	10	

Mt. McKinley, Denali National Park

Before we make a control chart, we need to know a few things about the distribution of rooms not made up by 3:30 P.M. Ms Tamara is aware from long experience that the distribution is symmetrical and bell-shaped. It is approximately normal, with mean $\mu = 19.3$ and standard deviation $\sigma = 4.7$. In addition, this distribution of x values is acceptable to the top administration of the Antlers Lodge.

SOLUTION: A control chart for a variable x is a plot of the observed x values (vertical scale) in time sequence order (the horizontal scale represents time). Place horizontal lines at

The mean $\mu = 19.3$

The control limits $\mu \pm 2\sigma = 19.3 \pm 2(4.7)$ or 9.90 and 28.70

The control limits $\mu \pm 3\sigma = 19.3 \pm 3(4.7)$ or 5.20 and 33.40

Then plot the data from Table 6-1. (See Figure 6-9.)   ◇

Once we have made a control chart, the main question is the following: As time goes on, is the x variable continuing in this same distribution, or is the distribution of x values changing? If the x distribution is continuing more or less the same, we say it is in *statistical control*. If it is not, we say it is *out of control*.

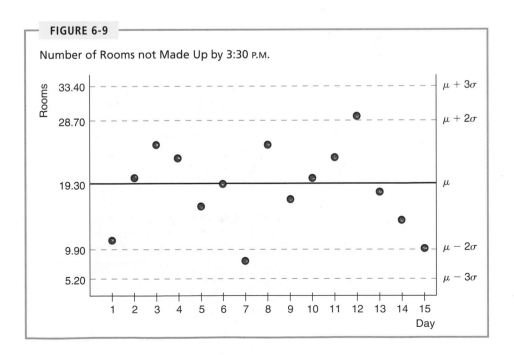

FIGURE 6-9

Number of Rooms not Made Up by 3:30 P.M.

Out-of-control warning signals

Many popular methods can set off a warning signal that a process is out of control. Remember, a random variable x is said to be *out of control* if successive time measurements of x indicate that it is no longer following the target probability distribution. We will assume that the target distribution is (approximately) normal and has (user-set) target values for μ and σ.

Three of the most popular warning signals are described next:

1. ***Out-of-Control Signal I:*** *One point falls beyond the 3σ level.* What is the probability signal I will be a false alarm? By the empirical rule, the probability that a point lies within 3σ of the mean is 0.997. The probability that signal I will give a false alarm is $1 - 0.997 = 0.003$. Remember, a false alarm means that the x distribution is really on the target distribution, and we simply have a very rare (probability of 0.003) event. (See Figure 6-10a.)

2. ***Out-of-Control Signal II:*** *A run of nine consecutive points on one side of the center line (the line at target value μ).* What is the probability that signal II is a false alarm? If the x distribution and the target distribution are the same, then there is a 50% chance that any x values will lie above or below the center line at μ. Because the samples are (time) independent, the probability of a run of nine points on one side of the center line is $(0.5)^9 = 0.002$. If we consider both sides, this probability becomes 0.004. Therefore, the probability that signal II is a false alarm is approximately 0.004. (See Figure 6-10b.)

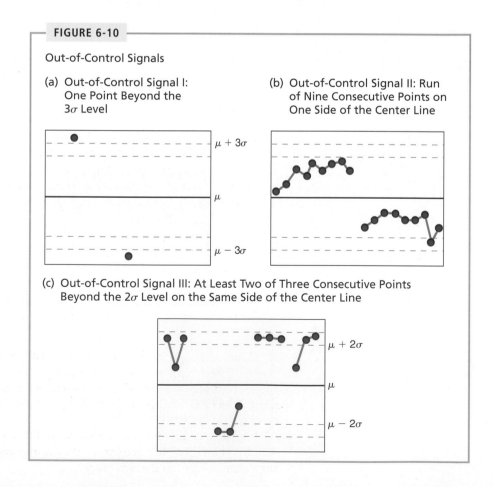

FIGURE 6-10

Out-of-Control Signals

(a) Out-of-Control Signal I: One Point Beyond the 3σ Level

(b) Out-of-Control Signal II: Run of Nine Consecutive Points on One Side of the Center Line

(c) Out-of-Control Signal III: At Least Two of Three Consecutive Points Beyond the 2σ Level on the Same Side of the Center Line

3. *Out-of-Control Signal III: At least two of three consecutive points lie beyond the 2σ level on the same side of the center line.* What is the probability that signal III will produce a false alarm? By the empirical rule, the probability that an *x* value will lie above the 2σ level is about 0.023. If we use the binomial probability distribution (with success being the point is above 2σ), then the probability of two or more successes out of three trials is

$$\frac{3!}{2!1!}(0.023)^2(0.977) + \frac{3!}{3!0!}(0.023)^3 \approx 0.002$$

If we take into account *both* above or below the center line, it follows that the probability that signal III is a false alarm is about 0.004. (See Figure 6-10c.)

◇ **SUMMARY**

Type of Warning Signal	Probability of a False Alarm
Type I: Point beyond 3σ	0.003
Type II: Run of nine consecutive points all below center line μ or all above center line μ	0.004
Type III: At least two out of three consecutive points beyond 2σ	0.004

◇

Remember, a control chart is only a warning device, and it is possible to get a false alarm. A false alarm happens when one (or more) of the out-of-control signals occurs, but the *x* distribution is really on the target or assigned distribution. In this case, we simply have a rare event (probability of 0.003 or 0.004). In practice, whenever a control chart indicates that a process is out of control, it is usually a good precaution to examine what is going on. If the process is out of control, corrective steps can be taken before things get a lot worse. The rare false alarm is a small price to pay if we can avert what might become real trouble.

From an intuitive point of view, signal I could be thought of as a blowup, something dramatically out of control. Signal II could be thought of as a slow drift out of control. Signal III is between a blowup and a slow drift.

EXAMPLE 3
Control chart

Ms Tamara of the Antlers Lodge examines the control chart for housekeeping. During the staff meeting, she makes recommendations about improving service, or if all is going well, she gives her staff a well-deserved "pat on the back." The most recent control chart for housekeeping is the one shown in Figure 6-11 on the next page. Look at this control chart to determine if the housekeeping process is out of control or not.

SOLUTION: The *x* values are more or less evenly distributed about the mean μ = 19.3. None of the points are outside the μ ± 3σ limit (i.e., above 33.40 or below 5.20 rooms). There is no run of nine consecutive points above or below μ. No two of three consecutive points are beyond the μ ± 2σ limit (i.e., above 28.7 or below 9.90 rooms).

It appears that the *x* distribution is "in control." At the staff meeting, Ms Tamara should tell her employees they are doing a reasonably good job and that they should keep up the fine work!   ◇

FIGURE 6-11

Number of Rooms not Made Up by 3:30 P.M.

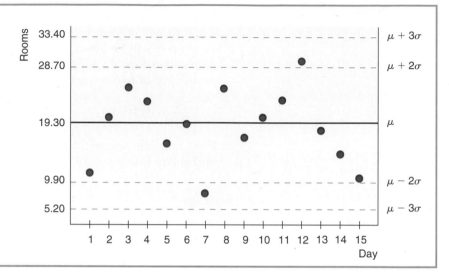

GUIDED EXERCISE 5

Control chart

Over the next 15-day period, let's suppose that housekeeping again reported the number of rooms not made up by 3:30 P.M. to Ms Tamara of the Antlers Lodge. The data in Table 6-2 show the results.

TABLE 6-2   Next 15-Day Report of Rooms not Made Up by 3:30 P.M.

Day	1	2	3	4	5	6	7	8
x = number of rooms	25	8	23	15	26	24	31	21
Day	9	10	11	12	13	14	15	
x = number of rooms	27	20	25	21	27	11	16	

(a) We assume that we are still working with the symmetrical, bell-shaped distribution of x values, with mean $\mu = 19.3$ and $\sigma = 4.7$. Compute the "control limits" of $\mu \pm 2\sigma$ and $\mu \pm 3\sigma$. Draw a control chart showing the solid line at the mean and the dashed lines at the control limits. Plot the data for the 15-day period.

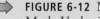 **FIGURE 6-12** Next 15-Day Report of Rooms not Made Up by 3:30 P.M.

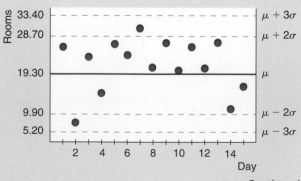

Continued

GUIDED EXERCISE 5 continued

(b) Interpret the control chart of part (a).

⟹ Days 5 to 13 are above $\mu = 19.3$. We have nine consecutive days on one side of the mean. This is a warning signal! It would appear that the mean μ is slowly drifting up beyond the target value of 19.3. The chart indicates that housekeeping is "out of control." Ms Tamara should take corrective measures at her next staff meeting.

(c) Over another 15-day period Ms Tamara obtained the data shown in Table 6-3 for housekeeping. Make a control chart using target values. $\mu = 19.3$ and $\sigma = 4.7$.

⟹ **FIGURE 6-13** 3rd Housekeeping Data Report

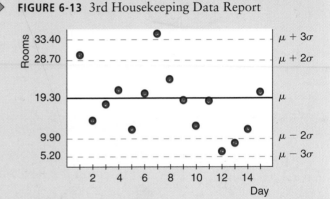

TABLE 6-3 3rd Housekeeping Data Report

Day	1	2	3	4	5	6	7	8
Number of rooms	29	14	18	21	11	20	35	24
Day	9	10	11	12	13	14	15	
Number of rooms	19	12	19	6	8	11	20	

(d) Interpret the control chart of part (c).

⟹ On day 7 we have a data value beyond $\mu + 3\sigma$ (i.e., above 33.40). On days 11, 12, and 13 we have two of three data values beyond $\mu - 2\sigma$. (i.e., below 9.90). The occurrences on both these periods are out-of-control warning signals. Ms Tamara might ask her staff about both these periods. There may be a lesson to be learned about day 7 when housekeeping apparently had a lot of trouble. Also, days 11, 12, and 13 were very good days. Perhaps a lesson could be learned about why things went so well.

VIEWP●INT

In Control? Out of Control?

If you care about quality, you also must care about control! Dr. Walter Shewhart invented control charts when he was working for Bell Laboratories. The great contribution of control charts is to separate variation into two sources: (1) random or chance causes (in control) and (2) special or assignable causes (out of control). A process is said to be in *statistical control* when it is no longer afflicted with special or assignable causes. The performance of a process that is in statistical control is predictable. Predictability and quality control tend to be closely associated.

(Source: Adapted from the classic text *Statistical Methods from the Viewpoint of Quality Control,* by W. A. Shewhart, with Foreword by W. E. Deming, Dover Publications.)

SECTION 6.1   PROBLEMS

1. *General: Curves* Which, if any, of the curves in Figure 6-14 look(s) like a normal curve? If a curve is not a normal curve, tell why.

2. *General: Normal Curves* Look at the normal curve in Figure 6-15, and find μ, $\mu + \sigma$, and σ.

3. *General: Normal Curves* Look at the two normal curves in Figures 6-16 and 6-17. Which has the larger standard deviation? What is the mean of the curve in Figure 6-16? What is the mean of the curve in Figure 6-17?

4. *General: Normal Curves* Sketch a normal curve
   (a) with mean 15 and standard deviation 2.
   (b) with mean 15 and standard deviation 3.
   (c) with mean 12 and standard deviation 2.
   (d) with mean 12 and standard deviation 3.
   (e) Consider two normal curves. If the first one has a larger mean than the second one, must it have a larger standard deviation than the second one as well? Explain your answer.

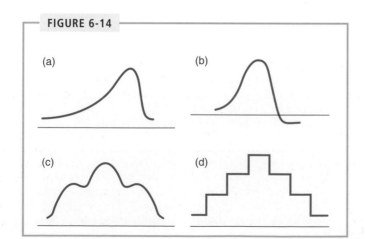

FIGURE 6-14

(a)   (b)

(c)   (d)

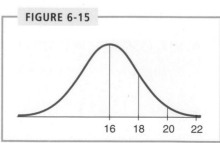

FIGURE 6-15

16   18   20   22

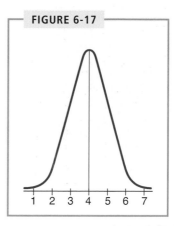

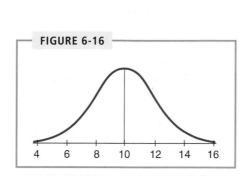

5. *General: Normal Curves* What percentage of the area under the normal curve lies
   (a) to the left of μ?
   (b) between $\mu - \sigma$ and $\mu + \sigma$?
   (c) between $\mu - 3\sigma$ and $\mu + 3\sigma$?

6. *General: Normal Curves* What percentage of the area under a normal curve lies
   (a) to the right of μ?
   (b) between $\mu - 2\sigma$ and $\mu + 2\sigma$?
   (c) to the right of $\mu + 3\sigma$?

7. *Distribution: Heights of Coeds* Assuming that the heights of college women are normally distributed, with mean 65 in. and standard deviation 2.5 in. (based on information from *Statistical Abstract of the United States,* 112th Edition), answer the
   following questions. (*Hint:* Use Problems 5 and 6 and Figure 6-5.)
   (a) What percentage of women are taller than 65 in.?
   (b) What percentage of women are shorter than 65 in.?
   (c) What percentage of women are between 62.5 in. and 67.5 in.?
   (d) What percentage of women are between 60 in. and 70 in.?

8. *Distribution: Rhode Island Red Chickens* The incubation time for Rhode Island
   Red chicks is normally distributed with mean of 21 days and standard deviation of
   approximately 1 day (based on information from *World Book Encyclopedia*). Look
   at Figure 6-5 and answer the following questions. If 1000 eggs are being incubated,
   how many chicks do we expect will hatch
   (a) in 19 to 23 days?
   (b) in 20 to 22 days?
   (c) in 21 days or fewer?
   (d) in 18 to 24 days? (Assume all eggs eventually hatch.)
   (*Note:* In this problem, let us agree to think of a single day or a succession of days
   as a continuous interval of time.)

9. *Archaeology: Tree Rings* At Burnt Mesa Pueblo, archaeological studies have used
   the method of tree-ring dating in an effort to determine when prehistoric people
   lived in the pueblo. Wood from several excavations gave a mean of (year) 1243
   with a standard deviation of 36 years (*Bandelier Archaeological Excavation
   Project: Summer 1989 Excavations at Burnt Mesa Pueblo,* edited by Kohler,
   Washington State University Department of Anthropology). The distribution of
   dates was more or less mound-shaped and symmetrical about the mean. Use the
   empirical rule to

(a) estimate a range of years centered about the mean in which about 68% of the data (tree-ring dates) will be found.

(b) estimate a range of years centered about the mean in which about 95% of the data (tree-ring dates) will be found.

(c) estimate a range of years centered about the mean in which almost all the data (tree-ring dates) will be found.

10. *Vending Machine: Soft Drinks* A vending machine automatically pours soft drinks into cups. The amount of soft drink dispensed into a cup is normally distributed with mean of 7.6 oz and standard deviation of 0.4 oz. Examine Figure 6-5 and answer the following questions.

(a) Estimate the probability that the machine will overflow an 8-oz cup.

(b) Estimate the probability that the machine will not overflow an 8-oz cup.

(c) The machine has just been loaded with 850 cups. How many of these do you expect will overflow when served?

11. *Pain Management: Laser Therapy* "Effect of Helium-Neon Laser Auriculotherapy on Experimental Pain Threshold" is the title of an article in the journal *Physical Therapy* (Vol. 70, No. 1, pp. 24–30). In this article, laser therapy was discussed as a useful alternative to drugs in pain management of chronically ill patients. To measure pain threshold, a machine was used that delivered low-voltage direct current to different parts of the body (wrist, neck, and back). The machine measured current in milliamperes (mA). The pretreatment experimental group in the study had an average threshold of pain (pain was first detectable) at $\mu = 3.15$ mA with standard deviation $\sigma = 1.45$ mA. Assume that the distribution of threshold pain so measured in milliamperes is symmetrical and more or less mound-shaped. Use the empirical rule to

(a) estimate a range of milliamperes centered about the mean in which about 68% of the experimental group will have a threshold of pain.

(b) estimate a range of milliamperes centered about the mean in which about 95% of the experimental group will have a threshold of pain.

12. *Control Charts: Yellowstone National Park* Yellowstone Park Medical Services (YPMS) provides emergency health care for park visitors. Such health care includes treatment for everything from indigestion and sunburn to more serious injuries. A recent issue of *Yellowstone Today* (National Park Service Publication) indicated that the average number of visitors treated each day by YPMS was 21.7. The estimated standard deviation was 4.2 (summer data). The distribution of numbers treated is approximately mound-shaped and symmetrical.

(a) For a 10-day summer period, the following data show the number of visitors treated each day by YPMS:

Day	1	2	3	4	5	6	7	8	9	10
Number treated	25	19	17	15	20	24	30	19	16	23

Make a control chart for the daily number of visitors treated by YPMS, and plot the data on the control chart. Do the data indicate that the number of visitors treated by YPMS is "in control"? Explain your answer.

(b) For another 10-day summer period, the following data were obtained:

Day	1	2	3	4	5	6	7	8	9	10
Number treated	20	15	12	21	24	28	32	36	35	37

Make a control chart, and plot the data on the chart. Do the data indicate that the number of visitors treated by YPMS is "in control" or "out of control"? Explain your answer. Identify all out-of-control signals by type (I, II, or III). If you were the park superintendent, do you think YPMS might need some (temporary) extra help? Explain.

13. *Control Charts: Bank Loans* Tri-County Bank is a small independent bank in central Wyoming. This is a rural bank that makes loans on items as small as horses and pickup trucks to items as large as ranch land. Total monthly loan requests are used by bank officials as an indicator of economic business conditions in this rural community. The mean monthly loan request for the past several years has been 615.1 (in thousands of dollars) with a standard deviation of 11.2 (in thousands of dollars). The distribution of loan requests is approximately mound-shaped and symmetrical.

(a) For 12 months, the following monthly loan requests (in thousands of dollars) were made to Tri-County Bank:

Month	1	2	3	4	5	6
Loan request	619.3	625.1	610.2	614.2	630.4	615.9

Month	7	8	9	10	11	12
Loan request	617.2	610.1	592.7	596.4	585.1	588.2

Make a control chart for the total monthly loan requests, and plot the preceding data on the control chart. From the control chart, would you say the local business economy is heating up or cooling down? Explain your answer by referring to any trend you may see on the control chart. Identify all out-of-control signals by type (I, II, or III).

(b) For another 12-month period, the following monthly loan requests (in thousands of dollars) were made to Tri-County Bank:

Month	1	2	3	4	5	6
Loan request	608.3	610.4	615.1	617.2	619.3	622.1

Month	7	8	9	10	11	12
Loan request	625.7	633.1	635.4	625.0	628.2	619.8

Make a control chart for the total monthly loan requests, and plot the preceding data on the control chart. From the control chart, would you say the local business economy is heating up, cooling down, or about normal? Explain your answer by referring to the control chart. Identify all out-of-control signals by type (I, II, or III).

14. *Control Charts: Motel Rooms* The manager of Motel 11 has 316 rooms in Palo Alto, California. From observation over a long time, she knows that on an average night 268 rooms will be rented. The long-term standard deviation is 12 rooms. This distribution is approximately mound-shaped and symmetrical.

(a) For 10 consecutive nights, the following number of rooms were rented each night:

Night	1	2	3	4	5	6
Number of rooms	234	258	265	271	283	267

Night	7	8	9	10
Number of rooms	290	286	263	240

Make a control chart for the number of rooms rented each night, and plot the preceding data on the control chart. Looking at the control chart, would you say the number of rooms rented during this 10-night period has been unusually low? unusually high? about what was expected? Explain your answer. Identify all out-of-control signals by type (I, II, or III).

(b) For another 10 consecutive nights, the following number of rooms were rented each night:

Night	1	2	3	4	5	6
Number of rooms	238	245	261	269	273	250

Night	7	8	9	10
Number of rooms	241	230	215	217

Make a control chart for the number of rooms rented each night, and plot the preceding data on the control chart. Would you say the room occupancy has been high? low? or about what was expected? Explain your answer. Identify all out-of-control signals by type (I, II, or III).

15. *Control Chart: Air Pollution* The visibility standard index (VSI) is a measure of Denver air pollution that is reported each day in the *Rocky Mountain News*. The index ranges from 0 (excellent air) to 200 (very bad air). During winter months, when air pollution is higher, the index has a mean of about 90 (rated as fair) with a standard deviation of approximately 30. Suppose that for 15 days the following VSI was reported each day:

Day	1	2	3	4	5	6	7	8	9
VSI	80	115	100	90	15	10	53	75	80

Day	10	11	12	13	14	15
VSI	110	165	160	120	140	195

Make a control chart for the VSI, and plot the preceding data on the control chart. Identify all out-of-control signals (high or low) that you find in the control chart by type (I, II, or III).

6.2   Standard Units and Areas Under the Standard Normal Distribution

FOCUS POINTS

✓ Given μ and σ, convert raw data to z scores.

✓ Given μ and σ, convert z scores to raw data.

✓ Graph the standard normal distribution, and find areas under the standard normal curve.

z Scores and Raw Scores

Normal distributions vary from one another in two ways: The mean μ may be located anywhere on the x axis, and the bell shape may be more or less spread according to the size of the standard deviation σ. The differences among the normal distributions cause difficulties when we try to compute the area under the curve in a specified interval of x values and, hence, the probability that a measurement will fall into that interval.

It would be a futile task to try to set up a table of areas under the normal curve for each different μ and σ combination. We need a way to standardize the distributions so that we can use *one* table of areas for *all* normal distributions. We achieve this standardization by considering how many standard deviations a measurements lies from the mean. In this way, we can compare a value in one normal distribution with a value in another different normal distribution. The next situation shows how this is done.

Suppose that Tina and Jack are in two different sections of the same course. Each section is quite large, and the scores on the midterm exams of each section follow a normal distribution. In Tina's section, the average (mean) was 64 and her score was 74. In Jack's section, the mean was 72 and his score was 82. Both Tina and Jack were pleased that their scores were each 10 points above the average of each respective section. However, the fact that each was 10 points above average does not really tell us how each did *with respect to the other students in the section*. In Figure 6-18, we see the normal distribution of grades for each section.

Tina's 74 was higher than most of the other scores in her section, while Jack's 82 is only an upper-middle score in his section. Tina's score is far better with respect to her class than Jack's score with respect to his class.

Standard score

The preceding situation demonstrates that it is not sufficient to know the difference between a measurement (x value) and the mean of a distribution. We need also to consider the spread of the curve, or the standard deviation. What we really want to know is the number of standard deviations between a measurement and the mean. This "distance" takes both μ and σ into account.

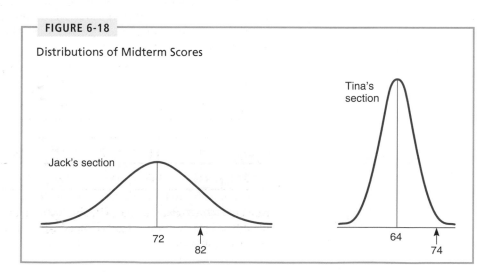

FIGURE 6-18

Distributions of Midterm Scores

Jack's section

72

82

Tina's section

64

74

We can use a simple formula to compute the number z of standard deviations between a measurement x and the mean μ of a normal distribution with standard deviation σ:

$$\begin{pmatrix} \text{Number of standard deviations} \\ \text{between the measurement and} \\ \text{the mean} \end{pmatrix} = \begin{pmatrix} \dfrac{\text{Difference between the}}{\text{measurement and the mean}} \\ \overline{\text{Standard deviation}} \end{pmatrix}$$

Written in symbols, this formula is

$$z = \frac{x - \mu}{\sigma}$$

z Score

> **Definition** The *z value* or *z score* tells us the number of standard deviations the original measurement is from the mean. The *z* value is in *standard units*.

The mean is a special value of a distribution. Let's see what happens when we convert $x = \mu$ to a z value:

$$z = \frac{x - \mu}{\sigma}$$

$$= \frac{\mu - \mu}{\sigma} \qquad \text{for } x = \mu$$

$$= 0$$

TABLE 6-4
x Values and Corresponding z Values

The mean of the original distribution is always zero, in standard units. This makes sense because the mean is zero standard variations from itself.

An x value in the original distribution that is *above* the mean μ has a corresponding z value that is *positive*. Again, this makes sense because a measurement above the mean would be a positive number of standard deviations from the mean. Likewise, an x value *below* the mean has a *negative z* value. (See Table 6-4.)

x Value in Original Distribution	Corresponding z Value or Standard Unit
$x = \mu$	$z = 0$
$x > \mu$	$z > 0$
$x < \mu$	$z < 0$

> **Note**
>
> Unless otherwise stated, in the remainder of the book we will take the word *average* to be either the sample arithmetic mean $\bar{x}$ or the population mean μ.

EXAMPLE 4

Standard score

A pizza parlor franchise specifies that the average (mean) amount of cheese on a large pizza should be 8 oz and the standard deviation only 0.5 oz. An inspector picks out a large pizza at random in one of the pizza parlors and finds that it is made with 6.9 oz of cheese. Assume that the amount of cheese on a pizza follows a normal distribution. If the amount of cheese is below the mean by more than *three* standard deviations, the parlor will be in danger of losing its franchise. (Remember, in a normal distribution we are unlikely to find measurements more than three standard deviations from the mean, since 99.7% of all measurements fall within three standard deviations of the mean.)

How many standard deviations from the mean is 6.9? Is the pizza parlor in danger of losing its franchise?

SOLUTION: Since we want to know the number of standard deviations from the mean, we want to convert 6.9 to standard z units.

$$z = \frac{x - \mu}{\sigma}$$

$$= \frac{6.9 - 8}{0.5}$$

$$= -2.20$$

Therefore, the amount of cheese on the selected pizza is only 2.20 standard deviations below the mean. Note that the fact that z is negative indicates that the amount of cheese was 2.20 standard deviations *below* the mean. The parlor will not lose its franchise based on this sample.   ◇

GUIDED EXERCISE 6

Standard score

A student has computed that it takes an average (mean) of 17 minutes with a standard deviation of 3 minutes to drive from home, park the car, and walk to an early morning class.

(a) One day it took the student 21 minutes to get to class. How many standard deviations from the average is that? Is the z value positive or negative. Explain why it should be either positive or negative.

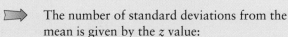   The number of standard deviations from the mean is given by the z value:

$$z = \frac{x - \mu}{\sigma} = \frac{21 - 17}{3} = 1.33$$

The z value is positive. We would expect a positive z value, since 21 minutes is *more* than the mean of 17.

(b) Another day it took only 12 minutes for the students to get to class. What is this measurements in standard units? Is the z value positive or negative? Why should it be positive or negative?

The measurement in standard units is

$$z = \frac{x - \mu}{\sigma} = \frac{12 - 17}{3} = -1.67$$

Here the z value is negative, as we should expect, because 12 minutes is less than the mean of 17 minutes.

(c) Another day it took 17 minutes for the students to go from home to class. What is the z value? Why should you expect this answer?

In this case, the value is

$$z = \frac{x - \mu}{\sigma} = \frac{17 - 17}{3} = 0.00$$

We expect this result because 17 minutes is the mean, and the z value of the mean is always zero.

Raw score

We have seen how to convert from x measurements to standard units z. We can easily reverse the process if we know μ and σ for the original distribution. For when we solve

$$z = \frac{x - \mu}{\sigma}$$

for x, we get

$$x = z\sigma + \mu$$

EXAMPLE 5

Raw score

In Example 4, we talked about the amount of cheese required by a franchise for a large pizza. Again, the mean amount of cheese required is 8 oz with a standard deviation of 0.5 oz. The franchise specifies that the minimum amount of cheese for a large pizza is three standard deviations below the mean. A pizza parlor can lose its franchise if the amount of cheese on a large pizza is less than the specified minimum. What is the minimum amount of cheese that can be placed on a large pizza according to the franchise?

SOLUTION: Here we need to convert $z = -3$ to information about x oz of cheese. We use the formula

$$x = z\sigma + \mu = -3(0.5) + 8 = 6.5 \text{ oz}$$

The franchise will not approve a large pizza with less than 6.5 oz of cheese.   ◊

In many testing situations, we hear the terms *raw score* and *z score*. The raw score is just the score in the original measuring units, and the z score is the score in standard units. Guided Exercise 7 illustrates these different units.

GUIDED EXERCISE 7

Raw score

Marulla's z score on a college entrance exam is 1.3. If the raw scores have a mean of 480 and a standard deviation of 70 points, what is her raw score?

 Here we are given z, σ and μ. We need to find the raw score x corresponding to the z score 1.3.

$$x = z\sigma + \mu$$
$$= 1.3(70) + 480$$
$$= 571$$

Standard normal distribution

If the original distribution of x values is normal, then the corresponding z values have a normal distribution as well. The z distribution has a mean of 0 and a standard deviation of 1. The normal curve with these properties has a special name.

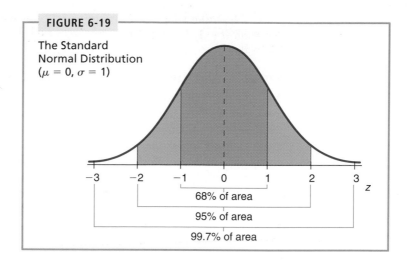

FIGURE 6-19

The Standard Normal Distribution ($\mu = 0$, $\sigma = 1$)

68% of area

95% of area

99.7% of area

Definition The *standard normal distribution* is a normal distribution with mean $\mu = 0$ and standard deviation $\sigma = 1$ (Figure 6-19).

Any normal distribution of x values can be converted to the standard normal distribution by converting all x values to their corresponding z values. Let's look at the graphic interpretation of this transformation in Figure 6-20.

FIGURE 6-20

The Transformation of a Normal Distribution to the Standard Normal Distribution

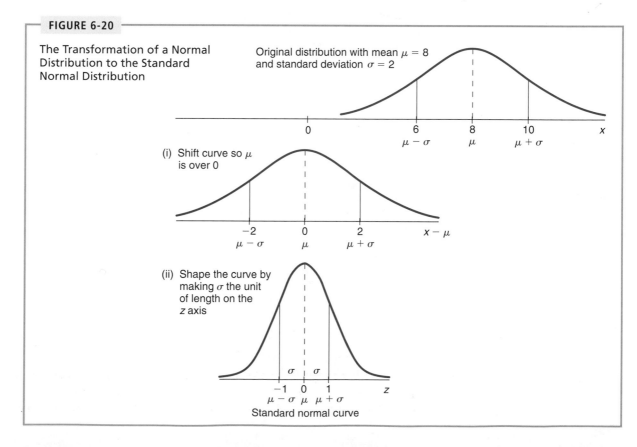

Original distribution with mean $\mu = 8$ and standard deviation $\sigma = 2$

(i) Shift curve so μ is over 0

(ii) Shape the curve by making σ the unit of length on the z axis

Standard normal curve

The resulting standard distribution will always have mean $\mu = 0$ and standard deviation $\sigma = 1$.

Areas Under the Standard Normal Curve

We have seen how to convert *any* normal distribution to the *standard* normal distribution. We can change any x value to a z value and back again. But what is the advantage of all this work? The advantage is that there are extensive tables that show the *area under the standard normal curve* for almost any interval along the z axis. The areas are important because they are equal to the *probability* that the measurement of an item selected at random falls in this interval. Thus, the *standard* normal distribution can be a tremendously helpful tool.

For instance, Sunshine Stereo guarantees their cassette decks for a period of 2 years. The company statistician has computed that the cassette deck life is normally distributed with a mean of 2.3 years and a standard deviation 0.4 year. What is the probability that a cassette deck will stop working during the guarantee period?

To answer questions of this type, we convert the given normal distribution to the standard normal distribution. Then we use a table to find the area over the interval in question and, hence, the probability an item selected at random will fall into that interval. Before we can carry out this plan, though, we must practice using Table 5 of Appendix II to find areas under the standard normal curve.

Using a Standard Normal Distribution Table

Using a table to find areas and probabilities associated with the standard normal distribution is a fairly straightforward activity. However, it is important to first observe the range of z values for which areas are given. This range is usually depicted in a picture that accompanies the table. Two commonly found styles of standard normal distribution tables are shown in Figure 6-21.

Middle-of-curve style table

Each style of table has its advantages. The middle-of-curve style is a compact table that fits on one page since it shows only z values on the right side of the

---

FIGURE 6-21

Styles of Standard Normal Distribution Tables

(a) Left-Tail: Gives cumulative area under the curve to the left of a specified z value

(b) Middle-of-Curve: Gives area under the curve between $z = 0$ and the specified z value

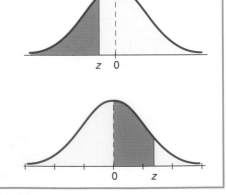

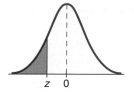

TABLE 6-5   Excerpt of Table 5 of Appendix II Showing Negative z Values

z	.00	.01	...	.07	.08	.09
−3.4	.0003	.0003	...	.0003	.0003	.0002
:						
−1.1	.1357	.1335	...	.1210	.1190	.1170
−1.0	.1587	.1562	...	.1423	.1401	.1379
−0.9	.1841	.1814	...	.1660	.1635	.1611
:						
−0.0	.5000	.4960	...	.4721	.4681	.4641

distribution. Using this type of table relies heavily on the inherent symmetry of the normal distribution and the fact that the area under the right half of the curve is 0.5. Such a table can be found in Appendix I along with brief instructions for use.

Left-tail style table

In this text, *we will use the left-tail style table.* This style table gives cumulative areas to the left of a specified z. Determining other areas under the curve utilizes the fact that the area under the entire curve is 1. Taking advantage of the symmetry of the normal distribution is also useful. The procedures you learn for using the left-tail style normal distribution table apply directly to cumulative normal distribution areas found on calculators and in computer software packages such as Excel and Minitab.

EXAMPLE 6

Standard normal distribution table

Use Table 5 of Appendix II to find the described areas under the standard normal curve.

(a) Find the area under the standard normal curve to the left of $z = -1.00$.

> **SOLUTION:** First, shade the area to be found on the standard normal distribution curve, as shown in Figure 6-22. Notice that the z value we are using is negative. This means we will look at the portion of Table 5 of Appendix II for which the z values are negative. In the upper-left corner of the table we see the letter z. The column under z gives us the units value and tenths for z. The other column headings indicate the hundredths value of z. Table entries give areas under the standard normal curve to the left of the listed z values. To find the area to the left of $z = -1.00$, we use the row headed by −1.0 and then move to the column headed by the hundredths position 0.00. This entry is shaded in Table 6-5. We see the area is 0.1587.

(b) Find the area to the left of $z = 1.18$ illustrated in Figure 6-23.

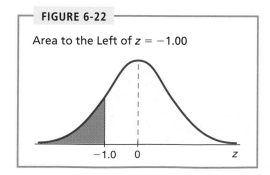

FIGURE 6-22

Area to the Left of $z = -1.00$

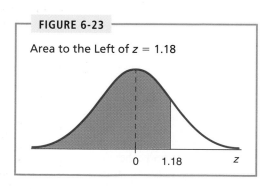

FIGURE 6-23

Area to the Left of $z = 1.18$

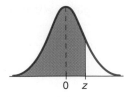

TABLE 6-6   Excerpt of Table 5 of Appendix II Showing Positive z Values

z	.00	.01	.02	...	.08	.09
0.0	.5000	.5040	.5080	...	.5319	.5359
:						
0.9	.8159	.8186	.8212	...	.8365	.8359
1.0	.8413	.8438	.8461	...	.8599	.8621
1.1	.8643	.8665	.8686	...	.8810	.8830
:						
3.4	.9997	.9997	.9997	...	.9997	.9998

SOLUTION: In this case, we are looking for an area to the left of a positive z value, so we look in the portion of Table 5 that shows positive z values. Again we first sketch the area to be found on a standard normal curve as shown in Figure 6-23 on the previous page. Look in the row headed by 1.1 and move to the column headed by .08. The desired area is shaded (see Table 6-6). We see that the area to the left of 1.18 is 0.8810.   ◊

GUIDED EXERCISE 8

Using the standard normal distribution table

Table 5, Areas of a Standard Normal Distribution, is located in Appendix II as well as in the endpapers of the text. Spend a little time studying the table, and then answer these questions.

(a) As z values increase, do the areas to the left of z increase?

⟹ Yes, as z values increase, we move to the right on the normal curve, and the areas increase.

(b) If a z value is negative, is the area to the left of z less than 0.5000?

⟹ Yes. Remember that a negative z value is on the left side of the standard normal distribution. The entire left half of the normal distribution has area 0.5, so any area to the left of $z = 0$ will be less than 0.5.

(c) If a z value is positive, is the area to the left of z greater than 0.5000?

⟹ Yes. Positive z values are on the right side of the standard normal distribution, and any area to the left of a positive z value includes the entire left half of the normal distribution.

Using Table 5 to find other areas   Table 5 gives areas under the standard normal distribution that are to the *left of a z value.* How do we find other areas under the standard normal curve?

Using left-tail style standard normal distribution table

1. For areas to the left of a specified z value, use the table entry directly.

2. For areas to the right of a specified z value, look up the table entry for z and subtract the area from 1.
   Note: Another way to find the same area is to use the symmetry of the normal curve and look up the table entry for $-z$.

3. For areas between two z values, z_1 and z_2 (where $z_2 > z_1$), *subtract* the table area for z_1 from the table area for z_2.

Figure 6-24 illustrates the procedure for using Table 5, Areas of a Standard Normal Distribution, to find any specified area under the standard normal distribution. Again, it is useful to sketch the area in question before you use Table 5.

◊ **COMMENT:** Notice that the z values shown in Table 5 of Appendix II are formatted to the hundredths position. It is convenient to *round or format z values to the hundredths position* before using the table. The areas are all given to 4 places after the decimal, so give your answers to 4 places after the decimal. ◊

FIGURE 6-24

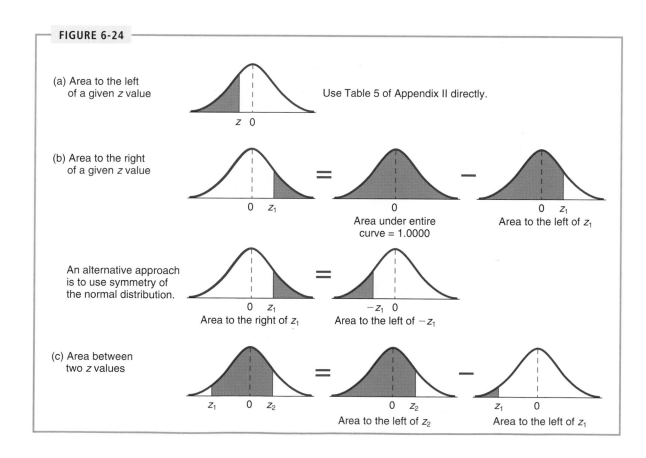

◇ **COMMENT:** The smallest z value shown in Table 5 is -3.49, while the largest value is 3.49. These values are, respectively, far to the left and far to the right on the standard normal distribution, with very little area beyond either value. We will follow the common convention of treating any area to the left of a z value smaller than -3.49 as 0.000. Similarly, we will consider any area to the right of a z value greater than 3.49 as 0.000. We understand that there is some area in these extreme tails. However, these areas are each less than 0.0002. Now let's get real about this! Some very specialized applications, beyond the scope of this book, do need to measure areas and corresponding probabilities in these extreme tails. But in most practical applications, *we follow the convention of treating the areas in the extreme tails as zero.* ◇

EXAMPLE 7

Using table to find areas

Use Table 5 of Appendix II to find the specified areas.

(a) Find the area between $z = 1.00$ and $z = 2.70$.

SOLUTION: First, sketch a diagram showing the area (see Figure 6-25). Because we are finding the area between two z values, we subtract corresponding table entries.

(Area between 1.00 and 2.70) = (Area left of 2.70) − (Area left of 1.00)

$$= 0.9965 - 0.8413$$

$$= 0.1552$$

(b) Find the area to the right of $z = 0.94$.

SOLUTION: First, sketch the area to be found (see Figure 6-26).

(Area to right of 0.94) = (Area under entire curve) − (Area to left of 0.94)

$$= 1.000 - 0.8264$$

$$= 0.1736$$

Alternatively,

(Area to the right of 0.94) = (Area to the left of −0.94)

$$= 0.1736$$ ◇

Probabilities associated with the standard normal distribution

We have practiced the skill of finding areas under the standard normal curve for various intervals along the z axis. This skill is important since *the probability*

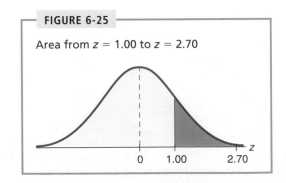

FIGURE 6-25

Area from $z = 1.00$ to $z = 2.70$

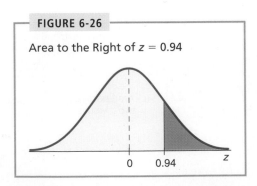

FIGURE 6-26

Area to the Right of $z = 0.94$

that z lies in an interval *is given by the area* under the standard normal curve above that interval.

Because the normal distribution is continuous, there is no area under the curve exactly over a specific z. Therefore, probabilities such as $P(z \geq z_1)$ are the same as $P(z > z_1)$. When dealing with probabilities or areas under a normal curve that are specified with inequalities, *strict inequality* symbols can be used *interchangeably* with *inequality-or-equal* symbols.

GUIDED EXERCISE 9

Probabilities associated with the standard normal distribution

Let z be a random variable with a standard normal distribution.

(a) $P(z \geq 1.15)$ refers to the probability that z values lie to the right of 1.15. Shade the corresponding area under the standard normal curve and find $P(z \geq 1.15)$.

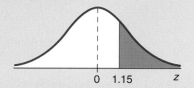 **FIGURE 6-27** Area to Be Found

$$P(z \geq 1.15) = 1.000 - P(z \leq 1.15) = 1.000 - 0.8749 = 0.1251$$

Alternatively,

$$P(z \geq 1.15) = P(z \leq -1.15) = 0.1251$$

(b) Find $P(-1.78 \leq z \leq 0.35)$. First, sketch the area under the standard normal curve corresponding to the area.

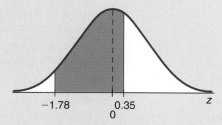 **FIGURE 6-28** Area to Be Found

$$P(-1.78 \leq z \leq 0.35) = P(z \leq 0.35) - P(z \leq -1.78)$$
$$= 0.6368 - 0.0375 = 0.5993$$

 TECH NOTE The TI-83Plus, Excel, and Minitab all provide cumulative areas under any normal distribution, including the standard normal. The Tech Note of Section 6.3 shows examples.

VIEWP◉INT | *Mighty Oaks from Little Acorns Grow!*

Just how big is that acorn? What if we compare it with other acorns? Is that oak tree taller than an average oak tree? How does it compare with other oak trees? What do you mean, this oak tree has a larger geographic range? Compared with what? Answers to questions such as these can be given only if we resort to *standardized statistical units*. Can you compare a single oak tree with an entire forest of oak trees? The answer is yes, if you use *standardized z scores*. For more information about sizes of acorns, oak trees, and geographic locations, visit the Brase/Brase statistics site at http://math.college.hmco.com/students and find the link to DASL, the Carnegie Mellon University Data and Story Library. From the DASL site, find Biology under Data Subjects, and select Acorns. Follow the links to Data Subjects, Biology, and Acorns.

SECTION 6.2 PROBLEMS

In these problems, assume that all the distributions are *normal*. In all problems in Chapter 6, *average* is always taken to be the arithmetic mean $\bar{x}$ or μ.

1. *Z Scores: First Aid Course* The college Physical Education Department offered an Advanced First Aid course last semester. The scores on the comprehensive final exam were normally distributed, and the z scores for some of the students are shown below:

Robert, 1.10	Jan, 1.70	Susan, −2.00
Joel, 0.00	John, −0.80	Linda, 1.60

   (a) Which of these students scored above the mean?
   (b) Which of these students scored on the mean?
   (c) Which of these students scored below the mean?
   (d) If the mean score was $\mu = 150$ with standard deviation $\sigma = 20$, what was the final exam score for each student?

2. *Z Scores: Teaching Duties* What do professors do with their time? They do research, teach classes, serve on academic committees, serve the student body (student advising, sponsor student clubs, attend student events), serve the community (consult, address civic groups), and a lot more! The specific answer depends on the individual professor and his or her special interests. Well, how much time does a professor spend on teaching activities? *The NEA Almanac of Higher Education,* published by the National Education Association, reports that the mean percentage of time professors spend on teaching activities is about $\mu = 51\%$ with standard deviation $\sigma = 25\%$. Find the standardized z value corresponding to the following professors' percentage of time allocated to teaching duties.
   (a) Dr. Taylor, 45% (b) Mr. Patterson, 72% (c) Dr. Smith, 75%
   (d) Ms Simms, 65% (e) Dr. Adams, 33% (f) Dr. Riley, 55%

3. *Z Scores: Honolulu Temperatures* Data collected over a period of years show that the average daily temperature in Honolulu is $\mu = 73°F$ with standard deviation

$\sigma = 5°F$ (U.S. Department of Commerce, Environmental Data Service). Convert each of the following intervals in °F to an interval of z values.
(a) $53°F < x < 93°F$ (b) $x < 65°F$ (c) $78°F < x$

Convert each of the following intervals of z values to intervals in °F.
(d) $1.75 < z$ (e) $z < -1.90$ (f) $-1.80 < z < 1.65$

4. *Z Scores: Fawns* Fawns between 1 and 5 months old in Mesa Verde National Park have a body weight that is approximately normally distributed with mean $\mu = 27.2$ kilograms and standard deviation $\sigma = 4.3$ kilograms (based on information from *The Mule Deer of Mesa Verde National Park,* by G. W. Mierau and J. L. Schmidt, Mesa Verde Museum Association). Let x be the weight of a fawn in kilograms. Convert each of the following x intervals to z intervals.
(a) $x < 30$ (b) $19 < x$ (c) $32 < x < 35$

Convert each of the following z intervals to x intervals.
(d) $-2.17 < z$ (e) $z < 1.28$ (f) $-1.99 < z < 1.44$
(g) If a fawn weighs 14 kilograms, would you say it is an unusually small animal? Explain using z values and Figure 6-19.
(h) If a fawn is unusually large, would you say that the z value for the weight of the fawn will be close to 0, -2, or 3? Explain.

5. *Z Scores: Deer Population* The fall deer population in Mesa Verde National Park is approximately normally distributed with mean 4400 deer and standard deviation 620 deer (see reference in Problem 4). Let x be the random variable that represents the size of the deer population in Mesa Verde National Park in the fall of a given year. Convert each of the following x intervals to z intervals.
(a) $3300 < x$ (b) $x < 5400$ (c) $3500 < x < 5300$

Convert each of the following z intervals to x intervals.
(d) $-1.12 < z < 2.43$ (e) $z < 1.96$ (f) $2.58 < z$
(g) If the fall deer population were 2800 deer, would that be considered an unusually low number? If the fall population were 6300, would that be considered an unusually high population? Explain using z values and Figure 6-19.

6. *Z Scores: White Blood Cell Count* Let x = white blood cell (WBC) count per cubic millimeter of whole blood. Then x has a distribution that is approximately normal with mean $\mu = 7500$ and standard deviation $\sigma = 1750$ (based on information from *Diagnostic Tests with Nursing Implications,* edited by S. Loeb, Springhouse Press). Convert each of the following x intervals to z intervals.
(a) $9000 < x$ (b) $x < 6000$ (c) $3500 < x < 4500$

Convert each of the following z intervals to x intervals.
(d) $z < 1.15$ (e) $2.19 < z$ (f) $0.25 < z < 1.25$
(g) If someone had a WBC count of 2500, would that be considered unusually high or low? Explain using z values and Figure 6-19.

7. *Z Scores: Red Blood Cell Count* Let x = red blood cell (RBC) count in millions per cubic millimeter of whole blood. For healthy females x has an approximately normal distribution with mean $\mu = 4.8$ and standard deviation $\sigma = 0.3$. (See reference in Problem 6.) Convert each of the following x intervals from laboratory tests to z intervals.
(a) $4.5 < x$ (b) $x < 4.2$ (c) $4.0 < x < 5.5$

Convert each of the following z intervals to x intervals.
(d) $z < -1.44$ (e) $1.28 < z$ (f) $-2.25 < z < -1.00$
(g) If a female had an RBC count of 5.9 or higher, would that be considered unusually high? Explain using z values and Figure 6-19.

8. *Normal Curve: Tree Rings* Tree-ring dates were used extensively in archaeological studies at Burnt Mesa Pueblo (*Bandelier Archaeological Excavation Project: Summer 1989 Excavations at Burnt Mesa Pueblo,* edited by Kohler, Washington State University Department of Anthropology). At one site on the mesa, tree-ring dates (for many samples) gave a mean date μ_1 = year 1272 with standard deviation σ_1 = 35 years. At a second, removed site, the tree-ring dates gave a mean of μ_2 = year 1122 with standard deviation σ_2 = 40 years. Assume that both sites had dates that were approximately normally distributed. In the first area, an object was found and dated as x_1 = year 1250. In the second area, another object was found and dated as x_2 = year 1234.

(a) Convert both x_1 and x_2 to z values, and locate both these values under the standard normal curve of Figure 6-19.

(b) Which of these two items is the more unusual as an archaeological find in its location?

In Problems 9–28, sketch the areas under the standard normal curve over the indicated intervals, and find the specified areas.

9. To the right of $z = 0$.   10. To the left of $z = 0$.

11. To the left of $z = -1.32$.   12. To the left of $z = -0.47$.

13. To the left of $z = 0.45$.   14. To the left of $z = 0.72$.

15. To the right of $z = 1.52$.   16. To the right of $z = 0.15$.

17. To the right of $z = -1.22$.   18. To the right of $z = -2.17$.

19. Between $z = 0$ and $z = 3.18$.   20. Between $z = 0$ and $z = 2.92$.

21. Between $z = 0$ and $z = -2.01$.   22. Between $z = 0$ and $z = -1.93$.

23. Between $z = -2.18$ and $z = 1.34$.   24. Between $z = -1.40$ and $z = 2.03$.

25. Between $z = 0.32$ and $z = 1.92$.   26. Between $z = 1.42$ and $z = 2.17$.

27. Between $z = -2.42$ and $z = -1.77$.   28. Between $z = -1.98$ and $z = -0.03$.

In Problems 29–48, let z be a random variable with a standard normal distribution. Find the indicated probability, and shade the corresponding area under the standard normal curve.

29. $P(z \le 0)$.   30. $P(z \ge 0)$.

31. $P(z \le -0.13)$.   32. $P(z \le -2.15)$.

33. $P(z \le 1.20)$.   34. $P(z \le 3.20)$.

35. $P(z \ge 1.35)$.   36. $P(z \ge 2.17)$.

37. $P(z \ge -1.20)$.   38. $P(z \ge -1.50)$.

39. $P(-1.20 \le z \le 2.64)$.   40. $P(-2.20 \le z \le 1.04)$.

41. $P(-2.18 \le z \le -0.42)$.   42. $P(-1.78 \le z \le -1.23)$.

43. $P(0 \le z \le 1.62)$.   44. $P(0 \le z \le 0.54)$.

45. $P(-0.82 \le z \le 0)$.   46. $P(-2.37 \le z \le 0)$.

47. $P(-0.45 \le z \le 2.73)$.   48. $P(-0.73 \le z \le 3.12)$.

6.3
Areas Under Any Normal Curve

FOCUS POINTS

✓ Compute the probability of "standardized events."

✓ Find a *z* score from a given normal probability (inverse normal).

✓ Use the inverse normal to solve guarantee problems.

Normal Distribution Areas

In many applied situations, the original normal curve is not the standard normal curve. Generally, there will not be a table of areas available for the original normal curve. This does not mean that we cannot find the probability that a measurement *x* will fall into an interval from *a* to *b*. What we must do is *convert* original measurements *x*, *a*, and *b* to *z* values.

EXAMPLE 8

Normal distribution probability

Let *x* have a normal distribution with $\mu = 10$ and $\sigma = 2$. Find the probability that an *x* value selected at random from this distribution is between 11 and 14. In symbols, find $P(11 \leq x \leq 14)$.

SOLUTION: Since probabilities correspond to areas under the distribution curve, we want to find the area under the *x* curve above the interval from $x = 11$ to $x = 14$. To do so, we will convert the *x* values to standard *z* values (see Figure 6-29) and then use Table 5 of Appendix II to find the corresponding area under the standard curve.

We use the formula

$$z = \frac{x - \mu}{\sigma}$$

to convert the given *x* interval to a *z* interval.

$$z_1 = \frac{11 - 10}{2} = 0.50 \qquad \text{(Use } x = 11, \ \mu = 10, \ \sigma = 2.\text{)}$$

$$z_2 = \frac{14 - 10}{2} = 2.00 \qquad \text{(Use } x = 14, \ \mu = 10, \ \sigma = 2.\text{)}$$

The corresponding areas under the *x* and *z* curves are shown in Figure 6-30 on the next page. From Figure 6-30 we see that

$$
\begin{aligned}
P(11 \leq x \leq 14) &= P(0.50 \leq z \leq 2.00) \\
&= P(z \leq 2.00) - P(z \leq 0.50) \\
&= 0.9772 - 0.6915 \qquad \text{(From Table 5, Appendix II)} \\
&= 0.2857
\end{aligned}
$$

FIGURE 6-29

The Interval $11 \leq x \leq 14$ Corresponds to the Interval $0.50 \leq z \leq 2.00$

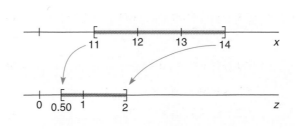

FIGURE 6-30

Corresponding Areas Under the *x* Curve and *z* Curve

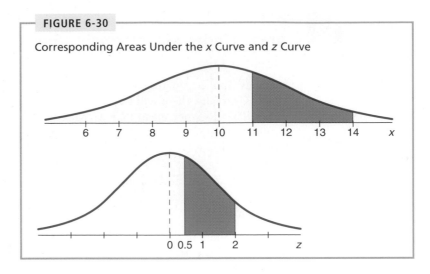

The probability is 0.2857 that an *x* value selected at random from a normal distribution with mean 10 and standard deviation 2 lies between 11 and 14.   ◊

GUIDED EXERCISE 10

Normal distribution probability

In Section 6.2, we talked about Sunshine Stereo cassette decks. The cassette deck life was normally distributed with a mean of 2.3 years and a standard deviation of 0.4 year. We wanted to know the probability that a cassette deck will break down during the guarantee period of 2 years.

(a) Let *x* represent the life of a cassette deck. The statement that the cassette deck breaks during the 2-year guarantee period means the life is less than 2 years, or $x \le 2$. Convert this to a statement about *z*.

$z = \dfrac{x - \mu}{\sigma} = \dfrac{2 - 2.3}{0.4} = -0.75$

So $x \le 2$ means $z \le -0.75$.

(b) Indicate the area to be found in Figure 6-31. Does this area correspond to the probability that $z \le -0.75$?

See Figure 6-32.
Yes, the shaded area does correspond to the probability that $z \le -0.75$.

FIGURE 6-31

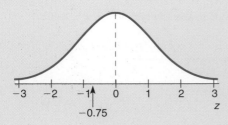

FIGURE 6-32 $z \le -0.75$

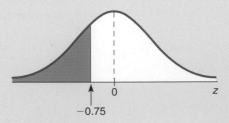

Continued

GUIDED EXERCISE 10 continued

(c) Use Table 5 of Appendix II to find $P(z \leq -0.75)$.   ➡   0.2266

(d) What is the probability that the cassette deck will break before the end of the guarantee period? [*Hint:* $P(x \leq 2) = P(z \leq -0.75)$.]   ➡

The probability is

$$P(x \leq 2) = P(z \leq -0.75)$$
$$\approx 0.23$$

This means the company will repair or replace about 23% of the cassette decks.

TECH NOTE   The TI-83Plus, Excel, and Minitab all provide areas under any normal distribution. Excel and Minitab give the left-tail area to the left of a specified x value. The TI-83Plus has you specify an interval from a lower bound to an upper bound and provides the area under the normal curve for that interval. For example, to solve Guided Exercise 10 regarding the probability a cassette deck will break during the guarantee period, we find $P(x \leq 2)$ for a normal distribution with $\mu = 2.3$ and $\sigma = 0.4$.

TI-83Plus Press **DISTR** Key, select 2:normalcdf (lower bound, upper bound, μ, σ) and press enter. Type in the specified values. For a left-tail area, use a lower bound setting at about 4 standard deviations below the mean. Likewise, for a right-tail area use an upper bound about 4 standard deviations above the mean. For our example, use a lower bound of $\mu - 4\sigma = 2.3 - 4(0.4) = 0.7$.

```
normalcdf(.7,2,2.3,
.4)
          .226955934
```

Excel Press the Paste Function (f_x) ➤ **Statistical** ➤ **NORMDIST.** Fill in the dialogue box, using True for cumulative.

=	=NORMDIST(2,2.3,0.4, TRUE)	
C	D	E
0.226627		

Minitab Use the menu selection **Calc** ➤ **Probability Distribution** ➤ **Normal.** Fill in the dialogue box, marking cumulative.

```
Cumulative Distribution Function
Normal with mean = 2.3 and
standard deviation = 0.4
      x      P(X <= x)
     2.0        0.2266
```

Finding *z* or *x*, given a probability

Inverse Normal Distribution

Sometimes we need to find z or x values that correspond to a given area under the normal curve. This situation arises when we want to specify a guarantee period so that a given percentage of the total products produced by a company lasts at least as long as the duration of the guarantee period. In such cases, we use the standard normal distribution table "in reverse." When we look up an area and find the corresponding z value, we are using the *inverse normal probability distribution*.

EXAMPLE 9

Find x, given probability

Magic Video Games, Inc., sells an expensive video computer games package. Because the package is so expensive, the company wants to advertise an impressive guarantee for the life expectancy of its computer control system. The guarantee policy will refund full purchase price if the computer fails during the guarantee period. The research department has done tests which show that the mean life for the computer is 30 months, with standard deviation of 4 months. The computer life is normally distributed. How long can the guarantee period be if management does not want to refund the purchase price on more than 7% of the Magic Video packages?

SOLUTION: Let us look at the distribution of lifetimes for the computer control system, and shade the portion of the distribution in which the computer lasts fewer months than the guarantee period. (See Figure 6-33.)

If a computer system lasts fewer months than the guarantee period, a full-price refund will have to be made. The lifetimes requiring a refund are in the shaded region in Figure 6-33. This region represents 7% of the total area under the curve.

We can use Table 5 of Appendix II to find the z value so that 7% of the total area under the *standard* normal curve lies to the left of the z value. Then we convert the z value to its corresponding x value to find the guarantee period.

We want to find the z value with 7% of the area under the standard normal curve to the left of z. Since we are given the area in a left tail, we can use Table 5 of Appendix II directly to find z. The area value is 0.0700. However, this area is not in our table, so we use the closest area, which is 0.0694, and the corresponding z value of $z = -1.48$ (see Table 6-7).

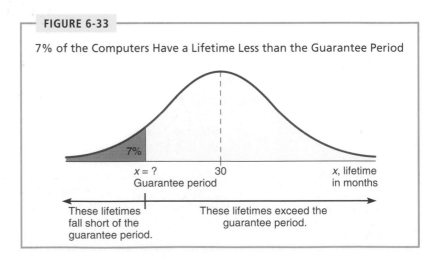

FIGURE 6-33

7% of the Computers Have a Lifetime Less than the Guarantee Period

TABLE 6-7   Excerpt from Table 5 of Appendix II

z	.00	...	.07		.08	.09
:						
−1.4	.0808		.0708		.0694	.0681
				↑		
				0.0700		

To translate this value back to an x value (in months), we use the formula

$$x = z\sigma + \mu$$
$$= -1.48(4) + 30 \qquad \text{(Use } \sigma = 4 \text{ months and } \mu = 30 \text{ months.)}$$
$$= 24.08 \text{ months}$$

The company can guarantee the Magic Video Games package for $x = 24$ months. For this guarantee period, they expect to refund the purchase price of no more than 7% of the video games packages. ◇

Example 9 had us find a z value corresponding to a given area to the left of z. What if the specified area is to the right of z or between $-z$ and z? Figure 6-34 shows us how to proceed.

◇ **COMMENT** When we use Table 5 in Appendix II to find a z value corresponding to a given area, we usually use the nearest area value rather than interpolation between values. However, when the area value given is exactly halfway

---

FIGURE 6-34

Inverse Normal: Use Table 5 of Appendix II to Find z Corresponding to a Given Area A ($0 < A < 1$)

(a) **Left-tail case:**
The given area A is to the left of z.

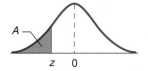

For the left-tail case, look up the number A in the body of the table and use the corresponding z value.

(b) **Right-tail case:**
The given area A is to the right of z.

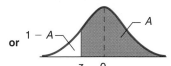

For the right-tail case, look up the number $1 - A$ in the body of the table and use the corresponding z value.

(c) **Center case:**
The given area A is symmetric and centered above $z = 0$. Half of A lies to the left and half lies to the right of $z = 0$.

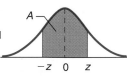

For the center case, look up the number $\dfrac{1 - A}{2}$ in the body of the table and use the corresponding $\pm z$ value.

between two area values of the table, we use the z value halfway between z values of the corresponding table areas. Example 10 demonstrates this procedure. However, this interpolation convention is not always used, especially if the area is changing slowly as it does in the tail ends of the distribution. *When the z value corresponding to an area is smaller than −2, the standard convention is to use the z value corresponding to the smaller area. Likewise, when the z value is larger than 2, the standard convention is to use the z values corresponding to the larger area.* We will see an example of this special case in Example 1 of Chapter 8. ◇

EXAMPLE 10

Find z

Find the z value so that 90% of the area under the standard normal curve lies between $-z$ and z.

SOLUTION: Sketch a picture showing the described area (see Figure 6-35).

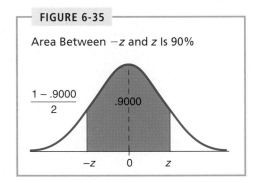

FIGURE 6-35

Area Between −z and z Is 90%

We find the corresponding area in the left tail.

$$(\text{Area left of } -z) = \frac{1 - 0.9000}{2}$$
$$= 0.0500$$

Looking in Table 6-8, we see that 0.0500 lies exactly between areas 0.0495 and 0.0505. The halfway value between $z = -1.65$ and $z = -1.64$ is $z = -1.645$. Therefore, we conclude that 90% of the area under the standard normal curve lies between the z values -1.645 and 1.645. ◇

TABLE 6-8   Excerpt from Table 5 in Appendix II

z	...	.04		.05
:				
−1.6		.0505		.0495
			0.0500 ↑	

GUIDED EXERCISE 11

Find z

Find the z value so that 3% of the area under the standard normal curve lies to the right of z.

(a) Draw a sketch of the standard normal distribution showing the described area.

➡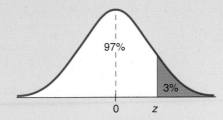

FIGURE 6-36 3% of Total Area Lies to the Right of z

(b) Find the area to the left of z.

➡ Area to the left of z = 1 − 0.0300 = 0.9700.

(c) Look up the area in Table 6-9 and find the corresponding z.

➡ The closest area is 0.9699. This area is to the left of z = 1.88.

TABLE 6-9   Excerpt from Table 5 in Appendix II

z	.00	.01	.02	.03	.04	.05	.06	.07	.08	.09
1.8	.9641	.9649	.9656	.9664	.9671	.9678	.9686	.9693	.9699	.9706
1.9	.9713	.9719	.9726	.9732	.9738	.9744	.9750	.9756	.9761	.9767

(d) Suppose the time to complete a test is normally distributed with $\mu = 40$ minutes and $\sigma = 5$ minutes. After how many minutes can we expect all but about 3% of the tests to be completed?

➡ We are looking for an x value so that 3% of the normal distribution lies to the right of x. In part (c), we found that 3% of the standard normal curve lies to the right of z = 1.88. We convert z = 1.88 to an x value.

$$x = z\sigma + \mu$$
$$= 1.88(5) + 40 = 49.4 \text{ minutes}$$

All but about 3% of the tests will be complete after 50 minutes.

(e) Use Table 6-10 to find a z value so that 3% of the area under the standard normal curve lies to the left of z.

➡ The closest area is 0.0301. This is the area to the left of z = −1.88.

TABLE 6-10   Excerpt from Table 5 in Appendix II

z	.00	.01	.02	.03	.04	.05	.06	.07	.08	.09
−1.9	.0287	.0281	.0274	.0268	.0262	.0256	.0250	.0244	.0239	.0233
−1.8	.0359	.0351	.0344	.0336	.0329	.0322	.0314	.0307	.0301	.0294

(f) Compare the z value of part (c) with the z value of part (e). Is there any relationship between the z values?

➡ One z value is the negative of the other. This result is expected because of the symmetry of the normal distribution.

TECH NOTE When we are given a z value and we find an area to the left of z, we are using a normal distribution function. When we are given an area to the left of z and we find the corresponding z, we are using an inverse normal distribution function. The TI-83Plus, Excel, and Minitab all have inverse normal distribution functions for any normal distribution. For instance, to find an x value from a normal distribution with mean 40 and standard deviation 5 such that 97% of the area lies to the left of x, use the described instructions.

TI-83Plus Press the **DISTR** key and select **3:invNorm(area,μ,σ)**.

```
invNorm(.97,40,5)
         49.40396805
```

Excel Press the Paste Function $\boxed{f_x}$ ➤ **Statistical** ➤ **NORMINV**. Fill in the dialogue box.

=	=NORMINV(0.97,40,5)	
C	D	
49.40395		

Minitab Use the menu selection **Calc** ➤ **Probability Distribution** ➤ **Normal**. Fill in the dialogue box, marking Inverse Cumulative.

```
Inverse Cumulative Distribution Function
Normal with mean = 40.000 and
  standard deviation = 5.00000
P(X <= x)            x
   0.9700        49.4040
```

VIEWPOINT

Want to Be an Archaeologist?

Each year about 4500 students work with professional archaeologists in scientific research at the Crow Canyon Archaeological Center, Cortez, Colorado. In fact, Crow Canyon was included in *The Princeton Review Guide to America's Top 100 Internships*. The nonprofit, multidisciplinary program at Crow Canyon enables students and laypeople with little or no background to get started in archaeological research. The only requirement is that you be interested in Native American culture and history. By the way, a knowledge of introductory statistics could come in handy in this internship. For more information about the program, visit the Brase/Brase statistics site at http://math.college.hmco.com/students and find the link to Crow Canyon.

SECTION 6.3   PROBLEMS

In Problems 1–10, assume that x has a normal distribution, with the specified mean and standard deviation. Find the indicated probabilities.

1. $P(3 \leq x \leq 6)$; $\mu = 4$; $\sigma = 2$
2. $P(10 \leq x \leq 26)$; $\mu = 15$; $\sigma = 4$
3. $P(50 \leq x \leq 70)$; $\mu = 40$; $\sigma = 15$
4. $P(7 \leq x \leq 9)$; $\mu = 5$; $\sigma = 1.2$
5. $P(8 \leq x \leq 12)$; $\mu = 15$; $\sigma = 3.2$
6. $P(40 \leq x \leq 47)$; $\mu = 50$; $\sigma = 15$
7. $P(x \geq 30)$; $\mu = 20$; $\sigma = 3.4$
8. $P(x \geq 120)$; $\mu = 100$; $\sigma = 15$
9. $P(x \geq 90)$; $\mu = 100$; $\sigma = 15$
10. $P(x \geq 2)$; $\mu = 3$; $\sigma = 0.25$

In Problems 11–20, find the z value described and sketch the area described.

11. Find z so that 6% of the standard normal curve lies to the left of z.

12. Find z so that 5.2% of the standard normal curve lies to the left of z.

13. Find z so that 55% of the standard normal curve lies to the left of z.

14. Find z so that 97.5% of the standard normal curve lies to the left of z.

15. Find z so that 8% of the standard normal curve lies to the right of z.

16. Find z so that 5% of the standard normal curve lies to the right of z.

17. Find z so that 82% of the standard normal curve lies to the right of z.

18. Find z so that 95% of the standard normal curve lies to the right of z.

19. Find the z value so that 98% of the standard normal curve lies between $-z$ and z.

20. Find the z value so that 95% of the standard normal curve lies between $-z$ and z.

21. *Medical: Blood Glucose* The level of blood glucose and diabetes are closely related. Let x be a random variable measured in milligrams of glucose per deciliter (1/10th of a liter) of blood. After a 12-hour fast, the random variable x will have a distribution that is approximately normal with mean $\mu = 85$ and standard deviation $\sigma = 25$ (*Diagnostic Tests with Nursing Implications,* edited by S. Loeb, Springhouse Press). *Note:* After 50 years of age, both the mean and standard deviation tend to increase. What is the probability that for an adult (under 50 years old) after a 12-hour fast
    (a) x is more than 60?
    (b) x is less than 110?
    (c) x is between 60 and 110?
    (d) x is greater than 140 (borderline diabetes starts at 140)?

22. *Medical: Blood Protoplasm* Porphyrin is a pigment in blood protoplasm and other body fluids that is significant in body energy and storage. Let x be a random variable that represents the number of milligrams of porphyrin per deciliter of blood. In healthy adults, x is approximately normally distributed with mean $\mu = 38$ and standard deviation $\sigma = 12$ (see reference in Problem 21). What is the probability that
    (a) x is less than 60?
    (b) x is greater than 16?
    (c) x is between 16 and 60?
    (d) x is more than 60? (This may indicate an infection, anemia, or another type of illness.)

23. *Education: SAT and ACT Scores* For a given population of high school seniors, the Scholastic Aptitude Test (SAT) in mathematics has a mean score of 500 with a standard deviation of 100. Another widely used test is American College Testing (ACT) exam. The mathematics portion of the ACT has a mean of 18 and a standard deviation of 6. (Visit the Brase/Brase statistics site at http://math.college.hmco.com/students and find the link to the College Board.) Both SAT and ACT scores are normally distributed. What is the probability that a randomly selected high school senior's score on the mathematics part of the SAT will be
    (a) more than 675?   (b) less than 450?   (c) between 450 and 675?

    What is the probability that a randomly selected high school senior's score on the mathematics part of the ACT will be
    (d) more than 28?   (e) more than 12?   (f) between 12 and 28?

24. *Education: SAT and ACT Scores* Please refer to SAT and ACT information from Problem 23.
    (a) Suppose that an engineering school honors program will accept only high school seniors with a mathematics SAT or ACT score in the top 10%. What is the minimum SAT score in mathematics for this program? What is the minimum ACT score in mathematics for this program?
    (b) Suppose that an engineering school will accept only high school seniors with a mathematics SAT or ACT score in the top 20%. What is the minimum SAT score in mathematics for this program? What is the minimum ACT score in mathematics for this program?
    (c) Suppose that an engineering school will accept only high school seniors with a mathematics SAT or ACT score in the top 60%. What is the minimum SAT score in mathematics for this program? What is the minimum ACT score in mathematics for this program?

25. *Archaeology: Hopi Village* Thickness measurements of ancient prehistoric Native American pot shards discovered in a Hopi village were approximately normally distributed with a mean of 5.1 mm and a standard deviation of 0.9 mm. (Source: *Homol'ovi*II: *Archaeology of an Ancestral Hopi Village, Arizona,* edited by E. C. Adams and K. A. Hays, University of Arizona Press.) For a randomly found shard, what is the probability that the thickness is
    (a) less than 3.0 mm?
    (b) more than 7.0 mm?
    (c) between 3.0 mm and 7.0 mm?

26. *Law Enforcement: Police Response Time* Police response time to an emergency call is the difference between the time the call is first received by the dispatcher and the time a patrol car radios that it has arrived at the scene (based on information from the *Denver Post*). Over a long period of time, it has been determined that the police response time has a normal distribution with a mean of 8.4 minutes and a standard deviation of 1.7 minutes. For a randomly received emergency call, what is the probability that the response time will be
    (a) between 5 and 10 minutes?
    (b) less than 5 minutes?
    (c) more than 10 minutes?

27. *Fuel Consumption: Boeing 747* Average fuel consumption for a Boeing 747 commercial jet in cruising position is 3213 gallons of jet fuel per hour (based on information from Canadian Pacific Air). Assume that the fuel consumption distribution is

mound-shaped and approximately normal with a standard deviation of 180 gallons of jet fuel per hour. When a 747 is in cruising position, what is the probability the fuel consumption is
(a) between 3000 and 3500 gallons per hour?
(b) less than 3000 gallons per hour?
(c) more than 3500 gallons per hour?

28. *Vermont Skiing: Temperatures* At a ski area in Vermont, the daytime high temperature is normally distributed during January, with a mean of 22°F and a standard deviation of 10°F (U.S. Department of Commerce, Environmental Data Services). You are planning to ski there this January. What is the probability that you will encounter daytime highs of
    (a) 42°F or higher?   (b) 15°F or lower?   (c) between 29°F and 40°F?

29. *Guarantee: Batteries* Quick Start Company makes 12-volt car batteries. After many years of product testing, the company knows that the average life of a Quick Start battery is normally distributed, with a mean of 45 months and a standard deviation of 8 months.
    (a) If Quick Start guarantees a full refund on any battery that fails within the 36-month period after purchase, what percentage of its batteries will the company expect to replace?
    (b) If Quick Start does not want to make refunds for more than 10% of its batteries under the full-refund guarantee policy, for how long should the company guarantee the batteries (to the nearest month)?

30. *Guarantee: Watches* Accrotime is a manufacturer of quartz crystal watches. Accrotime researchers have shown that the watches have an average life of 28 months before certain electronic components deteriorate, causing the watch to become unreliable. The standard deviation of watch lifetimes is 5 months, and the distribution of lifetimes is normal.
    (a) If Accrotime guarantees a full refund on any defective watch for 2 years after purchase, what percentage of total production will the company expect to replace?
    (b) If Accrotime does not want to make refunds on more than 12% of the watches it makes, how long should the guarantee period be (to the nearest month)?

31. *Consumer: TV Replacement Consumer Reports* gave information about the age at which various household products are replaced. For example, color TVs are replaced at an average age of $\mu = 8$ years after purchase, and the (95% of data) range was from 5 to 11 years. Thus, the range was $11 - 5 = 6$ years. Let x be the age (years) at which a color TV is replaced. Assume that x has a distribution that is approximately normal.
    (a) Read the empirical rule (Section 6.1); then explain why approximately four standard deviations will cover a 95% range of data values centered at the mean. Since the range is 6 years, explain why 1.5 years would be a good approximation for the standard deviation of x values.
    (b) What is the probability that someone will keep a color TV more than 5 years before replacement?
    (c) What is the probability that someone will keep a color TV fewer than 10 years before replacement?
    (d) Assume that the average life of a color TV is 8 years with a standard deviation of 1.5 years before it breaks. Suppose that a company guarantees color TVs and will replace a TV that breaks while under guarantee with a new one. However,

the company does not want to replace more than 10% of the TVs under guarantee. For how long should the guarantee be made (round to the nearest tenth of a year)?

32. *Consumer: Refrigerator Replacement* *Consumer Reports* indicated that the average life of a refrigerator before replacement is $\mu = 14$ years with a (95% of data) range from 9 to 19 years. Let x = age at which a refrigerator is replaced. Assume that x has a distribution that is approximately normal.
    (a) Find a good approximation for the standard deviation of x values. *Hint:* See Problem 31.
    (b) What is the probability that someone will keep a refrigerator fewer than 11 years before replacement?
    (c) What is the probability that someone will keep a refrigerator more than 18 years before replacement?
    (d) Assume that the average life of a refrigerator is 14 years with the standard deviation given in part (a) before it breaks. Suppose that a company guarantees refrigerators and will replace a refrigerator that breaks while under guarantee with a new one. However, the company does not want to replace more than 5% of the refrigerators under guarantee. For how long should the guarantee be made (round to the nearest tenth of a year)?

33. *Veterinary Science: Horses* The resting heart rate for an adult horse should average about $\mu = 46$ beats per minute with (95% of data) range from 22 to 70 beats per minute, based on information from *The Merck Veterinary Manual* (a classic reference used in most veterinary colleges). Let x be a random variable that represents the resting heart rate for an adult horse. Assume that x has a distribution that is approximately normal.
    (a) Estimate the standard deviation of the x distribution. *Hint:* See Problem 31.
    (b) What is the probability that the heart rate is less than 25 beats per minute?
    (c) What is the probability that the heart rate is greater than 60 beats per minute?
    (d) What is the probability that the heart rate is between 25 and 60 beats per minute?
    (e) A horse whose resting heart rate is in the upper 10% of the probability distribution of heart rates may have a secondary infection or illness that needs to be treated. What is the heart rate corresponding to the upper 10% cutoff point of the probability distribution?

34. *Veterinary Science: Kittens* How much should a healthy kitten weigh? A healthy 10-week-old (domestic) kitten should weigh an average of $\mu = 24.5$ oz with (95% of data) range from 14 to 35 oz. See reference in Problem 33. Let x be a random variable that represents the weight (in ounces) of a healthy 10-week-old kitten. Assume that x has a distribution that is approximately normal.
    (a) Estimate the standard deviation of the x distribution. *Hint:* See Problem 31.
    (b) What is the probability that a healthy 10-week-old kitten will weigh less than 14 oz?
    (c) What is the probability that a healthy 10-week-old kitten will weigh more than 33 oz?
    (d) What is the probability that a healthy 10-week-old kitten will weigh between 14 and 33 oz?
    (e) A kitten whose weight is in the bottom 10% of the probability distribution of weights is called *undernourished*. What is the cutoff point for the weight of an undernourished kitten?

35. *Insurance: Satellites* A relay microchip in a telecommunications satellite has a life expectancy that follows a normal distribution with a mean of 90 months and a

standard deviation of 3.7 months. When this computer-relay microchip malfunctions, the entire satellite is useless. A large London insurance company is going to insure the satellite for 50 million dollars. Assume that the only part of the satellite in question is the microchip. All other components will work indefinitely.

(a) For how many months should the satellite be insured to be 99% confident that it will last beyond the insurance date?

(b) If the satellite is insured for 84 months, what is the probability that it will malfunction before the insurance coverage ends?

(c) If the satellite is insured for 84 months, what is the expected loss to the insurance company?

(d) If the insurance company charges $3 million for 84 months of insurance, how much profit does the company expect to make?

36. *Focus Problem: Show Attendance* Solve the Focus Problem stated at the beginning of this chapter (page 271).

37. *Conditional Probability: Waiting Time* Suppose you want to eat lunch at a popular restaurant. The restaurant does not take reservations, so there is usually a waiting time before you can be seated. Let x represent the length of time to be seated. From past experience, you know that the mean waiting time is $\mu = 18$ minutes with $\sigma = 4$ minutes. You assume that the x distribution is approximately normal.

(a) What is the probability that the waiting time will *exceed* 20 minutes, given that it has *exceeded* 15 minutes? *Hint:* Let A = the event $x > 20$, and let B = the event $x > 15$. We want to compute $P(A, \text{given } B \text{ has occurred})$. Notice that the event A *and* B = the event $x > 20$. (Why?) Recall from Section 4.2 that

$$P(A \text{ and } B) = P(B) \cdot P(A, \text{ given } B \text{ has occurred})$$

so $P(x > 20) = P(x > 15) \cdot P(A, \text{ given } B \text{ has occurred})$

Use the normal distribution to compute $P(x > 20)$ and $P(x > 15)$. Use these values in the preceding equation, and solve for the requested probability $P(A, \text{ given } B \text{ has occurred})$.

(b) Use a method similar to that shown in part (a) to determine the probability that the waiting time will exceed 25 minutes given it has exceeded 18 minutes.

38. *Conditional Probability: Cycle Time* A cement truck delivers mixed cement to a large construction site. Let x represent the cycle time in minutes for the truck to leave the construction site, go back to the cement plant, fill up, and return to the construction site with another load of cement. From past experience, it is known that the mean cycle time is $\mu = 45$ minutes with $\sigma = 12$ minutes. The x distribution is approximately normal.

(a) What is the probability that the cycle time will *exceed* 60 minutes, given that it has exceeded 50 minutes? *Hint:* See Problem 37, part (a).

(b) What is the probability that the cycle time will exceed 55 minutes, given that it has exceeded 40 minutes?

39. *Budget: Maintenance* The weekly amount of money spent on cleaning, maintenance, and repairs at a large restaurant was observed over a long period of time to be approximately normally distributed with mean $\mu = \$615$ and standard deviation $\sigma = \$42$.

(a) If $646 is budgeted for next week, what is the probability the actual costs will exceed the budgeted amount?

(b) How much should be budgeted for weekly repairs, cleaning, and maintenance so that the probability the budgeted amount will be exceeded in a given week is only 0.10?

6.4
Normal Approximation to the Binomial Distribution

FOCUS POINTS

✓ State the assumptions needed for the normal approximation to the binomial.

✓ Compute μ and σ for the normal approximation.

✓ Use the continuity correction to convert a range of r values to a corresponding range of normal x values.

✓ Convert the x values to a range of standardized z scores and find desired probabilities.

Criteria $np > 5$ and $nq > 5$

The probability that a new vaccine will protect adults from cholera is known to be 0.85. It is administered to 300 adults who must enter an area where the disease is prevalent. What is the probability that more than 280 of these adults will be protected from cholera by the vaccine?

This question falls into the category of a binomial experiment with number of trials n equal to 300, the probability of success p equal to 0.85, and the number of successes r greater than 280. It is possible to use the formula for the binomial distribution to compute the probability that r is greater than 280. However, this approach would involve a number of tedious and long calculations. There is an easier way to do this problem, for under the conditions stated below, the normal distribution can be used to approximate the binomial distribution.

Again, let p be the probability of success and let q be the probability of failure in a single binomial trial. Let n be the number of trials in the binomial experiment. If n, p, and q are such that *both* $np > 5$ and $nq > 5$, then the normal probability distribution with $\mu = np$ and $\sigma = \sqrt{npq}$ will be a good *normal approximation to the binomial distribution*. As n gets larger, the approximation becomes better.

◇ **SUMMARY** Consider the binomial distribution with

n = number of trials

r = number of successes

p = probability of success

q = probability of failure = $1 - p$

If $np > 5$ and $nq > 5$, then r has a binomial distribution that is approximated by a *normal* distribution with

$$\mu = np$$

and

$$\sigma = \sqrt{npq} \quad ◇$$

EXAMPLE 11

Binomial distribution graphs

Graph the binomial distributions where $p = 0.25$, $q = 0.75$, and the number of trials is first $n = 3$, then $n = 10$, $n = 25$, and finally $n = 50$.

SOLUTION: The authors used the computer program ComputerStat to obtain the binomial distributions for the given values of p, q, and n. The results have been organized and graphed in Figures 6-37, 6-38, 6-39, and 6-40.

When $n = 3$, the outline of the histogram does not even begin to take the shape of a normal curve. But when $n = 10$, 25, or 50, it does begin to take a normal shape, indicated by the red curve.

From a theoretical point of view, the histograms in Figures 6-38, 6-39, and 6-40 would have bars for all values of r from $r = 0$ to $r = n$. However, in the construction of these histograms, the bars of height less than 0.001 unit have been omitted—that is, in this example, probabilities less than 0.001 have been rounded to 0.   ◇

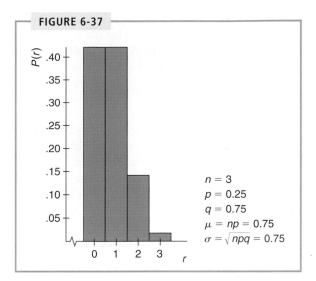

FIGURE 6-37

$n = 3$
$p = 0.25$
$q = 0.75$
$\mu = np = 0.75$
$\sigma = \sqrt{npq} = 0.75$

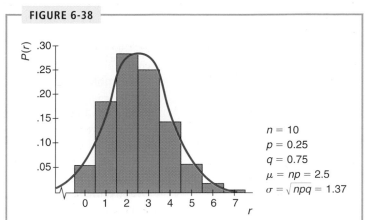

FIGURE 6-38

$n = 10$
$p = 0.25$
$q = 0.75$
$\mu = np = 2.5$
$\sigma = \sqrt{npq} = 1.37$

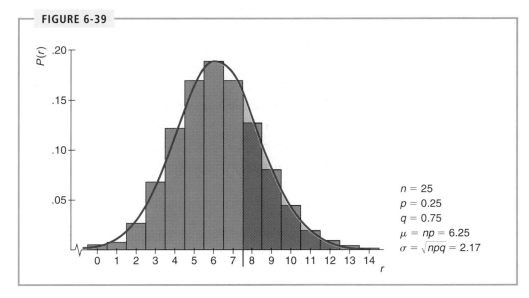

FIGURE 6-39

$n = 25$
$p = 0.25$
$q = 0.75$
$\mu = np = 6.25$
$\sigma = \sqrt{npq} = 2.17$

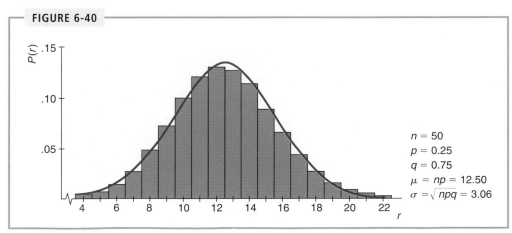

FIGURE 6-40

$n = 50$
$p = 0.25$
$q = 0.75$
$\mu = np = 12.50$
$\sigma = \sqrt{npq} = 3.06$

EXAMPLE 12

Normal approximation

The owner of a new apartment building must install 25 water heaters. From past experience in other apartment buildings, she knows that Quick Hot is a good brand. It is guaranteed for 5 years only, but from her past experience, she knows that the probability it will last 10 years is 0.25.

(a) What is the probability that 8 or more of the 25 water heaters will last at least 10 years?

SOLUTION: In this example, $n = 25$ and $p = 0.25$, so Figure 6-39 (on the preceding page) represents the probability distribution we will use. Let r be the binomial random variable corresponding to the number of successes out of $n = 25$ trials. We want to find $P(r \geq 8)$ by using the normal approximation. This probability is represented graphically (Figure 6-39) by the area of the bar over 8 and all bars to the right of the bar over 8.

Let x be a normal random variable corresponding to a normal distribution with $\mu = np = 25(0.25) = 6.25$ and $\sigma = \sqrt{npq} = \sqrt{25(0.25)0.75} \approx 2.17$. This normal curve is represented by the red line in Figure 6-39 on the preceding page. The area under the normal curve from $x = 7.5$ to the right is approximately the same as the area of the bars from the bar over $r = 8$ to the right. It is important to notice that we start with $x = 7.5$ because the bar over $r = 8$ really starts at $x = 7.5$.

The area of the bars and the area under the corresponding red (normal) curve are approximately equal, so we conclude that $P(r \geq 8)$ is approximately equal to $P(x \geq 7.5)$.

When we convert $x = 7.5$ to standard units, we get

$$z = \frac{x - \mu}{\sigma}$$

$$= \frac{7.5 - 6.25}{2.17}$$

$$= 0.58 \quad \text{(use } \mu = 6.25 \text{ and } \sigma = 2.17\text{)}$$

The probability we want is

$$P(x \geq 7.5) = P(z \geq 0.58) = 1 - P(z \leq 0.58)$$
$$= 1 - 0.7190$$
$$= 0.2810$$

(b) How does this result compare with the result we can obtain by using the formula for the binomial probability distribution with $n = 25$ and $p = 0.25$?

SOLUTION: Using the binomial distribution function on the TI-83Plus, the authors computed that $P(r \geq 8) \approx 0.2735$. This means that the probability is approximately 0.27 that 8 or more water heaters will last at least 10 years.

(c) How do the results of parts (a) and (b) compare?

SOLUTION: The error of approximation is the difference between the approximate normal value (0.2810) and the binomial value (0.2735). The error is only $0.2810 - 0.2735 = 0.0075$, which is negligible for most practical purposes.

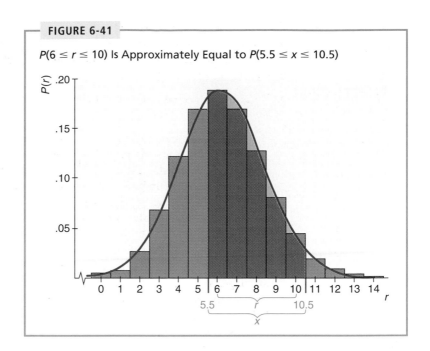

FIGURE 6-41

$P(6 \leq r \leq 10)$ Is Approximately Equal to $P(5.5 \leq x \leq 10.5)$

We knew in advance that the normal approximation to the binomial probability would be good, since $np = 25\,(0.25) = 6.25$ and $nq = 25(0.75) = 18.75$ are both greater than 5. These are the conditions that assure us that the normal approximation will be sufficiently close to the binomial probability for most practical purposes.   ◊

Continuity correction: converting r values to x values

Remember that when using the normal distribution to approximate the binomial, we are computing the areas under bars. The bar over r goes from $r - 0.5$ to $r + 0.5$. If r is a *left* endpoint of an interval, we *subtract* 0.5 to get the corresponding normal variable x. If r is a *right* endpoint of an interval, we *add* 0.5 to get the corresponding variable x. For instance, $P(6 \leq r \leq 10)$, where r is a binomial variable, is approximated by $P(5.5 \leq x \leq 10.5)$, where x is the corresponding normal variable (see Figure 6-41). Adding or subtracting 0.5 to or from r to obtain a corresponding range of normal values x is called the *continuity correction*. It is needed because we are approximating a discrete probability distribution by a continuous probability distribution.

◊ **COMMENT** Both the binomial and Poisson distributions are for *discrete* random variables. Therefore, adding or subtracting 0.5 to r was not necessary when we approximated the binomial distribution by the Poisson distribution (Section 5.4). However, the normal distribution is for a *continuous* random variable. In this case, adding or subtracting 0.5 to or from (as appropriate) r will improve the approximation of the normal to the binomial distribution.   ◊

GUIDED EXERCISE 12

Continuity correction

From many years of observation, a biologist knows the probability is only 0.65 that any given Arctic tern will survive the migration from its summer nesting area to its winter feeding grounds. A random sample of 500 Arctic terns were banded at their summer nesting area. Use the normal approximation to the binomial and the following steps to find the probability that between 310 and 340 of the banded Arctic terns will survive the migration. Let r be the number of surviving terns.

Arctic tern

(a) To approximate $P(310 \leq r \leq 340)$, we use the normal curve with $\mu =$ _____ and $\sigma =$ _____.

We use the normal curve with
$$\mu = np = 500(0.65) = 325$$
$$\sigma = \sqrt{npq} = \sqrt{500(0.65)(0.35)} = 10.67$$

(b) $P(310 \leq r \leq 340)$ is approximately equal to $P($ _____ $\leq x \leq$ _____ $)$, where x is a variable from the normal distribution described in part (a).

Since 310 is the left endpoint, we subtract 0.5, and since 340 is the right endpoint, we add 0.5. Consequently,
$$P(310 \leq r \leq 340) \approx P(309.5 \leq x \leq 340.5)$$

(c) Convert the condition $309.5 \leq x \leq 340.5$ to a condition in standard units.

Since $\mu = 325$ and $\sigma = 10.67$, the condition $309.5 \leq x \leq 340.5$ becomes
$$\frac{309.5 - 325}{10.67} \leq z \leq \frac{340.5 - 325}{10.67}$$

or

$$-1.45 \leq z \leq 1.45$$

(d) $P(310 \leq r \leq 340) = P(309.5 \leq x \leq 340.5)$
$\qquad = P(-1.45 \leq z \leq 1.45)$
$\qquad =$ _____

$P(-1.45 \leq z \leq 1.45) = P(z \leq 1.45) - P(z \leq -1.45)$
$\qquad\qquad = 0.9265 - 0.0735$
$\qquad\qquad = 0.8530$

(e) Will the normal distribution make a good approximation to the binomial for this problem? Explain your answer.

Since
$$np = 500(0.65) = 325$$
and
$$nq = 500(0.35) = 175$$

are both greater than 5, the normal distribution will be a good approximation to the binomial.

VIEWPOINT   *Sunspots, Tree Rings, and Statistics*

Ancient Chinese astronomers recorded extreme sunspot activity with a peak around 1200 A.D. Mesa Verde tree rings in the period between 1276 and 1299 were unusually narrow, indicating a drought and/or a severe cold spell in the region at that time. A cooling trend could have narrowed the window of frost-free days below the approximately 80 days needed for cultivation of aboriginal corn and beans. Is this the reason the ancient Anasazi dwellings in Mesa Verde were abandoned? Is there a connection with the extreme sunspot activity? Much research and statistical work continues to be done on this topic.

Reference: *Prehistoric Astronomy in the Southwest*, by J. McKim Malville and C. Putnam, Department of Astronomy, University of Colorado.

SECTION 6.4 PROBLEMS

Note: When we say *between a* and *b*, we mean every value from *a* to *b*, *including a* and *b*. Due to rounding, your answers might vary slightly from answers given in the text.

1. *Health: Lead Contamination* More than a decade ago high levels of lead in the blood put 88% of children at risk. A concerted effort was made to remove lead from the environment. Now, according to the *Third National Health and Nutrition Examination Survey (NHANES III)* conducted by the Centers for Disease Control, only 9% of children in the United States are at risk of high blood-lead levels.
   (a) In a random sample of 200 children taken more than a decade ago, what is the probability that 50 or more had high blood-lead levels?
   (b) In a random sample of 200 children taken now, what is the probability that 50 or more have high blood-lead levels?

2. *Insurance: Claims* Do you try to *pad* an insurance claim to cover your deductible? About 40% of all U.S. adults will try to pad their insurance claims! (Source: *Are You Normal?*, by Bernice Kanner, St. Martin's Press.) Suppose that you are the director of an insurance adjustment office. Your office has just received 128 insurance claims to be processed in the next few days. What is the probability that
   (a) half or more of the claims have been padded?
   (b) fewer than 45 of the claims have been padded?
   (c) from 40 to 64 of the claims have been padded?
   (d) more than 80 of the claims are *not* padded?

3. *Law Enforcement: Arrests* In Colorado Springs, a local newspaper ran a full page of nearly 100 mug shots of people the police wanted to arrest for serious crimes. Within 1 week, the police received enough information to locate and arrest about 17% of these "wanted" people (reported in *Rocky Mountain News*). If next month the newspaper runs a full page of 125 mug shots of fugitives, what is the probability that the police will receive enough information to locate and arrest (within 1 week)
   (a) at least 15 fugitives?
   (b) 28 or more fugitives?
   (c) between 15 and 28 fugitives?

(d) In the solution to this problem, what was n? p? q? Does it appear that both np and nq are larger than 5? Why is this an important consideration?

4. *Novels: Romance* USA Today reported that 11% of all books sold are of the romance genre. If a local bookstore sells 316 books on a given day, what is the probability that
   (a) fewer than 40 are romances?
   (b) at least 25 are romances?
   (c) between 25 and 40 are romances?
   (d) In the solution to this problem, what was n? p? q? Does it appear that both np and nq are larger than 5? Why is this an important consideration?

5. *Longevity: 90th Birthday* It is estimated that 3.5% of the general population will live past their 90th birthday (*Statistical Abstract of the United States,* 112th Edition). In a graduating class of 753 high school seniors, what is the probability that
   (a) 15 or more will live beyond their 90th birthday?
   (b) 30 or more will live beyond their 90th birthday?
   (c) between 25 and 35 will live beyond their 90th birthday?
   (d) more than 40 will live beyond their 90th birthday?

6. *Fishing: Billfish* Ocean fishing for billfish is very popular in the Cozumel Region of Mexico. In *World Record Game Fishes* (published by the International Game Fish Association), it was stated that in the Cozumel Region about 44% of strikes (while trolling) resulted in a catch. Suppose that on a given day a fleet of fishing boats got a total of 24 strikes. What is the probability that the number of fish caught was
   (a) 12 or fewer?
   (b) 5 or more?
   (c) between 5 and 12?
   (d) In the solution to this problem, what was n? p? q? Does it appear that both np and nq are larger than 5? Why is this an important consideration?

7. *Grocery Stores: New Products* The *Denver Post* stated that 80% of all new products introduced in grocery stores fail (are taken off the market) within 2 years. If a grocery store chain introduces 66 new products, what is the probability that within 2 years
   (a) 47 or more fail?
   (b) 58 or fewer fail?
   (c) 15 or more succeed?
   (d) fewer than 10 succeed?

8. *Crime: Murder* What are the chances that a person who is murdered actually knew the murderer? The answer to this question explains why a lot of police detective work begins with relatives and friends of the victim! About 64% of the people who are murdered actually knew the person who did the murder (*Chances: Risk and Odds in Everyday Life,* by James Burke). Suppose that a detective file in New Orleans has 63 current unsolved murders. What is the probability that
   (a) at least 35 of the victims knew their murderer?
   (b) at most 48 of the victims knew their murderer?
   (c) fewer than 30 victims did *not* know their murderer?
   (d) more than 20 victims did *not* know their murderer?

9. *Private Investigation: Finding People* Old Friends Information Service is a California company that finds addresses for people who have lost track of each other. Old Friends claims to be 70% successful in reuniting people (*Wall Street Journal*). In December, Old Friends had 430 requests for addresses of lost acquaintances. What is the probability that the number of addresses found was

(a) more than 280?

(b) at least 320?

(c) between 280 and 320?

(d) In the solution to this problem, what was n? p? q? Does it appear that both np and nq are larger than 5? Why is this an important consideration?

10. *Archaeology: Pottery* Santa Fe black on white is a style of pottery that occurs in about 61% of the pot shards found in the Bandelier National Monument area (*Bandelier Archaeological Excavation Project: Summer 1990 Excavations at Burnt Mesa Pueblo*, edited by Kohler, Washington State University). At one excavation site 8641 pot shards have been found that have not yet been cleaned and identified. What is the probability that

(a) fewer than 5200 are Santa Fe black on white?

(b) more than 5400 are Santa Fe black on white?

(c) between 5200 and 5400 are Santa Fe black on white?

(d) In the solution to this problem, what was n? p? q? Does it appear that both np and nq are larger than 5? Why is this an important consideration?

11. *Lawyers: Bar Exam* Over the years, it has been observed that of all the lawyers who take the state bar exam, only 57% pass (information from the National Conference on Bar Examiners, referenced in *The Book of Odds*, by Shook and Shook, Signet). Suppose that this year 850 lawyers are going to take the Ohio bar exam. What is the probability that

(a) 540 or more pass?

(b) 500 or fewer pass?

(c) between 485 and 525 pass?

12. *Coupons: Redemption* More than 200 billion grocery coupons are distributed each year for discounts exceeding $84 billion. However, according to a report in *USA Today*, only 3.2% of the coupons are redeemed. If a company distributes 5000 coupons, what is the probability that

(a) more than 100 are redeemed?

(b) fewer than 200 are redeemed?

(c) between 100 and 200 are redeemed?

13. *Supermarkets: Free Samples* Do you take the free samples offered in supermarkets? About 60% of all customers will take free samples. Furthermore, of those who take the free samples, about 37% will buy what they have sampled. (See reference in Problem 2.) Suppose that you set up a counter in a supermarket offering free samples of a new product. The day you were offering free samples, 317 customers passed by your counter.

(a) What is the probability that more than 180 will take your free sample?

(b) What is the probability that fewer than 200 will take your free sample?

(c) What is the probability that a customer will take a free sample *and* buy the product? *Hint:* Use the multiplication rule for *dependent* events. Notice that we are given the conditional probability $P(\text{buy, } given \text{ sample}) = 0.37$, while $P(\text{sample}) = 0.60$.

(d) What is the probability that between 60 and 80 customers will take the free sample *and* buy the product? *Hint:* Use the probability of success calculated in part (c).

14. *Ice Cream: Flavors* What's your favorite ice cream? For people who buy ice cream, the all-time favorite is still vanilla. About 25% of ice cream sales are vanilla. Chocolate accounts for only 9% of ice cream sales. (See reference in Problem 2.) Suppose that 175 customers go to a grocery store in Cheyenne, Wyoming, today to buy ice cream.

(a) What is the probability that 50 or more will buy vanilla?

(b) What is the probability that 12 or more will buy chocolate?

(c) A customer who buys ice cream is not limited to one container or one flavor. What is the probability that someone who is buying ice cream will buy chocolate or vanilla? *Hint:* Chocolate flavor and vanilla flavor are not mutually exclusive events. Assume that the choice to buy one flavor is independent of the choice to buy another flavor. Then use the multiplication rule for independent events together with the addition rule for events that are not mutually exclusive to compute the requested probability. (See Section 4.2.)

(d) What is the probability that between 50 and 60 customers buy chocolate or vanilla ice cream? *Hint:* Use the probability of success computed in part (c).

15. *Airline Flights: No-Shows* Based on long experience, an airline found that about 6% of the people making reservations on a flight from Miami to Denver do not show up for the flight. Suppose the airline overbooks this flight by selling 267 ticket reservations for an airplane with only 255 seats.

(a) What is the probability a person holding a reservation will show up for the flight?

(b) Let $n = 267$ represent the number of ticket reservations. Let r represent the number of people with reservations who show up for the flight. Which expression represents the probability that a seat will be available for everyone who shows up holding a reservation:

$$P(255 \leq r); \qquad P(r \leq 255); \qquad P(r \leq 267); \qquad P(r = 255)$$

(c) Use the normal approximation to the binomial distribution and part (b) to answer the question: What is the probability that a seat will be available for every person who shows up holding a reservation?

16. *General: Approximations* We have studied *two* approximations to the binomial, the normal approximation and the Poisson approximation (Section 5.4). Write a brief but complete essay in which you discuss and summarize the *conditions* under which each approximation would be used, the *formulas* involved, and *assumptions* made for each approximation. Give details and examples in your essay. How could you apply these statistical methods in your own everyday life?

SUMMARY

In this chapter, we have examined graphs of normal distributions, control charts, standard units, z scores, and areas under the standard normal curve.

It is important to remember that a normal distribution with mean μ and standard deviation σ can be transformed into the standard normal distribution with mean 0 and standard deviation 1 by the formula

$$z = \frac{x - \mu}{\sigma}$$

Values of z and Table 5 of Appendix II can be used to obtain the area under the standard normal curve to the left of z. Areas under the normal curve represent probabilities that a z value will fall into the interval over which the area lies.

Given a probability associated with an interval on the standard normal curve, we can use Table 5 of Appendix II to obtain the associated z values. When x follows a normal distribution, we can find the x values associated with a given probability by (1) finding the corresponding z values, and (2) converting to x values by using the formula

$$x = z\sigma + \mu$$

In the last section, we saw that we can use the normal distribution to approximate the binomial distribution if $np > 5$ and $nq > 5$, where n is the number of trials, p is the probability of success on a single trial, and $q = 1 - p$.

IMPORTANT WORDS & SYMBOLS

Section 6.1
Normal distributions
Normal curves
Upward cup and downward cup on normal curves
Symmetry of normal curves
Empirical rule
Control chart
Out-of-control signals

Section 6.2
z value or z score

Standard units
Standard normal distribution ($\mu = 0$ and $\sigma = 1$)
Raw score, x
Area under the standard normal curve

Section 6.3
Areas under any normal curve

Section 6.4
Normal approximation to the binomial distribution
Continuity correction

VIEWPOINT

Nenana Ice Classic

The Nenana Ice Classic is a betting pool offering a large cash prize to the lucky winner who can guess the time, to the nearest minute, of the ice breakup on the Tanana River at the town of Nenana, Alaska. Official breakup time is defined when the surging river dislodges a tripod on the ice. This breaks an attached line and stops a clock set to Yukon Standard Time. The event is so popular that the first state legislature of Alaska (1959) made the Nenana Ice Classic an official statewide lottery. Since 1918, the earliest breakup was April 20, 1940, at 3:27 P.M., and the latest recorded breakup was May 20, 1964, at 11:41 A.M. Want to make a statistical guess predicting when the ice will break up? Breakup times from 1918 to 1996 are recorded in *The Alaska Almanac,* published by Alaska Northwest Books, Anchorage.

CHAPTER REVIEW PROBLEMS

1. Given that z is the standard normal variable (with mean 0 and standard deviation 1), find
   (a) $P(0 \leq z \leq 1.75)$ (b) $P(-1.29 \leq z \leq 0)$
   (c) $P(1.03 \leq z \leq 1.21)$ (d) $P(z \geq 2.31)$
   (e) $P(z \leq -1.96)$ (f) $P(z \leq 1.00)$

2. Given that z is the standard normal variable (with mean 0 and standard deviation 1), find

(a) $P(0 \le z \le 0.75)$ (b) $P(-1.50 \le z \le 0)$
(c) $P(-2.67 \le z \le -1.74)$ (d) $P(z \ge 1.56)$
(e) $P(z \le -0.97)$ (f) $P(z \le 2.01)$

3. Given that x is a normal variable with mean $\mu = 47$ and standard deviation $\sigma = 6.2$, find
   (a) $P(x \le 60)$ (b) $P(x \ge 50)$ (c) $P(50 \le x \le 60)$

4. Given that x is a normal variable with mean $\mu = 110$ and standard deviation $\sigma = 12$, find
   (a) $P(x \le 120)$ (b) $P(x \ge 80)$ (c) $P(108 \le x \le 117)$

5. Find z so that 5% of the area under the standard normal curve lies to the right of z.

6. Find z so that 1% of the area under the standard normal curve lies to the left of z.

7. Find z so that 95% of the area under the standard normal curve lies between $-z$ and z.

8. Find z so that 99% of the area under the standard normal curve lies between $-z$ and z.

9. *Nursing: Exams* On a practical nursing licensing exam, the mean score is 79 and the standard deviation is 9 points.
   (a) What is the standardized score of a student with a raw score of 87?
   (b) What is the standardized score of a student with a raw score of 79?
   (c) Assuming the scores follow a normal distribution, what is the probability that a score selected at random is above 85?

10. *Aptitude Tests: Mechanical* On an auto mechanic aptitude test, the mean score is 270 points and the standard deviation is 35 points.
    (a) If a student has a standardized score of 1.9, how many points is that?
    (b) If a student has a standardized score of -0.25, how many points is that?
    (c) Assuming the scores follow a normal distribution, what is the probability that a student will get between 200 and 340 points?

11. *Recycling: Aluminum Cans* One environmental group did a study of recycling habits in a California community. It found that 70% of the aluminum cans sold in the area were recycled.
    (a) If 400 cans are sold today, what is the probability that 300 or more will be recycled?
    (b) Of the 400 cans sold, what is the probability that between 260 and 300 will be recycled?

12. *Guarantee: Disc Players* Future Electronics makes compact disc players. Their research department found that the life of the laser beam device is normally distributed with mean 5000 hours and standard deviation 450 hours.
    (a) Find the probability that the laser beam device will wear out in 5000 hours or sooner.
    (b) Future Electronics wants to place a guarantee on the players so that no more than 5% fail during the guarantee period. Because the laser pickup is the part most likely to wear out first, the guarantee period will be based on the life of the laser beam device. How many playing hours should the guarantee cover? (Round to the next playing hour.)

13. *Guarantee: Package Delivery* Express Courier Service has found that the delivery time for packages is normally distributed with mean 14 hours and standard deviation 2 hours.

(a) For a package selected at random, what is the probability that it will be delivered in 18 hours or less?

(b) What should be the guaranteed delivery time on all packages in order to be 95% sure that the package will be delivered before this time? (*Hint:* Note that 5% of the packages will be delivered at a time beyond the guarantee time period.)

14. *Control Chart: Landing Gear* Hydraulic pressure in the main cylinder of the landing gear of a commercial jet is very important for a safe landing. If the pressure is not high enough, the landing gear may not lower properly. If it is too high, the connectors in the hydraulic line may spring a leak.

In-flight landing tests show the actual pressure in the main cylinders is a variable with mean 819 pounds per square inch and standard deviation 23 pounds per square inch. Assume that these values for the mean and standard deviation are considered safe values by engineers.

(a) For nine consecutive test landings, the pressure in the main cylinder is recorded as follows:

Landing number	1	2	3	4	5	6	7	8	9
Pressure	870	855	830	815	847	836	825	810	792

Make a control chart for the pressure in the main cylinder of the hydraulic landing gear, and plot the data on the control chart. Looking at the control chart, would you say the pressure is "in control" or "out of control"? Explain your answer. Identify any out-of-control signals by type (I, II, or III).

(b) For 10 consecutive test landings, the pressure was recorded on another plane as follows:

Landing number	1	2	3	4	5	6	7	8	9	10
Pressure	865	850	841	820	815	789	801	765	730	725

Make a control chart and plot the data on the chart. Would you say the pressure is "in control" or not? Explain your answer. Identify any out-of-control signals by type (I, II, or III).

15. *Electronic Scanners: Errors* Instead of hearing the jingle of prices being rung up manually on cash registers, we now hear the beep of prices being scanned by electronic scanners. How accurate is price scanning? There are errors, and according to a *Denver Post* article, when the error occurs in the store's favor, it is larger than when it occurs in the customer's favor. An investigation of large discount stores by the Colorado state inspectors showed that the average error in the store's favor was $2.66. Assume that the distribution of scanner errors is more or less mound-shaped. If the standard deviation of scanner errors (in the store's favor) is $0.85, use the empirical rule to

(a) estimate a range of scanner errors centered about the mean in which 68% of the errors will lie.

(b) estimate a range of scanner errors centered about the mean in which 95% of the errors will lie.

(c) estimate a range of scanner errors centered about the mean in which almost all the errors will lie.

16. *Customer Complaints: Time* The Customer Service Center in a large New York department store has determined that the amount of time spent with a customer about a complaint is normally distributed with a mean of 9.3 minutes and a standard

deviation of 2.5 minutes. What is the probability that for a randomly chosen customer with a complaint the amount of time spent resolving the complaint will be
(a) shorter than 10 minutes?
(b) more than 5 minutes?
(c) between 8 and 15 minutes?

17. *Medical: Flight for Life* The Flight for Life emergency helicopter service is available for medical emergencies occurring from 15 to 90 miles from the hospital. Emergencies that occur closer to the hospital can be handled effectively by ambulance service. A long-term study of the service shows that the response time from receipt of the dispatch call to arrival at the scene of the emergency is normally distributed with a mean of 42 minutes and a standard deviation of 8 minutes. For a randomly received call, what is the probability that the response time will be
(a) between 30 and 45 minutes?
(b) shorter than 30 minutes?
(c) more than 60 minutes?

18. *Unlisted Phones: Sacramento* How easy is it to get into contact with a person in Sacramento, California? If you don't know the telephone number, it could be difficult. The data from Survey Sampling of Fairfield, Connecticut, reported in *American Demographics,* show that 68% of the telephone-owning households in Sacramento have unlisted numbers. For a random sample of 150 Sacramento households with telephones, what is the probability that
(a) 100 or more have unlisted numbers?
(b) fewer than 100 have unlisted numbers?
(c) between 50 and 65 (including 50 and 65) households have *listed* numbers?

19. *Medical: Blood Type* Blood type AB is found in only 3% of the population (*Textbook of Medical Physiology,* by A. Guyton, M.D.). If 250 people are chosen at random, what is the probability that
(a) 5 or more will have this blood type?
(b) between 5 and 10 will have this blood type?

DATA HIGHLIGHTS: GROUP PROJECTS

Break into small groups and discuss the following topics. Organize a brief outline in which you summarize the main points of your group discussion.

1. Examine Figure 6-42. Government documents and the Census Bureau show that the age at first marriage for U.S. citizens is approximately normally distributed for both men and women. For men, the average age is about 27 years, and for women, the average is about 24 years. For both sexes, the standard deviation is about 2.5 years.
   (a) If the distribution is symmetrical and mound-shaped (such as the normal distribution), why would you expect the median, mean, and mode to be equal?
   (b) Consider the age at first marriage in 1995. What is the probability that a man selected at random was over age 30 at the time of his first marriage? What is the probability that he was under age 20? What is the probability that he was between 20 and 30?
   (c) Consider the age at first marriage in 1995. What is the probability that a woman selected at random was over age 28 at the time of her first marriage? What is the probability that she was under age 18? What is the probability that she was between 18 and 28?

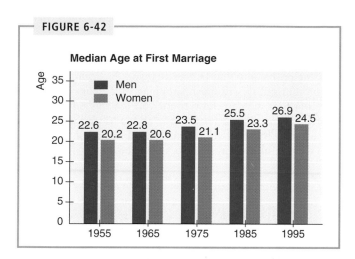

FIGURE 6-42

Median Age at First Marriage

FIGURE 6-43

USA SNAPSHOTS®
A look at statistics that shape our lives

Shopping by the clock
People spend an average of 69 minutes each visit to a mall or shopping center:

Time shopping

Less than 1 hour **52%**

More than 3 hours **4%**

2–3 hours **10%**

1–2 hours **34%**

Source: International Council of Shopping Centers

By Elys A. McLean, USA TODAY

Source: Copyright © 1992 USA TODAY. Reprinted with permission.

(d) At what age were only 10% of eligible men (who had never been married before) left? At what age were only 5% left?

(e) At what age were only 10% of eligible women (who had never been married before) left? At what age were only 5% left?

(f) The Census Bureau tracks data about marriages and age at first marriage. The *Statistical Abstract of the United States* is published each year and contains tables giving this information. For more information, visit the Brase/Brase statistics site at http://math.college.hmco.com/students and find the link to the Census Bureau. Look for marriage under the subject index, and follow the links to the table. Using either the Abstract or the Census Bureau web site, find the most recent information about median age at first marriage. Although the median age changes from year to year, the standard deviation usually does not change much. Under the assumption that the age at first marriage follows a distribution that is approximately normal, the mean age at first marriage equals the median age. Using standard deviation of about 2.5 years, repeat parts (b) through (e) for the most recent year for which median age at first marriage is available.

2. Examine Figure 6-43, "Shopping by the clock" (*USA Today*).

   (a) Notice that 52% of the people who go to a shopping center spend less than 1 hour. However, the figure also states that people spend an average of 69 minutes on each visit to a shopping center. How could *both* these claims be correct? Write a brief, complete essay in which you discuss mean, median, and mode in the context of symmetrical distribution. Also discuss mean, median, and mode for distributions that are skewed left, skewed right, or general distributions. Then answer the question: How could 52% of the people spend less than 1 hour in the shopping center while the average (we don't know which average was used) time spent was 69 minutes?

   (b) Ala Moana Shopping Center in Honolulu is sometimes advertised as the largest shopping center for 2500 miles and the best shopping center in the middle of the Pacific Ocean. There are many interesting Hawaiian, Asian, and other ethnic shops in Ala Moana, so this center is a favorite of tourists. Suppose that a tour group of 75 people has just arrived at the shopping center. The International Council of Shopping Centers indicates that 86% (52% plus 34%) of the people will spend less than 2 hours in a shopping center. Let us assume this applies to our tour group.

(i) For the tour group of 75 people, what is the expected number who finish shopping on or before 2 hours? What is the standard deviation?

(ii) For the tour group of 75 people, what is the probability that 55 or more will finish shopping in 2 hours or less?

(iii) For the tour group, what is the probability that 70 or more will finish shopping in 2 hours or less?

(iv) What is the probability that between 50 and 70 (including 50 and 70) people in the tour group will finish shopping in 2 hours or less?

(v) If you are a tour director, how could you use this information to plan an appropriate length of time for the stop at Ala Moana shopping center? You do not want to frustrate your group by allowing too little time for shopping, but you also do not want to spend too much time at the center.

LINKING CONCEPTS: WRITING PROJECTS

Discuss each of the following topics in class or review the topics on your own. Then write a brief but complete essay in which you summarize the main points. Please include formulas and graphs as appropriate.

1. If you look up the word *normal* in a dictionary, you will find it is synonymous with the words *standard* or *usual*. Consider the very wide and general applications of the normal probability distribution. Comment on why good synonyms for *normal probability distribution* might be the *standard probability distribution* or the *usual probability distribution*. List at least three random variables from everyday life for which you think the normal probability distribution could be applicable.

2. Why are standard *z* values so important? Is it true that *z* values have no units of measurement? Why would this be desirable for comparing data sets with *different* units of measurement? How can we assess differences in quality or performance by simply comparing *z* values under a standard normal curve? Examine the formula to compute standard *z* values. Notice it involves *both* the mean and standard deviation. Recall that in Chapter 3 we commented that the mean of a data collection was not entirely adequate to describe the data; you need the standard deviation as well. Discuss this topic again in the light of what you now know about normal distributions and standard *z* values.

3. If you look up the word *empirical* in a dictionary, you find it means relying on experiment and observation rather than on theory. Discuss the empirical rule in this context. The empirical rule certainly applies to the normal distribution, but does it also apply to a wide variety of other distributions that are not *exactly* (theoretically) normal? Discuss the terms *mound-shaped* and *symmetrical*. Draw several sketches of distributions that are mound-shaped *and* symmetrical. Draw sketches of distributions that are not mound-shaped or not symmetrical. To which distributions will the empirical rule apply?

4. Most companies that manufacture a product have a division of quality control or quality assurance. The purpose of the quality-control division is to make reasonably certain that the products manufactured are up to company standards. Write a brief essay in which you describe how the statistics you have learned so far could be applied to an industrial application (such as control charts and the Antlers Lodge example).

Using TECHNOLOGY

TI-83PLUS • EXCEL • MINITAB • COMPUTERSTAT

APPLICATION 1

How much money do people earn in a month? One way to answer this question is to look at government employees. The average earnings of a city government employee in the month of March are given for a sample of 40 large cities in the United States (*Statistical Abstract of the United States*, 120th Edition).

(a) To compare the earnings from one city to another, we will look at z values for each city. Use the sample mean and sample standard deviation to compute the z values. You may do this "by hand" using a calculator, or you may use a computer software package.

(b) Look at the z values for each salary. Which are above average? Which are below average?

(c) Which salaries are within one standard deviation of the mean?

City	Average Earnings ($) for March	City	Average Earnings ($) for March
Albuquerque	2494	Memphis	2973
Anchorage	3571	Miami	3654
Arlington	3069	Milwaukee	3636
Atlanta	2494	Mobile	3756
Baltimore	3229	New York	3694
Birmingham	3961	Newark	2507
Boston	3526	Norfolk	2577
Cincinnati	3399	Oakland	5084
Cleveland	2899	Philadelphia	3511
Dallas	3326	Phoenix	3909
Denver	3324	Portland	2899
Detroit	3301	Raleigh	3026
Honolulu	3479	San Antonio	2427
Houston	2813	San Diego	4072
Indianapolis	2756	San Francisco	4487
Kansas City	3023	San Jose	5227
Lincoln	2882	Seattle	4462
Las Vegas	4512	St. Louis	3033
Los Angeles	4534	Washington, D.C.	3725
Louisville	2674	Wichita	2928

Technology Hints: Standardizing Raw Scores

TI-83Plus

On the TI-83, enter the salary data in list L_1. Then use ➤ STAT ➤ CALC ➤ 1-Var Stats to compute the sample mean $\bar{x}$ and the standard deviation s. Then go back to ➤ STAT ➤ EDIT. Arrow up to the header label L_2. At the $L_2 =$ prompt, type in $(x - \bar{x})/s$ and press Enter.

You will find the $\bar{x}$ and s symbols under the ➤ **VARS** ➤ **Statistics** ... keys. The z values will then appear in list L_2.

Excel

In Excel, you need to compute the sample mean $\bar{x}$ and standard deviation s ahead of time. To do so, enter the salary data in column A. To compute $\bar{x}$, use the menu choices ➤ **Paste function** f_x ➤ **Statistical** ➤ **AVERAGE**. To compute s, use the menu choices ➤ **Paste function** f_x ➤ **Statistical** ➤ **STDEV**. Finally, to generate the z values, use the menu choices ➤ **Paste function** f_x ➤ **Statistical** ➤ **STANDARDIZED**. Remember to highlight the cell in which you want each value to appear before you use the commands.

Minitab

In Minitab, enter the salaries in column C1. To generate z values and store the results in column C2, use the following menu choices: ➤ **Calc** ➤ **Standardize**. In the dialogue box, select C1 for input column and C2 for the output column. Then choose the option "subtract mean and divide by standard deviation." Minitab will compute the sample mean $\bar{x}$ and standard deviation s for the salaries and then compute z using the formula $z = (x - \bar{x})/s$.

APPLICATION II

How can we determine if the data originated from a normal distribution? We can look at a stem-and-leaf plot or histogram of the data to check for general symmetry, skewness, clusters of data, or outliers. However, a more sensitive way to check that a distribution is normal is to look at a special graph called a *normal quantile plot* (or a variation of this plot called a normal probability plot in some software packages). It really is

not feasible to make a normal quantile plot by hand, but statistical software packages provide such plots. A simple version of the basic idea behind normal quantile plots involves the following process:

(a) Arrange the observed data values in order from smallest to largest, and determine the percentile occupied by each value. For instance, if there are 20 data values, the smallest datum is at the 5% point, the next smallest is at the 10% point.

(b) Find the z values that correspond to the percentile points. For instance, the z value that corresponds to the percentile 5% (i.e., % in the left tail of the distribution) is $z = -1.645$.

(c) Plot each data value x against the corresponding percentile z score. If the data are close to a normal distribution, the plotted points will lie close to a straight line. (If the data are close to a standard normal distribution, the points will lie close to the line $x = z$.)

The actual process that statistical software packages use to produce the z scores for the data is more complicated.

Interpreting Normal Quantile Plots

If the points of a normal quantile plot lie close to a straight line, the plot indicates that the data follow a normal distribution. Systematic deviations from a straight one or bulges in the plot indicate that the data distribution is not normal. Individual points off the line may be outliers.

Consider Figure 6-44. This figure shows Minitab-generated quantile plots for two data sets. The black dots show the normal quantile plot for the salary data of the first application. The red dots show the normal

FIGURE 6-44 Normal Quantile Plots

- Salary data for (city) government employees
- A random sample of 42 values from a theoretical normal distribution with the same mean and standard deviation as the salary data.

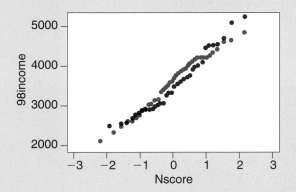

quantile plot for a random sample of 42 data drawn from a theoretical normal distribution with the same mean and standard deviation as the salary data ($\mu \approx 3421$, $\sigma \approx 709$).

(a) Do the black dots lie close to a straight line? Do the salaries appear to follow a normal distribution? Are there any outliers on the low or high side? Would you say any of the salaries are "out of line" for a normal distribution?

(b) Do the red dots lie close to a straight line? We know the red dots represent a sample drawn from a normal distribution. Is the normal quantile plot for the red dots consistent with this fact? Are there any outliers shown?

Technology Hints

Both the TI-83Plus and Minitab support normal quantile plots.

TI-83Plus

Enter the data. Press **STATPLOT** and select one of the plots. Highlight **ON.** Then highlight the 6th plot option. To get a plot similar to that of Figure 6-44, choose Y as the data axis.

Minitab

Minitab has several types of normal quantile plots that use different types of scales. To create a normal quantile plot similar to that of Figure 6-44, enter the data in column C1. Then use the menu choices **Calc ➤ Calculator.** In the dialogue box listing the functions, scroll to **Normal Scores.** Use **NSCOR(C1)** and store the results in column C2. Finally, use the menu choices **Graph ➤ Plot.** In the dialogue box, use C1 for variable "y" and C2 for variable "x."

7

Introduction to Sampling Distributions

No one wants to learn by mistakes, but we cannot learn enough from success to go beyond the state of the art. . . . Such is the nature not only of science and engineering, but of all human endeavors.

—Henry Petroski

Experience isn't what happens to you. It's what you make out of what happens to you.

—Aldous Huxley

**Aldous Huxley
(1894–1963)**

This British novelist and critic wrote about the theme of the human being confronted by the modern world.

For on-line student resources, visit **math.college.hmco.com/students** and follow the Statistics links to the Brase/Brase, *Understandable Statistics,* 7th edition web site.

Henry Petroski is a professor of engineering at Duke University, and Aldous Huxley was a well-known modern writer. Both quotes imply that experience, mistakes, information, and life itself are closely related. Life and uncertainty appear to be inseparable. Only those who are no longer living can escape chance happenings. Mistakes are bound to occur. However, not all mistakes are bad. The discovery of penicillin was a "mistake" when mold (penicillin) was accidentally introduced into a bacterial culture by Alexander Fleming in 1928.

Most of the really important decisions in life will involve incomplete information. In one lifetime, we simply cannot experience *everything*. Nor should we even want to! This is one reason why *experience by way of sampling* is so important. Statistics can help you have the experiences and yet maintain some control over mistakes. Remember, it is what you make out of experience (sample data) that is of real value.

In this chapter, we will study how information from samples relates to information about populations. We cannot be certain that the information from a sample reflects corresponding information about the entire population, but we can describe likely differences. Study this chapter and the following material carefully. We believe that your effort will be rewarded by helping you appreciate the joy and wonder of living in an uncertain universe.

PREVIEW QUESTIONS

◇ As humans, our experiences are finite and limited. Consequently, most of the important decisions in our lives are based on sample (incomplete) information. What do we mean by a probability sampling distribution? How will this help us?   (SECTION 7.1)

◇ There is an old saying: All roads lead to Rome. In statistics, we could recast this saying: All probability distributions average out to be normal distributions (as the sample size increases). How can we take advantage of this in our study of sampling distributions?   (SECTION 7.2)

◇ Many issues in life come down to success or failure. In most cases, we will not be successful all the time, so proportions of successes are very important. What is the probability sampling distribution for proportions?   (SECTION 7.3)

FOCUS PROBLEM

Impulse Buying

The Food Marketing Institute, Progressive Grocer, New Products News, and Point of Purchaser Advertising Institute are organizations that analyze supermarket sales. One of the interesting discoveries was that the average amount of impulse buying in a grocery store was very time-dependent. As reported in the *Denver Post*, "when you dilly dally in a store for 10 unplanned minutes, you can kiss nearly $20 goodbye." For this reason, it is in the best interest of the supermarket to keep you in the store longer. In the *Post* article, it was pointed out that long checkout lines (near end-aisle displays), "samplefest" events of tasting free samples, video kiosks, magazine and book sections, and so on help keep customers in the store longer. On average, a single customer who strays from his or her grocery list can plan on impulse spending of $20 for every 10 minutes spent wandering about in the supermarket.

Let x represent the dollar amount spent on supermarket impulse buying in a 10-minute (unplanned) shopping interval. Based on the *Post* article, the mean of the x distribution is about $20 and the (estimated) standard deviation is about $7.

335

(a) Consider a random sample of $n = 100$ customers, each of whom has 10 minutes of unplanned shopping time in a supermarket. From the central limit theorem, what can you say about the probability distribution of $\bar{x}$, the *average* amount spent by these customers due to impulse buying? Is the $\bar{x}$ distribution approximately normal? What are the mean and standard deviation of the $\bar{x}$ distribution? Is it necessary to make any assumption about the x distribution? Explain.

(b) What is the probability that $\bar{x}$ is between $18 and $22?

(c) Let us assume that x has a distribution that is approximately normal. What is the probability that x is between $18 and $22?

(d) In part (b), we used $\bar{x}$, the *average* amount spent, computed for 100 customers. In part (c), we used x, the amount spent by only *one* individual customer. The answers for parts (b) and (c) are very different. Why would this happen? In this example, $\bar{x}$ is a much more predictable or reliable statistic than x. Consider that almost all marketing strategies and sales pitches are designed for the *average* customer and *not* the *individual* customer. How does the central limit theorem tell us that the average customer is much more predictable than the individual customer?

7.1
Sampling Distributions

Let us begin with some common statistical terms. Most of these have been discussed before, but this is a good time to review them.

From a statistical point of view, a *population* can be thought of as a set of measurements (or counts), either existing or conceptual. We discussed populations at some length in Chapter 1. A *sample* is a subset of measurements from the population. For our purposes, the most important samples are *random samples*, which were discussed in Section 1.2.

A *parameter* is a numerical descriptive measure of a *population*. Examples of *population parameters* are the population mean μ, population variance σ^2, population standard deviation σ, and the population proportion of successes p in a binomial distribution.

A *statistic* is a numerical descriptive measure of a *sample* (usually a random sample). Examples of statistics are the sample mean $\bar{x}$, sample variance s^2, sample standard deviation s, and the sample estimate $\hat{p} = r/n$ (read $\hat{p}$ as "p hat") for the proportion of successes in a binomial distribution. Notice that for a given population a specified parameter is a *fixed* quantity, while the statistic might vary depending on which sample has been selected.

Some commonly used statistics and corresponding parameters		
Measure	*Statistic*	*Parameter*
Mean	$\bar{x}$ (x bar)	μ (mu)
Variance	s^2	σ^2 (sigma squared)
Standard deviation	s	σ (sigma)
Proportion	$\hat{p}$ (p hat)	p

TABLE 7-1   Length Measurements of Trout Caught by a Random Sample of 100 Children at the Pinedale Children's Pond

Sample	Length (to nearest inch)					$\bar{x}$ = Sample Mean	Sample	Length (to nearest inch)					$\bar{x}$ = Sample Mean
1	11	10	10	12	11	10.8	51	9	10	12	10	9	10.0
2	11	11	9	9	9	9.8	52	7	11	10	11	10	9.8
3	12	9	10	11	10	10.4	53	9	11	9	11	12	10.4
4	11	10	13	11	8	10.6	54	12	9	8	10	11	10.0
5	10	10	13	11	12	11.2	55	8	11	10	9	10	9.6
6	12	7	10	9	11	9.8	56	10	10	9	9	13	10.2
7	7	10	13	10	10	10.0	57	9	8	10	10	12	9.8
8	10	9	9	9	10	9.4	58	10	11	9	8	9	9.4
9	10	10	11	12	8	10.2	59	10	8	9	10	12	9.8
10	10	11	10	7	9	9.4	60	11	9	9	11	11	10.2
11	12	11	11	11	13	11.6	61	11	10	11	10	11	10.6
12	10	11	10	12	13	11.2	62	12	10	10	9	11	10.4
13	11	10	10	9	11	10.2	63	10	10	9	11	7	9.4
14	10	10	13	8	11	10.4	64	11	11	12	10	11	11.0
15	9	11	9	10	10	9.8	65	10	10	11	10	9	10.0
16	13	9	11	12	10	11.0	66	8	9	10	11	11	9.8
17	8	9	7	10	11	9.0	67	9	11	11	9	8	9.6
18	12	12	8	12	12	11.2	68	10	9	10	9	11	9.8
19	10	8	9	10	10	9.4	69	9	9	11	11	11	10.2
20	10	11	10	10	10	10.2	70	13	11	11	9	11	11.0
21	11	10	11	9	12	10.6	71	12	10	8	8	9	9.4
22	9	12	9	10	9	9.8	72	13	7	12	9	10	10.2
23	8	11	10	11	10	10.0	73	9	10	9	8	9	9.0
24	9	12	10	9	11	10.2	74	11	11	10	9	10	10.2
25	9	9	8	9	10	9.0	75	9	11	14	9	11	10.8
26	11	11	12	11	11	11.2	76	14	10	11	12	12	11.8
27	10	10	10	11	13	10.8	77	8	12	10	10	9	9.8
28	8	7	9	10	8	8.4	78	8	10	13	9	8	9.6
29	11	11	8	10	11	10.2	79	11	11	11	13	10	11.2
30	8	11	11	9	12	10.2	80	12	10	11	12	9	10.8
31	11	9	12	10	10	10.4	81	10	9	10	10	13	10.4
32	10	11	10	11	12	10.8	82	11	10	9	9	12	10.2
33	12	11	8	8	11	10.0	83	11	11	10	10	10	10.4
34	8	10	10	9	10	9.4	84	11	10	11	9	9	10.0
35	10	10	10	10	11	10.2	85	10	11	10	9	7	9.4
36	10	8	10	11	13	10.4	86	7	11	10	9	11	9.6
37	11	10	11	11	10	10.6	87	10	11	10	10	10	10.2
38	7	13	9	12	11	10.4	88	9	8	11	10	12	10.0
39	11	11	8	11	11	10.4	89	14	9	12	10	9	10.8
40	11	10	11	12	9	10.6	90	9	12	9	10	10	10.0
41	11	10	9	11	12	10.6	91	10	10	8	6	11	9.0
42	11	13	10	12	9	11.0	92	8	9	11	9	10	9.4
43	10	9	11	10	11	10.2	93	8	10	9	9	11	9.4
44	10	9	11	10	9	9.8	94	12	11	12	13	10	11.6
45	12	11	9	11	12	11.0	95	11	11	9	9	9	9.8
46	13	9	11	8	8	9.8	96	8	12	8	11	10	9.8
47	10	11	11	11	10	10.6	97	13	11	11	12	8	11.0
48	9	9	10	11	11	10.0	98	10	11	8	10	11	10.0
49	10	9	9	10	10	9.6	99	13	10	7	11	9	10.0
50	10	10	6	9	10	9.0	100	9	9	10	12	12	10.4

Often we do not have access to all the measurements of an entire population because of time, money, effort constraints, etc. So we must use measurements from a sample instead. In such cases, we will use a statistic (such as $\bar{x}$, s, or $\hat{p}$) to make *inferences* about corresponding population parameters (e.g., μ, σ, or p). The principal types of inferences we will make are

1. To *estimate* the value of a population parameter.

2. To formulate a *decision* about the value of a population parameter.

To evaluate the reliability of our inferences, we will need to know the probability distribution for the statistic we are using. Such a probability distribution is called a *sampling distribution*. Perhaps an example will help clarify this discussion.

Sampling distribution

EXAMPLE 1

Sampling distribution for x̄

Pinedale, Wisconsin, is a rural community with a children's fishing pond. Posted rules say that all fish under 6 inches must be returned to the pond, only children under 12 years old may fish, and a limit of five fish may be kept per day. Susan is a college student who was hired by the community last summer to make sure the rules were obeyed and to see that the children were safe from accidents. The pond contains only rainbow trout and has been well stocked for many years. Each child has no difficulty catching his or her limit of five trout.

As a project for her biometrics class, Susan kept a record of the lengths (to the nearest inch) of all trout caught last summer. Hundreds of children visited the pond and caught their limit of five trout, so Susan has a lot of data. To make Table 7-1 on the previous page, Susan selected 100 children at random and listed the length of each of the five trout caught by a child in the sample. Then, for each child, she listed the mean length of the five trout that child caught.

Now let us turn our attention to the following question: What is the average (mean) length of a trout taken from the Pinedale children's pond last summer?

SOLUTION: We can get an idea of the average length by looking at the far-right column of Table 7-1. But just looking at 100 of the $\bar{x}$ values doesn't tell us much. Let's organize our $\bar{x}$ values into a frequency table. We used a class width of 0.38 to make Table 7-2.

Note: Techniques of Section 2.2 dictate a class width of 0.4. However, this choice results in the tenth class being beyond the data. Consequently, we shortened the class width slightly and also started the first class with a value slightly lower than the smallest data value.

The far-right column of Table 7-2 contains relative frequencies $f/100$. Recall that the relative frequencies may be thought of as probabilities, so we effectively have a probability distribution. Because $\bar{x}$ represents the mean length of a trout (based on samples of five trout caught by each child), we estimate the probability of $\bar{x}$ falling into each class by using the relative frequencies. Figure 7-1 is a relative-frequency or probability distribution of the $\bar{x}$ values.

The bars of Figure 7-1 represent our estimated probabilities of $\bar{x}$ values based on the data of Table 7-1. The bell-shaped curve represents the theoretical probability distribution that would be obtained if the number of children (i.e., number of $\bar{x}$ values) were much larger.

Figure 7-1 represents a *probability sampling distribution* for the sample mean $\bar{x}$ of trout lengths based on random samples of size 5. We see that the distribution is mound-shaped and even somewhat bell-shaped. Irregularities are due to the small number of samples used (only 100 sample means) and the rather small sample size

TABLE 7-2   Frequency Table for 100 Values of $\bar{x}$

Class	Class Limits		f = Frequency	$f/100$ = Relative Frequency
	Lower	Upper		
1	8.39	8.76	1	0.01
2	8.77	9.14	5	0.05
3	9.15	9.52	10	0.10
4	9.53	9.90	19	0.19
5	9.91	10.28	27	0.27
6	10.29	10.66	18	0.18
7	10.67	11.04	12	0.12
8	11.05	11.42	5	0.05
9	11.43	11.80	3	0.03

(5 trout per child). These irregularities would become less obvious and even disappear if the sample of children became much larger, if we used a larger number of classes in Figure 7-1, and if the number of trout used in each sample became larger. In fact, the curve would eventually become a perfect bell-shaped curve. We will discuss this property at some length in the next section, which introduces the *central limit theorem.* ◇

There are other sampling distributions besides the $\bar{x}$ distribution. Section 7.3 shows the sampling distribution for $\hat{p}$. In the chapters ahead, we will see that other statistics have different sampling distributions. However, the $\bar{x}$ sampling distribution is very important. It will serve us well in our inferential work in Chapters 8 and 9 on estimation and testing.

Let us summarize the information about sampling distributions in the following exercise.

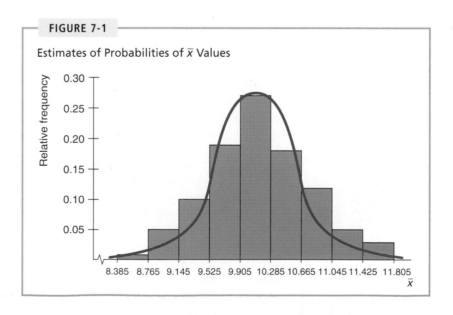

FIGURE 7-1

Estimates of Probabilities of $\bar{x}$ Values

GUIDED EXERCISE 1

Terminology

(a) What is a population parameter? Give an example.

⟹ A population parameter is a numerical descriptive measure of a population. Examples are μ, σ, and p. (There are many others.)

(b) What is a sample statistic? Give an example.

⟹ A sample statistic or a statistic is a numerical descriptive measure of a sample. Examples are $\bar{x}$, s, and $\hat{p}$.

(c) What is a sampling distribution?

⟹ A sampling distribution is a probability distribution for the sample statistic we are using.

(d) In Table 7-1, what makes up the members of the sample? What is the sample statistic corresponding to each sample? What is the sampling distribution? To which population parameter does this sampling distribution correspond?

⟹ There are 100 samples, each of which has five trout lengths. The first sample of five trout has lengths 11, 10, 10, 12, and 11. The sample statistic is the sample mean $\bar{x} = 10.8$. The sampling distribution is shown in Figure 7-1. This sampling distribution relates to the population mean μ of all lengths of trout taken from the Pinedale children's pond (i.e., trout over 6 inches long).

(e) Where will sampling distributions be used in our study of statistics?

⟹ Sampling distribution will be used for statistical inference. (Chapter 8 will concentrate on the method of inference called *estimation*. Chapter 9 will concentrate on a method of inference called *testing*.)

VIEWP◉INT   *"Chance Favors the Prepared Mind"*
—*Louis Pasteur*

It also has been said that a discovery is nothing more than an accident that meets a prepared mind. Sampling can be one of the best forms of preparation. In fact, sampling may be the primary way we humans venture into the unknown. Probability sampling distributions can provide new information for the sociologist, scientist, or economist. In addition, ordinary human sampling of life can help writers and artists develop preferences, style, and insight. Ansel Adams became famous for photographing lyrical, unforgettable landscapes such as "Moonrise, Hernandez, New Mexico." Adams claimed that he was a strong believer in the quote by Pasteur. In fact, he claims that the Hernandez photograph was just such a favored chance happening that his prepared mind readily grasped. During his lifetime, Adams made over $25 million from sales and royalties on the Hernandez photograph.

This is a good time to review several important concepts, some of which we have studied earlier. Please write out a careful but brief answer to each of the following questions.

1. What is a population? Give three examples.

2. What is a random sample from a population? (*Hint:* See Section 1.2.)

3. What is a population parameter? Give three examples.

4. What is a sample statistic? Give three examples.

5. What is the meaning of the term *statistical inference*? What types of inferences will we make about population parameters?

6. What is a sampling distribution?

7. How do frequency tables, relative frequencies, and histograms using relative frequencies help us understand sampling distributions?

8. How can relative frequencies be used to help us estimate probabilities occurring in sampling distributions?

9. Give an example of a specific sampling distribution we studied in this section. Outline other possible examples of sampling distributions from areas such as business administration, economics, finance, psychology, political science, sociology, biology, medical science, sports, engineering, chemistry, linguistics, and so on.

7.2
The Central Limit Theorem

FOCUS POINTS

✓ For a normal distribution, use μ and σ to construct the theoretical sampling distribution for the statistic $\bar{x}$.

✓ For large samples, use sample estimates to construct a good approximate sampling distribution for the statistic $\bar{x}$.

✓ Learn the statement and underlying meaning of the central limit theorem well enough to explain it to a friend who is intelligent, but (unfortunately) doesn't know much about statistics.

The $\bar{x}$ Distribution, Given x Is Normal

In Section 7.1, we began a study of the distribution of $\bar{x}$ values, where $\bar{x}$ was the (sample) mean length of five trout caught by children at the Pinedale children's fishing pond. Let's consider this example again in the light of a very important theorem of mathematical statistics.

◇ **THEOREM 7.1** Let x be a random variable with a *normal distribution* whose mean is μ and standard deviation is σ. Let $\bar{x}$ be the sample mean corresponding to random samples of size n taken from the x distribution. Then the following are true:

(a) The $\bar{x}$ distribution is a *normal distribution*.

(b) The mean of the $\bar{x}$ distribution is μ.

(c) The standard deviation of the $\bar{x}$ distribution is $\sigma/\sqrt{n}$.   ◇

We conclude from Theorem 7.1 that when x has a normal distribution, the $\bar{x}$ distribution will be normal *for any sample size n.* Furthermore, we can convert the $\bar{x}$ distribution to the standard normal z distribution using the following formulas.

$$\mu_{\bar{x}} = \mu$$

$$\sigma_{\bar{x}} = \frac{\sigma}{\sqrt{n}}$$

$$z = \frac{\bar{x} - \mu_{\bar{x}}}{\sigma_{\bar{x}}} = \frac{\bar{x} - \mu}{\sigma/\sqrt{n}}$$

where n is the sample size,
μ is the mean of the x distribution, and
σ is the standard deviation of the x distribution.

Theorem 7.1 is a wonderful theorem! It says that the $\bar{x}$ distribution will be normal provided the x distribution is normal. The sample size n could be 2, 3, 4, or any (fixed) sample size we wish. Furthermore, the mean of the $\bar{x}$ distribution is μ (same as for the x distribution), but the standard deviation is $\sigma/\sqrt{n}$ (which is, of course, smaller than σ). The next example illustrates Theorem 7.1.

EXAMPLE 2

*Probability regarding x;
regarding $\bar{x}$*

Suppose that a team of biologists has been studying the Pinedale children's fishing pond. Let x represent the length of a single trout taken at random from the pond. This group of biologists has determined that x has a normal distribution with mean $\mu = 10.2$ inches and standard deviation $\sigma = 1.4$ inches.

(a) What is the probability that a *single trout* taken at random from the pond is between 8 and 12 inches long?

SOLUTION: We use the methods of Chapter 6 with $\mu = 10.2$ and $\sigma = 1.4$ to get

$$z = \frac{x - \mu}{\sigma} = \frac{x - 10.2}{1.4}$$

Therefore,

$$P(8 < x < 12) = P\left(\frac{8 - 10.2}{1.4} < z < \frac{12 - 10.2}{1.4}\right)$$

$$= P(-1.57 < z < 1.29)$$

$$= 0.9015 - 0.0582 = 0.8433$$

Therefore, the probability is about 0.8433 that a *single* trout taken at random is between 8 and 12 inches long.

(b) What is the probability that the *mean length* $\bar{x}$ of five trout taken at random is between 8 and 12 inches?

SOLUTION: If we let $\mu_{\bar{x}}$ represent the mean of the distribution, then Theorem 7.1 part (b) tells us that

$$\mu_{\bar{x}} = \mu = 10.2$$

If $\sigma_{\bar{x}}$ represents the standard deviation of the $\bar{x}$ distribution, then Theorem 7.1 part (c) tells us that

$$\sigma_{\bar{x}} = \sigma/\sqrt{n} = 1.4/\sqrt{5} \approx 0.63$$

To create a standard z variable from $\bar{x}$, we subtract $\mu_{\bar{x}}$ and divide by $\sigma_{\bar{x}}$:

$$z = \frac{\bar{x} - \mu_{\bar{x}}}{\sigma_{\bar{x}}} = \frac{\bar{x} - \mu}{\sigma/\sqrt{n}} = \frac{\bar{x} - 10.2}{0.63}$$

To standardize the interval $8 < \bar{x} < 12$, we use 8 and then 12 in place of $\bar{x}$ in the preceding formula for z.

$$8 < \bar{x} < 12$$
$$\frac{8 - 10.2}{0.63} < z < \frac{12 - 10.2}{0.63}$$
$$-3.49 < z < 2.86$$

Theorem 7.1 part (a) tells us that $\bar{x}$ has a normal distribution. Therefore,

$$P(8 < \bar{x} < 12) = P(-3.49 < z < 2.86) = 0.9979 - 0.0002 = 0.9977$$

The probability is about 0.9977 that the mean length based on a sample size 5 is between 8 and 12 inches.

(c) Looking at the results of parts (a) and (b), we see that the probabilities (0.8433 and 0.9977) are quite different. Why is this the case?

SOLUTION: According to Theorem 7.1, both x and $\bar{x}$ have a normal distribution, and both have the same mean of 10.2 inches. The difference is in the standard deviation for x and $\bar{x}$. The standard deviation of the x distribution is $\sigma = 1.4$. The standard deviation of the $\bar{x}$ distribution is

$$\sigma_{\bar{x}} = \sigma/\sqrt{n} = 1.4/\sqrt{5} \approx 0.63$$

The standard deviation of $\bar{x}$ is less than half the standard deviation of x. Figure 7-2 shows the distribution of x and $\bar{x}$.

Looking at Figure 7-2(a) and (b), we see that both curves use the same scale on the horizontal axis. The means are the same, and the shaded area is above the interval from 8 to 12 on each graph. It becomes clear that the smaller standard deviation of the $\bar{x}$ distribution has the effect of gathering together much more of the total probability into the region over its mean. Therefore, the region from 8 to 12 has a much higher probability for the $\bar{x}$ distribution.   ◊

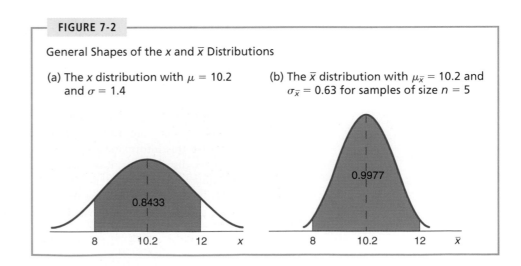

FIGURE 7-2

General Shapes of the x and $\bar{x}$ Distributions

(a) The x distribution with $\mu = 10.2$ and $\sigma = 1.4$

(b) The $\bar{x}$ distribution with $\mu_{\bar{x}} = 10.2$ and $\sigma_{\bar{x}} = 0.63$ for samples of size $n = 5$

Theorem 7.1 describes the distribution of a particular statistic: namely, the distribution of sample means $\bar{x}$. The standard deviation of a statistic is referred to as the *standard error* of that statistic. For the $\bar{x}$ sampling distribution, the standard error of $\bar{x}$ is $\sigma_{\bar{x}}$ or $\sigma/\sqrt{n}$. In other words, the *standard error of the mean* is $\sigma/\sqrt{n}$. The expression *standard error* appears commonly on printouts and refers to the standard deviation of the sampling distribution being used. (In Minitab, the expression SE MEAN refers to the standard error of the mean.)

The $\bar{x}$ Distribution, Given x Follows Any Distribution

Theorem 7.1 gives complete information about the $\bar{x}$ distribution, provided the original x distribution is known to be normal. What happens if we don't have information about the shape of the original x distribution? The *central limit theorem* tells us what to expect.

◊ **THEOREM 7.2 The Central Limit Theorem** If x possesses *any* distribution with mean μ and standard deviation σ, then the sample mean $\bar{x}$ based on a random sample of size n will have a distribution that approaches the distribution of a normal random variable with mean μ and standard deviation $\sigma/\sqrt{n}$ as n increases without limit. ◊

The central limit theorem is indeed surprising! It says that x can have *any* distribution whatever, but as the sample size gets larger and larger, the distribution of $\bar{x}$ will approach a *normal* distribution. From this relation, we begin to appreciate the scope and significance of the normal distribution.

In the central limit theorem, the degree to which the distribution of $\bar{x}$ values fit a normal distribution depends on both the selected value of n and the original distribution of x values. A natural question is: How large should the sample size be if we want to apply the central limit theorem? After a great deal of theoretical as well as empirical study, statisticians agree that if n is 30 or larger, the $\bar{x}$ distribution will appear to be normal and the central limit theorem will apply. However, this rule should not be applied blindly. If the x distribution is definitely not symmetrical about its mean, then the $\bar{x}$ distribution also will display a lack of symmetry. In such a case, a sample size larger than 30 may be required to get a reasonable approximation to the normal.

In practice, it is a good idea, when possible, to make a histogram of sample x values. If the histogram is approximately mound-shaped, and if it is more or less symmetrical, then we may be assured that, for all practical purposes, the $\bar{x}$ distribution will be well approximated by a normal distribution and the central limit theorem will apply when the sample size is 30 or larger. The main thing to remember is that in almost all practical applications, a sample size of 30 or more is adequate for the central limit theorem to hold. However, in a few rare applications, you may need a sample size larger than 30 to get reliable results.

Let's summarize this information for convenient reference:

For almost all x distributions, if we use a random sample of size 30 or larger, the $\bar{x}$ distribution will be approximately normal, and the larger the sample size becomes, the closer the $\bar{x}$ distribution gets to the normal. Furthermore, we may convert the $\bar{x}$ distribution to a standard normal distribution using the following formulas.

$$\mu_{\bar{x}} = \mu$$

$$\sigma_{\bar{x}} = \frac{\sigma}{\sqrt{n}}$$

$$z = \frac{\bar{x} - \mu_{\bar{x}}}{\sigma_{\bar{x}}} = \frac{\bar{x} - \mu}{\sigma/\sqrt{n}}$$

where n is the sample size ($n \geq 30$),
μ is the mean of the x distribution, and
σ is the standard deviation of the x distribution.

GUIDED EXERCISE 2

Central limit theorem

(a) Suppose x has a *normal* distribution with mean $\mu = 18$ and standard deviation $\sigma = 3$. If we draw random samples of size 5 from the x distribution and $\bar{x}$ represents the sample mean, what can you say about the $\bar{x}$ distribution? How could you standardize the $\bar{x}$ distribution?

 Since the x distribution is given to be *normal,* the $\bar{x}$ distribution also will be normal even though the sample size is much less than 30. The mean is $\mu_{\bar{x}} = \mu = 18$. The standard deviation is

$$\sigma_{\bar{x}} = \sigma/\sqrt{n} = 3/\sqrt{5} = 1.3$$

We could standardize $\bar{x}$ as follows:

$$z = \frac{\bar{x} - \mu}{\sigma/\sqrt{n}} = \frac{\bar{x} - 18}{1.3}$$

(b) Suppose we know that the x distribution has mean $\mu = 75$ and standard deviation $\sigma = 12$, but we have no information as to whether or not the x distribution is normal. If we draw samples of size 30 from the x distribution and $\bar{x}$ represents the sample mean, what can you say about the $\bar{x}$ distribution? How could you standardize the $\bar{x}$ distribution?

 Since the sample size is large enough, the $\bar{x}$ distribution will be approximately a normal distribution. The mean of the $\bar{x}$ distribution is

$$\mu_{\bar{x}} = \mu = 75$$

The standard deviation of the distribution is

$$\sigma_{\bar{x}} = \sigma/\sqrt{n} = 12/\sqrt{30} = 2.2$$

We could standardize $\bar{x}$ as follows:

$$z = \frac{\bar{x} - \mu}{\sigma/\sqrt{n}} = \frac{\bar{x} - 75}{2.2}$$

(c) Suppose you did not know that x had a normal distribution. Would you be justified in saying that the $\bar{x}$ distribution is approximately normal if the sample size was $n = 8$?

 No, the sample size should be 30 or larger if we don't know that x has a normal distribution.

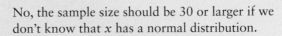

Now let's look at an example that demonstrates the use of the central limit theorem in a decision-making process.

EXAMPLE 3

Central limit theorem

A certain strain of bacteria occurs in all raw milk. Let x be the bacteria count per milliliter of milk. The health department has found that if the milk is not contaminated, then x has a distribution that is more or less mound-shaped and symmetrical. The mean of the x distribution is $\mu = 2500$, and the standard deviation is $\sigma = 300$. In a large commercial dairy, the health inspector takes 42 random samples of the milk produced each day. At the end of the day, the bacteria count in each of the 42 samples is averaged to obtain the sample mean bacteria count $\bar{x}$.

(a) Assuming that the milk is not contaminated, what is the distribution of $\bar{x}$?

SOLUTION: The sample size is $n = 42$. Since this value exceeds 30, the central limit theorem applies, and we know that $\bar{x}$ will be approximately normal with mean and standard deviation

$$\mu_{\bar{x}} = \mu = 2500$$
$$\sigma_{\bar{x}} = \sigma/\sqrt{n} = 300/\sqrt{42} \approx 46.3$$

(b) Assuming that the milk is not contaminated, what is the probability that the average bacteria count $\bar{x}$ for one day is between 2350 and 2650 bacteria per milliliter?

SOLUTION: We convert the interval

$$2350 \le \bar{x} \le 2650$$

to a corresponding interval on the standard z axis.

$$z = \frac{\bar{x} - \mu}{\sigma/\sqrt{n}} = \frac{\bar{x} - 2500}{46.3}$$

$\bar{x} = 2350$ converts to $z = \dfrac{2350 - 2500}{46.3} = -3.24$

$\bar{x} = 2650$ converts to $z = \dfrac{2650 - 2500}{46.3} = 3.24$

Therefore,

$$P(2350 \le \bar{x} \le 2650) = P(-3.24 \le z \le 3.24)$$
$$= 0.9994 - 0.0006$$
$$= 0.9988$$

The probability is 0.9988 that $\bar{x}$ is between 2350 and 2650.

(c) At the end of each day, the inspector must decide to accept or reject the accumulated milk that has been held in cold storage awaiting shipment. Suppose that the 42 samples taken by the inspector have a mean bacteria count $\bar{x}$ that is *not* between 2350 and 2650. If you were the inspector, what would be your comment on this situation?

SOLUTION: The probability that $\bar{x}$ is between 2350 and 2650 is very high. If the inspector finds that the average bacteria count for the 42 samples is not between 2350 and 2650, then it is reasonable to conclude that there is something wrong with the milk. If $\bar{x}$ is less than 2350, you might suspect someone added chemicals to the milk to artificially reduce the bacteria count. If $\bar{x}$ is above 2650, you might suspect some other kind of biologic contamination.   ◇

GUIDED EXERCISE 3

Probability regarding x̄

In mountain country, major highways sometimes use tunnels instead of long, winding roads over high passes. However, too many vehicles in a tunnel at the same time can cause a hazardous situation. Traffic engineers are studying a long tunnel in Colorado. If x represents the time for a vehicle to go through the tunnel, it is known that the x distribution has mean $\mu = 12.1$ minutes and standard deviation $\sigma = 3.8$ minutes under ordinary traffic conditions. From a histogram of x values, it was found that the x distribution is mound-shaped with some symmetry about the mean.

Engineers have calculated that, *on average*, vehicles should spend from 11 to 13 minutes in the tunnel. If the time is less than 11 minutes, traffic is moving too fast for safe travel in the tunnel. If the time is more than 13 minutes, there is a problem of bad air (too much carbon monoxide and other pollutants).

Under ordinary conditions, there are about 50 vehicles in the tunnel at one time. What is the probability that the mean time for 50 vehicles in the tunnel will be from 11 to 13 minutes?

We will answer this question in steps.

(a) Let $\bar{x}$ represent the sample mean based on samples of size 50. Describe the $\bar{x}$ distribution.

⇒ From the central limit theorem we expect the $\bar{x}$ distribution to be approximately normal with mean and standard deviation

$$\mu_{\bar{x}} = \mu = 12.1 \qquad \sigma_{\bar{x}} = \frac{\sigma}{\sqrt{n}} = \frac{3.8}{\sqrt{50}} \approx 0.54$$

(b) Find $P(11 < \bar{x} < 13)$.

⇒ We convert the interval

$$11 < \bar{x} < 13$$

to a standard z interval and use the standard normal probability table to find our answer. Since

$$z = \frac{\bar{x} - \mu}{\sigma/\sqrt{n}} = \frac{\bar{x} - 12.1}{0.54}$$

$\bar{x} = 11$ converts to $z = \dfrac{11 - 12.1}{0.54} = -2.04$

and $\bar{x} = 13$ converts to $z = \dfrac{13 - 12.1}{0.54} = 1.67$

Therefore,

$$P(11 < \bar{x} < 13) = P(-2.04 < z < 1.67)$$
$$= 0.9525 - 0.0207$$
$$= 0.9318$$

(c) Comment on your answer for part (b).

⇒ It would seem that about 93% of the time there should be no safety hazard for average traffic flow.

VIEWPOINT   *Chaos!*

Is there a different side to random sampling? Can sampling be used as a weapon? According to the *Wall Street Journal,* the answer could be yes! The acronym for *C*reate *H*avoc *A*round *O*ur *S*ystem is **CHAOS.** The Association of Flight Attendants (AFA) is a union that successfully used **CHAOS** against Alaska Airlines in 1994 as a negotiation tool. **CHAOS** involves a small sample of random strikes—a few flights at a time—instead of a mass walkout. The president of the AFA claims that by striking randomly, "we take control of the schedule." The entire schedule becomes unreliable, and that is something management cannot tolerate. In 1986, TWA flight attendants struck in a mass walkout, and all were permanently replaced! Using **CHAOS,** only a few jobs are put at risk, and these are usually not lost. It appears that random sampling can be used as a weapon.

SECTION 7.2   PROBLEMS

In these problems, the word *average* refers to the arithmetic mean $\bar{x}$ or μ, as appropriate.

1. *General* Suppose that x has a distribution with $\mu = 15$ and $\sigma = 14$.
   (a) If a random sample of size $n = 49$ is drawn, find $\mu_{\bar{x}}$, $\sigma_{\bar{x}}$, and $P(15 \le \bar{x} \le 17)$.
   (b) If a random sample of size $n = 64$ is drawn, find $\mu_{\bar{x}}$, $\sigma_{\bar{x}}$, and $P(15 \le \bar{x} \le 17)$.
   (c) Why should you expect the probability of part (b) to be higher than that of part (a)? (*Hint:* Consider the standard deviations in parts (a) and (b).)

2. *General* Suppose that x has a distribution with $\mu = 100$ and $\sigma = 48$.
   (a) If a random sample of size $n = 81$ is drawn, find $\mu_{\bar{x}}$, $\sigma_{\bar{x}}$, and $P(92 \le \bar{x} \le 100)$.
   (b) If a random sample of size $n = 121$ is drawn, find $\mu_{\bar{x}}$, $\sigma_{\bar{x}}$, and $P(92 \le \bar{x} \le 100)$.
   (c) Again, comment on the differences in the probabilities in parts (a) and (b). Why do you expect the differences?

3. *General* Suppose that x has a distribution with $\mu = 25$ and $\sigma = 3.5$.
   (a) If random samples of size $n = 9$ are selected, can we say anything about the $\bar{x}$ distribution of sample means?
   (b) If the original x distribution is normal, can we say anything about the $\bar{x}$ distribution from samples of size $n = 9$? Find $P(23 \le \bar{x} \le 26)$.

4. *General* Suppose that x has a distribution with $\mu = 72$ and $\sigma = 8$.
   (a) If random samples of size $n = 16$ are selected, can we say anything about the $\bar{x}$ distribution of sample means?
   (b) If the original x distribution is *normal,* can we say anything about the $\bar{x}$ distribution of random samples of size 16? Find $P(68 \le \bar{x} \le 73)$.

5. *Coal: Automatic Loader* Coal is carried from a mine in West Virginia to a power plant in New York in hopper cars on a long train. The automatic hopper car loader is set to put 75 tons of coal into each car. The actual weights of coal loaded into each car are *normally distributed* with mean $\mu = 75$ tons and standard deviation $\sigma = 0.8$ ton.
   (a) What is the probability that one car chosen at random will have less than 74.5 tons of coal?

(b) What is the probability that 20 cars chosen at random will have a mean load weight $\bar{x}$ of less than 74.5 tons of coal?

(c) Suppose that the weight of coal in one car was less than 74.5 tons. Would that fact make you suspect that the loader had slipped out of adjustment? Suppose the weight of coal in 20 cars selected at random had an average $\bar{x}$ less than 74.5 tons. Would that fact make you suspect that the loader had slipped out of adjustment? Why?

6. *Vital Statistics: Heights of Men* The heights of 18-year-old men are approximately *normally distributed,* with mean 68 inches and standard deviation 3 inches (based on information from *Statistical Abstract of the United States,* 112th Edition).

(a) What is the probability that an 18-year-old man selected at random is between 67 and 69 inches tall?

(b) If a random sample of nine 18-year-old men is selected, what is the probability that the mean height $\bar{x}$ is between 67 and 69 inches?

(c) Compare your answers for parts (a) and (b). Was the probability in part (b) much higher? Why would you expect this?

7. *Medical: Blood Glucose* Let x be a random variable that represents level of glucose in the blood (milligrams per deciliter of blood) after a 12-hour fast. Assume for people under 50 years old that x has a distribution that is approximately normal with mean $\mu = 85$ and an estimated standard deviation $\sigma = 25$ (based on information from *Diagnostic Tests with Nursing Applications,* edited by S. Loeb, Springhouse). A test result $x < 40$ is an indication of severe excess insulin, and medication is usually prescribed.

(a) What is the probability that on a single test $x < 40$?

(b) Suppose that a doctor uses the average $\bar{x}$ for two tests taken about a week apart. What can we say about the probability distribution of $\bar{x}$? *Hint:* See Theorem 7.1. What is the probability $\bar{x} < 40$?

(c) Repeat part (b) for $n = 3$ tests taken a week apart.

(d) Repeat part (b) for $n = 5$ tests taken a week apart.

(e) Compare your answers for parts (a), (b), (c), and (d). Did the probabilities decrease as n increased? Explain what this might say if you were a doctor or a nurse. If a patient had a test result of $\bar{x} < 40$ based on five tests, explain why either you are looking at an extremely rare event or (more likely) the person has a case of excess insulin.

8. *Medical: White Blood Cells* Let x be a random variable that represents white blood cell count per cubic milliliter of whole blood. Assume that x has a distribution that is approximately normal with mean $\mu = 7500$ and estimated standard deviation $\sigma = 1750$ (see reference in Problem 7). A test result of $x < 3500$ is an indication of leukopenia. This indicates bone marrow depression that may be the result of a viral infection.

(a) What is the probability that on a single test x is less than 3500?

(b) Suppose that a doctor uses the average $\bar{x}$ for two tests taken about a week apart. What can we say about the probability distribution of $\bar{x}$? What is the probability of $\bar{x} < 3500$?

(c) Repeat part (b) for $n = 3$ tests taken a week apart.

(d) Compare your answers for parts (a), (b), and (c). How did the probabilities change as n increased? If a person had $\bar{x} < 3500$ based on three tests, what conclusion would you draw as a doctor or a nurse?

9. *Wildlife: Deer* Let x be a random variable that represents weights in kilograms (kg) of healthy adult female deer (does) in December in Mesa Verde National Park. Then x has a distribution that is approximately normal with mean $\mu = 63.0$ kg and standard deviation $\sigma = 7.1$ kg (Source: *The Mule Deer of Mesa Verde National Park,*

by G. W. Mierau and J. L. Schmidt, Mesa Verde Museum Association). Suppose that a doe that weighs less than 54 kg is considered undernourished.

(a) What is the probability that a single doe captured at random in December (weighed and released) is undernourished?

(b) If the park has about 2200 does, what number do you expect to be undernourished in December?

(c) To estimate the health of the December doe population, park rangers use the rule that the average weight of $n = 50$ does should be more than 60 kg. If the average weight is less than 60 kg, it is thought that the entire population of does might be undernourished. What is the probability the average weight $\bar{x}$ for a random sample of 50 does is less than 60 kg (assume a healthy population)?

(d) Compute the probability that $\bar{x} < 64.2$ kg for 50 does (assume a healthy population). Suppose park rangers captured, weighed, and released 50 does in December, and the average weight was $\bar{x} = 64.2$ kg. Do you think the doe population is undernourished or not? Explain.

10. *Wildlife: Hummingbirds* *Selasphorus sasin* is the scientific name for what is commonly called "Allen's hummingbird." This beautiful hummingbird lives on the West Coast of the United States and is named for C. A. Allen (1841–1930), who studied these birds extensively. Let x be a random variable that represents the incubation time for the Allen hummingbird eggs. Based on information from *The Hummingbird Book*, by Donald and Lillian Stokes (Little, Brown and Company), the x distribution has a mean $\mu = 16$ days. Let us assume that the standard deviation is approximately $\sigma = 2$ days. The distribution of x values is more or less mound-shaped and symmetrical but not necessarily normal. Suppose that we have $n = 30$ eggs in an incubator. Let $\bar{x}$ be the average incubation time for these eggs.

(a) What can we say about the probability distribution of $\bar{x}$? Is it approximately normal? What are the mean and standard deviation?

(b) What is the probability that $\bar{x}$ is between 16 and 17 days?

(c) What is the probability that $\bar{x}$ is less than 15 days?

11. *Finance: Templeton Funds* Templeton World is a mutual fund that invests in both U.S. and foreign markets. Let x be a random variable that represents the monthly percentage return for the Templeton World fund. Based on information from the *Morningstar Guide to Mutual Funds* (available in most libraries), x has mean $\mu = 1.6\%$ and standard deviation $\sigma = 0.9\%$.

(a) Templeton World fund has over 250 stocks that combine together to give the overall monthly percentage return x. We can consider the monthly return of the stocks in the fund to be a sample from the population of monthly returns of all world stocks. Then we see that the overall monthly return x for Templeton World fund is itself an average return computed using all 250 stocks in the fund. Why would this indicate that x has an approximately normal distribution? Explain. *Hint:* See discussion after Theorem 7.2.

(b) After 6 months, what is the probability that the *average* monthly percentage return $\bar{x}$ will be between 1 and 2%? *Hint:* See Theorem 7.1, and assume that x has a normal distribution as based on part (a).

(c) After 2 years, what is the probability that $\bar{x}$ will be between 1 and 2%?

(d) Compare your answers for parts (b) and (c). Did the probability increase as n (number of months) increased? Why would this happen?

(e) If after 2 years the average monthly percentage return $\bar{x}$ was less than 1%, would that tend to shake your confidence in the statement that $\mu = 1.6\%$? Might you suspect that μ has slipped below 1.6%? Explain.

12. *Finance: Dean Witter Funds* Dean Witter European Growth is a mutual fund that specializes in stocks from the British Isles, continental Europe, and Scandinavia. The

fund has over 100 stocks. Let x be a random variable that represents the monthly percentage return for this fund. Based on information from *Morningstar* (see Problem 11), x has mean $\mu = 1.4\%$ and standard deviation $\sigma = 0.8\%$.

(a) Let's consider the monthly return of the stocks in the Dean Witter fund to be a sample from the population of monthly returns of all European stocks. Is it reasonable to assume that x (the average monthly return on the 100 stocks in the Dean Witter European Growth fund) has a distribution that is approximately normal? Explain. *Hint:* See Problem 11, part (a).

(b) After 9 months, what is the probability that the *average* monthly percentage return $\bar{x}$ will be between 1 and 2%? *Hint:* See Theorem 7.1 and the results of part (a).

(c) After 18 months, what is the probability that the *average* monthly percentage return $\bar{x}$ will be between 1 and 2%?

(d) Compare your answers for parts (b) and (c). Did the probability increase as n (number of months) increased? Why would this happen?

(e) If after 18 months the average monthly percentage return $\bar{x}$ is more than 2%, would that tend to shake your confidence in the statement that $\mu = 1.4\%$? If this happened, do you think the European stock market might be heating up? Explain.

13. *Finance: Invesco Funds High-yield bonds* is a polite term for "junk bonds." Good companies with good products can fall on bad times (poor management, too rapid expansion, wrong marketing approach, etc.). Many of these companies will bounce back to profitability and handsomely reward investors who picked up their "bad debt" when the company was in trouble. Invesco High Yield is a mutual fund that specializes in high-yield bonds. It has approximately 80 or more bonds at the B or below rating (S&P rating for junk bonds). Let x be a random variable that represents the annual percentage return for the Invesco High Yield fund. Based on information from *Morningstar* (see Problem 11), x has mean $\mu = 10.8\%$ and standard deviation $\sigma = 4.9\%$.

(a) Let's consider the annual yield of the bonds in the Invesco fund to be a sample from the population of annual yields of all high-yield bonds. Explain why it would be reasonable to assume that x (the average annual return of all bonds in the Invesco fund) has a distribution that is approximately normal. *Hint:* See Problem 11, part (a).

(b) Compute the probability that after 5 years $\bar{x}$ is less than 6%. After 5 years, if the average annual percentage return $\bar{x}$ is less than 6%, would that seem to indicate that μ is less than 10.8% and that the junk bond market is not as strong? *Hint:* See Theorem 7.1 and the results of part (a).

(c) Compute the probability that after 5 years $\bar{x}$ is greater than 16%. After 5 years, if the average annual percentage return $\bar{x}$ is greater than 16%, would that seem to indicate that μ is more than 10.8% and that the junk bond market is heating up?

14. *Security: Night Watchman* Arthur is a night watchman in a large warehouse. During one complete round of the warehouse, he must check in at 40 different checkpoints, all of which are about the same distance apart. At each checkpoint he inserts a key that tells a central computer the time he checked in. Let x be a random variable representing the length of time from one check-in to the next. While monitoring Arthur's check-in times for the past year, the computer found the mean of the x distribution to be $\mu = 6.4$ minutes with standard deviation $\sigma = 1.5$ minutes.

(a) In one complete round of the warehouse, there are 40 check-in time intervals. What is the probability that the average check-in time interval will be from 6 to 7 minutes?

(b) Answer part (a) for two complete rounds of the warehouse (i.e., use a sample of 80 check-in time intervals).

(c) If in two complete rounds of the warehouse the average check-in time interval is not between 6 and 7 minutes, do you think a second security guard should drop in for a look? Would this seem to indicate that Arthur is going too slowly or too fast for some unexplained reason? Assume that Arthur begins a round at a time randomly selected by the computer to foil burglars looking for a time schedule.

15. *General: Distribution of Sample Means*
(a) If we have a distribution of x values that is more or less mound-shaped and somewhat symmetrical, what is the size of sample needed to claim that the distribution of sample means $\bar{x}$ from random samples of that size is approximately normal?
(b) If the original distribution of x values is known to be normal, do we need to make any restriction about sample size in order to claim that the distribution of sample means $\bar{x}$ taken from random samples of a given size is normal?

16. *Focus Problem: Impulse Buying* Solve the Focus Problem at the beginning of this chapter.

17. *Grocery Store: Checkout Time* Let x be a random variable that represents checkout time (time spent in the actual checkout process) in minutes in the express lane of a large grocery. Based on a consumer survey, the mean of the x distribution is about $\mu = 2.7$ minutes with standard deviation $\sigma = 0.6$ minute. Assume that the express lane always has customers waiting to be checked out and that the distribution of x values is more or less symmetrical and mound-shaped. What is the probability that the *total* checkout time for the next 30 customers is less than 90 minutes? Let us solve this problem in steps.
(a) Let x_i (for $i = 1, 2, 3, \ldots, 30$) represent checkout time for each customer. For example, x_1 is the checkout time for the first customer, x_2 is the checkout time for the second customer, and so forth. Each x_i has a mean $\mu = 2.7$ minutes and standard deviation $\sigma = 0.6$ minute. Let $w = x_1 + x_2 + \cdots + x_{30}$. Explain why the problem is asking us to compute that the probability w is less than 90.
(b) Use a little algebra and explain why $w < 90$ is mathematically equivalent to $w/30 < 3$. Since w is the total of the thirty x values, then $w/30 = \bar{x}$. Therefore, the statement $\bar{x} < 3$ is equivalent to the statement $w < 90$. From this we conclude that the probabilities $P(\bar{x} < 3)$ and $P(w < 90)$ are equal.
(c) What does the central limit theorem say about the probability distribution of $\bar{x}$? Is it approximately normal? What are the mean and standard deviation of the $\bar{x}$ distribution?
(d) Use the result of part (c) to compute $P(\bar{x} < 3)$. What does this result tell you about $P(w < 90)$?

18. *Airport: Takeoff Time* The taxi and takeoff time for commercial jets is a random variable x with a mean of 8.5 minutes and a standard deviation of 2.5 minutes. Assume that the distribution of taxi and takeoff times is approximately normal. You may assume that the jets are lined up on a runway so that one taxies and takes off immediately after the other, and they take off one at a time on a given runway. What is the probability (*Hint:* See Problem 17) that for 36 jets on a given runway total taxi and takeoff time will be
(a) less than 320 minutes?
(b) more than 275 minutes?
(c) between 275 and 320 minutes?

19. *Archaeology: Stone Tools* Stone Age Native American dwellings were discovered at the Turquoise Ridge site near Fort Bliss, Texas. Let x be a random variable that represents grams (g) of chipped stone tools per cubic meter of excavated soil at

prehistoric house dwellings at Turquoise Ridge. The x distribution has approximate mean $\mu = 240$ and standard deviation $\sigma = 84$ (based on information from *Turquoise Ridge and Late Prehistoric Residential Mobility in the Desert Mogollon Region,* by M. E. Whalen, University of Utah Press). Suppose that 45 cubic meters of excavated soil from Turquoise Ridge house dwellings is scheduled to be dug up and examined. What is the probability (*Hint:* See Problem 17) that the total weight of chipped stone tools in this soil is

(a) less than 9500 g?

(b) more than 12,000 g?

(c) between 9500 and 12,000 g?

20. *Manufacturing: Assembly Line* It is important that each operation on an assembly line be completed in a predictable amount of time. In the assembly-line production of the Road Runner four-wheel-drive sport vehicle, the headlight installation process is designed to take an average of $\mu = 6.3$ minutes with standard deviation $\sigma = 1.2$ minutes for each vehicle.

(a) Assume that the assembly times follow a *normal* distribution and that the vehicles are lined up so that headlight units are installed on one vehicle right after another. What is the probability that the assembly time to install headlight units on 9 vehicles is less than 60 minutes? more than 65 minutes? (*Hint:* See Problem 17 and Theorem 7.1.)

(b) What is the probability that the assembly time to install headlight units on 50 vehicles is less than 342 minutes? Is the normal assumption of part (a) necessary for a sample size this large? Why?

7.3
Sampling Distributions for Proportions

In Section 6.4, we discussed the normal approximation to the binomial. There are many important situations where we prefer to work with the *proportion* of successes r/n rather than the actual *number* of successes r in binomial experiments. With this in mind, we make the following summary about the *sampling distribution* of the proportion $\hat{p} = r/n$.

Sampling distribution for the proportion $p = \dfrac{r}{n}$

Given n = number of binomial trials (fixed constant)

r = number of successes

p = probability of success on each trial

$q = 1 - p$ = probability of failure on each trial

If $np > 5$ and $nq > 5$, then the random variable $\hat{p} = r/n$ can be approximated by a normal random variable (x) with mean and standard deviation

$$\mu_{\hat{p}} = p \quad \text{and} \quad \sigma_{\hat{p}} = \sqrt{\frac{pq}{n}}$$

◇ **TERMINOLOGY** The *standard error* for $\hat{p}$ is the standard deviation $\sigma_{\hat{p}}$ of the $\hat{p}$ sampling distribution.   ◇

◇ **COMMENT** To obtain the information regarding the sampling distribution for the proportion $\hat{p} = r/n$, we consider the sampling distribution for r, the number of successes out of n binomial trials. In Section 6.4, we saw that when $np > 5$ and $nq > 5$, the r distribution is approximately normal with mean $\mu_r = np$ and $\sigma_r = \sqrt{npq}$. Notice that $\hat{p} = r/n$ is a linear function of r. This means that the $\hat{p}$ distribution is also approximately normal when np and nq are both greater than 5. In addition, from our work in Section 5.1 with linear functions of random variables, we know that $\mu_{\hat{p}} = \mu_r/n = np/n = p$ and $\sigma_{\hat{p}} = \sigma_r/n = \sqrt{npq}/n = \sqrt{pq/n}$. ◇

If $np > 5$ and $nq > 5$, then $\hat{p} = r/n$ can be approximated by a normal random variable, which we will call x. However, $\hat{p}$ is *discrete* while x is *continuous*. To adjust for this discrepancy, we apply an appropriate *continuity correction*. In Section 6.4, we noted that in the probability histogram for the binomial random variable r, the bar centered over r actually starts at $r - 0.5$ and ends at $r + 0.5$. See Figure 7-3(a).

When we shift our focus from r to $\hat{p} = r/n$, the bar centered at r/n actually starts at $(r - 0.5)/n$ and ends at $(r + 0.5)/n$ as shown in Figure 7-3(b).

This leads us to the conclusion that the appropriate continuity correction for $\hat{p}$ is to add or subtract $0.5/n$ to the endpoints of a $\hat{p}$ (discrete) interval to convert it to an x (continuous normal) interval.

Continuity correction for $\hat{p}$ intervals

(a) If r/n is the *right* endpoint of a $\hat{p}$ interval, we *add* $0.5/n$ to get the corresponding right endpoint of the x interval.

(b) If r/n is the *left* endpoint of a $\hat{p}$ interval, we *subtract* $0.5/n$ to get the corresponding left endpoint of the x interval.

FIGURE 7-3

Distribution of r and Corresponding Distribution of $\hat{p} = r/n$

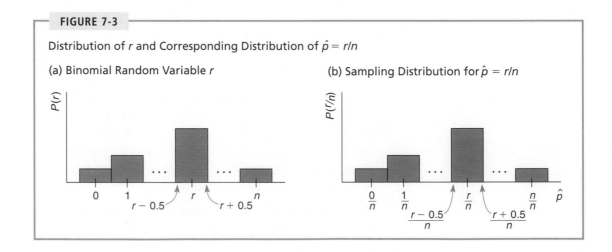

(a) Binomial Random Variable r

(b) Sampling Distribution for $\hat{p} = r/n$

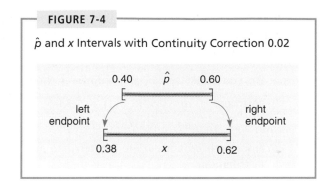

FIGURE 7-4

$\hat{p}$ and x Intervals with Continuity Correction 0.02

The next example illustrates the process of using the continuity correction to convert a $\hat{p}$ interval to an x (normal) interval.

EXAMPLE 4

Continuity correction

Suppose $n = 25$ and we have a $\hat{p}$ interval from $10/25 = 0.40$ to $15/25 = 0.60$. Use the continuity correction to convert this to an x interval.

SOLUTION: Since $n = 25$, then $0.5/n = 0.5/25 = 0.02$. This means that we subtract 0.02 from the left endpoint and add 0.02 to the right endpoint of the $\hat{p}$ interval (see Figure 7-4).

$\hat{p}$ interval: 0.40 to 0.60

x interval: $0.40 - 0.02$ to $0.60 + 0.02$ or 0.38 to 0.62   ◇

◇ **COMMENT** If n is large, the continuity correction for $\hat{p}$ won't change the x interval much. However, for smaller n values, it can make a difference.   ◇

EXAMPLE 5

Sampling distribution of $\hat{p}$

The annual crime rate in the Capital Hill neighborhood of Denver is 111 per 1000 residents. This means that 111 out of 1000 residents have been the victim of at least one crime. (Source: *Neighborhood Facts,* Piton Foundation). For more information, visit the Brase/Brase statistics site at http://math.college.hmco.com/students and find the link to the Piton Foundation. These crimes range from relatively minor (stolen hubcaps or purse snatching) to major crimes (murder). The Arms is an apartment building in this neighborhood that has 50 year-round residents. Suppose we view each of the $n = 50$ residents as a binomial trial. The random variable r (which takes on values 0, 1, 2, . . . , 50) represents the number of victims of at least one crime in the next year.

(a) What is the population probability p that a resident in the Capital Hill neighborhood will be the victim of a crime next year? What is the probability q that a resident will not be a victim?

SOLUTION: Using the Piton Foundation report, we take

$$p = 111/1000 = 0.111 \quad \text{and} \quad q = 1 - p = 0.889$$

(b) Consider the random variable

$$\hat{p} = \frac{r}{n} = \frac{r}{50}$$

Do you think we can approximate $\hat{p}$ with a normal distribution? Explain.

SOLUTION: $np = 50(0.111) = 5.55$

$nq = 50(0.889) = 44.45$

Since both np and nq are greater than 5, we can approximate $\hat{p}$ with a normal distribution.

(c) What are the mean and standard deviation for $\hat{p}$?

SOLUTION: $\mu_{\hat{p}} = p = 0.111$

$$\sigma_{\hat{p}} = \sqrt{\frac{pq}{n}}$$

$$= \sqrt{\frac{(0.111)(0.889)}{50}} \approx 0.044$$

(d) What is the probability that between 10% and 20% of the Arms residents will be victims of a crime next year?

SOLUTION: First we find the continuity correction so we can convert the $\hat{p}$ interval to an x interval. Since $n = 50$,

Continuity correction $= 0.5/n = 0.5/50 = 0.01$

We subtract 0.01 from the left $\hat{p}$ endpoint and add 0.01 to the right $\hat{p}$ endpoint (see Figure 7-5). The x interval is from 0.09 to 0.21. Therefore,

$$P(0.10 \le \hat{p} \le 0.20) \approx P(0.09 \le x \le 0.21)$$

$$\approx P\left(\frac{0.09 - 0.111}{0.044} \le z \le \frac{0.21 - 0.111}{0.044}\right)$$

$$\approx P(-0.48 \le z \le 2.25)$$

$$\approx 0.6722$$

There is about a 67% chance that between 10% and 20% of the Arms residents will be crime victims next year.   ◊

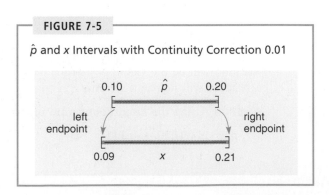

FIGURE 7-5

$\hat{p}$ and x Intervals with Continuity Correction 0.01

GUIDED
EXERCISE 4

Sampling distribution of $\hat{p}$

The general ethnic profile of Denver is about 42% minority and 58% Caucasian (Source: *Neighborhood Facts,* Piton Foundation). Suppose the city of Denver recently hired 56 new grounds and maintenance workers. It was claimed that the hiring practice is completely impartial without regard to ethnic background. However, only 27% of the new employees are minorities, and now there is a complaint. What is the probability that at most 27% of the new hires are minorities if the selection is impartial and the applicant pool reflects the ethnic profile of Denver?

(a) We take the point of view that each new hire is a binomial experiment with success being a minority hire. What are the values of n, p, q?

⟹ $n = 56$; $p = 0.42$; $q = 0.58$

(b) For the new hires, what is the sample proportion $\hat{p}$ of minority hires?

⟹ $\hat{p} = 0.27$

(c) Is the normal approximation for the distribution of $\hat{p}$ appropriate? Explain.

⟹ $np = 56(0.42) = 23.5$; nq
$= 56(0.58) = 32.5$

Both products are larger than 5, so the approximation is appropriate.

(d) Compute $\mu_{\hat{p}}$ and $\sigma_{\hat{p}}$.

⟹ $\mu_{\hat{p}} = p = 0.42$

$\sigma_{\hat{p}} = \sqrt{\dfrac{pq}{n}} = \sqrt{\dfrac{(0.42)(0.58)}{56}} = 0.066$

(e) Compute $P(\hat{p} \leq 0.27)$.

⟹ First find the continuity correction and convert $\hat{p} \leq 0.27$ to an x interval (see Figure 7-6).

FIGURE 7-6 $\hat{p}$ and x Intervals with Continuity Correction 0.009

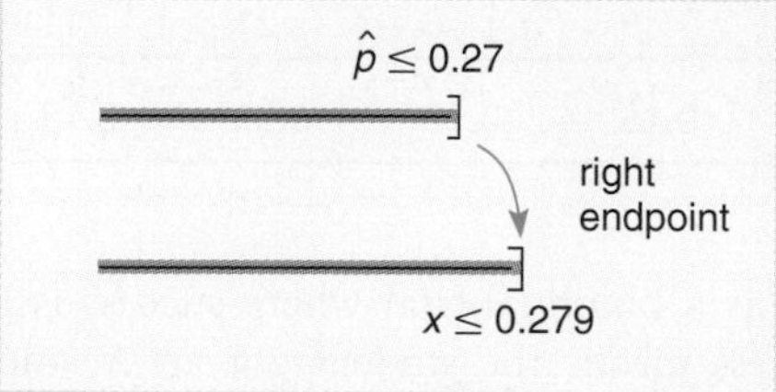

Continuity correction: $\dfrac{0.5}{56} \approx 0.009$

x interval: $x \leq 0.279$

$\hat{p} \leq 0.27$

right endpoint

$x \leq 0.279$

$P(\hat{p} \leq 0.27) \approx P(x \leq 0.279)$

$\approx P\left(z \leq \dfrac{0.279 - 0.42}{0.066}\right)$

$\approx P(z \leq -2.14)$

≈ 0.0162

(f) What is your conclusion?

⟹ Assuming all the conditions for binomial trials have been met, the probability is smaller than 2% that the proportion of minority hires would be 27% or less. It seems the hiring process might not be completely impartial or the applicant pool does not reflect the ethnic profile of Denver.

Control Charts for Proportions (*P*-Charts)

We conclude this section with an example of a control chart for proportions r/n. Such a chart is often called a *P-Chart*.

The control charts discussed in Section 6.1 were for *quantitative* data where the *size* of something is being measured. There are occasions where we prefer to examine a *quality* or *attribute* rather than just size. One way to do this is to use a binomial distribution in which success is defined as the quality or attribute we wish to study.

The basic idea for using *P*-Charts is to select samples of a fixed size n at regular time intervals and count the number of successes r from the n trials. We use the normal approximation for r/n and methods of Section 6.1 to plot control limits and r/n values, and to interpret results.

As in Section 6.1, we remind ourselves that control charts are used as warning devices tailored by a user for a particular need. Our assumptions and probability calculations need not be absolutely precise to achieve our purpose. For example, $\hat{p} = r/n$ need not follow a normal distribution exactly. A mound-shaped and more or less symmetric distribution for which the empirical rule applies will be sufficient.

EXAMPLE 6

P-Chart

Anatomy and Physiology is taught each semester. The course is required for several popular health-science majors, so it always fills up to its maximum of 60 students. The dean of the college asked the Biology Department to make a control chart for the proportion of A's given in the course each semester for the past 14 semesters. Using information from the registrar's office, the following data were obtained. Make a control chart and interpret the result.

Semester	1	2	3	4	5	6	7
r = no. of A's	9	12	8	15	6	7	13
$\hat{p}$ = r/60	0.15	0.20	0.13	0.25	0.10	0.12	0.22

Semester	8	9	10	11	12	13	14
r = no. of A's	7	11	9	8	21	11	10
$\hat{p}$ = r/60	0.12	0.18	0.15	0.13	0.35	0.18	0.17

SOLUTION: Let us view each student as a binomial trial where success is the quality or attribute we wish to study. Success means the student got an A, and failure is not getting an A. Since the class size is 60 each semester, the number of trials is $n = 60$.

(a) The first step is to use the data to estimate the overall proportion of successes. To do this, we pool the data for all 14 semesters, and use the symbol $\bar{p}$ (not to be confused with $\hat{p}$) to designate the pooled proportion of success.

$$\bar{p} = \frac{\text{Total number of A's from all 14 semesters}}{\text{Total number of students from all 14 semesters}}$$

$$\bar{p} = \frac{9 + 12 + 8 + \cdots + 10}{14(60)} = \frac{147}{840} = 0.175$$

Since the pooled estimate of the proportion of successes is $\bar{p} = 0.175$, the estimate for the proportion of failures is $\bar{q} = 1 - \bar{p} = 0.825$.

(b) For the random variable $\hat{p} = r/n$, we know the mean is $\mu_{\hat{p}} = p$, and the standard deviation is $\sigma_{\hat{p}} = \sqrt{pq/n}$. In our case, we don't have given values for p and q, so we use the pooled estimate $\bar{p} = 0.175$ and $\bar{q} = 0.825$. The number of trials is the class size $n = 60$. Therefore,

$$\mu_{\hat{p}} = p \approx \bar{p} = 0.175$$

$$\sigma_{\hat{p}} = \sqrt{\frac{pq}{n}} \approx \sqrt{\frac{\bar{p}\,\bar{q}}{n}} = \sqrt{\frac{(0.175)(0.825)}{60}} \approx 0.049$$

(c) Since a control chart is a *warning device*, it is not necessary that our probability calculations be absolutely precise. For instance, the empirical rule would substitute quite well for the normal distribution. However, it is a good idea to check that both $n\bar{p} = 60(0.175) = 10.5$ and $n\bar{q} = 60(0.825) = 49.5$ are larger than 5. This means the normal approximation should be reasonably good.

(d) Now we use the same basic methods of Section 6.1 to construct the control limits and control chart shown in Figure 7-7. The center line is at $\bar{p} = 0.175$.

$$\text{Control limits at } \bar{p} \pm 2\sqrt{\frac{\bar{p}\,\bar{q}}{n}} = 0.175 \pm 2(0.049) \text{ or } 0.077 \text{ and } 0.273$$

$$\text{Control limits at } \bar{p} \pm 3\sqrt{\frac{\bar{p}\,\bar{q}}{n}} = 0.175 \pm 3(0.049) \text{ or } 0.028 \text{ and } 0.322$$

(e) Interpretation: We use the three out-of-control signals discussed in Section 6.1.

Signal I—Beyond the 3σ level.

We see semester number 12 was above the 3σ level. That semester the class must have been very good indeed!

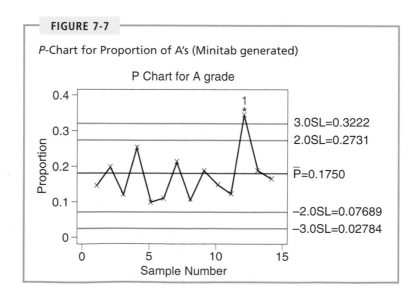

FIGURE 7-7

P-Chart for Proportion of A's (Minitab generated)

Signal II—Run of nine *consecutive* points on one side of center line.
Since this did not happen, there is no slow drift either up or down.

Signal III—At least 2 out of 3 *consecutive* points beyond the 2σ level (on the same side of center).
This out-of-control signal did not occur.

(f) Conclusion: The Biology Department can tell the dean that the proportion of A's given in Anatomy and Physiology is in statistical control with the exception of one unusually good class two semesters ago.   ◇

◇ **COMMENT** In some *P*-Charts the value of $\bar{p}$ may be near 0 or near 1. In this case, the control limits may drop below 0 or rise above 1. If this happens, we follow the usual convention of rounding negative control limits to 0 and rounding control limits above 1 to 1.   ◇

VIEW◉POINT   *Happy Memories! False Memories!!!*

"Memory isn't a record; it's an interpretation," says psychologist Mark Reinitz of the University of Puget Sound, Washington. Research indicates that about 68% of all people occasionally "fill in the blanks" in their memories. They claim to remember things that did not actually occur. In another study, it was found that 35% of visitors to Disneyland claim they shook hands with Bugs Bunny, who welcomed them at the entrance. However, Bugs Bunny is *not* a Disney character and was *not* at the Disneyland entrance.

Psychologists say memory is malleable for a reason. It helps us view ourselves in a more positive light. Statistical interpretation of proportions $\hat{p}$ are useful in this memory study, as well as all areas of natural science, business, linguistics, and social science. For more information about the memory study, see the July 2001 issue of the *Journal of Experimental Psychology*.

SECTION 7.3   PROBLEMS

1. *General Discussion* Suppose we have a binomial experiment in which success is defined to be a particular quality or attribute that interests us.
   (a) Examples of some attributes are an opinion about gun control (Problem 4), or identifying a special cell in bone marrow tissue (Problem 8). List at least three other attributes that could be used for success in a binomial trial.
   (b) Our binomial experiment has n trials. The probability of success is p, and the number of successes is the random variable r. In this section, we study the proportion $\hat{p} = r/n$. Under what conditions can the random variable $\hat{p}$ be approximated by a normal random variable? What are the formulas for $\mu_{\hat{p}}$ and $\sigma_{\hat{p}}$?

(c) Suppose $n = 33$ and $p = 0.21$. Can we approximate $\hat{p}$ by a normal distribution? Why? What are the values of $\mu_{\hat{p}}$ and $\sigma_{\hat{p}}$? What is the value of the continuity correction? Compute $P(0.15 \leq \hat{p} \leq 0.25)$.

(d) Suppose $n = 25$ and $p = 0.15$. Can we safely approximate $\hat{p}$ by a normal distribution? Why or why not?

(e) Suppose $n = 48$ and $p = 0.15$. Can we approximate $\hat{p}$ by a normal distribution? Why? What are the values of $\mu_{\hat{p}}$ and $\sigma_{\hat{p}}$? If a survey, experiment, or laboratory work gives us a $\hat{p}$ value of 0.22, what is the probability of getting a $\hat{p}$ value this high or higher?

2. *General Discussion* Suppose we have a binomial distribution with n trials and probability of success p. The random variable r is the number of successes in the n trials, and the random variable representing the proportion of successes is $\hat{p} = r/n$.
(a) $n = 50$; $p = 0.36$; Compute $P(0.30 \leq \hat{p} \leq 0.45)$.
(b) $n = 38$; $p = 0.25$; Compute the probability $\hat{p}$ will exceed 0.35.
(c) $n = 41$; $p = 0.09$; Can we approximate $\hat{p}$ by a normal distribution? Explain.

3. *Sociology: Criminal Justice* Courts sometimes make mistakes, but which do you believe is the worse mistake: convicting an innocent person or letting a guilty person go free? It turns out that about 60% of all Americans believe that convicting an innocent person is the worse mistake. (Source: *American Attitudes* by S. Mitchell, Sociology Department, Ithaca College.) Suppose you are taking a sociology class with 30 students enrolled. The question discussed today is: Do you agree with the statement that convicting an innocent person is worse than letting the guilty go free? What is the probability that the proportion of the class who agree is
(a) at least one half?
(b) at least two thirds?
(c) no more than one third?
(d) Is the normal approximation to the proportion $\hat{p} = r/n$ valid? Explain.

4. *Sociology: Gun Permits* Do you favor a law requiring a police permit to buy a gun? About 73% of American men and 86% of American women would favor such a law. (Source: See Problem 3.)
(a) A candidate for city council is speaking to a breakfast group of 38 men. The topic of gun permits comes up. What is the probability the majority of the audience (at least two thirds) will support gun permits? Assume the group is representative of all U.S. men.
(b) Answer part (a) if our candidate is speaking to a women's seminar with 45 women in the audience. Assume the group is representative of all U.S. women.
(c) Is the normal approximation to the proportion $\hat{p} = r/n$ valid in both applications? Explain.

5. *Who's Who: Misinformation* About 11% of Americans believe that Joan of Arc was Noah's wife. (Source: *Harper's Index*, Volume 3.) At a large freshman symposium, a college professor (jokingly) says that Joan of Arc was Noah's wife. Assume college freshmen are representative of the general American population regarding biblical knowledge.
(a) If the symposium was attended by 55 freshmen, what is the probability that up to 15% of the freshmen believe the professor's claim?
(b) What is the probability that between 10% and 15% (including 10% and 15%) of the freshmen believe the professor's claim?
(c) Is the normal approximation to the proportion $\hat{p} = r/n$ valid in this application? Explain.

6. *Grand Canyon: Boating Accidents* Thomas Myers is a staff physician at the clinic in Grand Canyon Village. Based on reports in recent years, Dr. Myers estimates that about 31% of the boating accidents on the Colorado River in Grand Canyon National Park occur at Crystal Rapids (mile 98). These range from small accidents (a few bruises) to major accidents (death). (Source: *Fateful Journey,* Myers, Becker, and Stevens.) Suppose there are 28 recently reported boating accidents in the park.
   (a) What is the probability that at least 25% of these accidents are from Crystal Rapids?
   (b) What is the probability that between 25% and 50% (including 25% and 50%) of these accidents are from Crystal Rapids?
   (c) Is the normal approximation to the proportion $\hat{p} = r/n$ valid in this application? Explain.

7. *Manufacturing: Defective Toys* A mechanical press is used to mold shapes for plastic toys. When the machine is adjusted and working well, it still produces about 6% defective toys. The toys are manufactured in lots of $n = 100$. Let r be a random variable representing the number of defective toys in a lot. Then $\hat{p} = r/n$ is the proportion of defective toys in a lot.
   (a) Explain why $\hat{p}$ can be approximated by a normal random variable. What are $\mu_{\hat{p}}$ and $\sigma_{\hat{p}}$?
   (b) Suppose a lot of 100 toys had a proportion 7% of defective toys. What is the probability a situation this bad or worse could occur? Compute $P(0.07 \leq \hat{p})$.
   (c) Suppose a lot of 100 toys had a proportion 11% of defective toys. What is the probability a situation this bad or worse could occur? Compute $P(0.11 \leq \hat{p})$. Do you think the machine might need an adjustment? Explain.

8. *Medical Tests: Leukemia* Healthy adult bone marrow contains about 56.5% neutrophils (a particular type of white blood cell). However, if this level is significantly reduced, it may be an early indicator of leukemia (Reference: *Diagnostic Tests with Nursing Implications,* edited by S. Loeb, Springhouse).
   (a) In a laboratory biopsy, a field of $n = 50$ bone marrow cells are observed under a microscope. A special dye is inserted, which only the neutrophils absorb. Then the number r of neutrophils in the field is counted. Although the field size $n = 50$ is fixed, the number of neutrophils r is a random variable. So the proportion $\hat{p} = r/n$ is also a random variable. Explain why $\hat{p}$ can be approximated by a normal random variable. What are $\mu_{\hat{p}}$ and $\sigma_{\hat{p}}$?
   (b) Suppose Jan had a bone marrow biopsy and $\hat{p}$ was observed to be 0.53. Assuming nothing is wrong (no leukemia), what is the probability of getting a biopsy this low or lower? Compute $P(\hat{p} \leq 0.53)$.
   (c) Suppose Meredith had a bone marrow biopsy and $\hat{p}$ was observed to be 0.41. Assuming nothing is wrong (no leukemia), what is the probability of getting a biopsy this low or lower? Compute $P(\hat{p} \leq 0.41)$.
   (d) Based on the probability estimates in parts (b) and (c), which do you think is the more serious case, Jan or Meredith? Explain.

9. *P-Chart: Property Crime* Lee is a cadet at the Honolulu Police Academy. He was asked to make a *P*-Chart for reported (minor) property crimes. Lee chose a small neighborhood with 92 families. Each family is viewed as a binomial trial. Success means the family was a victim of least one minor property crime in the past 3 months. Police reports gave the following data for the past 12 quarters (4 years). Assume the 92 families lived in the neighborhood all 4 years.

Quarter	1	2	3	4	5	6
r = no. of successes	11	14	18	23	19	15
$\hat{p} = r/92$	0.12	0.15	0.20	0.25	0.21	0.16

Quarter	7	8	9	10	11	12
r = no. of successes	12	16	13	22	24	19
$\hat{p} = r/92$	0.13	0.17	0.14	0.24	0.26	0.21

Make a *P*-Chart, and list any out-of-control signals by type (I, II, or III).

10. *P-Chart: Aluminum Cans* A high-speed metal stamp machine produces 12 oz. aluminum beverage cans. The cans are mass produced in lots of 110 cans for each square sheet of aluminum fed into the machine. However, some of the cans come out of the die stamp with folds and wrinkles. These are defective cans that must be recycled. Let us view each can as a binomial trial, where success is defined to mean the can is defective. So we have $n = 110$ trials (cans), and the random variable r is the number of defective cans. A test run of 15 consecutive aluminum sheets gave the following number r of defective cans.

Test sheet	1	2	3	4	5	6	7	8
r	8	11	6	9	12	8	7	11
$\hat{p} = r/110$	0.07	0.10	0.05	0.08	0.11	0.07	0.06	0.10

Test sheet	9	10	11	12	13	14	15
r	10	7	9	6	12	7	10
$\hat{p} = r/110$	0.09	0.06	0.08	0.05	0.11	0.06	0.09

Make a *P*-Chart and list any out-of-control signals by type (I, II, or III). Does it appear from the sequential test runs that the production process is in reasonable control? Explain.

11. *P-Chart: Temporary Work* Jobs for the homeless! A philanthropic foundation bought a used school bus that stops at homeless shelters early every weekday morning. The bus picks up people looking for temporary, unskilled, day jobs. The bus delivers these people to a work center. Later it picks them up after work. The bus can hold 75 people, and it fills up every morning. Not everyone finds work, so at 11 A.M. the bus goes to a soup kitchen where those not finding work that day volunteer their time. Let us view each person on the bus looking for work as a binomial trial. Success means he or she got a day job. The random variable r represents the number who got jobs. The foundation requested a *P*-Chart for the success ratios. For the past three weeks, we have the following data (see also next page).

Day	1	2	3	4	5	6	7	8
r	60	53	61	66	67	55	53	58
$\hat{p} = r/75$	0.80	0.71	0.81	0.88	0.89	0.73	0.71	0.77

Day	9	10	11	12	13	14	15
r	60	52	46	52	61	70	58
$\hat{p} = r/75$	0.80	0.69	0.61	0.69	0.81	0.93	0.77

Make a *P*-Chart, list any out-of-control signals, and interpret the results.

VIEWPOINT

Why Wait? Apply Now for a College Loan!

The cost of education is high. The cost of not having an education is higher! What can you expect? What about tuition and student fees? What about room and board? What is the total cost for 1 year at college? Perhaps some averages based on random samples of colleges could be useful. For more information, visit the Brase/Brase statistics site at http://math.college.hmco.com/students and find the link to the U.S. News site. Then select education. Search for the geographic regions of the colleges of interest.

SUMMARY

Sampling distributions give us the basis for inferential statistics. By studying the distribution of sample statistics, we can learn about a population parameter.

The central limit theorem describes the sampling distribution of sample means taken from samples of size *n*. It tells us that for increasing sample size *n*, the distribution of sample means $\bar{x}$ approaches a normal distribution with mean $\mu_{\bar{x}} = \mu$ and standard deviation $\sigma_{\bar{x}} = \sigma/\sqrt{n}$. The values of μ and σ are the population mean and standard deviation, respectively, of the original *x* distribution.

In the last section, we studied the proportion of successes $\hat{p} = r/n$ in binomial trials. When both *np* and *nq* are greater than 5, the sampling distribution for proportions is approximately normal with mean $\mu_{\hat{p}} = p$ and $\sigma_{\hat{p}} = \sqrt{pq/n}$. We use the normal distribution to compute the probability that $\hat{p}$ lies in a specified interval. However, we first make continuity corrections on the interval by adding $0.5/n$ to the right endpoint and subtracting $0.5/n$ from the left endpoint.

Finally, we looked at *P*-Charts, which are control charts for proportions.

IMPORTANT WORDS & SYMBOLS

Section 7.1
Population parameter
Statistic
Sampling distribution

Section 7.2
$\mu_{\bar{x}}$
$\sigma_{\bar{x}}$
Standard error of the mean
Central limit theorem

Section 7.3
Sampling distribution for $\hat{p}$
$\mu_{\hat{p}}$
$\sigma_{\hat{p}}$

Continuity correction, $0.5/n$
P-Chart

CHAPTER REVIEW PROBLEMS

1. *General Discussion* Let x be a random variable representing the amount of sleep each adult in New York City got last night. Consider a sampling distribution of sample means $\bar{x}$.
   (a) As the sample size becomes increasingly large, what distribution does the $\bar{x}$ distribution approach?
   (b) As the sample size becomes increasingly large, what value will the mean $\mu_{\bar{x}}$ of the $\bar{x}$ distribution approach?
   (c) What value will the standard deviation $\sigma_{\bar{x}}$ of the sampling distribution approach?
   (d) How do the two $\bar{x}$ distributions for sample size $n = 50$ and $n = 100$ compare?

2. *Normal Distributions: General Discussion* If x has a normal distribution with mean $\mu = 15$ and standard deviation $\sigma = 3$, describe the distribution of $\bar{x}$ values for sample size n, where $n = 4$, $n = 16$, $n = 100$. How do the $\bar{x}$ distributions compare for the various sample sizes?

3. *Job Interview: Length* The personnel office at a large electronics firm regularly schedules job interviews and maintains records of the interviews. From the past records, they have found that the length of a first interview is normally distributed with mean $\mu = 35$ minutes and standard deviation $\sigma = 7$ minutes.
   (a) What is the probability that a first interview will last 40 minutes or longer?
   (b) Nine first interviews are usually scheduled per day. What is the probability that the average length of time for the nine interviews will be 40 minutes or longer?

4. *Drugs: Effects* A new muscle relaxant is available. Researchers of the firm developing the relaxant have done studies that indicate that the time lapse between administration of the drug and beginning effects of the drug is normally distributed with mean $\mu = 38$ minutes and standard deviation $\sigma = 5$ minutes.
   (a) The drug is administered to one patient selected at random. What is the probability that the time it takes to go into effect is 35 minutes or less?
   (b) The drug is administered to a random sample of 10 patients. What is the probability that the average time before it is effective for all 10 patients is 35 minutes or less?
   (c) Comment on the differences of the results in parts (a) and (b).

5. *Psychology: IQ Scores* Assume that IQ scores are normally distributed with standard deviation of 15 points and mean of 100 points. If 100 people are chosen at random, what is the probability that the sample mean of IQ scores will not differ from the population mean by more than 2 points?

6. *Hatchery Fish: Length* A large tank of fish from a hatchery is being delivered to a lake. The hatchery claims that the mean length of fish in the tank is 15 inches, and the standard deviation is 2 inches. A random sample of 36 fish is taken from the tank. Let $\bar{x}$ be the mean sample length of these fish. What is the probability that $\bar{x}$ is within 0.5 inch of the claimed population mean?

7. *Light Bulbs: Hours of Life* A company that makes light bulbs claims that its bulbs have an average life of 750 hours with standard deviation of 20 hours. A random sample of 64 light bulbs is taken. Let $\bar{x}$ be the mean life of this sample.
   (a) What is the probability that $\bar{x} \geq 750$?
   (b) What is the probability that $745 \leq \bar{x} \leq 755$?

8. *Meteorology: Miami and Fairbanks* Let x be a random variable that represents daily high temperatures (degrees Fahrenheit) in January. The following information is based on a report from the U.S. Department of Commerce Environmental Data Services. For Miami, Florida, the mean of the x distribution is $\mu = 76$, and the standard deviation is approximately $\sigma = 1.9$. For Fairbanks, Alaska, the mean of the x distribution is $\mu = 0$ with approximate standard deviation $\sigma = 5.3$. Assume that x has a normal distribution.
   (a) For one day chosen at random in January, what is the probability that the high temperature in Miami will be less than 77°F? What is the probability the high temperature in Fairbanks will be less than 3°F?
   (b) If we choose $n = 7$ days in January, what can we say about the probability distribution of $\bar{x}$, the average high temperature? What is the probability that $\bar{x}$ is less than 77°F for Miami? less than 3°F for Fairbanks?
   (c) Suppose that we cannot assume that x has a normal distribution, but we can say that the distribution is approximately symmetrical and mound-shaped. In this case, what can we say about the $\bar{x}$ probability distribution? If we use all 31 days in January, what is the probability $\bar{x} < 77$°F in Miami? less than 3°F in Fairbanks?

9. *General Discussion* Suppose we have a binomial distribution with n trials and probability of success p. The random variable r is the number of successes in the n trials, and the random variable representing the proportion of successes is $\hat{p} = r/n$.
   (a) $n = 50$; $p = 0.22$; Compute $P(0.20 \leq \hat{p} \leq 0.25)$.
   (b) $n = 38$; $p = 0.27$; Compute the probability $\hat{p}$ will equal or exceed 0.35.
   (c) $n = 51$; $p = 0.05$; Can we approximate $\hat{p}$ by a normal distribution? Explain.

DATA HIGHLIGHTS: GROUP PROJECTS

Wild iris

Break into small groups and discuss the following topics. Organize a brief outline in which you summarize the main points of your group discussion.

Iris setosa is a beautiful wildflower that is found in such diverse places as Alaska, the Gulf of St. Lawrence, much of North America, and even in English meadows and parks. R. A. Fisher, with his colleague Dr. Edgar Anderson, studied these flowers extensively. Dr. Anderson described how he collected information on irises:

> I have studied such irises as I could get to see, in as great detail as possible, measuring iris standard after iris standard and iris fall after iris fall, sitting squat-legged with record book and ruler in mountain meadows, in cypress swamps, on lake beaches, and in English parks. [Anderson, E., "The Irises of the Gaspé Peninsula," *Bulletin, American Iris Society*, 59:2–5, 1935.]

The data in Table 7-3 were collected by Dr. Anderson and were published by his friend and colleague R. A. Fisher in a paper entitled "The Use of Multiple Measurements in Taxonomic Problems" (*Annals of Eugenics,* part II, 179–188, 1936). To find these data, visit the Brase/Brase statistics site at http://math.college.hmco.com/students and find the link to DASL, the Carnegie Mellon University Data and Story Library. From the DASL site, look under famous data sets.

TABLE 7-3 **Petal Length in Centimeters for**
Iris setosa

1.4	1.4	1.3	1.5	1.4
1.7	1.4	1.5	1.4	1.5
1.5	1.6	1.4	1.1	1.2
1.5	1.3	1.4	1.7	1.5
1.7	1.5	1	1.7	1.9
1.6	1.6	1.5	1.4	1.6
1.6	1.5	1.5	1.4	1.5
1.2	1.3	1.4	1.3	1.5
1.3	1.3	1.3	1.6	1.9
1.4	1.6	1.4	1.5	1.4

Let x be a random variable representing petal length. Using the TI-83 calculator, it was found that the sample mean is $\bar{x} = 1.46$ cm and the sample standard deviation is $s = 0.17$ cm. Figure 7-8 shows a histogram for the given data generated on a TI-83 calculator.

(a) Examine the histogram for petal lengths. Would you say that the distribution is approximately mound-shaped and symmetrical? Our sample has only 50 irises; if many thousands of irises had been used, do you think that the distribution would look even more like a normal curve? Let x be the petal length of *Iris setosa*. Research has shown that x has an approximately normal distribution with mean $\mu = 1.5$ cm and standard deviation $\sigma = 0.2$ cm.

(b) Use the empirical rule with $\mu = 1.5$ and $\sigma = 0.2$ to get an interval in which approximately 68% of the petal lengths will fall. Repeat this for 95% and 99.7%. Examine the raw data and compute the percentage of the raw data that actually falls into each of these intervals (the 68% interval, the 95% interval, and the 99.7% interval). Compare your computed percentages with those given by the empirical rule.

(c) Compute the probability that a petal length is between 1.3 and 1.6 cm. Compute the probability that the petal length is greater than 1.6 cm.

FIGURE 7-8

Petal Length (cm) *Iris setosa* (TI-83)

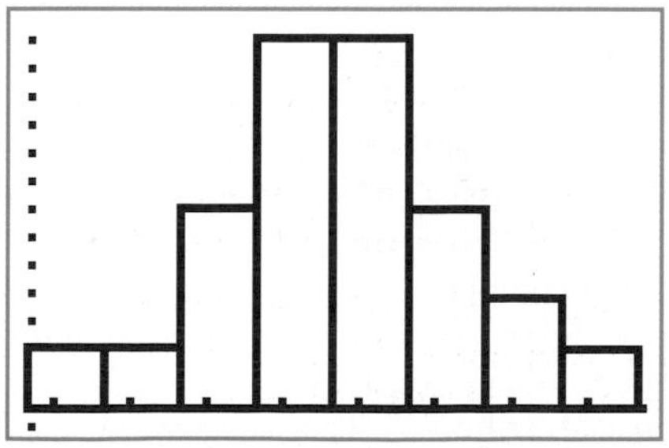

(d) Suppose that a random sample of 30 irises is obtained. Compute the probability that the average petal length for this sample is between 1.3 and 1.6 cm. Compute the probability that the average petal length is greater than 1.6 cm.

(e) Compare your answers for parts (c) and (d). Do you notice any differences? Why would these differences occur?

LINKING CONCEPTS: WRITING PROJECTS

Discuss each of the following topics in class, or review the topics on your own. Then write a brief but complete essay in which you summarize the main points. Please include formulas and graphs as appropriate.

1. Most people would agree that increased information should give better predictions. Discuss how sampling distributions actually enable better predictions by providing more information. Examine Theorem 7.1 again. Suppose that x is a random variable with a *normal* distribution. Then $\bar{x}$, the sample mean based on random samples of size n, also will have a normal distribution for *any* value of $n = 1, 2, 3, \ldots$.

    What happens to the standard deviation of the $\bar{x}$ distribution as n (the sample size) increases? Consider the following table for different values of n.

n	1	2	3	4	10	50	100
$\sigma/\sqrt{n}$	1σ	0.71σ	0.58σ	0.50σ	0.32σ	0.14σ	0.10σ

    In this case, "increased information" means a larger sample size n. Give a brief explanation why a large *standard deviation* will usually result in poor statistical predictions, whereas a small standard deviation usually results in much better predictions. Since the standard deviation of the sampling distribution $\bar{x}$ is $\sigma/\sqrt{n}$, we can decrease the standard deviation by increasing n. In fact, if we look at the preceding table, we see that if we use a sample of only size $n = 4$, we cut the standard deviation of $\bar{x}$ by 50% of the standard deviation σ of x. If we were to use a sample of size $n = 100$, we would cut the standard deviation of $\bar{x}$ to 10% of the standard deviation σ of x.

    Give the preceding discussion some thought and explain why you should get much better predictions for μ by using $\bar{x}$ from a sample of size n rather than by just using x. Write a brief essay in which you explain why sampling distributions are an important tool in statistics.

2. In a way, the central limit theorem can be thought of as a kind of "grand central station." It is a connecting hub or center for a great deal of statistical work. We will use it extensively in Chapters 8, 9, and 10. Put in a very elementary way, it says that as the sample size n increases, the mean $\bar{x}$ will always approach a normal distribution, no matter where the original x variable came from. For most people, it is the complete generality of the central limit theorem that is so awe inspiring: It applies to practically everything. List and discuss at least three variables from everyday life where you expect the variable x itself does *not* follow a normal or bell-shaped distribution. Then discuss what would happen to the sampling distribution $\bar{x}$ as the sample size increased. Sketch diagrams of the $\bar{x}$ distributions as the sample size n increases.

As we have seen in this chapter, the value of a sample statistic such as $\bar{x}$ varies from one sample to another. The central limit theorem describes the distribution of the sample statistic $\bar{x}$ when samples are sufficiently large.

We can use technology tools to generate samples of the same size from the same population. Then we can look at the statistic $\bar{x}$ for each sample, and the resulting $\bar{x}$ distribution.

Project Illustrating the Central Limit Theorem

Step 1: Generate random samples of specified size n from a population.

The random-number table allows us to sample from the uniform distribution of digits 0 through 9. Use either the random-number table or a random-number generator to generate 30 samples of size 10.

Step 2: Compute the sample mean $\bar{x}$ of the digits in each sample.

Step 3: Compute the sample mean of the means (i.e., $\bar{x}_{\bar{x}}$) as well as the standard deviation $s_{\bar{x}}$ of the sample means.

The population mean of the uniform distribution of digits from 0 through 9 is 4.5. How does $\bar{x}_{\bar{x}}$ compare to this value?

Step 4: Compare the sample distribution of $\bar{x}$ values to a normal distribution having the mean and standard deviation computed in Step 3.

(a) Use the values of $\bar{x}_{\bar{x}}$ and $s_{\bar{x}}$ computed in Step 3 to create the intervals shown in column 1 of Table 7-4.

(b) Tally the sample means computed in Step 2 to determine how many fall into each interval of column 2. Then compute the percent of data in each interval and record the results in column 3.

(c) The percentages listed in column 4 are those from a normal distribution (see Figure 6-5 showing the empirical rule). Compare the percentages in column 3 to those in column 4. How do the sample percentages compare with the hypothetical normal distribution?

Step 5: Create a histogram with six bars showing the sample means computed in Step 2.

TABLE 7-4 Frequency Table of Sample Means

1. Interval	2. Frequency	3. Percent	4. Hypothetical Normal Distribution
$\bar{x} - 3s$ to $\bar{x} - 2s$	Tally the sample means computed by the students in step 2 and place here.	Compute percents from column 2 and place here.	2 or 3%
$\bar{x} - 2s$ to $\bar{x} - s$			13 or 14%
$\bar{x} - s$ to $\bar{x}$			About 34%
$\bar{x}$ to $\bar{x} + s$			About 34%
$\bar{x} + s$ to $\bar{x} + 2s$			13 or 14%
$\bar{x} + 2s$ to $\bar{x} + 3s$			2 or 3%

Look at the histogram, and compare it to a normal distribution with the mean and standard deviation of the $\bar{x}$s (as computed in Step 3).

Step 6: Compare the results of this project to the central limit theorem.

Increase the sample size of Step 1 to 20, 30, and 40 and repeat Steps 1 to 5.

Computer Demonstration

The software package ComputerStat has the program Central Limit Theorem Demonstration. This program carries out the suggested project for samples from the uniform distribution. The user may select the sample size n to be any integer between 2 and 50. Figure 7-9 shows some displays from this program.

Technology Hints

The TI-83Plus, Excel, and Minitab all support the process of drawing random samples from a variety of distributions. Macros can be written in Excel and Minitab to repeat the six steps of the project.

TI-83Plus

You can generate random samples from uniform, normal, and binomial distributions. Press **MATH,** and select **PRB.** Selection **5:randInt(lower, upper, sample size m)** generates m random integers from the specified interval. Selection **6:randNorm(μ, σ, sample size m)** generates m random numbers from a normal distribution with mean μ and standard deviation σ. Selection **7:randBin(number of trials n, p, sample size m)** generates m random values (number of successes out of n

FIGURE 7-9 ComputerStat Central Limit Theorem Demonstration

(a) $n = 10$

(b) $n = 35$

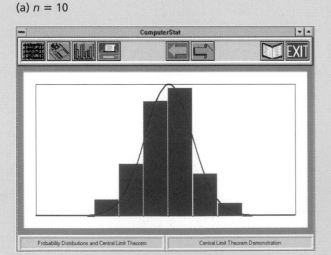

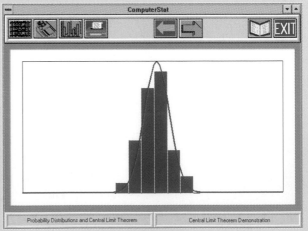

trials) for a binomial distribution with probability of success p on each trial. You can put these values in lists by using **Edit** under **Stat.** Highlight the list header, press Enter, and then select one of the options discussed.

Excel

Use the menu selection **Tools ➤ Data Analysis ➤ Random Number Generator.** The dialogue box provides choices for the population distribution, including uniform, binomial, and normal distributions. Fill in the required parameters and designate the location for the output.

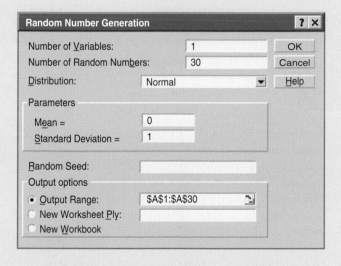

Minitab

Use the menu selections **Calc ➤ Random Data.** Then select the population distribution. The choices include uniform, binomial, and normal distributions. Fill in the dialogue box, where the number of rows indicates the number of data in the sample.

8 Estimation

We dance round in a ring and suppose,
But the Secret sits in the middle and knows.

—Robert Frost,
"The Secret Sits"*

Robert Lee Frost
(1874–1963)

This celebrated American poet drew poetic symbols largely from common experiences observed in his rural New England

For on-line student resources, visit
math.college.hmco.com/students and
follow the Statistics links to the Brase/Brase.
Understandable Statistics. 7th edition web site.

In Chapter 1 we said that statistics is the study of how to collect, organize, analyze, and interpret numerical data. That part concerned with analysis, interpretation, and forming conclusions about the source of the data is called *statistical inference*. Problems of statistical inference require us to draw a *sample* of observations from a larger *population*. A sample usually contains incomplete information, so in a sense we must "dance round in a ring and suppose." Nevertheless, conclusions about the population can be obtained from sample data by use of statistical estimates. This chapter introduces you to several widely used methods of estimation.

◇ How do we estimate the average outcome (mean) of a random variable? How much confidence should we place in such a process?   (SECTION 8.1)

◇ What famous statistician worked for an (equally famous) Irish brewing company? What has all this to do with small samples and Student's *t* distribution?   (SECTION 8.2)

◇ How about estimating the proportion *p* for success in a binomial experiment? How does the normal approximation fit into this process?   (SECTION 8.3)

◇ If we start out in the beginning design stage of a statistical project, how large a sample size would we plan to get?   (SECTION 8.4)

◇ So what is the big difference? Sometimes differences in life can be important. How do we estimate differences?   (SECTION 8.5)

FOCUS PROBLEM

The Trouble with Wood Ducks

The National Wildlife Federation published an article entitled "The Trouble with Wood Ducks" (*National Wildlife,* Vol. 31, No. 5). In this article, wood ducks are described as beautiful birds living in forested areas such as the Pacific Northwest and Southeast United States. Because of overhunting and habitat destruction, these birds were in danger of extinction. A federal ban on hunting wood ducks in 1918 helped save the species from extinction. Wood ducks like to nest in tree cavities. However, many such trees were disappearing due to heavy timber cutting. For a period of time it seemed that nesting boxes were the solution to disappearing trees. At first, the wood duck population grew, but after a few seasons, the population declined sharply. Good biology research combined with good statistics provided an answer to this disturbing phenomenon.

Cornell University professors of ecology Paul Sherman and Brad Semel found that the nesting boxes were placed too close to each other. Female wood ducks prefer a secluded nest that is a considerable distance from the next wood duck nest. In fact, female wood duck behavior changed when the nests were too close to each other. Some females would lay their eggs in another female's nest. The result was too many eggs in one nest. The biologists found that if there

373

were too many eggs in a nest, the proportion of eggs that hatched was considerably reduced. In the long run, this meant a decline in the population of wood ducks.

In their study, Sherman and Semel used two placements of nesting boxes. Group I had boxes that were well separated from each other and well hidden by available brush. Group II had boxes that were highly visible and grouped closely together.

In group I boxes, there were a total of 474 eggs, of which a field count showed that about 270 hatched. In group II boxes, there were a total of 805 eggs, of which a field count showed that, again, about 270 hatched.

The material in Chapter 8 will enable us to answer many questions about the hatch ratios of eggs from nests in the two groups.

(a) Find a point estimate $\hat{p}_1$ for p_1, the proportion of eggs that hatch in group I nest box placements. Find a 95% confidence interval for p_1.

(b) Find a point estimate $\hat{p}_2$ for p_2, the proportion of eggs that hatch in group II nest box placements. Find a 95% confidence interval for p_2.

(c) Find a 95% confidence interval for $p_1 - p_2$. Does the interval indicate that the proportion of eggs hatched from group I nest box placements is higher than, lower than, or not different from the proportion of eggs hatched from group II nest boxes?

(d) What conclusions about placement of nest boxes can be drawn? In the article, additional concerns are raised about the higher cost of placing and maintaining group I nest box placement. At issue is also the cost efficiency per successful wood duck hatch. Data in the article do not include information that would help us answer questions of *cost* efficiency. However, the data presented do help us answer questions about proportion of successful hatches in the two nest box configurations.

8.1
Estimating μ with Large Samples

FOCUS POINTS

✓ Explain the meaning of confidence level, error of estimate, and critical value.

✓ Find the critical value corresponding to a given confidence level.

✓ Compute confidence intervals for μ using large samples and interpret the results.

As we observed in Chapter 7, we often do not have access to all measurements of an entire population because of time, money, difficulty in finding population members, etc. Instead, we rely on information from a sample. We use the sample statistic to estimate the corresponding population parameter.

An estimate of a population parameter given by a single number is called a *point estimate* of that parameter. It should be no great surprise that we use $\bar{x}$ (the sample mean) as a point estimate for μ (the population mean) and s (the sample standard deviation) as a point estimate for σ (the population standard deviation). In this section, we will discuss estimates of μ and σ from *large samples* ($n \geq 30$).

Statistical theory and empirical results show that if a distribution is approximately mound-shaped and symmetrical, then when the sample size is 30 or larger, we are safe, for most practical purposes, if we estimate σ by s. The error resulting from taking the population standard deviation σ to be equal to the sample standard deviation s is negligible. However, when the sample size is less than 30, we will use special small sample methods, which we will study in Section 8.2.

For large samples of size $n \geq 30$,

$$\sigma \approx s$$

is a good estimate, for most practical purposes.

Basic terminology

Using $\bar{x}$ to estimate μ is not quite so simple, even when we have a large sample size. The *error of estimate* is the magnitude of the difference between the point estimate and the true parameter value. Using $\bar{x}$ as a *point estimate* for μ, the error of estimate is the magnitude of $\bar{x} - \mu$. If we use absolute-value notation, we can indicate the error of estimate for μ by the notation $|\bar{x} - \mu|$.

We cannot say exactly how close $\bar{x}$ is to μ when μ is unknown. Therefore, the exact error of estimate is unknown when the population parameter is unknown. Of course, μ is usually not known or there would be no need to estimate it. In this section, we will use the language of probability to give us an idea of the size of the error of estimate when we use $\bar{x}$ as a point estimate of μ.

First, we need to learn about *confidence levels*. The reliability of an estimate will be measured by the confidence level.

Finding the critical value

Suppose we want a confidence level of c (see Figure 8-1). Theoretically, you can choose c to be any value between 0 and 1, but usually c is equal to a number such as 0.90, 0.95, or 0.99. In each case, the value z_c is the number such that the area under the standard normal curve falling between $-z_c$ and z_c is equal to c. The value z_c is called the *critical value* for a confidence level of c.

The area under the normal curve from $-z_c$ to z_c is the probability that the standardized normal variable z lies in that interval. This means that

$$P(-z_c < z < z_c) = c$$

EXAMPLE 1

Find a critical value

Let us use Table 5 in Appendix II to find a number $z_{0.99}$ so that 99% of the area under the standard normal curve lies between $-z_{0.99}$ and $z_{0.99}$. That is, we will find $z_{0.99}$ so that

$$P(-z_{0.99} < z < z_{0.99}) = 0.99$$

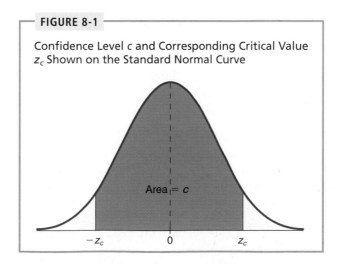

FIGURE 8-1

Confidence Level c and Corresponding Critical Value z_c Shown on the Standard Normal Curve

Area = c

$-z_c$ 0 z_c

SOLUTION: In Section 6.3, we saw how to find the z value when we were given an area between $-z$ and z. The first thing we did was to find the corresponding area to the left of $-z$. If A is the area between $-z$ and z, then $(1 - A)/2$ is the area to the left of z. In our case, the area between $-z$ and z is 0.99. The corresponding area in the left tail is $(1 - 0.99)/2 = 0.005$ (see Figure 8-2).

Next, we use Table 5 of Appendix II to find the z value corresponding to a left-tail area of 0.0050.

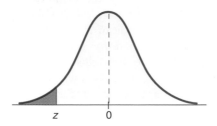

TABLE 8-1   Excerpt from Table 5 of Appendix II

z	.00	...	.07	.08	.09
-3.4	.0003		.0003	.0003	.0002
⋮					
-2.5	.0062		.0051	.0049	.0048
				↑ .0050	

From Table 8-1, we see that the desired area, 0.0050, is exactly halfway between areas corresponding to $z = -2.58$ and $z = -2.57$. Because the two area values are so close together, we use the more extreme z value -2.58 rather than interpolate. For $z_{0.99} = 2.58$, we have

$$P(-2.58 < z < 2.58) = 0.99 \qquad \diamond$$

The results of Example 1 will be used a great deal in our later work. For convenience, Table 8-2 gives some levels of confidence and corresponding critical values z_c. The same information is provided in Table 5(b) of Appendix II.

Error of estimate

An estimate is not very valuable unless we have some kind of measure of how "good" it is. The language of probability can give us an idea of the size of the error

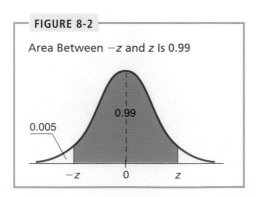

FIGURE 8-2

Area Between $-z$ and z Is 0.99

0.99

0.005

$-z$ 0 z

TABLE 8-2   Some Levels of Confidence and Their Corresponding Critical Values

Level of Confidence c	Critical Value z_c
0.70, or 70%	1.04
0.75, or 75%	1.15
0.80, or 80%	1.28
0.85, or 85%	1.44
0.90, or 90%	1.645
0.95, or 95%	1.96
0.98, or 98%	2.33
0.99, or 99%	2.58

of estimate caused by using the sample mean $\bar{x}$ as an estimate for the population mean.

Remember that $\bar{x}$ is a random variable. Each time we draw a sample of size n from a population, we can get a different value for $\bar{x}$. According to the central limit theorem, if the sample size is large, then $\bar{x}$ has a distribution that is approximately normal with mean $\mu_{\bar{x}} = \mu$, the population mean we are trying to estimate. The standard deviation is $\sigma_{\bar{x}} = \sigma/\sqrt{n}$.

This information, together with our work on confidence levels, leads us (as shown in the optional derivation on the next page) to the probability statement

$$P\left(-z_c \frac{\sigma}{\sqrt{n}} < \bar{x} - \mu < z_c \frac{\sigma}{\sqrt{n}}\right) = c \tag{1}$$

Equation (1) uses the language of probability to give us an idea of the size of the error of estimate for the corresponding confidence level c. In words, Equation (1) says that the probability is c that our point estimate $\bar{x}$ is within a distance $\pm z_c(\sigma/\sqrt{n})$ of the population mean μ. This relationship is shown in Figure 8-3.

In the following optional discussion, we derive Equation (1). If you prefer, you may jump ahead to the discussion about the error of estimate.

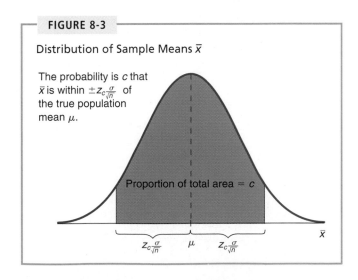

FIGURE 8-3

Distribution of Sample Means $\bar{x}$

The probability is c that $\bar{x}$ is within $\pm z_c \frac{\sigma}{\sqrt{n}}$ of the true population mean μ.

Proportion of total area $= c$

$z_c \frac{\sigma}{\sqrt{n}}$ μ $z_c \frac{\sigma}{\sqrt{n}}$ $\bar{x}$

Optional derivation of Equation (1)

For a c confidence level, we know

$$P(-z_c < z < z_c) = c \tag{2}$$

This statement gives us information about the size of z, but we want information about the size of $\bar{x} - \mu$. Is there a relationship between z and $\bar{x} - \mu$? The answer is yes since, by the central limit theorem, $\bar{x}$ has a distribution that is approximately normal with mean μ and standard deviation $\sigma/\sqrt{n}$. We can convert $\bar{x}$ to a standard z score by using the formula

$$z = \frac{\bar{x} - \mu}{\sigma/\sqrt{n}} \tag{3}$$

Substituting this expression for z in Equation (2) gives

$$P\left(-z_c < \frac{\bar{x} - \mu}{\sigma/\sqrt{n}} < z_c\right) = c \tag{4}$$

Multiplying all parts of the inequality in (4) by $\sigma/\sqrt{n}$ gives us

$$P\left(-z_c\frac{\sigma}{\sqrt{n}} < \bar{x} - \mu < z_c\frac{\sigma}{\sqrt{n}}\right) = c \tag{1}$$

Equation (1) is precisely the equation we set out to derive.

The *error of estimate* (or absolute error) using $\bar{x}$ as a point estimate for μ is $|\bar{x} - \mu|$. In most practical problems, μ is unknown, so the error of estimate is also unknown. However, Equation (1) allows us to compute an *error tolerance E*, which serves as a bound on the error of estimate. Using a $c\%$ level of confidence, we can say that the point estimate $\bar{x}$ differs from the population mean μ by a *maximal error tolerance* of

$$E = z_c\frac{\sigma}{\sqrt{n}}$$

Since $\sigma \approx s$ for large samples, we have

$$E \approx z_c\frac{s}{\sqrt{n}} \qquad \text{when } n \geq 30 \tag{5}$$

where E is the *maximal error tolerance* on the error of estimate for a given
   confidence level c (i.e., $|\bar{x} - \mu| < E$ with probability c)
   z_c is the critical value for the confidence level c (see Table 8-2)
   s is the sample standard deviation
   n is the sample size

Using Equations (1) and (5), we conclude that

$$P(-E < \bar{x} - \mu < E) = c \tag{6}$$

Equation (6) says that the probability is c that the difference between $\bar{x}$ and μ is no more than the maximal error tolerance E. If we use a little algebra on the inequality

$$-E < \bar{x} - \mu < E \tag{7}$$

for μ, we can rewrite it in the following mathematically equivalent way:

$$\bar{x} - E < \mu < \bar{x} + E \tag{8}$$

Confidence intervals (large sample)

Since (7) and (8) are mathematically equivalent, their probabilities are the same. Therefore, from (6), (7), and (8) we obtain

$$P(\bar{x} - E < \mu < \bar{x} + E) = c \tag{9}$$

Equation (9) says that there is a chance of c that the interval from $\bar{x} - E$ to $\bar{x} + E$ contains the population mean μ. We call this interval a *c confidence interval for μ*. We may get a different confidence interval for each different sample that is taken. Some intervals will contain the population mean μ and others will not. However, in the long run, the proportion of confidence intervals that contains μ is c.

◇ **SUMMARY** For large samples ($n \geq 30$) taken from a distribution that is approximately mound-shaped and symmetrical, and for which the population standard deviation σ is unknown, a *c confidence interval for the population mean μ* is given by the following:

> **c Confidence interval for μ (large samples)**
>
> $$\bar{x} - E < \mu < \bar{x} + E \tag{10}$$
>
> where $\bar{x}$ = sample mean
>
> $$E \approx z_c \frac{s}{\sqrt{n}}$$
>
> s = sample standard deviation
>
> c = confidence level ($0 < c < 1$)
>
> z_c = critical value for confidence level c
>   (See Table 8-2 for frequently used values.)
>
> n = sample size ($n \geq 30$)

◇

◇ **IMPORTANT COMMENT** What if you could assume

   (i) you were drawing your samples from a *normal* population *and*

   (ii) the population standard deviation σ was known to you?

Under these assumptions, a c confidence interval for the population mean would be $\bar{x} - E < \mu < \bar{x} + E$, where

$$E = z_c \frac{\sigma}{\sqrt{n}}$$

and the condition that we must have a large sample ($n \geq 30$) could be dropped.

In most practical situations, we don't know the population standard deviation σ. In these situations, we use Equation (10) with large samples ($n \geq 30$) to find a c confidence interval, or we use small samples ($n < 30$) and the methods of the next section to find a c confidence interval.

EXAMPLE 2

Confidence interval for μ

Julia enjoys jogging. She has been jogging over a period of several years, during which time her physical condition has remained constantly good. Usually, she jogs 2 miles per day. During the past year Julia has sometimes recorded her times required to run 2 miles. She has a sample of 90 of these times. For these 90 times the mean was $\bar{x} = 15.60$ minutes and the standard deviation was $s = 1.80$ minutes. Let μ be the mean jogging time for the entire distribution of Julia's 2-mile running times (taken over the past year). Find a 0.95 confidence interval for μ.

SOLUTION: The interval from $\bar{x} - E$ to $\bar{x} + E$ will be a 95% confidence interval for μ. In this case, $c = 0.95$, so $z_c = 1.96$ (see Table 8-2). The sample size $n = 90$ is large enough that we may approximate σ as $s = 1.80$ minutes. Therefore,

$$E \approx z_c \frac{s}{\sqrt{n}}$$

$$E = 1.96\left(\frac{1.80}{\sqrt{90}}\right)$$

$$E = 0.37$$

Using Equation (10), the given value of $\bar{x}$, and our computed value for E, we get the 95% confidence interval for μ.

$$\bar{x} - E < \mu < \bar{x} + E$$
$$15.60 - 0.37 < \mu < 15.60 + 0.37$$
$$15.23 < \mu < 15.97$$

We conclude with 95% confidence that the population mean μ of jogging times for Julia is between 15.23 and 15.97 minutes.   ◊

A few comments are in order about the general meaning of the term *confidence interval*. It is important to realize that the endpoints $\bar{x} \pm E$ are really statistical *variables*. Equation (9) says that we have a chance c of obtaining a sample such that the interval, once it is computed, will contain the parameter μ. Of course, after the confidence interval is numerically fixed, it either does or does not contain μ. So the probability is 1 or 0 that the interval, when it is fixed, will contain μ. A nontrivial probability statement can be made only about variables, not constants. Therefore, Equation (9) really says that if we repeat the experiment many times and get lots of confidence intervals (for the same sample size), then the proportion of all intervals that will turn out to contain the mean μ is c.

In Figure 8-4, the horizontal lines represent 0.90 confidence intervals for various samples of the same size from a distribution. Some of these intervals contain μ, and others do not. Since the intervals are 0.90 confidence intervals, about 90% of all such intervals should contain μ. For each sample, the interval goes from $\bar{x} - E$ to $\bar{x} + E$.

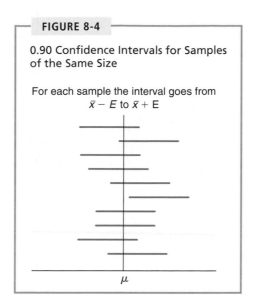

FIGURE 8-4

0.90 Confidence Intervals for Samples of the Same Size

For each sample the interval goes from $\bar{x} - E$ to $\bar{x} + E$

μ

◊ **COMMENT** Please see Using Technology at the end of this chapter for a computer demonstration of this discussion about confidence intervals.   ◊

GUIDED EXERCISE 1

Confidence interval for μ

Walter usually meets Julia at the track. He prefers to jog 3 miles. While Julia kept her record, he also kept one for his time required to jog 3 miles. For his 90 times, the mean was $\bar{x} = 22.50$ minutes and the standard deviation was $s = 2.40$ minutes. Let μ be the mean jogging time for the entire distribution of Walter's 3-mile running times over the past several years. Find a 0.99 confidence interval for μ.

(a) What is the value of $z_{0.99}$? (See Table 8-2.)   ⟹ $z_{0.99} = 2.58$

(b) Since the sample size is large, what can we use for σ?   ⟹ $\sigma \approx s = 2.40$

(c) What is the value of E?   ⟹ $E = z_c \dfrac{\sigma}{\sqrt{n}} \approx 2.58 \left(\dfrac{2.40}{\sqrt{90}} \right) = 0.65$

(d) What are the endpoints for a 0.99 confidence interval for μ?   ⟹   The endpoints are given by

$$\bar{x} - E \approx 22.50 - 0.65 = 21.85$$
$$\bar{x} + E \approx 22.50 + 0.65 = 23.15$$

GUIDED EXERCISE 2

Sample size

We have said that a sample of size 30 or larger is a large sample. In this section, we indicated two important reasons why our methods require large samples.

What are these reasons?   *Reason 1:* Our methods require $\bar{x}$ to have approximately a normal distribution. We know from the central limit theorem that this will be the case for large samples.

Reason 2: Unless we somehow know σ, our methods require us to approximate σ with the sample standard deviation s. This approximation will be good only if the sample size is large.

When we use samples to estimate the mean of a population, we generate a small error. However, samples are useful even when it is possible to survey the entire population because the use of a sample may yield savings of time or effort in collecting data.

TECH NOTE   The TI-83Plus, Excel, and Minitab all support confidence intervals for μ from large samples. The level of support varies according to the technology. When a confidence interval is given, the standard mathematical notation (lower value, upper value) is used. For instance, the notation (15.23, 15.97) means the interval from 15.23 to 15.97.

TI-83Plus   This calculator gives the most extensive support. The user can opt to enter raw data or just summary statistics. In each case, the value of σ (or the sample estimate) must be specified. Press the **STAT** key and select **TESTS**, use **7:ZInterval**. The T1-83Plus output shows the results for Example 2.

```
ZInterval
 Inpt:Data Stats
 σ:1.8
 x̄:15.6
 n:90
 C-Level:95
 Calculate
```

```
ZInterval
(15.228, 15.972)
x̄=15.6
n=90
```

Excel   Excel gives only the value of the maximal error of estimate E. Use the menu choice **Paste Function** $\boxed{f_x}$ ➤ **Statistics** ➤ **Confidence**(alpha, σ, n). In the dialogue box, the value of alpha is $1 -$ confidence level. The Excel output shows the value of E for Example 2.

=	=CONFIDENCE(0.05,1.8,90)		
	C	D	E
0.371876			

An alternate approach incorporating raw data (using the Student's *t* distribution presented in the next section) uses menu choices **Tools ➤ Data Analysis ➤ Describe Statistics**. Again, the value of E for the interval is given.

Minitab Raw data are required. Use the menu choice **Stat ➤ Basic Statistics ➤ 1-SampleZ**.

VIEWP⊙INT

Music and Techno Theft

Performing rights organizations ASCAP (American Society of Composers, Authors, and Publishers) and BMI (Broadcast Music, Inc.) collect royalties for song writers and music publishers. Radio, television, cable, nightclubs, restaurants, elevators, even beauty parlors play music that is copyrighted by a composer or publisher. The royalty payment for this music turns out to be more than a billion dollars a year (Source: *The Wall Street Journal*). How do ASCAP and BMI know who is playing what music? The answer is, they don't know! Instead of tracking *exactly* what gets played, they use random sampling and *confidence intervals*. For example, each radio station (there are more than 10,000 in the U.S.) has randomly chosen days of programming analyzed every year. The results are used to assess royalty fees. In fact, Deloitte & Touche (a financial services company) administers the sampling process.

Although the system is not perfect, it helps bring order into an otherwise chaotic accounting system. Such methods of "copyright policing" help prevent techno theft, ensuring that many song writers get a reasonable return for their creative work.

SECTION 8.1   PROBLEMS

Answers may vary slightly due to rounding.

1. *Chocolate Chip Cookies: Calories Consumer Reports* gave the following data about calories in a 30-gram serving of chocolate chip cookies. Both fresh-baked—such as Duncan Hines and Pillsbury—and packaged cookies—such as Pepperidge Farm and Nabisco—were included.

153	152	146	138	130	146	149	138	168
147	140	156	155	163	153	155	160	145
138	150	135	155	156	150	146	129	127
171	148	155	132	155	127	150	110	

(a) Use a calculator with mean and standard deviation keys to verify that the sample mean number of calories is $\bar{x} = 146.5$ with sample standard deviation $s = 12.7$ calories.

(b) We take the point of view that the preceding data are representative of the population of all chocolate chip cookies. Find an 80% confidence interval for the mean calories μ in a 30-gram serving of all chocolate chip cookies. Find the length of this interval.

(c) Repeat part (b) using a 90% confidence interval.

(d) Repeat part (b) using a 99% confidence interval.

(e) Compare the lengths from parts (b), (c), and (d). Comment on how these lengths change as c, the confidence level, increases.

2. *Aspen Trees: Base Circumference* In Roosevelt National Forest, the rangers took random samples of live aspen trees and measured the base circumference of each tree.

(a) The first sample had 30 trees with a mean circumference of $\bar{x} = 15.71$ inches and a sample standard deviation of $s = 4.63$ inches. Find a 95% confidence interval for the mean circumference of aspen trees from these data.

(b) The next sample had 90 trees with a mean of $\bar{x} = 15.58$ inches and a sample standard deviation of $s = 4.61$ inches. Again, find a 95% confidence interval from these data.

(c) The last sample had 300 trees with a mean of $\bar{x} = 15.59$ inches and a sample standard deviation of $s = 4.62$ inches. Again, find a 95% confidence interval from these data.

(d) Find the length of each interval of parts (a), (b), and (c). Comment on how these lengths change as the sample size increases.

3. *Profits: Banks* Jobs and productivity! How do banks rate? One way to answer this question is to examine annual profits per employee. *Forbes Top Companies,* edited by J. T. Davis (John Wiley & Sons), gave the following data about annual profits per employee (in units of one thousand dollars per employee) for representative companies in financial services. Companies such as Wells Fargo, First Bank System, Key Banks, and Norwest Banks were included.

42.9	43.8	48.2	60.6	54.9	55.1	52.9	54.9	42.5	33.0	33.6
36.9	27.0	47.1	33.8	28.1	28.5	29.1	36.5	36.1	26.9	27.8
28.8	29.3	31.5	31.7	31.1	38.0	32.0	31.7	32.9	23.1	54.9
43.8	36.9	31.9	25.5	23.2	29.8	22.3	26.5	26.7		

(a) Use a calculator with mean and sample standard deviation keys to verify that, for the preceding data, $\bar{x} \approx 36.0$ and $s \approx 10.2$.

(b) Let us say that the preceding data are representative of the entire sector of (successful) financial services corporations. Find a 75% confidence interval for μ, the average annual profit per employee for all successful banks.

(c) Let us say that you are the manager for a local bank with a large number of employees. Suppose the annual profits per employee are less than 30 thousand dollars per employee. Do you think that this might be somewhat low compared with other successful financial institutions? Explain by referring to the confidence interval you computed in part (b).

(d) Suppose the annual profits are more than 40 thousand dollars per employee. As manager of the bank, would you feel somewhat better? Explain by referring to the confidence interval you computed in part (b).

(e) Repeat parts (b), (c), and (d) for a 90% confidence level.

4. *Profits: Retail* Jobs and productivity! How do retail stores rate? One way to answer this question is to examine annual profits per employee. The following data give annual profits per employee (in units of one thousand dollars per employee) for companies in retail sales. See reference in Problem 3. Companies such as Gap,

$P(\bar{x} - E < M < \bar{x} + E)$

Nordstrom, Circuit City, Dillards, JCPenney, Sears, Wal-Mart, Office Depot, and Toys 'Я' Us are included.

4.4	6.5	4.2	8.9	8.7	8.1	6.1	6.0	2.6	2.9	8.1	−1.9
11.9	8.2	6.4	4.7	5.5	4.8	3.0	4.3	−6.0	1.5	2.9	4.8
−1.7	9.4	5.5	5.8	4.7	6.2	15.0	4.1	3.7	5.1	4.2	

(a) Use a calculator with mean and sample standard deviation keys to verify that, for the preceding data, $\bar{x} \approx 5.1$ and $s \approx 3.8$.

(b) Let us say that the preceding data are representative of the entire sector of retail sales companies. Find an 80% confidence interval for μ, the average annual profit per employee for retail sales.

(c) Let us say that you are the manager of a retail store with a large number of employees. Suppose the annual profits per employee are less than 3 thousand dollars per employee. Do you think that this might be low compared with other retail stores? Explain by referring to the confidence interval you computed in part (b).

(d) Suppose the annual profits are more than 6.5 thousand dollars per employee. As store manager, would you feel somewhat better? Explain by referring to the confidence interval you computed in part (b).

(e) Repeat parts (b), (c), and (d) for a 95% confidence interval.

5. *January Temperatures: Phoenix* The U.S. Department of Commerce Environmental Data Service gave the following information about average temperature (°F) in January in Phoenix, Arizona, for the past 40 years.

52.8	52.2	52.7	53.8	54.5	48.5	53.3	52.4
51.7	43.2	51.4	42.8	48.5	52.3	51.6	43.7
49.7	49.9	52.6	51.2	50.7	54.2	48.7	56.0
54.0	51.9	50.4	50.7	54.0	52.4	51.5	48.4
46.7	53.0	51.4	49.6	54.6	52.3	54.9	52.1

n: 40

(a) Use a calculator with mean and standard deviation keys to verify that the sample mean is $\bar{x} = 51.16°F$ and the sample standard deviation is $s = 3.04°F$.

(b) Find a 90% confidence interval for the January mean temperature in Phoenix.

(c) Find a 99% confidence interval for the January mean temperature in Phoenix.

(d) If someone told you that the earth was heating up and the average January temperature in Phoenix was now 53°F, what might you think about such a claim? Is it possible that a few more years of observation might be needed before such a claim could be made? Explain.

6. *Italian Restaurant: Lunch Tab* The Roman Arches is an Italian restaurant. The manager wants to estimate the average amount a customer spends on lunch Monday through Friday. A random sample of 115 customers' lunch tabs gave a mean of $\bar{x} = \$9.74$ with standard deviation $s = \$2.93$.

(a) Find a 95% confidence interval for the average amount spent on lunch by all customers.

(b) For a day when the Roman Arches has 115 lunch customers, use part (a) to estimate a range of dollar values for the total lunch income that day.

7. *Coyotes: Howls* A special feature of camping in the western United States is listening to coyotes howl at night. Just how long is one coyote howl? Since coyotes tend to hunt (mice, rabbits, etc.) in groups, it can seem that howls go on for a long time. This is the result of a collection of coyotes howling together. Individual coyotes tend

to howl for a shorter period of time. The howls may help individual members of the pack locate each other in the dark. The following data are based on information from *Coyotes: Biology, Behavior, and Management,* edited by M. Bekoff, University of Colorado (Academic Press).

(a) Based on a random sample of 102 coyotes, it was found that the mean duration of the howl was $\bar{x} = 1.2$ seconds with sample standard deviation $s = 0.4$ second. Compute a 99% confidence interval for the population mean duration μ of all coyote howls.

(b) For the 102 coyotes in the sample, the mean frequency of howls was $\bar{x} = 609$ Hz (hertz) with sample standard deviation $s = 248$ Hz. Compute a 95% confidence interval for the population mean frequency μ of all coyote howls.

8. *Ballooning: Air Temperature* How hot is the air in the top (crown) of a hot air balloon? Information from *Ballooning: The Complete Guide to Riding the Winds,* by Wirth and Young (Random House), claims that the air in the crown should be an average of 100°C for a balloon to be in a state of equilibrium. However, the temperature does not need to be exactly 100°C. What is a reasonable and safe range of temperatures? This may vary with the size and (decorative) shape of the balloon. All balloons have a temperature gauge in the crown. Suppose that 56 readings (for a balloon in equilibrium) gave a mean temperature $\bar{x} = 97$°C with sample standard deviation $s = 17$°C.

(a) Compute a 95% confidence interval for the average temperature for which this balloon will be in a steady-state equilibrium.

(b) If the average temperature in the crown of the balloon goes above the high end of your confidence interval, do you expect the balloon will go up or down? Explain.

9. *Medical: Sleep* How long does it take to fall asleep at night? This depends on what happened the night before. Alexander Borbely is a professor at the University of Zurich Medical School and director of the Sleep Laboratory. The following is adapted from the book, *Secrets of Sleep,* by Professor Borbely.

(a) Suppose that a random sample of 38 college students was kept awake all night and the next day (24 hours total). The mean time for this group to go to sleep the next night was $\bar{x} = 2.5$ minutes with standard deviation $s = 0.7$ minute. Compute a 90% confidence interval for the mean time of all such (sleep-deprived) students to fall asleep. What is the length of this interval?

(b) Suppose that a random sample of 38 college students had a normal (8-hour) sleep and a normal (16-hour) day. The mean time for this group to go to sleep the next night was $\bar{x} = 15.2$ minutes with sample standard deviation $s = 4.8$ minutes. Compute a 90% confidence interval for the mean time to go to sleep for all people in this (normal) group. What is the length of this interval?

(c) Suppose that a random sample of 38 college students stayed in bed at least 12 hours and then after another 12 hours went back to bed. The mean time for this group to fall asleep was $\bar{x} = 25.7$ minutes with sample standard deviation $s = 8.3$ minutes. Compute a 90% confidence interval for all people in this group to fall asleep. What is the length of this interval?

(d) Compare the lengths of the intervals in parts (a), (b), and (c). As the sample standard deviations got larger, did the intervals get longer? Why would you expect this from the method of calculation? Explain.

10. *Botany: Iris* Dr. Edgar Anderson was a botanist who collected vast amounts of data for several species of wild iris (see Data Highlights in Chapter 7). *Iris virginica,* a lovely wildflower spread over most of the American continent and much of Europe, is one of the species Dr. Anderson studied. His friend R. A. Fisher published the data in a paper entitled "The Use of Multiple Measurements in Taxonomic Problems," in *Annals of Eugenics,* 7 (part II):179–188, 1936. For a sample of 50 *Iris virginica,*

(a) the sample mean petal length was $\bar{x} = 5.55$ cm with sample standard deviation $s = 0.57$ cm. Compute an 85% confidence interval for the population mean petal length.

(b) the sample mean petal width was $\bar{x} = 2.03$ cm with sample standard deviation $s = 0.27$ cm. Compute a 90% confidence interval for the population mean petal width.

11. *Sociology: Courtship* A sociologist is studying the length of courtship before marriage in a rural district near Kyoto, Japan. A random sample of 56 middle-income families was interviewed. It was found that the average length of courtship was 3.4 years with sample standard deviation 1.2 years. Find an 85% confidence interval for the length of courtship for the population of all middle-income families in this district.

12. *Sociology: Age at Marriage* A sociologist by the name of N. Keyfitz studied the age at first marriage of women in the providence of Quèbec, Canada. The following is based on information from *American Journal of Sociology*, 53:470–480.
(a) In a large rural district of Quèbec far from any towns, a study of 75 families showed that the approximate sample mean age at which a woman married was $\bar{x} = 19.5$ years with sample standard deviation $s = 2.25$ years. Find a 95% confidence interval for the mean age of marriage for all married women in this district.
(b) In another study, only families from medium-sized and larger cities in Quèbec were used. This study of 89 families showed that the approximate mean age at which a woman married was $\bar{x} = 22.8$ years with sample standard deviation $s = 2.79$ years. Find a 99% confidence interval for the mean age of marriage for all married women in these communities.

13. *Marketing: Motel* Irv and Nancy are thinking about buying the Rockwood Motel located on Interstate 70. Before they make up their mind, they want to estimate the average number of vehicles that go by the motel each day in the summer. Fortunately, the highway department has been counting vehicles on I-70 near the motel. A random sample of 36 summer days shows an average of 16,000 cars per day with a standard deviation of 2400 cars. Find a 90% confidence interval for the mean number of cars per summer day going past the Rockwood Motel.

14. *Marketing: Motel* *The Wall Street Journal* stated that the average cost of a room (hotel or motel) in the Denver area was $111 per night. Suppose that you are a reporter for a local competing paper and you question the statement by *The Wall Street Journal*.
(a) Let us (hypothetically) say that you used the Yellow Pages to get a random sample of 50 hotels and motels in the Denver area. You called to get the price per night for a room. The sample average room rate was $\bar{x} = \$98.75$ with sample standard deviation $s = \$15.91$. Use this information to compute a 95% confidence interval for the mean room rate of all Denver area hotel and motel rooms.
(b) Is *The Wall Street Journal* figure of $111 in your confidence interval of part (a)? Do you suspect that *The Wall Street Journal* figure might be high or low? Explain.

15. *Medical: Vitamins and Cancer* In an article exploring blood serum levels of vitamins and lung cancer risks (*The New England Journal of Medicine*), the mean serum level of vitamin E in the control group was 11.9 mg/liter with standard deviation 4.30 mg/liter. There were 196 patients in the control group. (These patients were free of all cancer, except possibly skin cancer, in the subsequent 8 years.) Using this information, find a 95% confidence interval for the mean serum level of vitamin E in all persons similar to the control group.

16. *Medical: Vitamins and Cancer* In the same article as cited in Problem 15, 99 patients who were cancer-free at the time the blood was drawn later developed lung cancer. For these patients, the mean blood serum level of vitamin E was 10.5 mg/liter with standard deviation 3.2 mg/liter. Using this information, find a 95% confidence interval for the mean serum level of vitamin E in the population of all persons with similar lung cancer risks.

17. *Pro Football Players: Body Weight* How big are professional football players? Consider the following positions: defensive end, defensive guard, defensive tackle, offensive guard, offensive end, offensive tackle. A random sample of professional players in the given positions was taken from the San Diego Chargers, Kansas City Chiefs, Denver Broncos, Oakland Raiders, and Seattle Seahawks. The following list gives body weight in pounds and was obtained from *The Sports Encyclopedia, Pro Football* (11th Edition).

292	290	295	306	310	281	270	278	303	293
275	305	306	291	275	269	290	275	309	292
270	265	260	315	295	285	290	305	260	280
320	280	298	291	266	273	281	274	300	283

(a) Use a calculator with mean and sample standard deviation keys to verify that, for the preceding data, $\bar{x} \approx 287.4$ lb and $s \approx 15.5$ lb.

(b) Let us say that the preceding data are representative of the entire population of professional football players holding the designated positions. Compute an 85% confidence interval for μ, the population mean body weight of all professional football players in the designated positions.

(c) Let us say that you are the head coach for a professional football team. Suppose that a rookie end, tackle, or guard wants to join your team. This person weighs 200 lb. Do you think this weight might be somewhat low compared with other players in one of these positions? Explain by referring to the confidence interval you computed in part (b).

(d) As head coach, suppose that a rookie end, tackle, or guard wants to join your team. This person weighs 291 lb. Do you think this weight might be appropriate for a player in one of these positions? Explain by referring to the confidence interval you computed in part (b).

(e) Repeat parts (b), (c), and (d) for a 95% confidence interval.

18. *Pro Football Players: Age* How old are professional football players? Consider the following positions: defensive end, defensive guard, defensive tackle, offensive guard, offensive end, offensive tackle. A random sample of professional players in the given positions was taken from the San Diego Chargers, Kansas City Chiefs, Denver Broncos, Oakland Raiders, and Seattle Seahawks. The following information gives age in years and was obtained from *The Sports Encyclopedia, Pro Football* (11th Edition).

26	24	25	36	26	32	31	34
32	27	32	23	24	29	30	30
29	33	25	32	24	25	28	22
22	24	23	33	32	31	26	28
32	25	22	28	25	29	25	27

(a) Use a calculator with mean and sample standard deviation keys to verify that, for the preceding data, $\bar{x} \approx 27.8$ years and $s \approx 3.8$ years.

(b) Let us say that the preceding data are representative of the entire population of professional football players holding the designated positions. Compute an 80%

confidence interval for μ, the population mean age of all professional football players in the designated positions.

(c) Let us say that you are the head coach for a professional football team. Suppose that a rookie end, tackle, or guard wants to join your team. This person is 33 years old. Do you think that this age might be somewhat old for a rookie in a starting position? Explain by referring to the confidence interval you computed in part (b).

(d) Repeat parts (b) and (c) for a 99% confidence interval.

8.2
Estimating μ with Small Samples

FOCUS POINTS

✓ Learn about degrees of freedom and Student's *t* distribution.

✓ Find critical values using the degrees of freedom and confidence level.

✓ Compute confidence intervals for μ using small samples. Interpret the results.

Student's *t* distribution

For samples of size 30 or larger, we can approximate the population standard deviation σ by s, the sample standard deviation. Then we can use the central limit theorem to find bounds on the error of estimate and confidence intervals for μ.

There are many practical and important situations, however, where large samples are simply not available. Suppose an archaeologist discovers only seven fossil skeletons from a previously unknown species of miniature horse. Reconstructions of the skeletons of these seven miniature horses show their mean shoulder heights to be $\bar{x} = 46.1$ cm. Let μ be the mean shoulder height for this entire species of miniature horse. How can we find the maximal error of estimate $|\bar{x} - \mu|$? How can we find a confidence interval for μ? We will return to this problem later in this section.

To avoid the error involved in replacing σ by s—that is, approximating σ by s—when dealing with *small samples* ($n < 30$), we introduce a new variable called *Student's t variable*. The *t* variable and its corresponding distribution, called *Student's t distribution*, were discovered in 1908 by W. S. Gosset. He was employed as a statistician by Guinness brewing company, a company that frowned on the publication of research by its employees, so Gosset published his research under the pseudonym Student. Gosset was the first to recognize the importance of developing statistical methods for obtaining reliable information from small samples. It might be more fitting to call this *Gosset's t distribution*; however, in the literature of mathematical statistics it is known as *Student's t distribution*.

The *t* variable is defined by the following formula:

$$t = \frac{\bar{x} - \mu}{\dfrac{s}{\sqrt{n}}} \qquad (11)$$

where $\bar{x}$ is the mean of a random sample of n measurements, μ is the population mean of the x distribution, and s is the sample standard deviation.

◇ **COMMENT** You should note that our *t* variable is just like

$$z = \frac{\bar{x} - \mu}{\dfrac{\sigma}{\sqrt{n}}}$$

except that we replace σ with s. Unlike our methods for large samples, σ cannot be approximated by s when the sample size is less than 30 and we cannot use the normal distribution. However, we will be using the same methods as in Section 8.1 to find the maximal error of estimate and to find confidence intervals, but we use the Student's t distribution.   ◇

If many random samples of size n are drawn, then we get many t values from Equation (11). These t values can be organized into a frequency table, and a histogram can be drawn, thereby giving us an idea of the shape of the t distribution (for a given n).

Fortunately, all this work is not necessary because mathematical theorems can be used to obtain a formula for the t distribution. However, it is important to observe that these theorems say that the shape of the t distribution depends only on n, provided the basic variable x has a normal distribution. So *when we use the t distribution, we will assume that the x distribution is normal.*

Degrees of freedom

Table 6 in Appendix II gives values of the variable t corresponding to what we call the number of *degrees of freedom*, abbreviated *d.f.* For the methods used in this section, the number of degrees of freedom is given by the formula

$$d.f. = n - 1 \tag{12}$$

where *d.f.* stands for the degrees of freedom and n is the sample size being used.

Each choice for *d.f.* gives a different t distribution. However, for *d.f.* larger than about 30, the t distribution and the standard normal z distribution are almost the same.

The graph of a t distribution is always symmetrical about its mean, which (as for the z distribution) is 0. The main observable difference between a t distribution and the standard normal z distribution is that a t distribution has somewhat thicker tails.

Figure 8-5 shows a standard normal z distribution and Student's t distribution with *d.f.* = 3 and *d.f.* = 5.

Table 6 of Appendix II gives various t values for different degrees of freedom *d.f.* We will use this table to find *critical values* t_c for a c confidence level. In other words, we want to find t_c so that an area equal to c under the t distribution for a given number of degrees of freedom falls between $-t_c$ and t_c. In the language of probability, we want to find t_c so that

Using Table 6 to find critical values for confidence intervals

$$P(-t_c < t < t_c) = c$$

This probability corresponds to the area shaded in Figure 8-6.

Table 6 in Appendix II has been arranged so that c is one of the column headings, and the degrees of freedom *d.f.* are the row headings. To find t_c for any specific c, we find the column headed by that c value and read down until we reach the row headed by the appropriate number of degrees of freedom *d.f.* (You will notice two other column headings: α' and α''. We will use these later, but for the time being, ignore them.)

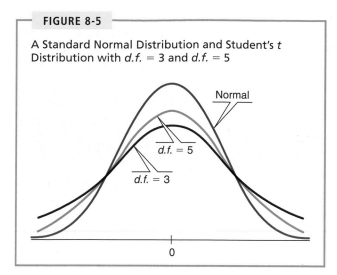

FIGURE 8-5

A Standard Normal Distribution and Student's t Distribution with $d.f. = 3$ and $d.f. = 5$

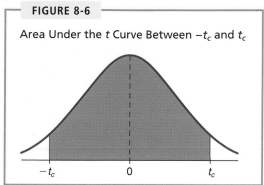

FIGURE 8-6

Area Under the t Curve Between $-t_c$ and t_c

EXAMPLE 3

Student's t distribution

Use Table 8-3 (an excerpt from Table 6 in Appendix II) to find the critical value t_c for a 0.99 confidence level for a t distribution with sample size $n = 5$.

SOLUTION:

(a) First, we find the column with c heading 0.99. This is the last column.

(b) Next, we compute the number of degrees of freedom:
$d.f. = n - 1 = 5 - 1 = 4.$

(c) We read down the column under the heading $c = 0.99$ until we reach the row headed by 4 (under $d.f.$). The entry is 4.604. Therefore, $t_{0.99} = 4.604$. ◇

TABLE 8-3   Student's t Distribution
(Excerpt from Table 6, Appendix II)

c	. . . 0.90	0.95	0.98	0.99
α'	—	—	—	—
α''	—	—	—	—
$d.f.$				
⋮				
3	. . . 2.353	3.182	4.541	5.841
4	. . . 2.132	2.776	3.747	4.604
⋮				
7	. . . 1.895	2.365	2.998	3.499
8	. . . 1.860	2.306	2.896	3.355

Source: Table 6, Appendix II, was generated by Minitab.

GUIDED EXERCISE 3

Student's t distribution table

Use Table 6 of Appendix II to find t_c for a 0.90 confidence level of a t distribution with sample size $n = 9$.

(a) We find the column headed by $c = $ _____. This is the _____ (first, second, third, fourth, fifth, sixth) column.

⟹ $c = 0.90$. This is the fourth column.

(b) The degrees of freedom are given by $d.f. = n - 1 = $ _____.

⟹ $d.f. = n - 1 = 9 - 1 = 8$

(c) Read down the column found in part (a) until you reach the entry in the row headed by $d.f. = 8$. The value of $t_{0.90}$ is _____ for a sample size of 9.

⟹ $t_{0.90} = 1.860$ for a sample size $n = 9$.

(d) Find t_c for a 0.95 confidence level of a t distribution with sample size $n = 9$.

⟹ $t_{0.95} = 2.306$ for a sample of size $n = 9$.

Maximal error of estimate

In Section 8.1, we found bounds $\pm E$ on the error of estimate for a c confidence level. Using the same basic approach, we arrive at the conclusion that

$$E = t_c \frac{s}{\sqrt{n}}$$

is the maximal error of estimate for a c confidence level with small samples (i.e., $|\bar{x} - \mu| < E$ with probability c). The analogue of Equation (1) in Section 8.1 is

$$P\left(-t_c \frac{s}{\sqrt{n}} < \bar{x} - \mu < t_c \frac{s}{\sqrt{n}}\right) = c \tag{13}$$

◇ **COMMENT** Comparing Equation (13) with Equation (1) in Section 8.1, it becomes evident that we are using the same basic method on the t distribution that we did on the z distribution. ◇

Likewise, for small samples from normal populations, Equation (9) of Section 8.1 becomes

$$P(\bar{x} - E < \mu < \bar{x} + E) = c \tag{14}$$

where $E = t_c(s/\sqrt{n})$. Let us organize what we have been doing in a convenient summary.

Confidence interval for μ (small sample)

◇ **SUMMARY** For small samples ($n < 30$) taken from a normal population where σ is unknown, a c confidence interval for the population mean μ is as follows:

> **c Confidence interval for μ (small sample)**
>
> $$\bar{x} - E < \mu < \bar{x} + E \tag{15}$$
>
> where $\bar{x}$ = sample mean
>
> $$E = t_c \frac{s}{\sqrt{n}}$$
>
> c = confidence level ($0 < c < 1$)
>
> t_c = critical value for confidence level c, and degrees of freedom $d.f. = n - 1$ taken from t distribution
>
> n = sample size (small samples, $n < 30$)
>
> s = sample standard deviation

◇ **COMMENT** In our applications of Student's t distribution, we have made the basic assumption that x has a normal distribution. However, the same methods apply even if x is only approximately normal. In fact, the main requirement for using the Student's t distribution is that the distribution of x values be reasonably symmetrical and mound-shaped. If this is the case, then the methods we employ with the t distribution can be considered valid for most practical applications.   ◇

EXAMPLE 4

Confidence interval for μ, small sample

Let's return to our archaeologist and the newly discovered (but extinct) species of miniature horse discussed at the beginning of this section. There are only seven known existing skeletons with shoulder heights (in centimeters) 45.3, 47.1, 44.2, 46.8, 46.5, 45.5, and 47.6. For this sample data, the mean is $\bar{x} = 46.14$ and the sampel standard deviation is $s = 1.19$. Let μ be the mean shoulder height (in centimeters) for this entire species of miniature horse, and assume that the population of shoulder heights is approximately normal.

Find a 99% confidence interval for μ, the mean shoulder height of the entire population of such horses.

SOLUTION: In this case, $n = 7$, so $d.f. = n - 1 = 7 - 1 = 6$. For $c = 0.99$, Table 6 in Appendix II gives $t_{0.99} = 3.707$ (for $d.f. = 6$). The sample standard deviation is $s = 1.19$.

$$E = t_c \frac{s}{\sqrt{n}} = (3.707)\frac{1.19}{\sqrt{7}} = 1.67$$

The 99% confidence interval is

$$\bar{x} - E < \mu < \bar{x} + E$$
$$46.14 - 1.67 < \mu < 46.14 + 1.67$$
$$44.5 < \mu < 47.8$$

◇

GUIDED EXERCISE 4

Confidence interval for μ, small sample

A company has a new process for manufacturing large artificial sapphires. The production of each gem is expensive, so the number available for examination is limited. In a trial run, 12 sapphires are produced. The mean weight for these 12 gems is $\bar{x} = 6.75$ carats, and the sample standard deviation is $s = 0.33$ carat. Let μ be the mean weight for the distribution of all sapphires produced by the new process.

(a) What is *d.f.* for this setting?

⟹ *d.f.* $= n - 1$ where n is the sample size. Since $n = 12$, *d.f.* $= 12 - 1 = 11$.

(b) Use Table 6 in Appendix II to find $t_{0.95}$.

⟹ Using Table 6 with *d.f.* $= 11$ and $c = 0.95$, we find $t_{0.95} = 2.201$.

(c) Find E.

⟹ $E = t_{0.95}\dfrac{s}{\sqrt{n}}$

$\quad = (2.201)\dfrac{0.33}{\sqrt{12}}$

$\quad = 0.21$

(d) Find a 95% confidence interval for μ.

⟹ $\bar{x} - E < \mu < \bar{x} + E$

$6.75 - 0.21 < \mu < 6.75 + 0.21$

$6.54 < \mu < 6.96$

(e) What assumption about the distribution of all sapphires had to be made to obtain these answers?

⟹ The population of artificial sapphire weights is approximately normal.

We have several formulas for confidence intervals for the population mean μ. How do we choose an appropriate one? We need to look at the sample size, the distribution of the original population, and whether or not the population standard deviation σ is known. There are essentially four cases for which we have the tools to find $c\%$ confidence intervals for the mean μ.

Summary: confidence intervals for the mean

Large-Sample Cases

$n \geq 30$

1. If σ is not known, then a $c\%$ confidence interval for μ is

$$\overline{x} - z_c \frac{s}{\sqrt{n}} < \mu < \overline{x} + z_c \frac{s}{\sqrt{n}}$$

2. If σ is known, then a $c\%$ confidence interval for μ is

$$\overline{x} - z_c \frac{\sigma}{\sqrt{n}} < \mu < \overline{x} + z_c \frac{\sigma}{\sqrt{n}}$$

Small-Sample Case

$n < 30$

If the population is approximately normal and σ is not known, then a $c\%$ confidence interval for μ is

$$\overline{x} - t_c \frac{s}{\sqrt{n}} < \mu < \overline{x} + t_c \frac{s}{\sqrt{n}} \qquad \text{Use } d.f. = n - 1.$$

For Any Sample Size

If the population *is normal* and σ is known, then for any sample size (large or small) a $c\%$ confidence interval for μ is

$$\overline{x} - z_c \frac{\sigma}{\sqrt{n}} < \mu < \overline{x} + z_c \frac{\sigma}{\sqrt{n}}$$

 TECH NOTE   The TI-83Plus, Excel, and Minitab support confidence intervals using the Student's t distribution.

TI-83Plus Press the **STAT** key, select **TESTS** and choose option **8:TInterval.** You may use either raw data in a list or summary statistics.

Excel Excel gives only the value of the maximal error of estimate E. You can easily construct the confidence interval by computing $\overline{x} - E$ and $\overline{x} + E$. Use the menu choices **Tools ➤ Data Analysis ➤ Describe Statistics.** In the dialogue box, check summary statistics and check confidence level for mean. Then set the desired confidence level. Under these menu choices, Excel uses the Student's t distribution for confidence intervals regardless of sample size.

Minitab Use the menu choices **Stat ➤ Basic Statistics ➤ 1-Sample t.** In the dialogue box, indicate the column that contains the raw data. The Minitab output shows the confidence interval for Example 4.

```
T Confidence Intervals
Variable    N     Mean      StDev     SE Mean      99.0 % CI
C1          7    46.143     1.190     0.450      (44.475, 47.810)
```

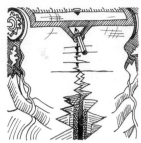

VIEWP⊙INT   *Earthquakes!*

California, Washington, Nevada, and even Yellowstone National Park all have earthquakes. Some earthquakes are severe! All earthquakes bring fear and anxiety to people living near the quake. Is San Francisco due for a really big quake like the 1906 major earthquake? How big are the sizes of recent earthquakes compared with really big earthquakes? What is the duration of an earthquake? How long is the time span between major earthquakes? One way to answer questions such as these is to use existing data to estimate confidence intervals on the average size, duration, and time interval between quakes. Recent data sets for computing such confidence intervals can be found at the National Earthquake Information Service of the U.S. Geological Survey web site. To access the site, visit the Brase/Brase statistics site at http://math.college.hmco.com/students and find the link to National Earthquake Information Service.

SECTION 8.2   PROBLEMS

In all the following problems, assume that the population of x values has an approximately normal distribution. Answers may vary slightly due to rounding.

1. Use Table 6 in Appendix II to find t_c for a 0.95 confidence level when the sample size is 18.

2. Use Table 6 in Appendix II to find t_c for a 0.99 confidence level when the sample size is 4.

3. Use Table 6 in Appendix II to find t_c for a 0.90 confidence level when the sample size is 22.

4. Use Table 6 in Appendix II to find the value of t_c for a 0.95 confidence level when the sample size is 12.

5. *Archaeology: Tree Rings* At Burnt Mesa Pueblo, the method of tree ring dating gave the following dates A.D. for an archaeological excavation site (*Bandelier Archaeological Excavation Project: Summer 1990 Excavations at Burnt Mesa Pueblo*, edited by Kohler, Washington State University):

   | 1189 | 1271 | 1267 | 1272 | 1268 | 1316 | 1275 | 1317 | 1275 |

   (a) Use a calculator with mean and standard deviation keys to verify that the sample mean date is $\bar{x} = 1272$ with sample standard deviation $s = 37$ years.
   (b) Find a 90% confidence interval for the mean of all tree ring dates from this archaeological site.

6. *Toys: Battery Life Consumer Reports* gave the following information about the life (hours) of size AA batteries in toys:

   | 2.3 | 5.1 | 5.4 | 4.2 | 6.1 | 5.7 | 5.5 | 1.3 |
   | 1.5 | 2.5 | 1.6 | 5.3 | 1.8 | 1.9 | 5.2 | 1.8 |

(a) Use a calculator with mean and standard deviation keys to verify that the sample mean of the data is $\bar{x} = 3.58$ hours with sample standard deviation $s = 1.85$ hours.

(b) Find a 95% confidence interval for the mean life μ hours for all brand-name AA batteries used in toys.

7. *Camping: Cost of a Tent* Just how expensive is camping equipment? How about a large-sized tent? A random sample of large tents listed in *Consumer Reports: Special Outdoor Issue* (Vol. 62, No. 6) gave the following prices ($). The tents listed include brand names such as L.L. Bean, Sears, Coleman, and Wal-Mart.

| 115 | 140 | 80 | 150 | 250 | 230 |
| 110 | 135 | 110 | 210 | 120 | 130 |

(a) Use a calculator with mean and sample standard deviation keys to verify that $\bar{x} \approx \$148.33$ and $s \approx 53.02$.

(b) Using the given data as representative of the population of all prices of large-sized tents, find a 90% confidence interval for the mean price μ of all such tents.

8. *Camping: Cost of a Sleeping Bag* How much does a sleeping bag cost? Let's say you want a sleeping bag that should keep you warm in temperatures from 20 to 45°F. A random sample of prices ($) for sleeping bags in this temperature range was taken from *Backpacker Magazine: Gear Guide* (Vol. 25, Issue 157, No. 2). Brand names include American Camper, Cabela's, Camp 7, Caribou, Cascade, and Coleman.

| 80 | 90 | 100 | 120 | 75 | 37 | 30 | 23 | 100 | 110 |
| 105 | 95 | 105 | 60 | 110 | 120 | 95 | 90 | 60 | 70 |

(a) Use a calculator with mean and sample standard deviation keys to verify that $\bar{x} \approx \$83.75$ and $s \approx \$28.97$.

(b) Using the given data as representative of the population of all prices of summer sleeping bags, find a 90% confidence interval for the mean price μ of all summer sleeping bags.

9. *Wildlife: Mountain Lions* How much do wild mountain lions weigh? *The 77th Annual Report of the New Mexico Department of Game and Fish*, edited by Bill Montoya, gave the following information. Adult wild mountain lions (18 months or older) captured and released for the first time in the San Andres Mountains gave the following weights (lb):

| 68 | 104 | 128 | 122 | 60 | 64 |

(a) Use a calculator with mean and sample standard deviation keys to verify that $\bar{x} \approx 91.0$ lb and $s \approx 30.7$ lb.

(b) Find a 75% confidence interval for the population average weight μ of all adult mountain lions in the specified region.

10. *Wildlife: Wolf Pups* The number of pups in wolf dens of the southwestern United States is recorded below for 16 wolf dens (*The Wolf in the Southwest: The Making of an Endangered Species*, edited by D. E. Brown, University of Arizona Press).

| 5 | 8 | 7 | 5 | 3 | 4 | 3 | 9 |
| 5 | 8 | 5 | 6 | 5 | 6 | 4 | 7 |

(a) Use a calculator with mean and standard deviation keys to verify that the sample mean is $\bar{x} = 5.63$ pups with sample standard deviation $s = 1.78$ pups.

(b) Compute an 85% confidence interval for the population mean number of wolf pups per den in the southwestern United States.

11. *Wildlife: Fawns* The shoulder height for a random sample of six fawns (less than 5 months old) in Mesa Verde National Park was $\bar{x} = 79.25$ cm with sample standard deviation $s = 5.33$ cm (*The Mule Deer of Mesa Verde National Park,* edited by G. W. Mierau and J. L. Schmidt, Mesa Verde Museum Association). Compute an 80% confidence interval for the mean shoulder height of the population of all fawns (less than 5 months old) in Mesa Verde National Park.

12. *French Fries: Calories* How many calories are there in 3 ounces of french fries? It depends on where you get them. *Good Cholesterol Bad Cholesterol,* by Roth and Streicher, gives the data from eight popular fast-food restaurants. The data are (in calories)

| 222 | 255 | 254 | 230 | 249 | 222 | 237 | 287 |

Use these data to create a 99% confidence interval for the mean calorie count in 3 ounces of french fries obtained from fast-food restaurants.

13. *Restaurant: Tips* André is head waiter at a famous gourmet restaurant in San Francisco. The Internal Revenue Service is doing an audit of his tax return this year. In particular, the IRS wants to know the average amount André gets for a tip. In an effort to satisfy the IRS, André took a random sample of eight credit card receipts, each of which indicated his tip. The results were

| $10.00 | $12.75 | $11.93 | $11.15 | $15.70 | $14.50 | $9.10 | $13.65 |

(a) Use a calculator to verify that the sample mean is $12.35 and the sample standard deviation is $2.25.

(b) Find a 90% confidence interval for the population mean of tips received by André.

14. *Franchise: Candy Store* Do you want to own your own candy store? Wow! With some interest in running your own business and a decent credit rating, you can probably get a bank loan on startup costs for franchises such as Candy Express, The Fudge Company, Karmel Corn, and Rocky Mountain Chocolate Factory. Startup costs (units in $1000) for a random sample of candy stores are given below (Source: *Entrepreneur Magazine,* Vol. 23, No. 10).

| 95 | 173 | 129 | 95 | 75 | 94 | 116 | 100 | 85 |

Use a calculator with mean and sample standard deviation keys to verify that $\bar{x} \approx 106.9$ thousand dollars and $s \approx 29.4$ thousand dollars. Find a 90% confidence interval for the population average startup costs μ for candy store franchises.

15. *Lincoln County Jail: Billy the Kid* Suppose that you are an anthropologist/historian doing research at an excavation site. How deep do you need to dig to locate artifacts of historical significance? The answer may depend on how long ago the events you hope to study occurred. If the events are relatively recent, you may not need to dig too deep. In 1881, the Lincoln County Cattle Baron Wars were in progress in the New Mexico Territory, and one special outlaw, Billy the Kid, killed two deputies in his escape from the Lincoln County Jail. In 1985, the anthropologist/historian Yvonne Oakes was commissioned by the Museum of New Mexico to study the now famous Lincoln Country Courthouse and Jail area (Source: *Archaeological Testing*

at Three Historic Sites at Lincoln State Monument, by Y. R. Oakes, Museum of New Mexico).

(a) The depths (in inches) below grade at which significant artifacts were found in a random sample of trenches are

10	10.5	14	14	4	4.5	5	12	12	16
9	6	8	9	10	9	11	11	12	

Use a calculator with mean and sample standard deviation keys to verify that $\bar{x} \approx 9.8$ inches and $s \approx 3.3$ inches. Compute a 90% confidence interval for the population mean depth μ at which significant artifacts will be found.

(b) The depth (in inches) to sterile ground (no more artifacts found) for seven excavation trenches are

17	19	21.5	16	14	16	16

Use a calculator with mean and sample standard deviation keys to verify that $\bar{x} \approx 17.1$ inches and $s \approx 2.5$ inches. Compute an 80% confidence interval for the population mean depth μ to sterile ground.

16. *College Tuition: Professors' Salaries* How much has college tuition gone up? How much have professors' salaries gone up? The National Education Association's *Almanac of Higher Education,* published by the NEA, gave the following information.

(a) A random sample of 12 East Coast colleges and universities (public and private) gave the following percentage tuition increases for a 2-year period:

3.9	2.3	3.7	9.4	4.8	3.9
6.7	6.6	4.6	4.3	3.1	3.1

Use a calculator with mean and sample standard deviation keys to verify that $\bar{x} \approx 4.70\%$ and $s \approx 1.98\%$. Compute a 90% confidence interval for the population mean percentage tuition increase μ of all East Coast colleges and universities.

(b) A random sample of 14 East Coast colleges and universities gave the following percentage increases in salaries for professors during the same 2-year period:

2.8	4.3	2.4	3.4	5.5	3.8	3.4
5.1	2.4	3.0	2.9	0.4	1.4	3.7

Use a calculator with mean and sample standard deviation keys to verify that $\bar{x} \approx 3.18\%$ and $s \approx 1.34\%$. Compute a 90% confidence interval for the population mean percentage salary increase μ of all East Coast college and university professors.

17. *Jobs: Annual Pay* What is the average annual pay for different jobs? The following problem is based on information taken from *State and Metropolitan Area Data Book,* 4th Edition (U.S. Department of Commerce).

(a) What if you have a factory job (manufacturing)? The following data are inflation adjusted and are from a random sample of reports from western and midwestern states. (Annual pay in units of $1000.)

37.0	34.6	36.1	44.6	32.3	24.8	27.1	28.5

Use a calculator with mean and sample standard deviation keys to verify that $\bar{x} \approx 33.1$ thousand dollars and $s \approx 6.4$ thousand dollars. Compute an 85% confidence interval for the population average μ of such income.

(b) What if you are a salesperson (retail trade)? The following data are inflation adjusted and are from a random sample of reports from eastern states. (Annual pay in units of $1000.)

| 21.8 | 19.1 | 17.3 | 18.6 | 24.6 | 22.5 | 23.8 | 18.9 | 20.8 |

Use a calculator with mean and sample standard deviation keys to verify that $\bar{x} \approx 20.8$ thousand dollars and $s \approx 2.5$ thousand dollars. Compute a 99% confidence interval for the population average μ of such income.

18. *Hospitals: Length of Stay* How long do patients stay in a hospital? What percentage of hospitals provide at least some charity care? The following problem is based on information taken from *State Health Care Data: Utilization, Spending, and Characteristics* (American Medical Association).

(a) How long do patients stay? Based on a random sample of hospital reports from eastern states, the following information was obtained (units in days):

| 7.2 | 7.9 | 7.0 | 6.9 | 9.9 | 7.5 | 6.6 | 7.3 |
| 7.3 | 6.8 | 7.1 | 6.8 | 7.3 | 6.9 | 6.8 |

Use a calculator with mean and sample standard deviation keys to verify that $\bar{x} \approx 7.3$ days and $s \approx 0.8$ days. Find a 99% confidence interval for the population average number of days μ patients spend in the hospital.

(b) What percentage of the hospitals provide (at least some) charity care? Based on a random sample of hospital reports from eastern states, the following information was obtained (units in percentage of hospitals providing at least some charity care):

| 57.1 | 56.2 | 53.0 | 66.1 | 59.0 | 64.7 | 70.1 | 64.7 | 53.5 | 78.2 |

Use a calculator with mean and sample standard deviation keys to verify that $\bar{x} \approx 62.3\%$ and $s \approx 8.0\%$. Find a 90% confidence interval for the population average μ of the percentage of hospitals providing at least some charity care.

19. *Vital Statistics: Heights* The distribution of the heights of 18-year-old men in the United States is approximately normal with mean 68 inches and standard deviation 3 inches (U.S. Census Bureau). In Minitab we can simulate the drawing of random samples of size 20 from this population (➤ Calc ➤ Random Data ➤ Normal with 20 rows from a distribution with mean 68 and standard deviation 3). Then we can have Minitab compute a 95% confidence interval and draw a boxplot of the data (➤ Stat ➤ Basic Statistics ➤ 1—Sample t, with boxplot selected in the graphs). The boxplots and confidence intervals for four different samples are shown in the accompanying figures. The four confidence intervals are

VARIABLE	N	MEAN	STDEV	SEMEAN	95.0 % CI
Sample 1	20	68.050	2.901	0.649	(66.692 , 69.407)
Sample 2	20	67.958	3.137	0.702	(66.490 , 69.426)
Sample 3	20	67.976	2.639	0.590	(66.741 , 69.211)
Sample 4	20	66.908	2.440	0.546	(65.766 , 68.050)

(a) Examine the figure [parts (a) to (d)]. How do the boxplots for the four samples differ? Why should you expect the boxplots to differ?

95% Confidence Intervals
for Mean Height of
18-Year-Old Men
(Sample size 20)

(a) **Boxplot of Sample 1**
(with 95% t-confidence interval for the mean)

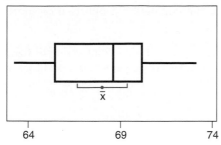

(b) **Boxplot of Sample 2**
(with 95% t-confidence interval for the mean)

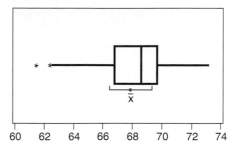

(c) **Boxplot of Sample 3**
(with 95% t-confidence interval for the mean)

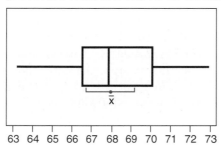

(d) **Boxplot of Sample 4**
(with 95% t-confidence interval for the mean)

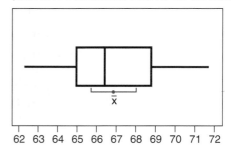

(b) Examine the 95% confidence intervals for the four samples shown in the above printout. Do the intervals differ in length? Do the intervals all contain the expected population mean of 68 inches? If we draw more samples, do you expect all of the resulting 95% confidence intervals to contain $\mu = 68$? Why or why not?

8.3
Estimating p in the Binomial Distribution

FOCUS POINTS

✓ Compute the maximal error tolerance for proportions using a given level of confidence.

✓ Compute confidence intervals for p and interpret the results.

✓ Interpret poll results and the margin of error.

The binomial distribution is completely determined by the number of trials n and the probability p of success in a single trial. For most experiments, the number of trials is chosen in advance. Then the distribution is completely determined by p. In this section, we will consider the problem of estimating p under the assumption that n has already been selected.

Again, we are employing what are called *large-sample methods*. We will assume that the normal curve is a good approximation to the binomial distribution, and when necessary, we will use sample estimates for the standard deviation. Empirical studies have shown that these methods are quite good provided that *both*

$$np > 5 \quad \text{and} \quad nq > 5 \quad \text{where } q = 1 - p$$

Basic criteria

Let r be the number of successes out of n trials in a binomial experiment. We will take the sample proportion of successes $\hat{p}$ (read "p hat") $= r/n$ as our *point estimate for p*, the population proportion of successes.

Point estimate for p

$$\hat{p} = \frac{r}{n}$$

Point estimate for q

$$\hat{q} = 1 - \hat{p}$$

For example, suppose that 800 students are selected at random from a student body of 20,000 and that they are each given shots to prevent a certain type of flu. These 800 students are then exposed to the flu, and 600 of them do not get the flu. What is the probability p that the shot will be successful for any single student selected at random from the entire population of 20,000 students? We estimate p for the entire student body by computing r/n from the sample of 800 students. The value $\hat{p} = r/n$ is 600/800, or 0.75. The value $\hat{p} = 0.75$ is then the *point estimate* for p.

Error of estimate

The difference between the actual value of p and the estimate $\hat{p}$ is the size of our error caused by using $\hat{p}$ as a point estimate for p. The magnitude of $\hat{p} - p$ is called the *error of estimate* for using $\hat{p} = r/n$ as a point estimate for p. In absolute value notation, the error of estimate is $|\hat{p} - p|$.

To compute the bounds for the error of estimate, we need some information about the distribution of $\hat{p} = r/n$ values for different samples of the same size n. It turns out that, for large samples, the distribution of $\hat{p}$ values is well approximated by a *normal curve* with

mean $\mu = p$ and standard error $\sigma = \sqrt{pq/n}$

Since the distribution of $\hat{p} = r/n$ is approximately normal, we use features of the standard normal distribution to find the bounds for the difference $\hat{p} - p$. Recall that z_c is the number such that an area equal to c under the standard normal curve falls between $-z_c$ and z_c. Then, in terms of the language of probability,

$$P\left(-z_c\sqrt{\frac{pq}{n}} < \hat{p} - p < z_c\sqrt{\frac{pq}{n}}\right) = c \tag{16}$$

Equation (16) says that the chance is c that the numerical difference between $\hat{p}$ and p is between $-z_c\sqrt{pq/n}$ and $z_c\sqrt{pq/n}$. With the c confidence level, our estimate $\hat{p}$ differs from p by no more than

$$E = z_c\sqrt{pq/n}$$

As in Section 8.1, we call E the *maximal error tolerance* of the error of estimate $|\hat{p} - p|$ for a confidence level c.

Optional derivation of Equation (16)

First, we need to show that $\hat{p} = r/n$ has a distribution that is approximately normal with $\mu = p$ and $\sigma = \sqrt{pq/n}$. From Section 6.4, we know that, for sufficiently large *n*, the binomial distribution can be approximated by a normal distribution with mean $\mu = np$ and standard deviation $\sigma = \sqrt{npq}$. If *r* is the number of successes out of *n* trials of a binomial experiment, then *r* is a binomial random variable with a binomial distribution. When we convert *r* to standard *z* units, we obtain

$$z = \frac{r - \mu}{\sigma} = \frac{r - np}{\sqrt{npq}}$$

For sufficiently large *n*, *r* will be approximately normally distributed, so *z* will be too.

If we divide both numerator and denominator of the last expression by *n*, the value of *z* will not change.

$$z = \frac{\dfrac{r - np}{n}}{\dfrac{\sqrt{npq}}{n}} \quad \text{simplified, we find} \quad z = \frac{\dfrac{r}{n} - p}{\sqrt{\dfrac{pq}{n}}} \tag{17}$$

Equation (17) tells us that the $\hat{p} = r/n$ distribution is approximated by a normal curve with $\mu = p$ and $\sigma = \sqrt{pq/n}$.

The probability is *c* that *z* lies in the interval between $-z_c$ and z_c because an area equal to *c* under the standard normal curve lies between $-z_c$ and z_c. Using the language of probability, we write

$$P(-z_c < z < z_c) = c$$

From Equation (17), we know that

$$z = \frac{\hat{p} - p}{\sqrt{\dfrac{pq}{n}}}$$

If we put this expression for *z* into the preceding equation, we obtain

$$P\left(-z_c < \frac{\hat{p} - p}{\sqrt{\dfrac{pq}{n}}} < z_c\right) = c$$

If we multiply all parts of the inequality by $\sqrt{pq/n}$, we obtain the equivalent statement

$$P\left(-z_c\sqrt{\frac{pq}{n}} < \hat{p} - p < z_c\sqrt{\frac{pq}{n}}\right) = c \tag{16}$$

Confidence interval for *p*

To find a *c* confidence interval for *p*, we will use *E* in place of the expression $z_c\sqrt{pq/n}$ in Equation (16). Then we get

$$P(-E < \hat{p} - p < E) = c \tag{18}$$

Some algebraic manipulation produces the mathematically equivalent statement

$$P(\hat{p} - E < p < \hat{p} + E) = c \tag{19}$$

Equation (19) says that the probability is c that p lies in the interval from $\hat{p} - E$ to $\hat{p} + E$. Therefore, the interval from $\hat{p} - E$ to $\hat{p} + E$ is the c confidence interval for p that we wanted to find.

There is one technical difficulty in computing the c confidence interval for p. The expression $E = z_c\sqrt{pq/n}$ requires that we know the values of p and q. In most situations, we will not know the actual values of p or q, so we will use our point estimates

$$p \approx \hat{p} \qquad \text{and} \qquad q = 1 - p \approx 1 - \hat{p}$$

to estimate E. These estimates are safe for most practical purposes, since we are dealing with large-sample theory ($np > 5$ and $nq > 5$).

For convenient reference, we'll summarize the information about c confidence intervals for p, the probability of success in a binomial distribution.

◇ **SUMMARY** Consider a binomial distribution where n = number of trials, r = number of successes out of the n trials, p = probability of success on each trial, and q = probability of failure on each trial.

If n, p, and q are such that

$$np > 5 \qquad \text{and} \qquad nq > 5$$

then a c confidence interval for p is

$$\hat{p} - E < p < \hat{p} + E$$

where $\hat{p} = \dfrac{r}{n}$

$$E \approx z_c\sqrt{\frac{\hat{p}(1 - \hat{p})}{n}}$$

z_c = critical value for confidence level c taken from a normal distribution (see Table 8-2).

EXAMPLE 5

Confidence interval for p

Let's return to our flu shot experiment described at the beginning of this section. Suppose that 800 students were selected at random from a student body of 20,000 and given shots to prevent a certain type of flu. All 800 students were exposed to the flu, and 600 of them did not get the flu. Let p represent the probability that the shot will be successful for any single student selected at random from the entire population of 20,000. Let q be the probability that the shot is not successful.

(a) What is the number of trials n? What is the value of r?

SOLUTION: Since each of the 800 students receiving the shot may be thought of as a trial, then $n = 800$, and $r = 600$ is the number of successful trials.

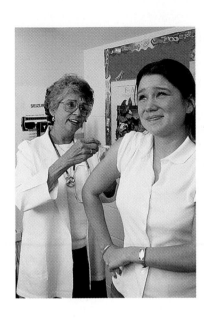

(b) What are the point estimates for p and q?

SOLUTION: We estimate p by the sample point estimate

$$\hat{p} = \frac{r}{n} = \frac{600}{800} = 0.75$$

We estimate q by

$$\hat{q} = 1 - \hat{p} = 1 - 0.75 = 0.25$$

(c) Would it seem that the number of trials is large enough to justify a normal approximation to the binomial?

SOLUTION: Since $n = 800$, $p \approx 0.75$, and $q \approx 0.25$, then

$$np \approx (800)(0.75) = 600 > 5 \qquad \text{and} \qquad nq \approx (800)(0.25) = 200 > 5$$

A normal approximation is certainly justified.

(d) Find a 99% confidence interval for p.

SOLUTION:

$$z_{0.99} = 2.58 \text{ (see Table 8-2)}$$

$$E \approx z_{0.99}\sqrt{\frac{\hat{p}(1-\hat{p})}{n}}$$

$$\approx 2.58\sqrt{\frac{(0.75)(0.25)}{800}}$$

$$\approx 0.0395$$

The 99% confidence interval is then

$$\hat{p} - E < p < \hat{p} + E$$
$$0.75 - 0.0395 < p < 0.75 + 0.0395$$
$$0.71 < p < 0.79$$

◇

Confidence interval for p

A random sample of 188 books purchased at a local bookstore showed that 66 of the books were murder mysteries. Let p represent the proportion of books sold by this store that are murder mysteries.

(a) What is a point estimate for p?

$$\Longrightarrow \quad \hat{p} = \frac{r}{n} = \frac{66}{188} = 0.35$$

Continued

GUIDED EXERCISE 5 continued

(b) Find a 90% confidence interval for p.

$$E = z_c \sqrt{\frac{\hat{p}(1 - \hat{p})}{n}}$$

$$= 1.645 \sqrt{\frac{(0.35)(1 - 0.35)}{188}} = 0.0572$$

The confidence interval is

$$\hat{p} - E < p < \hat{p} + E$$
$$0.35 - 0.0572 < p < 0.35 + 0.0572$$
$$0.29 < p < 0.41$$

(c) What is the meaning of the confidence interval you just computed?

If we had computed the interval for many different sets of 188 books, we would find that about 90% of the intervals actually contained p, the population proportion of mysteries. Consequently, we can be 90% confident that our interval is one of the ones that contains the unknown value p.

(d) To compute the confidence interval, we used a normal approximation. Does this seem justified?

$n = 188$;

$p \approx 0.35; q \approx 0.65$

Since $np \approx 65.8 > 5$ and $nq \approx 122.2 > 5$, the approximation is justified.

It is interesting to note that our sample point estimate $\hat{p} = r/n$ and the confidence interval for the population proportion p do not depend on the size of the population. In our bookstore example, it made no difference how many books the store sold. On the other hand, the size of the sample does affect the accuracy of a statistical estimate. In the next section, we will study the effect of sample size on the reliability of our estimate.

 TECH NOTE   Both the TI-83Plus and Minitab provide confidence intervals for proportions.

TI-83Plus Press the **STAT** key, select **TESTS,** and choose option **A:1-PropZInt.** The letter x represents the number of successes r. The TI-83Plus output shows the results for Guided Exercise 5.

```
1-PropZInt
(.29381,.40832)
p̂=.3510638298
n=188
```

Minitab Use menu selections **Stat ➤ Basic Statistics ➤ 1 Proportion.** In the dialogue box, select summarized data and fill in the number of trials and the number of successes. Under Options, select a confidence interval. Minitab uses the binomial distribution directly unless normal is checked. The Minitab output shows the results for Guided Exercise 5. Information from Chapter 9 material is also shown.

```
Test and Confidence Interval for One Proportion (Using Binomial)
Test of p = 0.5 vs p not = 0.5
                                                                Exact
Sample    X      N      Sample p         90.0 % CI            P-Value
  1      66    188     0.351064    (0.293222, 0.412466)       0.000
```

```
Test and Confidence Interval for One Proportion (Using Normal)
Test of p = 0.5 vs p not = 0.5
Sample   X    N    Sample p        90.0 % CI      Z-Value   P-Value
  1     66  188   0.351064    (0.293805, 0.408323)  -4.08    0.000
```

Margin of error

Newspapers frequently report the results of an opinion poll. In articles that give more information, a statement about the margin of error accompanies the poll results. The *margin of error* is the maximal error of estimate E for a confidence interval. Usually, a 95% confidence interval is assumed.

General interpretation of poll results

1. When a poll states the results of a survey, the proportion reported to respond in the designated manner is $\hat{p}$, the sample estimate of the population proportion.

2. The *margin of error* is the maximal error E of a 95% confidence interval for p.

3. A 95% confidence interval for the population proportion p is

   poll report $\hat{p}$ − margin of error $E < p <$ poll report $\hat{p}$ + margin of error E.

◊ **COMMENT:** Leslie Kish, a statistician at the University of Michigan, was the first to apply the term *margin of error*. He was a pioneer in the study of population sampling techniques. His book *Survey Sampling* is still widely used all around the world.   ◊

Some articles clarify the meaning of the margin of error further by saying that it is an error due to sampling. For instance, the following comments accompany results of a political poll reported in the October 29, 1993, issue of *The Wall Street Journal.*

How Poll Was Conducted

The Wall Street Journal/NBC news poll was based on nationwide telephone interviews of 1508 adults conducted last Friday through Tuesday by the polling organizations of Peter Hart and Robert Teeter.

The sample was drawn from 315 randomly selected geographic points in the continental U.S. Each region was represented in proportion to its population. Households were selected by a method that gave all telephone numbers, ... an equal chance of being included.

One adult, 18 years or older, was selected from each household by a procedure to provide the correct number of male and female respondents.

Chances are 19 of 20 that if all adults with telephones in the U.S. had been surveyed, the findings would differ from these poll results by no more than 2.6 percentage points in either direction.

GUIDED EXERCISE 6

Margin of error

Read the last paragraph of the article, "How Poll Was Conducted."

(a) What confidence level corresponds to the phrase "chances are 19 of 20 that if ..."

⟹ $\frac{19}{20} = 0.95$

A 95% confidence interval is being discussed.

(b) The article indicates that everyone in the sample was asked the question, "Which party, the Democratic Party or the Republican Party, do you think would do a better job handling ... education?" Possible responses were Democrats, neither, both, or Republicans. The poll reported that 32% of the respondents said "Democrats." Does 32% represent the sample statistic $\hat{p}$ or the population parameter p for the proportion of adults responding "Democrat"?

⟹ 32% represents a sample statistic $\hat{p}$ because 32% represents the percentage of the adults in the *sample* who responded "Democrats."

(c) Continue reading the last paragraph of the article. It goes on to state, "... if all adults with telephones in the U.S. had been surveyed, the findings would differ from these poll results by no more than 2.6 percentage points in either direction." Use this information together with parts (a) and (b) to find a 95% confidence interval for the proportion p of the specified population who would respond "Democrat" to the question.

⟹ The value 2.6 percentage points represents the margin of error. Since the margin of error is equivalent to E, the maximal error of estimate for a 95% confidence interval, the confidence interval is

$$32\% - 2.6\% < p < 32\% + 2.6\%$$
$$29.4\% < p < 34.6\%$$

The poll indicates that at the time of the poll, between 29.4% and 34.6% of the specified population think Democrats would do a better job handling education.

VIEWPOINT   *"Band-Aid Surgery"*

Faster recovery time and less pain! Sounds great. An alternate surgical technique called *laparoscopic* ("Band-Aid") *surgery* involves small incisions in which tiny video cameras and long surgical instruments are maneuvered. Instead of a 10-inch incision, surgeons might use four little stabs of about $\frac{1}{2}$-inch in length. However, not every such surgery is successful. An article in the Health Section of *The Wall Street Journal* recommends using a surgeon who has done at least 50 such surgeries. Then the prospective patient should ask about the *rate of conversion*, that is, the proportion p of times the surgeon has been forced by complications to switch in midoperation to conventional surgery. A confidence interval for the proportion p would be useful patient information!

SECTION 8.3   PROBLEMS

For all these problems, carry at least four digits after the decimal in your calculations. Answers may vary slightly due to rounding.

1. *Myers-Briggs: Actors* Isabel Myers was a pioneer in the study of personality types. The following information is taken from *A Guide to the Development and Use of the Myers-Briggs Type Indicator,* by Myers and McCaulley (Consulting Psychologists Press). In a random sample of 62 professional actors, it was found that 39 were extroverts.
   (a) Let p represent the proportion of all actors who are extroverts. Find a point estimate for p.
   (b) Find a 95% confidence interval for p. Give a brief interpretation of the meaning of the confidence interval you have found.
   (c) Do you think the conditions $np > 5$ and $nq > 5$ are satisfied in this problem? Explain why this would be an important consideration.

2. *Myers-Briggs: Judges* In a random sample of 519 judges, it was found that 285 were introverts (see reference of Problem 1).
   (a) Let p represent the proportion of all judges who are introverts. Find a point estimate for p.
   (b) Find a 99% confidence interval for p. Give a brief interpretation of the meaning of the confidence interval you have found.
   (c) Do you think the conditions $np > 5$ and $nq > 5$ are satisfied in this problem? Explain why this would be an important consideration.

3. *Navajo Lifestyle: Traditional Hogans* A random sample of 5222 permanent dwellings on the entire Navajo Indian Reservation showed that 1619 were traditional Navajo hogans (*Navajo Architecture: Forms, History, Distributions,* by Jett and Spencer, University of Arizona Press).
   (a) Let p be the proportion of all permanent dwellings on the entire Navajo Reservation that are traditional hogans. Find a point estimate for p.
   (b) Find a 99% confidence interval for p. Give a brief interpretation of the confidence interval.
   (c) Do you think that $np > 5$ and $nq > 5$ are satisfied for this problem? Explain why this would be an important consideration.

4. *Archaeology: Pottery* Santa Fe black on white is a type of pottery commonly found at archaeological excavations in Bandelier National Monument. At one excavation site a sample of 592 potsherds was found, of which 360 were identified as Santa Fe black on white (*Bandelier Archaeological Excavation Project: Summer 1990 Excavations at Burnt Mesa Pueblo and Casa del Rito,* edited by Kohler and Root, Washington State University).

   (a) Let p represent the population proportion of Santa Fe black on white potsherds at the excavation site. Find a point estimate for p.

   (b) Find a 95% confidence interval for p. Give a brief statement of the meaning of the confidence interval.

   (c) Do you think that the conditions $np > 5$ and $nq > 5$ are satisfied in this problem? Why would this be important?

5. *Health Care: Colorado Physicians* A random sample of 5792 physicians in Colorado showed that 3139 provided at least some charity care (i.e., treated poor people at no cost). These data are based on information from *State Health Care Data: Utilization, Spending, and Characteristics* (American Medical Association).

   (a) Let p represent the proportion of all Colorado physicians who provide some charity care. Find a point estimate for p.

   (b) Find a 99% confidence interval for p. Give a brief explanation of the meaning of your answer in the context of this problem.

   (c) Is the normal approximation to the binomial justified in this problem? Explain.

6. *Supermarkets: Broken Eggs* The manager of the dairy section of a large supermarket took a random sample of 250 egg cartons and found that 40 cartons had at least one broken egg.

   (a) Let p be the proportion of egg cartons with at least one broken egg out of the population of all egg cartons stocked by this store. Find a point estimate for p.

   (b) Find a 90% confidence interval for p. Give a brief interpretation of the meaning of your confidence interval.

   (c) Do you think that the conditions $np > 5$ and $nq > 5$ will be satisfied? Why is this important?

7. *Law Enforcement: Arrests* "Fugitive Task Force Runs 99 Photos in (Colorado) Springs Newspaper: Police Make 17 Arrests." This was a headline in the *Rocky Mountain News.*

   (a) Let p represent the population proportion of fugitives arrested after their photos are displayed in the newspaper. Find a point estimate for p.

   (b) Find an 85% confidence interval for p. Give a brief statement of the meaning of the confidence interval.

   (c) Do you think that the conditions $np > 5$ and $nq > 5$ are satisfied in this problem? Why would this be important?

8. *Law Enforcement: Escaped Convicts* Case studies showed that out of 10,351 convicts who escaped from U.S. prisons, only 7867 were recaptured (*The Book of Odds,* by Shook and Shook, Signet).

   (a) Let p represent the proportion of all escaped convicts who will eventually be recaptured. Find a point estimate for p.

   (b) Find a 99% confidence interval for p. Give a brief statement of the meaning of the confidence interval.

   (c) Is use of the normal approximation to the binomial justified in this problem? Explain.

9. *Fishing: Barbless Hooks* In a combined study of northern pike, cutthroat trout, rainbow trout, and lake trout, it was found that 26 out of 855 fish died when caught and released using barbless hooks on flies or lures. All hooks were removed from

the fish. (Source: *A National Symposium on Catch and Release Fishing*, Humboldt State University Press.)

(a) Let *p* represent the proportion of all pike and trout that die (i.e., *p* is the mortality rate) when caught and released using barbless hooks. Find a point estimate for *p*.

(b) Find a 99% confidence interval for *p*, and give a brief explanation of the meaning of the interval.

(c) Is the normal approximation of the binomial justified in this problem? Explain.

10. *Fishing: Barbed Hooks* A study of 200 rainbow trout caught on baited size 8 barbed hooks and released with the line cut at the hook (but the hook not removed from the fish) showed that 58 fish died. See reference in Problem 9.

(a) Let *p* represent the proportion of all rainbow trout that die when caught and released using the described method. Find a point estimate for *p*.

(b) Find a 95% confidence interval for *p*, and give a brief explanation of the meaning of the interval.

(c) Is the normal approximation of the binomial justified in this problem? Explain.

11. *Montana Open Range: Cattle* The U.S. Department of the Interior is checking cattle on the Windgate open range in Montana. A random sample of 900 cattle shows that 54 are undernourished.

(a) Let *p* represent the proportion of undernourished cattle on the Windgate range. Find a point estimate for *p*.

(b) Find a 0.99 confidence interval for *p*.

12. *Sports Events: Attendance* Attending sporting events is a popular source of entertainment. When 1000 people were surveyed, 590 said that getting together with friends was an important reason for attending a sporting event (*USA Today*).

(a) Let *p* be the proportion of all people attending a sporting event to be with friends. Find a point estimate for *p*.

(b) Find a 99% confidence interval for *p*.

13. *AM Radio: Audience* The Roper Organization conducted a poll of 2000 adults and found that 382 regularly listen to an AM radio station at home (*USA Today*).

(a) Let *p* represent the proportion of the adult population that regularly listens to an AM radio station at home. Find a point estimate for *p*.

(b) Find an 80% confidence interval for *p*.

14. *Physicians: Solo Practice* A random sample of 328 medical doctors showed that 171 had a solo practice. (Source: *Practice Patterns of General Internal Medicine*, American Medical Association.)

(a) Let *p* represent the proportion of all medical doctors who have a solo practice. Find a point estimate for *p*.

(b) Find a 95% confidence interval for *p*. Give a brief explanation of the meaning of the interval.

(c) As a news writer, how would you report the survey results regarding the percentage of medical doctors in solo practice? What is the margin of error based on a 95% confidence interval?

15. *Marketing: Customer Loyalty* In a marketing survey, a random sample of 730 women shoppers revealed that 628 remained loyal to their favorite supermarket during the past year (i.e., did not switch stores). (Source: *Trends in the United States: Consumer Attitudes and the Supermarket*, The Research Department, Food Marketing Institute.)

(a) Let *p* represent the proportion of all women shoppers who remain loyal to their favorite supermarket. Find a point estimate for *p*.

(b) Find a 95% confidence interval for p. Give a brief explanation of the meaning of the interval.

(c) As a news writer, how would you report the survey results regarding the percentage of supermarket shoppers who remained loyal to their favorite supermarket during the past year? What is the margin of error based on a 95% confidence interval?

16. *Marketing: Bargain Hunters* In a marketing survey, a random sample of 1001 supermarket shoppers revealed that 273 always stock up on an item when they find that item at a real bargain price. See reference in Problem 15.
    (a) Let p represent the proportion of all supermarket shoppers who always stock up on an item when they find a real bargain. Find a point estimate for p.
    (b) Find a 95% confidence interval for p. Give a brief explanation of the meaning of the interval.
    (c) As a news writer, how would you report the survey results on the percentage of supermarket shoppers who stock up on items when they find the item is a real bargain? What is the margin of error based on a 95% confidence interval?

17. *Lifestyle: Smoking* In a survey of 1000 large corporations, 250 said that, given a choice between a job candidate who smokes and an equally qualified nonsmoker, the nonsmoker will get the job (*USA Today*).
    (a) Let p represent the proportion of all corporations preferring a nonsmoking candidate. Find a point estimate for p.
    (b) Find a 0.95 confidence interval for p.
    (c) As a news writer, how would you report the survey results regarding the proportion of corporations who would hire the equally qualified nonsmoker. What is the margin of error based on a 95% confidence interval?

18. *Opinion Poll: Crime and Violence* A *New York Times*/CBS poll asked the question, "What do you think is the most important problem facing this country today?" Nineteen percent of the respondents answered "crime and violence." The margin of sampling error was plus or minus 3 percentage points. Following the convention that the margin of error is based on a 95% confidence interval, find a 95% confidence interval for the percentage of the population that would respond "crime and violence" to the question asked by the pollsters.

8.4
Choosing the Sample Size

FOCUS POINTS

✓ Compute the sample size to be used for estimating a mean μ.

✓ Compute the sample size to be used for estimating a proportion p when we have an estimate for p.

✓ Compute the sample size to be used for estimating a proportion p when we have no estimate for p.

Sample size for estimating μ

In the design stages of statistical research projects, it is a good idea to decide in advance on the confidence level you wish to use and to select the *maximum* error of estimate E you want for your project. How you choose to make these decisions depends on the requirements of the project and the practical nature of the problem. Whatever specifications you make, the next step is to determine the sample size. In this section, we will assume that the distribution of sample means $\bar{x}$ is approximately normal, and when necessary, we will approximate σ by the sample standard deviation s. These methods are technically justifiable since our samples will be of size at least 30.

Let's say that at a confidence level of c, we want our point estimate $\bar{x}$ for μ to be in error either way by less than some quantity E. In other words, E is the maximum error of estimate we can tolerate. Using the language of probability, we want the following to be true:

$$P(-E < \bar{x} - \mu < E) = c \tag{20}$$

This is essentially the same as Equation (1) of Section 8.1. Let's compare them.

$$P(-E < \bar{x} - \mu < E) = c \qquad (20)$$

$$P\left(-z_c \frac{\sigma}{\sqrt{n}} < \bar{x} - \mu < z_c \frac{\sigma}{\sqrt{n}}\right) = c \qquad (1)$$

From this comparison, we see that we want E to be

$$E = z_c \frac{\sigma}{\sqrt{n}}$$

Solving this equation for n, we get

$$n = \left(\frac{z_c \sigma}{E}\right)^2 \qquad (21)$$

To compute n from Equation (21), we must know the value of σ. If the value of σ is not previously known, we do a preliminary sampling to approximate it. For most practical purposes, a preliminary sample of size 30 or larger will give a sample standard deviation s, which we may use to approximate σ.

EXAMPLE 6

Sample size for estimating μ

A wildlife study is designed to find the mean weight of salmon caught by an Alaskan fishing company. As a preliminary study, a random sample of 50 freshly caught salmon is weighed. The sample standard deviation of the weights of these 50 fish is $s = 2.15$ lb. How large a sample should be taken to be 99% confident that the sample mean $\bar{x}$ is within 0.20 lb of the true mean weight μ?

SOLUTION: In this problem, $z_{0.99} = 2.58$ (see Table 8-2) and $E = 0.20$. The preliminary study of 50 fish is large enough to permit a good approximation of σ by $s = 2.15$. Therefore, Equation (21) becomes

$$n = \left(\frac{z_c \sigma}{E}\right)^2 \approx \left(\frac{(2.58)(2.15)}{0.20}\right)^2 = 769.2$$

NOTE In determining sample size, any fraction value of n is always rounded to the *next higher whole number.* We conclude that a sample size of 770 will be enough to satisfy the specifications. Of course, a sample size larger than 770 also works. ◇

Salmon moving upstream

EXAMPLE 7

Sample size for estimating μ

A certain company makes light fixtures on an assembly line. An efficiency expert wants to determine the mean time it takes an employee to assemble the switch on one of these fixtures. A preliminary study used a random sample of 45 observations and found that the sample standard deviation was $s = 78$ seconds. How many more observations are necessary for the efficiency expert to be 95% sure that the point estimate $\bar{x}$ will be "off" from the true mean μ by at most 15 seconds?

SOLUTION: In this example, we approximate σ by $s = 78$. We use $z_{0.95} = 1.96$ (see Table 8-2). The maximum error of estimate is specified to be $E = 15$ seconds. Equation (21) gives us

$$n = \left(\frac{z_c \sigma}{E}\right)^2 = \left(\frac{(1.96)(78)}{15}\right)^2 = 103.9$$

The efficiency expert should use a sample of minimum size 104. Since the preliminary study has 45 observations, an additional $104 - 45 = 59$ observations are necessary.   ◊

GUIDED EXERCISE 7

Sample size for estimating μ

A large state university has over 1800 faculty members. The dean of faculty wants to estimate the average teaching experience (in years) of the faculty members. A preliminary random sample of 60 faculty members yields a sample standard deviation of $s = 3.4$ years. The dean wants to be 99% confident that the sample mean $\bar{x}$ does not differ from the population mean by more than half a year. How large a sample should be used? Let's answer this question in parts.

(a) What value can we use to approximate σ? Why can we do this?

⟹ $s = 3.4$ years is a good approximation because a preliminary sample of 60 is fairly large.

(b) What is $z_{0.99}$? (*Hint:* See Table 8-2.)

⟹ $z_{0.99} = 2.58$

(c) What is E for this problem?

⟹ $E = 0.5$ year

(d) Which is the correct formula for n:

$$\left(\frac{z_c\sigma}{n}\right)^2, \quad \left(\frac{z_c\sigma}{E}\right)^2, \quad \text{or} \quad \left(\frac{z_cE}{\sigma}\right)^2$$

⟹ $n = \left(\dfrac{z_c\sigma}{E}\right)^2$

(e) Use the formula for n to find the minimum sample size. Should your answer be rounded up or down to a whole number?

⟹ $n = \left(\dfrac{(2.58)(3.4)}{0.5}\right)^2 = (17.54)^2 = 307.8$

Always round n up to the next whole number. Our final answer $n = 308$ is the minimum size.

Sample size for estimating $\hat{p} = r/n$

(If you omitted the binomial distribution, omit the rest of this section.)

Next, we will determine the minimum sample size when we use the sample proportion $\hat{p} = r/n$ as a point estimate for p in a binomial distribution. We will use the methods of normal approximation (large samples) discussed in Section 8.3. Suppose for a confidence level c we want the estimate $\hat{p} = r/n$ for p to be in error either way by less than some quantity E. Using the language of probability, we want the following to be true.

$$P(-E < \hat{p} - p < E) = c \tag{22}$$

Let's compare this with Equation (16) of Section 8.3. For convenience, they both are written together:

$$P(-E < \hat{p} - p < E) = c \tag{22}$$

$$P\left(-z_c\sqrt{\frac{pq}{n}} < \hat{p} - p < z_c\sqrt{\frac{pq}{n}}\right) = c \tag{16}$$

The comparison of the two equations gives a formula for E:

$$E = z_c \sqrt{\frac{pq}{n}}$$

Solving the last equation for n, we get

$$n = pq \left(\frac{z_c}{E} \right)^2$$

Since $q = 1 - p$, our equation for n can be written

$$n = p(1 - p) \left(\frac{z_c}{E} \right)^2 \tag{23}$$

Equation (23) cannot be used unless we already have a preliminary estimate for p. To get around this difficulty, we will use the equation $p(1 - p) = \frac{1}{4} - (p - \frac{1}{2})^2$. This is an algebraic identity that you are asked to verify in an optional exercise at the end of this section. In this exercise, you are also asked to use a little logical deduction to show that the maximum possible value of $p(1 - p)$ is $\frac{1}{4}$. Therefore, *when we have no preliminary estimate for p, we use the formula*

$$n = \frac{1}{4} \left(\frac{z_c}{E} \right)^2 \tag{24}$$

Since Equation (24) may make the sample size unnecessarily large, we can say the probability is *at least* (and possibly more than) c that the point estimate $\hat{p} = r/n$ for p will be in error either way by less than the quantity E.

EXAMPLE 8

Sample size for estimating p

A company is in the business of selling wholesale popcorn to grocery stores. The company buys directly from farmers. A buyer for the company is examining a large amount of corn from a certain farmer. Before the purchase is made, the buyer wants to estimate p, the probability that a kernel will pop.

Suppose that a random sample of n kernels is taken and r of these kernels pop. The buyer wants to be 95% sure that the point estimate $\hat{p} = r/n$ for p will be in error either way by less than 0.01.

(a) If no preliminary study is made to estimate p, how large a sample should the buyer use?

SOLUTION: In this case, we use Equation (24) with $z_{0.95} = 1.96$ (see Table 8-2) and $E = 0.01$.

$$n = \frac{1}{4} \left(\frac{z_c}{E} \right)^2 = \frac{1}{4} \left(\frac{1.96}{0.01} \right)^2 = 0.25(38,416) = 9604$$

We would need a sample of $n = 9604$ kernels.

(b) A preliminary study showed that p was approximately 0.86. If the buyer uses the results of the preliminary study, how large a sample should be used?

SOLUTION: In this case, we use Equation (23) with $p \approx 0.86$. Again, from Table 8-2, $z_{0.95} = 1.96$, and from the problem, $E = 0.01$.

$$n = p(1 - p)\left(\frac{z_c}{E}\right)^2 = (0.86)(0.14)\left(\frac{1.96}{0.01}\right)^2 = 4625.29$$

The sample size should be at least $n = 4626$ kernels. This sample is less than half the sample size necessary without the preliminary study.   ◇

GUIDED EXERCISE 8

Sample size for estimating p

In Indianapolis, the department of public health wants to estimate the proportion of children (grades 1–8) who require corrective lenses for their vision. A random sample of n children is taken, and r of these children are found to require corrective lenses. Let p be the true proportion of children requiring corrective lenses. The health department wants to be 99% sure that the point estimate $\hat{p} = r/n$ for p will be in error either way by less than 0.03.

(a) If no preliminary study is made to estimate p, how large a sample should the health department use? Let's answer this question in parts.

(i) Which formula shall we use: (23) or (24)?   ⟹   (i) We use Formula (24) because we do not have an estimate for p.

(ii) What is the value of E, and what is the value of z_c in this problem?   ⟹   (ii) $E = 0.03$ and $z_{0.99} = 2.58$ (see Table 8-2)

(iii) What is the value of n?   ⟹   (iii) $n = \dfrac{1}{4}\left(\dfrac{z_c}{E}\right)^2 = \dfrac{1}{4}\left(\dfrac{2.58}{0.03}\right)^2 = 1849$

So without a preliminary study to find p, we will need a sample size of at least $n = 1849$ children.

(b) A preliminary random sample of 100 children indicates that 23 require corrective lenses. Using the results of this preliminary study, how large a sample should the health department use? Again, let's answer this question in parts.

(i) Which formula should we use: (23) or (24)?   ⟹   We use Formula (23) because we have an estimate of p from a preliminary study.

(ii) What are the values of E and z_c for this problem?   ⟹ $E = 0.03$ and $z_{0.99} = 2.58$

(iii) What approximate value shall we use for p?   ⟹ $p \approx 0.23$

(iv) What is the value of n?   ⟹ $n = p(1 - p)\left(\dfrac{z_c}{E}\right)^2 = (0.23)(0.77)\left(\dfrac{2.58}{0.03}\right)^2 = 1309.83$

Therefore, the sample size should be at least 1310 children.

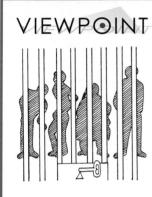

VIEWPOINT   *Profiles in Crime!*

What proportion of the U.S. population will eventually spend time in a prison? (*Answer:* About 5.1%.) What proportion of federal prison inmates have at least some college education? (*Answer:* About 28%.) If a person is released from prison, what is the probability that he or she will commit a felony and be returned to prison within 3 years? (*Answer:* About 41%.) What are the gender and age distributions of prison inmates? How accurate are these statistics? Sample size plays an important role in statistical accuracy. Methods of this section and data from the Bureau of Justice statistics can help you determine the implied level of accuracy. For more information, visit the Brase/Brase statistics site at http://math.college.hmco.com/students and find the link to the U.S. Justice Department.

SECTION 8.4   PROBLEMS

For each of the following problems,

(i) identify whether we are going to estimate a population mean μ or a population proportion p.

(ii) write down the appropriate sample size formula(s) for the problem, and then solve the problem.

1. *Botany: Yellowstone National Park* Test plots in Yellowstone National Park were used to study tree reproduction of lodgepole pine (*Yellowstone Vegetation,* by D. G. Despain, Roberts Rinehart). In plots of 50 square meters, the number x of new lodgepole saplings was counted. In a given region, based on long observation, the standard deviation of x values is estimated at $\sigma = 44$. How many 50-square-meter plots should be used in such a region to be 95% sure the sample mean $\bar{x}$ of lodgepole saplings is within 10 saplings of the population mean μ of lodgepole saplings in all 50-square-meter plots in this area?

2. *Botany: Yellowstone National Park* Root depth for grasses and shrubs in a type of soil known as glacial outwash was studied in Yellowstone National Park by D. G. Despain (see reference in Problem 1). Let x be a random variable representing root depth in this type of soil. It was found that the standard deviation of x values is approximately $\sigma = 8.94$ in. In a proposed study region of glacial outwash, how many plants should be carefully dug up and studied to be 90% sure that the sample mean root depth $\bar{x}$ is within 0.5 in. of the population mean root depth?

*3. *Marketing: McDonald's* Suppose that you live near a McDonald's restaurant and you want to estimate the proportion p of all people in your neighborhood who go to McDonald's on a typical day.

*Omit problems marked with an asterisk if you omitted the binomial distribution.

(a) If you had no preliminary estimate for p, how many people should you include in a random sample to be 85% sure that the point estimate $\hat{p}$ will be within a distance of 0.05 from p?

(b) Answer part (a) if you use the preliminary estimate that nationally about 1 in 20 Americans goes to McDonald's every day (*Fascinating McFacts*, provided by McDonald's Corporation).

4. *Pro Basketball: Heights of Players* A random sample of 41 basketball players from the "All-Time Player Directory" of *The Official NBA Basketball Encyclopedia* (Villard Books) gave a sample standard deviation for height of players $s = 3.32$ inches. How many more basketball players from the "All-Time Player Directory" should be included in the sample to be 95% sure that the sample mean $\bar{x}$ is within 0.75 inch of the population mean μ of all players listed in the NBA encyclopedia?

5. *Pro Basketball: Weights of Players* The NBA "All-Time Player Directory" also gives the weight of each basketball player. A random sample of 56 basketball players from the directory (see reference in Problem 4) gave a sample standard deviation $s = 26.58$ pounds for the weights of players. How many more basketball players from the "All-Time Player Directory" should be included in the sample to be 90% sure that the sample mean player weight $\bar{x}$ is within 4 pounds of the population mean μ?

*6. *Business: Phone Contact* How hard is it to reach a businessperson by phone? Let p be the proportion of calls to businesspeople for which the caller reaches the person being called on the *first* try.

(a) If you have no preliminary estimate for p, how many business phone calls should you include in a random sample to be 80% sure that the point estimate $\hat{p}$ will be within a distance of 0.03 from p?

(b) The *Book of Odds,* by Shook and Shook (Signet), reports that businesspeople can be reached by a single phone call approximately 17% of the time. Using this (national) estimate for p, answer part (a).

*7. *Campus Life: Coeds* What percentage of the campus student body is female? Let p be the proportion of women students on your campus.

(a) If no preliminary study is made to estimate p, how large a sample is needed to be 99% sure that a point estimate $\hat{p}$ will be within a distance of 0.05 from p?

(b) The *Statistical Abstract of the United States*, 112th Edition, indicates that approximately 54% of college students are females. Answer part (a) using this estimate for p.

*8. *Gasoline Stations: Self-Service* At many gasoline stations, customers have the option of using self-service pumps and receiving a discount instead of using full-service pumps. The question is: What proportion of customers in your neighborhood take advantage of this option and use the self-service pumps?

(a) If no preliminary study is made to estimate p, how large a sample of customers is necessary to be 90% sure that the point estimate $\hat{p}$ will be within a distance 0.08 of p?

(b) Nationally, about 81% of the gasoline customers use self-service pumps (Source: Amoco Oil Corporation). Answer part (a) for your neighborhood using this estimate for p.

9. *Anthropology: Pottery* A random sample of 83 reconstructed clay vessels from the Turquoise Ridge archaeological site indicated the standard deviation of diameters to

*Omit problems marked with an asterisk if you omitted the binomial distribution.

be approximately 5.5 cm (based on information from *Anthropological Paper Number 118*, University of Utah Press). How many more such clay vessels must be found and reconstructed to be 95% sure that the sample mean $\bar{x}$ of diameters is within 1 cm of the population mean μ of all such reconstructed clay vessels at this archaeological site?

10. *Wildlife: Bighorn Sheep* A random sample of 37 adult male desert bighorn sheep indicated the standard deviation of the sheep weights to be 15.8 lb (Source: *The Bighorn of Death Valley*, Fauna of the National Parks of the United States Monograph Number 6, U.S. Government Printing Office). How many more such adult male desert bighorn sheep should be included in the sample to be 90% sure that the sample mean weight $\bar{x}$ is within 2.5 lb of the population mean weight μ of all such bighorn sheep in this region?

*11. *Forestry: Pine Beetle* A ponderosa pine forest in Colorado has a pine beetle infestation. The beetles bore into a tree and carry a fungus that ultimately kills the tree. Let p be the proportion of trees in the forest that are infested.
   (a) If no preliminary sample is taken to estimate p, how large a sample is necessary to be 85% sure that a point estimate $\hat{p}$ will be within a distance of 0.06 from p?
   (b) A preliminary study of 58 trees showed that 19 were infested. How many *more* trees should be included in the sample to be 85% sure that a point estimate $\hat{p}$ will be within a distance of 0.06 from p?

12. *Hardware Store: Cash Flow* Gordon is thinking of buying a combination hardware/sporting goods store in Dove Creek, Montana. However, daily cash flow is an important consideration. A random sample of 40 business days from the past year was taken from the store records. For each day, the net income was determined. The sample standard deviation was found to be $57.19. How many *more* business days should be included in the sample to be 85% confident that the sample mean $\bar{x}$ of daily net incomes is within $10 of the population mean μ of daily net incomes?

*13. *Political Science: Capital Punishment* David is doing a research project in political science to determine the proportion p of voters in his district who favor capital punishment.
   (a) If no preliminary sample of voters is taken to estimate p, how large a sample is necessary to be 99% sure that a point estimate $\hat{p}$ will be within a distance of 0.01 from p?
   (b) The National Opinion Research Center at the University of Chicago found that approximately 67% of people in the United States favor capital punishment. Use this estimate to answer part (a).

14. *Sociology: Marriage Customs* A sociologist is studying marriage customs in a rural community in Denmark. A random sample of 35 women who have been married was used to determine the age of the woman at the time of her first marriage. The sample standard deviation of these ages was 2.3 years. The sociologist wants to estimate the population mean age of a woman at the time of her first marriage. How many *more* women should be included in the sample to be 95% confident that the sample mean $\bar{x}$ of ages is within 0.25 year of the population mean μ?

15. *Airline Reservations: Waiting Time* When customers phone an airline to make reservations, they usually find it irritating if they are kept on hold for a long time. In an effort to determine how long phone customers are kept on hold, one airline took a

---

*Omit problems marked with an asterisk if you omitted the binomial distribution.

random sample of 167 phone calls and determined the length of time (in minutes) each caller was kept on hold. The sample standard deviation was 3.8 minutes. How many more phone customers should be included in the sample to be 99% sure that the sample mean $\bar{x}$ of hold times is within 30 seconds of the population mean μ of hold times?

*16. *Small Business: Bankruptcy* The National Council of Small Businesses is interested in the proportion of small businesses that declared Chapter 11 bankruptcy last year. Since there are so many small businesses, the National Council intends to estimate the proportion from a random sample. Let p be the proportion of small businesses that declared Chapter 11 bankruptcy last year.
   (a) If no preliminary sample is taken to estimate p, how large a sample is necessary to be 95% sure that a point estimate $\hat{p}$ will be within a distance of 0.10 from p?
   (b) In a preliminary random sample of 38 small businesses, it was found that 6 had declared Chapter 11 bankruptcy. How many *more* small businesses should be included in the sample to be 95% sure that a point estimate $\hat{p}$ will be within a distance of 0.10 from p?

*17. *Women: Pickup Trucks* Let p be the proportion of all pickup truck owners in Cheyenne, Wyoming, who are women.
   (a) If no preliminary study is made to estimate p, how large a sample is necessary to be 90% sure that a point estimate $\hat{p}$ will be within a distance of 0.1 from p?
   (b) *USA Today* reported that nationally, approximately 24% of all pickup trucks are owned by women. Answer part (a) using this estimate for p.

*18. *Political Science: Voter Opinion* Linda Silbers is a social scientist studying voter opinion about a city bond proposal for a light-rail mass transit system in Denver.
   (a) If Linda wants to be 90% sure that her sample estimate of the proportion of voters who favor the bond is within 5% of the population percent p who favor the bond issue, how large a sample should she use?
   (b) If a preliminary study showed that p is approximately 73%, how large a sample is required?

*19. (a) Show that $p(1 - p) = 1/4 - (p - 1/2)^2$.
   (b) Why is $p(1 - p)$ never greater than 1/4?

*20. *Polling: Votes* You are working for a local polling firm. Your project involves interviewing a random sample of registered voters to determine the percentage favoring the use of state lottery proceeds for park improvements. The firm wants to ensure that the margin of error is no more than 3 percentage points either way. Assuming that there is no preliminary estimate for the percentage of registered voters favoring this use of lottery funds, what is the *minimum* number of respondents required? (*Hint:* Convert the margin of error to decimal and use the *New York Times*/CBS convention that the margin of error is based on a 95% confidence level.)

*21. *Election Results: Florida* A heavily populated county in Florida is told by the election commission to estimate the proportion p of popular vote count for the Democratic candidate for president.
   (a) The maximal error of estimate is to be 0.001 with a 99% confidence level. If no preliminary estimate for p can be agreed upon, how many ballots must be checked?
   (b) If $p = 0.5$ was used as a preliminary estimate for p, would the answer for part (a) change? Explain.

*Omit problems marked with an asterisk if you omitted the binomial distribution.

8.5
Estimating $\mu_1 - \mu_2$ and $p_1 - p_2$

FOCUS POINTS

✓ Compute confidence intervals for $\mu_1 - \mu_2$ using either large or small samples.

✓ Compute confidence intervals for $p_1 - p_2$ using the normal approximation.

✓ Interpret the meaning and implications of an all-positive, all-negative, or mixed confidence interval.

Independent samples and dependent samples

How can we tell if two populations are different? One way is to compare the difference in population means or the difference in population proportions. In this section, we will use samples from two populations to create confidence intervals for the difference between population parameters.

To make a statistical estimate about the difference between two population parameters, we need to have a sample from each population. Samples may be *independent* or *dependent* according to how they are selected. If the way we take a sample from one population is unrelated to the selection of sample data from the other population, the samples are called *independent*. However, if the samples are chosen in such a way that *each measurement* in one sample can be naturally *paired* with a measurement in the other sample, then the samples are called *dependent*.

Dependent samples and data pairs occur very naturally in "before and after" situations where the *same object* or item is *measured twice*. We will devote an entire section (9.6) to the study of dependent samples and paired data. However, in this section, we will confine our interest to independent samples.

Independent samples occur very naturally when we draw *two random samples*, one from the first population and one from the second population. Because *both* samples are random samples, there is no pairing of measurements between the two populations. All the examples of this section will involve independent random samples.

GUIDED EXERCISE 9

Distinguish between independent and dependent samples

For each experiment, categorize the sampling as independent or dependent, and explain your choice.

(a) In many medical experiments, a sample of subjects is randomly divided into two groups. One group is given a specific treatment, and the other group is given a placebo. After a certain period of time, both groups are measured for the same condition. Do the measurements from these two groups constitute independent or dependent samples?

⟹ Since the subjects are *randomly assigned* to the two treatment groups (one receives a treatment, the other a placebo), the resulting measurements would form independent samples.

(b) In an accountability study, a group of students in an English composition course is given a pretest. After the course, the same students are given a posttest over similar material. Are the two groups of scores independent or dependent?

⟹ Since the pretest scores and the posttest scores are from the same students, the samples are dependent. Each student has both a pretest score and a posttest score, so there is a natural pairing of data values.

The $\bar{x}_1 - \bar{x}_2$ sampling distribution

In this section, we will use probability distributions that arise from a difference of means (or proportions). How do we obtain such distributions? Suppose that we have two statistical variables x_1 and x_2, each with its own distribution. We take *independent* random samples of size n_1 from the x_1 distribution and of size n_2 from the x_2 distribution. Then we compute the respective means $\bar{x}_1$ and $\bar{x}_2$. Now consider the difference $\bar{x}_1 - \bar{x}_2$. This expression represents a difference of means. If we repeat this sampling process over and over, we will create lots of $\bar{x}_1 - \bar{x}_2$ values. Figure 8-7 illustrates the sampling distribution of $\bar{x}_1 - \bar{x}_2$.

The values of $\bar{x}_1 - \bar{x}_2$ that come from repeated (independent) sampling from populations 1 and 2 can be arranged in a relative-frequency table and a relative-frequency histogram (see Section 2.2). This would give us an experimental idea of the theoretical probability distribution of $\bar{x}_1 - \bar{x}_2$.

Fortunately, it is not necessary to carry out this lengthy process for each example. The results have been worked out mathematically. The next theorem presents the main results.

◇ **THEOREM 8.1** Let x_1 have a normal distribution with mean μ_1 and standard deviation σ_1. Let x_2 have a normal distribution with mean μ_2 and standard deviation σ_2. If we take independent random samples of size n_1 from the x_1 distribution and of size n_2 from the x_2 distribution, then the variable $\bar{x}_1 - \bar{x}_2$ has

1. a normal distribution

2. mean $\mu_1 - \mu_2$

3. standard deviation

$$\sqrt{\frac{\sigma_1^2}{n_1} + \frac{\sigma_2^2}{n_2}} \quad ◇$$

FIGURE 8-7

Sampling Distribution of $\bar{x}_1 - \bar{x}_2$

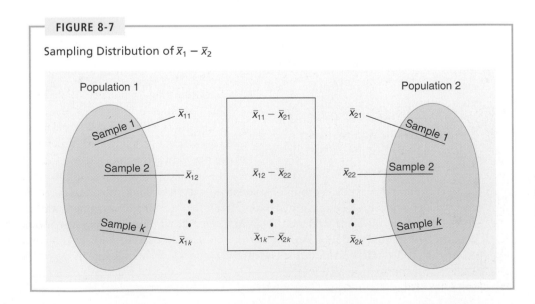

◇ **COMMENT** The theorem requires that x_1 and x_2 have *normal* distributions. However, if *both* n_1 and n_2 are 30 or larger (and the distributions are approximately mound-shaped and symmetrical), then the central limit theorem (Section 7.2) assures us that $\bar{x}_1$ and $\bar{x}_2$ are approximately normally distributed. In this case, the conclusions of the theorem are again valid even if the original x_1 and x_2 distributions are not exactly normal.

Furthermore, for most practical applications, if $n_1 \geq 30$ and $n_2 \geq 30$, then the approximation $s_1 \approx \sigma_1$ and $s_2 \approx \sigma_2$ give very usable results (where s_1 and s_2 are the sample standard deviations of populations 1 and 2, respectively).   ◇

Confidence Intervals for $\mu_1 - \mu_2$ (large independent samples)

If we use Theorem 8.1, then a discussion similar to that of Section 8.1 gives the following information.

---

c Confidence interval for $\mu_1 - \mu_2$ (large samples)

A c confidence interval for $\mu_1 - \mu_2$ is

$$(\bar{x}_1 - \bar{x}_2) - E < \mu_1 - \mu_2 < (\bar{x}_1 - \bar{x}_2) + E$$

where $\bar{x}_1$ = sample mean for population 1

s_1 = sample standard deviation for population 1

n_1 = sample size from population 1 ($n_1 \geq 30$)

$\bar{x}_2$ = sample mean for population 2

s_2 = sample standard deviation for population 2

n_2 = sample size from population 2 ($n_2 \geq 30$)

$$E \approx z_c \sqrt{\frac{s_1^2}{n_1} + \frac{s_2^2}{n_2}}$$

z_c = critical value for confidence level c (See Table 8-2 for frequently used values.)

c = confidence level ($0 < c < 1$)

EXAMPLE 9

Confidence interval for $\mu_1 - \mu_2$, large samples

Yellowstone National Park

In the summer of 1988, Yellowstone National Park had some major fires that destroyed large tracts of old timber near many famous trout streams. Fishermen were concerned about the long-term effect of the fire on these streams. However, biologists claimed that the new meadows that would spring up under dead trees would produce a lot more insects, which would in turn mean better fishing in the years ahead. Guide services registered with the park provided data about the daily catch for fishermen over many years. Ranger checks on the streams also provided data about the daily number of fish caught by fishermen. *Yellowstone Today* (a national park publication) indicated that the biologists' claim is basically correct and that Yellowstone anglers are delighted by their average increased catch.

Suppose that you are a biologist studying data from Yellowstone streams. A random sample of $n_1 = 167$ fishing reports in the years 1983 to 1988 (before the fire) showed that the average catch per day was $\bar{x}_1 = 5.2$

trout with sample standard deviation $s_1 = 1.9$. Then another random sample of $n_2 = 125$ fishing reports in the years 1990 to 1993 (after the fire) showed that the average catch per day was $\bar{x}_2 = 6.8$ trout with sample standard deviation $s_2 = 2.3$.

(a) For each sample, what is the population? Are the samples dependent or independent? Explain.

SOLUTION: The population for the first sample is the number of trout caught per day by fishermen before the fire. The population of the second sample is the number of trout caught per day after the fire. Both samples were random samples taken in their respective time periods. There was no effort to pair individual data values. Therefore, the samples can be thought of as independent samples.

(b) Compute a 95% confidence interval for $\mu_1 - \mu_2$, the difference of population means.

SOLUTION: Since $n_1 = 167$, $\bar{x}_1 = 5.2$, and $s_1 = 1.9$ and $n_2 = 125$, $\bar{x}_2 = 6.8$, $s_2 = 2.3$, and $z_{0.95} = 1.96$ (see Table 8-2), then

$$E \approx z_c \sqrt{\frac{s_1^2}{n_1} + \frac{s_2^2}{n_2}}$$

$$= 1.96 \sqrt{\frac{(1.9)^2}{167} + \frac{(2.3)^2}{125}}$$

$$\approx 1.96\sqrt{0.0639} = 0.4955$$

$$\approx 0.50$$

The 95% confidence interval is

$$(\bar{x}_1 - \bar{x}_2) - E < \mu_1 - \mu_2 < (\bar{x}_1 - \bar{x}_2) + E$$
$$(5.2 - 6.8) - 0.50 < \mu_1 - \mu_2 < (5.2 - 6.8) + 0.50$$
$$-2.10 < \mu_1 - \mu_2 < -1.10$$

(c) Explain the meaning of the confidence interval you computed in part (b).

SOLUTION: We are 95% confident that the interval -2.10 to -1.10 fish per day is one of the intervals containing the population difference $\mu_1 - \mu_2$, where μ_1 represents the population average daily catch before the fire and μ_2 represents the population average daily catch after the fire. Put another way, since the confidence interval contains only *negative values*, we can be 95% sure that $\mu_1 - \mu_2 < 0$. This means that we are 95% sure that $\mu_1 < \mu_2$. In words, we are 95% sure that the average catch before the fire is less than the average catch after the fire. ◊

Confidence interval for $\mu_1 - \mu_2$ (small independent samples)

Statistical methods for small samples and the difference between two means are very much like the methods we studied in Section 8.2 insofar as the Student's t distribution plays a key role.

Suppose that independent random samples are drawn from two populations that possess means μ_1 and μ_2. We assume that the parent population measurements have normal distributions or distributions that are approximately normal (i.e., mound-shaped and more or less symmetrical). We also assume that the standard deviations σ_1 and σ_2 for the two populations are equal. The condition $\sigma_1 = \sigma_2$ may seem very restrictive. However, in a vast number of practical applications, this condition is satisfied. In fact, our methods still apply even if the standard deviations are known to be only approximately equal. (See Section 11.4 for methods to test that $\sigma_1^2 = \sigma_2^2$.)

If we draw two independent random samples from our two populations, we can obtain the following information:

Sample 1 from Population of x_1 Values

n_1 = sample size

$\overline{x}_1$ = sample mean

s_1 = sample standard deviation

Sample 2 from Population of x_2 Values

n_2 = sample size

$\overline{x}_2$ = sample mean

s_2 = sample standard deviation

Pooled variance

We estimate the common standard deviation for the two populations by using a *pooled variance* of s_1^2 and s_2^2 values. The best estimate of the common variance of the x_1 and x_2 populations is given by the formula

$$s^2 = \frac{(n_1 - 1)s_1^2 + (n_2 - 1)s_2^2}{n_1 + n_2 - 2}$$

The best (*pooled*) estimate of the common standard deviation is then

$$s = \sqrt{\frac{(n_1 - 1)s_1^2 + (n_2 - 1)s_2^2}{n_1 + n_2 - 2}}$$

The next theorem presents the main result necessary to compute confidence intervals for $\mu_1 - \mu_2$ using *small samples*.

◇ **THEOREM 8.2** Let x_1 and x_2 have normal (or approximately normal) distributions with means μ_1 and μ_2 and standard deviations σ_1 and σ_2, respectively. We assume that σ_1 and σ_2 are (approximately) equal. Suppose that we take independent random samples of size n_1 and n_2 from the x_1 and x_2 populations. Let $\overline{x}_1$ and $\overline{x}_2$ and s_1 and s_2 be the sample means and standard deviations. If $n_1 < 30$ and/or $n_2 < 30$, then

$$t = \frac{(\overline{x}_1 - \overline{x}_2) - (\mu_1 - \mu_2)}{s\sqrt{\dfrac{1}{n_1} + \dfrac{1}{n_2}}}$$

has a Student's t distribution with degrees of freedom $d.f. = n_1 + n_2 - 2$. ◇

Using Theorem 8.2, it is not hard to obtain the following result.

> **c Confidence interval for $\mu_1 - \mu_2$ (small independent samples)**
>
> $$(\overline{x}_1 - \overline{x}_2) - E < \mu_1 - \mu_2 < (\overline{x}_1 - \overline{x}_2) + E$$
>
> where the samples are independent, and the standard deviations are approximately equal.
>
> *Population 1*
>
> $n_1 \ (< 30)$ = sample size
>
> $\overline{x}_1$ = sample mean
>
> s_1 = sample standard deviation
>
> *Population 2*
>
> $n_2 \ (< 30)$ = sample size
>
> $\overline{x}_2$ = sample mean
>
> s_2 = sample standard deviation
>
> $$s = \sqrt{\frac{(n_1 - 1)s_1^2 + (n_2 - 1)s_2^2}{n_1 + n_2 - 2}}$$
>
> $$E = t_c \, s \sqrt{\frac{1}{n_1} + \frac{1}{n_2}}$$
>
> c = confidence level, $0 < c < 1$
>
> t_c = critical value for confidence level c and degrees of freedom $d.f. = n_1 + n_2 - 2$ (See Table 6 in Appendix II.)

EXAMPLE 10

Confidence interval for $\mu_1 - \mu_2$, small samples

Alexander Borbely is a professor at the Medical School of the University of Zurich, where he is director of the Sleep Laboratory. Dr. Borbely and his colleagues are experts on sleep, dreams, and sleep disorders. In his book, *Secrets of Sleep*, Dr. Borbely discusses brain waves, which are measured in hertz, the number of oscillations per second. Rapid brain waves (wakefulness) are in the range of 16 to 25 hertz. Slow brain waves (sleep) are in the range of 4 to 8 hertz. During normal sleep, a person goes through several cycles (each cycle is about 90 minutes) of rapid to slow and back to rapid brain waves. During deep sleep, brain waves are at their lowest.

In his book, Professor Borbely comments that alcohol is a *poor* sleep aid. In one study, a number of subjects were given 1/2 liter of red wine before they went to sleep. The subjects fell asleep quickly but did not remain asleep the entire night. Toward morning, between 4 and 6 A.M., they tended to wake up and have trouble going back to sleep.

Suppose that a random sample of 29 college students was randomly divided into two groups. The first group of $n_1 = 15$ people was given 1/2 liter of red wine before going to sleep. The second group of $n_2 = 14$ people was given no alcohol before going to sleep. Everyone in both groups went to sleep at 11 P.M. The average brain wave activity (4 to 6 A.M.) was determined for each individual in the groups. The results follow:

Group 1 (x_1 values): $n_1 = 15$ (with alcohol)
Average brain wave activity in the hours 4 to 6 A.M.

16.0	19.6	19.9	20.9	20.3	20.1	16.4	20.6
20.1	22.3	18.8	19.1	17.4	21.1	22.1	

For group 1, we have the sample mean and standard deviation of

$$\overline{x}_1 = 19.65 \qquad \text{and} \qquad s_1 = 1.86$$

Group 2 (x_2 values): $n_2 = 14$ (no alcohol)
Average brain wave activity in the hours 4 to 6 A.M.

8.2	5.4	6.8	6.5	4.7	5.9	2.9
7.6	10.2	6.4	8.8	5.4	8.3	5.1

For group 2, we have the sample mean and standard deviation of

$$\bar{x}_2 = 6.59 \qquad \text{and} \qquad s_2 = 1.91$$

(a) Do you think that the samples are independent or dependent? Explain.

SOLUTION: Since the original random sample of 29 students was randomly divided into two groups, it is reasonable to say that the samples are independent.

(b) What assumptions are we making about the data?

SOLUTION: We are assuming that the populations of x_1 and x_2 values are each approximately normally distributed with approximately the same population standard deviations.

(c) Compute a 90% confidence interval for $\mu_1 - \mu_2$, the difference of population means.

SOLUTION: First, we need to find s, the pooled standard deviation:

$$s = \sqrt{\frac{(n_1 - 1)s_1^2 + (n_2 - 1)s_2^2}{n_1 + n_2 - 2}}$$

$$= \sqrt{\frac{(15 - 1)1.86^2 + (14 - 1)1.91^2}{15 + 14 - 2}}$$

$$= \sqrt{3.55} = 1.88$$

Next, we find the $t_{0.90}$ value and compute E. Since $d.f. = n_1 + n_2 - 2 = 15 + 14 - 2 = 27$, Table 6 in Appendix II gives $t_{0.90} = 1.703$. Then

$$E = t_c\, s\sqrt{\frac{1}{n_1} + \frac{1}{n_2}}$$

$$= 1.703(1.88)\sqrt{\frac{1}{15} + \frac{1}{14}}$$

$$= 1.1898 \approx 1.19$$

The c confidence interval is

$$(\bar{x}_1 - \bar{x}_2) - 1.19 < \mu_1 - \mu_2 < (\bar{x}_1 - \bar{x}_2) + 1.19$$
$$(19.65 - 6.59) - 1.19 < \mu_1 - \mu_2 < (19.65 - 6.59) + 1.19$$
$$11.87 < \mu_1 - \mu_2 < 14.25$$

Therefore, after further rounding, we have

$$11.9 \text{ hertz} < \mu_1 - \mu_2 < 14.3 \text{ hertz}$$

(d) Explain the meaning of the confidence interval you computed in part (c).

SOLUTION: μ_1 represents the population average brain wave activity for people who drink 1/2 liter of wine before sleeping. μ_2 represents the population average brain wave activity for people who take no alcohol before sleeping. Both periods of measurement are from 4 to 6 A.M. We are 90% confident the difference $\mu_1 - \mu_2$ is between 11.9 and 14.3 hertz (rounded values). It would seem reasonable to conclude that people who drink before sleeping might wake up in the early morning and have trouble going back to sleep. Since the confidence interval from 11.9 to 14.3 contains only *positive values*, we could express this by saying that we are 90% confident that $\mu_1 - \mu_2$ is *positive*. This means that $\mu_1 - \mu_2 > 0$. Thus we are 90% confident that $\mu_1 > \mu_2$ (that is, average brain wave activity from 4 to 6 A.M. for the group drinking wine is more than average brain wave activity for the group not drinking). ◇

Estimating the difference of proportions $p_1 - p_2$

We conclude this section with a discussion of confidence intervals for $p_1 - p_2$, the difference of two proportions from binomial probability distributions. The main result on this topic is the following theorem.

◇ **THEOREM 8.3** Consider two binomial probability distributions

Distribution 1	*Distribution 2*
n_1 = number of trials	n_2 = number of trials
r_1 = number of successes out of n_1 trials	r_2 = number of successes out of n_2 trials
p_1 = probability of success on each trial	p_2 = probability of success on each trial
$q_1 = 1 - p_1$ = probability of failure on each trial	$q_2 = 1 - p_2$ = probability of failure on each trial
$\hat{p}_1 = \dfrac{r_1}{n_1}$ = point estimate for p_1	$\hat{p}_2 = \dfrac{r_2}{n_2}$ = point estimate for p_2
$\hat{q}_1 = 1 - \dfrac{r_1}{n_1}$ = point estimate for q_1	$\hat{q}_2 = 1 - \dfrac{r_2}{n_2}$ = point estimate for q_2

For most practical applications, if the four quantities

$$n_1\hat{p}_1 \qquad n_1\hat{q}_1 \qquad n_2\hat{p}_2 \qquad n_2\hat{q}_2$$

are all larger than 5 (see Section 6.4), then the following statements are true about the random variable $\dfrac{r_1}{n_1} - \dfrac{r_2}{n_2}$:

1. $\dfrac{r_1}{n_1} - \dfrac{r_2}{n_2}$ has an approximately normal distribution.

2. The mean is $p_1 - p_2$.

3. The standard deviation is approximately

$$\hat{\sigma} = \sqrt{\frac{\hat{p}_1\hat{q}_1}{n_1} + \frac{\hat{p}_2\hat{q}_2}{n_2}} \quad ◇$$

Based on the preceding theorem, we can find confidence intervals for $p_1 - p_2$ in the following way:

> **c Confidence interval for $p_1 - p_2$ (large samples)**
>
> $$(\hat{p}_1 - \hat{p}_2) - E < p_1 - p_2 < (\hat{p}_1 - \hat{p}_2) + E$$
>
> where (using the notation of Theorem 8.3)
>
> $$E = z_c \sqrt{\frac{\hat{p}_1\hat{q}_1}{n_1} + \frac{\hat{p}_2\hat{q}_2}{n_2}}$$
>
> c = confidence level, $0 < c < 1$
>
> z_c = critical value for confidence level c (See Table 8-2 for frequently used values.)
>
> n_1 = number of trials in binomial experiment 1
>
> r_1 = number of successes in binomial experiment 1
>
> n_2 = number of trials in binomial experiment 2
>
> r_2 = number of successes in binomial experiment 2
>
> $$\hat{p}_1 = \frac{r_1}{n_1} \quad \text{and} \quad \hat{q}_1 = 1 - \frac{r_1}{n_1}$$
>
> $$\hat{p}_2 = \frac{r_2}{n_2} \quad \text{and} \quad \hat{q}_2 = 1 - \frac{r_2}{n_2}$$
>
> We assume that all four quantities
>
> $$n_1\hat{p}_1 \qquad n_1\hat{q}_1 \qquad n_2\hat{p}_2 \qquad n_2\hat{q}_2$$
>
> are greater than 5.

EXAMPLE 11

Confidence interval for $p_1 - p_2$

In his book, *Secrets of Sleep*, Professor Borbely describes research on dreams in the sleep laboratory at the University of Zurich Medical School. During normal sleep, there is a phase known as *REM* (rapid eye movement). For most people, REM sleep occurs about every 90 minutes or so, and it is thought that dreams occur just before or during the REM phase. Using electronic equipment in the sleep laboratory, it is possible to detect the REM phase in a sleeping person. If a person is wakened immediately after the REM phase, he or she usually can describe a dream that has just taken place. Based on a study of over 650 people in the Zurich sleep laboratory, it was found that about one-third of all dream reports contain feelings of fear, anxiety, or aggression. There is a conjecture that if a person is in a good mood when going to sleep, the proportion of "bad" dreams (fear, anxiety, aggression) might be reduced.

Suppose that two groups of subjects were randomly chosen for a sleep study. In group I, before going to sleep, the subjects spent 1 hour watching a comedy movie. In this group, there were a total of $n_1 = 175$ dreams recorded, of which $r_1 = 49$ were dreams with feelings of anxiety, fear, or aggression. In group II, the subjects did not watch a movie but simply went to sleep. In this group, there were a total of $n_2 = 180$ dreams recorded, of which $r_2 = 63$ were dreams with feelings of anxiety, fear, or aggression.

(a) Why could groups I and II be considered independent binomial distributions? Why do we have a "large-sample" situation?

SOLUTION: Since the two groups were chosen randomly, it is reasonable to assume that neither group's response would be related to the other. In both groups, each recorded dream could be thought of as a trial, with success being a dream with feelings of fear, anxiety, or aggression.

$$\hat{p}_1 = \frac{r_1}{n_1} = \frac{49}{175} = 0.28 \qquad \text{and} \qquad \hat{q}_1 = 1 - \hat{p}_1 = 0.72$$

$$\hat{p}_2 = \frac{r_2}{n_2} = \frac{63}{180} = 0.35 \qquad \text{and} \qquad \hat{q}_2 = 1 - \hat{p}_2 = 0.65$$

Since

$$n_1\hat{p}_1 = 49 > 5 \qquad n_1\hat{q}_1 = 126 > 5$$
$$n_2\hat{p}_2 = 63 > 5 \qquad n_2\hat{q}_2 = 117 > 5$$

then large-sample theory is appropriate.

(b) What is $p_1 - p_2$? Compute a 95% confidence interval for $p_1 - p_2$.

SOLUTION: p_1 is the population proportion of success (bad dreams) for all people who watch comedy movies before bed. Thus, p_1 can be thought of as the percentage of bad dreams for all people who are in a "good mood" when they go to bed. Likewise, p_2 is the percentage of bad dreams for the population of all people who just go to bed (no movie). The difference $p_1 - p_2$ is the population difference.

To find a confidence interval for $p_1 - p_2$, we need the values of z_c, $\hat{\sigma}$, and then E. From Table 8-2, we see that $z_{0.95} = 1.96$, so

$$\hat{\sigma} = \sqrt{\frac{\hat{p}_1\hat{q}_1}{n_1} + \frac{\hat{p}_2\hat{q}_2}{n_2}} = \sqrt{\frac{(0.28)(0.72)}{175} + \frac{(0.35)(0.65)}{180}}$$

$$= \sqrt{0.0024} = 0.0492$$

$$E = z_c\hat{\sigma} = 1.96(0.0492) \approx 0.096$$

$$(\hat{p}_1 - \hat{p}_2) - E < p_1 - p_2 < (\hat{p}_1 - \hat{p}_2) + E$$

$$(0.28 - 0.35) - 0.096 < p_1 - p_2 < (0.28 - 0.35) + 0.096$$

$$-0.166 < p_1 - p_2 < 0.026$$

(c) Explain the meaning of the confidence interval that you constructed in part (b).

SOLUTION: We are 95% sure that the percentage difference of "bad" dreams for group I and group II is between −16.6% and 2.6%. Since the interval −0.166 to 0.026 is not all negative (or all positive), we cannot say that $p_1 - p_2 < 0$ (or $p_1 - p_2 > 0$). Thus, at the 95% confidence level, we *cannot* conclude that $p_1 < p_2$ or $p_1 > p_2$. The comedy movies before bed help some people reduce the percentage of "bad" dreams, but at the 95% confidence level, we cannot say the *population difference* is reduced. ◇

◇ **COMMENT** Suppose that we construct a $c\%$ confidence interval for $\mu_1 - \mu_2$ (or $p_1 - p_2$). Then three cases arise:

1. The $c\%$ confidence interval contains only *negative values* (see Example 9). In this case, we conclude that $\mu_1 - \mu_2 < 0$ (or $p_1 - p_2 < 0$), and we are therefore $c\%$ confident that $\mu_1 < \mu_2$ (or $p_1 < p_2$).

2. The $c\%$ confidence interval contains only *positive values* (see Example 10). In this case, we conclude that $\mu_1 - \mu_2 > 0$ (or $p_1 - p_2 > 0$), and we can be $c\%$ confident that $\mu_1 > \mu_2$ (or $p_1 > p_2$).

3. The $c\%$ confidence interval contains *both positive and negative values*. In this case, we cannot at the $c\%$ confidence level conclude that either μ_1 or μ_2 (p_1 or p_2) is larger. However, if we *reduce* the confidence level c to a *smaller value*, then the confidence interval will, in general, be shorter (explain why). A shorter confidence interval *might* put us back into case 1 or case 2 above (again, explain why). ◊

GUIDED EXERCISE 10

Interpret a confidence interval

(a) A study reported a 90% confidence interval for the difference of means to be

$$10 < \mu_1 - \mu_2 < 20$$

For this interval, what can you conclude about the respective values of μ_1 and μ_2?

➡ At a 90% level of confidence, we can say that the difference $\mu_1 - \mu_2$ is positive, so $\mu_1 - \mu_2 > 0$ and $\mu_1 > \mu_2$.

(b) A study reported a 95% confidence interval for the difference of proportions to be

$$-0.32 < p_1 - p_2 < 0.16$$

From this interval, what can you conclude about the respective values of p_1 and p_2?

➡ At the 95% confidence level, we see that the difference of proportions ranges from negative to positive values. We cannot tell from this interval if p_1 is greater than p_2 or p_1 is less than p_2.

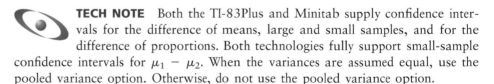 **TECH NOTE** Both the TI-83Plus and Minitab supply confidence intervals for the difference of means, large and small samples, and for the difference of proportions. Both technologies fully support small-sample confidence intervals for $\mu_1 - \mu_2$. When the variances are assumed equal, use the pooled variance option. Otherwise, do not use the pooled variance option.

TI-83Plus Use the **STAT** key and highlight **TESTS**. The choices **9:2-SampZInt, 0:2-SampTInt,** find confidence intervals for difference of means, large samples and small samples, respectively, while choice **B:2-PropZInt** provides confidence intervals for proportions. Large-sample $\mu_1 - \mu_2$ requires values for σ_1 and σ_2 or estimates for the values. For small-sample $\mu_1 - \mu_2$ intervals, select Pooled when the variances are approximately equal.

Minitab Use the menu choice **STAT ➤ Basic Statistics ➤ 2 sample t** or **2 proportions.** Minitab always uses the Student's t distribution for $\mu_1 - \mu_2$ confidence intervals. For large samples, do not check "assume equal variances." For small samples, check the "assume equal variances" option.

VIEWPOINT   *What's the Difference?*

Will two 15-minute piano lessons a week significantly improve a child's ana-lytical reasoning skills? Why piano? Why not computer keyboard instruction or maybe voice lessons? Professor Frances Rauscher, University of Wisconsin, and Professor Gordon Shaw, University of California at Irvine, claim that there is a difference! How could this be measured? A large number of piano students were given complicated tests of mental ability. Independent control groups of other students were given the same tests. Techniques involving the study of differences of means were used to draw the conclusion that students taking piano lessons did better on tests measuring analytical reasoning skills. (Reported in *The Denver Post.*)

SECTION 8.5   PROBLEMS

For each of the following problems,

(i) determine whether the requested confidence interval is for a difference of means or a difference of proportions.

(ii) if the confidence interval is for a difference of means, determine whether to use large-sample or small-sample technique.

Answers may vary slightly due to rounding.

1. *Pro Football and Basketball: Heights of Players* A random sample of 45 profes-sional football players indicated the mean height to be $\bar{x}_1 = 6.18$ feet with sample standard deviation $s_1 = 0.37$ feet. A random sample of 40 professional basketball players indicated the mean height to be $\bar{x}_2 = 6.45$ feet with sample standard devi-ation $s_2 = 0.31$ feet. (Sources: *Sports Encyclopedia of Pro Football,* 11th Edition, St. Martin's Press, and *The Official NBA Basketball Encyclopedia,* Villard Books.)
   (a) Assume that the preceding data are representative of professional football players and basketball players. Let μ_1 be the population mean height of all professional football players, and let μ_2 be the population mean height of all professional bas-ketball players. Find a 95% confidence interval for $\mu_1 - \mu_2$.
   (b) Examine the confidence interval and explain what it means in this context. Does the interval consist of numbers that are all positive? all negative? of different signs? At the 95% level of confidence, does it appear that the average height of football players tends to be shorter than the average height of basketball play-ers? Explain.
   (c) Redo parts (a) and (b) for a 99% confidence interval. Does your conclusion change? Explain.

2. *Pro Baseball and Basketball: Age of Players* A random sample of 38 professional baseball players indicated the mean age to be $\bar{x}_1 = 31.5$ years with sample stan-dard deviation $s_1 = 4.4$ years. A random sample of 41 professional basketball play-ers indicated the mean age to be $\bar{x}_2 = 30.3$ years with sample standard deviation $s_2 = 3.8$ years. (Sources: *The Baseball Encyclopedia,* 9th Edition, Macmillan, and *The Official NBA Basketball Encyclopedia,* Villard Books.)
   (a) Assume that the preceding data are representative of professional baseball play-ers and basketball players. Let μ_1 be the population mean age of all professional

baseball players, and let μ_2 be the population mean age of all professional basketball players. Find a 75% confidence interval for $\mu_1 - \mu_2$.

(b) Examine the confidence interval and explain what it means in this context. Does the interval consist of numbers that are all positive? all negative? of different signs? At the 75% level of confidence, does it appear that the average age of professional baseball players is older than that of basketball players? Explain.

(c) Redo parts (a) and (b) for a 90% confidence interval. Does your conclusion change? Explain.

3. *Industry Groups: Profits* Many companies take in a lot of money. How much of that money can they keep as profit? Do some industry groups have a better performance than others? One way to answer such questions is to study profit as a percentage of revenue. Annual profit as a percentage of revenue is shown below for a group of companies that manufacture electrical equipment. Names such as General Electric, Motorola, Whirlpool, and Maytag are included.

| 8.1 | 5.1 | 5.3 | 6.2 | 6.8 | 4.8 | 5.5 | 5.4 |
| 6.0 | 4.9 | 4.5 | 5.1 | 5.2 | 4.7 | 6.3 | |

For a group of food and drug stores, annual profit as a percentage of revenue is shown next. Names such as Kroger, Safeway, Albertson's, and Walgreen are included.

| 1.5 | 2.7 | 3.6 | 2.0 | 3.2 | 2.5 |
| 2.9 | 1.9 | 1.5 | 2.1 | 2.0 | 4.5 |

(Source: *Fortune 500,* Vol. 135, No. 8.)

(a) Assume that the preceding data are representative of each given industry group. Let x_1 represent annual profit as percentage of revenue for the electronic companies, and let x_2 represent profit as percentage of revenue for food and drug companies. Use a calculator with mean and sample standard deviation keys to verify that $\bar{x}_1 \approx 5.6$, $s_1 \approx 1.0$, $\bar{x}_2 \approx 2.5$, and $s_2 \approx 0.9$.

(b) Let μ_1 be the population mean for x_1 and μ_2 be the population mean for x_2. Find a 99% confidence interval for $\mu_1 - \mu_2$.

(c) Examine the confidence interval and explain what it means in this context. Does the interval consist of numbers that are all positive? all negative? of different signs? At the 99% level of confidence, does it appear that one industry group has a higher profit as a percentage of revenue?

4. *Industry Groups: Profits* It's what you keep that matters! What percentage of a company's revenue is profit? Let's consider the insurance industry and the health care industry. Annual profit as a percentage of revenue is shown below for a group of companies in the insurance industry. Names such as New York Life, Prudential, John Hancock, and Mutual of Omaha are included.

| 2.5 | 3.3 | 6.8 | 5.1 | 3.1 | 3.6 |
| 2.5 | 3.3 | 4.5 | 5.9 | 6.6 | 5.9 |

For a group of health care organizations, annual profit as a percentage of revenue is shown next. Names such as Humana, Columbia Health Care, Manor Care, and United Health Care are included.

| 7.6 | 3.5 | 6.3 | 4.8 | 4.2 | 3.3 | 3.8 | 5.3 | 3.2 | 4.2 |

(Source: *Fortune 500,* Vol. 135, No. 8.)

(a) Assume that the preceding data are representative of each given industry group. Let x_1 represent annual profit as percentage of revenue for the insurance companies, and let x_2 represent profit as percentage of revenue for health care companies. Use a calculator with mean and sample standard deviation keys to verify that $\bar{x}_1 \approx 4.4$, $s_1 \approx 1.6$, $\bar{x}_2 \approx 4.6$, and $s_2 \approx 1.4$.

(b) Let μ_1 be the population mean for x_1 and μ_2 be the population mean for x_2. Find a 90% confidence interval for $\mu_1 - \mu_2$.

(c) Examine the confidence interval and explain what it means in this context. Does the interval consist of numbers that are all positive? all negative? of different signs? At the 90% level of confidence, does it appear that one industry group has a higher profit as a percentage of revenue?

5. *Myers-Briggs: Marriage Counseling* Isabel Myers was a pioneer in the study of personality types. She identified four basic personality preferences that are described at length in the book *A Guide to the Development and Use of the Myers-Briggs Type Indicator,* by Myers and McCaulley (Consulting Psychologists Press). Marriage counselors know that couples who have none of the four preferences in common may have a stormy marriage. Myers took a random sample of 375 married couples and found that 289 had two or more personality preferences in common. In another random sample of 571 married couples, it was found that only 23 had no preferences in common. Let p_1 be the population proportion of all married couples who have two or more personality preferences in common. Let p_2 be the population proportion of all married couples who have no personality perferences in common.

(a) Find a 99% confidence interval for $p_1 - p_2$.

(b) Explain the meaning of the confidence interval of part (a) in the context of this problem. Does the confidence interval contain all positive, all negative, or both positive and negative numbers? What does this tell you (at the 99% confidence level) about the proportion of married couples with two or more personality preferences in common compared with the proportion of married couples sharing no personality preferences in common?

6. *Myers-Briggs: Marriage Counseling* Most married couples have two or three personality preferences in common (see reference in Problem 5). Myers used a random sample of 375 married couples and found that 132 had three preferences in common. Another random sample of 571 couples showed that 217 had two personality preferences in common. Let p_1 be the population proportion of all married couples who have three personality preferences in common. Let p_2 be the population proportion of all married couples who have two personality preferences in common.

(a) Find a 90% confidence interval for $p_1 - p_2$.

(b) Examine the confidence interval in part (a) and explain what it means in this context. Does the confidence interval contain all positive, all negative, or both positive and negative numbers? What does this tell you about the proportion of married couples with three personality preferences in common compared with the proportion of couples with two preferences in common (at the 90% confidence level)?

7. *Yellowstone National Park: Old Faithful Geyser* The U.S. Geological Survey compiled historical data about Old Faithful Geyser (Yellowstone National Park) from 1870 to 1987. Some of these data are published in the book *The Story of Old Faithful,* by G. D. Marler (Yellowstone Association Press). Let x_1 be a random variable that represents the time interval (minutes) between Old Faithful eruptions in the years 1948 to 1952. Based on 9340 observations, the sample mean interval was $\bar{x}_1 = 63.3$ minutes with sample standard deviation $s_1 = 9.17$ minutes. Let x_2 be a random variable that represents the time interval in minutes between Old Faithful eruptions in the years 1983 to 1987. Based on 25,111 observations, the sample mean time interval was $\bar{x}_2 = 72.1$ minutes with sample standard deviation $s_2 = 12.67$ minutes. Let μ_1 be the population mean of x_1 and μ_2 be the population mean of x_2.

(a) Compute a 99% confidence interval for $\mu_1 - \mu_2$.

(b) Comment on the meaning of the confidence interval in the context of this problem. Does the interval consist of positive numbers only? negative numbers only? a mix of positive and negative numbers? Does it appear (at the 99% confidence level) that a change in the interval length between eruptions has occurred? Many geologic experts believe that the distribution of eruption times of Old Faithful changed after the major earthquake that occurred in 1959.

8. *Psychology: Parental Sensitivity* "Parental Sensitivity to Infant Cues: Similarities and Differences Between Mothers and Fathers," by M. V. Graham (*Journal of Pediatric Nursing*, Vol. 8, No. 6), reports a study of parental empathy for sensitivity cues and baby temperament (higher scores mean more empathy). Let x_1 be a random variable that represents the score of a mother on an empathy test (as regards her baby). Let x_2 be the empathy score of a father. A random sample of 32 mothers gave a sample mean $\bar{x}_1 = 69.44$ with sample standard deviation $s_1 = 11.69$. Another random sample of 32 fathers gave $\bar{x}_2 = 59$ with $s_2 = 11.60$.

(a) Let μ_1 be the population mean of x_1 and μ_2 be the population mean of x_2. Find a 99% confidence interval for $\mu_1 - \mu_2$.

(b) Examine the confidence interval and explain what it means in this context. Does the confidence interval contain all positive, all negative, or both positive and negative numbers? What does this tell you about the relationship between average empathy scores for mothers compared with those for fathers at the 99% confidence level?

9. *Sociology: Canadian Families* A sociologist is studying the size of rural Canadian families. How large are families in rural Ontario, Canada? Does the level of family income make a difference? A random sample of low-income families in the region gave the following number of children in each family x_1:

5	9	3	4	2	6	8	3	6	5	0	4	10	3

Another random sample of high-income families in the region gave the following number of children in each family x_2:

7	3	4	4	4	1	6	4	1	5	3	2	1	10	7	5

(Source: *American Journal of Sociology*, Vol. 53, pp. 470–480.)

(a) Use a calculator with mean and sample standard deviation keys to verify that $\bar{x}_1 \approx 4.9$, $s_1 \approx 2.8$, $\bar{x}_2 \approx 4.2$, and $s_2 \approx 2.5$.

(b) Compute a 95% confidence interval for $\mu_1 - \mu_2$, where μ_1 is the population mean number of children in a low-income family and μ_2 is the population mean number of children in a high-income familiy.

(c) Examine the confidence interval and explain what it means in this context. Does the interval consist of numbers that are all positive? all negative? of different signs? At the 95% level of confidence, does there seem to be a difference in the average number of children for high- or low-income families?

10. *Wildlife: Wolves* David E. Brown is an expert in wildlife conservation. In his book, *The Wolf in the Southwest: The Making of an Endangered Species* (University of Arizona Press), he records the following weights of adult grey wolves from two regions in Old Mexico.

Chihuahua Region: x_1 variable in pounds

86	75	91	70	79
80	68	71	74	64

Durango Region: x_2 variable in pounds

68	72	79	68	77	89	62	55	68
68	59	63	66	58	54	71	59	67

(a) Use a calculator with mean and standard deviation keys to verify that $\bar{x}_1 =$ 75.80 pounds, $s_1 = 8.32$ pounds, $\bar{x}_2 = 66.83$ pounds, and $s_2 = 8.87$ pounds.

(b) Let μ_1 be the mean weight of the population of all grey wolves in the Chihuahua Region. Let μ_2 be the mean weight of the population of all grey wolves in the Durango Region. Find an 85% confidence interval for $\mu_1 - \mu_2$.

(c) Examine the confidence interval and explain what it means in this context. Does the interval consist of numbers that are all positive? all negative? of different signs? At the 85% level of confidence, what can you say about the comparison of the average weight of grey wolves in the Chihuahua Region with the average weight of those in the Durango Region?

11. *Navajo Culture: Traditional Hogans* S. C. Jett is a professor of geography at the University of California, Davis. He and a colleague, V. E. Spencer, are experts on modern Navajo culture and geography. The following information is taken from their book, *Navajo Architecture: Forms, History, Distributions* (University of Arizona Press). On the Navajo Reservation, a random sample of 210 permanent dwellings in the Fort Defiance Region showed that 65 were traditional Navajo hogans. In the Indian Wells Region, a random sample of 152 permanent dwellings showed that 18 were traditional hogans. Let p_1 be the population proportion of all traditional hogans in the Fort Defiance Region, and let p_2 be the population proportion of all traditional hogans in the Indian Wells Region.

(a) Find a 99% confidence interval for $p_1 - p_2$.

(b) Examine the confidence interval and comment on its meaning. Does it include numbers that are all positive? all negative? or mixed? What if it is hypothesized that Navajo who follow the traditional culture of their people tend to live in hogans. Comment on the confidence interval for $p_1 - p_2$ in this context.

12. *Archaeology: Culture Affiliation* "Unknown cultural affiliations and loss of identity at high elevations." These are words used to propose the hypothesis that archaeological sites tend to lose their identity as altitude extremes are reached. This idea is based on the notion that prehistoric people tended *not* to take trade wares to temporary settings and/or isolated areas (Source: *Prehistoric New Mexico: Background for Survey*, by D. E. Stuart and R. P. Gauthier, University of New Mexico Press). As elevation zones of prehistoric people (in what is now the state of New Mexico) increased, there seemed to be a loss of artifact identification. Consider the following information:

Elevation Zone	Number of Artifacts	Number Unidentified
7000–7500 ft	112	69
5000–5500 ft	140	26

Let p_1 be the population proportion of unidentified archaeological artifacts at the elevation zone 7000–7500 ft in the given archaeological area. Let p_2 be the population proportion of unidentified archaeological artifacts at an elevation zone 5000–5500 ft in the given archaeological area.

(a) Find a 99% confidence interval for $p_1 - p_2$.

(b) Explain the meaning of the confidence interval for part (a) in the context of this problem. Does the confidence interval contain all positive numbers? all negative numbers? both positive and negative numbers? What does this tell you (at the

99% confidence level) about the comparison of the population proportion of unidentified artifacts at high elevations (7000–7500 ft) with the population proportion of unidentified artifacts at lower elevations (5000–5500 ft)? How does this relate to the stated hypothesis?

13. *Wildlife: Deer* Mule deer in Colorado have been studied extensively by G. W. Mierau and J. L. Schmidt. In their book, *The Mule Deer of Mesa Verde National Park* (Mesa Verde Museum Association), they give the following information about weights of adult male deer in two Colorado regions. In the Cache la Poudre Region, a random sample of $n_1 = 51$ deer weighed an average of $\bar{x}_1 = 74.04$ kg with sample standard deviation $s_1 = 17.19$ kg. In the Mesa Verde Region, a random sample of $n_2 = 36$ deer gave a mean weight of $\bar{x}_2 = 94.53$ kg with sample standard deviation $s_2 = 19.66$ kg.

   (a) Let μ_1 be the population mean weight of all bucks in the Cache la Poudre Region. Let μ_2 be the population mean weight of all bucks in the Mesa Verde Region. Find a 95% confidence interval for $\mu_1 - \mu_2$.

   (b) Examine the confidence interval and comment on its meaning. Does it include numbers that are all positive? all negative? mixed? What conclusion can you draw (at the 95% level) about the average weight of bucks in the Cache la Poudre Region compared with those in the Mesa Verde Region? The bucks in Mesa Verde National Park are not hunted (tend to be older), and the browse is more abundant. How might these conditions account for the results shown by the confidence interval?

14. *Medical: Plasma Compress* At Community Hospital, the burn center is experimenting with a new plasma compress treatment. A random sample of $n_1 = 316$ patients with minor burns received the plasma compress treatment. Of these patients, it was found that 259 had no visible scars after treatment. Another random sample of $n_2 = 419$ patients with minor burns received no plasma compress treatment. For this group, it was found that 94 had no visible scars after treatment. Let p_1 be the population proportion of all patients with minor burns receiving the plasma compress treatment who have no visible scars. Let p_2 be the population proportion of all patients with minor burns not receiving the plasma compress treatment who have no visible scars.

   (a) Find a 95% confidence interval for $p_1 - p_2$.

   (b) Explain the meaning of the confidence interval found in part (a) in the context of the problem. Does the interval contain numbers that are all positive? all negative? both positive and negative? At the 95% level of confidence, does treatment with plasma compresses seem to make a difference in the proportion of visible scars from minor burns?

15. *Psychology: Self-Esteem* Female undergraduates in randomized groups of 15 took part in a self-esteem study ("There's More to Self-Esteem than Whether It Is High or Low: The Importance of Stability of Self-Esteem," by M. H. Kernis et al., *Journal of Personality and Social Psychology*, Vol. 65, No. 6). The study measured an index of self-esteem from the point of view of competence, social acceptance, and physical attractiveness. Let x_1, x_2, and x_3 be random variables representing the measure of self-esteem through x_1 (competence), x_2 (social acceptance), and x_3 (attractiveness). Higher index values mean a more positive influence on self-esteem.

Variable	Sample Size	$\bar{x}$ Mean	s Standard Deviation	Population Mean
x_1	15	19.84	3.07	μ_1
x_2	15	19.32	3.62	μ_2
x_3	15	17.88	3.74	μ_3

(a) Find an 85% confidence interval for $\mu_1 - \mu_2$.

(b) Find an 85% confidence interval for $\mu_1 - \mu_3$.

(c) Find an 85% confidence interval for $\mu_2 - \mu_3$.

(d) Comment on the meaning of each of the confidence intervals found in parts (a), (b), and (c). At the 85% confidence level, what can you say about the average differences in influence on self-esteem between competence and social acceptance? between competence and attractiveness? between social acceptance and attractiveness?

16. *Focus Problem: Wood Ducks* Solve the Focus Problem at the beginning of this chapter.

17. (a) Suppose that a 95% confidence interval for the difference of means (large sample) contains both positive and negative numbers. Will a 99% confidence interval based on the same data necessarily contain both positive and negative numbers? Explain. What about a 90% confidence interval? Explain.

(b) Suppose that a 95% confidence interval for the difference of proportions contains all positive numbers. Will a 99% confidence interval based on the same data necessarily contain all positive numbers as well? Explain. What about a 90% confidence interval? Explain.

18. *Difference of Means: Sample Size* What about the sample size? If we want a confidence interval with maximal error of estimate E and level of confidence c, then Section 8.4 shows us which formulas to apply for a *single* mean μ or a *single* proportion p.

(a) How about a *difference of means*? For large samples ($n_1 \geq 30$ and $n_2 \geq 30$), the error of estimate E for a $c\%$ confidence interval is

$$E = z_c \sqrt{\frac{\sigma_1^2}{n_1} + \frac{\sigma_2^2}{n_2}}$$

Let us make the simplifying assumption that we have *equal sample size n* so that $n = n_1 = n_2$ and $n \geq 30$. In this context, we get

$$E = z_c \sqrt{\frac{\sigma_1^2}{n} + \frac{\sigma_2^2}{n}} = \frac{z_c}{\sqrt{n}} \sqrt{\sigma_1^2 + \sigma_2^2}$$

Solve this equation for n and show that

$$n = \left(\frac{z_c}{E}\right)^2 (\sigma_1^2 + \sigma_2^2)$$

(b) In Problem 1 (football and basketball player heights), suppose we want to be 95% sure that our estimate $\bar{x}_1 - \bar{x}_2$ for the difference $\mu_1 - \mu_2$ has a maximum error of estimate $E = 0.05$ feet. How large should the sample size be (assuming equal sample size, i.e., $n = n_1 = n_2$)? Since we do not know σ_1 or σ_2, use s_1 and s_2, respectively, from the preliminary sample of Problem 1.

(c) In Problem 2 (baseball and basketball player ages), suppose we want to be 90% sure that our estimate $\bar{x}_1 - \bar{x}_2$ for the difference $\mu_1 - \mu_2$ has a maximum error of estimate $E = 0.5$ years. How large should the sample size be (assuming equal sample size, i.e., $n = n_1 = n_2$)? Since we do not know σ_1 or σ_2, use s_1 and s_2, respectively, from the preliminary sample of Problem 2.

19. *Difference of Proportions: Sample Size* What about the sample size n for confidence intervals for the difference of proportions $p_1 - p_2$? Let us make the following assumptions: *equal sample size $n = n_1 = n_2$* and *all four quantities $n_1\hat{p}_1$, $n_1\hat{q}_1$, $n_2\hat{p}_2$, and $n_2\hat{q}_2$ are greater than 5*. Those readers familiar with algebra can use the same procedure outlined in Problem 18 to show that if we have preliminary estimates $\hat{p}_1$

and $\hat{p}_2$ and a given maximal error of estimate E for a specified confidence level c, then the sample size n should be at least

$$n = \left(\frac{z_c}{E}\right)^2 (\hat{p}_1 \hat{q}_1 + \hat{p}_2 \hat{q}_2)$$

However, if we have no preliminary estimate for $\hat{p}_1$ and $\hat{p}_2$, theory similar to that used in Section 8.4 tells us the sample size n should be at least

$$n = \frac{1}{2}\left(\frac{z_c}{E}\right)^2$$

(a) In Problem 5 (Myers-Briggs personality type indicators in common for married couples), suppose we want to be 99% confident that our estimate $\hat{p}_1 - \hat{p}_2$ for the difference $p_1 - p_2$ has a maximum error of estimate $E = 0.04$. Use the preliminary estimates $\hat{p}_1 = 289/375$ for the proportion of couples sharing two personality traits and $\hat{p}_2 = 23/571$ for the proportion having no traits in common. How large should the sample size be (assuming equal sample size, i.e., $n = n_1 = n_2$)?

(b) Suppose that in Problem 5 we had no preliminary estimate for $\hat{p}_1$ and $\hat{p}_2$ and we want to be 95% confident that our estimate $\hat{p}_1 - \hat{p}_2$ for the difference $p_1 - p_2$ has a maximum error of estimate $E = 0.05$. How large should the sample size be (assuming equal sample size, i.e., $n = n_1 = n_2$)?

SUMMARY

We have studied point estimates and interval estimates. For point estimates, we found E, the maximal error of estimate, and for interval estimates, we found the interval endpoints. In each case, E or endpoints were determined by four factors: the confidence level c, the sample estimate for μ or p, the sample standard deviation s, and the sample size n.

For large samples ($n \geq 30$), we used the normal distribution, the central limit theorem, and sometimes the normal approximation to the binomial distribution. For small samples ($n < 30$), we made use of Student's t distribution. In applications to choose the sample size, we found formulas for n so that, with probability c, the sample estimate for μ or p is in error by less than a preassigned number E.

When we have two independent samples, it is often useful to find confidence intervals for the *difference* of means $\mu_1 - \mu_2$ or the difference of proportions $p_1 - p_2$. To generate confidence intervals for the difference of means, we used the normal distribution for large samples and the Student's t distribution for small samples. We used the normal approximation to the binomial to get confidence intervals for the difference of proportions.

IMPORTANT WORDS & SYMBOLS

Section 8.1
Large samples, $n \geq 30$
Error of estimate $|\bar{x} - \mu|$

Confidence level c
Critical values z_c
Point estimate for μ

VIEWPOINT

All Systems Go?

On January 28, 1986, the Space Shuttle *Challenger* caught fire and blew up only seconds after launch. A great deal of good engineering went into the design of the *Challenger*. However, when a system has several confidence levels operating at once, it can happen, in rare cases, that risks will increase rather than cancel out. (See Chapter Review Problem 17.) Diane Vaughn is a professor of sociology at Boston College and author of the book *The Challenger Launch Decision* (University of Chicago Press). Her book contains an excellent discussion of risks, the normalization of deviants, and cost/safety tradeoffs. Vaughn's book is described as "a remarkable and important analysis of how social structures can induce consequential errors in a decision process" (Robert K. Merton, Columbia University).

CHAPTER REVIEW PROBLEMS

For Problems 2–17 categorize each problem according to (a) parameter being estimated, proportion p, mean μ, difference of means $\mu_1 - \mu_2$, or difference of proportions $p_1 - p_2$ and (b) large sample or small sample. Then solve the problem.

1. In your own words, carefully explain the meaning of the following terms: point estimate, critical value, maximal error of estimate, confidence level, confidence interval, large samples, small samples.

2. *Auto Insurance: Claims* Anystate Auto Insurance Company took a random sample of 370 insurance claims paid out during a 1-year period. The average claim paid was $750 with a standard deviation of $150. Find 0.90 and 0.99 confidence intervals for the mean claim payment.

3. *Psychology: Closure* Three experiments investigating the relation between need for cognitive closure and persuasion were reported in "Motivated Resistance and Openness to Persuasion in the Presence or Absence of Prior Information," by A. W.

Kruglanski (*Journal of Personality and Social Psychology,* Vol. 65, No. 5, pp. 861–874). Part of the study involved administering a "need for closure scale" to a group of students enrolled in an introductory psychology course. The "need for closure scale" has scores ranging from 101 to 201. For the 73 students in the highest quartile of the distribution, the mean score was $\bar{x} = 178.70$ with sample standard deviation $s = 7.81$. These students were all classified as high on their need for closure. Assume that the 73 students represent a random sample of all students who are classified as high on their need for closure. Find a 95% confidence interval for the population mean score μ on the "need for closure scale" for all students with a high need for closure.

4. *Psychology: Closure* How large a sample is needed in Problem 3 if we wish to be 99% confident that the sample mean score is within 2 points of the population mean score for students who are high on the need for closure?

5. *Archaeology: Excavations* The Wind Mountain archaeological site is located in southwestern New Mexico. Wind Mountain was home to an ancient culture of prehistoric Native Americans called Anasazi. A random sample of excavations at Wind Mountain gave the following depth (in centimeters) from present-day surface grade to the location of significant archaeological artifacts (Source: *Mimbres Mogollon Archaeology,* by A. Woosley and A. McIntyre, University of New Mexico Press).

85	45	120	80	75	55	65	60
65	95	90	70	75	65	68	

(a) Use a calculator with mean and sample standard deviation keys to verify that $\bar{x} \approx 74.2$ cm and $s \approx 18.3$ cm.
(b) Compute a 95% confidence interval for the mean depth μ at which archaeological artifacts from the Wind Mountain excavation site can be found.

6. *Archaeology: Pottery* Sherds of clay vessels were put together to reconstruct rim diameters of the original ceramic vessel at the Wind Mountain archaeological site (see source in Problem 5). A random sample of ceramic vessels gave the following rim diameters (in centimeters):

15.9	13.4	22.1	12.7	13.1	19.6	11.7	13.5	17.7	18.1

(a) Use a calculator with mean and sample standard deviation keys to verify that $\bar{x} \approx 15.8$ cm and $s \approx 3.5$ cm.
(b) Compute an 80% confidence interval for the population mean μ of rim diameters for such ceramic vessels found at the Wind Mountain archaeological site.

7. *Telephone Interviews: Survey* The National Study of the Changing Work Force conducted an extensive survey of 2958 wage and salaried workers on issues ranging from relationship with their bosses to household chores. The data were gathered through hour-long telephone interviews with a nationally representative sample (*The Wall Street Journal*). In response to the question, "What does success mean to you?" 1538 responded, "Personal satisfaction from doing a good job." Let p be the population proportion of all wage and salaried workers who would respond the same way to the stated question. Find a 90% confidence interval for p.

8. *Telephone Interviews: Survey* How large a sample is needed in Problem 7 if we wish to be 95% confident that the sample percentage of those equating success with personal satisfaction is within 1% of the population percentage? (*Hint:* Use $p \approx 0.52$ as a preliminary estimate.)

9. *Archaeology: Pottery* Three circle, red on white is one distinctive pattern painted on ceramic vessels of the Anasazi period found at the Wind Mountain archaeological site (see source for Problem 5). At one excavation a sample of 167 potsherds indicated that 68 were of the three circle, red on white pattern.
   (a) Find a point estimate $\hat{p}$ for the proportion of all ceramic potsherds at this site that are of the three circle, red on white pattern.
   (b) Compute a 95% confidence interval for the population proportion p of all ceramic potsherds with this distinctive pattern found at the site.

10. *Archaeology: Pottery* Consider the three circle, red on white pattern discussed in Problem 9. How many ceramic potsherds must be found and identified if we are to be 95% confident that the sample proportion $\hat{p}$ of such potsherds is within 6% of the population proportion of three circle, red on white pattern found at this excavation site? (*Hint:* Use the results of Problem 9 as a preliminary estimate.)

11. *Professors: Salary* A random sample of 43 eastern colleges gave the following average percentage of salary increases for professors: $\bar{x}_1 = 3.6$ with sample standard deviation $s_1 = 1.8$. Another random sample of 40 western colleges gave the following average percentage of salary increases for professors: $\bar{x}_2 = 3.3$ with sample standard deviation $s_2 = 1.7$ (Source: *Academe: Bulletin of the American Association of University Professors,* Vol. 83, No. 2).
   (a) Let μ_1 represent the population mean percentage salary increase for professors in all eastern colleges. Let μ_2 represent the population mean percentage salary increase for professors in all western colleges. Compute a 90% confidence interval for $\mu_1 - \mu_2$.
   (b) Examine the confidence interval and explain what it means in this context. Does the interval consist of numbers that are all positive? all negative? of different signs? At the 90% level of confidence, does it appear that the average percentage salary increase for eastern college professors is higher than that for western college professors?

12. *Stocks: Retail and Utility* How profitable are different sectors of the stock market? One way to answer such a question is to examine profit as a percentage of stockholder equity. A random sample of 32 retail stocks such as Toys 'Я' Us, Best Buy, and Gap was studied for x_1, profit as a percentage of stockholder equity. These results were $\bar{x}_1 = 13.7$ with $s_1 = 4.1$. A random sample of 34 utility (gas and electric) stocks such as Boston Edison, Wisconsin Energy, and Texas Utilities was studied for x_2, profit as a percentage of stockholder equity. The results were $\bar{x}_2 = 10.1$ and $s_2 = 2.7$ (Source: *Fortune 500,* Vol. 135, No. 8).

   (a) Let μ_1 represent the population mean profit as a percentage of stockholder equity for retail stocks, and let μ_2 represent the population mean profit as a percentage of stockholder equity for utility stocks. Find a 95% confidence interval for $\mu_1 - \mu_2$.
   (b) Examine the confidence interval and explain what it means in this context. Does the interval consist of numbers that are all positive? all negative? of different signs? At the 95% level of confidence, does it appear that the profit as a percentage of stockholder equity for retail stocks is higher than for utility stocks?

13. *Wildlife: Wolves* A random sample of 18 adult male wolves from the Canadian Northwest Territories gave an average weight $\bar{x}_1 = 98$ lb with estimated sample standard deviation $s_1 = 6.5$ lb. Another sample of 24 adult male wolves from Alaska gave an average weight $\bar{x}_2 = 90$ lb with estimated sample standard deviation $s_2 = 7.3$ lb (Source: *The Wolf,* by L. D. Mech, University of Minnesota Press).

(a) Let μ_1 represent the population mean weight of adult male wolves from the Northwest Territories, and let μ_2 represent the population mean weight of adult male wolves from Alaska. Find a 75% confidence interval for $\mu_1 - \mu_2$.

(b) Examine the confidence interval and explain what it means in this context. Does the interval consist of numbers that are all positive? all negative? of different signs? At the 75% level of confidence, does it appear that the average weight of adult male wolves from the Northwest Territories is greater than that of the Alaska wolves?

14. *Wildlife: Wolves* A random sample of 17 wolf litters in Ontario, Canada, gave an average of $\bar{x}_1 = 4.9$ wolf pups per litter with estimated sample standard deviation $s_1 = 1.0$. Another random sample of 6 wolf litters in Finland gave an average of $\bar{x}_2 = 2.8$ wolf pups per litter with sample standard deviation $s_2 = 1.2$ (see source for Problem 13).

(a) Find an 85% confidence interval for $\mu_1 - \mu_2$, the difference in population mean litter size between Ontario and Finland.

(b) Examine the confidence interval and explain what it means in this context. Does the interval consist of numbers that are all positive? all negative? of different signs? at the 85% level of confidence, does it appear that the average litter size of wolf pups in Ontario is greater than the average litter size in Finland?

15. *Survey Response: Validity* The book *Survey Responses: An Evaluation of Their Validity,* by E. J. Wentland and K. Smith (Academic Press), includes studies reporting accuracy of answers to questions from surveys. A study by Locander et al. considered the question, "Are you a registered voter?" Accuracy of response was confirmed by a check of city voting records. Two methods of survey were used: a face-to-face interview and a telephone interview. A random sample of 93 people was asked the voter registration question face to face. Seventy-nine respondents gave accurate answers (as verified by city records). Another random sample of 83 people was asked the same question during a telephone interview. Seventy-four respondents gave accurate answers. Assume that the samples are representative of the general population.

(a) Let p_1 be the population proportion of all people who answer the voter registration question accurately during a face-to-face interview. Let p_2 be the population proportion of all people who answer the question accurately during a telephone interview. Find a 95% confidence interval for $p_1 - p_2$.

(b) Does the interval contain numbers that are all positive? all negative? mixed? Comment on the meaning of the confidence interval in the context of this problem. At the 95% level, do you detect any difference in the proportion of accurate responses from face-to-face interviews compared with the proportion of accurate responses from telephone interviews?

16. *Survey Response: Validity* Locander et al. (see reference in Problem 15) also studied the accuracy of responses on questions involving more sensitive material than voter registration. From public records, individuals were identified as having been charged with drunken driving not less than 6 months or more than 12 months from the starting date of the study. Two random samples from this group were studied. In the first sample of 30 individuals, the respondents were asked in a face-to-face interview if they had been charged with drunken driving in the last 12 months. Of these 30 people interviewed face to face, 16 answered the question accurately. The second random sample consisted of 46 people who had been charged with drunken driving. During a telephone interview, 25 of these responded accurately to the question asking if they had been charged with drunken driving during the past 12

months. Assume that the samples are representative of all people recently charged with drunken driving.

(a) Let p_1 represent the population proportion of all people with recent charges of drunken driving who respond accurately to a face-to-face interview asking if they have been charged with drunken driving during the past 12 months. Let p_2 represent the population proportion of people who respond accurately to the question when it is asked in a telephone interview. Find a 90% confidence interval for $p_1 - p_2$.

(b) Does the interval found in part (a) contain numbers that are all positive? all negative? mixed? Comment on the meaning of the confidence interval in the context of this problem. At the 90% level, do you detect any differences in the proportion of accurate responses to the question from face-to-face interviews as compared with the proportion of accurate responses from telephone interviews?

17. What happens if we want several confidence intervals to hold at the same time (concurrently). Do we still have the same level of confidence we had for *each* individual interval?

(a) Suppose we have two independent random variables x_1 and x_2 with respective population means μ_1 and μ_2. Let us say that we use sample data to construct two 80% confidence intervals.

Confidence Interval	Probability μ Is Included
$A_1 < \mu_1 < B_1$	0.80
$A_2 < \mu_2 < B_2$	0.80

Now what is the probability that *both* intervals hold together? Use methods of Section 4.2 to show that

$$P(A_1 < \mu_1 < B_1 \quad and \quad A_2 < \mu_2 < B_2) = 0.64$$

Hint: We are combining independent events. If the confidence is 64% that both intervals hold together, explain why the risk that at least one interval does not hold (i.e., fails) must be 36%.

(b) Suppose that we want *both* intervals to hold with 90% confidence (i.e., only 10% risk level). How much confidence c should each interval have to achieve this combined level of confidence? (Assume that each interval has the same confidence level c.)

Hint: $P(A_1 < \mu_1 < B_1 \quad and \quad A_2 < \mu_2 < B_2) = 0.90$
$P(A_1 < \mu_1 < B_1) \times P(A_2 < \mu_2 < B_2) = 0.90$
$c \times c = 0.90$

Now solve for c.

(c) If we want *both* intervals to hold at the 90% level of confidence, then the individual intervals must hold at a *higher* level of confidence. Write a brief but detailed explanation of how this could be of importance in a large, complex engineering design such as a rocket booster or a spacecraft.

DATA HIGHLIGHTS: GROUP PROJECTS

Break into small groups and discuss the following topics. Organize a brief outline in which you summarize the main points of your group discussion.

Digging clams

1. Garrison Bay is a small bay in Washington state. A popular recreational activity in the bay is clam digging. For several years, this harvest has been monitored and the size distribution of clams recorded. Data for lengths and widths of little neck clams (*Protothaca staminea*) were recorded by a method of systematic sampling in a study done by S. Scherba and V. F. Gallucci ("The Application of Systematic Sampling to a Study of Infaunal Variation in a Soft Substrate Intertidal Environment," *Fishery Bulletin* 74:937–948). The data in Tables 8-4 and 8-5 give lengths and widths for 35 little neck clams.
   (a) Use a calculator to compute the sample mean and the sample standard deviation for lengths and for widths. Compute the coefficient of variation for each.
   (b) Compute a 95% confidence interval for the population mean length of all Garrison Bay little neck clams.
   (c) How many more little neck clams would be needed in a sample if you wanted to be 95% sure the sample mean length is within a maximal error of 10 mm of the population mean length?
   (d) Compute a 95% confidence interval for the population mean width of all Garrison Bay little neck clams.
   (e) How many more little neck clams would be needed in a sample if you wanted to be 95% sure the sample mean width is within a maximal error of 10 mm of the population mean width?
   (f) The *same* 35 clams were used for measures of length and width. Are the sample measurements length and width independent or dependent? Why?

2. Examine Figure 8-8 on the next page, "Clocks roll back Sunday" (*USA Today*).
   (a) Of the 1024 adults surveyed, 66% were reported to favor Daylight Saving Time. How many people in the sample preferred Daylight Saving Time? Using the statistic $\hat{p} = 0.66$ and sample size $n = 1024$, find a 95% confidence interval for

TABLE 8-4   Length of Little Neck Clams (mm)

530	517	505	512	487	481	485	479	452	468
459	449	472	471	455	394	475	335	508	486
474	465	420	402	410	393	389	330	305	169
91	537	519	509	511					

TABLE 8-5   Width of Little Neck Clams (mm)

494	477	471	413	407	427	408	430	395	417
394	397	402	401	385	338	422	288	464	436
414	402	383	340	349	333	356	268	264	141
77	498	456	433	447					

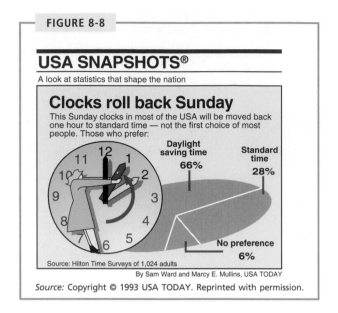

FIGURE 8-8

USA SNAPSHOTS®

A look at statistics that shape the nation

Clocks roll back Sunday

This Sunday clocks in most of the USA will be moved back one hour to standard time — not the first choice of most people. Those who prefer:

Daylight saving time
66%

Standard time
28%

No preference
6%

Source: Hilton Time Surveys of 1,024 adults

By Sam Ward and Marcy E. Mullins, USA TODAY

Source: Copyright © 1993 USA TODAY. Reprinted with permission.

the proportion of people p who favor Daylight Saving Time. How could you report this information in terms of a margin of error?

(b) Look at Figure 8-8 to find the sample statistic $\hat{p}$ for the proportion of people preferring standard time. Find a 95% confidence interval for the population proportion p of people who favor Standard Time. Report the same information in terms of a margin of error.

3. Examine Figure 8-9 "Coupons: Limited clipping" (*USA Today*).
   (a) Use Figure 8-9 to estimate the percentage of merchandise coupons that were redeemed. Also estimate the percentage of dollar value of the coupons that were redeemed. Are these numbers approximately equal?
   (b) Suppose that you are a marketing executive working for a national chain of toy stores. You wish to estimate the percentage of coupons that will be redeemed for the toy stores. How many coupons should you check to be 95% sure that the percentage of coupons redeemed is within 1% of the population proportion of all coupons redeemed for the toy store?
   (c) Use the results of part (a) as a preliminary estimate for p, the percentage of coupons that are redeemed, and redo part (b).
   (d) Suppose that you sent out 937 coupons and found that 27 were redeemed. Explain why you could be 95% confident that the proportion of such coupons redeemed in the future would be between 1.9% and 3.9%.
   (e) Suppose that the dollar value of a collection of coupons was $10,000. Use the data in Figure 8-9 to find the expected value and standard deviation of the dollar value of the redeemed coupons. What is the probability that between $225 and $275 (out of the $10,000) is redeemed?

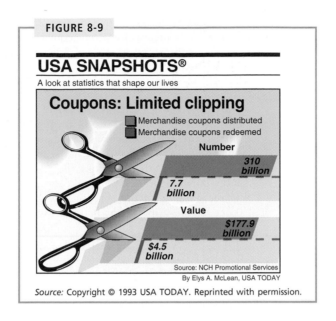

FIGURE 8-9

USA SNAPSHOTS®

A look at statistics that shape our lives

Coupons: Limited clipping

☐ Merchandise coupons distributed
☐ Merchandise coupons redeemed

Number

310 billion

7.7 billion

Value

$177.9 billion

$4.5 billion

Source: NCH Promotional Services

By Elys A. McLean, USA TODAY

Source: Copyright © 1993 USA TODAY. Reprinted with permission.

LINKING CONCEPTS: WRITING PROJECTS

Discuss each of the following topics in class or review the topics on your own. Then write a brief but complete essay in which you summarize the main points. Please include formulas and graphs as appropriate.

1. In this chapter, we have studied confidence intervals. Carefully read the following statements about confidence intervals:
   (a) Once the endpoints of the confidence interval are numerically fixed, then the parameter in question (either μ or p) does or does not fall inside the "fixed" interval.
   (b) A given fixed interval either does or does not contain the parameter μ or p; therefore, the probability is 1 or 0 that the parameter is in the interval.

   Next, read the following statements. Then discuss all four statements in the context of what we actually mean by a confidence interval.
   (c) Nontrivial probability statements can be made only about variables, not constants.
   (d) The confidence level c represents the proportion of all (fixed) intervals that would contain the parameter if we repeated the process many, many times.

2. Throughout Chapter 8, we have used the normal distribution, the central limit theorem, or the Student's t distribution.
   (a) Give a brief outline describing how confidence intervals for means use the central limit theorem or Student's t distribution in their basic construction.
   (b) Give a brief outline describing how the normal approximation to the binomial distribution is used in the construction of confidence intervals for a proportion p.
   (c) Give a brief outline describing how the sample size for a predetermined error tolerance and level of confidence is determined from the normal distribution or the central limit theorem.

3. When results of a survey or a poll are published, the sample size is usually given as well as the margin of error. For example, suppose the *Honolulu Star Bulletin*

reported that they surveyed 385 Honolulu residents and 78% said they favor mandatory jail sentences for people convicted of driving under the influence of drugs or alcohol (with margin of error of 3 percentage points in either direction). Usually the confidence level of the interval is not given, but it is standard practice to use the margin of error for a 95% confidence interval when no other confidence level is given.

(a) The paper reported a point estimate of 78% with margin of error of $\pm 3\%$. Write this information in the form of a confidence interval for p, the population proportion of residents favoring mandatory jail sentences for people convicted of driving under the influence. What is the assumed confidence level?

(b) The margin of error is simply the error due to using a sample instead of the entire population. It does not take into account the bias that might be introduced by the wording of the question, by the truthfulness of the respondents, or by other factors. Suppose the question was asked in this fashion: "Considering the devastating injuries suffered by innocent victims in auto accidents caused by drunken or drugged drivers, do you favor a mandatory jail sentence for those convicted of driving under the influence of drugs or alcohol?" Do you think the wording of the questions would influence the respondents? Do you think the population proportion of those favoring mandatory jail sentences is accurately represented by a confidence interval based on responses to such a question? Explain your answer.

Suppose the question had been: "Considering the existing overcrowding of our prisons, do you favor a mandatory jail sentence for people convicted of driving under the influence of drugs or alcohol?" Do you think the population proportion of those favoring mandatory jail sentences is accurately represented by a confidence interval based on responses to such a question? Explain your answer.

Using TECHNOLOGY

APPLICATION 1

Confidence Interval Demonstration

When we generate different random samples of the same size from a population, we discover that $\bar{x}$ varies from sample to sample. Likewise, different samples produce different confidence intervals for μ. The endpoints $\bar{x} \pm E$ of a confidence interval are statistical variables. A 90% confidence interval says that if we obtain lots of confidence intervals (for the same sample size), then the proportion of all intervals that will turn out to contain μ is 90%.

(a) Use the technology of your choice to generate 10 large random samples from a population with a known mean μ.

(b) Construct a 90% confidence interval for the mean for each sample.

(c) Examine the confidence intervals and note the percentage of the intervals that contain the population mean μ. We have 10 confidence intervals. Will exactly 90% of 10 intervals always contain μ? Explain. What if we have 1000 intervals?

APPLICATION 2

Finding a Confidence Interval for a Population Mean *Mu* (Small Sample)

Cryptanalysis, the science of breaking codes, makes extensive use of language patterns. The frequency of various letter combinations is an important part of the study. A letter combination consisting of a single letter is a monograph, while combinations consisting of two letters are called digraphs, and those with three letters are called trigraphs. In the English language the most frequent digraph is the letter combination TH.

The characteristic rate of a letter combination is a measurement of its rate of occurrence. To compute the characteristic rate, count the number of occurrences of a given letter combination and divide by the number of letters in the text. For instance, to estimate the characteristic rate of the digraph TH, you could select a newspaper text and pick a random starting place. From that place mark off 2000 letters and count the number of times that TH occurs. Then divide the number of occurrences by 2000.

The characteristic rate of a digraph can vary slightly depending on the style of the author; so to estimate an overall characteristic frequency, you want to consider several samples of newspaper text by different authors. Suppose you did this with a random sample of 15 articles and found the characteristic rate of the digraph TH in the articles. The results follow.

0.0275	0.0230	0.0300	0.0255
0.0280	0.0295	0.0265	0.0265
0.0240	0.0315	0.0250	0.0265
0.0290	0.0295	0.0275	

(a) Find a 95% confidence interval for the mean characteristic rate of the digraph TH.

(b) Repeat part (a) for a 90% confidence interval.

(c) Repeat part (a) for an 80% confidence interval.

(d) Repeat part (a) for a 70% confidence interval.

(e) Repeat part (a) for a 60% confidence interval.

(f) For each confidence interval in parts (a)–(e), compute the length of the given interval. Do you notice a relation between the confidence level and the length of the interval?

A good reference for cryptanalysis is a book by Sinkov:

Sinkov, Abraham, *Elementary Cryptanalysis,* New York: Random House, 1968

449

In the book, other common digraphs and trigraphs are given.

APPLICATION 3

Finding a Confidence Interval for a Population Proportion p

There must be nurses on duty in hospitals around the clock. Therefore, many nurses work various shifts. Tasto, Colligan, et al. did a study of the health consequences of shift work. Their results are published in the following government document:

> United States Department of Health, Education, and Welfare, NIOSH Technical Report, Tasto, Colligan, et al., *Health Consequences of Shift Work*. Washington: GPO, 1978, p. 25.

They used a large random sample of nurses on various shifts in 12 hospitals. Part of the report concerns the number of sick days nurses on various shifts take. A random sample of 315 day-shift nurses showed that 62 took no sick days during a six-month period. During that same period a random sample of 309 nurses on rotating duty showed that 51 took no sick days.

(a) We wish to estimate the proportion p of rotating-shift nurses who take no sick days in a six-month period. Find a $c\%$ confidence interval for p when $c = 98, 90, 85, 75, 60$.

(b) For each confidence interval in part (a), compute the length of the interval. Do you notice a relation between confidence level and length of the interval?

(c) We wish to estimate the proportion p of day-shift nurses who take no sick days in a six-month period. Find a $c\%$ confidence interval for p when $c = 99, 95, 80, 70, 60$. Is there a relation between confidence level and length of these intervals?

Technology Hints

TI-83Plus

The TI-83Plus generates random samples from uniform, normal, and binomial distributions. Press the **MATH** key and select **PRB**. Choice **5:randInt**(lower, upper, sample size n) generates random samples of size n from the integers between the specified lower and upper values. Choice **6:randNorm**(μ, σ, sample size n) generates random samples of size n from a normal distribution with specified mean and standard deviation. Choice **7:randBin**(number of trials, p, sample size) generates samples of the specified size from the designated binomial distribution. Under **STAT**, select **EDIT** and highlight the list name such as L1. At the = sign, use the **MATH** key to access the desired population distribution. Finally, use the **Zinterval** under the **TESTS** option of the **STAT** key to generate 90% confidence intervals.

Excel

Use the menu choices **Tools ➤ Data Analysis ➤ Random Number Generator.** In the dialogue box, the number of variables refers to the number of samples. The number of random numbers refers to the number of data in each sample. Select the population distribution (uniform, normal, binomial). The command **Paste function** ⨍ₓ **➤ Statistical ➤ Confidence**(1 − confidence level, σ, sample size) gives the maximal error of estimate E. To find a 90% confidence interval for each sample, use **Confidence**(0.10, σ, sample size) to find the maximal error of estimate E. Note that if you use the population standard deviation σ in the function, the value of E will be the same for all samples of the same size. Next, find the sample mean $\bar{x}$ for each sample (use **Paste function** ⨍ₓ **➤ Statistical ➤ Average**). Finally, construct the end-points $\bar{x} \pm E$ of the confidence interval for each sample.

Minitab

Minitab provides options for sampling from a variety of distributions. To generate random samples from a specific distribution, use the menu selection **Calc ➤ Random Data ➤** then select the population distribution. In the dialogue box, the *number of rows of data* represents the *sample size*. The *number of samples* corresponds to the number of columns selected for data storage. For example, c1−c10 in data storage produces 10 different random samples of the specified size. Use the menu selection **Stat ➤ Basic Statistics ➤ 1 sample *z*** to generate confidence intervals for the mean μ from each sample. In the variables box, list all the columns containing your samples. For instance, using c1−c10 in the variables list will produce confidence intervals for each of the 10 samples stored in columns c1 through c10.

The Minitab display shows 90% confidence intervals for 10 different random samples of size 50 taken from a normal distribution with $\mu = 30$ and $\sigma = 4$. Notice that, as expected, 9 out of 10 of the intervals contain $\mu = 30$.

Minitab Display:

```
Z Confidence Intervals (Samples from a Normal
Population with μ = 30 and σ = 4)
The assumed sigma = 4.00
Variable    N      Mean    StDev   SE Mean       90.0 % CI
C1         50     30.265   4.300    0.566    ( 29.334,  31.195)
C2         50     31.040   3.957    0.566    ( 30.109,  31.971)
C3         50     29.940   4.195    0.566    ( 29.010,  30.871)
C4         50     30.753   3.842    0.566    ( 29.823,  31.684)
C5         50     30.047   4.174    0.566    ( 29.116,  30.977)
C6         50     29.254   4.423    0.566    ( 28.324,  30.185)
C7         50     29.062   4.532    0.566    ( 28.131,  29.992)
C8         50     29.344   4.487    0.566    ( 28.414,  30.275)
C9         50     30.062   4.199    0.566    ( 29.131,  30.992)
C10        50     29.989   3.451    0.566    ( 29.058,  30.919)
```

ComputerStat

Select the menu item **Confidence Intervals,** and then choose **Confidence Interval Demonstration.** This program draws random samples of a designated size from a uniform distribution of numbers between 0 and 1. Then it creates 90% confidence intervals for the population mean μ. A graph of the intervals shows which ones actually contain μ.

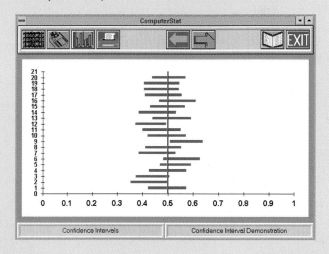

Hypothesis Testing

"Would you tell me, please, which way I ought to go from here?"
"That depends a good deal on where you want to get to," said the Cat.
"I don't much care where—" said Alice.
"Then it doesn't matter which way you go," said the Cat.

—Lewis Carroll
Alice's Adventures in Wonderland

Charles Lutwidge Dodgson (1832–1898)

Using the pseudonym Lewis Carroll, this English mathematician and author wrote *Alice's Adventures in Wonderland.*

Charles Dodgson was an English mathematician who loved to write children's stories in his free time. The dialogue above between Alice and the Cheshire Cat occurs in the masterpiece *Alice's Adventures in Wonderland,* written by Dodgson under the pen name Lewis Carroll. These lines relate to our study of hypothesis testing. Statistical tests cannot answer all of life's questions. They cannot always tell us "where to go," but after this decision is made on other grounds, they can help us find the best way to get there.

For on-line student resources, visit **math.college.hmco.com/students** and follow the Statistics links to the Brase/Brase, *Understandable Statistics,* 7th edition.

◊ Many of life's questions require a yes or no answer! When we must act on incomplete (sample) information, how shall we decide to accept or not accept a proposal?   (SECTION 9.1)

◊ How do we construct statistical tests for μ using large samples?   (SECTION 9.2)

◊ What is the attained level of significance (P value) of a statistical test? What does this have to do with performance reliability?   (SECTION 9.3)

◊ How do we construct statistical tests for μ using small samples?   (SECTION 9.4)

◊ How do we construct statistical tests for the proportion p of successes in a binomial experiment?   (SECTION 9.5)

◊ What are the advantages of pairing data values? How do we construct statistical tests for paired differences?   (SECTION 9.6)

◊ How do we construct statistical tests for differences of independent random variables?   (SECTION 9.7)

Business Opportunities and Start-Up Costs

Be your own boss! Be financially independent with your own business! These are dreams that are within reach if you are willing to do some work and occasionally take some calculated risks. The first step is to evaluate your financial and/or credit position. You do not need a lot of money to start a business, provided you have a reasonable credit rating. Next, you need to decide what kind of business interests you. Then you need to gather data and do a little homework. Some statistics should be helpful here. Where can you get data? Most local or national newspapers have a column called "Business Opportunities." For example, *The Wall Street Journal* has a section called "The Mart: International/National/Regional." Five days a week this section publishes various business opportunities. Most local newspapers—such as the *New York Times, Denver Post, Santa Fe New Mexican,* and *Honolulu Advertiser*—have sections that list business opportunities. If you want to look farther, *The Franchise Handbook,* published by *Enterprise Magazine,* contains over 1500 business start-up opportunities. For more information, visit the Brase/Brase statistics site at http://math.college.hmco.com/students and find the link to franchise businesses.

To be specific, let's say you want to start a pizza franchise. What should you expect to pay for the franchise fee? Franchise fees permit you to use the franchise trade name such as Pizza Factory, Papa John's Pizza, Domino's Pizza, and Rositi's Pizza. The fee usually includes basic training for operating the pizza franchise and making pizzas according to specifications, as well as marketing and advertising support. A random sample of 50 pizza franchises gave the following franchise fees (in thousands of dollars):

25	10	9.5	10	25	30	7.5	18.5	8	15
9.5	15	20	20	12	20	15	10	20	20
20	9.5	25	5	16	20	10	18.5	15	25
15	10	17	20	29	45	15	10	15	10
5.7	25	2	25	30	12	4.5	12	25	18

For these data, we have $n = 50$ data values with sample mean $\bar{x} = 16.58$ thousand dollars and $s = 8.11$ thousand dollars.

Suppose that you are taking out a small business loan, and the loan officer claims that the mean franchise costs for pizza businesses is only $13,900. However, in your loan application package, you are allowing more than $13,900 for the franchise fee. Using the preceding data (and in a gentle, persuasive way), could you effectively argue that the mean franchise costs for pizza businesses are in fact higher?

The sample data have a mean of $16,580, and this seems to contradict the claim that the mean franchise costs are only $13,900. The question is whether or not the sample mean $16,580 is *significantly* greater than $13,900. Put another way, how likely is it that the population mean franchise costs are really $13,900 and the preceding sample results are due only to chance fluctuations or sampling error? In this chapter, we will learn how to answer such questions.

9.1
Introduction to Hypothesis Testing

FOCUS POINTS

✓ Identify the null and alternate hypotheses in a statistical test.

✓ Recognize types of errors, level of significance, and power of a test.

✓ Use the hypotheses to identify an appropriate critical region.

✓ Understand the meaning and risk of "accepting" or "rejecting" a given hypothesis.

In Chapter 1, we emphasized the fact that a statistician's most important job is to draw inferences about populations based on samples taken from the population. Most statistical inference centers around the parameters of a population (often the mean or probability of success in a binomial trial). Methods for drawing inferences about parameters are of two types: Either we make decisions concerning the value of the parameter, or we actually estimate the value of the parameter. When we estimate the value (or location) of a parameter, we are using methods of estimation as studied in Chapter 8. Decisions concerning the value of a parameter are obtained by *hypothesis testing*, the topic we shall study in this chapter.

Students often ask which method should be used on a particular problem—that is, should the parameter be estimated, or should we test a *hypothesis* involving the parameter? The answer lies in the practical nature of the problem and the questions posed about it. Some people prefer to test theories concerning the parameters. Others prefer to express their inferences as estimates. Both estimation and hypothesis testing are found extensively in the literature of statistical applications.

Null hypothesis

Our first step is to establish a working hypothesis about the population parameter in question. This hypothesis is called the *null hypothesis,* denoted by the symbol H_0. The value specified in the null hypothesis is often a historical value, a claim, or a production specification. For instance, if the average height of a professional male basketball player was 6.5 feet 10 years ago, we might use a null hypothesis H_0: $\mu = 6.5$ feet for a study involving the average height of this year's professional male basketball players. If television networks claim that the average length of time devoted to commercials in a 60-minute program is 12 minutes, we would use H_0: $\mu = 12$ minutes as our null hypothesis in a study regarding the length of time devoted to commercials. Finally, if a repair shop claims that it should take an average of 25 minutes to install a new muffler on a passenger automobile, we would use H_0: $\mu = 25$ minutes as the null hypothesis for a study of how well the repair shop is conforming to specified times for a muffler installation.

Alternate hypothesis

Any hypothesis that differs from the null hypothesis is called an *alternate hypothesis*. An alternate hypothesis is constructed in such a way that it is the one to be accepted when the null hypothesis must be rejected. The alternate hypothesis is denoted by the symbol H_1. For instance, if we believe the average height of professional male basketball players is taller than it was 10 years ago, we would use an alternate hypothesis H_1: $\mu > 6.5$ feet with the null hypothesis H_0: $\mu = 6.5$ feet.

EXAMPLE 1

Null and alternate hypotheses

A car manufacturer advertises that its new subcompact models get 47 miles per gallon (mpg). Let μ be the mean of the mileage distribution for these cars. You assume that the dealer will not underrate the car, but you suspect that the mileage might be overrated.

(a) What shall we use for H_0?

We want to see if the dealer's claim $\mu = 47$ can be rejected. Therefore, our null hypothesis is simply that $\mu = 47$. We denote the null hypothesis as

$$H_0: \mu = 47$$

(b) What shall we use for H_1?

From experience with this dealer, we have every reason to believe that the advertised mileage is too high. If μ is not 47, we are sure it is less than 47. Therefore, the alternate hypothesis is

$$H_1: \mu < 47$$

◇

GUIDED EXERCISE 1

Null and alternate hypotheses

A company manufactures ball bearings for precision machines. The average diameter of a certain type of ball bearing should be 6.0 mm. To check that the average diameter is correct, the company formulates a statistical test.

(a) What should be used for H_0? (*Hint:* What is the company trying to test?)

⟹ If μ is the mean diameter of the ball bearings, the company wants to test $\mu = 6.0$ mm. Therefore, H_0: $\mu = 6.0$.

(b) What should be used for H_1? (*Hint:* An error either way, too small or too large, would be serious.)

⟹ An error either way could occur, and it would be serious. Therefore, H_1: $\mu \neq 6.0$ (μ is either smaller than or larger than 6.0).

Null and alternate hypotheses

A package delivery service claims it takes an average of 24 hours to send a package from New York to San Francisco. An independent consumer agency is doing a study to test the truth of this claim. Several complaints have led the agency to suspect that the delivery time is longer than 24 hours.

(a) What should be used for the null hypothesis?

⟹ The claim $\mu = 24$ hours is in question, so we take H_0: $\mu = 24$.

(b) Assuming that the delivery service does not underrate itself, what should be used for the alternate hypothesis?

⟹ If the delivery service does not underrate itself, then the only reasonable alternate hypothesis is H_1: $\mu > 24$.

Types of errors

If we *reject the null hypothesis when it is,* in fact, *true,* we have an error that is called a *type I error.* On the other hand, if we *accept the null hypothesis when it is,* in fact, *false,* we have made an error that is called a *type II error.* Table 9-1 indicates how these errors occur.

For tests of hypotheses to be well constructed, they must be designed to minimize possible errors of decision. (Usually, we do not know if an error has been made, and therefore, we can talk about only the probability of making an error.) Usually, for a given sample size, an attempt to reduce the probability of one type of error results in an increase in the probability of the other type of error. In practical applications, one type of error may be more serious than another. In such a case, careful attention is given to the more serious error. If we increase the sample size, it is possible to reduce both types of errors, but increasing the sample size may not be possible.

Level of significance

The probability with which we are willing to risk a type I error is called the *level of significance of a test.* The level of significance is denoted by the Greek letter α (pronounced "alpha"). In good statistical practice, α is specified in advance before any samples are drawn so that results will not influence the choice for the level of significance.

The *probability of making a type II error* is denoted by the Greek letter β (pronounced "beta"). Methods of hypothesis testing require us to choose α and β values to be as small as possible. In elementary statistical applications, we usually choose α first.

TABLE 9-1 Type I and Type II Errors

	Our Decision	
Truth of H_0	And if we do not reject H_0	And if we reject H_0
If H_0 is true	Correct decision; no error	Type I error
If H_0 is false	Type II error	Correct decision; no error

TABLE 9-2 Probabilities Associated with a Statistical Test

	Our Decision	
Truth of H_0	And if we accept H_0 as True	And if we reject H_0 as False
H_0 is true	Correct decision, with corresponding probability $1 - \alpha$	Type I error, with corresponding probability α called the *level of significance of the test*
H_0 is false	Type II error, with corresponding probability β	Correct decision, with corresponding probability $1 - \beta$ called the *power of the test*

Power of a test

The quantity $1 - \beta$ is called the *power of the test* and represents the probability of rejecting H_0 when it is in fact false. For a given level of significance, how much power can we expect from a test? The actual value of the power is usually difficult (and sometimes impossible) to obtain, since it requires us to know the H_1 distribution. However, we can make the following general comments:

1. The power of a statistical test increases as the level of significance α increases. A test performed at the $\alpha = 0.05$ level has more power than one at $\alpha = 0.01$. This means that the less stringent we make our significance level α, the more likely we will reject the null hypothesis when it is false.

2. Using a larger value of α will increase the power, but it also will increase the probability of a type I error. Despite this fact, most business executives, administrators, social scientists, and scientists use *small* α values. This choice reflects the conservative nature of administrators and scientists, who are usually more willing to make an error by failing to reject a claim (i.e., H_0) than to make an error by accepting another claim (i.e., H_1) that is false. Table 9-2 summarizes the probabilities of errors associated with a statistical test.

◇ **COMMENT** Since the calculation of the probability of a type II error is treated in advanced statistics courses, we will restrict our attention to the probability of a type I error. ◇

Meaning of accepting H_0

In most statistical applications, the level of significance is specified to be $\alpha = 0.05$ or $\alpha = 0.01$, although other values can be used. If $\alpha = 0.05$, then we say we are using a 5% level of significance. This means that in 100 similar situations, H_0 will be rejected 5 times, on average, when it should not have been rejected.

When we accept (or fail to reject) the null hypothesis, we should understand that we are *not proving the null hypothesis*. We are saying only that the sample evidence (data) is not strong enough to justify rejection of the null hypothesis. The word *accept* sometimes has a stronger meaning in common English usage than we are willing to give in our application of statistics. Therefore, we often use the expression *fail to reject* H_0 instead of accept H_0. *Fail to reject* the null hypothesis simply means the evidence in favor of rejection was not strong enough (see Table 9-3 on the following page). Often, in the case H_0 cannot be rejected, a confidence interval is used to estimate the parameter in question. The confidence interval gives the statistician a range of possible values for the parameter.

TABLE 9-3   Meaning of the Terms *Fail to Reject H_0* and *Reject H_0*

Term	Meaning
Fail to reject H_0	There is not enough evidence in the data (and the test being used) to justify a rejection of H_0. This means that we retain H_0 with the understanding that we have not proved it to be true beyond all doubt.
Reject H_0	There is enough evidence in the data (and the test employed) to justify rejection of H_0. This means that we choose the alternate hypothesis H_1 with the understanding that we have not proved H_1 to be true beyond all doubt.

EXAMPLE 2

Types of errors

Let's reconsider Example 1 regarding the average gas mileage of a compact car. The hypotheses for the test are

H_0: $\mu = 47$ mpg (manufacturer's claim)

H_1: $\mu < 47$ mpg (consumer researcher's suspicion)

(a) Suppose the level of significance is $\alpha = 0.05$. Describe a type I error and its probability.

SOLUTION: If, based on sample evidence, we reject the manufacturer's claim that $\mu = 47$ mpg when in fact the average number of miles per gallon achieved by the compact car is 47 mpg (or higher), we have committed a type I error. The probability of making such an error is as high as 5% because the level of significance is 5%.

(b) Describe a type II error.

SOLUTION: If, based on sample evidence, we fail to reject the manufacturer's claim of $\mu = 47$ mpg when, in fact, the alternate hypothesis H_1: $\mu < 47$ mpg is true, we have made a type II error. The probability of such an error is designated by the letter β. (The computation of β is beyond the scope of this text.)   ◇

GUIDED EXERCISE 3

Types of errors

Let's reconsider Guided Exercise 1 in which we were considering the manufacturing specifications for ball bearings. The hypotheses were

H_0: $\mu = 6.0$ mm (manufacturer's specification) H_1: $\mu \neq 6.0$ mm (cause for adjusting process)

(a) Suppose the manufacturer requires a 1% level of significance. Describe a type I error, its consequence and probability.

 A type I error is caused when sample evidence indicates that we should reject H_0 when, in fact, the average diameter of the ball bearings being produced is 6.0 mm. A type I error will cause a needless adjustment and delay of the manufacturing process. The probability of such an error is 1% because $\alpha = 0.01$.

Continued

GUIDED EXERCISE 3 continued

(b) Discuss a type II error and its consequences.

 A type II error occurs if the sample evidence tells us not to reject the null hypothesis H_0: $\mu = 6.0$ mm when, in fact, the average diameter of the ball bearing is either too large or too small to meet specifications. Such an error would mean that the production process would not be adjusted when it really needed to be adjusted. This could possibly result in a large production of ball bearings that do not meet specifications.

Critical regions

We use information from a sample to determine if we should reject or not reject the null hypothesis. However, we know that there will be variability among samples. For instance, if we test drive 40 of the compact automobiles discussed in Examples 1 and 2 and compute the average miles per gallon for the 40 automobiles, we know we are likely to find that $\bar{x}$ differs from the null-hypothesis claim H_0: $\mu = 47$ mpg. A difference is likely to occur regardless of the truth of H_0. The question is, How much less than 47 mpg can $\bar{x}$ be before we begin to suspect that the population mean mileage μ is less than 47, as stated in the alternate hypothesis? In other words, how much less can $\bar{x}$ be than 47 mpg before we reject H_0: $\mu = 47$ mpg and accept the alternate hypothesis H_1: $\mu < 47$ mpg?

The answer to the question regarding the relative size of $\bar{x}$ and μ, as stated in the null hypothesis, depends on the sample size, the sampling distribution of $\bar{x}$, the alternate hypothesis H_1, and the level of significance α. If the sample test statistic $\bar{x}$ is sufficiently different from the claim about μ made in the null hypothesis, we reject the null hypothesis. The values of $\bar{x}$ for which we reject H_0 are called the *critical region* of the $\bar{x}$ distribution. Depending on the alternate hypothesis, the critical region is located on the left side, the right side, or both sides of the $\bar{x}$ distribution. If the critical region is on the left side, we call the test a *left-tailed* test. Similarly, if the critical region is on the right side, the test is called a *right-tailed* test. If the critical region is on both sides, the test is called *two-tailed*.

H_1: $\mu < k$
Left-tailed test

H_1: $\mu > k$
Right-tailed test

H_1: $\mu \neq k$
Two-tailed test

Figure 9-1 on the next page shows the relationship of the critical region to the alternate hypothesis and the level of significance α.

In the next section, we will see how to use the level of significance α and the alternate hypothesis H_1 to compute the boundary for the critical region(s). Then we will see whether or not the sample test statistic falls in the critical region. If it does, we reject H_0. If it does not, we do not reject H_0.

FIGURE 9-1

Critical Regions for H_0: $\mu = k$ ($\bar{x}$ Values Are Under the Shaded Regions)

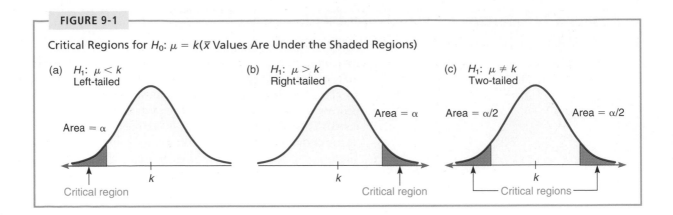

(a) H_1: $\mu < k$
    Left-tailed

Area = α

k

Critical region

(b) H_1: $\mu > k$
    Right-tailed

Area = α

k

Critical region

(c) H_1: $\mu \neq k$
    Two-tailed

Area = $\alpha/2$ Area = $\alpha/2$

k

Critical regions

Summary

In Chapter 8, we used *confidence intervals* to *estimate* the value of a parameter such as the population mean μ or probability of success p. Theory of estimation is very useful and certainly has a valuable place in statistical applications. However, confidence intervals by themselves don't answer every important type of question. For example, there are many situations in real life wherein the final decision must be a *simple yes or no*! Such a decision is based on the supposed value of a parameter, sample data, and a probability distribution involving the parameter.

Yes-or-no decisions are the essence of *hypothesis testing,* and this is the main topic of our chapter. When we make such yes-or-no decisions about the value of a statistical parameter, we realize that our decision could be wrong. Therefore, we also need to consider the types of errors and associated probabilities for such errors.

In this section, we have introduced some of the basic concepts of hypothesis testing. In particular, we have seen the null and alternate hypotheses, the level of significance, the types of errors possible, and the critical region. In the next section, we will see how to use information from a sample to conclude the test and determine whether or not to reject the null hypothesis.

◇ **COMMENT: NOTATION REGARDING THE NULL HYPOTHESIS** In statistical testing, the null hypothesis H_0 always contains the equals symbol. However, in the null hypothesis, some statistical software packages and texts also include the inequality symbol that is opposite that shown in the alternate hypothesis. For instance, if the alternate hypothesis is μ is less than 3 ($\mu < 3$), then the corresponding null hypothesis is sometimes written as μ is greater than or equal to 3 ($\mu \geq 3$). The mathematical construction of a statistical test uses the null hypothesis to assign a specific number (rather than a range of numbers) to the parameter μ in question. The null hypothesis establishes a single fixed value for μ so we are working with a single distribution having a specific mean. In this case, H_0 assigns $\mu = 3$. So when H_1: $\mu < 3$ is the alternate hypothesis, we follow the commonly used convention of writing the null hypothesis simply as H_0: $\mu = 3$. ◇

VIEWPOINT   *Lovers Take Heed!!!*

If you are going to whisper sweet nothings to your sweetheart, be sure to whisper in the *left* ear. Professor Sim of Sam Houston State University (Huntsville, Texas) found that emotionally loaded words had a higher recall rate when spoken into a person's left ear, not the right. Professor Sim presented his findings at the British Psychology Society European Congress. He told the Congress that his findings are consistent with the hypothesis that the brain's right hemisphere has more influence in the processing of emotional stimuli. The left ear is controlled by the right side of the brain. Sim's research involved statistical tests like the ones you will study in this chapter.

SECTION 9.1   PROBLEMS

1. Discuss each of the following topics in class or review the topics on your own. Then write a brief but complete essay in which you answer the following questions.
   (a) What is a null hypothesis H_0?
   (b) What is an alternate hypothesis H_1?
   (c) What is a type I error? a type II error?
   (d) What is the level of significance of a test? What is the probability of a type II error?

2. In a statistical test, we have a choice of a left-tailed critical region, right-tailed critical region, or two-tailed critical region. Is it the null hypothesis or the alternate hypothesis that determines which type of critical region is used? Explain your answer.

3. If we fail to reject (i.e., "accept") the null hypothesis, does this mean that we have *proven* it to be true beyond *all* doubt? Explain your answer.

4. If we reject the null hypothesis, does this mean that we have *proven* it to be false beyond *all* doubt? Explain your answer.

5. *Veterinary Science: Colts* The body weight of a healthy 3-month-old colt should be about $\mu = 60$ kg. (Source: *The Merck Veterinary Manual*, a standard reference manual used in most veterinary colleges.)
   (a) If you want to set up a statistical test to challenge the claim that $\mu = 60$ kg, what would you use for the null hypothesis H_0?
   (b) In Nevada, there are many herds of wild horses. Suppose that you want to test the claim that the average weight of a wild Nevada colt (3 months old) was less than 60 kg. What would you use for the alternate hypothesis H_1?
   (c) Suppose that you want to test the claim that the average weight of such a wild colt was greater than 60 kg. What would you use for the alternate hypothesis?
   (d) Suppose that you want to test the claim that the average weight of such a wild colt was *different* from 60 kg. What would you use for the alternate hypothesis?
   (e) For each of the tests in parts (b), (c), and (d), would the critical region be on the left, right, or both sides of the mean? Explain your answer in each case.

6. *Marketing: Shopping Time* How much customers buy is a direct result of how much time they spend in the store. A study of average shopping time in a large national houseware store gave the following information (Source: *Why We Buy: The Science*

of Shopping by P. Underhill):

> Women with female companion: 8.3 min.
> Women with male companion: 4.5 min.

Suppose you want to set up a statistical test to challenge the claim that a woman with a female friend spends an average of 8.3 minutes shopping in such a store.

(a) What would you use for the null and alternate hypotheses if you believe the average shopping time is less than 8.3 minutes? Is the critical region on the right, left, or on both sides of the mean?

(b) What would you use for the null and alternate hypotheses if you believe the average shopping time is different from 8.3 minutes? Is the critical region on the right, left, or on both sides of the mean?

Stores that sell mainly to women should figure out a way to engage the interest of men! Perhaps comfortable seats and a big TV with sports programs. Suppose such an entertainment center was installed and now you wish to challenge the claim that a woman with a male friend spends only 4.5 minutes shopping in a houseware store.

(c) What would you use for the null and alternate hypotheses if you believe the average shopping time is more than 4.5 minutes? Is the critical region on the right, left, or on both sides of the mean?

(d) What would you use for the null and alternate hypotheses if you believe the average shopping time is different from 4.5 minutes? Is the critical region on the right, left, or on both sides of the mean?

7. *Meteorology: Storms* *Weatherwise* magazine is published in association with the American Meteorological Society. Volume 46, Number 6 has a rating system to classify Nor'easter storms that frequently hit New England states and can cause much damage near the ocean coast. A *severe* storm has an average peak wave height of 16.4 feet for waves hitting the shore. Suppose that a Nor'easter is in progress at the severe storm class rating.

(a) Let us say we want to set up a statistical test to see if the wave action (i.e., height) is dying down or getting worse. What would be the null hypothesis regarding average wave height?

(b) If you wanted to test the hypothesis that the storm is getting worse, what would you use for the alternate hypothesis?

(c) If you wanted to test the hypothesis that the waves are dying down, what would you use for the alternate hypothesis?

(d) Suppose that you do not know if the storm is getting worse or dying out. You just want to test the hypothesis that the average wave height is *different* (either up or down) from the severe storm class rating. What would you use for the alternate hypothesis?

(e) For each of the tests in parts (b), (c), and (d) would the critical region be on the left, right, or both sides of the mean? Explain your answer in each case.

8. *Chrysler Concorde: Acceleration* *Consumer Reports* stated that the mean time for a Chrysler Concorde to go from 0 to 60 miles per hour was 8.7 seconds.

(a) If you want to set up a statistical test to challenge the claim of 8.7 seconds, what would you use for the null hypothesis?

(b) The town of Leadville, Colorado, has an elevation over 10,000 feet. Suppose that you wanted to test the claim that the average time to accelerate from 0 to 60 miles per hour is longer in Leadville (because of less oxygen). What would you use for the alternate hypothesis?

(c) Suppose that you made an engine modification and you think the average time to accelerate from 0 to 60 miles per hour is reduced. What would you use for the alternate hypothesis?

(d) For each of the tests in parts (b) and (c), would the critical region be on the left, right, or both sides of the mean? Explain your answer in each case.

9. *Pro Football: Western Division The Sports Encyclopedia: Pro Football* (11th Edition) indicated that the mean weight for football players in the Western division was approximately 288 pounds several years ago.
   (a) Suppose that we want to set up a statistical test that challenges the statement that the mean weight this year is 288 pounds. What would you use for the null hypothesis?
   (b) What would you use for the alternate hypothesis if you thought that the average weight was higher? lower? different (either way)? In each case, would the critical region be in the left, right, or both tails? Explain each answer.

9.2
Tests Involving the Mean μ (Large Samples)

FOCUS POINTS

✓ Review the general procedure for testing.
✓ Identify the components of large sample tests for μ.
✓ Compute critical values and critical regions for large sample tests of μ.
✓ Compute the sample test statistic.
✓ Understand the meaning of the terms "statistically significant" and "not statistically significant."

The general procedure for hypothesis testing involves several steps, some of which we examined in the preceding section:

1. Establish the null hypothesis H_0 and the alternate hypothesis H_1.

2. Use the level of significance α and the alternate hypothesis to determine the critical region.

In this section, we will complete the following steps:

3. Find the critical values that form the boundaries of the critical region.

4. Use the sample evidence to draw a conclusion regarding whether to reject the null hypothesis H_0 or not.

The basic concepts for hypothesis testing remain the same for all tests of hypotheses, regardless of the parameter in question (such as mean, proportion, difference of means, difference of proportions, standard deviation). However, the specific details will change, depending on the particular parameter being tested and the nature of the sampling distribution of the test statistic.

In this section, we will study the particular details of hypothesis tests about a population mean μ when our sample evidence comes from large samples ($n \geq 30$). We first establish the null and alternate hypotheses. For tests of means, Table 9-4 shows the hypotheses.

Null and alternate hypotheses

To decide whether or not to reject H_0, we look at sample evidence. From the sample, we compute $\bar{x}$. The question is: Is $\bar{x}$ far enough away from the value of μ

TABLE 9-4   The Null and Alternate Hypotheses

Null Hypothesis	Alternate Hypotheses and Type of Test		
Claim about μ or historical value of μ	You believe μ is less than value stated in H_0	You believe μ is more than value stated in H_0	You believe μ is different from value stated in H_0
H_0: $\mu = k$	H_1: $\mu < k$	H_1: $\mu > k$	H_1: $\mu \neq k$
	Left-tailed test	Right-tailed test	Two-tailed test

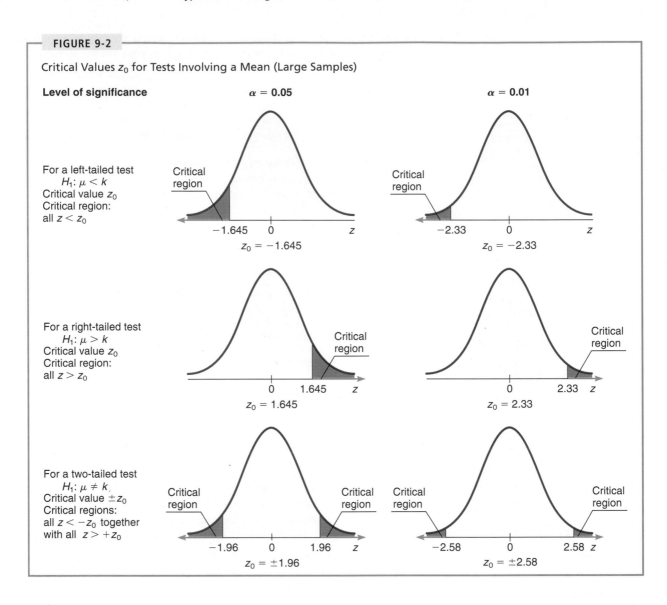

FIGURE 9-2

Critical Values z_0 for Tests Involving a Mean (Large Samples)

Critical values for testing μ (large samples)

Sample test statistic

hypothesized in H_0 to merit rejecting H_0? To answer this question, we look at the $\bar{x}$ distribution *under the assumption that H_0 is true.*

According to the central limit theorem, the $\bar{x}$ distribution based on samples of size n, where n is large ($n \geq 30$), is approximately *normal.* For this reason, our critical values are z values. Figure 9-2 gives the critical values for *one-tailed* and *two-tailed tests* for two commonly used levels of significance, $\alpha = 0.05$ and $\alpha = 0.01$. The critical values given in Figure 9-2 are based on the fact that the *area* above the critical region equals the level of significance. For instance, for a left-tailed test and $\alpha = 0.05$, the area above the critical region is 0.05. Using methods of Chapter 6 and the normal distribution table (Table 5, Appendix II), we see that $z = -1.645$ is the z value that "cuts off" 5% of the normal curve on the left.

To determine if the sample $\bar{x}$ is sufficiently far away from the mean proposed in the null hypothesis to merit rejecting H_0, we convert the sample value $\bar{x}$ to a z value and see if the sample z value falls in the critical region. If so, we reject H_0

and accept H_1. If not, we do not reject H_0. The conversion of $\bar{x}$ to z follows the formula

$$z = \frac{\bar{x} - \mu}{(\sigma/\sqrt{n})}$$

where μ = mean specified in H_0

σ = standard deviation of the x distribution

n = sample size being used

$\bar{x}$ = sample test statistic

$\diamond$ **COMMENT** *The methods used in this section will require $\bar{x}$ to be approximately normally distributed,* and we will approximate σ by the sample standard deviation s when necessary. To ensure that our methods yield accurate results, we will *require* our *sample size to be 30 or larger* ($n \geq 30$). Tests involving small samples ($n < 30$) will be presented in Section 9.4. $\diamond$

EXAMPLE 3

Hypothesis testing procedure

Statement of Problem: Suppose the St. Louis Zoo wishes to obtain eggs of a rare Mississippi river turtle. The zoo will hatch the eggs and raise the turtles as an exhibit of a rare and endangered species. Carol Wright is the staff biologist at the zoo who has been given the job of finding the eggs to be hatched. Turtles of the area bury their eggs in a nest in sandbanks along the river. Then the nest is abandoned, and the eggs hatch by themselves. A number of different species of turtles live in the region, and eggs from each species look much alike. Past research has shown that lengths of turtle eggs are normally distributed, and lengths of the rare turtle eggs have population mean $\mu = 7.50$ cm with standard deviation $\sigma = 1.5$ cm. The rare turtle egg is the only one with mean length 7.50 cm. The mean lengths of eggs from each of the other species are longer than 7.50 cm. The eggs of all the local turtle species have the same standard deviation $\sigma = 1.5$ cm for length.

After searching for some time, Carol found a nest with 36 eggs. The way the nest was constructed makes her suspect that it was made by the rare turtle. The mean length of this collection of 36 eggs is $\bar{x} = 7.74$ cm. Since $\bar{x} = 7.74$ cm is longer than the population mean $\mu = 7.50$ cm of the rare turtle, Carol is a little worried that the eggs may come from a species that lays larger eggs.

Let's use a statistical test to help the biologist make a decision. Pay close attention to this example; it contains principal features common to most statistical tests.

I. *Summary of Known Facts:* In any testing problem, it is a good idea to make a short summary of known facts before we start to construct the test.

(a) The distribution of lengths of eggs from the rare turtle is *normal*, with population mean $\mu = 7.50$ cm and standard deviation $\sigma = 1.5$ cm.

(b) The nest contains $n = 36$ eggs with an observed mean $\bar{x} = 7.74$ cm.

(c) Since the population is *normal* and μ and σ are known, Theorem 7.2 tells us that $\bar{x}$ is also normally distributed. The mean and standard deviation of the $\bar{x}$

distribution are

$$\mu_{\bar{x}} = \mu = 7.50 \text{ cm}$$

$$\sigma_{\bar{x}} = \frac{\sigma}{\sqrt{n}} = \frac{1.5}{\sqrt{36}} = 0.25 \text{ cm}$$

II. *Establishing H_0 and H_1:* The null hypothesis H_0 is set up for the primary purpose of seeing whether or not it can be rejected. The alternate hypothesis H_1 will be chosen when the null hypothesis must be rejected.

Let μ be the mean length of the population distribution from which our sample of 36 eggs is drawn. Our biologist suspects that the nest contains the rare turtle eggs; therefore, we will use the null hypothesis

$$H_0: \mu = 7.50$$

since the rare turtle eggs are known to have mean length $\mu = 7.50$ cm. However, all other local species lay eggs with a longer population mean. Therefore, the alternate hypothesis is

$$H_1: \mu > 7.50$$

III. *Choosing the Level of Significance α:* The null hypothesis says that the eggs in the nest are from the rare turtle. A type I error means that we reject the nest as being formed by the rare turtle when it was, in fact, formed by the rare turtle. A type II error means that we accept the nest as coming from the rare turtle when it really came from a common species.

A type I error could be serious; we don't want to reject the eggs if they are from the rare turtle. A type II error is not too serious. If the eggs are from a common species, the turtles can be released after they hatch. Naturally, the zoo doesn't want to incubate the wrong eggs, but it would not be too serious if it did.

Although other levels of significance may be used, most researchers use either $\alpha = 0.05$ or $\alpha = 0.01$. Our biologist wants to use a level of significance $\alpha = 0.01$. This means that she is willing to risk a type I error with a probability of $\alpha = 0.01$.

IV. *Graphical Model for the Test:* What decision procedure shall we use to test our hypothesis? In a way, the logic of our decision process is similar to that used in a typical courtroom setting, but our methods are mathematical rather than legal. In a courtroom setting, the person charged with a crime is initially considered innocent (null hypothesis). If evidence (data) presented in court can sufficiently discredit the person's innocence, he or she is then judged to be guilty (alternate hypothesis). In a similar way, we will consider the null hypothesis to be *true* until there is enough data (mathematical evidence) to discredit it at the $\alpha = 0.01$ level of significance.

Figure 9-3 shows the graphical model for a test with hypotheses

$$H_0: \mu = 7.50 \qquad H_1: \mu > 7.50$$

and a large sample. Let us examine the details of Figure 9-3.

(a) First, we determine the *sampling distribution*. By the central limit theorem (Theorem 7.2), the *sample test statistic* $\bar{x}$ follows a *normal* distribution. We convert the $\bar{x}$ distribution to the standard normal distribution by the formula

$$z = \frac{\bar{x} - \mu}{(\sigma/\sqrt{n})}$$

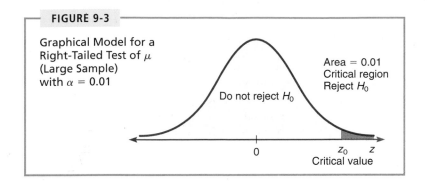

FIGURE 9-3

Graphical Model for a
Right-Tailed Test of μ
(Large Sample)
with $\alpha = 0.01$

Do not reject H_0

Area = 0.01
Critical region
Reject H_0

0

z_0 z
Critical value

where μ is given in H_0, σ is given or estimated from the sample (large sample), and n is the sample size. In our case, $\mu = 7.50$, $\sigma = 1.5$, and $n = 36$. Figure 9-3 is the standard normal distribution.

(b) Next, we find the critical region. Since H_1: $\mu > 7.50$ claims that μ is greater than 7.50, the *critical region* for this problem is a *right tail* of the standard normal distribution. This tail has an area of $\alpha = 0.01$. If the z value corresponding to the sample statistic $\bar{x} = 7.74$ falls in the critical region, we say that there is enough evidence to discredit the null hypothesis, and we *reject H_0* at the $\alpha = 0.01$ level of significance. If our observed value falls outside the critical region, we conclude that the evidence is not strong enough to reject H_0, so in this case, we *fail to reject H_0*. Since H_1: $\mu > 7.50$ claims that μ is greater than 7.50, then an observed sample statistic $\bar{x}$ *far enough* to the *right* of $\mu = 7.50$ discredits the null hypothesis and supports the alternate hypothesis. In other examples, we will encounter left-tailed and two-tailed tests.

(c) The value z_0 is called the *critical value for the test*. It is the separation point for the critical region. If the z value of the sample statistic $\bar{x} = 7.74$ falls to the *left* of z_0, we *fail to reject H_0*. If it falls to the *right* of z_0, we *reject H_0*. Once we know z_0 and convert the sample statistic $\bar{x}$ to z, we can quickly finish the test.

V. *Finding the Critical Value, Critical Region, and z Value for $\bar{x}$*: Figure 9-2 gives critical values z_0 to use for tests with sampling distributions that are normal. We find these values, as we did in Chapter 6, by using Table 5, "Areas of a Standard Normal Distribution," in Appendix II. In this problem, we have a right-tailed test with level of significance $\alpha = 0.01$. We want to find the value z_0 so that 1% of the standard normal curve lies to the right of z_0. This is equivalent to finding the z_0 value so that 99% of the area lies to the left of z_0. Table 5 shows that $z_0 = 2.33$. The critical region is all values of z greater than or equal to $z_0 = 2.33$ (see Figure 9-4 on the next page).

The sample "evidence" we have is $\bar{x} = 7.74$ based on a nest of 36 eggs. We convert $\bar{x}$ to z using the formula in part IV:

$$z = \frac{\bar{x} - \mu}{(\sigma/\sqrt{n})} = \frac{7.74 - 7.50}{(1.5/\sqrt{36})} = 0.96$$

Now we have all the information we need to complete the test.

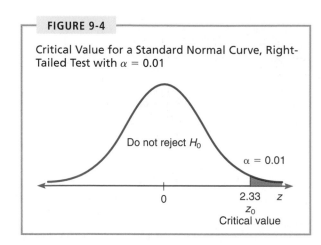

FIGURE 9-4

Critical Value for a Standard Normal Curve, Right-Tailed Test with $\alpha = 0.01$

Do not reject H_0

$\alpha = 0.01$

0   2.33 z
           z_0
        Critical value

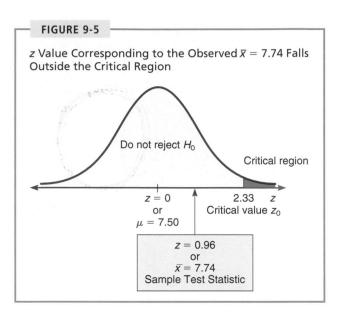

FIGURE 9-5

z Value Corresponding to the Observed $\bar{x} = 7.74$ Falls Outside the Critical Region

Do not reject H_0

Critical region

$z = 0$ 2.33 z
or   Critical value z_0
$\mu = 7.50$

$z = 0.96$
or
$\bar{x} = 7.74$
Sample Test Statistic

VI. *Conclusion:* Earlier we pointed out a similarity between the logic of a statistical test and proceedings in a court of law. We are now at the point where all the sample evidence has been presented, and we are awaiting a decision. Who makes the decision? In a court, the jury or judge would make the decision, but in our statistical test, it is Mother Nature who makes the final decision. The observed value $\bar{x} = 7.74$ cm of length for turtle eggs taken from the turtle nest corresponds to the z value 0.96. This z value does not lie in the critical region (see Figure 9-5).

We conclude that there is not enough evidence to discredit the null hypothesis at the $\alpha = 0.01$ level of significance. Therefore, we do not reject H_0: $\mu = 7.50$, and we still think that the nest could have been made by the rare endangered species of turtle. The biologist should collect the eggs and take them to the zoo for incubation. It is important to remember that we do not claim to have *proven* that the eggs are from the rare turtle; all we say is that there is not enough evidence to reject H_0, and therefore, we "accept" it.

VII. *Meaning of α and β:* Remember that β is the probability of accepting H_0 when it is false. In our setting, β is the probability that the zoo incubates and hatches eggs from a common turtle species. To calculate β requires knowledge of the H_1 distribution. Since there are several species of common turtles in the area and we don't know which species made the nest, we really don't know the H_1 distribution. However, a type II error is not too serious because if the wrong eggs (from the common turtle) are hatched, there is little harm done. The baby turtles simply can be released near the place the eggs were found.

The level of significance α is the probability of rejecting H_0 when it is true. This is the probability of concluding that the eggs are not from the rare turtle species when they, in fact, are. It would be a serious mistake to reject the nest of the rare turtle. To guard against such a mistake, we have taken a relatively small α value of 0.01.   ◇

GUIDED EXERCISE 4

Testing μ (large sample)

A research meteorologist has been studying wind patterns over the Pacific Ocean. Based on these studies, a new route is proposed for commercial airlines going from San Francisco to Honolulu. The new route is intended to take advantage of existing wind patterns to reduce flying time. It is known that for the old route the distribution of flying times for a large four-engine jet has mean $\mu = 5.25$ hours with standard deviation $\sigma = 0.6$ hour. Thirty-six flights on the new route have yielded a mean flying time of $\bar{x} = 4.90$ hours. Does this indicate that the average flying time for the new route is less than 5.25 hours? Use a 5% level of significance.

(a) What is H_0?

⟹ The average time for the old route is $\mu = 5.25$ hours, so we will set up the null hypothesis H_0: $\mu = 5.25$. In words, we are saying that the mean flying time on the new route is the same as that on the old route.

(b) What is H_1?

⟹ We want to see if the average flying time for the new route is less than 5.25 hours, so the alternate hypothesis is H_1: $\mu < 5.25$.

(c) What type of critical region must be used?

⟹ Since the $<$ symbol is used in the alternate hypothesis, Figure 9-1 tells us to use a left-tailed test.

(d) What is the critical value?

⟹ By Figure 9-2 (or using Table 5 in Appendix II), the critical value for a left-tailed test with $\alpha = 0.05$ is $z_0 = -1.645$.

(e) What is the z value corresponding to the sample test statistic $\bar{x} = 4.90$?

⟹ We use the formula from the central limit theorem with $\mu = 5.25$ (the value specified in H_0), $\sigma = 0.6$, and $n = 36$. Therefore,

$$z = \frac{\bar{x} - \mu}{(\sigma/\sqrt{n})} = \frac{4.90 - 5.25}{(0.6/\sqrt{36})} = -3.50$$

(f) Do we reject or fail to reject H_0?

⟹ Look at the critical region of Figure 9-6. The sample z value corresponding to the sample test statistic $\bar{x} = 4.90$ falls in the critical region. We reject H_0, and choose H_1 at the $\alpha = 0.05$ level. This means that we conclude that the flying time on the new route is less than the flying time on the old route.

FIGURE 9-6 Critical Region, $\alpha = 0.05$

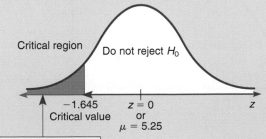

EXAMPLE 4

Testing μ (large sample)

A large company has branch offices in several major cities of the world. From time to time, it is necessary for company employees to move their families from one city to another. From long experience, the company knows that its employees move on the average of once every 8.50 years. However, trends for the past few years have led people to think that a change might have occurred. To determine if such a change has occurred, a random sample was taken of 48 employees (from the entire company). The employees were asked to provide either the number of years since the company asked them to move or the number of years employed by the company if they had never been asked to move. For this sample of 48 employees, the mean time was $\bar{x} = 7.91$ years with sample standard deviation 3.62 years.

Let us see if we can reject the hypothesis H_0: $\mu = 8.50$ at the $\alpha = 0.05$ level of significance. Since we have no way of knowing if the average moving time (μ) has increased or decreased, the alternate hypothesis will simply be H_1: $\mu \neq 8.50$.

(a) Do we use a right-tailed, left-tailed, or two-tailed test? Find the critical values.

SOLUTION: Since the alternate hypothesis uses the $\neq$ symbol, we use a two-tailed test. Since the sample size $n = 48$ is large, the sampling distribution $\bar{x}$ is approximately normal by the central limit theorem. Therefore, we can use Figure 9-2 to find the critical values. For $\alpha = 0.05$, Figure 9-2 tells us that the critical values are $z_0 = \pm 1.96$.

(b) Convert the sample test statistic $\bar{x} = 7.91$ to a z value. Show the location of the critical region and sample test statistic on the standard normal distribution.

SOLUTION: We convert $\bar{x} = 7.91$ years to a z value using $\mu = 8.50$ from H_0, $n = 48$, and estimating σ by $s = 3.62$:

$$z = \frac{\bar{x} - \mu}{(\sigma/\sqrt{n})} = \frac{7.91 - 8.50}{(3.62/\sqrt{48})} \approx -1.13$$

● **CALCULATOR NOTE** It is best to do the entire calculation for z internally on your calculator. For instance, on the TI-83, key in

$$(7.91 - 8.50) \div (3.62 \div \sqrt{(48)}) \qquad \text{ENTER}$$

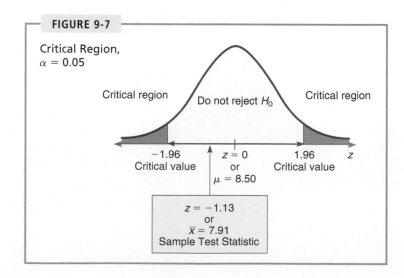

FIGURE 9-7

Critical Region, $\alpha = 0.05$

Critical region   Do not reject H_0 Critical region

-1.96 $z = 0$ 1.96 z
Critical value   or   Critical value
         $\mu = 8.50$

$z = -1.13$
or
$\bar{x} = 7.91$
Sample Test Statistic

Then round z to 2 places after the decimal. If you calculate the denominator $3.62/\sqrt{48}$ separately, be sure to carry at least four places after the decimal. Then divide the quantity $(7.91 - 8.50)$ by that number.

(c) Do we reject or fail to reject H_0 at the 5% level of significance?

> **SOLUTION:** The sample test statistic does not fall in the critical region (see Figure 9-7). Therefore, we *fail to reject* H_0. It appears that employees still move on the average of once every 8.50 years.   ◇

GUIDED EXERCISE 5

Testing μ (large sample)

A machine makes twist-off caps for bottles. The machine is adjusted to make caps of diameter 1.85 cm. Production records show that when the machine is so adjusted, it will make caps with mean diameter 1.85 cm and with standard deviation $\sigma = 0.05$ cm. During production, an inspector checks the diameters of caps to see if the machine has slipped out of adjustment. A random sample of 64 caps is taken. If the mean diameter for this sample is $\bar{x} = 1.87$ cm, does this indicate that the machine has slipped out of adjustment and the average diameter of caps is no longer $\mu = 1.85$ cm? (Use a 1% level of significance.)

(a) What is the null hypothesis?

> H_0: $\mu = 1.85$
>
> Be sure to use the hypothesized parameter, *not* the sample statistic $\bar{x}$ in the null hypothesis.

(b) An error either way would be serious. Therefore, we want to test the null hypothesis against the hypothesis that the mean diameter is *not* 1.85 cm. State the alternate hypothesis.

> H_1: $\mu \neq 1.85$

(c) Should we use a one- or two-tailed test? What are the critical values?

> Because the alternate hypothesis uses the $\neq$ symbol, we use a two-tailed test. Since the sample size $n = 64$ is large, the central limit theorem tells us that the sampling distribution of $\bar{x}$ is approximately normal. Therefore, we can use Figure 9-2 or Table 5 in Appendix II to find the critical values. For $\alpha = 0.01$, $z_0 = \pm 2.58$.

(d) What is the value of the sample test statistic? Convert it to a z value.

> Because we are testing μ, the sample test statistic is $\bar{x}$. The sample of 64 bottle caps yielded a sample mean diameter $\bar{x} = 1.87$ cm. To convert $\bar{x}$ to z, we use $\mu = 1.85$ from H_0, $n = 64$, and $\sigma = 0.05$:
>
> $$z = \frac{\bar{x} - \mu}{(\sigma/\sqrt{n})} = \frac{1.87 - 1.85}{(0.05/\sqrt{64})} = 3.20$$

Continued

GUIDED EXERCISE 5 continued

(e) Show the critical region and the z value of the sample test statistic on the normal curve. Does the sample test statistic fall in the critical region or not? Do we reject or fail to reject H_0? At the 1% level of significance, can we say that the machine needs adjustment?

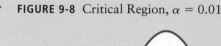

FIGURE 9-8 Critical Region, $\alpha = 0.01$

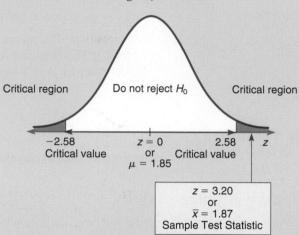

We see that the z value corresponding to the sample test statistic $\bar{x} = 1.87$ falls in the critical region. This means that we reject H_0 and conclude that at the 1% level of significance the machine needs adjustment.

Statistical significance

In a sense, the sample estimate of the parameter represents the evidence favoring rejection of H_0. For this reason, statisticians say the results of a random sample are *statistically significant* if the estimated parameter falls in the critical region of a test. In Guided Exercise 5, the random sample of 64 caps gave a sample test statistic $\bar{x} = 1.87$. Since the corresponding z value was in the critical region, we say that the results of the sample were statistically significant—that is, there was sufficient evidence to reject H_0 at the specified level of significance.

Statistical significance

If we reject H_0, we say that the data are *statistically significant*.
If we do not reject H_0, we say that the data are *not statistically significant*.

◇ **COMMENT: Using the Student's t Distribution to Test μ Even with Large Samples** When samples are large ($n \geq 30$), the central limit theorem assures us that the $\bar{x}$ distribution is approximately normal. In addition, we use the common convention that when we have large samples ($n \geq 30$), we can estimate σ by the sample standard deviation s. This convention allows us to use the normal distribution to test μ for large samples even when we don't know σ for the x distribution. However, when n is large and we don't know σ, we can use the Student's t distribution with $n - 1$ degrees of freedom for the $\bar{x}$ distribution. Using the Student's t distribution avoids the errors inherent in estimating σ by s.

With more widespread access to technology, such as the TI-83Plus calculator or statistical software packages, it is becoming more common to use the Student's t distribution to test μ when we have large samples, but don't know σ. Using the Student's t distribution moves the critical region farther into the tail(s) of the $\bar{x}$ distribution, and generates slightly larger P values (discussed in Section 9.3). Generally speaking, test conclusions for large sample tests of μ are the same whether we use the normal distribution with σ approximated by s or use the Student's t distribution with $n-1$ degrees of freedom.   ◇

TECH NOTE   In conducting hypothesis tests for μ using the normal distribution, the TI-83Plus calculator and Minitab both provide the z value for the sample test statistic $\bar{x}$. To conclude the test, compare the z value of the test statistic with the critical value(s) for the specified level of significance and type of test.

TI-83Plus You can select to enter raw data or summary statistics. Enter the value of μ_0 used in the null hypothesis H_0: $\mu = \mu_0$. Select the symbol used in the alternate hypothesis ($\neq\mu_0$, $<\mu_0$, $>\mu_0$). The value for sigma is also required. Using data from Guided Exercise 5 regarding the diameter of twist-off caps, we have the following displays. Press **Stat**, select **Tests**, use option **1:z-Test**.

```
Z-Test
 Inpt:Data Stats
 µo:1.85
 σ:.05
 x̄:1.87
 n:64
 µ:≠µo <µo >µo
 Calculate Draw
```

```
Z-Test
 µ≠1.85
 z=3.2
 p=.0013744042
 x̄=1.87
 n=64
```

Minitab Enter the raw data from a sample. Use the menu selections **Stat ➤ Basic Stat ➤ 1-Sample Z**. The z value of the sample mean $\bar{x}$ is included in the display. We will see such a display in the next section.

VIEWPOINT   *Money, Money, Money!*

What is the statistical profile of U.S. millionaires? Are they flashy spenders? Are they stingy or generous with friends? Do they drive big cars and wear expensive clothes? Do millionaires work or inherit their wealth? How many millionaires are there in the U.S.? What's the buying power of a million dollars today compared with 10 or 20 years ago? For a good statistical profile of U.S. millionaires, see *The Millionaire Next Door*, by T. J. Stanley and W. D. Danko. The book points out that most U.S. millionaires *earned* what they have, behave like ordinary people, and are not flashy or arrogant. A million dollars has a great deal less buying power than it used to have. Anyone who works hard and is willing to take some risks has a good chance of becoming a millionaire. A good understanding of statistics helps.

SECTION 9.2   PROBLEMS

For Problems 1 through 12, please provide the requested information.

(a) What is the null hypothesis? What is the alternate hypothesis? Will we use a left-tailed, right-tailed, or two-tailed test? What is the level of significance?

(b) What sampling distribution will we use? What is the critical value z_0 (or critical values $\pm z_0$)?

(c) Sketch the critical region and show the critical value (or critical values).

(d) Calculate the z value corresponding to the sample statistic $\bar{x}$ and show its location on the sketch of part (c).

(e) Based on your answers for parts (a) to (d), shall we reject or fail to reject (i.e., "accept") the null hypothesis at the given level of significance α? Explain your conclusion in the context of the problem.

(f) Are the data statistically significant?

1. *Meteorology: Storms* *Weatherwise* is a magazine published in association with the American Meteorological Society. In one issue, there is a rating system to classify Nor'easter storms that frequently hit New England states and can cause much damage near the ocean coasts. A *severe* storm has an average peak wave height of 16.4 feet for waves hitting the shore. Suppose that a Nor'easter is in progress at the severe storm class rating. Peak wave heights are usually measured from land (using binoculars) off fixed cement piers. Suppose that a reading of 36 peak waves showed an average wave height of $\bar{x} = 15.1$ feet with sample standard deviation $s = 3.2$ feet. Does this information indicate that the storm is (perhaps temporarily) retreating from its severe rating? Use $\alpha = 0.01$.

2. *Ford Taurus: Assembly Time* Let x be a random variable that represents assembly time for the Ford Taurus. *The Wall Street Journal* reported that the average assembly time for the Ford Taurus is $\mu = 38$ hours. A modification to the assembly procedure has been made. It is thought that the average assembly time may be reduced because of this modification. A random sample of 47 new Ford Taurus automobiles coming off the assembly line showed the average assembly time to be $\bar{x} = 37.5$ hours with sample standard deviation $s = 1.2$ hours. Does this indicate that the average assembly time has been reduced? Use $\alpha = 0.01$.

3. *Phone Calls: Priority List* Message mania! According to *The Wall Street Journal*, a professional employee working in a large company receives an average of $\mu = 31.8$ calls per day. Most of the calls are from other employees in the company. Because of the large number of calls, employees find themselves distracted and are unable to concentrate when they return to their tasks. In an effort to reduce distraction caused by such interruptions, one company established a "priority list" that all employees were to use before making phone calls. One month after the new priorities were put into use, a random sample of 63 employees showed they were receiving an average of $\bar{x} = 28.5$ calls per day with sample standard deviation $s = 10.7$ calls per day. Use a 1% level of significance to test the claim that there has been a change (either way) in the average number of calls per day received per employee.

4. *Fashion Design: Display Windows* Judy Povich is a fashion design artist who designs the display windows in front of a large clothing store in New York City. Electronic counters at the entrances total the number of people entering the store each business day. Before Judy was hired by the store, the mean number of people entering the store each day was 3218. However, since Judy has started working, it is thought that this number has increased. A random sample of 42 business days after Judy began work gave an average $\bar{x} = 3392$ people entering the store each

day. The sample standard deviation was $s = 287$ people. Does this indicate that the average number of people entering the store each day has increased? Use a 1% level of significance.

5. *Cocktail Hostess: Tips* Maureen is a cocktail hostess in a very exclusive private club. The Internal Revenue Service is auditing her tax return this year. Maureen claims that her average tip last year was $4.75. To support this claim, she sent the IRS a random sample of 52 credit card receipts showing her bar tips. When the IRS got the receipts, they computed the sample average and found it to be $\bar{x} = \$5.25$ with sample standard deviation $s = \$1.15$. Do these receipts indicate that the average tip Maureen received last year was more than $4.75? Use a 1% level of significance.

6. *Farmer's Co-op: Alfalfa* Let x be a random variable representing the percentage of protein content for early bloom alfalfa hay. The average percentage protein content of such early bloom alfalfa should be $\mu = 17.2\%$. (Source: *The Merck Veterinary Manual,* a manual commonly used in veterinary colleges.) A farmer's co-op is thinking of buying a large amount of baled hay but suspects that the hay is from a later summer cutting with lower protein content. A small amount of hay was removed from each bale of a random sample of 50 bales. The average protein content from the samples was determined by the local agriculture college to be $\bar{x} = 15.8\%$ with sample standard deviation $s = 5.3\%$. Use a 5% level of significance to test the claim that this hay has lower average protein content than early bloom alfalfa.

7. *Dodge Intrepid: Acceleration* *Consumer Reports* indicated that the mean acceleration time (0 to 60 miles per hour) for the Dodge Intrepid was 10.2 seconds. In most tests of this type, regular unleaded gasoline is used. Suppose that 41 such tests were made using premium unleaded gasoline (octane over 91), and the sample mean acceleration time (0 to 60 miles per hour) was $\bar{x} = 9.7$ seconds with sample standard deviation $s = 2.1$ seconds. Does this indicate that premium gasoline tends to reduce average acceleration time (0 to 60 miles per hour)? Use $\alpha = 0.05$.

8. *Mercury Sable: Braking Distance* *Consumer Reports* also indicated that the mean braking distance (from 60 miles per hour) on wet pavement for the Mercury Sable was 159 feet. Suppose that Sables equipped with tires having a new tread designed to grip the road better on wet pavement were used in 45 tests (braking from 60 miles per hour). The sample mean braking distance was $\bar{x} = 148$ feet with sample standard deviation $s = 23.5$ feet. Does this information indicate that the population mean braking distance for Mercury Sables on wet pavement is reduced for the new tire tread? Use $\alpha = 0.01$.

9. *Vitamins: Oxygen* Let x be a random variable that represents milliliters of oxygen per deciliter of whole blood. For healthy adults, the population mean of x is $\mu = 19.0$ milliliters of oxygen per deciliter. (Source: *The Merck Manual,* a commonly used reference in most medical schools and nursing programs.) A company that sells vitamins claims that its multivitamin complex will increase the oxygen capacity of the blood. A random sample of 48 adults took the vitamins for 6 months. After blood tests, it was found that the sample mean was $\bar{x} = 20.7$ milliliters of oxygen per deciliter with sample standard deviation $s = 9.9$. Use a 1% level of significance to test the claim that the average oxygen capacity has been increased.

10. *Major League Baseball: Salary* The average annual salary of major league baseball players is now $1,789,556 (Source: *USA Today*). Suppose this year a random sample of 35 major league players in Florida had an average annual salary of $\bar{x} = \$1,621,726$ with sample standard deviation $s = \$591,218$. Use a 1% level of

significance to test the claim that the average salary for all Florida players is different from the national average.

11. *Blood Plasma: pH* Let x be a random variable that represents the pH of arterial plasma (i.e., the acidity of the blood). For healthy adults, the mean of the x distribution is $\mu = 7.4$ pH. (See source for Problem 9.) A new drug for arthritis has been developed. However, it is thought this drug might change blood pH. A random sample of 33 patients with arthritis took the drug for 3 months. Blood tests showed $\bar{x} = 8.1$ pH with sample standard deviation $s = 1.9$ pH. Use a 5% level of significance to test the claim that the drug has changed (either way) the mean pH of the blood.

12. *Air Force: Mess Hall* An Air Force base mess hall has received a shipment of 10,000 gallon-size cans of cherries. The supplier claims that the average amount of liquid is 0.25 gallon per can. A government inspector took a random sample of 100 cans and found the average liquid content to be 0.28 gallon per can with a standard deviation of 0.10. Does this indicate that the supplier's claim is too low? (Use a 5% level of significance.)

13. *Focus Problem: Pizza Franchise* Solve the Chapter 9 Focus Problem. Use $\alpha = 0.05$.

14. Suppose that you did some statistical research on a given topic.
    (a) If you had a null hypothesis that you hoped to *reject* based on your sample data, would you be better off using a one-tailed test (either left or right tail as appropriate) or a two-tailed test? Explain your answer.
    (b) Answer part (a) if you had a null hypothesis you hoped *not to reject* based on your sample data. Explain your answer. *Hint:* Examine Figure 9-2 and the size of the critical values for each test.
    (c) If a report states that certain data were used to reject (or fail to reject) a given hypothesis, would it be a good idea to know what type of test (one-tailed or two-tailed) was used? Explain your answer.

15. Compare similarities of statistical testing with legal methods used in a U.S. court setting. Then discuss the following topics in class or consider the topics on your own. Please write a brief but complete essay in which you answer the following questions.
    (a) In a court setting, the person charged with a crime is initially considered to be innocent. The claim of innocence is maintained until the jury returns with a decision. Explain how the claim of innocence could be taken to be the null hypothesis. Do we assume that the null hypothesis is true throughout the testing procedure? What would the alternate hypothesis be in a court setting?
    (b) The court claims that a person is innocent if the evidence against the person is not adequate to find him or her guilty. This does not mean, however, that the court has necessarily *proven* the person to be innocent. It simply means that the evidence against the person was not adequate for the jury to find him or her guilty. How does this situation compare with a statistical test for which the conclusion is "do not reject" (i.e., accept) the null hypothesis? What would be a type II error in this context?
    (c) If the evidence against a person is adequate for the jury to find him or her guilty, then the court claims that the person is guilty. Remember, this does not mean that the court has necessarily *proven* the person to be guilty. It simply means that the evidence against the person was strong enough to find him or her guilty. How does this situation compare with a statistical test for which the conclusion is "reject" the null hypothesis? What would be a type I error in this context?
    (d) In a court setting, the final decision as to whether the person charged is innocent or guilty is made at the end of the trial, usually by a jury of impartial people. In hypothesis testing, the final decision to reject or not reject the null

hypothesis is made at the end of the test by using information or data from an (impartial) random sample. Discuss these similarities between statistical hypothesis testing and a court setting.

(e) We hope that you are able to use this discussion to increase your understanding of statistical testing by comparing it with something that is a well-known part of our American way of life. However, all analogies have weak points. It is important not to take the analogy between statistical hypothesis testing and legal court methods too far. For instance, the judge does not set a level of significance and tell the jury to determine a verdict that is wrong only 5% or 1% of the time. Discuss some of these weak points in the analogy between the court setting and hypothesis testing.

16. Is there a relationship between confidence intervals and two-tailed hypothesis tests? Let *c* be the level of confidence used to construct a confidence interval from sample data. Let α be the level of significance for a two-tailed hypothesis test. The following statement applies to hypothesis tests of the mean (for both large and small samples).

> For a two-tailed hypothesis test with level of significance α and null hypothesis H_0: $\mu = k$, we *reject* H_0 whenever *k* falls *outside* the $c = 1 - \alpha$ confidence interval for μ based on the sample data. When *k* falls within the $c = 1 - \alpha$ confidence interval, we do not reject H_0.

(A corresponding relationship between confidence intervals and two-tailed hypothesis tests also is valid for other parameters such as p, $\mu_1 - \mu_2$, or $p_1 - p_2$, which we will study in Sections 9.5 and 9.7.) Whenever the value of *k* given in the null hypothesis falls *outside* the $c = 1 - \alpha$ confidence interval for the parameter, we *reject* H_0. For example, consider a two-tailed hypothesis test with $\alpha = 0.01$ and

$$H_0\text{: } \mu = 20 \qquad H_1\text{: } \mu \neq 20$$

A random sample of size 36 has a sample mean $\bar{x} = 22$ with sample standard deviation $s = 4$.

(a) What is the value of $c = 1 - \alpha$? Using the methods of Chapter 8, construct a $1 - \alpha$ confidence interval for μ from the sample data. What is the value of μ given in the null hypothesis (i.e., what is *k*)? Is this value in the confidence interval? Do we reject or fail to reject H_0 based on this information?

(b) Using methods of Chapter 9, find the critical region for the hypothesis test. Do we reject or fail to reject H_0? Compare your result to that of part (a).

17. Change the null hypothesis of Problem 16 to H_0: $\mu = 21$. Repeat parts (a) and (b).

9.3
The *P* Value in Hypothesis Testing

FOCUS POINTS

✓ Identify the meaning of *P* value for a statistical test.

✓ Compute *P* values for large sample tests of μ.

✓ Draw test conclusions based on the *P* value and the level of significance.

The level of significance α in a test of hypothesis is the probability of making a type I error; that is, α is the probability of rejecting the null hypothesis when it is true.

The experimenter sets α before beginning the statistical test. Then this level of significance is used to determine whether or not H_0 is rejected. However, as we shall see in the next example, the same sample data may lead to two different test conclusions, depending on the level of significance used.

EXAMPLE 5

Test conclusion depends
on α

Viva, a credit card company, lowered its annual interest rate by 1%. Records from before the rate change showed the average outstanding balance on credit card accounts to be $\mu = \$576$. The managers believe that reducing interest rates spurs greater use of credit cards, which results in higher outstanding balances. To test this claim, a random sample of 36 accounts was examined 6 months after the interest-rate reduction. The average outstanding balance was $\bar{x} = \$615$ with standard deviation $s = \$120$. Test the managers' claim at the 0.05 and 0.01 levels of significance.

SOLUTION: For both levels of significance, we use the same null and alternate hypotheses:

$$H_0: \mu = 576 \qquad H_1: \mu > 576$$

The sample test statistic $\bar{x} = 615$ is also the same for both cases. Using $\mu = 576$ from the null hypothesis, $n = 36$, and $\sigma \approx s = 120$, we find the z value corresponding to the sample test statistic $\bar{x} = 615$:

$$z = \frac{\bar{x} - \mu}{(\sigma/\sqrt{n})} = \frac{615 - 576}{(120/\sqrt{36})} = 1.95$$

In both cases, we have a right-tailed test. Figure 9-9 shows the test conclusion for each of the two levels of significance.   ◇

As we see in Example 5, smaller levels of significance move the critical value for the rejection region farther from the mean μ. A natural question arises: What is the smallest level of significance at which the sample data will tell us to reject H_0? The answer is the *P value* associated with the observed sample statistic. The *P* value is also called the *probability of chance* or the *attained level of significance*.

FIGURE 9-9

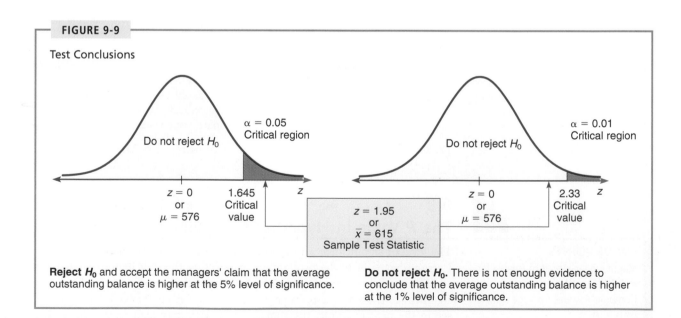

Test Conclusions

Reject H_0 and accept the managers' claim that the average outstanding balance is higher at the 5% level of significance.

Do not reject H_0. There is not enough evidence to conclude that the average outstanding balance is higher at the 1% level of significance.

For the distribution described by the null hypothesis, the *P value* is the smallest level of significance for which the observed sample statistic tells us to reject H_0. Consequently, if

P value $\leq \alpha$, then we reject H_0

P value $> \alpha$, then we do not reject H_0

FIGURE 9-10

P Values, Level of Significance α, and Decision Procedure for Different Tests of the Mean

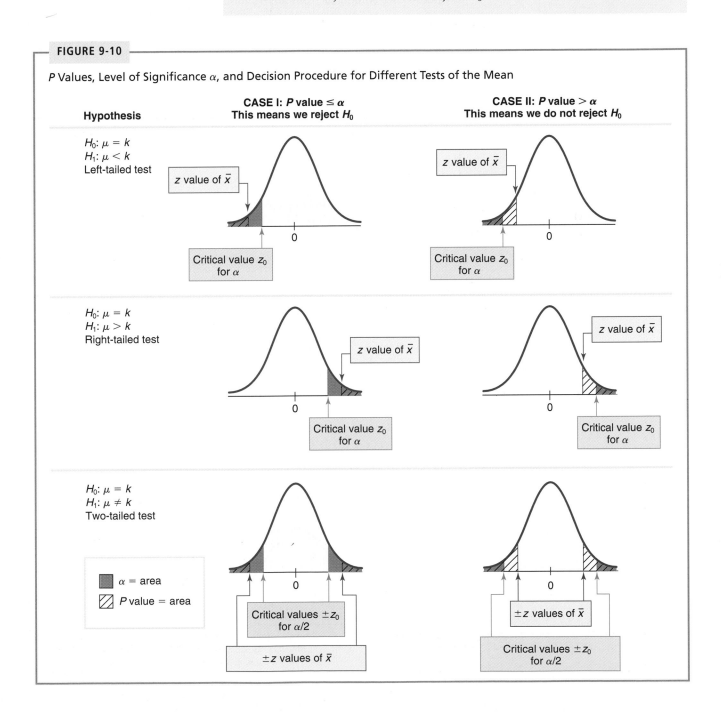

| | **CASE I: *P* value $\leq \alpha$**
This means we reject H_0 | **CASE II: *P* value $> \alpha$**
This means we do not reject H_0 |

Hypothesis

$H_0: \mu = k$
$H_1: \mu < k$
Left-tailed test

$H_0: \mu = k$
$H_1: \mu > k$
Right-tailed test

$H_0: \mu = k$
$H_1: \mu \neq k$
Two-tailed test

α = area

P value = area

Finding the P value for tests of μ

How do we go about computing the P value? We need to express the P value in terms of probability. Let's use the model of a right-tailed test of the mean. For a right-tailed test of the mean, the P value is simply the probability that the sample mean from any random sample of the same size will be greater than or equal to the observed sample mean $\bar{x}$. In symbols, for right-tailed tests of μ, we have

P value $= P(\bar{x}$ computed from any random sample of size $n \geq$ observed $\bar{x})$

P values are the *areas* in the tail or tails of a probability distribution beyond the observed sample statistic. Figure 9-10 on the preceding page shows the P values for right-, left-, and two-tailed tests of the mean. The P values are the hatched areas of the figure.

EXAMPLE 6

P value for right-tailed test

Compute the P value for the hypothesis test of Example 5. Recall that the problem involved the claim that the outstanding balance on Viva credit cards increased after the annual interest decreased. We had

$H_0: \mu = 576$
$H_1: \mu > 576$

Observed sample data: $n = 36$, $\bar{x} = \$615$, and $s = \$120$.
For what levels of significance α do we reject H_0?

SOLUTION:

(a) To find the P value, we first convert $\bar{x}$ to z using $\mu = 576$ given in the null hypothesis, standard deviation $\sigma \approx s = 120$, and $n = 36$ in the formula

$$z = \frac{\bar{x} - \mu}{(\sigma/\sqrt{n})} = \frac{615 - 576}{(120/\sqrt{36})} = 1.95$$

(b) Sketch the P value on a diagram. Because the alternate hypothesis specifies a *right-tailed* test, the P value is the area to the *right* of the observed sample statistic (see Figure 9-11).

(c) Compute the P value. Using techniques of Chapter 6 and Table 5 in Appendix II, we find

$$P \text{ value} = P(\bar{x} \geq 615) = P(z \geq 1.95)$$
$$= 1 - 0.9744 = 0.0256$$

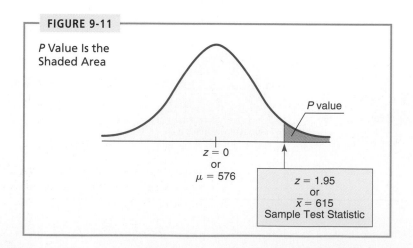

FIGURE 9-11

P Value Is the Shaded Area

P value

$z = 0$
or
$\mu = 576$

$z = 1.95$
or
$\bar{x} = 615$
Sample Test Statistic

(d) *Conclusion:* Since the *P* value is the *smallest* level of significance for which the sample data tell us to reject H_0, we reject H_0 for any $\alpha \geq 0.0256$. For $\alpha < 0.0256$, we fail to reject H_0.   ◇

GUIDED EXERCISE 6

P value for right-tailed test

Last year the average age of students attending Fremont College was 21.3 years. This semester, in order to meet the needs of older students, more classes were scheduled during evening and weekend hours. Has the average age of the students at Fremont College increased this semester? To answer this question, a random sample of 64 students enrolled this semester was studied. The average age of these students was $\bar{x} = 22.1$ years with sample standard deviation $s = 2.7$ years. Use the *P* value of the sample statistic $\bar{x}$ to test the hypothesis that the average age of students this semester is higher than it was last year. Use $\alpha = 0.01$.

(a) What are the null and alternate hypotheses?

→ H_0: $\mu = 21.3$

H_1: $\mu > 21.3$

(b) Convert $\bar{x}$ to z.

→   Using $\mu = 21.3$ from H_0, $n = 64$, and $\sigma \approx s = 2.7$, we get

$$z = \frac{\bar{x} - \mu}{(\sigma/\sqrt{n})} = \frac{22.1 - 21.3}{(2.7/\sqrt{64})} = 2.37$$

(c) Sketch a diagram showing the *P* value of the sample test statistic $\bar{x} = 22.1$.

→   **FIGURE 9-12** *P* Value Is the Shaded Area

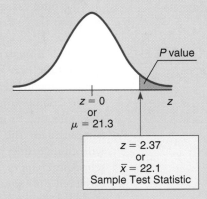

(d) Use Table 5 in Appendix II to compute the *P* value.

→   *P* value = $P(\bar{x} \geq 22.1) = P(z \geq 2.37)$

= $1 - 0.9911 = 0.0089$

(e) Compare the *P* value to $\alpha = 0.01$. Do we reject H_0 or not? Has the average age at the college increased?

→   The *P* value = 0.0089 is the smallest level of significance for which we reject H_0. Since

P value = $0.0089 < 0.01 = \alpha$

we reject H_0 for $\alpha = 0.01$. At the 1% level of significance, it seems that the average age of the students enrolled this semester has increased.

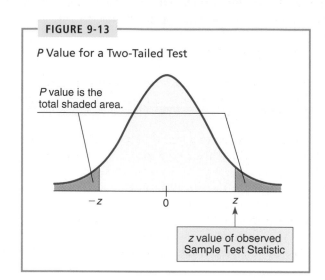

FIGURE 9-13

P Value for a Two-Tailed Test

P value is the total shaded area.

z value of observed Sample Test Statistic

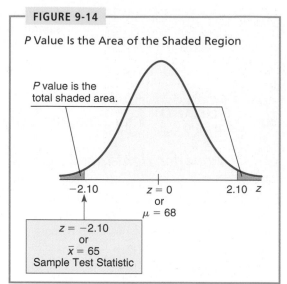

FIGURE 9-14

P Value Is the Area of the Shaded Region

P value is the total shaded area.

-2.10 $z = 0$ or $\mu = 68$ 2.10 z

$z = -2.10$ or $\bar{x} = 65$ Sample Test Statistic

For a distribution described by the null hypothesis, the P value is the probability of obtaining any sample statistic as *far away* as, or even *farther* from μ than, the given sample test statistic $\bar{x}$. In the case of a *two-tailed* test, this means that we must consider *both* tails of the distribution. For instance, the P value associated with a sample z value is the *total* of the area to the *left of* $-z$ added to the area to the *right of z* (see Figure 9-13).

The next example shows how to compute the P value for a *two-tailed* test of the mean (large sample size).

EXAMPLE 7

P value for two-tailed test

Whitehall Construction Company figures the cost of a project based on an average idle time of 68 minutes per worker per 8-hour shift. Idle time does not include lunch breaks but can be caused by delays in material delivery, delays in the completion of a prior stage of the project, worker injury, inadequate crew size due to absenteeism, and so forth. A new work schedule has just been completed by the construction foreman. It is not known whether this schedule will change idle time or not (increase or decrease). To determine if the new schedule *changes* the average idle time, a random sample of 49 workers was observed during an 8-hour shift. The average idle time was $\bar{x} = 65$ minutes with standard deviation $s = 10$ minutes. Use a 5% level of significance. Does the new schedule *change* idle time? Use the P value of the sample statistic to make the decision.

SOLUTION:

(a) The null and alternate hypotheses are

H_0: $\mu = 68$ minutes

H_1: $\mu \neq 68$ minutes

Statistical significance based on *P* values

Interpreting results from *P* values generated on a computer

(b) Next, we convert the sample test statistic $\bar{x} = 65$ to a *z* value using $n = 49$, $\mu = 68$ from H_0, and $s = 10$ as the approximation for σ:

$$z = \frac{\bar{x} - \mu}{(\sigma/\sqrt{n})} = \frac{65 - 68}{(10/\sqrt{49})} = -2.10$$

(c) Show the *P* value on a diagram. The alternate hypothesis corresponds to a *two-tailed* test, so the *P* value is the area to the *left* of $z = -2.10$ plus the area to the *right* of $z = 2.10$ (see Figure 9-14).

(d) Compute the *P* value. By symmetry, the *P* value is *twice* the area in the tail to the *left of* -2.10. Using Table 5 in Appendix II, we see that

$$P \text{ value} = 2(0.0179) = 0.0358$$

(e) Conclude the test. *The P value is the smallest level of significance for which we reject H_0.* This means that we reject H_0 for all $\alpha \geq 0.0358$. In particular, we reject H_0 for $\alpha = 0.05$, since 0.05 is greater than 0.0358. (Note that we fail to reject H_0 for $\alpha = 0.01$ because the *P* value 0.0358 is greater than 0.01.)   ◇

In hypothesis testing, we use sample data to draw a conclusion about the null hypothesis. If the sample data results are quite different from the claim stated in H_0, then we suspect that the difference is due to some effect other than just random chance. In this case, we reject H_0, and we say that the data are *statistically significant*. If we do not reject the null hypothesis H_0, we say that the data are *not statistically significant.*

The advantage of knowing the *P* value is that we know *all* levels of significance for which the observed sample statistic tells us to reject H_0. Many research journals require authors to include the *P* value of the observed sample statistic. Then readers will have more information and will know the test conclusion for any preset level of significance.

Again, let us caution you to establish the level of significance α before doing the hypothesis test. The level of significance reflects the probability level at which you are willing to risk a type I error. Also, the accuracy and reliability of your measurement instruments might affect your choice of α. Thus, select α first, then use a computer or table to find the *P* value of your test statistic, and finally, draw the appropriate conclusion.

Hypothesis testing using *P* values from a computer

1. Establish the level of significance α of the test.

2. Determine the null and alternate hypotheses, H_0 and H_1.

3. Enter information about the observed sample into the computer and look for the *P* value of the observed sample statistic in this computer output.

4. Compare your level of significance with the *P* value.

   If $P \text{ value} \leq \alpha$, reject H_0.

   If $P \text{ value} > \alpha$, do not reject H_0.

Summary

The P value (probability of chance) is the area of the sampling distribution that lies beyond the observed sample statistic. It tells us the probability that a sample statistic will be more extreme than the observed sample statistic. We compute the P values as follows:

1. When H_1 indicates a right-tailed test,

    P value = area to the right of the observed sample statistic

    P value = P(sample statistic > observed sample statistic)

2. When H_1 indicates a left-tailed test,

    P value = area to the left of the observed sample statistic

    P value = P(sample statistic < observed sample statistic)

3. When H_1 indicates a two-tailed test,

    P value = sum of the areas in the two tails

    (a) In the case that the observed sample statistic falls in the right half of a symmetric curve, then

    P value = $2P$(sample statistic > observed sample statistic)

    (b) In the case that the observed sample statistic falls in the left half of a symmetric curve, then

    P value = $2P$(sample statistic < observed sample statistic)

To conclude the test, we compare the level of significance α with the P value.

1. If the P value is less than or equal to α, reject H_0.

2. If the P value is greater than α, do not reject H_0.

TECH NOTE   When doing hypothesis testing using statistical software or a calculator similar to the TI-83Plus, the P value corresponding to the sample statistic is generally provided. The user compares it with a specified level of significance α to conclude the test.

TI-83Plus The displays use the data of Guided Exercise 6 regarding the average age of students attending Fremont College. Press **STAT**, select **TESTS**, option **1:Z-Test**, then **Draw.**

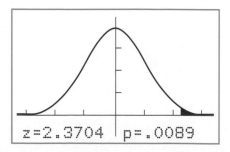

Excel The **ZTEST** function claims to return the *P* value for a two-tailed test. However, it seems to give the area in the right tail bounded by the sample statistic $\bar{x}$. With some care, you can figure out the appropriate *P* value from the given area.

Minitab Enter the raw data in a column. Use menu choices **Stat ➤ Basic Statistics ➤ 1-Sample Z**. Section Problem 10 shows a Minitab display.

VIEWP●INT   *Predator or Prey?*

Consider animals such as the arctic fox, gray wolf, desert lion, and South American jaguar. Each animal is a predator. What are the total sleep (hours per day), maximum life span (years), and overall danger index from other animals? Now consider prey such as rabbits, deer, wild horses, and the Brazilian tapir (a wild pig). Is there a statistically significant difference in average sleep, life span, and danger index? What about other variables such as the ratio of brain weight to body weight or the sleep exposure index (sleeping in a well-protected den or out in the open)? How did prehistoric humans fit into this picture? Scientists have collected a lot of data, and a great deal of statistical work has been done regarding such questions. For more information, visit the Brase/Brase statistics site at http://math.college.hmco.com/students and find the link to the StatLib site hosted by the Department of Statistics at Carnegie Mellon University. Follow links to Sleep in the Datasets Archive.

SECTION 9.3   PROBLEMS

For Problems 1 through 9, please do the following:
(a) Convert the sample mean $\bar{x}$ to a sample *z* value. (*Hint:* Due to the large sample size, approximate σ with *s*.)
(b) Sketch a graph of the *z* distribution. Show the location of the area corresponding to the *P* value for this test.
(c) Compute the *P* value for this test.
(d) Is the data significant at the specified level of significance α?

1. Given H_0: $\mu = 5$ and H_1: $\mu > 5$ with observed sample mean $\bar{x} = 6.1$, sample standard deviation $s = 2.5$, and sample size $n = 40$. Use $\alpha = 0.01$.

2. Given H_0: $\mu = 53.1$ and H_1: $\mu < 53.1$ with observed sample mean $\bar{x} = 52.7$, sample standard deviation $s = 4.5$, and sample size $n = 41$. Use $\alpha = 0.01$.

3. Given H_0: $\mu = 21.7$ and H_1: $\mu \neq 21.7$ with observed sample mean $\bar{x} = 20.5$, sample standard deviation $s = 6.8$, and sample size $n = 45$. Use $\alpha = 0.05$.

4. Given H_0: $\mu = 18.7$ and H_1: $\mu \neq 18.7$ with observed sample mean $\bar{x} = 19.1$, sample standard deviation $s = 5.2$, and sample size $n = 32$. Use $\alpha = 0.05$.

5. *Wildlife: Coyotes* A random sample of 68 adult coyotes in a region of northern Minnesota showed the average age to be $\bar{x} = 2.05$ years with sample standard deviation $s = 0.82$ years (based on information from the book *Coyotes: Biology, Behavior and Management*, by M. Bekoff, Academic Press). However, it is thought that

the overall population mean age of coyotes is $\mu = 1.75$. Does the sample data indicate that coyotes in this region of northern Minnesota tend to live longer than an average of 1.75 years? Use $\alpha = 0.01$.

6. *Archaeology: Pueblos* An archaeology professor claims that the average floor space of Mesa Verde Anasazi kivas is $\mu = 12$ square meters. However, in the book *Architecture of Social Integration in Prehistoric Pueblos* (by W. Lipe and M. Hegmon, Crow Canyon Archaeological Center Press), it is stated that a random sample of 56 Mesa Verde Anasazi kivas had a sample mean area of $\bar{x} = 12.3$ square meters with sample standard deviation $s = 3.4$ square meters. Do these data indicate that the mean floor area of all such Mesa Verde kivas is different from $\mu = 12$ square meters? Use $\alpha = 0.01$.

7. *Fishing: Trout* Pyramid Lake is on the Paiute Indian Reservation in Nevada. The lake is famous for cutthroat trout. Suppose a friend tells you that the average length of a trout caught in Pyramid Lake is $\mu = 19$ inches. However, the January 1995 Creel Survey (published by the Pyramid Lake Paiute Tribe Fisheries Association) reported that of a random sample of 73 fish caught, the mean length was $\bar{x} = 18.7$ inches with estimated standard deviation $s = 3.2$ inches. Do these data indicate that the average length of a trout caught in Pyramid Lake is less than $\mu = 19$ inches? Use $\alpha = 0.05$.

8. *Franchise: Flower Shop Franchise and Business Opportunities Annual Report* has over 1500 listings of small business start-up opportunities. Each of these involves a start-up cost (money needed to get the business going). In most cases, bank financing is readily available if you have a reasonable credit rating. One category of small business is the flower and gift shop business. For this type of business, the mean start-up costs (across the United States) amount to $61,400. However, start-up costs may vary from one region to the next. Suppose that you are interested in starting a flower and gift shop in San Antonio, Texas. Local newspaper advertising and business sections can be used to obtain a sample of start-up costs for this type of business. Suppose that a sample of 34 small flower and gift shops in the San Antonio region gave a sample mean start-up cost of $\bar{x} = \$55,200$ with sample standard deviation $s = \$18,800$. Does this indicate that the population average start-up cost in this region is lower than the national average? Use $\alpha = 0.05$.

9. *Automobile: Cost* Based on information from *The Statistical Abstract of the United States* (116th Edition), the average daily cost of owning and operating an automobile is $15.35. This includes depreciation, finance charges, general maintenance, gasoline, insurance, and 12,000 miles driven annually. A random sample of 34 college students who own cars found the average cost per day to be $11.85 with sample standard deviation of $6.21. Test the claim that college students' average daily ownership expenses are less than the national average. Use a 5% level of significance.

10. *Automobile: Mileage* How do new automobiles stack up mileage-wise? A random sample of 100 autos from *Consumer Review CAR Buyer's Guide* showed city and highway mileage. The autos tested included sport, luxury, compact, sedan, van, and sport utility models.
    (a) Is the city mileage less than 19.5 mpg for the cars? Let's look at a Minitab (**Stat ➤ Basic Statistics ➤ 1-Sample Z**) display of the sample results.

```
Z-Test
Test of mu = 19.500 vs mu < 19.500
The assumed sigma = 4.48
VARIABLE    N      MEAN    STDEV    SE MEAN     Z        P
City       100    18.750   4.480    0.448    -1.67    0.047
```

Refer to the Minitab display. Write out the null and alternate hypotheses. What is the value of the sample mean? What is the P value? At what levels of significance may we reject the null hypothesis and conclude that the avérage mileage is less than 19.5 mpg?

(b) Is the city mileage different from 19.5 mpg for these cars? Look at the next Minitab display.

```
Z-Test
Test of mu = 19.500 vs mu not = 19.500
The assumed sigma = 4.48
VARIABLE     N      MEAN    STDEV    SE MEAN      Z        P
City        100    18.750   4.480    0.448     -1.67    0.094
```

Refer to the Minitab display. Write out the null and alternate hypotheses. What is the value of the sample mean? What is the P value? At what levels of significance may we reject the null hypothesis and conclude that the average mileage is different than 19.5 mpg?

(c) How does the P value of a one-tailed test compare with the P value of a two-tailed test for the same sample data and same null hypothesis?

9.4
Tests Involving the Mean μ (Small Samples)

FOCUS POINTS

✓ Identify the components of a small-sample test for μ.

✓ Use degrees of freedom and the t table to find critical values.

✓ Compute the sample test statistic.

✓ Estimate the P value for a small sample test of μ.

✓ Draw conclusions using level of significance α and the P value.

Sometimes it is not practical or even possible to obtain large samples. Cost, available time, and other factors may require us to work with *small samples*. For our purposes, we will say that a sample size less than 30 is a small sample. In Section 8.2, we found that the t distribution is suitable for small samples in which the sampled population has a normal distribution. However, it can be shown that our applications of the t distribution are still appropriate for populations that are not normal but possess a "mound-shaped" and symmetric probability distribution. Because such distributions commonly occur, the t distribution is very useful in applied work.

When doing hypothesis tests of μ with small samples, we will use essentially the same methods we used for large samples. The main difference is that when σ is not known, the sampling distribution $\bar{x}$ follows a *Student's t distribution with degrees of freedom d.f. = n − 1* instead of a normal distribution.

When we draw a random sample of size n from a population that has a normal distribution (or at least a mound-shaped symmetric distribution) with mean μ, then the t values, that is,

$$t = \frac{\bar{x} - \mu}{(s/\sqrt{n})}$$

where $\bar{x}$ = sample mean

n = sample size

s = sample standard deviation

follow a Student's t distribution with degrees of freedom d.f. = n − 1.

Critical values from a t distribution

In Table 6, "Student's t Distribution," of Appendix II, you see the column headings c, α', and α''. (Up to now we have ignored the α' and α'' headings.) In this table, c represents the level of confidence, α' is the significance level of a one-tailed

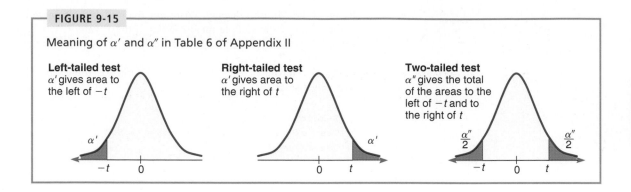

FIGURE 9-15

Meaning of α' and α'' in Table 6 of Appendix II

Left-tailed test
α' gives area to the left of $-t$

Right-tailed test
α' gives area to the right of t

Two-tailed test
α'' gives the total of the areas to the left of $-t$ and to the right of t

test (α' is the area to the right of t or, equivalently, to the left of $-t$), and α'' is the significance level for a two-tailed test (α'' is the area beyond $-t$ and t, so, in fact, $\alpha'' = 2\alpha'$). Figure 9-15 illustrates these different values.

To find the critical value(s) t_0, we look in the column headed by the level of significance α' for a *one-tailed test* or α'' for a *two-tailed test*. The critical value t_0 is located in the row headed by degrees of freedom $d.f. = n - 1$.

EXAMPLE 8

Using the Student's t distribution table

Use Table 6 in Appendix II to find the critical value(s) t_0 and critical region for the described tests of μ with sample size $n = 11$. (Assume that the sample came from a population with a distribution that is mound-shaped and symmetric.)

(a) Suppose that we have a *right-tailed* test of μ with level of significance 0.01.

SOLUTION: Since the test is a *one-tailed* test, we use the column headed by $\alpha' = 0.01$. The degrees of freedom are $d.f. = 11 - 1 = 10$, so we use the row headed by 10. The critical value is $t_0 = 2.764$ (see Figure 9-16).

(b) We have a *two-tailed* test of μ with level of significance 0.01.

SOLUTION: The test is a *two-tailed* test, so we look in the column under $\alpha'' = 0.01$ and the row headed by $d.f. = 11 - 1 = 10$. The critical values are $\pm t_0 = \pm 3.169$ (see Figure 9-17).   ◊

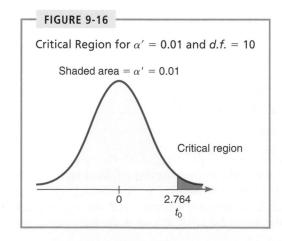

FIGURE 9-16

Critical Region for $\alpha' = 0.01$ and $d.f. = 10$

Shaded area = $\alpha' = 0.01$

Critical region

0 2.764
      t_0

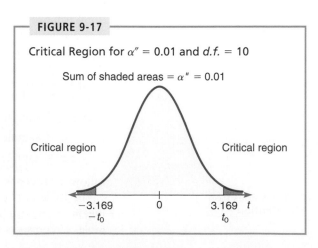

FIGURE 9-17

Critical Region for $\alpha'' = 0.01$ and $d.f. = 10$

Sum of shaded areas = $\alpha'' = 0.01$

Critical region   Critical region

-3.169 0 3.169 t
$-t_0$ t_0

GUIDED EXERCISE 7

Using the Student's t distribution table

Use Table 6 of Appendix II to find the critical value(s) t_0 and critical region for the test described based on a sample of size $n = 8$.

(a) A *left-tailed* test of μ with level of significance 0.05.

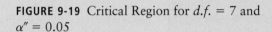 In this case, we use the column headed by $\alpha' = 0.05$ and the row headed by $d.f. = n - 1 = 8 - 1 = 7$. This gives us $t = 1.895$. For a left-tailed test, we use the symmetry of the distribution to get $t_0 = -1.895$ (see Figure 9-18).

(b) A *two-tailed* test of μ with level of significance 0.05.

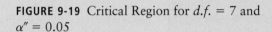 Use the column headed by $\alpha'' = 0.05$ and the row headed by $d.f. = n - 1 = 7$. By the symmetry of the curve, the critical values are $\pm t_0 = \pm 2.365$ (see Figure 9-19).

FIGURE 9-18 Critical Region for $d.f. = 7$ and $\alpha' = 0.05$

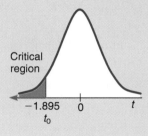

FIGURE 9-19 Critical Region for $d.f. = 7$ and $\alpha'' = 0.05$

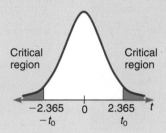

Sample test statistic

Once we have the critical value(s) t_0 and critical region, we convert the sample test statistic $\bar{x}$ to a t value using the formula

$$t = \frac{\bar{x} - \mu}{(s/\sqrt{n})} \text{ with } d.f. = n - 1$$

where $\bar{x}$ = sample mean and μ is specified in H_0

n = sample size

s = sample standard deviation

To conclude the test, we locate the t value of the sample statistic on a diagram showing the critical region. If the sample t value falls in the critical region, we reject H_0. If the sample t value falls outside the critical region, we fail to reject H_0.

The next example demonstrates the traditional method of using critical regions to conclude a test of a mean using small samples.

EXAMPLE 9

Testing μ (small sample)

A company manufactures large rocket engines used to project satellites into space. The government buys the rockets, and the contract specifies that these engines are to use an average of 5500 pounds of rocket fuel the first 15 seconds of operation. The company claims that its engines fit specifications. To test the claim, an inspector randomly selects six such engines from the warehouse. These six engines are fired 15 seconds each, and the fuel consumption for each engine is measured. For all six engines, the mean fuel consumption is $\bar{x} = 5690$ pounds and the standard deviation is $s = 250$ pounds. Is the claim justified at the 5% level of significance?

SOLUTION:

(a) We want to see if we can reject the hypothesis $\mu = 5500$ pounds, so the null hypothesis is

$$H_0: \mu = 5500$$

A substantial difference either way from 5500 pounds could be important, so the alternate hypothesis is

$$H_1: \mu \neq 5500$$

(b) Next, we find the critical values. Since our sample is small, we use Table 6 in Appendix II. For a *two-tailed* test with level of significance 0.05, we use the $\alpha'' = 0.05$ column and the row headed by $d.f. = n - 1 = 6 - 1 = 5$. This gives us $\pm t_0 = \pm 2.571$.

(c) Using $\mu = 5500$ from H_0, $s = 250$, and $n = 6$, we convert the sample test statistic $\bar{x} = 5690$ to a t value:

$$t = \frac{\bar{x} - \mu}{(s/\sqrt{n})} = \frac{5690 - 5500}{(250/\sqrt{6})} = 1.862$$

FIGURE 9-20

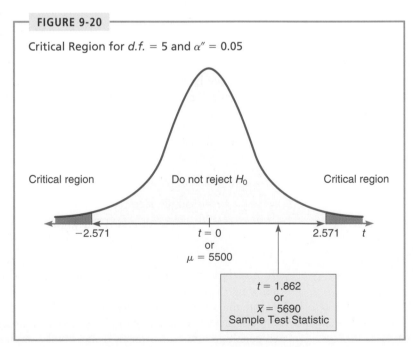

Critical Region for $d.f. = 5$ and $\alpha'' = 0.05$

(d) On a diagram show the critical regions and the location of the sample test statistic. (See Figure 9-20.) We see that the t value of the sample test statistic falls outside the critical region. Therefore, we cannot reject H_0. The data do not present sufficient evidence to indicate that the average fuel consumption for the first 15 seconds of operation is different from $\mu = 5500$ pounds.   ◇

GUIDED EXERCISE 8

Testing μ (small sample)

Suppose in Example 9 that a random sample of eight engines was used and the mean fuel consumption was $\bar{x} = 5880$ pounds (for the first 15 seconds). Again, the standard deviation was 250 pounds. Use a 1% level of significance to test the claim that the population average fuel consumption *exceeds* 5500 pounds.

(a) We will use $H_0: \mu = 5500$. State H_1.

→ $H_1: \mu > 5500$ because we want to test the claim that the average fuel consumption *exceeds* 5500 pounds.

(b) Find the critical value t_0.

→ Because the sample size is small, we use Table 6 of Appendix II. For a one-tailed test, we look in the column headed by $\alpha' = 0.01$ and the row headed by $d.f. = 8 - 1 = 7$. The critical value is $t_0 = 2.998$.

(c) Convert the sample test statistic $\bar{x}$ to a t value.

→ We use $\mu = 5500$, $n = 8$, and $s = 250$:

$$t = \frac{\bar{x} - \mu}{(s/\sqrt{n})} = \frac{5880 - 5500}{(250/\sqrt{8})} = 4.299$$

(d) Sketch the critical region and show the location of the sample test statistic on the diagram. Do we reject H_0 or not at the 1% level of significance?

→ Since the sample test statistic falls in the critical region, we reject H_0. There is evidence that the average fuel consumption exceeds 5500 pounds during the first 15 seconds of operation. (See Figure 9-21.)

FIGURE 9-21 Critical Region for $d.f. = 7$ and $\alpha' = 0.01$

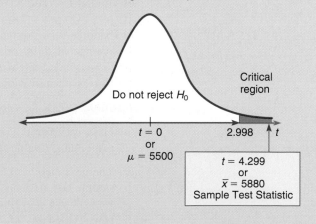

FIGURE 9-22

P Values Corresponding to α' and α''

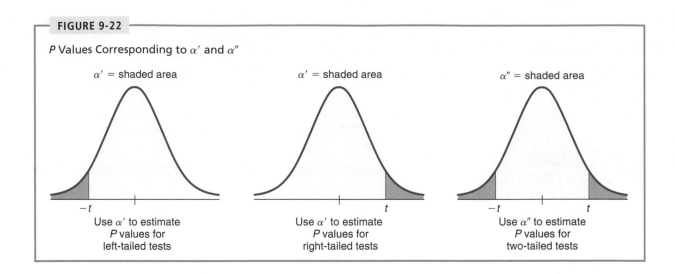

Using *P* values for tests of μ, small samples

Recall that the *P* value for a statistical test is the probability of getting a sample statistic as far (or even farther) into the tails of the sampling distribution as the observed sample statistic. The smaller the *P* value, the stronger the evidence is to reject H_0.

In the case of small sample tests of μ, the sample distribution $\bar{x}$ follows a Student's *t* distribution. Therefore, we will use Table 6 of Appendix II to estimate the *P* values. However, the areas given in Table 6 are limited. Consequently, when we use Table 6 to estimate *P* values for $\bar{x}$, we usually find an *interval containing the P value* rather than a single number for the *P* value.

◇ **NOTE REGARDING TABLE 6 OF APPENDIX II** In Table 6, α' represents the area in *one tail* beyond *t*. The α'' values represent the area in the *two tails*. Notice that in each column $\alpha'' = 2\alpha'$. Consequently, we use α' values as endpoints of the *P*-value intervals for *one-tailed tests* and α'' values as endpoints of the *P*-value intervals for *two-tailed tests* (see Figure 9-22). ◇

Let's find the *P* values associated with the sample test statistic of Example 9 and of Guided Exercise 8.

EXAMPLE 10

P-value interval for two-tailed test

In Example 9, we tested the rocket manufacturer's claim that its rockets consumed an average of 5500 pounds of fuel in the first 15 seconds of operation. The test was a two-tailed test with sample statistic $\bar{x} = 5690$, $s = 250$, and $n = 6$. Find the *P* value associated with $\bar{x} = 5690$. What does the *P* value tell us?

SOLUTION:

(a) First, we convert $\bar{x} = 5690$ to a *t* value. This was done in Example 9. The corresponding *t* value is 1.862.

(b) The next step is to find an interval containing the *P* value corresponding to the sample test statistic $t = 1.862$ using a two-tailed test. Our test is two-tailed, so we will use α'' values. To find the associated *P*-value interval, we look in the row headed by *d.f.* $= n - 1 = 6 - 1 = 5$. We find that the sample *t* value 1.862 falls between 1.699 and 2.015. The corresponding *P* value then lies between corresponding α'' values 0.150 and 0.100 (see Table 9-5).

TABLE 9-5   Excerpt from Student's t Distribution (Table 6, Appendix II)

One-tailed test P value	α'	...		
✓ Two-tailed test P value	α''	...	0.150	0.100 ···
d.f.	5	...	1.699	2.015 ···
			↑	
			Sample $t = 1.862$	

Put the adjacent α'' values in increasing order to form the interval

$$0.100 < P \text{ value} < 0.150$$

(c) We see that the P value may be almost as large as 0.150. For $\alpha = 0.05$, we see that the P value is greater than α, so we do not reject H_0. This result is consistent with the conclusion of Example 9.   ◇

The next example shows how to find an interval containing the P value for a one-tailed test using the Student's t distribution.

EXAMPLE 11

P-value interval for one-tailed test

In Guided Exercise 8, we again looked at the fuel consumption of rockets. This time the sample mean consumption for the first 15 seconds of flight was $\bar{x} = 5880$ pounds with $s = 250$ pounds and $n = 8$ rockets. We used a right-tailed test with H_0: $\mu = 5500$ and H_1: $\mu > 5500$. Find the P value associated with the sample statistic $\bar{x}$. What does the P value tell you?

SOLUTION:

(a) Again, our first step is to convert the sample statistic $\bar{x} = 5880$ to a t value. We did this in Guided Exercise 8 and got $t = 4.299$.

(b) This test is a one-tailed test, so we use α' values to create the P-value interval. To find the associated P value, we look in the row headed by $d.f. = n - 1 = 8 - 1 = 7$ and find that the sample t statistic $t = 4.299$ falls to the *right* of 3.499 (Table 9-6).

Notice that reading from left to right, the P values decrease. Therefore, the P value corresponding to the sample t value is smaller than that corresponding to $t = 3.499$. Therefore,

$$P \text{ value} < 0.005$$

This relation is shown in Figure 9-23 on the following page.

(c) Since the P value is less than the level of significance $\alpha = 0.01$, it is significant, and we reject H_0. This result is consistent with the conclusion of Guided Exercise 8.   ◇

TABLE 9-6   Excerpt from Student's t Distribution (Table 6, Appendix II)

✓ One-tailed test P value	α'	...	0.010	0.005
Two-tailed test P value	α''	...		
d.f.	7	...	2.998	3.499
				↑
				Sample $t = 4.299$

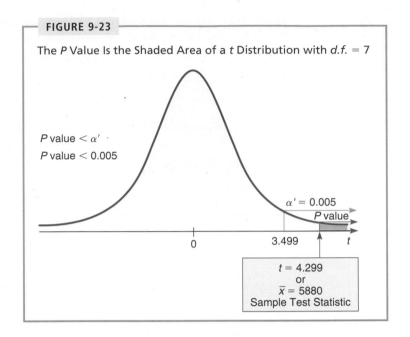

FIGURE 9-23

The *P* Value Is the Shaded Area of a *t* Distribution with *d.f.* = 7

P value $< \alpha'$
P value < 0.005

$\alpha' = 0.005$

P value

0   3.499   *t*

$t = 4.299$
or
$\bar{x} = 5880$
Sample Test Statistic

 TECH NOTE Both the TI-83Plus and Minitab support testing a single mean using the Student's *t* distribution. The output for each technology is similar to the output for testing a single mean using the normal distribution. The technologies give the specific *P* value for the sample test statistic rather than an interval of *P* values.

TI-83Plus Press **STAT**, select **TESTS**, use option **2:T-Test**.

Minitab Menu selection: **Stats ➤ Basic Statistics ➤ 1-Sample t**.

VIEWP⊙INT

A merry heart doeth good like a medicine.
—Proverbs 17:22

This Old Testament quote summarizes much of the research done by the nonprofit group Rx Laughter. To find a link to the Rx Laughter Web site, visit the Brase/Brase statistics site at http://math.college.hmco.com/students and look up Rx Laughter. Lonnie Zeltzer is a professor of pediatrics at the UCLA Medical School who uses laughter to help children endure painful treatments like chemotherapy and dialysis. Professor Zeltzer works with descendants of famous comedians such as Charlie Chaplin, Harpo Marx, and Lou Costello in an unusual partnership of show business and medicine. Chris Costello, daughter of Lou Costello and a board member of Rx Laughter, claims the research done by the group is a wonderful way to bring the medical community up to speed with children's need for pain management.

Many scientists are exploring the relation between humor and health. The *Journal of the American Medical Association* contains numerous articles about laughter and its effect on stress hormones and the immune system. In all studies of this type, statistics turns out to be an extremely important research tool.

SECTION 9.4   PROBLEMS

In Problems 1–6, use Table 6 of Appendix II to find the critical value(s) t_0 for the described test of the mean (small sample).

1. Sample size $n = 9$, left-tailed test, level of significance 5%.

2. Sample size $n = 13$, right-tailed test, level of significance 1%.

3. Sample size $n = 24$, two-tailed test, level of significance 1%.

4. Sample size $n = 18$, left-tailed test, level of significance 5%.

5. Sample size $n = 12$, two-tailed test, level of significance 5%.

6. Sample size $n = 29$, right-tailed test, level of significance 1%.

For each of the following problems, please provide the requested information.
(a) What is the null hypothesis? What is the alternate hypothesis? Will we use a left-tailed, right-tailed, or two-tailed test? What is the level of significance?
(b) What sampling distribution will we use? What is the critical value t_0 (or critical values $\pm t_0$)?
(c) Sketch the critical region and show the critical value (or critical values).
(d) Calculate the t value corresponding to the sample statistic $\bar{x}$ and show its location on the sketch of part (c).
(e) Find the P value (or interval containing the P value) for your test and explain what the P value means in this context.
(f) Based on your answers for parts (a) to (e), shall we reject or fail to reject (i.e., "accept") the null hypothesis based on the given level of significance? Explain your conclusion in the context of the problem.

7. *Medical: Red Blood Cell Count* Let x be a random variable that represents red blood cell (RBC) count in millions per cubic millimeter of whole blood. Then x has a distribution that is approximately normal, and for the population of healthy female adults, the mean of the x distribution is about 4.8 (based on information from *Diagnostic Tests with Nursing Implications*, Springhouse Corporation). Suppose that a female patient has taken six laboratory blood tests over the past several months and the RBC count data sent to the patient's doctor were

| 3.5 | 4.2 | 4.5 | 4.6 | 3.7 | 3.9 |

(a) Use a calculator with sample mean and sample standard deviation keys to verify that $\bar{x} = 4.07$ and $s = 0.44$.
(b) Do the given data indicate that the population mean RBC count for this patient is lower than 4.8? Use $\alpha = 0.05$.

8. *Medical: Hemoglobin Count* Let x be a random variable that represents hemoglobin count (HC) in grams per 100 milliliters of whole blood. Then x has a distribution that is approximately normal with population mean about 14 for healthy adult women (see reference in Problem 7). Suppose that a female patient has taken 12 laboratory blood tests during the past year. The HC data sent to the patient's doctor were

| 19 | 23 | 15 | 21 | 18 | 16 |
| 14 | 20 | 19 | 16 | 18 | 21 |

(a) Use a calculator with sample mean and sample standard deviation keys to verify that $\bar{x} = 18.33$ and $s = 2.71$.

(b) Does this information indicate that the population average HC for this patient is higher than 14? Use $\alpha = 0.01$.

9. *Hummingbirds: Incubation Selasphorus platycerus* is the scientific name of the broad-tailed hummingbird that is commonly found throughout the western United States. It is known that the mean incubation time for eggs of this bird is approximately 16.5 days (based on information from *The Hummingbird Book*, by D. Stokes and L. Stokes, Little, Brown and Company). Assume that the incubation times are approximately normally distributed. However, at higher elevations (above 8000 feet), it is thought that the average incubation time might be different from 16.5 days. A number of broad-tailed hummingbird nests were located in mountain country above 8000 feet, and incubation times for 18 eggs gave the following information (in days):

15	19	16	16	18	17	21	18	17
16	17	18	19	16	15	21	23	19

(a) Use a calculator with mean and standard deviation keys to verify that $\bar{x} = 17.83$ days and $s = 2.20$ days.
(b) Does this information indicate that the population average incubation time above 8000 feet elevation is different (either more or less) from 16.5 days? Use $\alpha = 0.05$.

10. *Fishing: Atlantic Salmon* Homser Lake, Oregon, has an Atlantic salmon catch and release program that has been very successful. The average fisherman's catch has been $\mu = 8.8$ Atlantic salmon per day. (Source: *National Symposium on Catch and Release Fishing*, Humboldt State University.) Suppose that a new quota system restricting the number of fishermen has been put into effect this season. A random sample of fishermen gave the following catch per day:

12	6	11	12	5	0	2
7	8	7	6	3	12	12

(a) Use a calculator with mean and sample standard deviation keys to verify $\bar{x} = 7.36$ and sample standard deviation $s = 4.03$.
(b) Assuming that the catch per day has an approximately normal distribution, use a 5% level of significance to test the claim that the population average catch per day is now different from 8.8.

11. *Archaeology: Tree Rings* Tree-ring dating from archaeological excavation sites is used in conjunction with other chronologic evidence to estimate occupation dates of prehistoric Indian ruins in southwestern United States. It is thought that Burnt Mesa Pueblo was occupied around 1300 A.D. (based on evidence from potsherds and stone tools). The following data give tree-ring dates (A.D.) from adjacent archaeological sites (*Bandelier Archaeological Excavation Project: Summer 1990 Excavations at Burnt Mesa Pueblo*, edited by T. Kohler, Washington State University Department of Anthropology, 1992):

1189	1267	1268	1275	1275
1271	1272	1316	1317	1230

(a) Use a calculator with mean and standard deviation keys to verify that $\bar{x} = 1268$ and $s = 37.29$ years.
(b) Assuming that the tree-ring dates in this excavation area follow a distribution that is approximately normal, does this information indicate the population mean of tree-ring dates in the area is different (either high or low) from 1300 A.D.? Use a 1% level of significance.

12. *Ski Patrol: Avalanches* Snow avalanches can be a real problem for travelers in western United States and Canada. A very common type of avalanche is called the slab avalanche. These have been studied extensively by David McClung, a professor of civil engineering at the University of British Columbia. Slab avalanches studied in Canada had an average thickness of $\mu = 67$ cm. (Source: *Avalanche Handbook*, by D. McClung and P. Schaerer). The ski patrol at Vail, Colorado, is studying slab avalanches in their region. A random sample of avalanches in spring gave the following thicknesses (cm):

59	51	76	38	65	54	49	62
68	55	64	67	63	74	65	79

(a) Use a calculator with mean and standard deviation keys to verify that $\bar{x} = 61.8$ cm and $s = 10.6$ cm.

(b) Assume the slab thickness has an approximately normal distribution. Use a 1% level of significance to test the claim that the mean slab thickness in the Vail region is different from that in Canada.

13. *Longevity: Honolulu* *USA Today* reported that the state with the longest mean life span is Hawaii, where the population mean life span is 77 years. A random sample of 20 obituary notices in the *Honolulu Advertizer* gave the following information about life span (in years) of Honolulu residents:

72	68	81	93	56	19	78	94	83	84
77	69	85	97	75	71	86	47	66	88

(a) Use a calculator with mean and standard deviation keys to verify that $\bar{x} = 74.45$ years and $s = 18.09$ years.

(b) Assuming that life span in Honolulu is approximately normally distributed, does this information indicate that the population mean life span for Honolulu residents is less than 77 years? Use a 5% level of significance.

14. *Veterinary Science: Lions* The heart rate of a healthy lion is approximately normally distributed with mean $\mu = 40$ beats per minute. (Source: *The Merck Veterinary Manual*, a reference used in most veterinary colleges.) A heart rate that is too slow or too fast can indicate a health problem. A veterinarian has removed an abscessed tooth from a young, healthy zoo lion. As the animal slowly starts to come out of the anesthetic, the heart rate is taken and recorded for half an hour (beats per minute):

30	37	43	38	35	36

(a) Use a calculator with mean and standard deviation keys to verify $\bar{x} = 36.5$ and $s = 4.2$.

(b) Use a 5% level of significance to test the claim that the population average heart rate of the lion is different (either way) from 40 beats per minute.

15. *Shopping Time: Housewares* How much customers buy is a direct result of how much time they spend in the store. The mean shopping time for a woman accompanied by children in national houseware stores is 7.3 minutes. (Source: *Why We Buy: The Science of Shopping* by P. Underhill.) A retail research team is studying shopping habits in the Cherry Creek Mall (Denver). A random sample of women shoppers with children in a large houseware store gave the following shopping times (minutes):

7.7	8.1	8.2	9.0	5.8	9.3	8.4	6.9	12.1	9.4
8.1	6.2	7.3	7.9	8.2	8.5	7.2	6.3	9.1	8.8

(a) Use a calculator with mean and standard deviation keys to verify that $\bar{x} = 8.1$ min and $s = 1.4$ min.

(b) Assume shopping time follows an approximately normal distribution. Use a 5% level of significance to test the claim that the average shopping time for women with children in the Cherry Creek Mall is higher than the national average for this type of store.

9.5
Tests Involving a Proportion

Tests for a single proportion

Many situations arise that call for tests of proportions or percentages rather than means. For instance, a college registrar may want to determine if the proportion of students wanting 3-week intensive courses has increased.

How can we make such a test? In this section, we will study tests involving proportions (i.e., percentages or proportions). Such tests are similar to those in Sections 9.2 and 9.4. The main difference is that we are working with a distribution of proportions.

Throughout this section, we will assume that the situations we are dealing with satisfy the conditions underlying the binomial distribution. In particular, we will let r be a binomial random variable. This means that r is the number of successes out of n independent binomial trials (for the definition of binomial trial, see Section 5.2). We will use $\hat{p} = r/n$ as our estimate for p, the probability of success on each trial. The letter q again represents the probability of failure on each trial, and so $q = 1 - p$. We also assume that the samples are large (i.e., $np > 5$ and $nq > 5$).

For large samples, the distribution of $\hat{p} = r/n$ values is well approximated by a *normal curve* with mean μ and standard deviation σ as follows:

$$\mu = p$$

$$\sigma = \sqrt{\frac{pq}{n}}$$

The null and alternate hypotheses for tests of proportions are

H_0: $p = k$	H_0: $p = k$	H_0: $p = k$
H_1: $p < k$	H_1: $p > k$	H_1: $p \neq k$

depending on what is asked for in the problem. Notice that since p is a probability, the value k must be between 0 and 1.

Critical values for testing p ($np > 5$ and $nq > 5$)

Since the $\hat{p}$ distribution is approximately normal when n is sufficiently large, we will use the same *critical values* z_0 for our tests as those we used for testing the mean with large samples (Section 9.2). Table 9-7 shows these values.

Sample test statistic

For tests of proportions, we need to convert our sample test statistic $\hat{p}$ to a z value. Then we can compare the sample z value with a specified critical value z_0 to determine

TABLE 9-7   Critical Values for Testing a Proportion

Level of Significance	$\alpha = 0.05$	$\alpha = 0.01$
Critical value z_0 for a left-tailed test	−1.645	−2.33
Critical value z_0 for a right-tailed test	1.645	2.33
Critical values $\pm z_0$ for a two-tailed test	±1.96	±2.58

whether or not to reject H_0. The $\hat{p}$ distribution is approximately normal with mean p and standard deviation $\sqrt{pq/n}$. Therefore, the conversion of $\hat{p}$ to z follows the formula

$$z = \frac{\hat{p} - p}{\sqrt{\dfrac{pq}{n}}}$$

where $\hat{p} = r/n$ is the sample test statistic

$\quad\quad n$ = number of trials

$\quad\quad p$ = proportion specified in H_0

$\quad\quad q = 1 - p$

EXAMPLE 12

Testing p

A team of eye surgeons has developed a new technique for a risky eye operation to restore the sight of people blinded from a certain disease. Under the old method, it is known that only 30% of the patients who undergo this operation recover their eyesight.

Suppose that surgeons in various hospitals have performed a total of 225 operations using the new method and that 88 have been successful (the patients fully recovered their sight). Can we justify the claim that the new method is better than the old one? (Use a 1% level of significance.)

SOLUTION:

(a) Let p be the probability that a patient fully recovers his or her eyesight. The null hypothesis is that p is still 0.30, even for the new method of operation. Therefore,

$\quad\quad H_0$: $p = 0.30$

(b) The alternate hypothesis is that the new method has improved the patient's chances for eyesight recovery. Therefore,

$\quad\quad H_1$: $p > 0.30$

(c) Since the alternate hypothesis is H_1: $p > 0.30$, we use a right-tailed test. For $\alpha = 0.01$, Table 9-7 shows a critical value of $z_0 = 2.33$.

(d) Next, we find the sample test statistic $\hat{p}$ and convert it to a z value:

$$\hat{p} = \frac{r}{n} = \frac{88}{225} \approx 0.39$$

The z value corresponding to $\hat{p}$ is

$$z = \frac{\hat{p} - p}{\sqrt{\dfrac{pq}{n}}} = \frac{0.39 - 0.30}{\sqrt{\dfrac{0.30(0.70)}{225}}} = 2.95$$

In the formula, the value for p came from the null hypothesis. H_0 specified that $p = 0.30$, so $q = 1 - 0.30 = 0.70$.

● **CALCULATOR NOTE**   If you evaluate the denominator separately, carry at least four digits after the decimal. Since your calculator automatically carries at least eight or nine digits, it is best to do the entire calculation internally on your calculator.

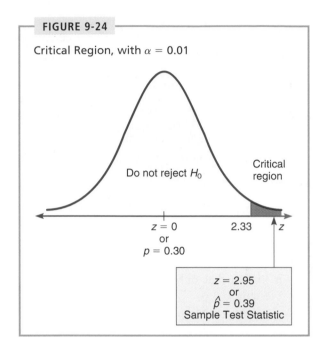

FIGURE 9-24

Critical Region, with $\alpha = 0.01$

Do not reject H_0

Critical region

$z = 0$
or
$p = 0.30$

2.33

z

$z = 2.95$
or
$\hat{p} = 0.39$
Sample Test Statistic

(e) Finally, we draw a normal distribution showing the critical region and the location of the z value corresponding to the sample test statistic $\hat{p}$ (Figure 9-24). Since the sample z value falls inside the critical region, we reject H_0 and conclude that the new method seems to have improved a patient's chances for eyesight recovery.   ◇

GUIDED EXERCISE 9

Testing p

A botanist has produced a new variety of hybrid wheat that is better able to withstand drought than other varieties. The botanist knows that for the parent plants the proportion of seeds germinating is 80%. The proportion of seeds germinating for the hybrid variety is unknown, but the botanist claims that it is 80%. To test this claim, 400 seeds from the hybrid plant are tested, and it is found that 312 germinated. Use a 5% level of significance to test the claim that the proportion germinating for the hybrid is 80%.

(a) Let p be the proportion of hybrid seeds that will germinate. What is the null hypothesis?

⟹ H_0: $p = 0.80$

(b) Because we have no prior knowledge about germination proportion for the hybrid plant, what would be a good choice for the alternate hypothesis?

⟹ H_1: $p \neq 0.80$

Continued

GUIDED EXERCISE 9 continued

(c) Should our test be a one-tailed or a two-tailed test? For $\alpha = 0.05$, what is (are) the critical value(s)?

⟹ The test is a two-tailed test. By Table 9-7, the critical values are $z_0 = \pm 1.96$.

(d) Calculate the sample test statistic $\hat{p}$.

⟹ The number of trials is $n = 400$, and the number of successes is $r = 312$. Thus,

$$\hat{p} = \frac{r}{n} = \frac{312}{400} = 0.78$$

(e) Next, we convert the sample test statistic $\hat{p} = 0.78$ to a z value. Based on our choice for H_0, what value should we use for p in our formula? Since $q = 1 - p$, what value should we use for q? Using these values for p and q, convert $\hat{p}$ to a z value.

⟹ According to H_0, $p = 0.80$. Then $q = 1 - p = 0.20$. Using these values in the following formula gives

$$z = \frac{\hat{p} - p}{\sqrt{\dfrac{pq}{n}}}$$

$$= \frac{0.78 - 0.80}{\sqrt{\dfrac{0.80(0.20)}{400}}}$$

$$= -1.00$$

⬤ **CALCULATOR NOTE**   If you evaluate the denominator separately, be sure to carry at least 4 digits after the decimal.

(f) On the normal curve, show the critical regions and the z value corresponding to the sample test statistic $\hat{p}$. Do we reject or fail to reject H_0?

⟹ The critical region is shown in Figure 9-25. Since the sample statistic does not fall in the critical region, we fail to reject H_0. At the 5% level of significance, there is not enough evidence to say the botanist is wrong.

FIGURE 9-25 Critical Region, $\alpha = 0.05$

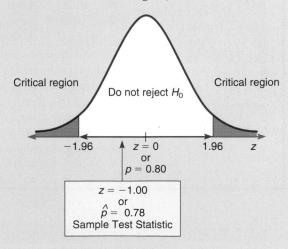

P values for tests of proportions

Since the $\hat{p}$ sampling distribution is approximately normal, we use Table 5, "Areas of a Standard Normal Distribution," in Appendix II to find *P* values.

EXAMPLE 13

P value for testing p

Let's solve Guided Exercise 9 using the *P*-value approach. In that problem, 312 of 400 seeds from a hybrid wheat variety germinated. For the parent plants, the proportion of germinating seeds was 80%. Use a 5% level of significance to test the claim that the population proportion of germinating seeds from the hybrid wheat is different from that of the parent plants.

(a) First, we find the sample statistic $\hat{p}$ and the corresponding z value. This was done in Guided Exercise 9, where we found that $\hat{p} = 0.78$ with corresponding $z = -1.00$.

(b) Next, we find the *P* value associated with $z = -1.00$ and a two-tailed test. Using Table 5 (Appendix II) together with techniques of Chapter 6, we see that $P(z \leq -1.00) = 0.1587$. The test is a two-tailed z test, so we double the area found in the left tail to get *P* value = 0.3174. Since this *P* value is greater than the level of significance, $\alpha = 0.05$, we do not reject the null hypothesis, and we conclude that there is not enough evidence to show that the germination rate for the hybrid wheat is different from that of the parent plants.   ◇

 TECH NOTE Both the TI-83Plus and Minitab support tests of proportions. The output for both technologies includes the sample proportion $\hat{p}$ and the *P* value of $\hat{p}$. Minitab also includes the z value corresponding to $\hat{p}$.

TI-83Plus Press **STAT**, select **TESTS**, use option **5:1-PropZTest**. The value of p_0 is from the null hypothesis H_0: $p = p_0$. The number of successes is the value for x.

Minitab Menu Selections: **Stat ➤ Basic Statistics ➤ 1 Proportion**. Under options set the test proportion as the value in H_0. Choose to use the normal distribution.

VIEWPOINT

Who Did What?

Art, music, literature, and science share a common need to classify things: Who painted that picture? Who composed that music? Who wrote that document? Who should get that patent? In statistics, such questions are called *classification problems.* For example, the *Federalist Papers* were published anonymously in 1787–1788 by Alexander Hamilton, John Jay, and James Madison. But who wrote what? That question is addressed by F. Mosteller (Harvard University) and D. Wallace (University of Chicago) in the book *Statistics: A Guide to the Unknown,* edited by J. M. Tanur. Other scholars have studied authorship regarding Plato's *Republic* and Plato's *Dialogues,* including the *Symposium.* For more information on this topic, see the source in Problems 7 and 8 in this exercise set.

SECTION 9.5   PROBLEMS

For each of the following problems, please provide the requested information.
(a) What is the null hypothesis? What is the alternate hypothesis? Will we use a left-tailed, right-tailed, or two-tailed test? What is the level of significance?

(b) What sampling distribution will we use? What is the critical value z_0 (or critical values $\pm z_0$)?

(c) Sketch the critical region and show the critical value (or critical values).

(d) Compute the z value corresponding to the sample statistic $\hat{p}$ and show its location on the sketch of part (c).

(e) Find the P value for your test and explain what the P value means in this context.

(f) Based on your answers for parts (a) to (e), shall we reject or fail to reject (i.e., "accept") the null hypothesis for the given level of significance α? Explain your conclusion in the context of the problem.

1. *Sociology: Crime Rate* Is the national crime rate really going down? Some sociologists say yes! They say that the reason for the decline in crime rates for the 1980s and 1990s is demographics. It seems that the population is aging, and older people commit fewer crimes. According to the FBI and the Justice Department, 70% of all arrests are of males aged 15 to 34 years. (Source: *True Odds*, by J. Walsh, Merritt Publishing.) Suppose that you are a sociologist in Rock Springs, Wyoming, and a random sample of police files showed that of 32 arrests last month, 24 were of males aged 15 to 34 years. Use a 1% level of significance to test the claim that the population proportion of such arrests in Rock Springs is different from 70%.

2. *College Athletics: Graduation Rate* Women athletes at the University of Colorado, Boulder, have a long-term graduation rate of 67% (Source: *The Chronicle of Higher Education*). Over the past several years, a random sample of 38 women athletes at the school showed that 21 eventually graduated. Does this indicate the population proportion of women athletes who graduate from the University of Colorado, Boulder, is now less than 67%? Use a 5% level of significance.

3. *Highway Accidents: DUI* The U.S. Department of Transportation, National Highway Traffic Safety Administration, reported that 77% of all fatally injured automobile drivers were intoxicated. A random sample of 27 records of automobile driver fatalities in Kit Carson County, Colorado, showed that 15 involved an intoxicated driver. Do these data indicate that the population proportion of driver fatalities related to alcohol is less than 77% in Kit Carson County? Use $\alpha = 0.01$.

4. *Highway Accidents: Airbags* Do drivers with air bags take greater risks? Researchers at Virginia Commonwealth University studied fatal automobile accidents between cars in which one has an air bag and the other does not. They found that in 73% of all such fatal crashes the driver of the air bag vehicle was responsible for the accident. (Source: See Problem 1.) Suppose that you obtain police records of fatal automobile crashes (involving a vehicle with an air bag and one without) near Fargo, North Dakota. A random sample of 41 such accidents showed that 33 were the fault of the driver of the air bag vehicle. Does this indicate that the population proportion of such accidents in which the driver of the air bag vehicle was at fault is higher than 73% in the Fargo district? Use $\alpha = 0.05$.

5. *Wildlife: Wolves* The following is based on information from *The Wolf in the Southwest: The Making of an Endangered Species*, by David E. Brown (University of Arizona Press). Before 1918, the proportion of female wolves in the general population of all southwestern wolves was about 50%. However, after 1918, southwestern cattle ranchers began a widespread effort to destroy wolves. In a recent sample of 34 wolves, there were only 10 females. One theory is that male wolves tend to return sooner than females to their old territory, where their predecessors were exterminated. Do these data indicate that the population proportion of female wolves is now less than 50% in the region? Use $\alpha = 0.01$.

6. *Fishing: Northern Pike* Athabasca Fishing Lodge is located on Lake Athabasca in northern Canada. In one of its recent brochures, the lodge advertises that 75% of

.75

its guests catch northern pike over 20 pounds. Suppose that last summer 64 out of a random sample of 83 guests did, in fact, catch northern pikes weighing over 20 pounds. Does this indicate that the population proportion of guests who catch pikes over 20 pounds is different from 75% (either higher or lower)? Use $\alpha = 0.05$.

7. *Plato's* Republic: *Syllable Patterns* Prose rhythm is characterized as the occurrence of five-syllable sequences in long passages of text. This characterization may be used to assess the similarity among passages of text and sometimes the identity of authors. The following information is based on an article by D. Wishart and S. V. Leach appearing in *Computer Studies of the Humanities and Verbal Behavior* (Vol. 3, pp. 90–99). Syllables were categorized as long or short. On analyzing Plato's *Republic*, Wishart and Leach found that about 26.1% of the five-syllable sequences are of the type in which two are short and three are long. Suppose that Greek archaeologists have found an ancient manuscript dating back to Plato's time (about 427–347 B.C.). A random sample of 317 five-syllable sequences from the newly discovered manuscript showed that 61 are of the type two short and three long combination. Do the data indicate that the population proportion of this type of five-syllable sequence is different (either way) from the text of Plato's *Republic*? Use $\alpha = 0.01$.

8. *Plato's* Dialogues: *Prose Rhythm Symposium* is part of a larger work referred to as Plato's *Dialogues*. Wishart and Leach (see source in Problem 7) found that about 21.4% of five-syllable sequences in *Symposium* are of the type in which four are short and one is long. Suppose that an antiquities store in Athens has a very old manuscript that the owner claims to be part of Plato's *Dialogues*. A random sample of 493 five-syllable sequences from this manuscript showed that 136 were of the type four short and one long. Do the data indicate that the population proportion of this type five-syllable sequence is higher than that found in Plato's *Symposium*? Use $\alpha = 0.01$.

9. *Consumers: Product Loyalty USA Today* reported that about 47% of the general consumer population in the United States is loyal to the automobile manufacturer of their choice. Suppose that Chevrolet did a study of a random sample of 1006 Chevrolet owners and found that 490 said that they would buy another Chevrolet. Does this indicate that the population proportion of consumers loyal to Chevrolet is more than 47%? Use $\alpha = 0.01$.

10. *Supermarket: Prices Harper's Index* reported that 80% of all supermarket prices end in the digit 9 or 5. Suppose that you check a random sample of 115 items in a supermarket and find that 88 have prices that end in 9 or 5. Does this indicate that less than 80% of the prices in the store end in the digits 9 or 5? Use $\alpha = 0.05$.

11. *Medical: Hypertension* This problem is based on information taken from *The Merck Manual* (a reference manual used in most medical and nursing schools). Hypertension is defined as a blood pressure over 140 mm Hg systolic and/or over 90 mm Hg diastolic. Hypertension, if not corrected, can cause long-term health problems. In the college-age population (18–24 years), about 9.2% have hypertension. Suppose that a blood donor program is occurring in a college dormitory this week (final exams week). Before each student gives blood, the nurse takes a blood pressure reading. Of 196 donors, it was found that 29 have hypertension. Do these data indicate that the population proportion of students with hypertension during final exams week is higher than 9.2%? Use a 5% level of significance.

12. *Medical: Hypertension* Diltiazem is a commonly prescribed drug for hypertension (see source in Problem 11). However, diltiazem causes headaches in about

12% of patients using the drug. It is hypothesized that regular exercise might help reduce the headaches. If a random sample of 209 patients using diltiazem exercised regularly and only 16 had headaches, would this indicate a reduction in the population proportion of patients having headaches? Use a 1% level of significance.

13. *Myers-Briggs: Extroverts* Are most student government leaders extroverts? According to Myers-Briggs estimates, about 82% of college student government leaders are extroverts. (Source: *Myers-Briggs Type Indicator Atlas of Type Tables.*) Suppose that a Myers-Briggs personality preference test was given to a random sample of 73 student government leaders attending a large national leadership conference and that 56 were found to be extroverts. Does this indicate that the population proportion of extroverts among college student government leaders is different (either way) from 82%? Use $\alpha = 0.01$.

14. *American Attitudes: NAFTA* Generally speaking, would you say that America benefits from being a member of NAFTA? Nationally, about 28% of the U.S. population believes NAFTA benefits America (Source: *American Attitudes* by S. Mitchell, Sociology Department, Ithaca College). A random sample of 48 interstate truck drivers showed that 19 believe NAFTA benefits America. Does this indicate the population proportion of interstate truckers who believe NAFTA benefits America is higher than 28%? Use a 5% level of significance.

15. *Careers: College Professors* If you could start your career all over again, would you still choose to be a college professor? About 76% of all U.S. college professors respond yes, they would choose to be a college professor again (Source: *The Chronicle of Higher Education*). A random sample of 59 college professors in Colorado showed that 47 claim they would choose college teaching again. Does this indicate the proportion of professors in Colorado who would choose the career again is different from the national rate of 76%? Use a 1% level of significance.

9.6
Tests Involving Paired Differences (Dependent Samples)

FOCUS POINTS
- ✓ Identify paired data and dependent samples.
- ✓ Explain the advantages of paired data tests.
- ✓ Compute differences and the sample test statistic.
- ✓ Use degrees of freedom and the *t* table to find critical values.
- ✓ Estimate the *P* value and conclude the test.

Creating data pairs

Many statistical applications use *paired data* samples to draw conclusions about the difference between two population means. Data *pairs* occur very naturally in "before and after" situations, where the *same* object or item is measured both before and after a treatment. Applied problems in social science, natural science, and business administration frequently involve a study of matching pairs. Psychological studies of identical twins; biological studies of plant growth on plots of land matched for soil type, moisture, and sun; and business studies on sales of matched inventories are examples of paired data studies.

When working with paired data, it is very important to have a definite and uniform method of creating data pairs that clearly utilizes a natural matching of characteristics. The next example and exercise demonstrate this feature.

EXAMPLE 14

Paired data

A shoe manufacturer claims that among the general population of adults in the United States, the average length of the left foot is longer than that of the right. To compare the average length of the left foot with that of the right, we can take a random sample of 15 U.S. adults and measure the length of the left foot and then the length of the right foot for each person in the sample. Is there a natural way of pairing the measurements? How many pairs will we have?

SOLUTION: In this case, we can pair each left foot measurement with the same person's right foot measurement. The person serves as the "matching link" between the two distributions. We will have 15 pairs of measurements. ◇

GUIDED EXERCISE 10

Paired data

A psychologist has developed a series of exercises called the Instrumental Enrichment (IE) Program, which he claims to be useful in overcoming cognitive deficiencies in mentally retarded children. To test the program, extensive statistical tests are being conducted. In one experiment, a random sample of 10-year-old students with IQ scores below 80 was selected. An IQ test was given to these students before they spent 2 years in an IE Program, and an IQ test was given to the same students after the program.

(a) On what basis can you pair the IQ scores?

⟹ Take the "before and after" IQ scores of each individual student.

(b) If there were 20 students in the sample, how many data pairs would you have?

⟹ Twenty data pairs. Note that there would be 40 IQ scores, but only 20 pairs.

◇ **COMMENT** To compare two populations, we cannot always employ paired data tests, but when we can, what are the advantages? Using matched or paired data often can reduce the danger of introducing extraneous or uncontrollable factors into our sample measurements because the matched or paired data have essentially the *same* characteristics except for the *one* characteristic that is being measured. Furthermore, it can be shown that pairing data has the theoretical effect of reducing measurement variability (i.e., variance), which increases the accuracy of statistical conclusions. ◇

When we wish to compare the means of two samples, the first item to be determined is whether or not there is a natural pairing between the data in the two samples. Again, data pairs are created from "before and after" situations, or from matching data by using studies of the same object, or by a process of taking measurements of closely matched items.

Testing the differences d

When testing *paired* data, we take the difference d of the data pairs *first* and look at the mean difference $\bar{d}$. Then we use a test on $\bar{d}$. Theorem 9.1 provides the basis for our work with paired data.

Theorem 9.1

Consider a random sample of n data pairs. Suppose the differences d between the first and second members of each data pair are (approximately) normally distributed with population mean μ_d. Then the t values

$$t = \frac{\bar{d} - \mu_d}{s_d / \sqrt{n}}$$

where $\bar{d}$ is the sample mean of the d values, n is the number of data pairs, and

$$s_d = \sqrt{\frac{\Sigma(d - \bar{d})^2}{n - 1}}$$

is the sample standard deviation of the d values, follow a Student's t distribution with degrees of freedom $d.f. = n - 1$.

Hypotheses for testing the mean of the differences

When testing the mean of the differences of paired data values, the null hypothesis is that there is no difference among the pairs. That is, the mean of the differences μ_d is zero.

$$H_0: \mu_d = 0$$

The alternate hypothesis depends on the problem and can be

$H_1: \mu_d < 0$	$H_1: \mu_d > 0$	$H_1: \mu_d \neq 0$
(left-tailed)	(right-tailed)	(two-tailed)

Critical values

Since the $\bar{d}$ distribution follows a Student's t distribution with degrees of freedom $d.f. = n - 1$, our critical values will be found in Table 6, "Student's t Distribution," in Appendix II. The methods to find the critical values are the same as those of Section 9.4 when we used the Student's t distribution to find critical values for tests of the mean with small samples. If the test is a *one-tailed* test, we find the critical t value in the column headed by $\alpha' =$ level of significance. If the test is a *two-tailed* test, we find the critical value in the column headed by $\alpha'' =$ level of significance. In each case, we use the row containing the degrees of freedom $n - 1$, where n is the *number of data pairs*.

Sample test statistic

For paired difference tests, we make our decision regarding H_0 according to the evidence of the sample mean $\bar{d}$ of the differences of measurements. By Theorem 9.1, we convert the sample test statistic $\bar{d}$ to a t value using the formula

$$t = \frac{\bar{d} - \mu_d}{(s_d / \sqrt{n})} \text{ with } d.f. = n - 1$$

where $s_d =$ sample standard deviation of the differences d

$\quad\quad n =$ number of data pairs

$\quad\quad \mu_d = 0$ as specified in H_0

EXAMPLE 15

Paired difference test

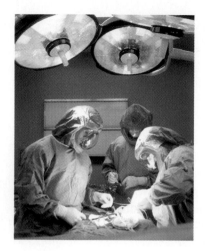

A team of heart surgeons at Saint Ann's Hospital knows that many patients who undergo corrective heart surgery have a dangerous buildup of anxiety before their scheduled operations. The staff psychiatrist at the hospital has started a new counseling program intended to reduce this anxiety. A test of anxiety is given to patients who know they must undergo heart surgery. Then each patient participates in a series of counseling sessions with the staff psychiatrist. At the end of the counseling sessions, each patient is retested to determine anxiety level. Table 9-8 indicates the results for a random sample of nine patients. Higher scores mean higher levels of anxiety.

From the given data, can we conclude that the counseling sessions reduce anxiety? Use a 0.01 level of significance.

SOLUTION: Before we answer this question, let us notice two important points: (1) we have a *random sample* of nine patients, and (2) we have a *pair* of measurements taken on the same patient before and after counseling sessions. In our problem, the sample size is $n = 9$ pairs (i.e., patients), and the d values are found in the fourth column of Table 9-8.

(a) First, we need to find the sample mean $\overline{d}$ and sample standard deviation s_d of the d values. Using the formulas for $\overline{d}$ and for s_d or a calculator with mean and standard deviation keys, we find

$$\overline{d} = 33.33$$

and

$$s_d = 22.92$$

(b) Next, we set the hypotheses. In our problem, we want to test the claim that the counseling sessions reduce anxiety. This means that the anxiety level before counseling is expected to be higher than the anxiety level after counseling. In symbols, $d = B - A$ should tend to be positive, and the

TABLE 9-8

Patient	B Score before Counseling	A Score after Counseling	d = B − A Difference
Jan	121	76	45
Tom	93	93	0
Diane	105	64	41
Barbara	115	117	−2
Mike	130	82	48
Bill	98	80	18
Frank	142	79	63
Carol	118	67	51
Alice	125	89	36

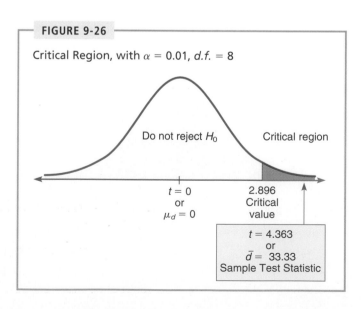

FIGURE 9-26

Critical Region, with $\alpha = 0.01$, $d.f. = 8$

Do not reject H_0 Critical region

$t = 0$ or $\mu_d = 0$

2.896 Critical value

$t = 4.363$ or $\overline{d} = 33.33$ Sample Test Statistic

population mean of differences μ_d also should be positive. Therefore, we have the hypotheses

$$H_0: \mu_d = 0$$
$$H_1: \mu_d > 0$$

(c) Find the critical values. Critical values are found using the Student's t distribution. We have a right-tailed test, with level of significance 0.01, so we look in the column headed by $\alpha' = 0.01$. The degrees of freedom are $d.f. = n - 1 = 9 - 1 = 8$. In Table 6 in Appendix II, we find the critical value is $t_0 = 2.896$.

(d) Convert the sample test statistic $\overline{d}$ to a t value. Using $n = 9$, $\overline{d} = 33.33$, $s_d = 22.92$, and $\mu_d = 0$ from H_0, we get

$$t = \frac{\overline{d} - \mu_d}{(s_d/\sqrt{n})} = \frac{33.33 - 0}{(22.92/\sqrt{9})}$$
$$= 4.363$$

(e) Finally, sketch the critical region, show the sample statistic on the sketch, and decide whether or not to reject H_0. Figure 9-26 shows a Student's t distribution with 8 degrees of freedom. We see that the t value corresponding to the observed sample statistic $\overline{d} = 33.33$ falls in the critical region. We reject H_0, select H_1, and conclude that the counseling sessions do reduce anxiety (at the 0.01 level of significance).

P-value approach

(f) To find an interval containing the P value associated with the sample test statistic $t = 4.363$, we use techniques shown in Section 9.4. In Table 6 (Appendix II), use the row with $d.f. = 8$ to find the location of $t = 4.363$. Then, since this is a one-tailed test, use α' values to create the interval containing the corresponding P value (see Table 9-9).

Reading from left to right, we notice that as the t values increase, the P values decrease. Therefore,

$$P \text{ value} < 0.005$$

This means that we reject H_0 for all levels of significance α greater than or equal to 0.005. In particular, we reject H_0 for $\alpha = 0.01$.   ◇

The problem we have just solved was a paired difference problem of the "before and after" type. The next guided exercise demonstrates a paired difference problem of the "matched pair" type.

TABLE 9-9   Excerpt from Student's t Distribution Table (Table 6, Appendix II)

✓ P value for one-tailed test	α'	...	0.010	0.005
$d.f.$	8	...	2.896	3.355

↑
Sample $t = 4.363$

GUIDED EXERCISE 11

Paired difference test

Do educational toys make a difference in age at which a child learns to read? To study this question, researchers designed an experiment in which one group of preschool children spent 2 hours each day (for 6 months) in a room well supplied with "educational" toys such as alphabet blocks, puzzles, ABC readers, coloring books featuring letters, and so forth. A control group of children spent 2 hours a day for 6 months in a "noneducational" toy room. It was anticipated that IQ differences and home environment might be uncontrollable factors unless identical twins could be used. Therefore, six pairs of identical twins of preschool age were randomly selected. From each pair, one member was randomly selected to participate in the experimental (i.e., educational toy room) group and the other in the control (i.e., noneducational toy room) group. For each twin the data item recorded is the age in months when the child began reading at the primary level (Table 9-10).

TABLE 9-10 Reading Ages for Identical Twins in Months

Twin Pair	Experimental Group B = Reading Age	Control Group A = Reading Age	Difference $d = B - A$
1	58	60	
2	61	64	
3	53	52	
4	60	65	
5	71	75	
6	62	63	

(a) Compute the entries in the $d = B - A$ column of Table 9-10.

Pair	$d = B - A$
1	−2
2	−3
3	1
4	−5
5	−4
6	−1

(b) Using formulas for the mean and sample standard deviation or a calculator with mean and sample standard deviation keys, compute $\bar{d}$ and s_d.

$\bar{d} = -2.33$
$s_d = 2.16$

(c) What is the null hypothesis?

$H_0: \mu_d = 0$

(d) To test the claim that the experimental group learned to read at a *different age* (either younger or older), what should the alternate hypothesis be?

$H_1: \mu_d \neq 0$

(e) What is the value of the degrees of freedom (*d.f.*)? For a 5% level of significance, find the critical values.

$d.f. = n - 1 = 6 - 1 = 5$

Use Table 6 in Appendix II. Since we have a two-tailed test, use the column headed by $\alpha'' = 0.05$. The critical values are $\pm t_0 = \pm 2.571$.

Continued

GUIDED EXERCISE 11 continued

(f) Convert the sample test statistic $\bar{d}$ to a t value.

Using $\mu_d = 0$ from H_0, $\bar{d} = -2.33$, $n = 6$, and $s_d = 2.16$, we get

$$t = \frac{\bar{d} - \mu_d}{(s_d/\sqrt{n})} = \frac{-2.33 - 0}{(2.16/\sqrt{6})} = -2.642$$

(g) Sketch the critical regions and place the t value corresponding to the sample test statistic $\bar{d}$ on the sketch. Do we reject or fail to reject H_0 at the 5% level of significance?

See Figure 9-27. Since the sample statistic falls in the critical region, we reject H_0 at the 5% level of significance.

(h) Do the results of this experiment indicate that there is a difference in reading age when a preschool child is exposed to educational toys?

This experiment indicates that at the 5% level of significance, educational toys make a difference.

(i) Find an interval containing the P value of the sample test statistic, and conclude the test using the P value.

In Table 6 of Appendix II, use the row with $d.f. = 5$ and find the location of the positive value corresponding to the sample test statistic $t = -2.642$ (see Table 9-11). Because this is a two-tailed test, use α'' values to form the P-value interval.

FIGURE 9-27 Critical Region, with $\alpha = 0.05$, $d.f. = 5$

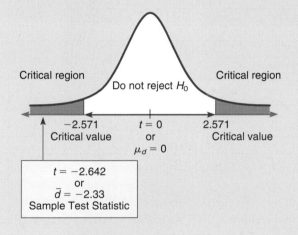

TABLE 9-11   Excerpt from Table 6 (Appendix II)

α''	0.050	0.020
$d.f.$ 5	2.571	3.365

↑
Sample $t = 2.642$

The corresponding P value is in the interval

$0.020 < P$ value < 0.050

This says that we reject H_0 for all levels of significance greater than or equal to 0.05. In particular, we reject H_0 for $\alpha = 0.05$.

TECH NOTE Both Excel and Minitab support paired difference tests directly. On the TI-83Plus calculator, construct a column of differences and then do a t test on the data in that column. For each technology, be sure to relate the alternate hypothesis to the "before and after" assignments. All the displays show the results for the data of Guided Exercise 11.

TI-83Plus Enter the "before" data in column L1 and the "after" data in column L2. Highlight L3, type L1 − L2 and press enter. The column L3 now contains the $B - A$ differences. To conduct the test, press **STAT**, select **TESTS**, and use option **2:T-Test**. Note that the letter x is used in place of d.

```
L1        L2        L3      3          T-Test
 58        60        -2                 μ≠0
 61        64        -3                 t=-2.645751311
 53        52         1                 p=.0456591238
 60        65        -5                 x̄=-2.333333333
 71        75        -4                 Sx=2.160246899
 62        63        -1                 n=6
L3(7)  =
```

Excel Enter the data in two columns. Use the menu choices **Tools ➤ Data Analysis ➤ t-Test: Paired Two-Sample for Means.** Fill in the dialogue box with hypothesized mean difference of 0. Set alpha.

B	C	D
t-Test: Paired Two Sample for Means		
	Variable 1	*Variable 2*
Mean	60.83333	63.16666667
Variance	34.96667	55.76666667
Observations	6	6
Pearson Correlation	0.974519	
Hypothesized Mean Difference	0	
df	5	
t Stat	−2.64575	
P(T<=t) one-tail	0.02283	
t Critical one-tail	2.015049	
P(T<=t) two-tail	0.045659	
t Critical two-tail	2.570578	

Minitab Enter the data in two columns. Use the menu selection **Stat ➤ Basic Statistics ➤ Paired t.** Under Options, set the null and alternate hypotheses.

```
Paired T-Test and Confidence Interval
Paired T for B — A
              N      Mean    St Dev   SE Mean
B             6     60.83     5.91      2.41
A             6     63.17     7.47      3.05
Difference    6     -2.333    2.160     0.882
95% CI for mean difference: (-4.601, -0.066)
T-Test of mean difference = 0 (vs not = 0):
     T-Value = -2.65 P-Value = 0.046
```

VIEWPOINT   *DUI*

DUI usually means "driving under the influence" of *alcohol,* but driving under the influence of *sleep loss* can be just as dangerous. Researchers in Australia have found that after staying awake for 24 hours straight, a person will be about as impaired as if he had enough alcohol to be legally drunk in most U.S. states (Source: *Rocky Mountain News*). Using driver simulation exams and statistical tests (paired difference tests) found in this section, it is possible to show that the null hypothesis H_0: $\mu_d = 0$ cannot be rejected. Or put another way, the average level of impairment for a given individual from alcohol (at the DUI level) is about the same as the average level of impairment for sleep loss (24 hours without sleep).

SECTION 9.6   PROBLEMS

For each of the following problems, please provide the requested information.
(a) What is the null hypothesis? What is the alternate hypothesis? Will we use a left-tailed, right-tailed, or two-tailed test? What is the level of significance?
(b) What sampling distribution will we use? What is the critical value t_0 (or critical values $\pm t_0$)?
(c) Sketch the critical region and show the critical value (or critical values).
(d) Calculate the t value corresponding to the sample statistic $\bar{d}$ and show its location on the sketch of part (c).
(e) Find the P value (or interval containing the P value) for your test and explain what the P value means in this context.
(f) Based on your answers for parts (a) to (e), shall we reject or fail to reject (i.e., "accept") the null hypothesis for the given level of significance α? Explain your conclusion in the context of the problem.

1. *Business: CEO Raises* Are America's top chief executive officers (CEOs) really worth all that money? One way to answer this question is to look at row *B,* the annual company percentage increase in revenue, versus row *A,* the CEO's annual percentage salary increase in that same company. (Source: *Forbes,* Vol. 159, No. 10.) A random sample of companies such as John Deere & Co., General Electric, Union Carbide, and Dow Chemical yields the following data:

B: Percent for company	24	23	25	18	6	4	21	37
A: Percent for CEO	21	25	20	14	−4	19	15	30

Do these data indicate that the population mean percentage increase in corporate revenue (row *B*) is different from the population mean percentage increase in CEO's salary? Use a 5% level of significance.

2. *Fishing: Shore or Boat?* Is fishing better from a boat or from the shore? Pyramid Lake is on the Paiute Indian Reservation in Nevada. Presidents, movie stars, and

people who just want to catch fish go to Pyramid Lake for really large cutthroat trout. Let row B represent hours per fish fishing from the shore, and let row A represent hours per fish using a boat. The following data are paired by month from October through April. (Source: *Pyramid Lake Fisheries*, Paiute Reservation, Nevada.)

	Oct.	Nov.	Dec.	Jan.	Feb.	March	April
B: Shore	1.6	1.8	2.0	3.2	3.9	3.6	3.3
A: Boat	1.5	1.4	1.6	2.2	3.3	3.0	3.8

Use a 1% level of significance to test if there is a difference in the population mean hours per fish using a boat compared with fishing from the shore.

3. *Ecology: Rocky Mountain National Park* The following is based on information taken from *Winter Wind Studies in Rocky Mountain National Park*, by D. E. Glidden (Rocky Mountain Nature Association). At five weather stations on Trail Ridge Road in Rocky Mountain National Park, the peak wind gusts (miles per hour) in January and April are recorded below.

Weather Station	1	2	3	4	5
January	139	122	126	64	78
April	104	113	100	88	61

Does this information indicate that the peak wind gusts are higher in January than in April? Use $\alpha = 0.01$.

4. *Wildlife: Highways* The western United States has a number of four-lane interstate highways that cut through long tracts of wilderness. To prevent car accidents with wild animals, the highways are bordered on both sides with 12-foot-high woven wire fences. Although the fences prevent accidents, they also disturb the winter migration pattern of many animals. To compensate for this disturbance, the highways have frequent wilderness underpasses designed for exclusive use by deer, elk, and other animals.

In Colorado, there is a large group of deer that spend their summer months in a region on one side of a highway and survive the winter months in a lower region on the other side. To determine if the highway has disturbed deer migration to the winter feeding area, the following data were gathered on a random sample of 10 wilderness districts in the winter feeding area. Row B represents the average January deer count in a 5-year period before the highway was built, and row A represents the average January deer count for a 5-year period after the highway was built. The highway department claims that the January population has not changed. Test this against the claim that the January population has dropped. Use a 1% level of significance. Units used in the table are hundreds of deer.

Wilderness District	1	2	3	4	5	6	7	8	9	10
B: Before highway	10.3	7.2	12.9	5.8	17.4	9.9	20.5	16.2	18.9	11.6
A: After highway	9.1	8.4	11.6	5.8	16.1	10.2	22.7	14.3	21.8	9.9

5. *Wildlife: Wolves* In environmental studies, sex ratios are of great importance. Wolf society, packs, and ecology have been studied extensively at different locations in the U.S. and foreign countries. Sex ratios for 8 study sites in Northern Europe are shown below (based on *The Wolf* by L. D. Mech, University of Minnesota Press).

Gender Study of Large Wolf Packs

Location of Wolf Pack	% Males (Winter)	% Males (Summer)
Finland	72	53
Finland	47	51
Finland	89	72
Lapland	55	48
Lapland	64	55
Russia	50	50
Russia	41	50
Russia	55	45

It is hypothesized that in winter "loner" males (not present in summer packs) join the pack to increase survival rate. Use a 5% level of significance to test the claim the average percentage of males in a wolf pack is higher in winter.

6. *Temperature: Miami vs. Honolulu* The following information is taken from the U.S. Department of Commerce Environmental Data Service. In the table, the months January to December are listed along with the average temperature in degrees Fahrenheit for Miami and Honolulu.

	Jan.	Feb.	March	April	May	June
Miami	67.5	68.0	71.3	74.9	78.0	80.9
Honolulu	74.4	72.6	73.3	74.7	76.2	78.0

	July	August	Sept.	Oct.	Nov.	Dec.
Miami	82.2	82.7	81.6	77.8	72.3	68.5
Honolulu	79.1	79.8	79.5	78.4	76.1	73.7

Do these data indicate that the average temperature in Miami is different (either higher or lower) from Honolulu? Use $\alpha = 0.01$.

7. *Navajo Reservation: Hogans* The following data are based on information taken from the book *Navajo Architecture: Forms, History, Distributions*, by S. C. Jett and V. E. Spencer (University of Arizona Press). A survey of houses and traditional hogans was made in a number of different regions in the modern Navajo Indian Reservation. The following table is the result of a random sample of eight regions on the Navajo Reservation.

Area on Navajo Reservation	Number of Inhabited Houses	Number of Inhabited Hogans
Bitter Springs	18	13
Rainbow Lodge	16	14
Kayenta	68	46
Red Mesa	9	32
Black Mesa	11	15
Canyon de Chelly	28	47
Cedar Point	50	17
Burnt Water	50	18

Does this information indicate that the population mean number of inhabited houses is greater than that of hogans on the Navajo Reservation? Use a 5% level of significance.

8. *Archaeology: Stone Tools* The following is based on information taken from *Bandelier Archaeological Excavation Project: Summer 1990 Excavations at Burnt Mesa Pueblo and Casa del Rito,* edited by T. A. Kohler (Washington State University, Department of Anthropology). The artifact frequency for an excavation of a kiva in Bandelier National Monument gave the following information:

Stratum	Flaked Stone Tools	Nonflaked Stone Tools
1	7	3
2	3	2
3	10	12
4	1	8
5	4	18
6	38	33
7	51	38

Does this information indicate that there tend to be more flaked stone tools than nonflaked stone tools at this excavation site? Use a 5% level of significance.

9. *Archaeology: Pot Sherds* On the same stratum of an excavated block of rooms at Bandelier National Monument, the following information was obtained about sherds of service ware in two different subareas (see reference in Problem 8):

Service Ware	Subarea 1	Subarea 2
Socorro black on white	10	4
Santa Fe black on white	42	39
Galisteo black on white	15	21
Puerco black on red	6	9
Wingate black on red	11	6

Does this information indicate that there is a difference (either higher or lower) for the population mean number of service ware sherds in subarea 1 as compared with subarea 2? Use a 5% level of significance.

10. *Psychology: Babies* Many mothers say that they can recognize the cries of their own babies and that they can distinguish between a cry of pain and a hunger cry. A psychologist studied this phenomenon using the following experiment. A random sample of mothers listened to a tape-recorded set of five cries from different babies, one of which was their own. They had to decide which was their baby. Each mother heard 20 such sets, in which 10 were cries of hungry babies and 10 were cries produced by a slight pin prick on a foot. The results are shown in the following table, where row *B* is the correct number of identifications (out of 10) for a hunger cry, and row *A* is the correct number of identifications (out of 10) for a pain cry. The psychologist claims that the mothers are more successful in picking out their own babies when a hunger cry is involved, since the mothers have more experience with that situation. Test this claim at the 5% level of significance.

Mother	1	2	3	4	5	6	7	8
B: Hunger cry	6	6	6	5	3	7	9	2
A: Pain cry	5	4	7	3	2	6	4	3

11. *Medical: Heart Rate* The following problem is based on information taken from *Diagnostic Tests with Nursing Implications* (Springhouse). A patient who had undergone a triple coronary artery bypass was tested in the following way. The patient took his pulse at rest and then on a treadmill exercised at a 10% grade at 1.7 miles per hour for 2 minutes. This procedure was repeated on six different days. The pulse rate 6 minutes after the test and the pulse rate before the test are shown in the following table (beats per minute):

Test	1	2	3	4	5	6
Pulse before	69	72	75	73	70	74
Pulse after	85	79	83	84	87	78

Do these data indicate that the population mean heart rate 6 minutes after the test is higher than before the test? Use a 5% level of significance.

12. *Medical: Blood Pressure* The patient described in Problem 11 also had his systolic blood pressure measured before and after the treadmill tests. The results for the tests are shown in the following table:

Test	1	2	3	4	5	6
Before	135	120	138	127	122	133
After	146	132	144	122	121	130

Does this indicate that the population mean systolic blood pressure is different (either higher or lower) before and 6 minutes after the treadmill test? Use a 5% level of significance.

13. *Golf: Tournaments* Do professional golfers play better in their first round? Let row *B* represent the score in the fourth (and final) round, and let row *A* represent the score in the first round of a professional golf tournament. A random sample of

finalists in the British Open gave the following data for their first and last rounds in the tournament. (Source: *Golf Almanac*.)

B: Last	73	68	73	71	71	72	68	68	74
A: First	66	70	64	71	65	71	71	71	71

Do the data indicate that the population mean score on the last round is higher than that on the first? Use a 5% level of significance.

14. *Psychology: Training Rats* The following data are based on information from the Regis University Psychology Department. In an effort to determine if rats perform certain tasks more quickly if offered larger rewards, the following experiment was performed. On day 1, a group of 3 rats was given a reward of one food pellet each time they ran a maze. A second group of 3 rats was given a reward of five food pellets each time they ran the maze. On day 2, the groups were reversed, so the first group now got five food pellets for running the maze and the second group got only one pellet for running the same maze. The average times in seconds for each rat to run the maze 30 times are shown in the following table:

Rat	A	B	C	D	E	F
Time with one food pellet	3.6	4.2	2.9	3.1	3.5	3.9
Time with five food pellets	3.0	3.7	3.0	3.3	2.8	3.0

Do these data indicate that rats receiving larger rewards tend to run the maze in less time? Use a 5% level of significance.

15. *Psychology: Training Rats* The same experimental design discussed in Problem 14 also was used to test rats trained to climb a sequence of short ladders. Times in seconds for eight rats to do this task are shown in the following table:

Rat	A	B	C	D	E	F	G	H
Time 1 pellet	12.5	13.7	11.4	12.1	11.0	10.4	14.6	12.3
Time 5 pellets	11.1	12.0	12.2	10.6	11.5	10.5	12.9	11.0

Do these data indicate that rats receiving larger rewards tend to perform the ladder climb in less time? Use a 5% level of significance.

16. *Careers: Professors* The following is based on information taken from *Academe: Bulletin of the American Association of University Professors* (Vol. 79, No. 2). A random sample of nine small and medium-sized colleges and universities in the western United States gave the following information about numbers of male and female assistant professors in each institution:

Institution	A	B	C	D	E	F	G	H	I
Male assistant professors	17	59	7	41	22	25	20	23	14
Female assistant professors	16	55	8	36	30	34	15	22	11

Do these data indicate that there is a difference (either higher or lower) in the population mean number of male versus female assistant professors in small and medium-sized colleges in the western United States? Use a 5% level of significance.

9.7   Testing Differences of Two Means or Two Proportions (Independent Samples)

FOCUS POINTS

✓ Identify independent samples and sampling distributions.

✓ Compute the sample test statistic, critical value, and P value for $\mu_1 - \mu_2$ (large samples).

✓ Compute the sample test statistic, critical value, and P value for $\mu_1 - \mu_2$ (small samples).

✓ Compute the sample test statistic, critical values, and P value for $p_1 - p_2$.

✓ Use the level of significance α and the P value to conclude the test.

Independent Samples

Many practical applications of statistics involve a comparison of two population means or two population proportions. In Section 9.6, we considered tests of differences of means for *dependent samples*. With dependent samples, we could pair the data and then consider the difference of data measurements d. In this section, we will turn our attention to tests of differences of means from *independent samples*. We will see new techniques for testing the difference of means from *independent samples*.

First, let's consider independent samples.

> **Definition** We say that two sampling distributions are *independent* if there is no relation whatsoever between specific values of the two distributions.

EXAMPLE 16

Independent sample

A teacher wishes to compare the effectiveness of two teaching methods. Students are randomly divided into two groups: The first group is taught by method 1; the second group, by method 2. At the end of the course, a comprehensive exam is given to all students, and the mean score $\bar{x}_1$ for group 1 is compared with the mean score $\bar{x}_2$ for group 2. Are the samples independent or dependent?

SOLUTION: Because the students were *randomly* divided into two groups, it is reasonable to say that the $\bar{x}_1$ distribution is independent of the $\bar{x}_2$ distribution.   ◇

EXAMPLE 17

Dependent sample

In Section 9.6, we considered a situation in which a shoe manufacturer claims that for the general population of adult U.S. citizens, the average length of the left foot is longer than the average length of the right foot. To study this claim, the manufacturer gathers data in this fashion: Sixty adult U.S. citizens are drawn at random, and for these 60 people, both their left and right feet are measured. Let $\bar{x}_1$ be the mean length of the left feet and $\bar{x}_2$ be the mean length of the right feet.

Are the $\bar{x}_1$ and $\bar{x}_2$ distributions independent for this method of collecting data?

SOLUTION: In this method, there is only *one* random sample of people drawn, and both the left and right feet are measured from this sample. The length of a person's left foot is usually related to the length of the right foot, so in this case the $\bar{x}_1$ and $\bar{x}_2$ distributions are *not* independent. In fact, we could pair the data and consider the distribution of the differences, left foot length minus right foot length. Then we would use the techniques of paired difference tests as found in Section 9.6.   ◇

GUIDED EXERCISE 12

Independent sample

Suppose the shoe manufacturer of Example 17 gathers data in the following way: Sixty adult U.S. citizens are drawn at random and their left feet are measured; then another 60 adult U.S. citizens are drawn at random and their right feet are measured. Again, $\bar{x}_1$ is the mean of the left foot measurements and $\bar{x}_2$ is the mean of the right foot measurements.

Are the $\bar{x}_1$ and $\bar{x}_2$ distributions independent for this method of collecting data?

For this method of gathering data, two random samples are drawn: one for the left foot measurements and one for the right foot measurements. The first sample is not related to the second sample. The $\bar{x}_1$ and $\bar{x}_2$ distributions are independent.

Testing Difference of Means for Large, Independent Samples

Properties of $\bar{x}_1 - \bar{x}_2$ distribution, large sample size

In this portion, we will use distributions that arise from a difference of means from independent samples. How do we obtain such distributions? If we have two statistical variables x_1 and x_2, each with its own distribution, we take independent random samples of size n_1 from the x_1 distribution and size n_2 from the x_2 distribution. Then we can compute the respective means $\bar{x}_1$ and $\bar{x}_2$. Consider the difference $\bar{x}_1 - \bar{x}_2$. This represents a difference of means. If we repeat the sampling process over and over, we will come up with lots of $\bar{x}_1 - \bar{x}_2$ values. These values can be arranged in a frequency table, and we can make a histogram for the distribution of $\bar{x}_1 - \bar{x}_2$ values. This would give us an experimental idea of the theoretical distribution of $\bar{x}_1 - \bar{x}_2$.

Fortunately, it is not necessary to carry out this lengthy process for each example. The results have already been worked out mathematically. The next theorem presents the main results for large, independent samples.

Theorem 9.2

Let x_1 have a normal distribution with mean μ_1 and standard deviation σ_1. Let x_2 have a normal distribution with mean μ_2 and standard deviation σ_2. If we take independent random samples of size n_1 from the x_1 distribution and of size n_2 from the x_2 distribution, then the variable $\bar{x}_1 - \bar{x}_2$ has

1. A normal distribution

2. Mean $\mu_1 - \mu_2$

3. Standard deviation

$$\sqrt{\frac{\sigma_1^2}{n_1} + \frac{\sigma_2^2}{n_2}}$$

TABLE 9-12 Alternate Hypotheses and Type of Test: Difference of Two Means

H_1			Type of Test
$H_1: \mu_1 - \mu_2 < 0$	or equivalently	$H_1: \mu_1 < \mu_2$	Left-tailed test
$H_1: \mu_1 - \mu_2 > 0$	or equivalently	$H_1: \mu_1 > \mu_2$	Right-tailed test
$H_1: \mu_1 - \mu_2 \neq 0$	or equivalently	$H_1: \mu_1 \neq \mu_2$	Two-tailed test

◇ **COMMENT** The theorem requires that x_1 and x_2 have normal distributions. However, if both n_1 and n_2 are 30 or larger, then for most practical applications, the central limit theorem assures us that $\bar{x}_1$ and $\bar{x}_2$ are approximately normally distributed. In this case, the conclusions of the theorem are again valid even if the original x_1 and x_2 distributions were not normal. ◇

Hypotheses for testing difference of means

When testing the difference of means, it is customary to use the null hypothesis

$$H_0: \mu_1 - \mu_2 = 0 \text{ or equivalently, } H_0: \mu_1 = \mu_2$$

As mentioned in Section 9.1, the null hypothesis is set up to see if it can be rejected. When testing the difference of means, we first set up the hypothesis H_0 that there is no difference. The alternate hypothesis could then be any of the ones listed in Table 9-12. The alternate hypothesis and consequent type of test used depend on the particular problem. Note that μ_1 is always listed first.

Critical values

Since the $\bar{x}_1 - \bar{x}_2$ distribution is *normal,* we use the usual critical values z_0 for normal distributions given in Figure 9-2 of Section 9.2. For convenience, we list the values again in Table 9-13 on the following page.

Sample test statistic

For tests of difference of means $\mu_1 - \mu_2$, independent samples, we make a decision regarding H_0 based on the sample evidence $\bar{x}_1 - \bar{x}_2$. For large samples, Theorem 9.2 tells us that the $\bar{x}_1 - \bar{x}_2$ distribution is normal with mean $\mu_1 - \mu_2$ and standard deviation $\sqrt{\sigma_1^2/n_1 + \sigma_2^2/n_2}$. We use this information to convert the sample $\bar{x}_1 - \bar{x}_2$ value to a standard z value using the formula

$$z = \frac{(\bar{x}_1 - \bar{x}_2) - (\mu_1 - \mu_2)}{\sqrt{\dfrac{\sigma_1^2}{n_1} + \dfrac{\sigma_2^2}{n_2}}}$$

where $\mu_1 - \mu_2 = 0$, as stated in the null hypothesis

$\bar{x}_1 =$ sample mean of x_1 data

$\bar{x}_2 =$ sample mean of x_2 data

$\sigma_1 =$ standard deviation of the x_1 distribution

$\sigma_2 =$ standard deviation of the x_2 distribution

$n_1 =$ sample size from the x_1 distribution

$n_2 =$ sample size from the x_2 distribution

TABLE 9-13   Critical Values for Difference of Means Tests, Independent Samples, Large Sample Size

Level of Significance	$\alpha = 0.05$	$\alpha = 0.01$
Critical value z_0 for a left-tailed test	-1.645	-2.33
Critical value z_0 for a right-tailed test	1.645	2.33
Critical values $\pm z_0$ for a two-tailed test	± 1.96	± 2.58

EXAMPLE 18

Testing the difference of means (large samples)

A consumer group is testing camp stoves. To test the heating capacity of a stove, they measure the time required to bring 2 quarts of water from 50°F to boiling (at sea level).

Two competing models are under consideration. Thirty-six stoves of each model were tested, and the following results were obtained.

Model 1: mean time $\bar{x}_1 = 11.4$ min; standard deviation $s_1 = 2.5$ min

Model 2: mean time $\bar{x}_2 = 9.9$ min; standard deviation $s_2 = 3.0$ min

Is there any difference between the performances of these two models? (Use a 5% level of significance.) Also, find the P value for the sample test statistic.

SOLUTION:

(a) The problem is whether the observed difference between the sample means is significant or not at the 0.05 level. Let μ_1 and μ_2 be the mean of the distribution of times for models 1 and 2, respectively. We set up the null hypothesis to say that there is no difference:

$$H_0: \mu_1 = \mu_2 \quad \text{or} \quad H_0: \mu_1 - \mu_2 = 0$$

The alternate hypothesis says that there is a difference:

$$H_1: \mu_1 \neq \mu_2 \quad \text{or} \quad H_1: \mu_1 - \mu_2 \neq 0$$

(b) Since the alternate hypothesis is $H_1: \mu_1 \neq \mu_2$, we use a two-tailed test. By Table 9-13, we see that the critical values are $\pm z_0 = \pm 1.96$ for $\alpha = 0.05$.

(c) The next step is to compute the sample test statistic $\bar{x}_1 - \bar{x}_2$ and then convert it to a z value.

We are given the values $\bar{x}_1 = 11.4$ and $\bar{x}_2 = 9.9$. Therefore, the sample test statistic is $\bar{x}_1 - \bar{x}_2 = 11.4 - 9.9 = 1.5$. To convert this value to z, we use the values $\sigma_1 \approx s_1 = 2.5$, $\sigma_2 \approx s_2 = 3.0$, $n_1 = 36$, and $n_2 = 36$ in the formula. From the null hypothesis, we use $\mu_1 - \mu_2 = 0$ as well. Then

$$z = \frac{(\bar{x}_1 - \bar{x}_2) - (\mu_1 - \mu_2)}{\sqrt{\dfrac{\sigma_1^2}{n_1} + \dfrac{\sigma_2^2}{n_2}}} = \frac{1.5 - 0}{\sqrt{\dfrac{2.5^2}{36} + \dfrac{3.0^2}{36}}} = 2.30$$

(d) Figure 9-28 shows the critical region and the location of the sample test statistic. We see that the sample test statistic $\bar{x}_1 - \bar{x}_2$ falls in the critical region. Therefore, we reject H_0 at the 5% level of significance. The mean times for the two stoves to boil water are statistically different.

P value

(e) To find the P value of the sample test statistic $\bar{x}_1 - \bar{x}_2 = 1.5$, we first find the corresponding z value. By part (c), the z value is 2.30. Since this is a *two-tailed* test, the P value represents the total of the areas to the right of $z = 2.30$ and

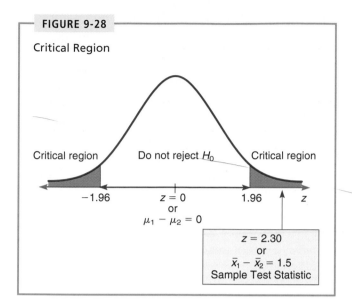

FIGURE 9-28

Critical Region

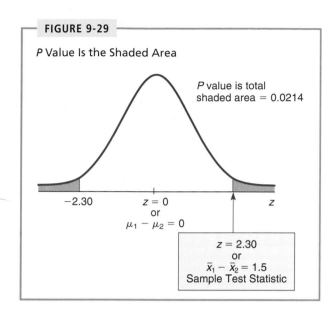

FIGURE 9-29

P Value Is the Shaded Area

to the left of $z = -2.30$ (see Figure 9-29). We use Table 5 in Appendix II to find the P value. Since we are using a *two-tailed* test, we *double* the area for one tail found in Table 5:

$$P \text{ value} = 2(0.0107) = 0.0214$$

This means that we reject H_0 for any $\alpha \geq 0.0214$. In particular, we reject H_0 for $\alpha = 0.05$, but we fail to reject H_0 for $\alpha = 0.01$ because 0.01 is not greater than 0.0214. ◇

GUIDED EXERCISE 13

Testing the difference of means (large samples)

Let us return to Example 16 at the beginning of this section. A teacher wishes to compare the effectiveness of two teaching methods. Students are randomly divided into two groups. The first group is taught by method 1; the second group, by method 2. At the end of the course, a comprehensive exam is given to all students.

The first group consists of $n_1 = 49$ students with a mean score of $\bar{x}_1 = 74.8$ points and standard deviation $s_1 = 14$ points. The second group has $n_2 = 50$ students with a mean score of $\bar{x}_2 = 81.3$ points and standard deviation $s_2 = 15$ points. The teacher claims that the second method will increase the mean score on the comprehensive exam. Is this claim justified at the 5% level of significance?

Let μ_1 and μ_2 be the mean score of the distribution of all scores using method 1 and method 2, respectively.

Continued

GUIDED EXERCISE 13 continued

(a) What is the null hypothesis?

➡ $H_0: \mu_1 = \mu_2$ or $H_0: \mu_1 - \mu_2 = 0$

(b) To examine the validity of the teacher's claim, what should the alternate hypothesis be?

➡ $H_1: \mu_1 < \mu_2$ (the second method gives a higher average score) or $H_1: \mu_1 - \mu_2 < 0$.

(c) Find the critical value z_0.

➡ By Table 9-13, we see that for a left-tailed test where $\alpha = 0.05$, $z_0 = -1.645$.

(d) Compute the sample test statistic $\bar{x}_1 - \bar{x}_2$.

➡ $\bar{x}_1 - \bar{x}_2 = 74.8 - 81.3 = -6.5$

(e) Convert $\bar{x}_1 - \bar{x}_2 = -6.5$ to a z value.

➡ By the null hypothesis, $\mu_1 - \mu_2 = 0$. From the problem, we have $s_1 = 14$ and $s_2 = 15$. Since the samples are both large, we can estimate σ_1 and σ_2 by these values, respectively. Also, $n_1 = 49$ and $n_2 = 50$. Putting all these values in the formula gives

$$z = \frac{(\bar{x}_1 - \bar{x}_2) - (\mu_1 - \mu_2)}{\sqrt{\dfrac{\sigma_1^2}{n_1} + \dfrac{\sigma_2^2}{n_2}}} = \frac{-6.5 - 0}{\sqrt{\dfrac{14^2}{49} + \dfrac{15^2}{50}}} = -2.23$$

(f) Show the z value of the sample test statistic and the critical region on a diagram, and then conclude the test. Is the claim that method 2 is better justified at the 5% level of significance?

➡ See Figure 9-30. Since the z value of the sample test statistic falls in the critical region, we reject H_0, select H_1, and conclude that method 2 is better at the 5% level of significance.

(g) Find the P value for the sample test statistic $\bar{x}_1 - \bar{x}_2$. What does the P value tell us?

➡ First, we convert $\bar{x}_1 - \bar{x}_2 = -6.5$ to a z value. In part (e), we found that the corresponding z value is -2.23. Since the test is a *left-tailed* test, the P value is the area to the left of $z = -2.23$ (see Figure 9-31). From Table 5 in Appendix II we have

P value $= 0.0129$

This tells us to reject H_0 for any $\alpha \geq 0.0129$.

FIGURE 9-31 P Value Is the Shaded Area

FIGURE 9-30 Critical Region

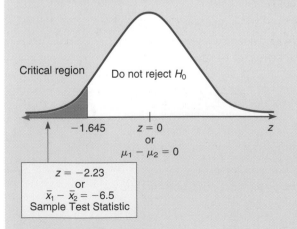

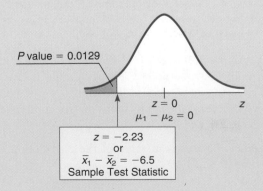

Testing Difference of Means for Small Samples

Statistical methods involving small samples and the difference between two means are much like the methods for large samples. However, for small samples, we will use the Student's t distribution for critical values instead of the normal distribution.

Independent random samples of size n_1 and n_2, respectively, are drawn from two populations that possess means μ_1 and μ_2. Again, *we assume that the parent populations have normal distributions, and we also assume that the standard deviations σ_1 and σ_2 for the two populations are equal*. The condition $\sigma_1 = \sigma_2$ may seem quite restrictive. However, in a great many practical applications, this condition is satisfied. Furthermore, our methods still apply even if the standard deviations are known to be only approximately equal. (See Section 11.4 for a method to test if variances or corresponding standard deviations from two populations are equal.)

Suppose we draw two independent random samples, one from the x_1 population and one from the x_2 population. Say the sample from the x_1 population is of size n_1 and has sample standard deviation s_1. Likewise for the x_2 population, the sample size is n_2 and the sample standard deviation is s_2. We estimate the common standard deviation for the two populations by using a *pooled variance* of the s_1^2 and s_2^2 values. Research shows that the best estimate of the common variance of the x_1 and x_2 populations is given by the formula

$$s^2 = \frac{(n_1 - 1)s_1^2 + (n_2 - 1)s_2^2}{n_1 + n_2 - 2}$$

Pooled standard deviation

The best estimate of the common or *pooled standard deviation* is then

$$s = \sqrt{\frac{(n_1 - 1)s_1^2 + (n_2 - 1)s_2^2}{n_1 + n_2 - 2}}$$

The null hypothesis is usually set up to see if it can be rejected. When testing the difference of means, we first set up the hypothesis H_0 that there is no difference. That is, we take the null hypothesis to be

$$H_0: \mu_1 - \mu_2 = 0 \qquad \text{or, equivalently,} \qquad H_0: \mu_1 = \mu_2$$

Table 9-12 on page 521 lists the possibilities for the alternate hypothesis and the type of test to be used with the alternate hypothesis.

Let $\bar{x}_1$ and $\bar{x}_2$ be the sample means for our two random samples from the x_1 and x_2 populations. Then, under the assumption that $\sigma_1 = \sigma_2$ and under the null hypothesis $H_0: \mu_1 - \mu_2 = 0$, it is possible to show that

$$t = \frac{\bar{x}_1 - \bar{x}_2}{s\sqrt{\dfrac{1}{n_1} + \dfrac{1}{n_2}}}$$

has a t distribution with degrees of freedom:

$$d.f. = n_1 + n_2 - 2$$

Using TECHNOLOGY

SIMULATION

Recall that the level of significance α is the probability of mistakenly rejecting a true null hypothesis. If $\alpha = 0.05$, then we expect to mistakenly reject a true null hypothesis about 5% of the time. The following simulation conducted with Minitab demonstrates this phenomenon.

We draw 40 random samples of size 50 from a population that is normally distributed with mean $\mu = 30$ and $\sigma = 2.5$. The display shows the results of a hypothesis test with

$$H_0: \mu = 30 \qquad H_1: \mu > 30$$

for each of the 40 samples labeled C1 through C40. Because each of the 40 samples are drawn from a population with mean $\mu = 30$, the null hypothesis $H_0: \mu = 30$ is true for the test based on each sample. However, as the display shows, for some samples we reject the true null hypothesis.

(a) How many of the 40 samples have a sample mean $\bar{x}$ above $\mu = 30$? below $\mu = 30$?

(b) Look at the P value of the sample statistic $\bar{x}$ in each of the 40 samples. How many P values are less than or equal to $\alpha = 0.05$? What percent of the P values are less than or equal to α? What percent of the samples have us reject H_0 when, in fact, each of the samples was drawn from a normal distribution with $\mu = 30$ as hypothesized in the null hypothesis.

(c) If you have access to computer or calculator technology that creates random samples from a normal distribution with a specified mean and standard deviation, repeat this simulation. Do you expect to get the same results? Why or why not?

Minitab Display: Random samples of size 50 from a Normal Population with $\mu = 30$ and $\sigma = 2.5$

```
Z-Test
Test of mu = 30.000 vs. mu > 30.000
The assumed sigma = 2.50
Variable   N    Mean    StDev   SE Mean      Z      P
C1        50   30.002   2.776   0.354     0.01   0.50
C2        50   30.120   2.511   0.354     0.34   0.37
C3        50   30.032   2.721   0.354     0.09   0.46
C4        50   30.504   2.138   0.354     1.43   0.077
C5        50   29.901   2.496   0.354    -0.28   0.61
C6        50   30.059   2.836   0.354     0.17   0.43
C7        50   30.443   2.519   0.354     1.25   0.11
C8        50   29.775   2.530   0.354    -0.64   0.74
C9        50   30.188   2.204   0.354     0.53   0.30
C10       50   29.907   2.302   0.354    -0.26   0.60
C11       50   30.036   2.762   0.354     0.10   0.46
C12       50   30.656   2.399   0.354     1.86   0.032
C13       50   30.158   2.884   0.354     0.45   0.33
C14       50   29.830   3.129   0.354    -0.48   0.68
C15       50   30.308   2.241   0.354     0.87   0.19
C16       50   29.751   2.165   0.354    -0.70   0.76
C17       50   29.833   2.358   0.354    -0.47   0.68
C18       50   29.741   2.836   0.354    -0.73   0.77
C19       50   30.441   2.194   0.354     1.25   0.11
C20       50   29.820   2.156   0.354    -0.51   0.69
C21       50   29.611   2.360   0.354    -1.10   0.86
C22       50   30.569   2.659   0.354     1.61   0.054
C23       50   30.294   2.302   0.354     0.83   0.20
C24       50   29.978   2.298   0.354    -0.06   0.53
C25       50   29.836   2.438   0.354    -0.46   0.68
C26       50   30.102   2.322   0.354     0.29   0.39
C27       50   30.066   2.266   0.354     0.19   0.43
C28       50   29.071   2.219   0.354    -2.63   1.00
C29       50   30.597   2.426   0.354     1.69   0.046
C30       50   30.092   2.296   0.354     0.26   0.40
C31       50   29.803   2.495   0.354    -0.56   0.71
C32       50   29.546   2.335   0.354    -1.28   0.90
C33       50   29.702   1.902   0.354    -0.84   0.80
C34       50   29.233   2.657   0.354    -2.17   0.98
C35       50   30.097   2.472   0.354     0.28   0.39
C36       50   29.733   2.588   0.354    -0.76   0.78
C37       50   30.379   2.976   0.354     1.07   0.14
C38       50   29.424   2.827   0.354    -1.63   0.95
C39       50   30.288   2.396   0.354     0.81   0.21
C40       50   30.195   3.051   0.354     0.55   0.29
```

Technology Hints

TI-83Plus

Press **STAT**, select **EDIT**. Highlight the list name such as L1. Then press **MATH**, select **PRB**, highlight **6:randNorm(μ, σ, sample size)**. Press enter. Fill in the values of $\mu = 30$, $\sigma = 2.5$, and sample size = 50. Press enter. Now list L1 contains a random sample from the normal distribution specified.

To test the hypothesis H_0: $\mu = 30$ against H_1: $\mu > 30$, press **STAT**, select **TESTS**, and use option **1:Z-Test**. Fill in the value 30 for μ_0, 2.5 for σ, and $> \mu_0$. The output provides the value of the sample statistic $\bar{x}$, its corresponding z value, and the P value of the sample statistic.

Excel

To draw random samples from a normal distribution with $\mu = 30$ and $\sigma = 2.5$, use the menu choice **Tools ➤ Data Analysis ➤ Random Number Generator**. In the dialogue box, the number of variables is the number of samples. Fill in the rest of the dialogue box as shown.

To conduct a hypothesis test of H_0: $\mu = 30$ against H_1: $\mu > 30$, use the command **ZTEST(data range, 30, 2.5)**. This command is described as returning the two-tailed P value of the z test. However, it appears to give the P value of a right-tailed test. Use the ZTEST command with caution and check your results against the table results.

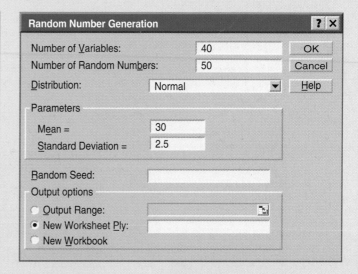

Minitab

To generate random samples from a normal distribution, use the menu choices **Calc ➤ Random Data ➤ Normal**. In the dialogue box, the number of rows refers to the sample size. Use 50 rows. Then designate the columns for the samples. Using C1–C40 will generate 40 random samples and put the samples in columns C1 through C40.

To test the hypothesis H_0: $\mu = 30$ against H_1: $\mu > 30$, use the menu choices **Stat ➤ Basic Statistics ➤ 1-SampleZ**. Use columns C1–C40 as the variables. Fill in 30 for the test mean, use greater than for the alternate hypothesis, and use 2.5 for sigma.

10

Regression and Correlation

When it is not in our power to determine what is true, we ought to follow what is most probable.

—René Descartes

**René Descartes
(1596–1650)**
French mathematician and philosopher.

For on-line student resources, visit **math.college.hmco.com/students** and follow the Statistics links to the Brase/Brase, *Understandable Statistics*, 7th edition web site.

It is important to realize that statistics and probability do not deal in the realm of certainty. If there is any realm of human knowledge where genuine certainty exists, you may be sure that our statistical methods are not needed there. In most human endeavors, and in almost all of the natural world around us, the element of chance happenings cannot be avoided. When we cannot expect something with true certainty, we must rely on probability to be our guide. In this chapter we will study regression, correlation, and forecasting. One of the tools we use is a scatter plot. René Decartes was the first mathematician to systematically use rectangular coordinate plots. For this reason, such a coordinate axis is called a Cartesian axis.

◇ How can we use a scatter plot to visually estimate the degree of linear correlation of random variables?   (SECTION 10.1)

◇ How is the least-squares line mathematically determined? How do we compute confidence intervals for predictions using the least-squares line?   (SECTION 10.2)

◇ What is the mathematical definition of the correlation coefficient, and how is it computed? (SECTION 10.3)

◇ How do we determine (i) if a sample correlation coefficient is statistically significant (ii) if a sample slope for a least-squares line is significant? (SECTION 10.4)

◇ How do we construct a confidence interval for the slope β of the population least-squares line? What does such an interval tell us about the expected change in y based on a change in x? (SECTION 10.4)

◇ What if we have more than just two random variables? How do we construct a linear regression model for three, four, or more random variables?   (SECTION 10.5)

Changing Populations and Crime Rate

Is the crime rate higher in neighborhoods where people might not know each other very well? Is there a relationship between crime rate and population change? If so, can we make predictions based on such a relationship? Is the relationship statistically significant? Is it possible to predict crime rate from population changes?

Denver is a city that has had a lot of growth and consequently a lot of population change in recent years. Sociologists studying population changes and crime rate could find a wealth of information in Denver statistics. Let x be a random variable representing percentage change in neighborhood population in the past few years, and let y be a random variable representing crime rate (crimes per 1,000 population). A random sample of six Denver neighborhoods gave the following information (Source: *Neighborhood Facts*, The Piton Foundation.) To find out more about the Piton Foundation, visit the Brase/Brase statistics site at http://math.college.hmco.com/students and find the link to the Piton Foundation.

x	29	2	11	17	7	6
y	173	35	132	127	69	53

Using information presented in this chapter, you will be able to make the following analysis:

(a) Draw a scatter diagram to display the data.

(b) Compute the least-squares line and graph it on the scatter diagram.

(c) If a neighborhood had a 12% change in population in the past few years, could you predict the crime rate from the least-squares line?

(d) Find an 80% confidence interval for the prediction of part (c).

(e) What is the sample correlation coefficient? Is it statistically significant at the 1% level?

(f) Does a high correlation guarantee a "cause-and-effect" situation, or is it simply an indication of a mathematical relationship between variables?

(g) Test the claim that the slope β of the population least-squares line is positive at the 1% level of significance.

(h) Find an 80% confidence interval for the population slope β, and interpret the interval in the context of the problem.

FOCUS POINTS

✓ Make a scatter diagram.

✓ Visually estimate the location of the "best-fitting" line for a scatter diagram.

✓ Visually estimate the degree of linear correlation.

10.1
Introduction to Paired Data and Scatter Diagrams

The study of correlation and regression of two variables usually begins with a table and/or a graph of *paired data values*. Example 1 shows what we mean.

EXAMPLE 1

Scatter diagram

For her botany class project, Jan wonders if there is a connection between the average weight of watermelons a vine produces and the root depth of the vine. Jan suspects that vines with deeper roots have a better water supply and also larger average melons. From a large watermelon field, 15 vines are chosen at random. At the end of 8 weeks, the watermelons are removed from each vine, and the average weight of watermelons from each vine is determined. Each plant is carefully dug up and its root depth (or length) is measured. Table 10-1 shows the results.

(a) For each plant, there is an ordered pair (x, y) of data values. If we plot the pairs (x, y) as points on a coordinate system, we obtain the graph of Figure 10-1. Figure 10-1 is called a *scatter diagram* for the paired data values of Table 10-1.

Scatter diagrams

(b) By inspecting Figure 10-1, we see that to some extent, larger values of x tend to be associated with larger y values, and smaller x values tend to be associated with smaller y values. Roughly speaking, the general trend seems to be reasonably well represented by the line segment shown in Figure 10-1.   ◊

Of course, it is possible to draw many curves in Figure 10-1, but the straight line is the simplest and most widely used curve for elementary studies of paired

TABLE 10-1   Results of a Botany Experiment

x = Root Depth (nearest inch)	y = Mean Weight of Watermelon (nearest pound)
26	20
14	10
18	13
10	9
26	19
21	17
7	8
26	15
13	9
19	13
17	12
13	7
16	9
28	17
23	14

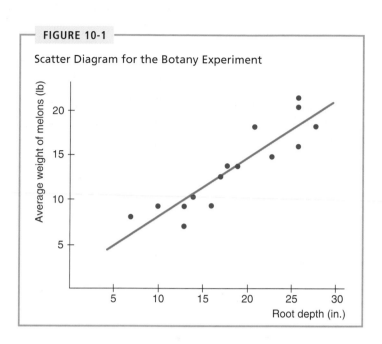

FIGURE 10-1

Scatter Diagram for the Botany Experiment

Introduction to linear correlation

data. We can draw many lines in Figure 10-1, but in some sense, the "best" line should be the one that comes closest to each of the points of the scatter diagram. To single out one line as the "best-fitting line," we must find a mathematical criterion for this line and a formula representing the line. This will be done in Section 10.2 by the *method of least squares.*

Another problem precedes that of finding the "best-fitting line." That is the problem of determining how well the points of the scatter diagram are suited for fitting *any* line. Certainly, if the points are a very poor fit to *any* line, there is little use in trying to find the "best" line. This problem will be dealt with in Section 10.3 by use of the Pearson product-moment coefficient of correlation.

If the points of a scatter diagram are located so that *no* line is realistically a "good" fit, we then say that the points possess *no linear correlation.* We see some examples of scatter diagrams for which there is no linear correlation in Figure 10-2.

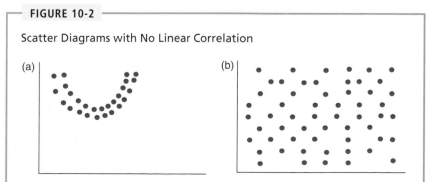

FIGURE 10-2

Scatter Diagrams with No Linear Correlation

(a)   (b)

GUIDED EXERCISE 1

Scatter diagram

A large industrial plant has seven divisions that do the same type of work. A safety inspector visits each division of 20 workers quarterly. The number x of work-hours devoted to safety training and the number y of work-hours lost due to industry-related accidents are recorded for each separate division in Table 10-2.

TABLE 10-2 Safety Report

Division	x	y
1	10.0	80
2	19.5	65
3	30.0	68
4	45.0	55
5	50.0	35
6	65.0	10
7	80.0	12

(a) Make a scatter diagram for these pairs. Use the x values on the horizontal axis and the y values on the vertical one.

 FIGURE 10-3 Scatter Diagram for Safety Report

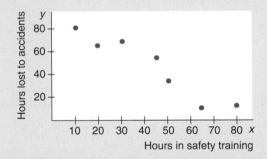

(b) As the number of hours spent on safety training increases, what happens to the number of hours lost due to industry-related accidents?

In general, as the number of hours in safety training goes up, the number of hours lost due to accidents goes down.

(c) Does a line fit the data reasonably well?

A line fits reasonably well.

(d) Draw a line that you think "fits best."

Use a downward-sloping line that lies close to the points. Later, you will see the equation of a line that is a "best fit."

If the points seem close to a straight line, we say the linear correlation is low to moderate, depending on how close the points lie to a line. If all the points do, in fact, lie on a line, then we have *perfect linear correlation*. In Figure 10-4, we see some diagrams with perfect linear correlation. In statistical applications, perfect linear correlation almost never occurs.

 TECH NOTE The TI-83Plus, Excel, and Minitab all produce scatter plots. For each technology, enter the x values in one column and the corresponding y values in another column. The displays show the data from

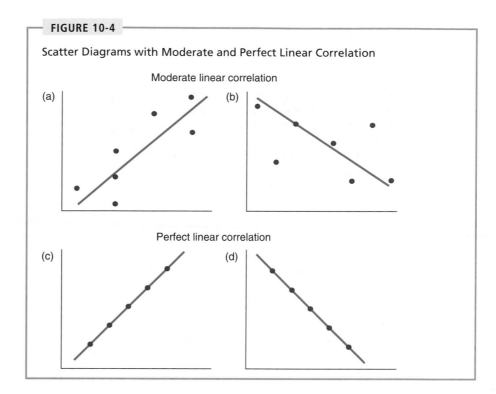

FIGURE 10-4

Scatter Diagrams with Moderate and Perfect Linear Correlation

Moderate linear correlation

(a)   (b)

Perfect linear correlation

(c)   (d)

GUIDED EXERCISE 2

Scatter diagram and linear correlation

Examine the scatter diagrams in Figure 10-5 and then answer the following questions.

FIGURE 10-5 Scatter Diagrams

(a) 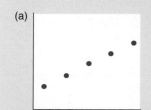   (b) 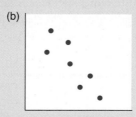   (c)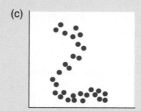

(a) Which diagram has no linear correlation?

⟹ Figure 10-5(c) has no linear correlation. No straight-line fit should be attempted.

(b) Which has perfect linear correlation?

⟹ Figure 10-5(a) has perfect linear correlation and can be fitted exactly by a straight line.

(c) Which can be reasonably fitted by a straight line?

⟹ Figure 10-5(b) can be reasonably fitted by a straight line.

Guided Exercise 2 regarding the safety training and hours lost because of accidents. Notice that the scatter plots do not necessarily show the origin.

TI-83Plus Use **Stat Plot** and choose the first type. Use option **9: ZoomStat** under **Zoom.** To check the scale, look at the settings displayed under **Window.**

Excel Use the menu choices **Chart wizard ➤ Scatter Diagram.** Dialogue box choices permit you to label the axes and title the chart. Changing the size of the diagram box changes the scale on the axes.

Minitab Use the menu selections **Stat ➤ Regression ➤ Fitted Line Plot.** The best-fit line is automatically plotted on the scatter diagram.

TI-83Plus Display

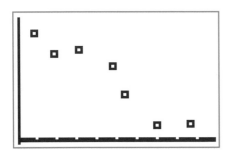

Excel Display

Minitab Display

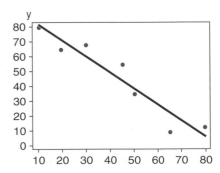

VIEWP◉INT   *Hawaii Island Hopping!*

Suppose you want to go camping in Hawaii. Yes! Hawaii has both state and federal parks where you can enjoy camping on the beach or in the mountains. However, you will probably need to rent a car to get to the different campgrounds. How much will the car rental cost? That depends on the islands you visit. For car rental data and regression statistics you can compute regarding costs on different Hawaiian Islands, visit the Brase/Brase statistics site at http://math.college.hmco.com/students and find the link to Hawaiian Islands.

SECTION 10.1 PROBLEMS

For Problems 1–6, look at the scatter diagrams and state which of the following conditions you think is true for each diagram:
(a) High linear correlation
(b) Moderate or low linear correlation
(c) No linear correlation

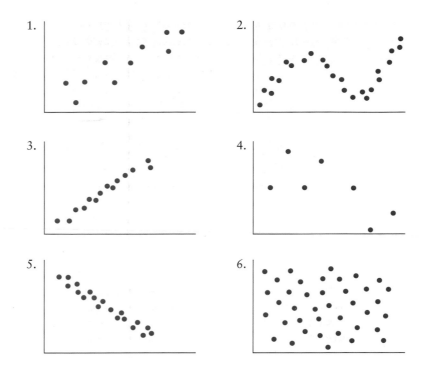

7. *Cars: Negotiating Price* Suppose you are interested in buying a new Pontiac Grand Am. You are standing on the sales lot looking at a model with different options. The list price is on the vehicle. As a salesperson approaches, you wonder what the dealer invoice price is for this model with its options. The following data are based on information taken from *Consumer Guide* (vol. 677). Let x be the list price (in thousands of dollars) for a random selection of Pontiac Grand Am vehicles of different models and options. Let y be the dealer invoice (in thousands of dollars) for the given vehicle.

x	15.8	16.5	19.2	20.7	21.7	22.0
y	14.6	15.1	16.7	18.9	19.8	21.1

(a) Draw a scatter diagram for the given data.
(b) Draw a straight line that you think best fits the data.
(c) Would you say the correlation is low, moderate, or high?

8. *Trucks: Negotiating Price* Suppose you are interested in buying a new Dodge Ram 1500 pickup truck. You are standing on the sales lot looking at a model with different options. The list price is on the vehicle. As a salesperson approaches, you wonder what the dealer invoice price is for this model with its options. The following data are based on information taken from *Consumer Guide* (vol. 677). Let x be the list price (in thousands of dollars) for a random selection of Dodge Ram 1500 pickups of different models and options. Let y be the dealer invoice (in thousands of dollars) for the given vehicle.

x	15.3	16.2	17.4	20.9	24.1	25.3
y	13.2	15.2	16.6	19.1	20.0	23.7

(a) Draw a scatter diagram for the given data.
(b) Draw a straight line that you think best fits the data.
(c) Would you say the correlation is low, moderate, or high?

9. *Veterinary Science: Shetland Ponies* How much should a healthy Shetland pony weigh? Let x be the age of the pony (in months), and let y be the average weight of the pony (in kilograms). The following information is based on data taken from *The Merck Veterinary Manual* (a reference used in most veterinary colleges).

x	3	6	12	18	24
y	60	95	140	170	185

(a) Draw a scatter diagram for the given data.
(b) Draw a straight line that you think best fits the data.
(c) Would you say the correlation is low, moderate, or high?

10. *Health Insurance: Administrative Cost* The following data are based on information from *Domestic Affairs*. Let x be the average number of employees in a group health insurance plan, and let y be the average administrative cost as a percentage of claims.

x	3	7	15	35	75
y	40	35	30	25	18

(a) Draw a scatter diagram for the given data.
(b) Draw a straight line that you think best fits the data.
(c) Would you say the correlation is low, moderate, or high?

11. *Economics: Wages and Prices* The following data are based on information taken from *The Economist*. Let x be the percentage change in wages, and let y be the percentage change in consumer prices for a recent year in Australia, Austria, Canada, France, Italy, Spain, and the United States.

x	3.3	4.1	1.9	2.6	3.9	6.3	2.5
y	2.2	3.5	1.9	2.1	4.0	4.9	2.7

(a) Draw a scatter diagram for the given data.
(b) Draw a straight line that you think best fits the data.
(c) Would you say the correlation is low, moderate, or high?

12. *Geology: Earthquakes* Is the magnitude of an earthquake related to the depth below the surface at which the quake occurs? Let x be the magnitude of an earthquake (on the Richter scale), and let y be the depth (in kilometers) of the quake below the surface at the epicenter. The following is based on information taken from National Earthquake Information Service of the U.S. Geological survey. Additional data may be found by visiting the Brase/Brase statistics site at http://math.college.hmco.com/students and finding the link to earthquakes.

x	2.9	4.2	3.3	4.5	2.6	3.2	3.4
y	5.0	10.0	11.2	10.0	7.9	3.9	5.5

(a) Draw a scatter diagram for the given data.
(b) Draw a straight line that you think best fits the data.
(c) Would you say the correlation is low, moderate, or high?

13. *Archaeology: Pottery* Wind Mountain archaeological site is located in southwest New Mexico. Ancient, prehistoric pottery vessels are usually found as sherds (broken pieces) and carefully reconstructed if enough sherds can be found. For reconstructed (or even rare unbroken) pottery vessels, let x be the body diameter (in centimeters), and let y be the height (in centimeters) of the vessel. The following data are based on information taken from *Mimbres Mogollon Archaeology*, by A. I. Woosley and A. J. McIntyre (University of New Mexico Press).

x	7.3	31.0	18.4	6.5	4.9	2.6	19.5	9.2	23.7
y	5.5	28.5	19.7	5.0	5.7	2.1	11.5	5.0	11.6

(a) Draw a scatter diagram for the given data.
(b) Draw a straight line that you think best fits the data.
(c) Would you say the correlation is low, moderate, or high?

14. *Physiology: Children* The following problem is based on information taken from the pediatrics section of *The Merck Manual* (a commonly used reference in medical schools and nursing programs). Let x be the body weight of a child (in kilograms), and let y be the metabolic rate of the child (100 kcal/24 h).

x	3.0	5.0	9.0	11.0	15.0	17.0	19.0	21.0
y	1.4	2.7	5.0	6.0	7.1	7.8	8.3	8.8

(a) Draw a scatter diagram for the given data.
(b) Draw a straight line that you think best fits the data.
(c) Would you say the correlation is low, moderate, or high?

15. *Graph: Effect of Scale* The initial visual impact of a scatter diagram depends on the scales used on the x and y axes. Consider the following data:

x	1	2	3	4	5	6
y	1	4	6	3	6	7

(a) Make a scatter diagram using the same scale on both the x and y axes (i.e., make sure the unit lengths on the two axes are equal).
(b) Make a scatter diagram using a scale on the y axis that is twice as long as that on the x axis.
(c) Make a scatter diagram using a scale on the y axis that is half as long as that on the x axis.
(d) On each of the three graphs, draw the straight line that you think best fits the data points. How does the slope (or direction) of the three lines appear to change? (*Note:* The actual slopes will be the same; they just appear different because of the choice of scale factors.)

10.2
Linear Regression and Confidence Bounds for Prediction

FOCUS POINTS

✓ Learn about explanatory variables, response variables, and the least-squares criterion.

✓ Use sample data to compute the least-squares line. Graph the least-squares line.

✓ Use sample data to compute the standard error of estimate and confidence intervals for least-squares predictions.

Anyone who has been outdoors on a summer evening has probably heard crickets. Did you know that it is possible to use the cricket as a thermometer? Crickets tend to chirp more frequently as temperatures increase. A Harvard physics professor made a detailed study of this phenomenon. Using sophisticated equipment, Professor George W. Pierce studied the striped ground cricket and compiled the data in Table 10-3.

Do the data indicate a linear relation between chirping frequency and temperature? Is there a way we can use the data to predict the temperature that corresponds to a chirp frequency that is not listed in the table? For instance, how can we use the data to predict the temperature for $x = 19$ chirps per second? Let us first make a scatter diagram (Figure 10-6) for the data of Table 10-3.

Looking at the scatter diagram of Figure 10-6, we ask two questions:

1. Can we find a relationship between x and y?

2. If so, how strong is the relationship?

The first step in answering these questions is to try to express the relationship as a mathematical equation. There are many possible equations, but the simplest and most widely used is the linear equation, or the equation of a straight line. Because we will be using this line to predict the y values (temperature) from the x values (chirps per second), we call x the *explanatory variable* and y the *response variable*.

Our job is to find the "best" linear equation representing the points of the scatter diagram. For our criterion of best-fitting line, we use the *least-squares criterion*, which says that the line we fit to the data points must be such that *the sum of the*

Explanatory variable
Response variable

Least-squares criterion

TABLE 10-3 Chirping Frequency and Temperature for the Striped Ground Cricket

x (chirps/s)	y (temp., °F)
20.0	88.6
16.0	71.6
19.8	93.3
18.4	84.3
17.1	80.6
15.5	75.2
14.7	69.7
17.1	82.0
15.4	69.4
16.2	83.3
15.0	79.6
17.2	82.6
16.0	80.6
17.0	83.5
14.4	76.3

Source: Reprinted by permission of the publisher from *The Songs of Insects* by George W. Pierce, Cambridge, Mass.: Harvard University Press, Copyright © 1948 by the President and Fellows of Harvard College.

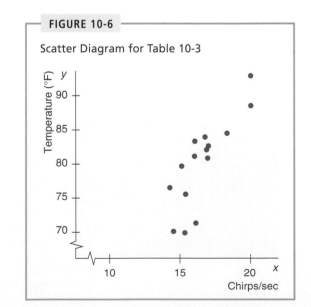

FIGURE 10-6

Scatter Diagram for Table 10-3

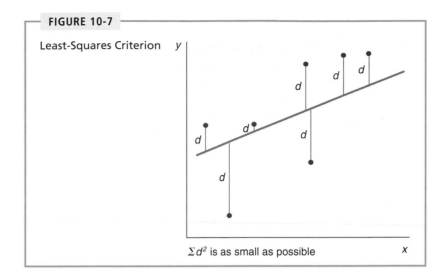

FIGURE 10-7

Least-Squares Criterion

Σd^2 is as small as possible

squares of the vertical distances from the points to the line be made as small as possible. The least-squares criterion is illustrated in Figure 10-7.

In Figure 10-7, d represents the difference between the y coordinate of the data point and the corresponding y coordinate on the line. Thus, if the data point lies above the line, d is positive, but if the data point is below the line, d is negative. As a result, the sum of the d values can be small even if the points are widely spread in the scatter diagram. However, the squares d^2 cannot be negative. By minimizing the sum of the squares, we are, in effect, not allowing positive and negative d values to "cancel out" one another in the sum. It is in this way that we can meet the least-squares criterion of minimizing the sum of the squares of the vertical distances between the points and the line over *all* points in the scatter diagram.

Least-squares line
    Techniques of calculus can be applied to show that the line that meets the least-squares criterion is as follows:

Least-squares line

$$y = a + bx \tag{1}$$

where $b = \dfrac{SS_{xy}}{SS_x}$ b is the slope   (2)

    $a = \bar{y} - b\bar{x}$ a is the y-intercept   (3)

and $\bar{y}$ = mean of y values in scatter diagram
    $\bar{x}$ = mean of x values in scatter diagram

$$SS_{xy} = \Sigma xy - \frac{(\Sigma x)(\Sigma y)}{n} \tag{4}$$

$$SS_x = \Sigma x^2 - \frac{(\Sigma x)^2}{n} \tag{5}$$

    n = number of points in a scatter diagram

In Formulas (4) and (5), the sums are taken over all x or y values in the scatter diagram.

◇ **COMMENT** In other mathematics courses, the slope-intercept form of the equation of a line is usually given as $y = mx + b$, where m refers to the slope of the line and b to the y-coordinate of the y-intercept. In statistics, it is standard to use the letter b to designate the slope of the least-squares line and the letter a to designate the y-coordinate of the intercept. Be attentive! ◇

◇ **COMMENT** The notation SS_x is the same expression used in the computation for the standard deviation of x values (see Section 3.2). Recall that the formula for SS_x is simply a faster way of computing $\Sigma(x - \bar{x})^2$. Likewise, the formula for SS_{xy} is a more efficient way to compute $\Sigma(x - \bar{x})(y - \bar{y})$. ◇

Using the formulas to find the values of b and a

Most calculators supporting two-variable statistics provide the values of the slope b and the y-intercept a directly. However, computing these values from formulas shows how the raw data are related to the equation of the least-squares line. The following discussion shows you how to use the formulas to find the values of b and a. If you are using your calculator to find the values of b and a directly, then you may omit the discussion regarding the use of the formulas. Go to the margin header "Using values a and b to construct the equation of the least-squares line" on the following page.

● **COMPUTATION NOTES**

1. The formulas used to compute the slope and y intercept of the least-squares line are sensitive to rounding. Answers to the exercises at the end of each section and at the end of the chapter are computer-generated, so you may expect slight differences in your answers, depending on how you round intermediate steps.

2. Do not confuse Σx^2 and $(\Sigma x)^2$. For Σx^2, we *first square* each x value and then find the total sum of these values. For $(\Sigma x)^2$, we *first sum* the x values and then square the total. Note that these sums are directly available on most calculators supporting two-variable statistics.

3. On calculators with a statistics mode, you can compute the standard deviation s and then the variance s^2 of the x values. Then

$$SS_x = s^2(n - 1)$$

The simplest way to find a and b is to organize your work into a table. In Table 10-4, we use the data relating rate of cricket chirps to temperature to obtain the sums needed for Equations (4) and (5). We will then use these sums to compute the formula for the least-squares line.

Using the formulas to find the equation of the least-squares line

From Table 10-4, we have

$$SS_x = \Sigma x^2 - \frac{(\Sigma x)^2}{n} = 4200.6 - \frac{(249.8)^2}{15} = 40.6$$

$$SS_{xy} = \Sigma xy - \frac{(\Sigma x)(\Sigma y)}{n} = 20{,}127.5 - \frac{(249.8)(1200.6)}{15} = 133.5$$

We also find

$$\bar{x} = \frac{\Sigma x}{n} = \frac{249.8}{15} = 16.7 \qquad \text{and} \qquad \bar{y} = \frac{\Sigma y}{n} = \frac{1200.6}{15} = 80.0$$

TABLE 10-4   Sums for Computing $\bar{x}$, $\bar{y}$, SS_x, and SS_{xy}

x (chirps/s)	y (°F)	x^2	xy
20.0	88.6	400.0	1,772.0
16.0	71.6	256.0	1,145.6
19.8	93.3	392.0	1,847.3
18.4	84.3	338.6	1,551.1
17.1	80.6	292.4	1,378.3
15.5	75.2	240.3	1,165.6
14.7	69.7	216.1	1,024.6
17.1	82.0	292.4	1,402.2
15.4	69.4	237.2	1,068.8
16.2	83.3	262.4	1,349.5
15.0	79.6	225.0	1,194.0
17.2	82.6	295.8	1,420.7
16.0	80.6	256.0	1,289.6
17.0	83.5	289.0	1,419.5
14.4	76.3	207.4	1,098.7
$\Sigma x = 249.8$	$\Sigma y = 1{,}200.6$	$\Sigma x^2 = 4{,}200.6$	$\Sigma xy = 20{,}127.5$

Therefore, using Equations (2) and (3), we find a and b in the equation of the least-squares line.

Slope: $b = \dfrac{SS_{xy}}{SS_x} = \dfrac{133.5}{40.6} = 3.3$

y intercept: $a = \bar{y} - b\bar{x} = (80.0) - (3.3)(16.7) = 24.9$

We conclude that the least-squares line for the data of Table 10-3 is

$$y = a + bx \tag{6}$$
$$y = 24.9 + 3.3x$$

◇ **COMMENT** *Meaning of slope.* In the equation $y = a + bx$, the slope b tells us how many units y changes for each unit change in x. In our cricket example,

$$y = 24.9 + 3.3x$$

The slope 3.3 tells us that if the number of cricket chirps per second changes by 1 chirp, then the corresponding temperature changes by 3.3 degrees Fahrenheit. For instance, if the number of cricket chirps per second increases by 2, then we expect an increase in the corresponding temperature of $2b$ or $2(3.3)$ or 6.6 degrees. ◇

To graph the least-squares line (6), we have several options available. The slope–intercept method of college algebra is probably the quickest. The slope is $b = 3.3$, and the y intercept is $a = 24.9$. However, if you don't remember this method, it is almost as easy to plot two points and connect them with a straight line. For

x values, we usually use any two values in the range of x data values. Corresponding y values are computed from the equation of the least-squares line.

The value of $\bar{x}$ will always be in the range of x values. When we use $x = \bar{x}$ in Equations (1) and (3), we see that the corresponding y value is $\bar{y}$.

> Therefore, the point $(\bar{x}, \bar{y})$ will always be on the least-squares line.

Because we have already computed these values, the point $(\bar{x}, \bar{y})$ is a convenient choice for one of the two points we use to graph the least-squares line. From Table 10-3, we see that $x = 20$ is also in the range of x values. We compute the corresponding y value by using the equation of the least-squares line.

x	$y = 24.9 + 3.3x$
When we choose $x = \bar{x} = 16.7$	$y = 24.9 + 3.3(16.7) = 80.0 = \bar{y}$
When we choose $x = 20.0$	$y = 24.9 + 3.3(20.0) = 90.9$

The line going through the points (16.7, 80.0) and (20.0, 90.9) is the least-squares line for the scatter diagram of Figure 10-6. This line is shown in Figure 10-8.

Predicting y for a specified x

Now suppose we find a striped ground cricket and, with a listening device, discover that it chirps at the rate of 19.0 chirps per second. What should we predict for the temperature? We could read the y value above $x = 19.0$ from the least-squares line graphed in Figure 10-8. But a more accurate estimate can be obtained by using the value $x = 19.0$ in the equation of the least-squares line and computing the corresponding y.

$y = 24.9 + 3.3x$ equation of least-squares line

$y = 24.9 + 3.3(19.0)$ using 19.0 in place of x

$y = 87.6°\text{F}$ evaluating y

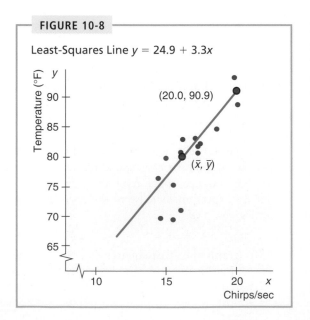

FIGURE 10-8

Least-Squares Line $y = 24.9 + 3.3x$

Rounded to the nearest whole number, we should predict the temperature to be 88°F. Of course, this is just a prediction, and we would be quite happy if the temperature turned out to be relatively close to our prediction. This brings up the natural question: How *good* are predictions based on the least-squares line? This is a fairly difficult question, and much of the answer requires advanced mathematics; however, a partial answer will be given later in this section.

GUIDED EXERCISE 3

Least-squares line

The Quick Sell car dealership has been using 1-minute spot ads on a local TV station. The ads always occur during the evening hours and advertise the different models and price ranges of cars on the lot that week. During a 10-week period, the Quick Sell dealer kept a weekly record of the number x of TV ads versus the number y of cars sold. The results are given in Table 10-5.

The manager decided that Quick Sell can afford only 12 ads per week. At that level of advertisement, how many cars can Quick Sell expect to sell each week? We'll answer this question in several steps.

TABLE 10-5

x	y
6	15
20	31
0	10
14	16
25	28
16	20
28	40
18	25
10	12
8	15

(a) Draw a scatter diagram for the data.

⟹ The scatter diagram is shown in Figure 10-9 on the following page.

(b) Look at Equations (1) to (5) pertaining to the least-squares line (page 561). Two of the quantities we need to find b are (Σx) and (Σxy). List the others.

⟹ We also need n, (Σy), (Σx^2), and $(\Sigma x)^2$.

(c) Complete Table 10-6(a).

⟹ The missing table entries are shown in Table 10-6(b).

TABLE 10-6(a)

x	y	x^2	xy
6	15	36	90
20	31	400	620
0	10	0	0
14	16	196	224
25	28	625	700
16	20	256	320
28	40	——	——
18	25	——	——
10	12	——	——
8	15	64	120
$\Sigma x = 145$	$\Sigma y = 212$	$\Sigma x^2 =$ ——	$\Sigma xy =$ ——

TABLE 10-6(b)

x^2	xy
$(28)^2 = 784$	$28(40) = 1120$
$(18)^2 = 324$	$18(25) = 450$
$(10)^2 = 100$	$10(12) = 120$
$\Sigma x^2 = 2785$	$\Sigma xy = 3764$

Continued

GUIDED EXERCISE 3 continued

FIGURE 10-9 Scatter Diagram and Least-Squares Line for Table 10-5

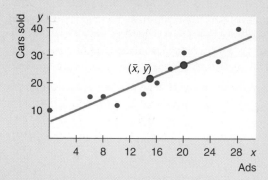

(d) Use Table 10-6(a) to compute SS_x, SS_{xy}, $\bar{x}$, and $\bar{y}$.

$$SS_x = \Sigma x^2 - \frac{(\Sigma x)^2}{n} = 2785 - \frac{(145)^2}{10} = 682.5$$

$$SS_{xy} = \Sigma xy - \frac{(\Sigma x)(\Sigma y)}{n} = 3764 - \frac{(145)(212)}{10}$$

$$= 690.0$$

$$\bar{x} = \frac{\Sigma x}{n} = \frac{145}{10} = 14.5, \qquad \bar{y} = \frac{\Sigma y}{n} = \frac{212}{10} = 21.2$$

(e) Compute a and b in the formula

$$y = a + bx$$

for the least-squares line. What is the equation of the least-squares line?

$$b = \frac{SS_{xy}}{SS_x} = \frac{690.0}{682.5} \approx 1.01$$

$$a = \bar{y} - b\bar{x} = 21.2 - 1.01(14.5) = 6.56$$

The equation for the least-squares line is

$$y = 6.56 + 1.01x$$

(f) Plot the least-squares line on your scatter diagram.

The least-squares line goes through the point $(\bar{x}, \bar{y})$ = (14.5, 21.2). To get another point on the line, select a value for x and compute the corresponding y value using the equation $y = 6.56 + 1.01x$. For $x = 20$, we get $y = 6.56 + 1.01(20) = 26.8$, so the point (20, 26.8) is also on the line. The least-squares line is shown in Figure 10-9.

(g) Read the y value for $x = 12$ from your graph. Then use the equation of the least-squares line to calculate y when $x = 12$. How many cars can the manager expect to sell if 12 ads per week are aired on TV?

The graph gives $y \approx 19$. From the equation we get

$$y = 6.56 + 1.01x$$

$$= 6.56 + 1.01(12) \quad \text{using 12 in place of } x$$

$$= 18.68$$

To the nearest whole number, the manager can expect to sell 19 cars when 12 ads are aired on TV each week.

TECH NOTE   When we have more data pairs, it is convenient to use a technology tool such as the TI-83Plus, Excel, or Minitab to find the equation of the least-squares line. The displays show results for data of Guided Exercise 3 regarding car sales and ads.

TI-83Plus Press **STAT**, choose **Calculate**, use option **8:LinReg(a+bx)**. For a graph showing the scatter plot and the least-squares line, press the **STAT PLOT** key, turn on a plot, and highlight the first type. Then press the **Y=** key. To enter the equation of the least-squares line, press **VARS**, select **5:Statistics**, highlight **EQ**, and then **1:RegEQ**. Press **ENTER**. Finally, press **ZOOM** and choose **9:ZoomStat**.

Excel   There are several ways to find the equation of the least-squares line in Excel. One way is to make a scatter plot using the menu choices **Chart wizard ➤ Scatter Diagram**. When the diagram is complete, **right click** on one of the points on the diagram, select **trendline**, and under options check to display equation of line.

TI-83Plus Display

Excel Display

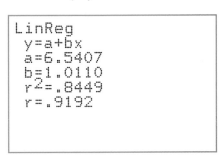

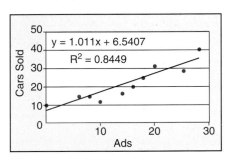

Minitab   There are a number of ways to generate the least-squares line. One way is to use the menu selection **Stat ➤ Regression ➤ Fitted Line Plot**. The least-squares equation is shown with the diagram.

Interpolation

We have used the least-squares line to predict y values for x values that were *between* x values observed in the experiment. Predicting y values for x values that are between x values of points in the scatter diagram is called *interpolation*. The least-squares lines can be used for interpolation. Predicting y values for an x value beyond the range of observed x values is a complex problem that is

Extrapolation

not treated in this book. Prediction beyond the range of observations is called *extrapolation*.

The *least-squares line*

$$y = a + bx$$

was developed with y as the response variable and x as the explanatory variable. This model can be used only to predict y values from specified x values. If you wish to begin with y values and predict corresponding x values, you must use a different formula. Such a formula would be developed using a model with x as the response variable and y as the explanatory variable. For our purposes, we'll always arrange to predict y values from given x values.

Standard Error of Estimate

Sometimes a scatter diagram clearly indicates the existence of a linear relationship between x and y, but it can happen that the points are widely scattered around the least-squares line. We need a method (besides just looking) for measuring the spread of a set of points about the least-squares line. There are three common methods of measuring the spread. One method uses the *standard error of estimate*. The others, to be studied in the next section, use the *coefficient of correlation* and the *coefficient of determination*.

For the standard error of estimate, we use a measure of spread that is in some ways like the standard deviation of measurements of a single variable. Let

$$y_p = a + bx$$

be the predicted value of y from the least-squares line. Then $y - y_p$ is the difference between the y value of the *data point* (x, y) shown on the scatter diagram (Figure 10-10) and the y value of the point on the *least-squares line* with the same x value. The quantity $y - y_p$ is known as the *residual*. To avoid the difficulty of having some positive and some negative values, we square the quantity $(y - y_p)$. Then we sum the squares and, for technical reasons, divide this sum by $n - 2$. Finally, we take the square root to obtain the *standard error of estimate*, denoted by S_e.

Residual

$$\text{Standard error of estimate} = S_e = \sqrt{\frac{\Sigma(y - y_p)^2}{n - 2}} \qquad (7)$$

where $n \geq 3$

Note: To compute the standard error of estimate, we require that there be at least three points on the scatter diagram. If we had only two points, the line would be a perfect fit, since two points determine a line. In such a case, there would be no need to compute S_e.

The nearer the scatter points lie to the least-squares line, the smaller S_e will be. In fact, if $S_e = 0$, it follows that each $y - y_p$ is also zero. This means that all the

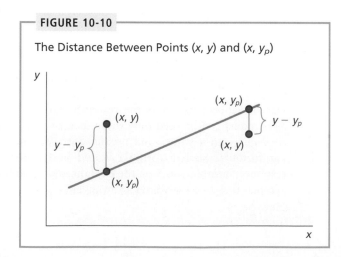

FIGURE 10-10

The Distance Between Points (x, y) and (x, y_p)

scatter points lie *on* the least-squares line if $S_e = 0$. The larger S_e becomes, the more scattered the points are.

The formula for the standard error of estimate is reminiscent of the formula for the standard deviation. It, too, is a measure of dispersion. However, the standard deviation involves differences of data values from a mean, whereas the standard error of estimate involves the differences between experimental and predicted y values for a given x.

The actual computation of S_e using Equation (7) is quite long because the formula requires us to use the least-squares line equation to compute a predicted value y_p for *each* x value in the data pairs. There is a computational formula that we strongly recommend that you use. However, as with all the computation formulas, be careful about rounding. This formula is sensitive to rounding, and you should carry as many digits as seems reasonable for your problem. Answers will vary, depending on rounding used. We give the formula here and follow it with an example of its use.

Formula to calculate S_e

$$S_e = \sqrt{\frac{SS_y - bSS_{xy}}{n - 2}}$$

(8)

where $SS_y = \Sigma y^2 - \dfrac{(\Sigma y)^2}{n}$ and $SS_{xy} = \Sigma xy - \dfrac{(\Sigma x)(\Sigma y)}{n}$

$SS_x = \Sigma x^2 - \dfrac{(\Sigma x)^2}{n}$ and $b = \dfrac{SS_{xy}}{SS_x}$

n = number of points in scatter diagram

Use caution in rounding.

With a considerable amount of algebra, Equations (7) and (8) can be shown to be mathematically equivalent. Equation (7) shows the strong similarity between the standard error of estimate and standard deviation. Equation (8) is a shortcut calculation formula because it involves few subtractions and uses quantities SS_x, b, and SS_{xy} that are also used to determine the least-squares line. The sums Σx, Σy, Σx^2, Σy^2, and Σxy, are provided directly on most calculators that support two-variable statistics.

In the next example, we show you how to compute the standard error of estimate using the computation formula. Then, in the following example and guided exercise, we will show you how to use S_e to create confidence intervals for the y value corresponding to a given x value.

EXAMPLE 2

Least-squares line and S_e

June and Jim are partners in the chemistry lab. Their assignment is to determine how much copper sulfate ($CuSO_4$) will dissolve in water at 10, 20, 30, 40, 50, 60, and 70°C. Their lab results are shown in Table 10-7 on the next page, where y is the weight in grams of copper sulfate that will dissolve in 100 g of water at x°C.

Sketch a scatter diagram, find the equation of the least-squares line, and compute S_e.

TABLE 10-7   Lab Results (x = °C, y = amount of CuSO$_4$)

x	y
10	17
20	21
30	25
40	28
50	33
60	40
70	49

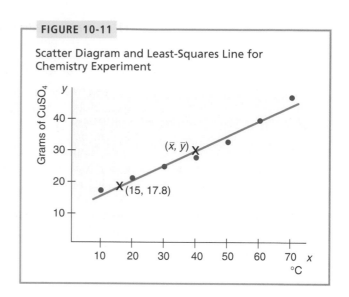

FIGURE 10-11

Scatter Diagram and Least-Squares Line for Chemistry Experiment

SOLUTION: Figure 10-11 includes a scatter diagram for the data of Table 10-7. To find the equation of the least-squares line and the value of S_e, we set up a computational table (Table 10-8).

$$SS_x = \Sigma x^2 - \frac{(\Sigma x)^2}{n} = 14{,}000 - \frac{(280)^2}{7} = 2800$$

$$SS_{xy} = \Sigma xy - \frac{(\Sigma x)(\Sigma y)}{n} = 9940 - \frac{(280)(213)}{7} = 1420$$

$$SS_y = \Sigma y^2 - \frac{(\Sigma y)^2}{n} = 7229 - \frac{(213)^2}{7} = 747.714$$

$$b = \frac{SS_{xy}}{SS_x} = \frac{1420}{2800} = 0.507143 \approx 0.51$$

$$\bar{x} = \frac{280}{7} = 40 \quad \text{and} \quad \bar{y} = \frac{213}{7} = 30.43$$

$$a = \bar{y} - b\bar{x} = \frac{213}{7} - (0.507143)\left(\frac{280}{7}\right) \approx 10.14$$

TABLE 10-8   Computational Table

x	y	x^2	y^2	xy
10	17	100	289	170
20	21	400	441	420
30	25	900	625	750
40	28	1,600	784	1,120
50	33	2,500	1,089	1,650
60	40	3,600	1,600	2,400
70	49	4,900	2,401	3,430
$\Sigma x = 280$	$\Sigma y = 213$	$\Sigma x^2 = 14{,}000$	$\Sigma y^2 = 7{,}229$	$\Sigma xy = 9{,}940$

The equation of the least-squares line is

$$y = a + bx$$
$$y = 10.14 + 0.51x$$

The graph of the least-squares line is shown in Figure 10-11. Notice that it passes through the point $(\bar{x}, \bar{y}) = (40, 30.4)$. Another point on the line can be found by using $x = 15$ in the equation of the line $y = 10.14 + 0.51x$. When we use 15 in place of x, we obtain $y = 10.14 + 0.51(15) = 17.8$. The point $(15, 17.8)$ is the other point we used to graph the least-squares line in Figure 10-11.

The standard error of estimate is computed using the computational formula

$$S_e = \sqrt{\frac{SS_y - bSS_{xy}}{n-2}}$$
$$= \sqrt{\frac{747.714 - (0.507143)(1420)}{5}}$$
$$\approx 2.35$$ ◇

TECH NOTE Although many calculators that support two-variable statistics and linear regression do not provide the value of the standard error of estimate S_e directly, they do provide the sums required for the calculation of S_e. The TI-83Plus, Excel, and Minitab all provide the value of S_e.

TI-83Plus The value for S_e is given as s under **STAT, TEST**, option **E: Lin Reg TTest.**

Excel Use the paste function (f_x), select **Statistical,** and choose the function STEYX.

Minitab Use menu choices **Stat ➤ Regression ➤ Regression.** The value for S_e is given as s in the display.

Confidence Intervals for y

The least-squares line gives us a predicted value y_p for a specified x value. However, we used sample data to get the equation of the line. The line derived from the population of all data pairs is likely to have a slightly different slope, which we designate by the symbol β for population slope, and a slightly different y intercept, which we designate by the symbol α for population intercept. In addition, there is some random error ϵ, so the true y value would be

$$y = \alpha + \beta x + \epsilon$$

Because of the random variable ϵ, for each x value there is a corresponding distribution of y values. The methods of linear regression were developed so that the distribution of y values for a given x is centered on the population regression line. Furthermore, the distributions of y values corresponding to each x value all have the same standard deviation, estimated by the standard error of estimate S_e.

Using all this background, the theory tells us that for a specific x, a *c confidence interval for y* is given by the next formula.

> ### c Confidence interval for y at a specific value of x
>
> $$y_p - E \leq y \leq y_p + E$$
>
> where $E = t_c S_e \sqrt{1 + \dfrac{1}{n} + \dfrac{(x - \bar{x})^2}{SS_x}}$
>
> y_p = the predicted value of y from the least-squares line for the specified x value
>
> t_c = critical value from the Student's t distribution for a c confidence level using $n - 2$ degrees of freedom
>
> S_e = the standard error of estimate [see Equation (8)]
>
> $SS_x = \Sigma x^2 - \dfrac{(\Sigma x)^2}{n}$
>
> n = number of data pairs

The formulas involved in the computation of a c confidence interval look complicated. However, they involve quantities we have already computed or values we can easily look up in tables. The next example illustrates this point.

EXAMPLE 3

Confidence interval for prediction

Using the data of Table 10-7, find a 95% confidence interval for the amount of copper sulfate that will dissolve in 100 g of water at 45°C.

SOLUTION: First, we need to find y_p for $x = 45°C$. We use the equation of the least-squares line that we found in Example 2.

$$y = 10.14 + 0.51x \qquad \text{from Example 2}$$
$$y_p = 10.14 + 0.51(45) \qquad \text{using 45 in place of } x$$
$$y_p \approx 33$$

A 95% confidence interval is then

$$33 - E \leq y \leq 33 + E$$

where $E = t_{0.95} S_e \sqrt{1 + \dfrac{1}{n} + \dfrac{(x - \bar{x})^2}{SS_x}}$

Using $n - 2 = 7 - 2 = 5$ degrees of freedom, we find from Table 6 in Appendix II that $t_{0.95} = 2.571$. We computed S_e, SS_x, and $\bar{x}$ in Example 2. Therefore,

$$E = (2.571)(2.35)\sqrt{1 + \dfrac{1}{7} + \dfrac{(45 - 40)^2}{2800}} = (2.571)(2.35)\sqrt{1.15179} \approx 6.5$$

A 95% confidence interval for y is

$$33 - 6.5 \leq y \leq 33 + 6.5$$
$$26.5 \leq y \leq 39.5$$

This means that we are 95% sure that the actual amount of copper sulfate that will dissolve in 100 g of water at 45°C is between 26.5 and 39.5 g. The interval is fairly wide but would decrease with more sample data.   ◇

GUIDED EXERCISE 4

Confidence interval for prediction

Let's use the data of Example 2 to compute a 95% confidence interval for y = amount of copper sulfate that will dissolve at $x = 15°C$.

(a) From Example 2, we have

$$y = 10.14 + 0.51x$$

Evaluate y_p for $x = 15$.

➡ $y_p = 10.14 + 0.51x$
$\qquad = 10.14 + 0.51(15)$
$\qquad \approx 17.8$

(b) The bound E on the error of estimate is

$$E = t_c S_e \sqrt{1 + \frac{1}{n} + \frac{(x - \bar{x})^2}{SS_x}}$$

From Example 2, we know that $S_e = 2.35$, $SS_x = 2800$, and $\bar{x} = 40$. Recall that there were $n = 7$ data pairs. Find $t_{0.95}$ and compute E.

➡ $t_{0.95} = 2.571$

$$E = (2.571)(2.35)\sqrt{1 + \frac{1}{7} + \frac{(15-40)^2}{2800}}$$

$\qquad = (2.571)(2.35)\sqrt{1.366071}$
$\qquad \approx 7.1$

(c) Find the 95% confidence interval for y.

$$y_p - E \leq y \leq y_p + E$$

➡ The confidence interval is

$$17.8 - 7.1 \leq y \leq 17.8 + 7.1$$
$$10.7 \leq y \leq 24.9$$

As we compare the results of Guided Exercise 4 and Example 3, we notice that the 95% confidence interval of y values for $x = 15°C$ is 7.1 units above and below the least-squares line, while the 95% confidence interval of y values for $x = 45°C$ is only 6.5 units above and below the least-squares line. This comparison reflects

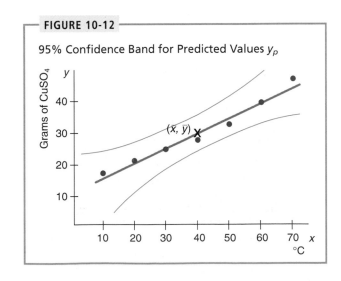

FIGURE 10-12

95% Confidence Band for Predicted Values y_p

the general property that confidence intervals for y are narrower the nearer we are to the mean $\bar{x}$ of the x values. As we move near the extremes of the x distribution, the confidence intervals for y become wider. This is another reason that we should not try to use the least-squares line to predict y values for x values beyond the data extremes of the sample x distribution.

If we were to compute a 95% confidence interval for all x values in the range of the sample x values, the confidence interval band would curve away from the least-squares line, as shown in Figure 10-12 on the preceding page.

TECH NOTE   Minitab provides confidence intervals for predictions. Use the menu selection **Stat ➤ Regression ➤ Regression.** Under options, enter the observed x value and set the confidence level. In the output, the confidence interval for predictions is designated by %PI.

VIEWPOINT   *It's Freezing!*

Can we use average temperatures in January to predict how bad the rest of the winter will be? Can you predict the number of days with freezing temperatures for the entire calendar year using conditions in January? How good would such a forecast be for predicting growing season or number of frost-free days? Methods of this section can help you answer such questions. For more information, visit the Brase/Brase statistics site at http://math.college.hmco.com/students and find the link to temperatures.

SECTION 10.2   PROBLEMS

Please do the following as parts a–d for Problems 1–15.
(a) Draw a scatter diagram for the data.
(b) Find $\bar{x}$, $\bar{y}$, and b. Then find the equation of the least-squares line.
(c) Graph the least-squares line on your scatter diagram. Be sure to use the point $(\bar{x}, \bar{y})$ as one of the points on the line.
(d) Find the standard error of estimate S_e.

Answers may vary slightly due to rounding.

1. *Assembly Line: Defects* The number of workers on an assembly line varies due to the level of absenteeism on any given day. In a random sample of production output from several days of work, the following data were obtained, where $x =$ number of workers absent from the assembly line and $y =$ number of defects coming off the line.

x	3	5	0	2	1
y	16	20	9	12	10

Complete parts (a) through (d).

(e) On a day when four workers are absent from the assembly line, what would the least-squares line predict for the number of defects coming off the line?

(f) Find a 95% confidence interval for the number of defects when four workers are absent.

2. *Ranching: Cattle* You are the foreman of the Bar-S cattle ranch in Colorado. A neighboring ranch has calves for sale, and you are going to buy some calves to add to the Bar-S herd. How much should a healthy calf weigh? Let x be the age of the calf (in weeks), and let y be the weight of the calf (in kilograms). The following information is based on data taken from *The Merck Veterinary Manual* (a reference used by many ranchers).

x	1	3	10	16	26	36
y	42	50	75	100	150	200

Complete parts (a) through (d).

(e) The calves you want to buy are 12 weeks old. What does the least-squares line predict for a healthy weight?

(f) Find a 90% confidence interval for the forecast y value of part (e).

3. *Weight of Car: Miles per Gallon* Do heavier cars really use more gasoline? Suppose that a car is chosen at random. Let x be the weight of the car (in hundreds of pounds), and let y be the miles per gallon (mpg). The following information is based on data taken from *Consumer Reports* (vol. 62, no. 4).

x	27	44	32	47	23	40	34	52
y	30	19	24	13	29	17	21	14

Complete parts (a) through (d).

(e) Suppose that a car weighs $x = 38$ (hundred pounds). What does the least-squares line forecast for y = miles per gallon?

(f) Find an 80% confidence interval for the forecast of part (e).

4. *Basketball: Fouls* Data for this problem are based on information from *STATS Basketball Scoreboard*. It is thought that basketball teams that make too many fouls in a game tend to lose the game even if they otherwise play well. Let x be the number of fouls more than (i.e., over and above) the opposing team. Let y be the percentage of times the team with the larger number of fouls wins the game.

x	0	2	5	6
y	50	45	33	26

Complete parts (a) through (d).

(e) If a team had $x = 4$ fouls over and above the opposing team, what does the least-squares equation forecast for y?

(f) Find an 80% confidence interval for the forecast of part (e).

5. *College: Income* The following data are based on information from the book *Life in America's Small Cities* (by G. S. Thomas, Prometheus Books). Let x be the percentage of those 25 years or older with 4 or more years of college. Let y be the per capita income in thousands of dollars. Five small cities in South Carolina

(Greenwood, Hilton Head Island, Myrtle Beach, Orangeburg, and Sumpter) reported the following information regarding the x and y variables:

x	13.8	21.9	12.5	12.7	11.5
y	9.0	10.8	8.8	6.9	7.2

Complete parts (a) through (d).
(e) In a small city in South Carolina where $x = 20$ percent of the population 25 years or older who have had 4 or more years of college, what would the least-squares equation forecast for y = per capita income (in thousands of dollars) in this community?
(f) Find an 80% confidence interval for your forecast y value of part (e).

6. *High School Dropouts: Income* Five small cities in California (El Centro, Eureka, Hanford, Madera, and San Luis Obispo–Atascadero) reported the following information. Let x be the percentage of 16- to 19-year-olds not in school and not high school graduates. Let y be the per capita income in thousands of dollars. The following information was obtained (see reference in Problem 5):

x	16.2	9.9	19.5	19.7	9.8
y	7.2	8.8	7.9	8.1	10.3

Complete parts (a) through (d).
(e) In a small city in California where $x = 17$, what would the least-squares equation forecast for y = per capita income (in thousands of dollars) in this community?
(f) Find a 75% confidence interval for the forecast y value of part (e).

7. *Income: Sales* Five small cities in Kentucky (Bowling Green, Madisonville, Paducah, Radcliff–Elizabethtown, and Richmond) reported the following information about the random variables x = per capita income and y = per capita retail sales (both in thousands of dollars). See reference in Problem 5.

x	9.0	8.9	9.9	8.2	7.6
y	5.1	4.5	6.2	3.7	4.2

Complete parts (a) through (d).
(e) Suppose that you plan to open a retail store in a small city in Kentucky where the per capita income is $x = 9.5$ (thousand dollars). What does the least-squares equation forecast for y = per capita retail sales (in thousands of dollars)?
(f) Find an 80% confidence interval for the forecast y value of part (e).

8. *Pickup Truck: Price* Suppose that you are interested in buying a new GMC Sierra 1500 pickup truck at a super weekend sale. You see a list price of $23,500 on one particular model with some extra options, and you wonder what the dealer invoice price (cost for the dealer) is for this truck so that you can compare the sale price with the invoice price and maybe negotiate an even better deal. The following information is based on data from *Consumer Guide* (vol. 677). Let x be the sale list price (in thousands of dollars) for a random sample of GMC Sierra pickups of various models and options. Let y be the dealer invoice price (in thousands of dollars) for the given truck.

x	16.7	22.5	25.9	28.3	32.2
y	15.1	19.0	21.0	24.8	28.2

Complete parts (a) through (d).
(e) The truck that interests you has a list price of 23.5 thousand dollars. What does the least-squares line forecast for the y = dealer invoice price (in thousands of dollars)?
(f) Find a 75% confidence interval for the forecast y value of part (e).

9. *Pickup Truck: Price* Suppose that you are interested in buying a new Chevrolet Silverado 1500 pickup truck at a super weekend sale. You see a list price of $22,900 on one particular model with some extra options, and you wonder what the dealer invoice price (cost for the dealer) is for this truck so that you can compare the sale price with the invoice price and maybe negotiate an even better deal. The following information is based on data from *Consumer Guide* (vol. 677). Let x be the sale list price (in thousands of dollars) for a random sample of Chevrolet Silverado pickups of various models and options. Let y be the dealer invoice price (in thousands of dollars) for the given truck.

x	16.0	20.5	23.2	25.3	28.3	35.1
y	14.2	17.9	18.2	24.0	24.8	29.0

Complete parts (a) through (d).
(e) The truck that interests you has a list price of 22.9 thousand dollars. What does the least-squares line forecast for the y = dealer invoice price (in thousands of dollars)?
(f) Find an 80% confidence interval for the forecast y value of part (e).

10. *Research: Patents* The following data are based on information from the *Harvard Business Review* (vol. 72, no. 1). Let x be the number of different research programs, and let y be the mean number of patents per program. As in any business, a company can spread itself too thin. For example, too many research programs might lead to a decline in overall research productivity. The following data are for a collection of pharmaceutical companies and their research programs:

x	10	12	14	16	18	20
y	1.8	1.7	1.5	1.4	1.0	0.7

Complete parts (a) through (d).
(e) Suppose that a pharmaceutical company had 15 different research programs. What does the least-squares equation forecast for y = mean number of patents per program?
(f) Find an 85% confidence interval for the forecast y value of part (e).

11. *Archaeology: Artifacts* Data for this problem are based on information taken from *Prehistoric New Mexico: Background for Survey* (by D. E. Stuart and R. P. Gauthier, University of New Mexico Press). It is thought that prehistoric Indians did not take their best tools, pottery, and household items when they visited higher elevations for their summer camps. It is hypothesized that archaeological sites tend to lose their cultural identity and specific cultural affiliation as the elevation of the site increases. Let x be the elevation (in thousands of feet) for an archaeological site in

the southwestern United States. Let y be the percentage of unidentified artifacts (no specific cultural affiliation) at a given elevation. The following data were obtained for a collection of archaeological sites in New Mexico:

x	5.25	5.75	6.25	6.75	7.25
y	19	13	33	37	62

Complete parts (a) through (d).

(e) At an archaeological site with elevation 6.5 (thousand feet), what does the least-squares equation forecast for y = percentage of culturally unidentified artifacts?

(f) Find a 75% confidence interval for the forecast y value of part (e).

12. *Psychiatry: Irrelevant Responses* A child psychiatrist is studying the mental development of children. A random sample of nine children was given a standard set of questions appropriate to the age of each child. The number of irrelevant responses to the questions was recorded for each child. In the following data, x = age of child in years and y = number of irrelevant responses:

x	2	3	4	5	7	9	10	11	12
y	15	15	12	13	11	10	8	6	5

Complete parts (a) through (d).

(e) If a child is 9.5 years old, what does the least-squares line predict for the number of irrelevant responses?

(f) Find a 99% confidence interval for the number of irrelevant responses for a child who is 9.5 years old.

13. *Climatology: Frost* Data for this problem are from *Climatology Report No. 77-3* (by J. F. Benci and T. B. McKee, Department of Atmospheric Science, Colorado State University). Let x be the elevation (in thousands of feet), and let y be the average number of frost-free days in a year. For Denver, Gunnison, Aspen, Crested Butte, and Dillon, Colorado, the following data were obtained:

x	5.3	7.7	7.9	8.9	9.8
y	162	63	73	49	21

Complete parts (a) through (d).

(e) Colorado Springs is at an elevation of 6 (thousand feet). What does the least-squares equation forecast for the average number of frost-free days per year in Colorado Springs?

(f) Find an 85% confidence interval for the forecast y value of part (e).

14. *Medical: Metabolic Rate* This problem is based on information taken from the pediatrics section of *The Merck Manual* (a commonly used reference in medical schools and nursing programs). Let x be the body weight of a child (in kilograms), and let y be the metabolic rate of the child (100 kcal/24 h).

x	3	5	9	11	15	17	19	21
y	1.4	2.7	5.0	6.0	7.1	7.8	8.3	8.8

Complete parts (a) through (d).

(e) Suppose that a child weighs 16 kg. What does the least-squares line forecast for y = metabolic rate of the child?

(f) Find a 75% confidence interval for the forecast y of part (e).

15. *Physics: CO_2* The following data are taken from the *Handbook of Physics and Chemistry* (CRC Publishing Company). Here x = water temperature in degrees Celsius and y = weight of carbon dioxide in grams that will dissolve in 100 g of water at 1 atmosphere of pressure for the corresponding temperature.

x	3	6	9	12	15
y	0.298	0.268	0.240	0.224	0.210

Complete parts (a) through (d).

(e) If the temperature is 10°C, what does the least-squares line predict for the weight of carbon dioxide that will dissolve in 100 g of water?

(f) Find a 90% confidence interval for your prediction of part (e).

16. *Least-Squares Equation: Exchange x and y*

(a) Suppose that you are given the following x, y data pairs:

x	1	3	4
y	2	1	6

Show that the least-squares equation for these data is $y = 1.071x + 0.143$ (where we round to three digits after the decimal).

(b) Now suppose that you are given these x, y data pairs:

x	2	1	6
y	1	3	4

Show that the least-squares equation for these data is $y = 0.357x + 1.595$ (where we round to three digits after the decimal).

(c) In the data for parts (a) and (b), did we simply exchange the x and y values of each data pair?

(d) Solve $y = 0.143 + 1.071x$ for x. Do you get the least-squares equation of part (b) with the symbols x and y exchanged?

(e) In general, suppose that we have the least-squares equation $y = a + bx$ for a set of data pairs x and y. If we solve this equation for x, will we *necessarily* get the least-squares equation for the set of data pairs y, x (with x and y exchanged)? Explain using parts (a) through (d).

17. *Least-Squares Model: Residual Plot* There are several ways to assess how well a least-squares line serves as a model for the data. One measure based on residuals is the standard error of estimate S_e. Another is a graphic tool called a *residual plot*. Let x be the given explanatory variable and y the corresponding response variable of the data. Let the symbol y_p represent the predicted value for x determined by the least-squares line. To make a residual plot, we put the x values in order on the horizontal axis and plot the corresponding residuals $y - y_p$ in the vertical direction. Because for a least-squares model the mean of the residuals is always zero, we dash in a horizontal line at zero. The accompanying figure shows a residual plot for the data of Guided Exercise 3, in which the relationship between the number of ads run per week and the number of cars sold that week was explored. To make the

residual plot, first compute all the residuals. Remember that x and y are the given data values, and y_p is computed from the least-squares line $y_p \approx 6.56 + 1.01x$.

Residual				Residual			
x	y	y_p	$y - y_p$	x	y	y_p	$y - y_p$
6	15	12.6	2.4	16	20	22.7	−2.7
20	31	26.8	4.2	28	40	34.8	5.2
0	10	6.6	3.4	18	25	24.7	0.3
14	16	20.7	−4.7	10	12	16.7	−4.7
25	28	31.8	−3.8	8	15	14.6	0.4

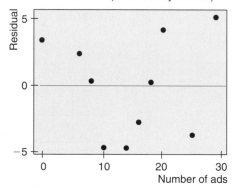

Residual Plot (Produced by Minitab)

(a) If the least-squares line provides a reasonable model for the data, the pattern of points in the plot will seem random and unstructured around the horizontal line at 0. Is this the case for the residual plot?

(b) If a point on the residual plot seems far outside the pattern of other points, it might reflect an unusual data point (x, y), called an *outlier*. Such points may have quite an influence on the least-squares model. Do there appear to be any outliers in the data for the residual plot?

18. *Residual Plot: Miles per Gallon* Consider the data of Problem 3.
   (a) Make a residual plot for the least-squares model.
   (b) Use the residual plot to comment about the appropriateness of the least-squares model for these data. See Problem 17.

10.3
The Linear Correlation Coefficient

If we are given a set of data pairs, we know how to find the equation of the line that "best" predicts y from a given x. This equation is called the *least-squares regression line of y on x*. However, we need more information about how well the line "fits" the data. In the preceding section, we saw that the standard error of estimate gives us some measure of how well our predicted values computed from the regression line fit actual experimental data values. The standard error of estimate has the same units of measurement as our y values, and as such, its value is, to a certain extent, dependent on the unit of measure selected for y. If we compare two different sets of data pairs, we cannot necessarily use the standard error of estimate to say that the regression line from one data set is "better" than the regression line from another data set. Finally, if we exchange the order in our data pairs, we get the regression line of x on y and obtain a different value for the standard error of estimate. In other words, the value of the standard error of estimate changes according to which variable is the response variable.

We need a unitless measurement to describe the strength of the linear association that exists between two variables regardless of which is listed first. Such a measure is the *correlation coefficient r*. The full name for r is the *Pearson product–moment correlation coefficient,* named in honor of the English statistician Karl Pearson (1857–1936), who is credited with formulating r. We'll develop the defining formula for r and then give a more convenient computation formula.

Correlation coefficient r

Development of Formula for *r*

If there is a *positive* linear relation between variables x and y, then high values of x are paired with high values of y, and low values of x are paired with low values of y. [See Figure 10-13(a).] In the case of *negative* linear correlation, high values of x are paired with low values of y, and low values of x are paired with high values of y. This relation is pictured in Figure 10-13(b). If there is little or no *linear correlation* between x and y, however, then we will find both high and low x values sometimes paired with high y values and sometimes paired with low y values. This relation is shown in Figure 10-13(c).

These observations lead us to the development of the formula for the correlation coefficient r. Taking *high* to mean "above the mean," we can express the relationships pictured in Figure 10-13 by considering the products

$$(x - \bar{x})(y - \bar{y})$$

If both x and y are high, both factors will be positive, and the product will be positive as well. The sign of this product will depend on the relative values of x and y compared with their respective means.

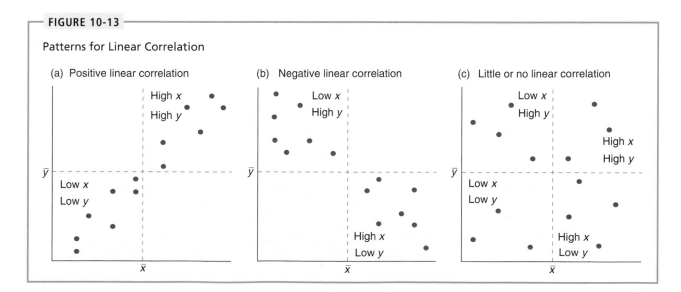

FIGURE 10-13

Patterns for Linear Correlation

(a) Positive linear correlation   (b) Negative linear correlation   (c) Little or no linear correlation

$$(x - \bar{x})(y - \bar{y}) \begin{cases} \text{is positive if } x \text{ and } y \text{ are both "high"} \\ \text{is positive if } x \text{ and } y \text{ are both "low"} \\ \\ \text{is negative if } x \text{ is "low," but } y \text{ is "high"} \\ \text{is negative if } x \text{ is "high," but } y \text{ is "low"} \end{cases}$$

In the case of positive linear correlation, most of the products $(x - \bar{x})(y - \bar{y})$ will be positive and so will the sum over all the data pairs

$$\Sigma(x - \bar{x})(y - \bar{y})$$

For negative linear correlation, the products will tend to be negative, so the sum also will be negative. On the other hand, in the case of little, if any, linear correlation, the sum will tend to be zero.

One trouble with the preceding sum is that it will be larger or smaller, depending on the units of x and y. Because we want r to be unitless, we standardize both x and y of a data pair by dividing each factor $(x - \bar{x})$ by the sample standard deviation s_x and each factor $(y - \bar{y})$ by s_y. Finally, we take an average of all the products. For technical reasons, we take the average by dividing by $n - 1$ instead of by n. This process leads us to the desired measurement, r.

$$r = \frac{1}{n - 1} \Sigma \frac{(y - \bar{y})}{s_y} \cdot \frac{(x - \bar{x})}{s_x}$$

Computation Formula for r

The defining formula for r is awkward to use because of all the subtractions. As before, we can simplify the formula and produce one that is much easier to use. In fact, it should be no surprise that the sums utilized in computing the standard error of estimate and the slope of the least-squares line are also used in the computational formula for r.

Formula to calculate correlation coefficient, r

$$r = \frac{SS_{xy}}{\sqrt{SS_x SS_y}} \tag{9}$$

where $\quad SS_{xy} = \Sigma xy - \dfrac{(\Sigma x)(\Sigma y)}{n} \qquad$ Note: $SS_{xy} = \Sigma(x - \bar{x})(y - \bar{y})$

$$SS_x = \Sigma x^2 - \frac{(\Sigma x)^2}{n} \qquad \text{and} \qquad SS_y = \Sigma y^2 - \frac{(\Sigma y)^2}{n}$$

n = number of data pairs in scatter diagram

This formula for r is sensitive to rounding. As in other formulas involving SS_x, SS_{xy}, and SS_y, you want to carry as many digits as is reasonable for your problem until the last step. Again, depending on the rounding process used, answers will vary slightly. The answers for the end-of-section and end-of-chapter exercises were computer-generated, so your answers might differ from them slightly and still be essentially correct.

TABLE 10-9   Some Facts About the Correlation Coefficient

If r Is	Then	The Scatter Diagram Might Look Something Like
0	There is no linear relation for the points of the scatter diagram.	
1 or −1	There is a perfect linear relation between x and y values; all points lie on the least-squares line.	$r = -1$ $r = 1$
Between 0 and 1 $(0 < r < 1)$	The x and y values have a *positive correlation.* By this, we mean that *large x* values are associated with *large y* values, and *small x* values are associated with *small y* values.	As we go from left to right, the least-squares line goes *up.*
Between −1 and 0 $(-1 < r < 0)$	The x and y values have a *negative correlation.* By this, we mean *large x* values are associated with *small y* values, and *small x* values are associated with *large y* values.	As we go from left to right, the least-squares line goes *down.*

Let us delay an example showing how to compute r until we know a little more about the meaning of the correlation coefficient. It can be shown mathematically that r is always a number between +1 and −1 ($-1 \leq r \leq +1$). Table 10-9 gives a quick summary of some basic facts about r. Guided Exercise 5 on the next page shows relationships between scatter diagrams and respective values of r.

Now let's actually compute r for some data.

EXAMPLE 4

Computing r

Most of us have heard someone say that more intelligent people tend to do better in school. Is this always true? Experienced teachers know that it is only partially true. Students with higher IQs (intelligence quotients) often do better schoolwork, but factors other than IQ can affect academic success. However, let's see if there is a correlation between IQ and cumulative grade averages (CGA). The principal of Delta High School chose 12 students from the senior class at random and compiled the data in Table 10-10.

TABLE 10-10   IQ versus CGA (on a Four-Point Scale) of 12 High School Seniors

IQ, x	117	92	102	115	87	76	107	108	121	91	113	98
CGA, y	3.7	2.6	3.3	2.2	2.4	1.8	2.8	3.2	3.8	3.0	4.0	3.5

GUIDED EXERCISE 5

Scatter diagrams and r

Match the appropriate statement about r to each scatter diagram in Figure 10-14.

(1) $r = 0$.   (2) $r = 1$.   (3) $r = -1$.

(4) r is between 0 and 1.   (5) r is between -1 and 0.

FIGURE 10-14 Scatter Diagrams

(a)   (b)   (c)

(a) $r = 1$ because all the points are on the line and the line goes up from left to right.

(b) r is between -1 and 0 because the points are fairly close to the line, and as we read from left to right, the least-squares line goes down.

(c) $r = 0$ because there is no apparent linear relation among the points.

(a) First, make a scatter diagram, and determine if r is positive, close to 0, or negative.

SOLUTION: The scatter diagram indicates that r is positive. See Figure 10-15.

(b) Compute r. The solution shows the use of the computation formula to compute r. Alternatively, you may read the value of r directly from your calculator or computer display.

SOLUTION: To find r, we must compute Σx, Σy, Σx^2, Σy^2, and Σxy. The values for $(\Sigma x)^2$ and $(\Sigma y)^2$ can be obtained from Σx and Σy. It is easiest to organize our work into a table of five columns (Table 10-11). The first two columns are just a repetition of Table 10-10.

To use Equation (9) to calculate r, we will first compute SS_{xy}, SS_x, and SS_y.

$$SS_{xy} = \Sigma xy - \frac{(\Sigma x)(\Sigma y)}{n} = 3780.3 - \frac{(1227)(36.3)}{12} = 68.63$$

$$SS_x = \Sigma x^2 - \frac{(\Sigma x)^2}{n} = 127{,}535 - \frac{(1227)^2}{12} = 2074.25$$

$$SS_y = \Sigma y^2 - \frac{(\Sigma y)^2}{n} = 114.9 - \frac{(36.3)^2}{12} = 5.09$$

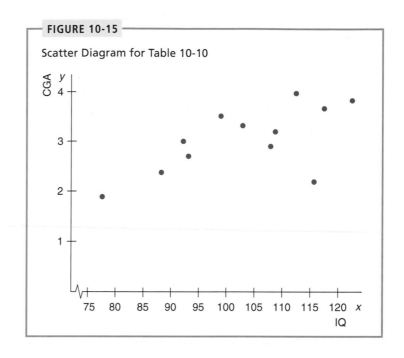

FIGURE 10-15

Scatter Diagram for Table 10-10

Therefore, the correlation coefficient is

$$r = \frac{SS_{xy}}{\sqrt{SS_x SS_y}} = \frac{68.63}{\sqrt{(2074.25)(5.09)}} = 0.6679 \approx 0.67$$

Our correlation coefficient is $r \approx 0.67$. Let's make sure that this answer agrees with what we expect from a quick glance at the scatter diagram (Figure 10-15). In Figure 10-15, the general trend is upward as we read from left to right, so we would

TABLE 10-11   Information Necessary to Compute r

x(IQ)	y(CGA)	x^2	y^2	xy
117	3.7	13,689	13.7	432.9
92	2.6	8,464	6.8	239.2
102	3.3	10,404	10.9	336.6
115	2.2	13,225	4.8	253.0
87	2.4	7,569	5.8	208.8
76	1.8	5,776	3.2	136.8
107	2.8	11,449	7.8	299.6
108	3.2	11,664	10.2	345.6
121	3.8	14,641	14.4	459.8
91	3.0	8,281	9.0	273.0
113	4.0	12,769	16.0	452.0
98	3.5	9,604	12.3	343.0
$\Sigma x = 1{,}227$	$\Sigma y = 36.3$	$\Sigma x^2 = 127{,}535$	$\Sigma y^2 = 114.9$	$\Sigma xy = 3{,}780.3$
$(\Sigma x)^2 = 1{,}505{,}529$	$(\Sigma y)^2 = 1{,}317.7$			

expect a positive value for r. The value 0.67 is in the expected range—that is, it is between 0 and 1.   ◇

It is quite a task to compute r for even 12 data pairs. The use of columns as in Example 4 is extremely helpful. Your value for r should always be between -1 and 1, inclusive. Use a scatter diagram to get a rough idea of the value of r. If your computed value of r is outside the allowable range, or if it disagrees quite a bit with the scatter diagram, recheck your calculations. Be sure you distinguish between expressions such as (Σx^2) and $(\Sigma x)^2$. Negligible rounding errors may occur, depending on how you (or your calculator) round.

GUIDED EXERCISE 6

Computing r

In one of the Boston city parks, there has been a problem with muggings in the summer months. A police cadet took a random sample of 10 days (out of the 90-day summer) and compiled the following data. For each day, x represents the number of police officers on duty in the park and y represents the number of reported muggings on that day.

x	10	15	16	1	4	6	18	12	14	7
y	5	2	1	9	7	8	1	5	3	6

(a) Construct a scatter diagram of x and y values.   Figure 10-16 shows the scatter diagram.

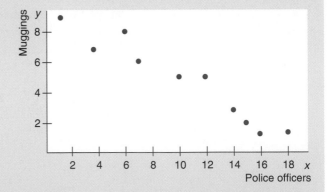

FIGURE 10-16 Scatter Diagram for Number of Police Officers versus Number of Muggings

(b) From the scatter diagram, do you think the computed value of r will be positive, negative, or zero? Explain. r will be negative. The general trend is that large x values are associated with small y values and vice versa. From left to right, the least-squares line goes down.

Continued

GUIDED EXERCISE 6 continued

(c) Complete **TABLE 10-12.**

x	y	x^2	y^2	xy
10	5	100	25	50
15	2	225	4	30
16	1	256	1	16
1	9	1	81	9
4	7	16	49	28
6	8	___	___	___
18	1	___	___	___
12	5	___	___	___
14	3	___	___	___
7	6	49	36	42

$\Sigma x = 103$ $\Sigma y = 47$ $\Sigma x^2 =$ ___ $\Sigma y^2 =$ ___ $\Sigma xy =$ ___

$(\Sigma x)^2 =$ ___ $(\Sigma y)^2 =$ ___

  TABLE 10-13 Completion of Table 10-12.

x	y	x^2	y^2	xy
6	8	36	64	48
18	1	324	1	18
12	5	144	25	60
14	3	196	9	42

$\Sigma x^2 = 1{,}347$ $\Sigma y^2 = 295$ $\Sigma xy = 343$

$(\Sigma x)^2 = 10{,}609$ $(\Sigma y)^2 = 2209$

(d) Compute SS_{xy}, SS_x, SS_y, and then r.
Alternatively, find the value of r directly by using a calculator or computer software.

$$SS_{xy} = \Sigma xy - \frac{(\Sigma x)(\Sigma y)}{n}$$

$$= 343 - \frac{(103)(47)}{10} = -141.1$$

$$SS_x = \Sigma x^2 - \frac{(\Sigma x)^2}{n}$$

$$= 1347 - \frac{(103)^2}{10} = 286.1$$

$$SS_y = \Sigma y^2 - \frac{(\Sigma y)^2}{n}$$

$$= 295 - \frac{(47)^2}{10} = 74.1$$

$$r = \frac{SS_{xy}}{\sqrt{SS_x SS_y}}$$

$$= \frac{-141.1}{\sqrt{(286.1)(74.1)}} = -0.9691 \approx -0.97$$

TECH NOTE Most calculators that support two-variable statistics provide the value of the correlation coefficient r directly. Statistical software provides r, r^2, or both.

TI-83Plus First use **CATALOGUE**, find, **DiagnosticOn**, and press **Enter** twice. Then when you use **STAT**, **CALC**, option 8:**LinReg(a+bx)**, the values of r and r^2 will be given. (Data from Example 4)

Excel Use the menu selection **Paste function** ⌐ f_x ⌐ ➤ **Statistical** ➤ **Correl.**

Minitab Use the menu selection **Stat** ➤ **Basic Statistics** ➤ **Correlation.**

```
LinReg
y=a+bx
a=-.3579
b=.0331
r²=.4415
r=.6645
```

The correlation coefficient can be thought of as another measure of how "good" the least-squares line fits the data points of the scatter diagram. (Recall that the standard error of estimate S_e in Section 10.2 is another such measure.) The closer r is to $+1$ or -1, the better the least-squares line "fits" the data. Values of r close to 0 indicate a poor "fit."

Usually, our scatter diagram does not contain *all* possible data points that could be gathered. Most scatter diagrams represent only a *random sample* of data pairs taken from a very large population of possible pairs. Because r is computed by Equation (9) on the basis of a random sample of (x, y) pairs, we expect the values of r to vary from one sample to the next (much as sample means $\bar{x}$ varied from sample to sample). This brings up the question of the *significance* of r. Or put another way, what are the chances that our random sample of data pairs indicates a high correlation when, in fact, the population x and y values are not so strongly correlated? Right now let us just say the significance of r is a separate issue that is left to the next section.

◇ **COMMENT** As we use computing formulas for the slope of the least-squares line, for r, and for standard deviations s_x and s_y, we see many of the same sums used. There is, in fact, a relationship between the correlation coefficient r and the slope of the least-squares line b. In instances when we know r, s_x, and s_y, we can use the following formula to compute b.

$$b = r\left(\frac{s_y}{s_x}\right) \quad ◇$$

Coefficient of Determination

There is another way to answer the question, How good is the least-squares line as an instrument of regression? The *coefficient of determination* r^2 is the square of the sample correlation coefficient r.

Suppose we have a scatter diagram and corresponding least-squares line as shown in Figure 10-17.

Let us take the point of view that $\bar{y}$ is a kind of baseline for the y values. If you were given an x value, and if you were completely ignorant of regression and correlation but you wanted to predict a value of y corresponding to the given x, a reasonable guess for y would be the mean $\bar{y}$. However, since we do know how to construct the least-squares regression line, we can calculate $y_p = a + bx$, the predicted value corresponding to x. Now in most cases the predicted value y_p on the least-squares line will not be the same as the actual data value y. We will measure deviations (or differences) from the baseline $\bar{y}$. (See Figure 10-17.)

Total deviation $= y - \bar{y}$

Explained deviation $= y_p - \bar{y}$

Unexplained deviation $= y - y_p$ (also known as the *residual*)

The total deviation $y - \bar{y}$ is a measure of how far y is from the baseline $\bar{y}$. This can be broken into two parts: the explained deviation $y_p - \bar{y}$ tells us how far the estimated y value "should" be from the baseline $\bar{y}$. (The "explanation" of this part of the deviation is the least-squares line, so to speak.) The unexplained deviation $y - y_p$ tells us how far our data value y is "off." This amount is called *unexplained* because it is due to random chance and other factors that the least-squares line cannot account for.

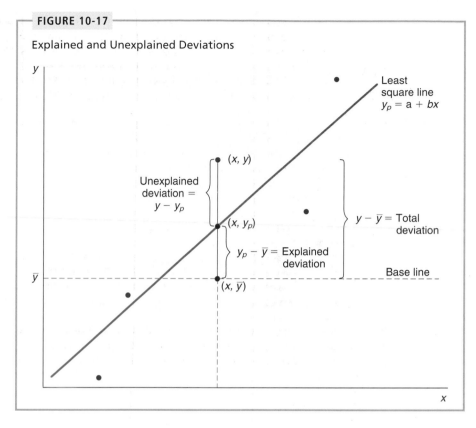

FIGURE 10-17

Explained and Unexplained Deviations

$$(y - \bar{y}) \qquad = \qquad (y_p - \bar{y}) \qquad + \qquad (y - y_p)$$

$$\begin{pmatrix} \text{Total} \\ \text{deviation} \end{pmatrix} \qquad = \qquad \begin{pmatrix} \text{Explained} \\ \text{deviation} \end{pmatrix} \qquad + \qquad \begin{pmatrix} \text{Unexplained} \\ \text{deviation} \end{pmatrix}$$

At this point, we wish to include all the data pairs and we wish to deal only with nonnegative values (so positive and negative deviations won't cancel out). Therefore, we construct the following equation for the sum of squares. This equation can be derived using some lengthy algebraic manipulations, which we omit.

$$\Sigma(y - \bar{y})^2 \qquad = \qquad \Sigma(y_p - \bar{y})^2 \qquad + \qquad \Sigma(y - y_p)^2$$

$$\begin{pmatrix} \text{Total} \\ \text{variation} \end{pmatrix} \qquad = \qquad \begin{pmatrix} \text{Explained} \\ \text{variation} \end{pmatrix} \qquad + \qquad \begin{pmatrix} \text{Unexplained} \\ \text{variation} \end{pmatrix}$$

Note that the sum of *squares* is taken over all data points and is then referred to as *variation* (not deviation).

The preceding concepts are connected together in the following important statement (whose proof we omit):

If r is the correlation coefficient [see Equation (9)], then it can be shown that

$$r^2 = \frac{\Sigma(y_p - \bar{y})^2}{\Sigma(y - \bar{y})^2} = \frac{\text{Explained variation}}{\text{Total variation}}$$

r^2 is called the *coefficient of determination*.

Let us make some comments about the coefficient of determination.

1. The calculation of r is described in Equation (9). The calculation of the coefficient of determination is then very easy; it is simply r^2.

2. The ratio of explained variation over total variation is that fractional amount of total variation in y that can be explained by using the least-squares line and the x variable.

In other words, the coefficient of determination r^2 is a measure of the proportion of variation in y that is explained by the regression line using x as the predicting variable. If $r = 0.90$, then $r^2 = 0.81$ is the coefficient of determination, and we can say that about 81% of the (variation) behavior of the y variable can be

GUIDED EXERCISE 7

Coefficient of determination

(a) In Example 4, we found the correlation coefficient r of the relationship between IQ and cumulative grade averages. In that case, r was 0.67. How would you describe the strength of this relationship?

⟹ The correlation coefficient $r = 0.67$ is moderate but not extremely high. It seems that other factors besides IQ are significant in determining a cumulative grade point average.

(b) Compute the coefficient of determination for the data of Example 4 and comment on the meaning of this number.

⟹ Since $r = 0.67$, then $r^2 = 0.449$ is the value of the coefficient of determination. This says that 44.9% of the variation of y = CGA can be explained by the least-squares line and x = IQ. The remaining $100 - 44.9 = 55.1\%$ of the variation of y is due to random chance or other variables besides x that influence y (possibly, amount of time spent studying!).

(c) In Guided Exercise 6, dealing with the relation between the number of police officers in the park and the number of muggings, we found r to be -0.97. How would you describe the strength of this relationship? Do you think the city is justified in asking for more police officers to be assigned to park duty?

⟹ $r = -0.97$ is a high correlation. The relation between the number of police officers in the park and the number of muggings in the park is a strong and dependable negative correlation. The authors feel that the city would be wise to hire more police officers to patrol the park. But many other aspects of the situation must be considered. Perhaps more crimes would be prevented by putting those officers elsewhere.

(d) Compute the coefficient of determination for the data of Guided Exercise 6 and comment on the meaning of this number.

⟹ Since $r = -0.97$, then $r^2 = 0.941$ is the coefficient of determination. About 94.1% of the variation of y can be explained by the least-squares line and the x variable. The remaining $100 - 94.1 = 5.9\%$ of the variation of y is due to random chance or other variables besides x that influence y.

explained by the corresponding (variation) behavior of the x variable if we use the equation of the least-squares line. The remaining 19% of the (variation) behavior of the y variable is due to random chance or to the presence of other variables besides the x that may influence y. Guided Exercise 7 illustrates these concepts.

The correlation coefficient is a mathematical tool for measuring the strength of the linear relationship between two variables. As such, it makes no implication about cause or effect. Just because two variables tend to increase or decrease together does not mean a change in one is *causing* a change in the other. A strong correlation between x and y is sometimes due to other (either known or unknown) confounding variables.

Causation

EXAMPLE 5

Causation

Over a period of years, a certain town observed that the correlation between x, the number of people attending churches, and y, the number of people in the city jail, was $r = 0.90$.

Does going to church cause people to go to jail? We hope not. During this period, there was a steady increase in population. Therefore, it is not too surprising that both the number of people attending churches and the number of people in jail increased together. The high correlation between x and y is due to the common effect of the increase in the general population.   ◇

VIEWP◉INT

Low on Credit, High on Cost!!!

How do you measure automobile insurance risk? One way is to use a little statistics and customer credit rating. Insurers say statistics show that drivers who have a history of bad credit are more likely to be in serious car accidents. According to a high-level executive at Allstate Insurance Company, financial instability is an extremely powerful predictor of future insurance losses. In short, there seems to be a strong correlation between bad credit ratings and auto insurance claims. Consequently, insurance companies want to charge higher premiums to customers with bad credit ratings. Consumer advocates object strongly because they say bad credit *does not cause* automobile accidents. More than 20 states prohibit or restrict the use credit ratings to determine auto insurance premiums. Insurance companies respond by saying your best defense is to pay your bills on time!

SECTION 10.3   PROBLEMS

1. Over the past 10 years, there has been a high positive correlation between the number of South Dakota safety inspection stickers issued and the number of South Dakota traffic accidents.
   (a) Do safety inspection stickers cause traffic accidents?
   (b) What third factor might cause traffic accidents and the number of safety stickers to increase together?

2. There is a high positive correlation in the United States between teachers' salaries and annual consumption of liquor.

(a) Do you think increasing teachers' salaries has caused increased liquor consumption?

(b) As teachers' salaries have been going up, most other salaries have been going up, too. To some extent, this means an upward trend in buying power for everyone. How might this explain the high correlation between teachers' salaries and liquor consumption?

3. Over the past 30 years in the United States, there has been a strong negative correlation between number of infant deaths at birth and number of people over age 65.

(a) Is the fact that people are living longer causing a decrease in infant mortalities at birth?

(b) What third factor might be decreasing infant mortalities and at the same time increasing life span?

4. Over the past few years, there has been a strong positive correlation between the annual consumption of diet soda pop and the number of traffic accidents.

(a) Do you think that an increasing consumption of diet pop has led to more traffic accidents?

(b) What third factor or factors might be causing both the annual consumption of diet pop and the number of traffic accidents to increase together?

For each of the Problems 5–14, please do the following:
(a) Draw a scatter diagram for the data.
(b) From the scatter diagram, would you estimate the correlation coefficient r to be closest to 1, 0, or -1?
(c) Compute the correlation coefficient r and the coefficient of determination r^2. What percentage of the variation in y can be *explained* by the corresponding variation in x using the least-squares line? What percentage of the variation is *unexplained?*
Answers may vary slightly due to rounding.

5. *Economics: Entry-Level Jobs* An economist is studying the job market in Denver area neighborhoods. Let x represent the total number of jobs in a given neighborhood, and let y represent the number of entry-level jobs in the same neighborhood. A sample of 6 Denver neighborhoods gave the following information (units in 100s of jobs).

x	16	33	50	28	50	25
y	2	3	6	5	9	3

Source: Neighborhood Facts, The Piton Foundation. To find out more, visit the Brase/Brase statistics site at http://math.college.hmco.com/students and find the link to the Piton Foundation.

Complete parts (a) through (c) for these data, and comment on the meaning of r and r^2 in the context of this problem.

6. *Violent Crimes: Prisons* Does prison really deter violent crime? Let x represent percent change in the rate of violent crime and y represent percent change in the rate of imprisonment in the general U.S. population. For 7 recent years the following data have been obtained (Source: *The Crime Drop in America,* edited by Blumstein and Wallman, Cambridge University Press).

x	6.1	5.7	3.9	5.2	6.2	6.5	11.1
y	−1.4	−4.1	−7.0	−4.0	3.6	−0.1	−4.4

Complete parts (a) through (c) for these data, and comment on the meaning of r and r^2 in the context of this problem.

7. *Education: Violent Crime* The following data are based on information from the book *Life in America's Small Cities* (by G. S. Thomas, Prometheus Books). Let x be the percentage of 16- to 19-year-olds not in school and not high school graduates. Let y be the reported violent crimes per 1000 residents. Six small cities in Arkansas (Blytheville, El Dorado, Hot Springs, Jonesboro, Rogers, and Russellville) reported the following information about x and y:

x	24.2	19.0	18.2	14.9	19.0	17.5
y	13.0	4.4	9.3	1.3	0.8	3.6

Complete parts (a) through (c) for these data, and comment on the meaning of r and r^2 in the context of this problem.

8. *Income: Death Rate* Let x be per capita income in thousands of dollars. Let y be death rate per 1000 residents. Six small cities in Oregon (Albany, Bend, Corvallis, Grants Pass, Klamath Falls, and Roseburg) gave the following information about x and y values (see reference in Problem 7):

x	8.6	9.3	10.1	8.0	8.3	8.7
y	8.4	7.6	5.4	10.6	8.3	9.3

Complete parts (a) through (c) for these data, and comment on the meaning of r and r^2 in the context of this problem.

9. *Income: Medical Care* Let x be per capita income in thousands of dollars. Let y be number of medical doctors per 10,000 residents. Six small cities in Oregon (see reference and list of cities in Problem 8) gave the following information about x and y values:

x	8.6	9.3	10.1	8.0	8.3	8.7
y	9.6	18.5	20.9	10.2	11.4	13.1

Complete parts (a) through (c) for these data, and comment on the meaning of r and r^2 in the context of this problem.

10. *Education: Death Rate* Let x be the percentage of 16- to 19-year-olds not in school and not high school graduates. Let y be the death rate per 1000 residents. Five small cities in California (El Centro, Eureka, Hanford, Madera, and San Luis Obispo–Atascadero) gave the following information about x and y values (see reference in Problem 7):

x	16.2	9.9	19.5	19.7	9.8
y	7.7	8.8	7.0	8.1	8.4

Complete parts (a) through (c) for these data, and comment on the meaning of r and r^2 in the context of this problem.

11. *Auto Accidents: Age* Data for this problem are based on information taken from *The Wall Street Journal*. Let x be the age in years of a licensed automobile driver. Let y be the percentage of all fatal accidents (for a given age) due to speeding. For example, the first data pair indicates that 36% of all fatal accidents of 17-year-olds are due to speeding.

x	17	27	37	47	57	67	77
y	36	25	20	12	10	7	5

Complete parts (a) through (c) for these data, and comment on the meaning of r and r^2 in the context of this problem.

12. *Auto Accidents: Age* Let x be the age of a licensed driver in years. Let y be the percentage of all fatal accidents (for a given age) due to failure to yield the right of way. For example, the first data pair says that 5% of all fatal accidents of 37-year-olds are due to failure to yield the right of way. *The Wall Street Journal* article referenced in Problem 11 reported the following data:

x	37	47	57	67	77	87
y	5	8	10	16	30	43

Complete parts (a) through (c) for these data, and comment on the meaning of r and r^2 in the context of this problem.

13. *Criminology: Skeletons* The state criminology laboratory must sometimes estimate a person's height from partial skeletal remains. How strong a correlation is there between body height and bone size? A random sample of eight adult male x-rays gave the following information, where x = length of femur (thigh bone) and y = body height.

x (in.)	17.5	20	21	19	15.5	18.5	16	18
y (in.)	50	80	78	73	63	71	64	71

Complete parts (a) through (c) for these data, and comment on the meaning of r and r^2 in the context of this problem.

14. *Meteorology: Cyclones* Can a low barometer reading be used to predict maximum wind speed of an approaching tropical cyclone? Data for this problem are based on information taken from *Weatherwise* (vol. 46, no. 1), a publication of the American Meteorological Society. For a random sample of tropical cyclones, let x be the lowest pressure (in millibars) as a cyclone approaches, and let y be the maximum wind speed (in miles per hour) of the cyclone.

x	1004	975	992	935	985	932
y	40	100	65	145	80	150

Complete parts (a) through (c) for these data, and comment on the meaning of r and r^2 in the context of this problem.

15. Examine the computation formula for r, the sample correlation coefficient [Equation (9) of this section].

    (a) In the formula for SS_{xy}, if we exchange the symbols x and y, do we get a different result or do we get the same (i.e., equivalent) result? Explain.

    (b) In the formula for r [see Equation (9) of this section], if we exchange the symbols x and y, do we get a different result or do we get the same (equivalent) result? Explain.

    (c) If we have a set of x and y data values and we exchange each corresponding x and y value to get a new data set, should the sample correlation coefficient be the same for both sets of data? Explain.

    (d) Compute the sample correlation coefficient r for each of the following data sets and show that they are the same.

x	1	3	4	x	2	1	6
y	2	1	6	y	1	3	4

    What can you say about the least-squares equation for each data set? Are the equations algebraically equivalent? *Hint:* See Problem 16 of Section 10.2.

16. *Dependent Variables: Investing* In Section 5.1, we studied linear combinations of *independent* random variables. What happens if the variables are not independent? A lot of mathematics can be used to prove the following:

    Let x and y be random variables with means μ_x, μ_y, and variances σ_x^2, σ_y^2 and population correlation coefficient ρ (the Greek letter rho). Let a and b be any constants and let $w = ax + by$. Then

$$\mu_w = a\mu_x + b\mu_y$$
$$\sigma_w^2 = a^2\sigma_x^2 + b^2\sigma_y^2 + 2ab\sigma_x\sigma_y\rho$$

    In this formula, ρ is the population correlation coefficient theoretically computed using the population of all (x, y) data pairs. The expression $\sigma_x\sigma_y\rho$ is called the *covariance* of x and y. If x and y are independent, then $\rho = 0$ and the formula for σ_w^2 reduces to the appropriate formula for independent variables (see Section 5.1). In most real-world applications the population parameters are not known so we use sample estimates with the understanding that our conclusions are also estimates.

    Do you have to be rich to invest in bonds and real estate? No, mutual fund shares are available to you even if you aren't rich. Let x represent annual percentage return (after expenses) on the Vanguard Total Bond Index Fund, and let y represent annual percentage return on Fidelity Real Estate Investment Fund. Over a long period of time, we have the following population estimates (based on *Morningstar Mutual Fund Report*).

$$\mu_x \approx 7.32 \qquad \sigma_x \approx 6.59 \qquad \mu_y \approx 13.19 \qquad \sigma_y \approx 18.56 \qquad \rho \approx 0.424$$

    (a) Do you think the variables x and y are independent? Explain.

    (b) Suppose you decide to put 60% of your investment in bonds and 40% in real estate. This means we will use a weighted average $w = 0.6x + 0.4y$. Estimate your expected percentage return μ_w and risk σ_w.

    (c) Repeat part (b) if $w = 0.4x + 0.6y$.

    (d) Compare your results for parts (c) and (d). Which investment has the higher expected return? Which has the greater risk as measured by σ_w?

10.4
Inferences Concerning Regression Parameters

Learn more, earn more! We have probably all heard this platitude. The question is whether or not there is some truth in the statement. Do college graduates have an improved chance at a better income? Is there a trend in the general population to support the "learn more, earn more" statement?

Consider the following variables: x = percentage of the population 25 or older with at least four years of college and y = percentage *growth* in per capita income over the past seven years. A random sample of six communities in Ohio gave the following information (based on *Life in America's Small Cities,* by G. S. Thomas):

If we use what we learned in Sections 10.2 and 10.3, we can compute the correlation coefficient r and the least-squares line $y = a + bx$ using the data of Table 10-14.

TABLE 10-14 Education and Income Growth Percentages

x	9.9	11.4	8.1	14.7	8.5	12.6
y	37.1	43.0	33.4	47.1	26.5	40.2

However, r is only a *sample* correlation coefficient and $y = a + bx$ is only a "sample-based" least-squares line. What if we used *all* possible data pairs (x, y) from *all* U.S. cities, not just the six towns in Ohio? If you did this seemingly impossible task, you would have the *population* of all (x, y) pairs.

From this population of (x, y) pairs you could (in theory) compute the *population correlation coefficient*, which we call ρ (Greek letter rho that is pronounced like "row"). You could also compute the least-squares line for the entire population, which we denote as $y = \alpha + \beta x$ using more Greek letters α (alpha) and β (beta) (see Table 10-15).

To make inferences regarding the population correlation coefficient ρ and the slope β of the population least-squares line, we need to be sure that

Assumptions for inferences concerning linear regression

(a) The set (x, y) of ordered pairs is a *random sample* from the population of all possible such (x, y) pairs.

(b) For each fixed value of x, the y values have a normal distribution.

We *assume these conditions are met for all inferences presented in this section.*

Testing the Correlation Coefficient

The first topic we want to study is the statistical significance of the sample correlation coefficient r. To do this, we construct a statistical test of ρ, the population correlation coefficient. The test will be based on the following theorem.

TABLE 10-15 Sample Statistics and Corresponding Population Parameters

Sample		Population
r	$\rightarrow$	ρ
$y = a + bx$	$\rightarrow$	$y = \alpha + \beta x$

◊ **THEOREM 10.1** Let r be the sample correlation coefficient computed using data pairs (x, y). We use the null hypothesis

H_0: x and y have no linear correlation, so $\rho = 0$

The alternate hypothesis may be

H_1: $\rho > 0$ or H_1: $\rho < 0$ or H_1: $\rho \neq 0$

The conversion of r to a Student's t distribution is

$$t = \frac{r \sqrt{n - 2}}{\sqrt{1 - r^2}} \quad \text{with } d.f. = n - 2$$

where n is the number of sample (x, y) data pairs $(n \geq 3)$.   ◊

EXAMPLE 6

Testing ρ

Let's return to our data from Ohio regarding the percentage of the population with at least four years of college and the percentage of growth in per capita income (Table 10-14). We'll develop a test for the population correlation coefficient ρ.

SOLUTION: First, we compute the sample correlation coefficient r. Using a calculator, statistical software, or "by-hand" calculation from Section 10.3, we find

$r \approx 0.887$

Now we test the correlation coefficient ρ. Remember x represents percentage college "graduates" and y represents percentage salary increases in the general population. We suspect the population correlation is positive, $\rho > 0$. Let's use a 1% level of significance:

H_0: $\rho = 0$ (no linear correlation)

H_1: $\rho > 0$ (positive linear correlation)

Convert the sample test statistic $r = 0.887$ to t using $n = 6$.

$$t = \frac{r \sqrt{n - 2}}{\sqrt{1 - r^2}} = \frac{0.887\sqrt{6 - 2}}{\sqrt{1 - 0.887^2}} \approx 3.84 \quad \text{with } d.f. = n - 2 = 6 - 2 = 4$$

From the Student's t distribution table, we find a critical value (for a right-tailed test with $\alpha' = 0.01$) of $t_0 = 3.747$. This is shown in Figure 10-18.

Using the Student's t table, we see that $0.005 < P$ value < 0.01

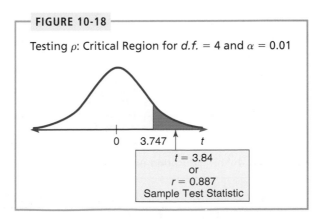

FIGURE 10-18

Testing ρ: Critical Region for $d.f. = 4$ and $\alpha = 0.01$

0   3.747 t

$t = 3.84$
or
$r = 0.887$
Sample Test Statistic

Both the critical value analysis and the P value of the sample test statistic lead us to reject $H_0: \rho = 0$ and conclude that x and y have a positive correlation. Note that just because x and y are positively correlated, we have not shown that an increase in education *causes* an increase in earning. ◇

Testing the Slope β

In many real-world applications, the slope β is very important, so our next topic is to develop a statistical test for β. The test is based on the following theorem.

◇ **THEOREM 10.2** Let b be the slope of the sample-based least-squares line $y = a + bx$. Let β be the slope of the population-based least-squares line $y = \alpha + \beta x$. Let S_e be the standard error of estimate (from Section 10.2).

We use the null hypothesis

$H_0: \beta = 0$. The population slope is zero.

The alternate hypothesis may be

$$H_1: \beta > 0 \qquad \text{or} \qquad H_1: \beta < 0 \qquad \text{or} \qquad H_1: \beta \neq 0$$

The conversion of b to a Student's t distribution is

$$t = \frac{b}{\dfrac{S_e}{\sqrt{SS_x}}} \qquad \text{with } d.f. = n - 2$$

where n is the number of sample (x, y) data pairs ($n \geq 3$).

$$SS_x = \Sigma x^2 - \frac{(\Sigma x)^2}{n} \qquad \text{and} \qquad SS_y = \Sigma y^2 - \frac{(\Sigma y)^2}{n}$$

$$SS_{xy} = \Sigma xy - \frac{(\Sigma x)(\Sigma y)}{n} \qquad \text{and} \qquad S_e = \sqrt{\frac{SS_y - bSS_{xy}}{n - 2}} \quad ◇$$

◇ **COMMENT** The expression $\dfrac{S_e}{\sqrt{SS_x}}$ is called the *standard error for b*. ◇

EXAMPLE 7

Testing β

Let's return to our sample data from Ohio and develop a test for β. Recall that x is the percentage of the population 25 or older with at least four years of college and y is the percentage growth in per capita income over the past seven years.

SOLUTION: The first step is to find the value of b, the sample slope of the least-squares line. Using a calculator, computer software, or "by-hand" calculations of Section 10.2, we find that

$$b \approx 2.55$$

The next step is to get the values for S_e and SS_x. Although the TI-83Plus calculator provides the value for S_e, many calculators do not. We will use the computation formula given in Section 10.2.

TABLE 10-16   Education and Income Growth Percentages

x	9.9	11.4	8.1	14.7	8.5	12.6
y	37.1	43.0	33.4	47.1	26.5	40.2

$$S_e = \sqrt{\frac{SS_y - bSS_{xy}}{n - 2}}$$

Using the x and y values in Table 10-16, find the following sums:

$$\Sigma x = 65.20; \ \Sigma y = 227.30; \ \Sigma x^2 = 740.68; \ \Sigma y^2 = 8877.67; \ \Sigma xy = 2552.17$$

Using $n = 6$, calculate the sum of squares.

$$SS_x = \Sigma x^2 - \frac{(\Sigma x)^2}{n} \approx 32.17; \quad SS_y = \Sigma y^2 - \frac{(\Sigma y)^2}{n} \approx 266.79;$$

$$SS_{xy} = \Sigma xy - \frac{(\Sigma x)(\Sigma y)}{n} \approx 82.18$$

Using the formula for S_e and the values just computed, we get

$$S_e = 3.78$$

Now we set up the test for β. We suspect that incomes increase when education increases. Therefore we expect the slope β is positive. Let us use a 1% level of significance.

$$H_0\text{: } \beta = 0 \qquad H_1\text{: } \beta > 0$$

Convert the sample test statistic $b = 2.55$ to a t value:

$$t = \frac{b}{\left(\dfrac{S_e}{\sqrt{SS_x}}\right)} = \frac{2.55}{\left(\dfrac{3.78}{\sqrt{32.17}}\right)} \approx 3.83$$

From the Student's t distribution table, we find a critical value (for a right-tailed test with $\alpha' = 0.01$) of $t_0 = 3.747$.

Using the Student's t table, we see that $0.005 < P$ value < 0.01.

Both the critical value analysis and the P value of the sample test statistic lead us to reject $H_0\text{: } \beta = 0$ and conclude that β is positive at the 1% level of significance. ◊

◊ **COMMENT** Recall that the *slope* of the least-squares line tells us how many units we expect y to change for each unit change in x. Using the value of b as an estimate for β, we expect that for each percentage point increase in the number of people having at least 4 years of college, there should be about a

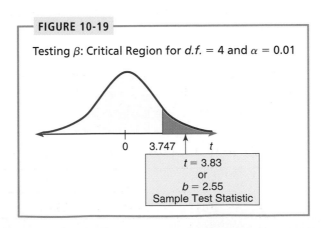

FIGURE 10-19

Testing β: Critical Region for *d.f.* = 4 and $\alpha = 0.01$

0   3.747 t

$t = 3.83$
or
$b = 2.55$
Sample Test Statistic

2.55 percent increase in the growth of per capita income. However, be careful not to extrapolate too far outside the range of x values. ◊

◊ **COMMENT** *CALCULATION HINTS.* Most calculators that support two-variable statistics give values for the correlation coefficient r and the slope of the least-squares line b directly. Although many such calculators do not provide the value of the standard error of estimate S_e, they do provide the sums Σx, Σx^2, Σy, Σy^2, and Σxy. In addition, note that $SS_x = s_x^2(n-1)$ where s_x is the sample standard deviation of the x values. Similarly, $SS_y = s_y^2(n-1)$ where s_y is the sample standard deviation of the y values. ◊

Confidence Intervals for the Slope β

The slope β measures the rate at which y changes per unit change in x (on the population level). In many real-world applications, it is important to estimate this rate of change β. Therefore, our next topic is confidence intervals for β.

◊ **THEOREM 10.3** Let b be the slope of the sample-based least-squares line $y = a + bx$, and let β be the slope of the population-based least-squares line $y = \alpha + \beta x$. Let S_e be the standard error of estimate (see Section 10.2). Then we can convert b to a Student's t distribution using

$$t = \frac{b - \beta}{\dfrac{S_e}{\sqrt{SS_x}}} \qquad \text{with } d.f. = n - 2$$

where n is the number of sample (x, y) data pairs ($n \geq 3$). ◊

Theorem 10.3 and methods similar to those of Chapter 8 can be used to obtain the following summary.

c Confidence interval for β

$$b - E < \beta < b + E$$

where b is the slope of the sample least-squares line.

$$SS_x = \Sigma x^2 - \frac{(\Sigma x)^2}{n} \qquad \text{and} \qquad SS_y = \Sigma y^2 - \frac{(\Sigma y)^2}{n}$$

$$SS_{xy} = \Sigma xy - \frac{\Sigma x \Sigma y}{n} \qquad \text{and} \qquad S_e = \sqrt{\frac{SS_y - bSS_{xy}}{n-2}}$$

$$E = t_c \frac{S_e}{\sqrt{SS_x}}$$

c = confidence level ($0 < c < 1$)

t_c = critical value for confidence level c and degrees of freedom $d.f. = n - 2$

n = number of sample (x, y) data pairs ($n \geq 3$)

EXAMPLE 8

Confidence interval for β

Plate tectonics and the spread of the ocean floor are very important in modern studies of earthquakes and earth science in general. The following data give

x = age of volcanic islands in the Indian Ocean (units 10^6 years)

y = distance of the island from the center of the midoceanic ridge (units 100 Km)

(Based on *Physical Geography* by C. A. M. King, Oxford, England.)

x	120	83	60	50	35	30	20	17
y	30	16	15.5	14.5	22	18	12	0

The population-based least-squares line is $y = \alpha + \beta x$. In this case, β, the slope, can be used to estimate the rate at which the ocean floor is moving. Let us construct a 75% confidence interval for β.

(a) Starting from raw data (x, y) values, the first step is simple but tedious. In short, you may verify (if you wish) that

$$SS_x \approx 8674.88$$
$$SS_y \approx 510.50$$
$$SS_{xy} \approx 1493.00$$

(b) The next step is to compute b and S_e. Using a calculator, statistical software, or the formulas, we get

$$b \approx 0.17$$

Since $n = 8$, we get

$$S_e = \sqrt{\frac{SS_y - bSS_{xy}}{n - 2}} \approx 6.50$$

(c) We are given $c = 0.75$. Since $n = 8$, we use degrees of freedom $d.f. = n - 2 = 8 - 2 = 6$. This gives us a t value

$$t_{0.75} = 1.273$$

Therefore,

$$E = t_c \frac{S_e}{\sqrt{SS_x}} \approx 1.273\left(\frac{6.50}{\sqrt{8674.88}}\right) \approx 0.089 \approx 0.09$$

(d) The 75% confidence interval for β is

$$b - E < \beta < b + E$$
$$0.17 - 0.09 < \beta < 0.17 + 0.09$$
$$0.08 < \beta < 0.26$$

(e) What does this mean?

Recall the units involved (x in 100 Km and y in 10^6 yr). It appears that, in this part of the world, we can be 75% confident the ocean floor is moving at a rate between 8 mm and 26 mm per year.   ◇

GUIDED EXERCISE 8

Inference for ρ and β

How fast do puppies grow? That depends on the puppy. How about male wolf pups in the Helsinki Zoo (Finland)? Let x = age in weeks and y = weight in kilograms for a random sample of male wolf pups. The following data are based on the article *Studies of the Wolf in Finland Canis lupus L (Ann. Zool. Fenn.* 2: 215–259) by E. Pulliainen, University of Helsinki.

x	8	10	14	20	28	40	45
y	7	13	17	23	30	34	35

(a) Verify the following values.

$SS_x \approx 1279.71$; $SS_y \approx 705.43$; $SS_{xy} \approx 911.14$;
$r \approx 0.959$; $b \approx 0.712$; $S_e \approx 3.368$

▷ Find the sums Σx, Σy, Σx^2, Σy^2, Σxy from your calculator or by hand, and then use the formulas for SS_x, SS_y, and SS_{xy}. Find the values for r, b, and S_e on your calculator or use the formulas of Sections 10.2 and 10.3.

(b) Use a 1% level of significance to test the claim that $\rho > 0$.

▷ Convert the sample correlation coefficient $r \approx 0.959$ to a t value.

$$t = \frac{r\sqrt{n-2}}{\sqrt{1-r^2}} \approx \frac{0.959\sqrt{5}}{\sqrt{1-0.959^2}} \approx 7.56$$

$H_0: \rho = 0$; $H_1: \rho > 0$; $d.f. = 5$; $t_0 = 3.365$. Reject H_0 and conclude there is a positive correlation between age and weight.

(c) Use a 1% level of significance to test the claim that $\beta > 0$.

▷ Convert $b \approx 0.712$ to a t value.

$$t = \frac{b}{\dfrac{S_e}{\sqrt{SS_x}}} \approx \frac{0.712}{\dfrac{3.368}{\sqrt{1279.71}}} \approx 7.56$$

$H_0: \beta = 0$; $H_1: \beta > 0$; $d.f. = 5$; $t_0 = 3.365$. Reject H_0 and conclude the slope of the least-squares line is positive.

(d) Compute an 80% confidence interval for β and interpret the results in the context of this problem.

▷ $d.f. = 5$; $t_c = 1.476$; $b \approx 0.712$. The confidence interval is

$$b - E < \beta < b + E$$

where

$$E = t_c \frac{S_e}{\sqrt{SS_x}} \approx 1.476\left(\frac{3.368}{\sqrt{1269.71}}\right) \approx 0.139$$

The interval is from 0.57 to 0.85 kg. For each week's change in age, we are 80% confident that the weight change is between 0.57 and 0.85 kg.

TECH NOTE It turns out that the t value corresponding to the correlation coefficient r is the same as the t value corresponding to b, the slope of the least-squares line (see Problem 8 at the end of this section). Consequently, the two tests H_0: $\rho = 0$ and H_0: $\beta = 0$ (with similar corresponding alternate hypotheses) have the same conclusions. The TI-83Plus uses this fact explicitly. Minitab and Excel show the two-tailed P value for the slope b of the least-squares line. Excel also shows confidence intervals for β. The displays show data from Guided Exercise 8 regarding the age and weight of wolf pups.

TI-83Plus Under **STAT**, select **TEST** and use option **E:LinRegTTest.**

```
LinRegTTest
  y=a+bx
  β>0 and ρ>0
  t=7.5632
  p=3.2037E-4
  df=5.0000
↓a=5.9317
```

```
LinRegTTest
  y=a+bx
  β>0 and ρ>0
↑b=.7120
  s=3.3676
  r²=.9196
  r=.9590
```

Note that the value of S_e is given as s.

Excel Use menu selection **Tools ➤ Data Analysis ➤ Regression.**

Regression Statistics	
Multiple R	0.958966516
R Square	0.919616778
Adjusted R Square	0.903540133
Standard Error	3.367628886
Observations	7

⟵ Value of S_e

	Coefficients	Standard Error	t Stat	P-value	Lower 95%	Upper 95%
Intercept	5.931681179	2.558126184	2.318760198	0.068158803	-0.644180779	12.50754314
X Variable 1	0.711989283	0.094138596	7.563202697	0.000640746	0.469998714	0.953979853

        b for two-tailed test

Minitab Use menu selection **Stat ➤ Regression ➤ Regression.** The value of S_e is S; P is the P value of a two-tailed test. For a one-tailed test, divide the P value by 2.

```
Regression Analysis
The regression equation is
y = 5.93 + 0.712 x
Predictor        Coef        StDev          T          P
Constant        5.932        2.558       2.32      0.068
x             0.71199      0.09414       7.56      0.001
S = 3.368      R-Sq = 92.0%      R-Sq(adj) = 90.4%
```

Answers to these problems may vary slightly due to rounding.

1. *Salary: Tuition* Is tuition correlated to professors' salaries? If you pay more tuition, does your professor get a higher salary? Let x = average annual tuition and fees for 2-year public colleges (units $100). Let y = average annual full-time faculty salary (units $1,000). A random sample of 8 states gave the following information (Source: *The Chronicle of Higher Education*, Vol. XLVI, Number 1):

x	13.4	8.2	9.4	3.8	14.5	12.5	23.9	8.2
y	37.9	50.5	33.3	56.5	37.9	42.5	43.9	41.6

(a) Please verify that $SS_x = 252.40$; $SS_y = 387.63$; $SS_{xy} = -118.06$; $r \approx -0.377$; $b \approx -0.468$; $S_e \approx 7.443$.

(b) Use a 5% level of significance to test the claim that $\rho < 0$. Include an estimate for the P value.

(c) Use a 5% level of significance to test the claim that $\beta \neq 0$. Include an estimate for the P value.

(d) Compute a 75% confidence interval for β and interpret the meaning in the context of this problem.

2. *Salary: Tuition* Is tuition correlated to professors' salaries? If you pay more tuition, does your professor get a higher salary? Let x = average annual tuition and fees for 4-year public colleges (units $100). Let y = average annual full-time faculty salary (units $1,000). A random sample of 9 states gave the following information (Source: *The Chronicle of Higher Education*, Vol. XLVI, Number 1):

x	24.9	20.6	24.5	27.1	26.2	19.1	40.1	22.7	33.9
y	42.7	47.8	42.3	62.7	48.3	50.7	50.3	45.0	49.1

(a) Please verify that $SS_x = 350.70$; $SS_y = 294.30$; $SS_{xy} = 60.17$; $r \approx 0.187$; $b \approx 0.172$; $S_e \approx 6.369$.

(b) Use a 5% level of significance to test the claim that $\rho \neq 0$. Include an estimate for the P value.

(c) Use a 5% level of significance to test the claim that $\beta \neq 0$. Include an estimate for the P value.

(d) Compute an 80% confidence interval for β and interpret the meaning in the context of this problem.

3. *Scuba Diving: Depth* What is the optimal time for a scuba diver to be on the bottom? That depends on the depth of the dive. The U.S. Navy has done a lot of research on this topic. The Navy defines the "optimal time" to be the time at each depth for the best balance between length of work period and decompression time after surfacing. Let x = depth of dive in meters, and let y = optimal time in hours. A random sample of divers gave the following data (based on information taken from *Medical Physiology* by A. C. Guyton, M.D.)

x	14.1	24.3	30.2	38.3	51.3	20.5	22.7
y	2.58	2.08	1.58	1.03	0.75	2.38	2.20

(a) Please verify that $SS_x = 940.89$; $SS_y = 2.93$; $SS_{xy} = -51.23$; $r \approx -0.976$; $b \approx -0.054$; $S_e \approx 0.166$.

(b) Use a 1% level of significance to test the claim that $\rho < 0$. Include an estimate for the P value.

(c) Use a 1% level of significance to test the claim that $\beta < 0$. Include an estimate for the P value.

(d) Compute a 90% confidence interval for β, and interpret the meaning of the result in the context of this problem.

4. *Physiology: Oxygen* Aviation and high-altitude physiology is a specialty in the study of medicine. Let x = partial pressure of oxygen in the alveoli (air cells in the lungs) when breathing naturally available air. Let y = partial pressure when breathing pure oxygen. The (x, y) data pairs correspond to elevations from 10,000 feet up to 30,000 feet in 5,000-foot intervals for a random sample of volunteers. Although the medical data were collected using airplanes, it will apply equally well to Mt. Everest climbers (summit 29,028 feet).

x	6.7	5.1	4.2	3.3	2.1 (units: mm Hg/10)
y	43.6	32.9	26.2	16.2	13.9 (units: mm Hg/10)

(based on information taken from *Medical Physiology* by A. C. Guyton, M.D.)

(a) Please verify that $SS_x = 12.25$; $SS_y = 598.29$; $SS_{xy} = 84.22$; $r \approx 0.984$; $b \approx 6.876$; $S_e \approx 2.532$.

(b) Use a 1% level of significance to test the claim that the population correlation coefficient is positive. Include an estimate for the P value.

(c) Use a 1% level of significance to test the claim that the slope of the population least-squares line is positive. Include an estimate for the P value.

(d) Compute a 95% confidence interval for β. For each unit change in oxygen pressure breathing only available air, what do you expect to be the corresponding change if you were breathing pure oxygen?

5. *New Car: Negotiating Price* Suppose you are interested in buying a new Toyota Corolla. You are standing on the sales lot looking at a model with different options. The list price is on the vehicle. As a salesperson approaches, you wonder what the dealer invoice price is for this model with its options. The following data are based on information taken from *Consumer Guide* (vol. 677). Let x be the list price (in thousands of dollars) for a random selection of Toyota Corollas of different models and options. Let y be the dealer invoice (in thousands of dollars) for the given vehicle.

x	12.6	13.0	12.8	13.6	13.4	14.2
y	11.6	12.0	11.5	12.2	12.0	12.8

(a) Please verify that $SS_x = 1.733$; $SS_y = 1.088$; $SS_{xy} = 1.313$; $r \approx 0.956$; $b \approx 0.758$; $S_e \approx 0.1527$.

(b) Use a 1% level of significance to test the claim that $\rho > 0$. Include an estimate for the P value.

(c) Use a 1% level of significance to test the claim that $\beta > 0$. Include an estimate for the P value.

(d) Compute a 90% confidence interval for β, and interpret the meaning of the result in the context of this problem.

6. *New Car: Negotiating Price* Suppose you are interested in buying a new Lincoln Navigator or Town Car. You are standing on the sales lot looking at a model with different options. The list price is on the vehicle. As a salesperson approaches, you wonder what the dealer invoice price is for this model with its options. The following data are based on information taken from *Consumer Guide* (vol. 677). Let x be the list price (in thousands of dollars) for a random selection of these cars of different models and options. Let y be the dealer invoice (in thousands of dollars) for the given vehicle.

x	32.1	33.5	36.1	44.0	47.8
y	29.8	31.1	32.0	42.1	42.2

(a) Please verify that $SS_x = 188.26$; $SS_y = 152.53$; $SS_{xy} = 165.55$; $r \approx 0.977$; $b \approx 0.879$; $S_e \approx 1.522$.

(b) Use a 1% level of significance to test the claim that $\rho > 0$. Include an estimate for the P value.

(c) Use a 1% level of significance to test the claim that $\beta > 0$. Include an estimate for the P value.

(d) Compute a 90% confidence interval for β, and interpret the meaning of the result in the context of this problem.

7. (a) Suppose $n = 6$ and the sample correlation coefficient is $r = 0.90$. Is r significant at the 1% level of significance (based on a two-tailed test)?

(b) Suppose $n = 10$ and the sample correlation coefficient is $r = 0.90$. Is r significant at the 1% level of significance (based on a two-tailed test)?

(c) Explain why the test results of parts (a) and (b) are different even though the sample correlation coefficient $r = 0.90$ in both parts. Does it appear that sample size plays an important role in determining the significance of a correlation coefficient? Explain.

8. In testing both ρ and β, we use a Student's t distribution with degrees of freedom $d.f. = n - 2$. The formulas to convert r and b to t are, respectively,

$$t = \frac{r\sqrt{n-2}}{\sqrt{1-r^2}} \qquad \text{and} \qquad t = \frac{b\sqrt{SS_x}}{S_e}$$

(a) Look at Examples 6 and 7 of this section. The sample correlation coefficient r and sample slope b of the least-squares line are based on the same data. Now look at the t values corresponding to r and to b. Are they approximately equal? Look at other problems (1 through 6) where you computed r, b, and respective t values. Are the respective t values equal?

(b) Just by looking at the two formulas, it is not obvious that the t values for r and b are equal. However, they are the same. Prove that

$$t = \frac{r\sqrt{n-2}}{\sqrt{1-r^2}} = \frac{b\sqrt{SS_x}}{S_e}$$

Hint: Work on the right side. Use the relationship

$b = r(s_y/s_x) = r(\sqrt{SS_y}/\sqrt{SS_x})$ together with the formulas

$S_e = \sqrt{(SS_y - bSS_{xy})/(n-2)}$ $\qquad$ and $\qquad$ $r = SS_{xy}/\sqrt{SS_x SS_y}$

10.5
Multiple Regression

FOCUS POINTS

✔ Learn about the advantages of multiple regression.

✔ Learn the basic ingredients that go into a multiple regression model.

✔ Discuss standard error for computed coefficients and coefficient of multiple determination.

✔ Test coefficients in the model for statistical significance.

✔ Compute confidence intervals for predictions.

Advantage of Multiple Regression

There are many examples in statistics where one variable can be predicted very accurately in terms of another *single* variable. However, predictions usually improve if we consider additional relevant information. For example, the sugar content y of golden delicious apples taken from an apple orchard in Colorado could be predicted from x_1 = number of days in growing season. If we also included information regarding x_2 = soil quality rating and x_3 = amount of available water, then we would expect our prediction of y = sugar content to be more accurate.

Likewise, the annual net income y of a new franchise auto parts store could be predicted using only x_1 = population size of sales district. However, we would probably get a better prediction of y values if we included the explanatory variables x_2 = size of store inventory, x_3 = dollar amount spent on advertising in local newspapers, and x_4 = number of competing stores in the sales district.

For most statistical applications, we gain a definite advantage in the reliability of our predictions if we include more *relevant* data and corresponding (relevant) random variables in the computation of our predictions. In this section, we will give you an idea of how this can be done by methods of *multiple regression*. You should be aware that an in-depth study of multiple regression requires the use of advanced mathematics. However, if you are willing to let the computer be a "friend who gives you useful information," then you will learn a great deal about multiple regression in this section. We will let the computer do most of the calculating work while we interpret the results.

Basic Terminology and Notation

In statistics, the most commonly used mathematical formulas for expressing linear relationships among more than two variables are *equations* of the form

$$y = b_0 + b_1x_1 + b_2x_2 + \cdots + b_kx_k \tag{10}$$

Here y is the variable that we want to predict or to forecast. We will employ the usual terminology and call y the *response variable*. The k variables x_1, x_2, ... , x_k are specified variables on which the predictions are going to be based. Once again, we will employ the popular terminology and call x_1, x_2, ... , x_k the *explanatory variables*. This terminology is easy to remember if you just think of the explanatory variables x_1, x_2, ... , x_k as "explaining" the response y.

In Equation (10), b_0, b_1, b_2, ... , b_k are numerical constants (called *coefficients*) that must be mathematically determined from given data. The numerical values of these coefficients are obtained from the *least-squares criterion*, which we will discuss after the following exercise.

GUIDED EXERCISE 9

Components of multiple regression equation

An industrial psychologist in a hospital supply company is studying the following variables for a random sample of company employees:

x_1 = number of years an employee has been with the company

x_2 = job training level (0 = lowest level and 5 = highest level)

x_3 = interpersonal skills (0 = lowest level and 10 = highest level)

y = job performance rating from supervisor (1 = lowest rating, 20 = highest rating)

The psychologist wants to predict y using x_1, x_2, and x_3 together in a least-squares equation.

(a) Identify the response variable and the explanatory variables.

The response variable is what we want to predict. This is y, job performance. The explanatory variables are years of experience x_1, training level x_2, and interpersonal skills x_3. In a sense, these variables "explain" the response variable.

(b) After collecting data, the psychologist used a computer with appropriate software to obtain the least-squares linear equation

$$y = 1 + 0.2x_1 + 2.3x_2 + 0.7x_3$$

Identify the constant term and each of the coefficients with its corresponding variable.

The constant term is 1.

Explanatory Variable	Coefficient
x_1	0.2
x_2	2.3
x_3	0.7

(c) Use the equation to predict the job performance rating of an employee with 3 years of experience, a training level of 4, and an interpersonal skill rating of 2.

Substituting $x_1 = 3$, $x_2 = 4$, $x_3 = 2$ in the least-squares equation and multiplying by the respective coefficients, we obtain the predicted job performance rating of

$$y = 1 + 0.2(3) + 2.3(4) + 0.7(2) = 12.2$$

Of course, the *predicted* value for job performance might differ from the actual rating given by the supervisor.

Theory for the least-squares criterion (optional)

This material is a little sophisticated, so you may wish to skip ahead to the discussion of regression models and computers and omit the following explanation of basic theory.

In multiple regression, the least-squares criterion says the following sum (over all data points),

$$\Sigma[y_i - (b_0 + b_1x_{1i} + b_2x_{2i} + \cdots + b_kx_{ki})]^2 \tag{11}$$

must be made as small as possible. In this formula,

$$y_i = i\text{th data value for } y$$
$$x_{1i} = i\text{th data value for } x_1$$
$$x_{2i} = i\text{th data value for } x_2$$
$$\vdots$$
$$x_{ki} = i\text{th data value for } x_k$$

Recall that Equation (10) gives the predicted y value; therefore,

$$y_i - (b_0 + b_1x_{1i} + b_2x_{2i} + \cdots + b_kx_{ki}) \tag{12}$$

represents the *difference* between the *observed* y value (that is, y_i) and the *predicted* y value based on the data values $x_{1i}, x_{2i}, \ldots, x_{ki}$. When we square this difference and total the result over all data points and choose the values of $b_0, b_1, b_2, \ldots, b_k$ to minimize the sum [i.e., minimize Equation (11)], then we are satisfying the least-squares criterion.

◊ **COMMENT** The algebraic expression in Equation (12) is very important. In fact, it has a special name in the theory of regression. It is called a *residual*. The residual is simply the difference between the actual data value and the predicted value of the response variable based on given data values for the explanatory variables. Advanced topics in the theory of regression will study residuals in great detail. Such a detailed treatment is beyond the scope of this text. However, from the discussion presented so far, we see that the method of least squares chooses the values of the coefficients b_i to make the sum of the squares of the residuals as small as possible. ◊

After a good deal of mathematics has been done (involving a considerable amount of calculus), the least-squares criterion can be reduced to solving a system of linear equations. These are usually called *normal equations* (not to be confused with the normal distribution).

In the simplest case, where there are only *two* explanatory variables x_1 and x_2 and we want to fit the equation

$$y = b_0 + b_1x_1 + b_2x_2$$

to given data, there are three normal equations that must be solved for b_0, b_1, and b_2. These normal equations are

$$\Sigma y_i = nb_0 + b_1(\Sigma x_{1i}) + b_2(\Sigma x_{2i})$$
$$\Sigma x_{1i}y_i = b_0(\Sigma x_{1i}) + b_1(\Sigma x_{1i}^2) + b_2(\Sigma x_{1i}x_{2i}) \tag{13}$$
$$\Sigma x_{2i}y_i = b_0(\Sigma x_{2i}) + b_1(\Sigma x_{1i}x_{2i}) + b_2(\Sigma x_{2i}^2)$$

In the system of Equations (13), n represents the number of data points and x_{1i}, x_{2i}, and y_i all represent given data values.

Therefore, the only unknowns are the coefficients b_0, b_1, and b_2; we can use the system of Equations (13) to solve for these unknowns. This is the procedure that lets us obtain the least-squares regression equation in Equation (10) when we have only *two* explanatory variables.

As you can see, all this is rather complicated, and the more explanatory variables $x_1, x_2, \ldots, x_k$ we have, the more involved the calculations become. In the general case, if you have k explanatory variables, there will be $k + 1$ normal equations that must be solved for the coefficients $b_0, b_1, b_2, \ldots, b_k$.

Regression Models and Computers

As we can see from the preceding optional discussion, the work required to find an equation satisfying the least-squares criterion is tremendous and can be very complex. Today, such work is conveniently left to computers. In this text, we use two computer software packages that specialize in statistical applications.

Minitab is a widely used statistical software package. It fully supports multiple regression. Excel has a multiple regression component that performs much of the multiple regression analysis. We will use Minitab in our example. Many other software packages, including ComputerStat, support multiple regression and have outputs similar to Minitab.

In this section, we will often refer to a *regression model*. What do we mean by this? We mean a mathematical package that consists of the following ingredients:

Ingredients of the regression model

1. A collection of random variables, *one* of which has been identified as the response variable, with *any or all* of the remaining variables being identified as explanatory variables.

2. Associated with a given application will be a collection of numerical data values for each of the variables of part 1.

3. Using the numerical data values, the least-squares criterion, and the declared response and explanatory variables, a *least-squares equation* (also called a *regression equation*) will be constructed. In Section 10.2, we were able to construct the least-squares equation using only a hand calculator. However, in multiple regression, we will use a computer to construct the least-squares equation.

4. The model usually includes additional information about the variables used, the coefficients and regression equation, and a measure of "goodness of fit" of the regression equation to the data values. In modern practice, this information usually comes to you in the form of computer displays.

5. Finally, the regression model allows you to supply given values of the explanatory variables for the purpose of predicting or forecasting the corresponding value of the response variable. You also should be able to construct a $c\%$ confidence interval for your least-squares prediction. In multiple regression, this will be done by the computer at your request.

The next example demonstrates computer applications of a typical multiple regression problem. In the context of the example, we will introduce some of the basic techniques of multiple regression.

Example Utilizing Minitab

EXAMPLE 9

Multiple regression

Antelope are beautiful and graceful animals that live on the high plains of the western United States. Thunder Basin National Grasslands in Wyoming is home to hundreds of antelope. The Bureau of Land Management (BLM) has been studying the Thunder Basin antelope population for the past 8 years. The variables used are

TABLE 10-17   Data for Thunder Basin Antelope Study

Year	x_1	x_2	x_3	x_4
1	2.9	9.2	13.2	2
2	2.4	8.7	11.5	3
3	2.0	7.2	10.8	4
4	2.3	8.5	12.3	2
5	3.2	9.6	12.6	3
6	1.9	6.8	10.6	5
7	3.4	9.7	14.1	1
8	2.1	7.9	11.2	3

$x_1 =$ spring fawn count (in hundreds of fawns)

$x_2 =$ size of adult antelope population (in hundreds)

$x_3 =$ annual precipitation (in inches)

$x_4 =$ winter severity index (1 = mild and 5 = extremely severe) (This is an index based on temperature and wind chill factors.)

The data obtained in the study over the 8-year period are shown in Table 10-17.

Summary Statistics for Each Variable

It is a good idea to first look at the summary statistics for each variable. Figure 10-20 shows the Minitab display of the summary statistics.

Menu selection: **Stat ➤ Basic Statistic ➤ Display Descriptive Statistics**

FIGURE 10-20

Minitab Display of Summary Statistics for Each Variable

```
Descriptive Statistics
Variable     N       Mean     Median     TrMean      StDev    SE Mean
x1           8      2.525      2.350      2.525      0.570      0.202
x2           8      8.450      8.600      8.450      1.076      0.380
x3           8     12.037     11.900     12.037      1.229      0.435
x4           8      2.875      3.000      2.875      1.246      0.441
Variable   Minimum    Maximum        Q1         Q3
x1           1.900      3.400      2.025      3.125
x2           6.800      9.700      7.375      9.500
x3          10.600     14.100     10.900     13.050
x4           1.000      5.000      2.000      3.750
```

This type of information can be very useful because it tells you basic information about the variables you are studying. Sample means and sample standard deviations with Student's t distribution are essential ingredients for estimating or testing population means (Chapters 8 and 9).

For example, if μ_2 represents the *population mean* of x_2 (adult antelope population), then by using the methods of Section 8.2 we can quickly estimate a 90% confidence interval for μ_2:

$$7.729 < \mu_2 < 9.171$$

Since our units are in hundreds, this means that we can be 90% sure the *population mean* μ_2 of adult antelope in the Thunder Basin Grasslands is between 773 and 917.

Correlation Between Variables

It is also useful to examine how the variables relate to each other. Figure 10-21 on the following page shows the correlation coefficients r between each of the two variables. A natural question arises: Which of the variables are closely related to each other, and which are not as closely related? Recall (from Section 10.3) that if the correlation coefficient is near 1 or -1, then the corresponding variables have a lot in common. If the correlation coefficient is near zero, the variables have much less influence on each other.

Menu selection: **Stat ➤ Basic Statistics ➤ Correlation**

```
Correlations  (Pearson)
              x1           x2           x3
x2         0.939
x3         0.924        0.903
x4        -0.739       -0.836       -0.901
```

Look at Figure 10-21. Which of the variables has the greatest influence on x_1? The correlation coefficient between x_1 and x_2 is $r = 0.939$, with a corresponding coefficient of determination of $r^2 \approx 0.88$. This means that if we consider only x_1 and x_2 (and none of the other variables), then about 88% of the variation of x_1 could be explained by the corresponding variation of x_2 (by itself). Similarly, if we consider only x_1 and x_3, we see the correlation coefficient $r = 0.924$, with a corresponding coefficient of determination of $r^2 \approx 0.85$. About 85% of the variation of x_1 could be explained by the corresponding variation of x_3. The variable x_4 has much less influence on x_1 because the correlation coefficient between these two variables is $r = -0.739$, with corresponding coefficient of determination $r^2 \approx 0.55$, or only 55%.

These relationships are very reasonable in the context of our problem. It is common sense that the number of spring fawns x_1 is strongly related to x_2, the size of the adult antelope population. Furthermore, the spring fawn count x_1 is very much influenced by available food for the fawn (and its mother). Thunder Basin National Grasslands is a semiarid region, and available food (grass) is almost completely determined by annual precipitation x_3. Antelope are naturally strong and hardy animals. Therefore, the temperature and wind chill index x_4 will have much less effect on the adult does and corresponding number of spring fawns provided there is plenty of available food.

Least-Squares Equation

Figure 10-22 shows a display that gives an expression for the actual least-squares equation and a lot of information about the equation. To get this display or a similar display, the user needs to declare which variable is the response variable and which are the explanatory variables. For Figure 10-22, we designated x_1 as the response variable. This means that x_1 is the variable we choose to predict. We also designated variables x_2, x_3, and x_4 as explanatory variables. This means that x_2, x_3, and x_4 will be used *together* to predict x_1. There is a lot of flexibility here. We could have designated any *one* of the variables x_1, x_2, x_3, x_4 as the response variable and *any or all* of the remaining variables as explanatory variables. So there are a lot of possible regression models the computer can construct for you, depending on the type of information you want. In this example, we want to predict x_1 (spring fawn count) by using x_2 (adult population), x_3 (annual precipitation), and x_4 (winter index) *together*.

Menu selection: **Stat ➤ Regression ➤ Regression.** In the dialogue box, select x_1 as the response and x_2, x_3, x_4 as the predictors.

```
Regression Analysis
The regression equation is
x1 = -5.92 + 0.338 x2 + 0.402 x3 + 0.263 x4
Predictor       Coef       StDev        T         P
Constant      -5.922       1.256     -4.72     0.009
x2            0.33822     0.09947     3.40     0.027
x3            0.4015      0.1099      3.65     0.022
x4            0.26295     0.08514     3.09     0.037
S = 0.1209   R-Sq = 97.4%    R-Sq(adj) = 95.5%
```

The least-squares regression equation is given near the top of the display. Then more information is given about the constant and coefficients. The parts of the equation are

$$x_1 = -5.92 + 0.338x_2 + 0.402x_3 + 0.263x_4 \tag{14}$$

response variable   constant   coefficient of associated explanatory variable

◇ **COMMENT** In the case of a simple regression model, where we have only one explanatory variable, the coefficient of that variable is the *slope* of the least-squares line. This slope (or coefficient) represents the change in the response variable per unit change in the explanatory variable. In a multiple regression model such as Equation (14), the coefficients also can be thought of as a slope—*provided* we hold the other variables as arbitrary and fixed constants. For example, the coefficient of x_2 in Equation (14) is $b_2 = 0.338$. This means that if x_3 (precipitation) and x_4 (winter index) are taken into account but held constant, then $b_2 = 0.338$ represents the change in x_1 (spring fawn count) per unit change in x_2 (adult antelope count). Since our units are in hundreds, this would say that if x_3 and x_4 are taken into account as arbitrary but fixed values, then an increase of 100 adult antelope would give an expected increase of 33.8 or 34 spring fawns. ◇

A natural question arises: How good a fit is the least-squares regression Equation (14) for our given data?

Coefficient of multiple determination

One way to answer this question is to examine the *coefficient of multiple determination*. The coefficient of multiple determination is a direct generalization of the concept of coefficient of determination (between *two* variables) as discussed in Section 10.3, and it has essentially the same meaning. The coefficient of multiple determination is given in the display of Figure 10-22 as a percent. We see R-Sq = 97.4%. This means that about 97.4% of the variation of the response variable x_1 can be explained from the least-squares Equation (14) and the corresponding *joint* variation of the variables x_2, x_3, and x_4 taken together. The remaining 100% − 97.4% = 2.6% of the variation in x_1 is due to random chance or possibly the presence of other variables not included in this regression equation. (We will discuss the *standard error* associated with each coefficient later in this section.)

Predictions

Let's use the current regression model to predict the response variable x_1. Recall that in Section 10.2 we first made predictions from the least-squares line and then constructed a confidence interval for our prediction. Although the exact details are beyond the scope of this text, this process can be generalized to multiple regression. The calculations are very tedious, but that's why we use a computer!

Suppose we ask the following question: In a year when $x_2 = 8.2$ (hundreds of adult antelope) and $x_3 = 11.7$ (inches of precipitation) and $x_4 = 3$ (winter index), what do we predict for x_1 (spring fawn count)? Furthermore, let's suppose we want an 85% confidence interval for our prediction.

To answer this question, we look at Figure 10-23, which shows the Minitab prediction result for x_1 from the specified values of x_2, x_3, x_4.

Menu selection: **Stat ➤ Regression ➤ Regression**. In the dialogue box, select Options. List the new observations for x_2, x_3, x_4 in order, separated by spaces. Specify the confidence level. Be sure that Fit Intercept is checked.

FIGURE 10-23

Minitab Display Showing the Predicted Value of x_1

```
Predicted  Values
Fit        StDev Fit       85.0% CI            85.0% PI
2.3378        0.0472    (2.2539,  2.4217)    (2.1069,  2.5687)
```

The value for Fit is 2.3378. That is the predicted value for x_1. The 85% confidence interval for the prediction is designated as 85% PI. We see that the interval for x_1 (rounded to two digits after the decimal) is $2.11 \le x_1 \le 2.57$. This means that we are 85% confident the number of spring fawns will be in the range from 211 to 257.

Please note that this is *not* a confidence interval for the population mean of x_1. Rather, we have constructed a confidence interval for the *actual value* of x_1 under the conditions when $x_2 = 8.2$, $x_3 = 11.7$, and $x_4 = 3$.   ◊

◊ **COMMENT** Extrapolation much beyond the data range in a multiple regression model in any of the variables can produce results that might be meaningless and unrealistic. Many computer software packages warn about computing a confidence interval for a prediction when some of the values of the explanatory variables are beyond the data range in either direction.   ◊

Testing a Coefficient for Significance

In applications of multiple regression, it is possible to have many different variables. Occasionally, you might suspect that one of the explanatory variables x_i is not very useful as a tool for predicting the response variable. It simply may not influence the response variable much at all. To decide whether or not this is the case, we construct a test for the significance of the coefficient of x_i in the least-squares equation.

Recall that the general least-squares equation is

$$y = b_0 + b_1 x_1 + b_2 x_2 + \cdots + b_k x_k \tag{15}$$

where y = response variable

x_i = explanatory variable for $i = 1, 2, \ldots, k$

b_i = numerical coefficient for $i = 0, 1, 2, \ldots, k$

Equation (15) was constructed from given data. Usually, the data are only a small subset of all possible data that could have been collected.

Let us suppose (in theory) that we used *all possible data* that could ever be obtained for our regression problem and that we constructed the regression equation using the entire population of all possible data. Then we would get the *theoretical* regression equation

$$y = \beta_0 + \beta_1 x_1 + \beta_2 x_2 + \cdots + \beta_k x_k \qquad (16)$$

where y and x_i are as in Equation (15), but β_i is the *theoretical* coefficient of x_i.

Now look back at the regression analysis in Figure 10-22. Beside the constant and each coefficient there is a number in the StDev column. This is the *standard error* corresponding to that coefficient. The standard error can be thought of as similar to a standard deviation that corresponds to the coefficient. The calculation of the number is beyond the scope of this text, but it is available on computer printouts, and we will use it to construct our test.

Let us call S_i the standard error for coefficient x_i (S_0 is the standard error for the constant). Under very basic and general assumptions, it can be proved that

$$t = \frac{b_i - \beta_i}{S_i} \qquad (17)$$

has a Student's t distribution with degrees of freedom $d.f. = n - k - 1$, where n = number of data points and k = number of explanatory variables in the least-squares equation.

Now let us return to the question: Is x_i useful as an explanatory variable in the least-squares equation?

The answer is that it is *not* useful if $\beta_i = 0$. In this case, the (theoretical) coefficient of x_i would be zero and x_i would contribute nothing to the least-squares equation. However, if $\beta_i \neq 0$, then the explanatory variable x_i does contribute information in the least-squares equation.

Consider the following hypotheses,

$$H_0: \beta_i = 0 \qquad \text{and} \qquad H_1: \beta_i \neq 0$$

If we accept H_0, we conclude that $\beta_i = 0$ and x_i probably should be dropped as an explanatory variable in the least-squares equation. If we accept H_1, we conclude that $\beta_i \neq 0$ and x_i should be included as an explanatory variable in the least-squares equation.

EXAMPLE 10

Test a coefficient

We'll use the data and printouts of Example 9 and test the significance of x_3 as an explanatory variable using $\alpha = 0.05$ as level of significance.

$$H_0: \beta_3 = 0 \qquad \text{and} \qquad H_1: \beta_3 \neq 0$$

Since the sample coefficients follow a Student's t distribution, we look in Table 6 of Appendix II for the critical values. The degrees of freedom are $d.f. = 8 - 3 - 1 = 4$ (since $n = 8$ and $k = 3$). The critical values for a two-tailed test with $\alpha = 0.05$ are $t = \pm 2.776$.

To find the t value corresponding to b_3, we use Equation (17) and the null hypothesis $H_0: \beta_3 = 0$. This gives us the equation

$$t = \frac{b_3}{S_3} \qquad (18)$$

In the regression analysis shown in Figure 10-22, we see a t-ratio value for the constant and each coefficient. This t ratio is exactly the value of $t = b_i/S_i$. This is the t ratio corresponding to the sample test statistic. For the coefficient of x_3, we see

$$t \text{ ratio} \approx 3.652$$

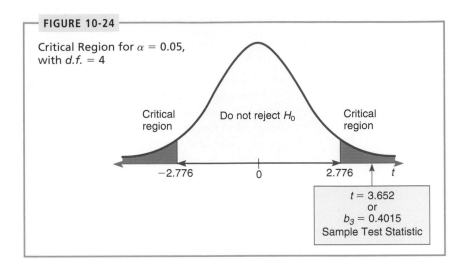

FIGURE 10-24

Critical Region for $\alpha = 0.05$, with $d.f. = 4$

Critical region

Do not reject H_0

Critical region

-2.776 0 2.776 t

$t = 3.652$
or
$b_3 = 0.4015$
Sample Test Statistic

The sample test statistic falls in the critical region (see Figure 10-24). Therefore, we reject the null hypothesis H_0: $\beta_3 = 0$ and accept H_1: $\beta_3 \neq 0$.

We conclude at the 5% level of significance that x_3 (annual precipitation) should be included as an explanatory variable in the least-squares equation. Notice that Figure 10-22 also gives the P value for each ratio, so we can conclude the test using P values directly. Using the P values, we see that x_2 and x_3 are also significant at the 5% level.   ◊

Confidence Intervals for Coefficients

Equation (17) also gives us the basis of finding *confidence intervals* for β_i. A $c\%$ confidence interval for β_i will be

$$b_i - tS_i < \beta_i < b_i + tS_i$$

where $d.f. = n - k - 1$, t is selected according to the specified confidence level, b_i is the numerical value of the coefficient from Figure 10-24, S_i is the numerical value of the standard error from Figure 10-24, n is the number of data points, and k is the number of explanatory variables in the least-squares equation.

EXAMPLE 11

Confidence interval for a coefficient

Suppose we want to compute a 90% confidence interval for β_2, the coefficient of x_2. From Figure 10-24 we have (rounding to three digits after the decimal)

$$b_2 = 0.338, \qquad S_2 = 0.099, \qquad \text{and} \qquad d.f. = 4$$

From the t table (Table 6, Appendix II), we find $t = 2.132$, so

$$b_2 - tS_2 < \beta_2 < b_2 + tS_2$$
$$0.338 - 2.132(0.099) < \beta_2 < 0.338 + 2.132(0.099)$$
$$0.127 < \beta_2 < 0.549 \qquad\qquad ◊$$

Excel Displays

Excel gives information very similar to that supplied by Minitab. The least-squares equation is not explicitly displayed. However, the intercept (constant) and coefficients of the variables are shown with the corresponding standard errors and t ratios

FIGURE 10-25

Excel Display of Regression Analysis

Regression Statistics						
Multiple R	0.987060478					
R Square	0.974288388					
Adjusted R Square	0.955004679					
Standard Error	0.120927579					
Observations	8					
ANOVA						
	df	*SS*	*MS*	*F*	*Significance F*	
Regression	3	2.216506083	0.738835361	50.5239104	0.001228863	
Residual	4	0.058493917	0.014623479			
Total	7	2.275				
	Coefficients	*Standard Error*	*t Stat*	*P-value*	*Lower 95%*	*Upper 95%*
Intercept	-5.922011616	1.255623292	-4.716391972	0.009196085	-9.40818798	-2.435835251
x2	0.338217487	0.099470083	3.400193085	0.027272474	0.062043691	0.614391283
x3	0.401503945	0.109900277	3.653347874	0.021707246	0.096371226	0.706636664
x4	0.262946128	0.085136028	3.088541172	0.036626194	0.02657013	0.499322125

with P values. Excel shows the confidence interval for each coefficient. However, there is no built-in function to provide predicted values or confidence intervals for predicted values. Note that as in the Minitab regression analysis, we will not make use of the ANOVA information in the Excel display.

Menu selection: **Tools ➤ Data Analysis ➤ Regression.** Note that when you enter data into the worksheet, all the explanatory variables must be together in a block.

VIEWP◉INT   *Synoptic Climatology*

Synoptic means "giving a summary from the same basic point of view." In this case, the point of view is Niwot Ridge, high above the timberline in the Rocky Mountains. Vegetation, water, temperature, and wind all affect the delicate balance of this alpine environment. How do these elements of nature interact to sustain life in such a harsh land? One answer can be found by collecting data at the location and using multiple regression to study the interaction of variables. For more information, visit the Brase/Brase statistics site at http://math.college.hmco.com/students and find the link to Niwot ridge climate study.

SECTION 10.5   PROBLEMS

1. Given the linear regression equation

$$x_1 = 1.6 + 3.5x_2 - 7.9x_3 + 2.0x_4$$

(a) Which variable is the response variable? Which variables are the explanatory variables?

(b) Which number is the constant term? List the coefficients with their corresponding explanatory variables.

(c) If $x_2 = 2$, $x_3 = 1$, and $x_4 = 5$, what is the predicted value for x_1?

(d) Explain how each coefficient can be thought of as a "slope" under certain conditions. Suppose that x_3 and x_4 were held at fixed but arbitrary values and x_2 increased by one unit. What would be the corresponding change in x_1? Suppose x_2 were increased by two units. What would be the expected change in x_1? Suppose x_2 decreased by four units. What would be the expected change in x_1?

(e) Suppose that $n = 12$ data points were used to construct the given regression equation and that the standard error for the coefficient of x_2 is 0.419. Construct a 90% confidence interval for the coefficient of x_2.

(f) Using the information of part (e) and level of significance 5%, test the claim that the coefficient of x_2 is different from zero. Explain how the conclusion of this test would affect the regression equation.

2. Given the linear regression equation

$$x_3 = -16.5 + 4.0x_1 + 9.2x_4 - 1.1x_7$$

(a) Which variable is the response variable? Which variables are the explanatory variables?

(b) Which number is the constant term? List the coefficients with their corresponding explanatory variables.

(c) If $x_1 = 10$, $x_4 = -1$, and $x_7 = 2$, what is the predicted value for x_3?

(d) Explain how each coefficient can be thought of as a "slope." Suppose that x_1 and x_7 were held as fixed but arbitrary values. If x_4 increased by one unit, what would we expect the corresponding change in x_3 to be? If x_4 increased by three units, what would be the corresponding expected change in x_3? If x_4 decreased by two units, what do we expect for the corresponding change in x_3?

(e) Suppose that $n = 15$ data points were used to construct the given regression equation and that the standard error for the coefficient of x_4 was 0.921. Construct a 90% confidence interval for the coefficient of x_4.

(f) Using the information of part (e) and level of significance 1%, test the claim that the coefficient of x_4 is different from zero. Explain how the conclusion has a bearing on the regression equation.

For Problems 3–6, use appropriate multiple regression software of your choice and enter the data. Note that the CD-ROM that comes with this text has the data in formats for Excel, Minitab portable files, and ASCII files. The data are also in ComputerStat.

3. *Medical: Blood Pressure* The systolic blood pressure of individuals is thought to be related to both age and weight. For a random sample of 11 men, the following data were obtained:

Systolic Blood Pressure x_1	Age (years) x_2	Weight (pounds) x_3	Systolic Blood Pressure x_1	Age (years) x_2	Weight (pounds) x_3
132	52	173	137	54	188
143	59	184	149	61	188
153	67	194	159	65	207
162	73	211	128	46	167
154	64	196	166	72	217
168	74	220			

(a) Generate summary statistics, including the mean and standard deviation of each variable. Compute the coefficient of variation (see Section 3.2) for each variable. Relative to its mean, which variable has the greatest spread of data values? Which variable has the smallest spread of data values relative to its mean?

(b) For each pair of variables, generate the correlation coefficient r. Compute the corresponding coefficient of determination r^2. Which variable (other than x_1) has the greatest influence (by itself) on x_1? Would you say both variables x_2 and x_3 show a strong influence on x_1? Explain your answer. What percent of the variation of x_1 can be explained by the corresponding variation in x_2? Answer the same question for x_3.

(c) Perform a regression analysis with x_1 as the response variable. Use x_2 and x_3 as explanatory variables. Look at the coefficient of multiple determination. What percentage of the variation in x_1 can be explained by the corresponding variation in x_2 and x_3 *taken together*?

(d) Look at the coefficients of the regression equation. Write out the regression equation. Explain how each coefficient can be thought of as a slope. If age were held fixed, but a person put on 10 pounds, what would you expect for the corresponding change in systolic blood pressure? If a person kept the same weight but got 10 years older, what would you expect for the corresponding change in systolic blood pressure?

(e) Test each coefficient to determine if it is zero or not zero. Use level of significance 5%. Why would the outcome of each test help us determine whether or not a given variable should be used in the regression model?

(f) Find a 90% confidence interval for each coefficient.

(g) Suppose that Michael is 68 years old and weighs 192 pounds. Predict his systolic blood pressure, and find a 90% confidence range for your prediction (if your software produces prediction intervals).

4. *Education: Exam Scores* Professor Gill has taught General Psychology for many years. During the semester, she gives three multiple-choice exams each worth 100 points. At the end of the course, Dr. Gill gives a comprehensive final worth 200 points. Let x_1, x_2, and x_3 represent a student's scores on exams 1, 2, and 3, respectively. Let x_4 represent the student's score on the final exam. Last semester Dr. Gill had 25 students in class and the student exam scores are shown below.

x_1	x_2	x_3	x_4	x_1	x_2	x_3	x_4	x_1	x_2	x_3	x_4
73	80	75	152	79	70	88	164	81	90	93	183
93	88	93	185	69	70	73	141	88	92	86	177
89	91	90	180	70	65	74	141	78	83	77	159
96	98	100	196	93	95	91	184	82	86	90	177
73	66	70	142	79	80	73	152	86	82	89	175
53	46	55	101	70	73	78	148	78	83	85	175
69	74	77	149	93	89	96	192	76	83	71	149
47	56	60	115	78	75	68	147	96	93	95	192
87	79	90	175								

Since Professor Gill has not changed the course much from last semester to the present semester, the preceding data should be useful for constructing a regression model that describes this semester as well.

(a) Generate summary statistics, including the mean and standard deviation of each variable. Compute the coefficient of variation (see Section 3.2) for each variable. Relative to its mean, would you say each exam had about the same spread of scores? Most professors do not wish to give an exam that is extremely easy or extremely hard. Would you say all of the exams were about the same level of difficulty? (Consider both means and spread of test scores.)

(b) For each pair of variables, generate the correlation coefficient r. Compute the corresponding coefficient of determination r^2. Of the three exams 1, 2, and 3, which do you think had the most influence on the final exam 4? Although one exam had more influence on the final exam, did the other two exams still have a lot of influence on the final? Explain each answer.

(c) Perform a regression analysis with x_4 as the response variable. Use x_1, x_2, and x_3 as explanatory variables. Look at the coefficient of multiple determination. What percentage of the variation in x_4 can be explained by the corresponding variation in x_1, x_2, x_3 taken together?

(d) Write out the regression equation. Explain how each coefficient can be thought of as a slope. If a student were to study "extra hard" for exam 3 and increase his or her score on that exam by 10 points, what corresponding change would you expect on the final exam? (Assume that exams 1 and 2 remain "fixed" in their scores.)

(e) Test each coefficient in the regression equation to determine if it is zero or not zero. Use level of significance 5%. Why would the outcome of each hypothesis test help us decide whether or not a given variable should be used in the regression equation?

(f) Find a 90% confidence interval for each coefficient.

(g) This semester Susan has scores 68, 72, and 75 on the exams 1, 2, and 3, respectively. Make a prediction for Susan's score on the final exam and find a 90% confidence interval for your prediction (if your software supports prediction intervals).

5. *Entertainment: Movies* A motion picture industry analyst is studying movies based on epic novels. The following data were obtained for 10 Hollywood movies made in the past five years. Each movie was based on an epic novel. In these data, x_1 = first year box office receipts of the movie, x_2 = total production costs of the movie, x_3 = total promotional costs of the movie, x_4 = total book sales prior to movie release. All units are in millions of dollars.

x_1	x_2	x_3	x_4	x_1	x_2	x_3	x_4
85.1	8.5	5.1	4.7	30.3	3.5	1.2	3.5
106.3	12.9	5.8	8.8	79.4	9.2	3.7	9.7
50.2	5.2	2.1	15.1	91.0	9.0	7.6	5.9
130.6	10.7	8.4	12.2	135.4	15.1	7.7	20.8
54.8	3.1	2.9	10.6	89.3	10.2	4.5	7.9

(a) Generate summary statistics, including the mean and standard deviation of each variable. Compute the coefficient of variation (see Section 3.2) for each variable. Relative to its mean, which variable has the largest spread of data values? Why would a variable with a large coefficient of variation be expected to change a lot relative to its average value? Although x_1 has the largest standard deviation, it has the smallest coefficient of variation. How does the mean of x_1 help explain this?

(b) For each pair of variables, generate the correlation coefficient r. Compute the corresponding coefficient of determination r^2. Which of the three variables x_2, x_3, x_4 has the *least* influence on box office receipts? What percent of the variation of box office receipts can be attributed to the corresponding variation in production costs?

(c) Perform a regression analysis with x_1 as the response variable. Use x_2, x_3, and x_4 as explanatory variables. Look at the coefficient of multiple determination. What percentage of the variation in x_1 can be explained by the corresponding variation in x_2, x_3, and x_4 taken together?

(d) Write out the regression equation. Explain how each coefficient can be thought of as a slope. If x_2 (production costs) and x_4 (book sales) were held fixed but x_3 (promotional costs) were increased by one million dollars, what would you expect for the corresponding change in x_1 (box office receipts)?

(e) Test each coefficient in the regression equation to determine if it is zero or not zero. Use level of significance 5%. Explain why book sales x_4 are probably not contributing much information in the regression model to forecast box office receipts x_1.

(f) Find a 90% confidence interval for each coefficient.

(g) Suppose that a new movie (based on an epic novel) has just been released. Production costs were $x_2 = 11.4$ million; promotion costs were $x_3 = 4.7$ million; book sales were $x_4 = 8.1$ million. Make a prediction for $x_1 =$ first year box office receipts and find an 85% confidence interval for your prediction (if your software supports prediction intervals).

(h) Construct a new regression model with x_3 as the response variable and x_1, x_2, and x_4 as explanatory variables. Suppose that Hollywood is planning a new epic movie with projected box office sales $x_1 = 100$ million and production costs $x_2 = 12$ million. The book on which the movie is based sold $x_4 = 9.2$ million. Forecast the dollar amount (in millions) that should be budgeted for x_3 promotion costs and find an 80% confidence interval for your prediction.

6. *Franchise Business: Market Analysis* All Greens is a franchise store that sells house plants and lawn and garden supplies. Although All Greens is a franchise, each store is owned and managed by private individuals. Some friends have asked you to go into business with them to open a new All Greens store in the suburbs of San Diego. The national franchise headquarters sent you the following information at your request. These data are about 27 All Greens stores in California. Each of the 27 stores has been doing very well, and you would like to use the information to help set up your own new store. The variables for which we have data are

$x_1 =$ annual net sales in thousands of dollars

$x_2 =$ number of square feet of floor display in store in thousands of square feet

$x_3 =$ value of store inventory in thousands of dollars

$x_4 =$ amount spent on local advertising in thousands of dollars

$x_5 =$ size of sales district in thousands of families

$x_6 =$ number of competing or similar stores in sales district

A sales district was defined to be the region within a radius of 5 miles of an All Greens store.

x_1	x_2	x_3	x_4	x_5	x_6	x_1	x_2	x_3	x_4	x_5	x_6
231	3	294	8.2	8.2	11	65	1.2	168	4.7	3.3	11
156	2.2	232	6.9	4.1	12	98	1.6	151	4.6	2.7	10
10	0.5	149	3	4.3	15	398	4.3	342	5.5	16.0	4
519	5.5	600	12	16.1	1	161	2.6	196	7.2	6.3	13
437	4.4	567	10.6	14.1	5	397	3.8	453	10.4	13.9	7
487	4.8	571	11.8	12.7	4	497	5.3	518	11.5	16.3	1
299	3.1	512	8.1	10.1	10	528	5.6	615	12.3	16.0	0
195	2.5	347	7.7	8.4	12	99	0.8	278	2.8	6.5	14
20	1.2	212	3.3	2.1	15	0.5	1.1	142	3.1	1.6	12
68	0.6	102	4.9	4.7	8	347	3.6	461	9.6	11.3	6
570	5.4	788	17.4	12.3	1	341	3.5	382	9.8	11.5	5
428	4.2	577	10.5	14.0	7	507	5.1	590	12.0	15.7	0
464	4.7	535	11.3	15.0	3	400	8.6	517	7.0	12.0	8
15	0.6	163	2.5	2.5	14						

(a) Generate summary statistics, including the mean and standard deviation of each variable. Compute the coefficient of variation (see Section 3.2) for each variable. Relative to its mean, which variable has the largest spread of data values? Which variable has the least spread of data values relative to its mean?

(b) For each pair of variables, generate the correlation coefficient r. For all pairs involving x_1, compute the corresponding coefficient of determination r^2. Which variable has greatest influence on annual net sales? Which variable has the least influence on annual net sales?

(c) Perform a regression analysis with x_1 as the response variable. Use x_2, x_3, x_4, x_5, and x_6 as explanatory variables. Look at the coefficient of multiple determination. What percentage of the variation in x_1 can be explained by the corresponding variations in x_2, x_3, x_4, x_5, and x_6 taken together?

(d) Write out the regression equation. If two new competing stores moved into the sales district, but the other explanatory variables did not change, what do you expect for the corresponding change in annual net sales? Explain your answer. If you increase the local advertising by a thousand dollars, but the other explanatory variables did not change, what do you expect for the corresponding change in annual net sales? Explain.

(e) Test each coefficient to determine if it is or is not zero. Use level of significance 5%.

(f) Suppose you and your business associates rent a store, get a bank loan to start up your business, and do a little research about the size of your sales district and number of competing stores in the district. If $x_2 = 2.8$, $x_3 = 250$, $x_4 = 3.1$, $x_5 = 7.3$, and $x_6 = 2$, use the computer to forecast $x_1 = $ annual net sales and find an 80% confidence interval for your forecast (if your software produces prediction intervals).

(g) Construct a new regression model with x_4 as the response variable and x_1, x_2, x_3, x_5, and x_6 as explanatory variables. Suppose an All Greens store in Sonoma, California, wanted to estimate a range of advertising costs appropriate to its store. If it spends too little on advertising, it will not reach enough customers. However, it does not want to overspend on advertising for this type and size of store. At this store, $x_1 = 163$, $x_2 = 2.4$, $x_3 = 188$, $x_5 = 6.6$, and $x_6 = 10$. Use these data to predict x_4 (advertising cost) and find an 80% confidence interval for your prediction. At the 80% confidence level, what range of advertising costs do you think is appropriate for this store?

● **NOTE** In the Using Technology section at the end of this chapter, you will find a "mini case study" of seven important variables from the economy of the United States for the years 1976 to 1987. A reader interested in applications of multiple regression and the United States economy is referred to this material.

SUMMARY

We can often get information about a population of y values by attempting to predict them from known values of another population x. If a random sample of data pairs (x, y) are obtained from an overall population of such data pairs, we may use the equation of the least-squares line

$$y = a + bx$$

to predict y values from given x values. In this equation, a and b are computed from Equations (2) to (5) of Section 10.2. When we use the least-squares line to predict y values from x values, y is the *response variable* and x is the *explanatory variable*.

If data pairs (x, y) are gathered, the extent to which x and y values are related can be measured by the Pearson product–moment correlation coefficient r given in Equation (9) of Section 10.3. The quantity r^2 is called the *coefficient of determination*. We use this value to measure the proportion of variation in y explained by the least-squares line.

The significance of the correlation coefficient is tested using the null hypothesis

H_0: $\rho = 0$ (i.e., there is no correlation)

against an appropriate alternate hypothesis (either $\rho \neq 0$, $\rho > 0$, or $\rho < 0$).

Scatter diagrams of (x, y) values are useful to help determine visually if there is any linear relation between the x and y values and, if so, how strong the relation might be.

The slope of the least-squares line is especially important because it tells how many units y changes for every unit change in x. We test the slope β of the population least-squares line using the null hypothesis

H_0: $\beta = 0$ (y does not change when x changes)

against an appropriate alternate hypothesis (either $\beta \neq 0$, $\beta > 0$, or $\beta < 0$). In addition, confidence intervals for β are useful.

Techniques of multiple regression (with computer assistance) help us analyze the linear relation among several variables.

IMPORTANT WORDS & SYMBOLS

Section 10.1
Paired data values
Scatter diagram
No linear correlation
Perfect linear correlation

Section 10.2
Explanatory variable
Response variable
Least-squares criterion
Regression

Method of least squares
Interpolation and extrapolation
Least-squares line $y = a + bx$
Residual
Standard error of estimate S_e
Confidence interval for y

Section 10.3
Correlation coefficient r (later called sample correlation coefficient)
Linear correlation

Positive, negative, and zero correlation
Coefficient of determination

Section 10.4
Population correlation coefficient ρ
A significant sample correlation coefficient r
Population slope β

Section 10.5
Multiple regression
Coefficient of multiple determination

VIEWP●INT   *Living Arrangements*

Male, female, married, single, living alone, living with friends or relatives—all these categories are of interest to the U.S. Census Bureau. In addition to these categories, there are others, such as age, income, and health needs. How strongly correlated are these variables? Can we use one or more of these variables to predict the others? How good is such a prediction? Methods of this chapter can help you answer such questions. For more information regarding such data, visit the Brase/Brase statistics site at http://math.college.hmco.com/students and find the link to Census Bureau.

CHAPTER REVIEW PROBLEMS

When solving chapter problems involving the standard error of estimate, testing the correlation coefficient, or testing β or confidence intervals for β, make the assumption that x and y are normally distributed random variables. Answers may vary slightly due to rounding.

1. *Desert Ecology: Wildlife* Bighorn sheep are beautiful wild animals found throughout the western United States. Data for this problem are based on information taken from *The Desert Bighorn*, edited by Monson and Sumner (University of Arizona Press). Let x be the age of a bighorn sheep (in years), and let y be the mortality rate (percent that die) in this age group. For example, $x = 1$, $y = 14$ means that 14% of the bighorn sheep between 1 and 2 years old died. A random sample of Arizona bighorn sheep gave the following information:

x	1	2	3	4	5
y	14	18.9	14.4	19.6	20.0

(a) Draw a scatter diagram.
(b) Find the equation of the least-squares line.
(c) Find r. Find the coefficient of determination r^2. Explain what these measures mean in the context of the problem.
(d) Test the claim that the population correlation coefficient is positive at the 1% level of significance.

(e) Test the claim that the slope β of the population least-squares line is positive at the 1% level of significance. (Note that $S_e \approx 2.468$ and $SS_x = 10$.)

2. *Sociology: Job Changes* A sociologist is interested in the relation between $x = $ number of job changes and $y = $ annual salary (in thousands of dollars) for people living in the Nashville area. A random sample of 10 people employed in Nashville provided the following information:

x (Number of job changes)	4	7	5	6	1	5	9	10	10	3
y (Salary in $1000)	33	37	34	32	32	38	43	37	40	33

(a) Draw a scatter diagram for the data.
(b) Find $\bar{x}$, $\bar{y}$, and b and then the equation of the least-squares line.
(c) Graph the least-squares line on your scatter diagram.
(d) If someone had $x = 2$ job changes, what does the least-squares line predict for y, the annual salary?
(e) Verify that the standard error of estimate $S_e \approx 2.56$.
(f) Find a 90% confidence interval for the annual salary of an individual with two job changes.
(g) Looking at the scatter diagram and least-squares line, do you think the correlation coefficient will be positive, negative, or zero?
(h) Find r. Find the coefficient of determination.
(i) Test the claim that the population correlation coefficient is positive (use a 5% level of significance).
(j) Test the claim that the slope β of the population least-squares line is positive at the 5% level of significance. (Note that $SS_x \approx 82$.)

3. *Medical: Fat Babies* Modern medical practice tells us not to encourage babies to become too fat. Is there a positive correlation between the weight x of a 1-year-old baby and the weight y of a mature adult (30 years old)? A random sample of medical files produced the following information for 14 females:

x (lb)	21	25	23	24	20	15	25	21	17	24	26	22	18	19
y (lb)	125	125	120	125	130	120	145	130	130	130	130	140	110	115

(a) Draw a scatter diagram for the data.
(b) Find $\bar{x}$, $\bar{y}$, and b and the equation of the least-squares line.
(c) Graph the least-squares line on your scatter diagram.
(d) If a female baby weighs 20 lb at 1 year, what would you predict she would weigh at 30 years of age?
(e) Verify that the standard error of estimate $S_e \approx 8.38$.
(f) Find a 95% confidence interval for the weight at age 30 of a female who weighed 20 lb at 1 year of age.
(g) Looking at the scatter diagram, do you think the correlation coefficient is positive, negative, or zero?
(h) Find r. Find the coefficient of determination. Interpret the meaning of these measurements in the context of the problem.
(i) Test the claim that the population correlation coefficient is positive (use a 1% level of significance).
(j) Test the claim that the slope β of the population least-squares line is positive at the 1% level of significance. (Note that $SS_x \approx 143.4$.)

4. *Sales: Insurance* Dorothy Kelly sells life insurance for the Prudence Insurance Company. She sells insurance by making visits to her clients' homes. Dorothy believes that the number of sales should depend, to some degree, on the number of visits made. For the past several years, she kept careful records of the number of visits (x) she made each week and the number of people (y) who bought insurance that week. For a random sample of 15 such weeks, the x and y values follow:

x	11	19	16	13	28	5	20	14	22	7	15	29	8	25	16
y	3	11	8	5	8	2	5	6	8	3	5	10	6	10	7

(a) Draw a scatter diagram for the data.
(b) Find $\bar{x}$, $\bar{y}$, and b and the equation of the least-squares line.
(c) Graph the least-squares line on your scatter diagram.
(d) On a week in which Dorothy made 18 visits, how many people would you predict would buy insurance from her?
(e) Verify that the standard error of estimate $S_e \approx 1.73$.
(f) Find a 90% confidence interval for the number of sales Dorothy would make in a week in which she made 18 visits.
(g) Find the correlation coefficient and the coefficient of determination.
(h) Test the claim that the population correlation coefficient is positive (use a 1% level of significance).
(i) Test the claim that the slope β of the population least-squares line is positive at the 1% level of significance. (Note that $SS_{\hat{x}} \approx 755.7$.)

5. *Marketing: Coupons* Each box of Healthy Crunch breakfast cereal contains a coupon entitling you to a free package of garden seeds. At the Healthy Crunch home office, they use the weight of incoming mail to determine how many of their employees are to be assigned to collecting coupons and mailing out seed packages on a given day. (Healthy Crunch has a policy of answering all its mail the day it is received.)

Let x = weight of incoming mail and y = number of employees required to process the mail in one working day. A random sample of 8 days gave the following data:

x (lb)	11	20	16	6	12	18	23	25
y (Number of employees)	6	10	9	5	8	14	13	16

(a) Draw a scatter diagram for the data.
(b) Find $\bar{x}$, $\bar{y}$, and b and the equation of the least-squares line.
(c) Graph the least-squares line on your scatter diagram.
(d) If Healthy Crunch receives 15 lb of mail, how many employees should be assigned mail duty?
(e) Verify that the standard error of estimate $S_e \approx 1.73$.
(f) Find a 95% confidence interval for the number of employees required to process mail for 15 lb of mail.
(g) Find r. Find the coefficient of determination. Interpret the meaning of these measurements in the context of the problem.
(h) Test the claim that the population correlation coefficient is positive (use a 1% level of significance).

   (i) Test the claim that the slope β of the population least-squares line is positive at the 1% level of significance. (Note that $SS_x \approx 289.9$.)

   (j) Find an 80% confidence interval for β, and interpret the meaning in the context of this problem.

6. *Focus Problem: Changing Population and Crime Rate* Solve the Focus Problem found at the beginning of this chapter.

DATA HIGHLIGHTS: GROUP PROJECTS

Break into small groups and discuss the following topics. Organize a brief outline in which you summarize the main points of your group discussion.

  Scatter diagrams! Are they really useful? Scatter diagrams give a first impression of the data relationship and help us assess whether a linear relation provides a reasonable model for the data. In addition, we can spot *influential points*. A data point with an extreme x value can heavily influence the position of the least-squares line. In this project, we look at data sets with an influential point.

x	1	4	5	9	10	15
y	3	7	6	10	12	4

(a) Compute r and b, the slope of the least-squares line. Find the equation of the least-squares line, and sketch the line on the scatter diagram.

(b) Notice the point boxed in blue in Figure 10-26. Does it seem to lie away from the linear pattern determined by the other points? The coordinates of that point are (15, 4). Is it an influential point? Remove that point from the model and recompute r, b, and the equation of the least-squares line. Sketch this least-squares line on the diagram. How does the removal of the influential point affect the values of r, b, and the position of the least-squares line?

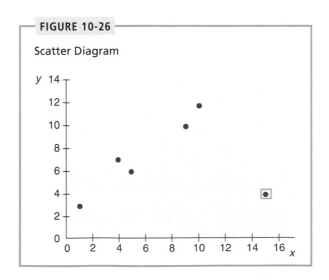

FIGURE 10-26

Scatter Diagram

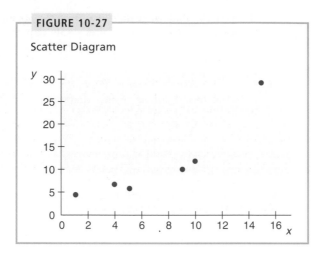

FIGURE 10-27

Scatter Diagram

(c) Consider the scatter diagram of Figure 10-27. Is there an influential point? If you remove the influential point, will the slope of the new least-squares line be larger or smaller than the slope of the line from the original data? Will the correlation coefficient be larger or smaller?

LINKING CONCEPTS: WRITING PROJECTS

Discuss each of the following topics in class or review the topics on your own. Then write a brief but complete essay in which you summarize the main points. Please include formulas and graphs as appropriate.

1. What do we mean when we say two variables have a strong positive (or negative) linear correlation? What would a scatter diagram for these variables look like? Is it possible that two variables could be strongly related somehow, but have a low *linear* correlation? Explain and draw a scatter diagram to demonstrate your point.

2. What do we mean by the least-squares criterion? Give a very general description of how the least-squares criterion is involved in the construction of the least-squares line. Why do we say the least-squares line is the "best-fitting" line for the data set?

3. In this chapter, we discussed three measures for "goodness of fit" of the least-squares line to given data. These measures were standard error of estimate, correlation coefficient, and coefficient of determination. Discuss the ways these measurements are different and the ways they are similar to each other. Be sure to include a discussion of explained variation, unexplained variation, and total variation in your answer. Draw a sketch and include appropriate formulas.

4. Look at the formula for confidence bounds for least-squares predictions. Which of the following conditions do you think will result in a *shorter* confidence interval for a prediction.
   (a) Larger or smaller values for the standard error of estimate
   (b) Larger or smaller number of data pairs
   (c) A value of x near $\bar{x}$ or a value of x far away from $\bar{x}$

Why would a shorter confidence interval for a prediction be more desirable than a longer interval?

5. If you did not cover Section 10.5, Multiple Regression, then omit this problem.

   For many applications in statistics, more data lead to more accurate results. In multiple regression, we have more variables (and data) than we have in most simple regression problems. Why will this usually lead to more accurate predictions? Will additional variables *always* lead to more accurate predictions? Explain your answer. Discuss the coefficient of multiple determination and its meaning in the context of multiple regression. How do we know if an explanatory variable has a statistically significant influence on the response variable? What do we mean by a regression model?

6. Use the Internet or go to the library and find a magazine or journal article in your field of major interest where the content of this chapter could be applied. List the variables used, method of data collection, and the general type of information and conclusions drawn.

Using TECHNOLOGY

Simple Linear Regression (one explanatory variable)

APPLICATION 1

The data in this section are taken from this reference:

King, Cuchlaine A. M. *Physical Geography.* Oxford: Basil Blackwell, 1980, 77–86, 196–206. Reprinted with permission of Basil Blackwell Limited, Oxford, England.

Throughout the world, natural ocean beaches are beautiful sights to see. If you have visited natural beaches, you may have noticed that when the gradient or dropoff is steep, the grains of sand tend to be larger. In fact, a manmade beach with the "wrong" size granules of sand tends to be washed away and eventually replaced when the proper size grain is selected by the action of the ocean and the gradient of the bottom. Since manmade beaches are expensive, grain size is an important consideration.

In the data that follow, x = median diameter (in millimeters) of granules of sand, and y = gradient of beach slope in degrees on natural ocean beaches.

x	y
0.17	0.63
0.19	0.70
0.22	0.82
0.235	0.88
0.235	1.15
0.30	1.50
0.35	4.40
0.42	7.30
0.85	11.30

1. Find the sample mean and standard deviation for x and for y.

2. Make a scatter plot. Would you expect a moderately high correlation and good fit for the least-squares line?

3. Find the equation of the least-squares line, and graph the line on the scatter plot.

4. Find the correlation coefficient r and the coefficient of determination r^2. Is r significant at the 1% level of significance (two-tailed test)?

5. Test that $\beta > 0$ at the 1% level of significance. Find the standard error of estimate S_e and form an 80% confidence interval for β. As the diameter of granules of sand changes by 0.10 mm, how much does the gradient of beach slope change?

6. Suppose that you have a truckload of sifted sand in which the median size of granules is 0.38 mm. If you want to put this sand on a beach and you don't want the sand to wash away, then what does the least-squares line predict for the angle of the beach? *Note:* Heavy storms that produce abnormal waves may also wash out the sand. However, in the long run, the size of sand granules that remain on the beach or that are brought back to the beach by long-term wave action are determined to a large extent by the angle at which the beach drops off. What range of angles should the beach have if we want to be 90% confident that we are matching the size of our sand granules (0.38 mm) to the proper angle of the beach?

7. Suppose we now have a truckload of sifted sand in which the median size of the granules is 0.45 mm. Repeat Problem 6.

Technology Hints (Simple Regression)

TI-83Plus

Be sure to set **DiagnosticOn** (under **Catalogue**).

(a) Scatter diagram: Use **STAT PLOT**, select the first type, use **ZOOM** option **9:ZoomStat**.

(b) Least-squares line and r: Use **STAT, CALC**, option **8:LinReg(a + bx)**.

(c) Graph least-squares line and predict: Press **Y=**. Then under **VARS**, select **5:Statistics**, then select **EQ**, and finally select item **1:RegEQ**. Press enter. This sequence of steps will automatically set Y_1 = your regression equation. Press **GRAPH**. To find a predicted value, when the graph is showing, press the **CALC** key and select item **1:Value**. Enter the x value and the corresponding y value will appear.

(d) Testing ρ and β, value for S_e: Use **STAT, TEST**, option **E:LinRegTTest**. The value of S_e is in the display as s.

(e) Confidence intervals for β or predictions: Use formulas from Section 10.4.

Excel

(a) Scatter plot, least-squares line, r^2: Use **Chart wizard**. Select **scatter plot**. Once plot is displayed, *right* click on any data point. Select **trend line**. Under options, check display line and display r^2.

(b) Prediction: Use paste function $\boxed{f_x}$ ➤ **Statistical** ➤ **Forecast**.

(c) Coefficient r: Use $\boxed{f_x}$ ➤ **Statistical** ➤ **Correl**.

(d) Testing β and confidence intervals for β: Use menu selection **Tools** ➤ **Data Analysis** ➤ **Regression**.

(e) Confidence interval for prediction: Use formulas from Section 10.4.

Minitab

(a) Scatter plot, least-squares line, r^2, S_e: Use menu selection **Stat** ➤ **Regression** ➤ **Fitted line plot**. The value for S_e is displayed as the value of s.

(b) Coefficient r: Use menu selection **Stat** ➤ **Basic Statistics** ➤ **Correlation**.

(c) Testing β, predictions, confidence interval for predictions: Use menu selection **Stat** ➤ **Regression** ➤ **Regression**.

(d) Confidence interval for β: Use formulas from Section 10.4.

ComputerStat

(a) Scatter plot, least-squares line, r, r^2, S_e, prediction, confidence interval for prediction, testing ρ: Use the menu item **Linear Regression and Correlation**.

(b) Confidence intervals for β: Use formulas from Section 10.4.

Multiple Regression

APPLICATION 2

Data values in the following study are taken from *Statistical Abstract of the United States*, U.S. Department of Commerce, 103rd and 109th Editions (see Table 10-18).

All data values represent annual averages as determined by the U.S. Department of Commerce.

1. Construct a regression model with

   Response variable: x_3 (foreign investments)

   Explanatory variables: x_5 (GNP), x_6 (U.S. dollar), and x_7 (consumer credit)

631

TABLE 10-18 Economic Data 1976–1987 (on the data disk)

Year	x_1	x_2	x_3	x_4	x_5	x_6	x_7
1976	10.9	7.61	31	974.9	1,718	1.757	234.4
1977	12.0	7.42	35	894.6	1,918	1.649	263.8
1978	12.5	8.41	42	820.2	2,164	1.532	308.3
1979	17.7	9.44	54	844.4	2,418	1.380	347.5
1980	28.1	11.46	83	891.4	2,732	1.215	349.4
1981	35.6	13.91	109	932.9	3,053	1.098	366.6
1982	31.8	13.00	125	884.4	3,166	1.035	381.1
1983	29.0	11.11	137	1,190.3	3,406	1.000	430.4
1984	28.6	12.44	165	1,178.5	3,772	0.961	511.8
1985	26.8	10.62	185	1,328.2	4,015	0.928	592.4
1986	14.6	7.68	209	1,792.8	4,240	0.913	646.1
1987	17.9	8.38	244	2,276.0	4,527	0.880	685.5

We will use the following notation:

x_1 = price of a barrel of crude oil in dollars per barrel

x_2 = percent interest on 10-year U.S. Treasury notes

x_3 = total foreign investments in U.S. in billions of dollars

x_4 = Dow Jones Industrial Average (DJIA)

x_5 = Gross National Product, GNP, in billions of dollars

x_6 = purchasing power of U.S. dollar with base 1983 corresponding to $1.000

x_7 = consumer credit (i.e., consumer debt) in billions of dollars

What is the coefficient of multiple determination?

(a) Use level of significance 1% and test each coefficient for significance (two-tailed test).

(b) Examine the coefficients of the regression equation. Then explain why you think the following statement is true or false: If the purchasing power of the U.S. dollar did not change and the GNP did not change, then an increase in consumer credit would likely be accompanied by a reduction in foreign investments.

(c) Suppose $x_3 = 3500$, $x_6 = 0.975$, and $x_7 = 450$. Predict the level of foreign investment. Find a 90% confidence interval for your prediction.

2. Construct a new regression model with

Response variable: x_4 (DJIA)

Explanatory variables: x_3 (foreign investments), x_5 (GNP), and x_7 (consumer credit)

What is the coefficient of multiple determination?

(a) Use level of significance 5% and test each coefficient for significance (two-tailed test).

(b) Examine the coefficients of the regression equation; then explain why you think the following statement is true or false: If the GNP and consumer credit didn't change but foreign investments increased, the DJIA should show a strong increase.

(c) Suppose $x_3 = 210$, $x_5 = 4260$, and $x_7 = 650$. Predict the DJIA and find an 85% confidence interval for your prediction.

3. Construct a new regression model with

Response variable: x_7 (consumer credit)

Explanatory variables: x_3 (foreign investments), x_5 (GNP), and x_6 (U.S. dollar)

What is the coefficient of multiple determination?

(a) Use level of significance 1% and test each coefficient for significance (two-tailed test).

(b) Examine the coefficients of the regression equation; then explain why you think each of the following statements is true or false: If both GNP and purchasing power of the U.S. dollar didn't change, then an increase in foreign investments would likely be accompanied by a reduction in consumer credit. If both foreign investments and purchasing power of the U.S. dollar remained fixed, then an increase in GNP would likely be accompanied by an increase in consumer credit.

(c) Suppose $x_3 = 88$, $x_5 = 2750$, and $x_6 = 1.250$. Predict consumer credit, and find an 80% confidence interval for your prediction.

Technology Hints (Multiple Regression)

TI-83Plus

Does not support multiple regression.

Excel

Use menu selection **Tools ➤ Data Analysis ➤ Regression**. On the spreadsheet, the columns containing the explanatory variables need to be adjacent.

Minitab

Use menu selection Stat ➤ Regression ➤ Regression.

ComputerStat

Use menu selection **Linear regression and correlation**.

11

Chi-Square and F Distributions

"So what!"

—Anonymous

"Girl with Black Eye" by Norman Rockwell (1894–1978)

Norman Rockwell painted everyday people and situations. In this cover for the *Saturday Evening Post* (May 23, 1953), a young lady is about to have a conference with her school principal. So what!

We have all heard the exclamation, "So what!" Philologists (people who study cultural linguistics) tell us that this expression is a shortened version of "So what is the difference!" They also tell us that there are similar popular or slang expressions about differences in all languages and cultures. It is human nature to challenge the claim that something is better, worse, or just simply different. In this chapter, we will focus on this very human theme by studying a variety of topics regarding questions of whether or not differences exist between two population variances or among several population means.

PART I: Inferences using the chi-square distribution

11.1 Chi-Square: Tests of Independence

11.2 Chi-Square: Goodness of Fit

11.3 Testing and Estimating a Single Variance or Standard Deviation

PART II: Inferences using the *F* distribution

11.4 Testing Two Variances

11.5 One-Way ANOVA: Comparing Several Sample Means

11.6 Introduction to Two-Way ANOVA

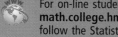

For on-line student resources, visit **math.college.hmco.com/students** and follow the Statistics links to the Brase/Brase, *Understandable Statistics,* 7th web site.

◇ How do we decide if random variables are dependent or independent?   (SECTION 11.1)

◇ How do we decide if two distributions are not only dependent, but actually the same distribution? (SECTION 11.2)

◇ How do we compute confidence intervals and tests for σ? (SECTION 11.3)

◇ How do we test two variances σ_1^2 and σ_2^2?   (SECTION 11.4)

◇ What is one-way ANOVA? Where is it used?   (SECTION 11.5)

◇ What about two-way ANOVA? Where is it used? (SECTION 11.6)

Mesa Verde National Park

Archaeological excavation site

FOCUS PROBLEM

Stone Age Tools and Archaeology

Archaeologists at Washington State University did an extensive summer excavation at Burnt Mesa Pueblo in Bandelier National Monument. Their work is published in the book *Bandelier Archaeological Excavation Project: Summer 1990 Excavations at Burnt Mesa Pueblo and Casa del Rito*, edited by T. A. Kohler.

One question the archaeologists asked was: Is raw material used by prehistoric Indians for stone tool manufacture independent of the archaeological excavation site? Two different excavation sites at Burnt Mesa Pueblo gave the information in the following table.

Use a chi-square test with 1% level of significance to test the claim that raw material used for construction of stone tools and excavation site are independent.

Stone Tool Construction Material, Burnt Mesa Pueblo

Material	Site A	Site B	Row Total
Basalt	3,657	1,238	4,895
Obsidian	497	68	565
Pedernal chert	3,606	232	3,838
Other	357	36	393
Column total	8,117	1,574	9,691

PART I: INFERENCES USING THE CHI-SQUARE DISTRIBUTION

Overview of the Chi-Square Distribution

So far, we have used several probability distributions for hypothesis testing and confidence intervals, with the most frequently used being the normal distribution and the Student's t distribution. In this chapter, we will use two other probability distributions, namely, the chi-square (where *chi* is pronounced like the first two letters in the word *kite*) and the F distribution. In Part I, we will see applications of the chi-square distribution, whereas in Part II, we will see some important applications of the F distribution.

Chi is a Greek letter denoted by the symbol χ, so chi-square is denoted by the symbol χ^2. Because the distribution is of chi-*square* values, the χ^2 values begin at 0 and then are all positive. The graph of the χ^2 distribution is not symmetrical, and like the Student's t distribution, it depends on the number of degrees of freedom. Figure 11-1 shows the χ^2 distribution for several degrees of freedom (*d.f.*).

As the degrees of freedom increase, the graph of the chi-square distribution becomes more bell-like and begins to look more and more symmetric. Notice that the mode, or high point, of the graph with n degrees of freedom occurs over $n - 2$ (for $n \geq 3$).

We use Table 7 of Appendix II to find critical values of chi-square distributions for which a designated area α falls to the right of the critical value. Table 11-1 gives an excerpt from Table 7. Notice that the row headers are degrees of freedom, and the column headers are areas α in the *right tail* of the distribution. For instance, according to the table, for a χ^2 distribution with 3 degrees of freedom, $\alpha = 0.995$ is the area falling to the right of $\chi^2 = 0.072$. For a χ^2 distribution with 4 degrees of freedom, $\alpha = 0.010$ is the area falling to the right of $\chi^2 = 13.28$.

In the next three sections, we will see how to apply the chi-square distribution to different applications.

FIGURE 11-1

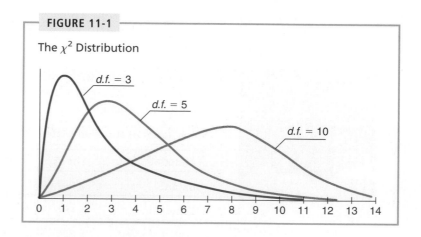

The χ^2 Distribution

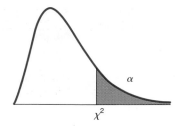

TABLE 11-1   Excerpt from Table 7 (Appendix II): The χ^2 Distribution

$d.f. \setminus \alpha$	0.995	0.990	0.975	...	0.010	0.005
⋮	⋮	⋮	⋮		⋮	⋮
3	0.072	0.115	0.216		11.34	12.84
4	0.207	0.297	0.484		13.28	14.86

11.1
Chi-Square: Tests of Independence

FOCUS POINTS

✓ Set up a test to investigate independence of random variables.

✓ Use contingency tables to compute the sample χ^2 statistic.

✓ Complete the test and estimate the P value.

Innovative Machines Incorporated has developed two new letter arrangements for computer keyboards. The company wishes to see if there is any relationship between the arrangement of letters on the keyboard and the number of hours it takes a new typing student to learn to type at 20 words per minute. Or, from another point of view, is the time it takes a student to learn to type *independent* of the arrangement of the letters on a keyboard?

To answer questions of this type, we test the hypotheses

H_0: Keyboard arrangement and learning times *are independent.*

H_1: Keyboard arrangement and learning times *are not independent.*

Chi-square distribution

In problems of this sort, we are testing the *independence* of two factors. The probability distribution we use to make the decision is the *chi-square distribution.* Recall from the overview of the chi-square distribution that *chi* is pronounced like the first two letters of the word *kite* and is a Greek letter denoted by the symbol χ, so chi-square is denoted by χ^2.

The first task for Innovative Machines is to gather data. Suppose that the company took a random sample of 300 beginning typing students and randomly assigned them to learn to type on one of three keyboards. The learning times for this sample are shown in Table 11-2.

Contingency table

Table 11-2 is called a *contingency table.* The *shaded boxes* that contain observed frequencies are called *cells.* The row and column totals are not considered to be cells. This contingency table is of size 3×3 (read, "three-by-three") because there are three rows of cells and three columns. When giving the size of a contingency table, we always list the number of *rows first.*

TABLE 11-2   Keyboard versus Time to Learn to Type at 20 wpm

Keyboard	21–40 h	41–60 h	61–80 h	Row Total
A	#1 25	#2 30	#3 25	80
B	#4 30	#5 71	#6 19	120
Standard	#7 35	#8 49	#9 16	100
Column total	90	150	60	300 Sample size

GUIDED EXERCISE 1

Size of contingency table

Give the size of the contingency tables in Figures 11-2 and 11-3. Also, count the number of cells in each table. (Remember, each cell is a shaded box.)

(a) **FIGURE 11-2** Contingency Table

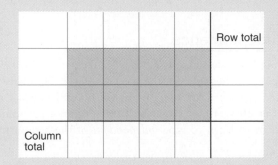

⟹ There are two rows and four columns, so this is a 2×4 table. There are eight cells.

(b) **FIGURE 11-3** Contingency Table

⟹ Here we have three rows and two columns, so this is a 3×2 table with six cells.

Expected frequency

We are testing the null hypothesis that the keyboard arrangement and the time it takes a student to learn to type are *independent*. We use this hypothesis to determine the *expected frequency* of each cell.

For instance, to compute the expected frequency of cell 1 in Table 11-2, we observe that cell 1 consists of all the students in the sample who learned to type on keyboard A and who mastered the skill at the 20-words-per-minute level in 21 to 40 hours. By the assumption (null hypothesis) that the two events are independent, we use the multiplication law to obtain the probability that a student is in cell 1.

$$P(\text{cell 1}) = P(\text{keyboard A } and \text{ skill in 21–40 h})$$
$$= P(\text{keyboard A}) \cdot P(\text{skill in 21–40 h})$$

Because there are 300 students in the sample and 80 used keyboard A,

$$P(\text{keyboard A}) = \frac{80}{300}$$

Also, 90 of the 300 students learned to type in 21–40 hours, so

$$P(\text{skill in 21–40 h}) = \frac{90}{300}$$

Using these two probabilities and the assumption of independence,

$$P(\text{keyboard A } and \text{ skill in 21–40 h}) = \frac{80}{300} \cdot \frac{90}{300}$$

Finally, because there are 300 students in the sample, we have the *expected frequency E* for cell 1.

$$E = P(\text{student in cell 1}) \cdot (\text{no. of students in sample})$$

$$= \frac{80}{300} \cdot \frac{90}{300} \cdot 300 = \frac{80 \cdot 90}{300} = 24$$

We can repeat this process for each cell. However, the last step yields an easier formula for the expected frequency E.

Formula for expected frequency *E*

$$E = \frac{(\text{Row total})(\text{Column total})}{\text{Sample size}}$$

Note: If the expected value is not a whole number, do *not* round it to the nearest whole number.

Let's use this formula in Example 1 to find the expected frequency for cell 2.

EXAMPLE 1

Expected frequency

Find the expected frequency for cell 2 of contingency Table 11-2.

SOLUTION: Cell 2 is in row 1 and column 2. The *row total* is 80, and the *column total* is 150. The size of the sample is still 300.

$$E = \frac{(\text{Row total})(\text{Column total})}{\text{Sample size}}$$

$$= \frac{(80)(150)}{300} = 40 \qquad \Diamond$$

GUIDED EXERCISE 2

Expected frequency

Table 11-3 contains the *observed frequencies O* and *expected frequencies E* for the contingency table giving keyboard arrangement and number of hours it takes a student to learn to type at 20 words per minute. Fill in the missing expected frequencies.

Continued

GUIDED EXERCISE 2 continued

TABLE 11-3 Complete Contingency Table of Keyboard Arrangement and Time to Learn to Type

Keyboard	21–40 h	41–60 h	61–80 h	Row Total
A	#1 $O = 25$ $E = 24$	#2 $O = 30$ $E = 40$	#3 $O = 25$ $E = \underline{\quad}$	80
B	#4 $O = 30$ $E = 36$	#5 $O = 71$ $E = \underline{\quad}$	#6 $O = 19$ $E = \underline{\quad}$	120
Standard	#7 $O = 35$ $E = \underline{\quad}$	#8 $O = 49$ $E = 50$	#9 $O = 16$ $E = 20$	100
Column Total	90	150	60	300 Sample Size

For cell 3, we have

$$E = \frac{(80)(60)}{300} = 16$$

For cell 5, we have

$$E = \frac{(120)(150)}{300} = 60$$

For cell 6, we have

$$E = \frac{(120)(60)}{300} = 24$$

For cell 7, we have

$$E = \frac{(100)(90)}{300} = 30$$

Computing the sample test statistic χ^2

Now we are ready to compute the sample statistic χ^2 for the typing students. The χ^2 value is a measure of the sum of differences between observed frequency O and expected frequency E in each cell. These differences are listed in Table 11-4.

As we see, if we sum the differences between the observed frequencies and the expected frequencies of the cells, we get the value zero. This total certainly does not reflect the fact that there were differences between the observed and expected frequencies. To obtain a measure whose sum does reflect the magnitude of the differences, we square the differences and work with the quantities $(O - E)^2$. But instead of using the terms $(O - E)^2$, we use the values $(O - E)^2/E$. The reason we use this expression is that a small difference between the observed and expected frequency is not nearly as important if the expected frequency is large as it is if the

TABLE 11-4 Difference Between the Observed and Expected Frequencies

Cell	Observed O	Expected E	Difference $(O - E)$
1	25	24	1
2	30	40	−10
3	25	16	9
4	30	36	−6
5	71	60	11
6	19	24	−5
7	35	30	5
8	49	50	−1
9	16	20	−4
			$\Sigma(O - E) = 0$

expected frequency is small. For instance, for both cells 1 and 8, the squared difference $(O - E)^2$ is 1. However, this difference is more meaningful in cell 1, where the expected frequency is 24, than it is in cell 8, where the expected frequency is 50. When we divide the quantity $(O - E)^2$ by E, we take the size of the difference with respect to the size of the expected value. We use the sum of these values to form the sample statistic χ^2:

$$\chi^2 = \Sigma \frac{(O - E)^2}{E}$$

where the sum is over all cells in the contingency table.

◇ **COMMENT** If you look up the word *irony* in a dictionary, you will find one of its meanings is described as "the difference between actual (or observed) results and expected results." Because irony is so prevalent in much of our human experience, it is not surprising that statisticians have incorporated a related chi-square distribution into their work.   ◇

GUIDED EXERCISE 3

Sample χ^2

(a) Complete Table 11-5.

⇨ The last two rows of Table 11-5 are

TABLE 11-5   Data of Table 11-4

Cell	O	E	O − E	$(O - E)^2$	$(O - E)^2/E$
1	25	24	1	1	0.04
2	30	40	−10	100	2.50
3	25	16	9	81	5.06
4	30	36	−6	36	1.00
5	71	60	11	121	2.02
6	19	24	−5	25	1.04
7	35	30	5	25	0.83
8	49	50	___	___	___
9	16	20	___	___	___

$$\Sigma \frac{(O - E)^2}{E} = \underline{\hspace{2cm}}$$

Cell	O	E	O − E	$(O - E)^2$	$(O - E)^2/E$
8	49	50	−1	1	0.02
9	16	20	−4	16	0.80

$$\Sigma \frac{(O - E)^2}{E} = \text{total of last column} = 13.31$$

(b) Compute the statistic χ^2 for this sample.

⇨ Since $\chi^2 = \Sigma \dfrac{(O - E)^2}{E}$, then $\chi^2 = 13.31$.

Notice that when the observed frequency and the expected frequency are very close, the quantity $(O - E)^2$ is close to zero, and so the statistic χ^2 is near zero. As the difference increases, the statistic χ^2 also increases. To determine how large the statistic can be before we must reject the null hypothesis of independence, we find a *critical value* χ_α^2 in Table 7 of Appendix II for the specified level of significance α and the number of degrees of freedom in the sample.

Critical value χ_α^2

As we saw in the chi-square overview at the beginning of this chapter, the chi-square distribution changes as the degrees of freedom change. To find a critical value χ_α^2, we need to know the degrees of freedom in the sample as well as the level of significance designated. To test independence, the degrees of freedom $d.f.$ of a sample are determined by the following formula:

Degrees of freedom for test of independence

Degrees of freedom = (Number of rows − 1) · (Number of columns − 1)

or $\qquad\qquad d.f. = (R - 1)(C - 1)$

where R = number of cell rows

$\qquad C$ = number of cell columns

GUIDED EXERCISE 4

Degrees of freedom

Determine the number of degrees of freedom in the example of keyboard arrangements (see Table 11-2). Recall that the contingency table had three rows and three columns.

$\Rightarrow \quad d.f. = (R - 1)(C - 1)$

$\qquad\quad = (3 - 1)(3 - 1) = (2)(2) = 4$

To test the hypothesis that the letter arrangement on a keyboard and the time it takes to learn to type at 20 words per minute are independent at the $\alpha = 0.05$ level of significance, we look up the critical value $\chi_{0.05}^2$ in Table 7 of Appendix II. For $d.f. = 4$ and $\alpha = 0.05$, we see $\chi_{0.05}^2 = 9.49$. When we compare the sample statistic $\chi^2 = 13.31$ with the critical value $\chi_{0.05}^2$, we see that the sample statistic is larger. Because it is larger, we *reject* the null hypothesis of independence and conclude that keyboard arrangement and learning time are *not* independent (Figure 11-4).

◇ **COMMENT** For tests of independence, we always use a *right-tailed* test on the chi-square distribution. This is because we are testing to see if the χ^2 measure of the difference between the observed and expected frequencies is too large to be due to chance alone. ◇

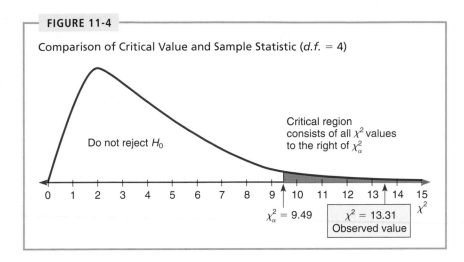

FIGURE 11-4

Comparison of Critical Value and Sample Statistic ($d.f. = 4$)

Do not reject H_0

Critical region consists of all x^2 values to the right of χ^2_α

$\chi^2_\alpha = 9.49$

$\chi^2 = 13.31$
Observed value

Summary

Let's summarize how we use the chi-square distribution to test the independence of two variables.

STEP 1: Set up the hypotheses:

H_0: The variables *are* independent.

H_1: The variables *are not* independent.

STEP 2: Compute the expected frequency for each cell in the contingency table by use of the formula

$$E = \text{expected frequency} = \frac{(\text{Row total})(\text{Column total})}{\text{Sample size}}$$

STEP 3: Compute the statistic χ^2 for the sample:

$$\chi^2 = \Sigma \frac{(O - E)^2}{E}$$

where O is the observed frequency, E is the expected frequency, and the sum Σ is over all cells.

STEP 4: Find the critical value χ^2_α in Table 7 of Appendix II. Use the level of significance α and the number of degrees of freedom $d.f.$ to find the critical value.

$$d.f. = (R - 1)(C - 1)$$

where R is the number of rows, and C is the number of columns of cells in the contingency table. The critical region consists of all values of χ^2 to the *right* of the critical value χ^2_α.

STEP 5: Compare the sample statistic χ^2 of Step 3 with the critical value χ^2_α of Step 4. If the sample statistic is *larger,* reject the null hypothesis of independence. Otherwise, do not reject the null hypothesis.

◊ **NOTE** We compare the sample statistic χ^2 with the critical value χ^2_α. But the distribution of sample statistics is only approximately the same as the theoretical distribution whose critical values χ^2_α are found in Table 7 of Appendix II. To safely use critical values χ^2_α, we must be sure that all the cells have an *expected frequency* larger than or equal to 5. If this condition is not met, the sample size should be increased.   ◊

GUIDED EXERCISE 5

Testing independence

Super Vending Machines Company is to install soda pop machines in elementary schools and high schools. The market analysts wish to know if flavor preference and school level are independent. A random sample of 200 students was taken. Their school level and soda pop preference are given in Table 11-6. Is independence indicated at the $\alpha = 0.01$ level of significance?

Step 1: State the null and alternate hypotheses.

⟹ H_0: School level and soda pop preference are independent.

H_1: School level and soda pop preference are not independent.

Step 2: Complete the contingency Table 11-6 by filling in the required expected frequencies.

⟹ The expected frequency

for cell 5 is $\dfrac{(40)(80)}{200} = 16$

for cell 6 is $\dfrac{(40)(120)}{200} = 24$

for cell 7 is $\dfrac{(20)(80)}{200} = 8$

for cell 8 is $\dfrac{(20)(120)}{200} = 12$

TABLE 11-6   School Level and Soda Pop Preference

Soda Pop	High School	Elementary School	Row Total
Kula Kola	$O = 33$ #1 $E = 36$	$O = 57$ #2 $E = 54$	90
Mountain Mist	$O = 30$ #3 $E = 20$	$O = 20$ #4 $E = 30$	50
Jungle Grape	$O = 5$ #5 $E = $ ___	$O = 35$ #6 $E = $ ___	40
Diet Pop	$O = 12$ #7 $E = $ ___	$O = 8$ #8 $E = $ ___	20
Column Total	80	120	200 Sample Size

Note: In this example, the expected frequencies are all whole numbers. If the expected frequency has a decimal part such as 8.45, do *not* round the value to the nearest whole number; rather, give the expected frequency as the decimal number.

Continued

GUIDED EXERCISE 5 continued

Step 3: Fill in Table 11-7 and use the table to find the sample statistic χ^2.

The last three rows of Table 11-7 read as shown here:

TABLE 11-7   Computational Table for χ^2

Cell	O	E	$O - E$	$(O - E)^2$	$(O - E)^2/E$
1	33	36	−3	9	0.25
2	57	54	3	9	0.17
3	30	20	10	100	5.00
4	20	30	−10	100	3.33
5	5	16	−11	121	7.56
6	35	24	11	_____	_____
7	12	8	_____	_____	_____
8	8	12	_____	_____	_____

Cell	O	E	$O - E$	$(O - E)^2$	$(O - E)^2/E$
6	35	24	11	121	5.04
7	12	8	4	16	2.00
8	8	12	−4	16	1.33

χ^2 = total of last column

$$= \Sigma \frac{(O - E)^2}{E} = 24.68$$

Step 4: What is the size of the contingency table? Use the number of rows and columns to determine the number of degrees of freedom. For $\alpha = 0.01$, use Table 7 of Appendix II to find the critical value $\chi^2_{0.01}$.

The contingency table is of size 4×2. Since there are four rows and two columns,

$$d.f. = (4 - 1)(2 - 1) = 3$$

For $\alpha = 0.01$, the critical value χ^2_α is 11.34.

Step 5: Do we reject or fail to reject the null hypothesis that school level and soda pop flavor preference are independent?

Because the statistic χ^2 is larger than the critical value χ^2_α, we reject the null hypothesis of independence and conclude that school level and soda pop preference are dependent.

P values

◇ **COMMENT** *P* values for tests of independence can be estimated from Table 7 of Appendix II. For instance, to estimate the *P* value for the sample statistic $\chi^2 = 24.68$ found in Guided Exercise 5, we look in the row headed by the degrees of freedom for the contingency table, *d.f.* = 3. We see that the sample statistic $\chi^2 = 24.68$ falls to the right of the row entry 12.84. This means that the *P* value is smaller than the column header 0.005 corresponding to the entry 12.84. Therefore,

P value < 0.005

Consequently, we reject H_0 for all $\alpha \geq 0.005$. In particular, we reject H_0 for $\alpha = 0.01$. This result is consistent with the conclusion stated in Guided Exercise 5. ◇

TECH NOTE The TI-83Plus, Excel, and Minitab all support chi-square tests of independence. In each case, the observed data are entered in the format of the contingency table.

TI-83Plus Enter the observed data into a matrix. Set the dimension of matrix **[B]** to match that of the matrix of observed values. Expected values will be placed in matrix **[B]**. Press **STAT, TESTS,** and option **C:χ^2-Test.** The output gives the sample χ^2 with the *P* value.

Excel Enter the table of observed values. Use formulas of the section to compute expected values. Enter the corresponding table of observed values. Finally, use **paste function** $\boxed{f_x}$ ➤ **Statistical** ➤ **Chitest.** Excel returns the *P* value of the sample χ^2.

Minitab Enter the contingency table of expected values. Use menu selection **Stat** ➤ **Tables** ➤ **Chi-Square Test.** The output shows the contingency table with expected values, the sample χ^2 with *P* value.

VIEWPOINT

Loyalty! Going, Going, Gone!

Was there a time in the past when people worked for the same company all their lives, regularly purchased the same brand names, always voted for candidates from the same political party, and loyally cheered for the same sports team? One way to look at this question is to consider tests of *statistical independence.* Is customer loyalty independent of company profits? Can a company maintain its productivity independent of loyal workers? Can politicians do whatever they please independent of the voters back home? Americans may be ready to act on a pent-up desire to restore a sense of loyalty in their lives. For more information, see *American Demographics,* vol. 19, no. 9.

SECTION 11.1 PROBLEMS

For each of the problems, please do the following:
(a) State the null and alternate hypotheses.
(b) Find the value of the chi-square statistic from the sample.
(c) Find the degrees of freedom and the appropriate critical chi-square value.
(d) Sketch the critical region and locate your sample chi-square value and critical chi-square value on the sketch.
(e) Decide whether you should reject or fail to reject the null hypothesis using the given level of significance α. Interpret the results in the context of the problem.

Use the expected values *E* to the hundredths place.

1. *Psychology: Myers-Briggs* The following table shows the Myers-Briggs personality preference and professions for a random sample of 2408 people in the listed professions (*Atlas of Type Tables,* by Macdaid, McCaulley, and Kainz). E refers to extroverted, and I refers to introverted.

Occupation	Personality Preference Type		Row Total
	E	I	
Clergy (all denominations)	308	226	534
M.D.	667	936	1603
Lawyer	112	159	271
Column total	1087	1321	2408

Use the chi-square test to determine if the listed occupations and personality preferences are independent at the 0.01 level of significance.

2. *Psychology: Myers-Briggs* The following table shows the Myers-Briggs personality preference and professions for a random sample of 2408 people in the listed professions (*Atlas of Type Tables,* by Macdaid, McCaulley, and Kainz). T refers to thinking, and F refers to feeling.

Occupation	Personality Preference Type		Row Total
	T	F	
Clergy (all denominations)	114	420	534
M.D.	785	818	1603
Lawyer	176	95	271
Column total	1075	1333	2408

Use the chi-square test to determine if the listed occupations and personality preferences are independent at the 0.01 level of significance.

3. *Archaeology: Pottery* The following table shows site type and type of pottery for a random sample of 628 sherds at a location in Sand Canyon Archaeological Project, Colorado (*The Sand Canyon Archaeological Project,* edited by Lipe).

Site Type	Pottery Type			Row Total
	Mesa Verde Black on White	McElmo Black on White	Mancos Black on White	
Mesa Top	75	61	53	189
Cliff-Talus	81	70	62	213
Canyon Bench	92	68	66	226
Column total	248	199	181	628

Use a chi-square test to determine if site type and pottery type are independent at the 0.05 level of significance.

4. *Archaeology: Pottery* The following table shows ceremonial ranking and type of pottery sherds for a random sample of 1404 sherds at a location in the Sand Canyon Archaeological Project, Colorado (*The Architecture of Social Integration in Prehistoric Pueblos,* edited by Lipe and Hegmon).

Ceremonial Ranking	Cooking Jar Sherds	Decorated Jar Sherds (Noncooking)	Row Total
A	242	26	268
B	658	45	703
C	371	62	433
Column total	1271	133	1404

Use a chi-square test to determine if ceremonial ranking and pottery type are independent at the 0.05 level of significance.

5. *Ecology: Buffalo* The following table shows age distribution and location of a random sample of 166 buffalo in Yellowstone National Park (based on information from *The Bison of Yellowstone National Park,* National Park Service Scientific Monograph Series).

Age	Lamar District	Nez Perce District	Firehole District	Row Total
Calf	13	13	15	41
Yearling	10	11	12	33
Adult	34	28	30	92
Column total	57	52	57	166

Use a chi-square test to determine if age distribution and location are independent at the 0.05 level of significance.

6. *Psychology: Myers-Briggs* The following table shows the Myers-Briggs personality preference and area of study for a random sample of 519 college students (*Applications of the Myers-Briggs Type Indicator in Higher Education,* edited by Provost and Anchors). In the table IN refers to introvert, intuitive; EN refers to extrovert, intuitive; IS refers to introvert, sensing; ES refers to extrovert, sensing.

Myers-Briggs Preference	Arts & Science	Business	Allied Health	Row Total
IN	64	15	17	96
EN	82	42	30	154
IS	68	35	12	115
ES	75	42	37	154
Column total	289	134	96	519

Use a chi-square test to determine if Myers-Briggs preference type is independent of area of study at the 0.01 level of significance.

7. *Sociology: Movie Preference* Mr. Acosta, a sociologist, is doing a study to see if there is a relationship between the age of a young adult (18 to 35 years old) and the type of movie preferred. A random sample of 93 adults revealed the following data. Test if age and type of movie preferred are independent at the 0.05 level.

| Movie | Person's Age | | | Row Total |
	18–23 yr	24–29 yr	30–35 yr	
Drama	8	15	11	34
Science fiction	12	10	8	30
Comedy	9	8	12	29
Column total	29	33	31	93

8. *Sociology: Ethnic Groups* After a large fund drive to help the Boston City Library, the following information was obtained from a random sample of contributors to

the library fund. Using a 1% level of significance, test the claim that the amount contributed to the library fund is independent of ethnic group.

Ethnic Group	Number of People Making Contribution					Row Total
	$1–50	$51–100	$101–150	$151–200	Over $200	
A	310	715	201	105	42	1373
B	619	511	312	97	22	1561
C	402	624	217	88	35	1366
D	544	571	309	79	29	1532
Column total	1875	2421	1039	369	128	5832

9. *Marketing: Movies* Blue Bird Consolidated Theaters has more than 600 theaters located across the country. Each theater has four separate screens, and a customer can choose from one of four different movies. The president of Blue Bird Consolidated wants to know if a variety of shows (spy, comedy, horror, children's) or a coordinated bill (all spy, all comedy, all horror, all children's) has any effect on the total ticket sales at a theater. The president randomly assigned 47 theaters to use a variety of shows and 53 other theaters to use a coordinated bill of shows. For all theaters, total ticket sales for one week were recorded. Using the following data and a 5% level of significance, test the claim that total ticket sales are independent of the four shows being varied or coordinated.

Type of Billing	Ticket Sales for One Week				Row Total
	Less than 1000	1000 to 2000	2001 to 3000	More than 3000	
Variety	10	12	18	7	47
Coordinated	6	16	22	9	53
Column total	16	28	40	16	100

10. *Political Affiliation: Funding* A random sample of senators and representatives in Washington, D.C., gave the following information about party affiliation and number of dollars spent on federal projects in their home districts. Using a 1% level of significance, test the claim that federal spending level in home districts is independent of party affiliation.

Party	Dollars Spent on Federal Projects in Home Districts			Row Total
	Less than 5 Billion	5 to 10 Billion	More than 10 Billion	
Democrat	8	15	22	45
Republican	12	19	16	47
Column total	20	34	38	92

11. *Focus Problem: Archaeology* Solve the Focus Problem at the beginning of the chapter.

11.2
Chi-Square: Goodness of Fit

Hypotheses

Last year the labor union bargaining agents listed five items and asked each employee to mark the *one* most important to her or him. The items and corresponding percentage of favorable responses are shown in Table 11-8. The bargaining agents need to determine if the distribution of responses *now* "fits" last year's distribution or if it is different.

In questions of this type, we are asking if a population follows a specified distribution. In other words, we are testing the hypotheses

H_0: The population fits the given distribution.

H_1: The population has a different distribution.

Computing sample χ^2

We use the chi-square distribution to test "goodness-of-fit" hypotheses.

Just as with tests of independence, we compute the sample statistic:

$$\chi^2 = \Sigma \frac{(O - E)^2}{E} \text{ with degrees of freedom} = n - 1$$

where E = expected frequency

O = observed frequency

$\dfrac{(O - E)^2}{E}$ is summed for each item in the distribution

n = number of items in the distribution

Then we compare it with an appropriate critical value χ^2_α from Table 7 of Appendix II. In the case of a *goodness-of-fit test*, we use the null hypothesis to compute the expected values. Let's look at the bargaining item problem to see how this is done.

In the bargaining item problem, the two hypotheses are

H_0: The present distribution of responses is the same as last year's.

H_1: The present distribution of responses is different.

TABLE 11-8   **Bargaining Items (last year)**

Item	Percentage of Favorable Responses
Vacation time	4%
Salary	65%
Safety regulations	13%
Health and retirement benefits	12%
Overtime policy and pay	6%

TABLE 11-9   Observed and Expected Frequencies for Bargaining Items

Item	O	E	$(O - E)^2$	$(O - E)^2/E$
Vacation time	30	4% of 500 = 20	100	5.00
Salary	290	65% of 500 = 325	1225	3.77
Safety	70	13% of 500 = 65	25	0.38
Health and retirement	70	12% of 500 = 60	100	1.67
Overtime	40	6% of 500 = 30	100	3.33
	$\Sigma O = 500$	$\Sigma E = 500$		$\Sigma \dfrac{(O - E)^2}{E} = 14.15$

The null hypothesis tells us that the expected frequencies of the present response distribution should follow the percentages indicated in last year's survey. To test this hypothesis, a random sample of 500 employees was taken. If the null hypothesis is true, then there should be 4%, or 20 responses, out of the 500 rating vacation time as the most important bargaining issue. Table 11-9 gives the other expected values and all the information necessary to compute the sample statistic χ^2. We see that the sample statistic is

$$\chi^2 = \Sigma \frac{(O - E)^2}{E} = 14.15$$

Again, larger values of the sample statistic χ^2 indicate greater differences between the proposed probability distribution and the one followed by the sample. The critical value χ^2_α tells us how large the sample statistic can be before we reject the null hypothesis that the population does follow the distribution proposed in that hypothesis.

To find the critical value χ^2_α, we need to know the level of significance α and the number of degrees of freedom $d.f.$ In the case of a goodness-of-fit test, the degrees of freedom are found by the following formula:

Critical value χ^2_α

Degrees of freedom for goodness-of-fit test

$d.f. =$ (number of E entries) $- 1$

Notice that when we compute the expected values E, we must use the null hypothesis to compute all but the last one. To compute the last one, we can subtract the previous expected values from the sample size. For instance, for the bargaining issues, we could have found the number of responses for overtime policy by adding up the other expected values and subtracting that sum from the sample size 500. We would again get an expected value of 30 responses. The degrees of freedom, then, is the number of E values that *must* be computed by using the null hypothesis.

For the bargaining issues, we have

$d.f. = 5 - 1 = 4$

To test the hypothesis at the 0.05 level of significance, we find the critical value $\chi^2_{0.05}$ for four degrees of freedom in Table 7 of Appendix II. The critical value, 9.49, and the sample statistic, 14.15, are shown in Figure 11-5 on the next page. As shown in

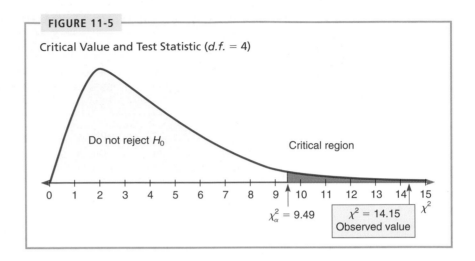

FIGURE 11-5

Critical Value and Test Statistic (*d.f.* = 4)

Do not reject H_0

Critical region

0 1 2 3 4 5 6 7 8 9 10 11 12 13 14 15 χ^2

$\chi_\alpha^2 = 9.49$

$\chi^2 = 14.15$
Observed value

the figure, the sample statistic χ^2 is in the critical region. Note that the critical region is always to the right of the critical value χ_α^2. Because the sample statistic is in the critical region, we reject the null hypothesis and conclude that the distribution of responses to the bargaining issues now is different from the distribution of last year.

One important application of goodness-of-fit tests is to genetic theories. Such an application is shown in Guided Exercise 6.

GUIDED EXERCISE 6

Goodness-of-fit test

According to genetics theory, red-green colorblindness in humans is a recessive sex-linked characteristic. In this case, the gene is carried on the X chromosome only. We will denote an X chromosome with the gene by X_c and one without the gene by X_n. Women have two X chromosomes, and they will be red-green colorblind only if both chromosomes have the gene, designated $X_c X_c$. A woman can have normal vision but still carry the colorblind gene if only one of the chromosomes has the gene, designated $X_c X_n$. A man carries an X and Y chromosome; if the X chromosome carries the colorblind gene ($X_c Y$), the man is colorblind.

According to genetics theory, if a man with normal vision ($X_n Y$), and a woman carrier ($X_c X_n$) have a child, the probabilities that the child will have red-green colorblindness, have normal vision and not carry the gene, or have normal vision and carry the gene are given by the *equally likely* events in Table 11-10.

TABLE 11-10
Red-Green Colorblindness

Mother	Father	
	X_n	Y
X_c	$X_c X_n$	$X_c Y$
X_n	$X_n X_n$	$X_n Y$

Continued

GUIDED EXERCISE 6 continued

$P(\text{child has normal vision and is not a carrier}) = P(X_n Y) + P(X_n X_n) = \dfrac{1}{2}$

$P(\text{child has normal vision and is a carrier}) = P(X_c X_n) = \dfrac{1}{4}$

$P(\text{child is red-green colorblind}) = P(X_c Y) = \dfrac{1}{4}$

To test this genetics theory, Genetics Labs took a random sample of 200 children whose mothers were carriers of the colorblind gene and whose fathers had normal vision. The results are in Table 11-11. We wish to test the hypothesis that the population follows the distribution predicted by the genetics theory (see Table 11-10).

(a) State the null and alternate hypotheses.

➡ H_0: The population fits the distribution predicted by genetics theory.

H_1: The population does not fit the distribution predicted by genetics theory.

(b) Fill in the rest of Table 11-11 and use the table to compute the sample statistic χ^2.

TABLE 11-11   Colorblindness Sample

Event	O	E	$(O - E)^2$	$(O - E)^2/E$
Red-green colorblind	35	50	225	4.50
Normal vision, noncarrier	105	_____	_____	_____
Normal vision, carrier	60	_____	_____	_____

➡ **TABLE 11-12   Completion of Table 11-11**

Event	O	E	$(O - E)^2$	$(O - E)^2/E$
Red-green colorblind	35	50	225	4.50
Normal vision, noncarrier	105	100	25	0.25
Normal vision, carrier	60	50	100	2.00

The sample statistic is $\chi^2 = \Sigma \dfrac{(O - E)^2}{E} = 6.75.$

(c) There are three expected frequencies listed in Table 11-11. Use this information to compute the degrees of freedom.

➡ $d.f. = (\text{no. of } E \text{ values}) - 1$

$= 3 - 1$

$= 2$

(d) Find the critical value $\chi^2_{0.01}$ for a 0.01 level of significance. Do we reject the hypothesis that the population follows the distribution predicted by genetics theory or not?

➡ From Table 7 of Appendix II, we see that for $d.f. = 2$ and level of significance 0.01, the critical value is $\chi^2_{0.01} = 9.21$. Because the sample statistic $\chi^2 = 6.75$ is less than the critical value, we do not reject the null hypothesis that the population follows the distribution predicted by genetics theory.

P values

◊ **COMMENT** To estimate *P* values for goodness-of-fit tests, we use Table 7 of Appendix II. As an example, let's find an interval containing the *P* value of the sample chi-square statistic $\chi^2 = 6.75$ of Guided Exercise 6. The degrees of freedom for this test are *d.f.* = 2. Therefore, we look in the row headed by *d.f.* = 2 in Table 7. Notice that the sample statistic $\chi^2 = 6.75$ falls between the row entries 5.99 and 7.38. Therefore, the *P* value falls between the corresponding column headers 0.050 and 0.025.

$$0.025 < P \text{ value} < 0.050$$

This means that we reject H_0 for all $\alpha \geq 0.050$. In particular, we reject H_0 for $\alpha = 0.05$, but we fail to reject H_0 for $\alpha = 0.01$. This result is consistent with the test conclusion in Guided Exercise 6.   ◊

VIEWPOINT      *Run! Run! Run!*

What description would you use for marathon runners? How about age distribution? Body weight? Length of stride? Heart rate? Blood pressure? What countries do these runners come from? What are their best running times? Make your own estimated distribution for these variables, and then consider a goodness-of-fit test for your distribution compared with available data. For more information on marathon runners, visit the Brase/Brase statistics site at http://math.college.hmco.com/students and find links to the Honolulu marathon site and to the *Runners World* site.

SECTION 11.2   PROBLEMS

For each of the problems, please do the following:
(a) State the null and alternate hypotheses.
(b) Find the value of the chi-square statistic from the sample.
(c) Find the degrees of freedom and the appropriate critical chi-square value.
(d) Sketch the critical region and locate your sample chi-square value and critical chi-square value on the sketch.
(e) Decide whether you should reject or fail to reject the null hypothesis.

1. *Census: Age* The age distribution of the Canadian population and the age distribution of a random sample of 455 residents in the Indian community of Red Lake (Northwest Territories) are shown below (based on *U.S. Bureau of the Census, International Data Base*).

Age (years)	Percent of Canadian Population	Observed Number in Red Lake Village
Under 5	7.2%	47
5 to 14	13.6%	75
15 to 64	67.1%	288
65 and older	12.1%	45

Use a 5% level of significance to test the claim that the age distribution of the general Canadian population fits the age distribution of Red Lake Village.

2. *Census: Type of Household* The type of household for the U.S. population and a random sample of 411 households of the community of Dove Creek, Montana, are shown (based on *Statistical Abstract of the United States*).

Type of Household	Percent of U.S. Households	Observed Number of Households in Dove Creek
Married, with children	26%	102
Married, no children	29%	112
Single parent	9%	33
One person	25%	96
Other (e.g., roommates, siblings)	11%	68

Use a 5% level of significance to test the claim that the distribution of U.S. households fits the Dove Creek distribution.

3. *Archaeology: Stone Tools* The type of raw material used to construct stone tools found at the archaeological site Casa del Rito is shown below (*Bandelier Archaeological Excavation Project*, edited by Kohler and Root). A random sample of 1486 stone tools was obtained from a current excavation site.

Raw Material	Regional Percent of Stone Tools	Observed Number of Tools at Current Excavation Site
Basalt	61.3%	906
Obsidian	10.6%	162
Welded tuff	11.4%	168
Pedernal chert	13.1%	197
Other	3.6%	53

Use a 1% level of significance to test the claim that the regional distribution of raw materials fits the distribution at the current excavation site.

4. *Ecology: Deer* The type of browse favored by deer is shown in the following table (*The Mule Deer of Mesa Verde National Park*, edited by Mierau and Schmidt). Using binoculars, volunteers observed feeding habits of a random sample of 320 deer.

Type of Browse	Plant Composition in Study Area	Observed Number of Deer Feeding on This Plant
Sage brush	32%	102
Rabbit brush	38.7%	125
Salt brush	12%	43
Service berry	9.3%	27
Other	8%	23

Use a 5% level of significance to test the claim that the natural distribution of browse fits the deer feeding pattern.

5. *Meteorology: Colorado* The following problem is based on information from the *National Oceanic and Atmospheric Administration (NOAA) Environmental Data Service*. Let x be a random variable that represents the average daily temperature (degrees

Fahrenheit) in July in the town of Kit Carson, Colorado. The x distribution has a mean μ approximately 75°F and standard deviation σ approximately 8°F. A 20-year study (620 July days) gave the entries in the right column of the following table.

I Region under Normal Curve	II $x°F$	III Expected % from Normal Curve	IV Observed Number of Days in 20 Years
$\mu - 3\sigma \le x < \mu - 2\sigma$	$51 \le x < 59$	2.35%	16
$\mu - 2\sigma \le x < \mu - \sigma$	$59 \le x < 67$	13.5%	78
$\mu - \sigma \le x < \mu$	$67 \le x < 75$	34%	212
$\mu \le x < \mu + \sigma$	$75 \le x < 83$	34%	221
$\mu + \sigma \le x < \mu + 2\sigma$	$83 \le x < 91$	13.5%	81
$\mu + 2\sigma \le x < \mu + 3\sigma$	$91 \le x < 99$	2.35%	12

(i) Remember that $\mu = 75$ and $\sigma = 8$. Examine Figure 6-5 in Chapter 6. Write a brief explanation for columns I, II, and III in the context of this problem.

(ii) Use a 1% level of significance to test the claim that the average daily July temperature follows a normal distribution with $\mu = 75$ and $\sigma = 8$.

6. *Meteorology: Hawaii* Let x be a random variable that represents the average daily temperature (degrees Fahrenheit) in January at the town of Hana, Maui. The x variable has a mean μ approximately 68°F and standard deviation σ approximately 4°F (see reference in Problem 5). A 20-year study (620 January days) gave the entries in the right column of the following table.

I Region under Normal Curve	II $x°F$	III Expected % from Normal Curve	IV Observed Number of Days in 20 Years
$\mu - 3\sigma \le x < \mu - 2\sigma$	$56 \le x < 60$	2.35%	14
$\mu - 2\sigma \le x < \mu - \sigma$	$60 \le x < 64$	13.5%	86
$\mu - \sigma \le x < \mu$	$64 \le x < 68$	34%	207
$\mu \le x < \mu + \sigma$	$68 \le x < 72$	34%	215
$\mu + \sigma \le x < \mu + 2\sigma$	$72 \le x < 76$	13.5%	83
$\mu + 2\sigma \le x < \mu + 3\sigma$	$76 \le x < 80$	2.35%	15

(i) Remember that $\mu = 68$ and $\sigma = 4$. Examine Figure 6-5 in Chapter 6. Write a brief explanation for columns I, II, and III in the context of this problem.

(ii) Use a 1% level of significance to test the claim that the average daily January temperature follows a normal distribution with $\mu = 68$ and $\sigma = 4$.

7. *Ecology: Fish* The Fish and Game Department stocked Lake Lulu with fish in the following proportions: 30% catfish, 15% bass, 40% bluegill, and 15% pike. Five years later they sampled the lake to see if the distribution of fish had changed. They found the 500 fish in the sample were distributed as follows:

Catfish	Bass	Bluegill	Pike
120	85	220	75

In the 5-year interval, did the distribution of fish change at the 0.05 level?

8. *Library: Book Circulation* The director of library services at Fairmont College did a survey of types of books (by subject) in the circulation library. Then she used library records to take a random sample of 4217 books checked out last term and classified the books in the sample by subject. The results are shown below.

Subject Area	Percent of Books in Circulation Library on This Subject	Number of Books in Sample on This Subject
Business	32%	1210
Humanities	25%	956
Natural science	20%	940
Social science	15%	814
All other subjects	8%	297

Using a 5% level of significance, test the claim that the subject distribution of books in the library fits the distribution of books checked out by students.

9. *Census: California* The accuracy of a census report on a city in southern California was questioned by some government officials. A random sample of 1215 people living in the city was used to check the report, and the results are shown here:

Ethnic Origin	Census Percent	Sample Result
Black	10%	127
Asian	3%	40
Anglo	38%	480
Latino/Latina	41%	502
Native American	6%	56
All others	2%	10

Using a 1% level of significance, test the claim that census distribution and sample distribution agree.

10. *Marketing: Compact Discs* Snoop Incorporated is a firm that does market surveys. The Rollum Sound Company hired Snoop to study the age distribution of people who buy compact discs. To check the Snoop report, Rollum used a random sample of 519 customers and obtained the following data:

Customer Age (years)	Percent of Customers from Snoop Report	Number of Customers in Sample
Less than 14	12%	88
14–18	29%	135
19–23	11%	52
24–28	10%	40
29–33	14%	76
More than 33	24%	128

Using a 1% level of significance, test the claim that the distribution of customer ages in the Snoop report agrees with the sample report.

11.3   Testing and Estimating a Single Variance or Standard Deviation

Testing σ^2

Many problems arise that require us to make decisions about variability. In this section, we will study two kinds of problems: (1) we will test hypotheses about the variance (or standard deviation) of a population, and (2) we will find confidence intervals for the variance (or standard deviation) of a population. It is customary to talk about variance instead of standard deviation because our techniques employ the sample variance rather than the standard deviation. Of course, the standard deviation is just the square root of the variance, so any discussion about variance is easily converted to a similar discussion about standard deviation.

Let us consider a specific example in which we might wish to test a hypothesis about the variance. Almost everyone has had to wait in line. In a grocery store, bank, post office, or registration center, there are usually several checkout or service areas. Frequently, each service area has its own independent line. However, many businesses and government offices are adopting a "single-line" procedure.

In a single-line procedure there is only one waiting line for everyone. As any service area becomes available, the next person in line gets served. The old independent-lines procedure has a line at each service center. An incoming customer simply picks the shortest line and hopes it will move quickly. In either procedure, the number of clerks and the rate at which they work is the same, so the average waiting time is the *same*. What is the advantage of the single-line procedure? The difference is in the *attitudes* of people who wait in the lines. A lengthy waiting line will be more acceptable if the variability of waiting times is smaller, even though the average waiting time is the same. When the variability is small, the inconvenience of waiting (although it might not be reduced) does become more predictable. This means impatience is reduced and people are happier.

To test the hypothesis that variability is less in a single-line process, we use the chi-square distribution. The next theorem tells us how to use the sample and population variance to compute values of χ^2.

Theorem 11.1

If we have a normal population with variance σ^2 and a random sample of n measurements is taken from this population with sample variance s^2, then

$$\chi^2 = \frac{(n-1)s^2}{\sigma^2}$$

has a chi-square distribution with degrees of freedom $d.f. = n - 1$.

Recall that the chi-square distribution is *not* symmetrical and that there are different chi-square distributions for different degrees of freedom. Table 7 in Appendix II gives chi-square values for which the area α is to the *right* of the given chi-square value.

FIGURE 11-6

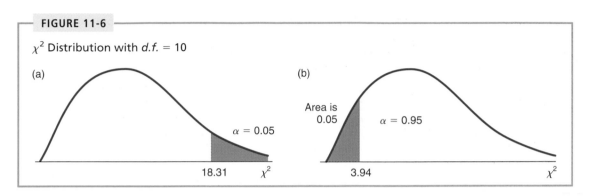

χ^2 Distribution with *d.f.* = 10

(a) $\alpha = 0.05$ 18.31 χ^2

(b) Area is 0.05 $\alpha = 0.95$ 3.94 χ^2

EXAMPLE 2

χ^2 distribution

(a) Find the χ^2 value so that the area to the right of χ^2 is 0.05 when *d.f.* = 10.

   SOLUTION: Since the area to the *right* of χ^2 is to be 0.05, we look in the $\alpha = 0.050$ column and the row with *d.f.* = 10. $\chi^2 = 18.31$ (see Figure 11-6a).

(b) Find the χ^2 value so that the area to the *left* of χ^2 is 0.05 when *d.f.* = 10.

   SOLUTION: When the area to the left of χ^2 is 0.05, the corresponding area to the *right* is $1 - 0.05 = 0.95$, so we look in the $\alpha = 0.950$ column and the row with *d.f.* = 10. We find $\chi^2 = 3.94$ (see Figure 11-6b). ◇

GUIDED EXERCISE 7

χ^2 distribution

(a) Find the χ^2 value so that 1% of the area under the χ^2 curve is to the right of χ^2 when *d.f.* = 20.

➡ The χ^2 value is in the column under $\alpha = 0.010$ and the row with *d.f.* = 20. $\chi^2 = 37.57$ (see Figure 11.7).

(b) Find the χ^2 value so that 1% of the area under the curve is to the *left* of χ^2 when *d.f.* = 20.

   (i) First find the corresponding area α to the *right* of the desired χ^2 value.

   (ii) Then look up the χ^2 value with *d.f.* = 20 and $\alpha = 0.99$.

➡ The area to the right of the desired χ^2 value is

$1 - 0.01 = 0.99$.

The χ^2 value is in the column under $\alpha = 0.990$ and the row with *d.f.* = 20. $\chi^2 = 8.26$ (see Figure 11-8).

FIGURE 11-7 χ^2 Distribution with *d.f.* = 20

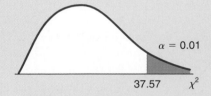

$\alpha = 0.01$ 37.57 χ^2

FIGURE 11-8 χ^2 Distribution with *d.f.* = 20

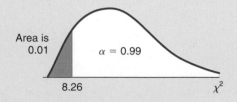

Area is 0.01 $\alpha = 0.99$ 8.26 χ^2

Now let's use Theorem 11.1 and our knowledge of the chi-square distribution to determine if a single-line procedure has less variance of waiting times than independent lines.

EXAMPLE 3

Testing the variance

A large discount hardware store in San Antonio has been using the independent-lines procedure to check out customers. After long observation, the manager knows that the standard deviation of waiting times is 7 minutes. The manager decided to introduce the single-line procedure on a trial basis to see if a reduction in waiting time variability would occur. A random sample of 25 customers was monitored, and their waiting times for checkout were determined. The sample standard deviation was $s = 4$ minutes. We will use a 5% level of significance to test the claim that the variance of waiting times has been reduced.

As a null hypothesis, we assume that the variance in waiting times is the same as that of the former independent-lines procedure. The alternate hypothesis is that the variance for the single-line procedure is less than that for the independent lines. If we let σ be the standard deviation of waiting times for the single-line procedure, then σ^2 is the variance, and we have

$$H_0: \sigma^2 = 49 \qquad \text{(use } 7^2 = 49\text{)}$$
$$H_1: \sigma^2 < 49$$

We use the chi-square distribution to test the hypotheses. Assuming that the waiting times are normally distributed, we compute our observed value of χ^2 by using Theorem 11.1. Since

$$n = 25$$
$$s = 4 \quad \text{so} \quad s^2 = 16 \qquad \text{(observed)}$$
$$\sigma = 7 \quad \text{so} \quad \sigma^2 = 49 \qquad \text{(from } H_0: \sigma^2 = 49\text{)}$$
$$\chi^2 = \frac{(n-1)s^2}{\sigma^2} = \frac{(25-1)16}{49} = 7.8 \qquad \text{(by Theorem 11.1)}$$

The critical value is obtained from Table 7 of Appendix II using the degrees of freedom and level of significance. By the alternate hypothesis $H_1: \sigma^2 < 49$, we want a *left* tail with area 0.05. Therefore, the area in the corresponding right tail is $\alpha = 1 - 0.05 = 0.95$. Since $d.f. = n - 1 = 25 - 1 = 24$, the desired critical value is $\chi^2_{0.95} = 13.85$ (see Figure 11-9).

Because the observed value of $\chi^2 = 7.8$ is in the critical or rejection region, we reject $H_0: \sigma^2 = 49$ and accept $H_1: \sigma^2 < 49$. The variance of the single-line procedure is less than 49. ◇

Checkout lines

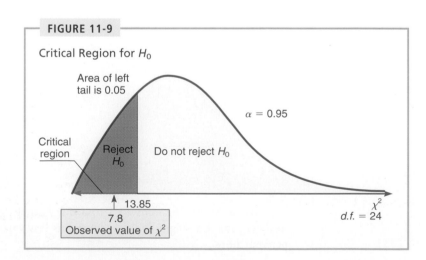

FIGURE 11-9

Critical Region for H_0

Area of left tail is 0.05

$\alpha = 0.95$

Critical region

Reject H_0

Do not reject H_0

13.85

χ^2

$d.f. = 24$

7.8
Observed value of χ^2

GUIDED EXERCISE 8

Testing the variance

Certain industrial machines require overhaul when wear on their parts introduces too much variability to pass inspection. A government official is visiting a dentist's office to inspect the operation of an x-ray machine. If the machine emits too little radiation, clear photographs cannot be obtained. However, too much radiation can be harmful to the patient. Government regulations specify an average emission of 60 millirads with standard deviation σ of 12 millirads, and the machine has been set for these readings. After examining the machine, the inspector is satisfied that the average emission is still 60 millirads. However, there is wear on certain mechanical parts. To test variability, the inspector takes a random sample of 30 x-ray emissions and finds the sample standard deviation to be $s = 15$ millirads. Does this support the claim that the variance is too high (i.e., the machine should be overhauled)? Use a 1% level of significance.

Let σ be the (population) standard deviation of emissions (in millirads) of the machine in its present condition.

(a) Which of the following shall we use for the null hypothesis? Explain.

$H_0: \sigma^2 = 12$ $H_0: \sigma^2 = 144$ $H_0: \sigma^2 > 144$

➡ $H_0: \sigma^2 = 144$. We use $\sigma = 12$ and the initial claim that the variance is still what it should be according to specifications.

(b) Which of the following shall we use for the alternate hypothesis? Explain.

$H_1: \sigma^2 > 12$ $H_1: \sigma^2 \neq 144$ $H_1: \sigma^2 > 144$

➡ $H_1: \sigma^2 > 144$. We want to test the claim that the variance is too large.

(c) What is the observed value of χ^2?

➡ $\chi^2 = \dfrac{(n-1)s^2}{\sigma^2} = \dfrac{(30-1)15^2}{144} = 45.3$

(since $n = 30$, $s = 15$, and by H_0, $\sigma^2 = 144$)

(d) What are the degrees of freedom? Are we using a left-, right-, or two-tailed test? Use Table 7 in Appendix II to find the chi-square critical value.

➡ $d.f. = n - 1 = 30 - 1 = 29$. Since H_1 is $\sigma^2 > 144$, we use a right-tailed test. The problem calls for $\alpha = 0.01$ and an area of 0.01 is to the right of $\chi^2 = 49.59$ when $d.f. = 29$. The critical value is $\chi^2_{0.01} = 49.59$.

(e) Sketch the critical region on a chi-square curve. Locate the observed chi-square value on the sketch.

➡ **FIGURE 11-10** Critical Region with $d.f. = 29$

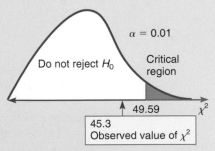

(f) Do we reject H_0 or not? Should the inspector recommend that the machine be overhauled?

➡ Because the observed chi-square value 45.3 is not in the critical region, we fail to reject H_0 and conclude that the machine does not need an overhaul at this time.

Confidence Interval for σ^2

Sometimes it is important to have a confidence interval for the variance or standard deviation. Let us look at another example.

Mr. Wilson is a truck farmer in California who makes his living on a large single-vegetable crop of green beans. Because of modern machinery being used, the entire crop must be harvested at once. Therefore, it is important to plant a variety of green beans that matures all at once. This means that Mr. Wilson wants a small standard deviation between maturing times of individual plants. A seed company is trying to develop a new variety of green beans with a small standard deviation of maturing times. To test their new variety, Mr. Wilson planted 30 of the new seeds and carefully observed the number of days required for each plant to arrive at its peak of maturity. The maturing times for these plants had a sample standard deviation of $s = 3.4$ days. How can we find a 95% confidence interval for the population standard deviation of maturing times of this variety of green bean? The answer to this question is based on the following theorem.

Theorem 11.2

Let a random sample of size n be taken from a normal population with population standard deviation σ, and let c be a chosen confidence level $(0 < c < 1)$. Then

$$P\left(\frac{(n-1)s^2}{\chi_U^2} < \sigma^2 < \frac{(n-1)s^2}{\chi_L^2}\right) = c$$

and

$$P\left(\sqrt{\frac{(n-1)s^2}{\chi_U^2}} < \sigma < \sqrt{\frac{(n-1)s^2}{\chi_L^2}}\right) = c$$

where n = sample size

$$s = \sqrt{\frac{\Sigma(x - \bar{x})^2}{n-1}} \text{ is the sample standard deviation}$$

χ_U^2 = chi-square value from Table 7 of Appendix II using
  d.f. = $n - 1$ and $\alpha = (1 - c)/2$

χ_L^2 = chi-square value from Table 7 of Appendix II using
  d.f. = $n - 1$ and $\alpha = (1 + c)/2$

From Figure 11-11 we see that a c confidence level on a chi-square distribution with equal probability in each tail does not center the middle of the corresponding interval under the peak of the curve. This is to be expected because a chi-square curve is skewed to the right.

Let us summarize Theorem 11.2 in the following way: A c confidence interval for σ^2 is

$$\frac{(n-1)s^2}{\chi_U^2} < \sigma^2 < \frac{(n-1)s^2}{\chi_L^2} \tag{1}$$

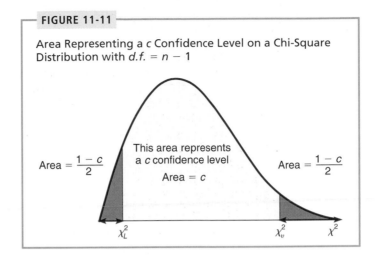

FIGURE 11-11

Area Representing a c Confidence Level on a Chi-Square Distribution with $d.f. = n - 1$

Area $= \dfrac{1-c}{2}$

This area represents a c confidence level

Area $= c$

Area $= \dfrac{1-c}{2}$

χ_L^2 χ_v^2 χ^2

and a c confidence interval for σ is

$$\sqrt{\frac{(n-1)s^2}{\chi_U^2}} < \sigma < \sqrt{\frac{(n-1)s^2}{\chi_L^2}}$$

Now let us finish our example regarding the variance of maturing times for green beans.

EXAMPLE 4

Confidence interval for σ^2 and σ

A random sample of $n = 30$ plants has a sample standard deviation of $s = 3.4$ days for maturity. Find a 95% confidence interval for the population variance σ^2.

SOLUTION: To find the confidence interval, we use the following values:

$c = 0.95$ confidence level

$n = 30$ sample size

$d.f. = n - 1 = 30 - 1 = 29$ degrees of freedom

$s = 3.4$ sample standard deviation

To find the value of χ_U^2, we use Table 7 in Appendix II with $d.f. = 29$ and $\alpha = (1 - c)/2 = (1 - 0.95)/2 = 0.025$. From Table 7, we get

$\chi_U^2 = 45.72$

To find χ_L^2, we use Table 7 with $d.f. = 29$ and $\alpha = (1 + c)/2 = (1 + 0.95)/2 = 0.975$. From Table 7, we get

$\chi_L^2 = 16.05$

Formula (1) tells us that our desired 95% confidence interval for σ^2 is

$$\frac{(n-1)s^2}{\chi_U^2} < \sigma^2 < \frac{(n-1)s^2}{\chi_L^2}$$

$$\frac{(30-1)(3.4)^2}{45.72} < \sigma^2 < \frac{(30-1)(3.4)^2}{16.05}$$

$$7.33 < \sigma^2 < 20.89$$

To find a 95% confidence interval for σ, we simply take square roots; therefore, a 95% confidence interval for σ is

$$\sqrt{7.33} < \sigma < \sqrt{20.89}$$
$$2.71 < \sigma < 4.57$$

◇

GUIDED EXERCISE 9

Confidence interval for σ^2 and σ

A few miles off the Kona coast of the island of Hawaii a research vessel lies anchored. This ship makes electrical energy from the solar temperature differential of (warm) surface water versus (cool) deep water. The basic idea is that the warm water is flushed over coils to vaporize a special fluid. The vapor is under pressure and drives electrical turbines. Then some electricity is used to pump up cold water to cool the vapor back to a liquid, and the process is repeated. Even though some electricity is used to pump up the cold water, there is plenty left to supply a moderate-sized Hawaiian town. The subtropic sun always warms up surface water to a reliable temperature, but ocean currents can change the temperature of the deep, cooler water. If the deep-water temperature is too variable, the power plant cannot operate efficiently or possibly not operate at all. To estimate the variability of deep ocean water temperatures, a random sample of 25 near-bottom readings gave a sample standard deviation of 7.3°C.

Find a 99% confidence interval for the variance σ^2 and standard deviation σ of deep-water temperatures.

(a) Determine the following values: $c =$ _____; $n =$ _____; $d.f. =$ _____; $s =$ _____.

⟹ $c = 0.99$; $n = 25$; $d.f. = 24$; $s = 7.3$

(b) What is the value of χ_U^2?_____ of χ_L^2?_____

⟹ We use Table 7 of Appendix II with $d.f. = 24$.

For χ_U^2, $\alpha = (1 - 0.99)/2 = 0.005$
$$\chi_U^2 = 45.56$$

For χ_L^2, $\alpha = (1 + 0.99)/2 = 0.995$
$$\chi_L^2 = 9.89$$

(c) Find a 99% confidence interval for σ^2.

⟹ $$\frac{(n-1)s^2}{\chi_U^2} < \sigma^2 < \frac{(n-1)s^2}{\chi_L^2}$$

$$\frac{(24)(7.3)^2}{45.56} < \sigma^2 < \frac{24(7.3)^2}{9.89}$$

$$28.07 < \sigma^2 < 129.32$$

(d) Find a 99% confidence interval for σ.

⟹ $$\sqrt{28.07} < \sqrt{\sigma^2} < \sqrt{129.32}$$

$$5.30 < \sigma < 11.37$$

VIEWP●INT   *Adoption—A Good Choice!*

Cuckoos are birds that are known to lay their eggs in the nests of other (host) birds. The host birds then hatch the eggs and adopt the cuckoo chicks as their own. Birds such as the meadow pipit, tree pipit, hedge sparrow, robin, and wren have all played host to cuckoo eggs and adopted their chicks. L. H. C. Tippett (1902–1985) was a pioneer in the field of statistical quality control who collected data on cuckoo eggs found in the nests of other birds. For more information and data from Tippett's study, visit the Brase/Brase statistics site at http://math.college.hmco.com/students and find a link to DASL, the Carnegie Mellon University Data and Story Library. Find Biology under Data Subjects, and then select the Cuckoo Egg Length Data file.

SECTION 11.3   PROBLEMS

In each of the problems, please do the following:
(a) State the null and alternate hypotheses.
(b) Find the degrees of freedom and appropriate critical value or critical values.
(c) Find the appropriate chi-square value using the sample standard deviation.
(d) Sketch the critical region and show the critical chi-square value and the value of part (c).
(e) Decide whether to reject or fail to reject the null hypothesis at the given level of significance. Interpret the results in the context of the problem.
(f) Find the requested confidence interval for the population variance or population standard deviation. Interpret the results in the context of the problem.

In each of the following problems, assume a normal population distribution.

1. *Archaeology: Chaco Canyon* The following problem is based on information from *Archaeological Surveys of Chaco Canyon, New Mexico,* by A. Hayes, D. Brugge, and W. Judge, University of New Mexico Press. A *transect* is an archaeological study area that is 1/5 mile wide and 1 mile long. A *site* in a transect is the location of a significant archaeological find. Let x represent the number of sites per transect. In a section of Chaco Canyon, a large number of transects showed that x has a population variance $\sigma^2 = 42.3$. In a different section of Chaco Canyon, a random sample of 23 transects gave a sample variance $s^2 = 46.1$ for the number of sites per transect. Use a 5% level of significance to test the claim that the variance in the new section is greater than 42.3. Find a 95% confidence interval for the population variance.

2. *Sociology: Marriage* The following problem is based on information from an article by N. Keyfitz in *The American Journal of Sociology* (vol. 53, pp. 470–480). Let x = age in years of a rural Quebec woman at the time of her first marriage. In the year 1941, the population variance of x was approximately $\sigma^2 = 5.1$. Suppose that a recent study of age of first marriage for a random sample of 41 women in rural Quebec gave a sample variance $s^2 = 3.3$. Use a 5% level of significance to test the claim that the current variance is less than 5.1. Find a 90% confidence interval for the population variance.

3. *Mountain Climbing: Accidents* The following problem is based on information taken from *Accidents in North American Mountaineering* (jointly published by The American Alpine Club and The Alpine Club of Canada). Let x represent the number of mountain climbers killed each year. The long-term variance of x is approximately $\sigma^2 = 136.2$. Suppose that for the past 8 years the variance has been $s^2 = 115.1$. Use a 1% level of significance to test the claim that the recent variance for number of mountain-climber deaths is less than 136.2. Find a 90% confidence interval for the population variance.

4. *Professors: Salaries* The following problem is based on information taken from *Academe, Bulletin of the American Association of University Professors*. Let x represent the average annual salary of college and university professors (in thousands of dollars) in the United States. For all colleges and universities in the United States, the population variance of x is approximately $\sigma^2 = 47.1$. However, a random sample of 15 colleges and universities in Kansas showed that x has a sample variance $s^2 = 83.2$. Use a 5% level of significance to test the claim that the variance for colleges and universities in Kansas is greater than 47.1. Find a 95% confidence interval for the population variance.

5. *Medical: Clinical Test* A new kind of typhoid shot is being developed by a medical research team. The old typhoid shot was known to protect the population for a mean of 36 months with a standard deviation of 3 months. To test the variability of the new shot, a random sample of 24 people was given the new shot. Regular blood tests showed that the sample standard deviation of protection times was 1.9 months. Using a 0.05 level of significance, test the claim that the new typhoid shot has a smaller variance of protection times. Find a 90% confidence interval for the population standard deviation.

6. *Veterinary Science: Tranquilizer* Jim Mead is a veterinarian who visits a Vermont farm to examine prize bulls. In order to examine a bull, Jim first gives the animal a tranquilizer shot. The effect of the shot is supposed to last an average of 65 minutes, and it usually does. However, Jim sometimes gets chased out of the pasture by a bull that recovers too soon, and other times he becomes worried about prize bulls that take too long to recover. By reading journals, Jim found that the tranquilizer should have a mean duration of 65 minutes with standard deviation of 15 minutes. A random sample of 10 of Jim's bulls had a mean tranquilized duration time of close to 65 minutes but a standard deviation of 24 minutes. At the 1% level of significance, is Jim justified in the claim that the variance is larger than that stated in his journal? Find a 95% confidence interval for the population standard deviation.

7. *Engineering: Jet Engines* The fan blades on commercial jet engines must be replaced when wear on these parts indicates too much variability to pass inspection. If a single fan blade broke during operation, it could severely endanger a flight. A large engine contains thousands of fan blades, and safety regulations require that variability measurements on the population of all blades not exceed $\sigma^2 = 0.15$ mm^2. An engine inspector took a random sample of 61 fan blades from an engine. She measured each blade and found a sample variance of 0.27 mm^2. Using a 0.01 level of significance, is the inspector justified in claiming that all the engine fan blades must be replaced? Find a 90% confidence interval for the population standard deviation.

8. *Law: Bar Exam* A factor in determining the usefulness of an examination as a measure of demonstrated ability is the amount of spread that occurs in the grades. If the spread or variation of examination scores is very small, it usually means that the examination was either too hard or too easy. However, if the variance of scores is moderately large, then there is a definite difference in scores between "better," "average," and "poorer" students. A group of attorneys in a Midwest state has been given the task of making up this year's bar examination for the state. The examination has 500 total possible points, and from the history of past examinations, it is known that a

standard deviation of around 75 points is desirable. Of course, too large or too small a standard deviation is not good. The attorneys want to test their examination to see how good it is. A preliminary version of the examination (with slight modification to protect the integrity of the real examination) is given to a random sample of 24 newly graduated law students. Their scores give a sample standard deviation of 72 points.

(a) Using a 0.01 level of significance, test the claim that the population standard deviation for the new examination is 75 against the claim that the population standard deviation is different from 75.

(b) Find a 99% confidence interval for the population variance.

(c) Find a 99% confidence interval for the population standard deviation.

9. *Engineering: Solar Batteries* A set of solar batteries is used in a research satellite. The satellite can run on only one battery, but it runs best if more than one battery is used. The variance σ^2 of lifetimes of these batteries affects the useful lifetime of the satellite before it goes dead. If the variance is too small, all the batteries will tend to die at once. Why? If the variance is too large, the batteries are simply not dependable. Why? Engineers have determined a variance of $\sigma^2 = 15$ months (squared) is most desirable for these batteries. A random sample of 22 batteries gave a sample variance of 14.3 months (squared).

(a) Using a 0.05 level of significance, test the claim that $\sigma^2 = 15$ against the claim that σ^2 is different from 15.

(b) Find a 90% confidence interval for the population variance σ^2.

(c) Find a 90% confidence interval for the population standard deviation σ.

PART II: INFERENCES USING THE *F* DISTRIBUTION*

11.4
Testing Two Variances

FOCUS POINTS

✓ Set up a test for two variances σ_1^2 and σ_2^2.

✓ Use sample variances to compute the sample *F* statistic.

✓ Use the *F* distribution to complete the test and estimate a *P* value.

In this section, we present a method for testing two variances (or equivalently, two standard deviations). We use independent random samples from two populations to test the claim that the population variances are equal. The concept of variation among data is very important, so there will be many possible applications in science, industry, business administration, social science, and so on.

In Section 11.3, we tested a *single* variance. The main mathematical tool we used was the chi-square probability distribution. In this section, the main tool will be the *F* probability distribution. This distribution was discovered by the English statistician Sir Ronald Fisher (1890–1962).

Let us begin by stating what we need to assume for a test of two population variances. We assume that

Basic assumptions

- The two populations are independent of each other. Recall from Section 9.7 that two sampling distributions are *independent* if there is no relation whatsoever between specific values of the two distributions.

- The two populations each have a *normal* probability distribution. This is important because the test we will use is sensitive to changes away from normality.

Setup for test

Now that we know the basic assumptions, let's consider the setup.

---

*Section 11.4, "Testing Two Variances," and Section 11.5, "One-Way ANOVA: Comparing Several Sample Means," are self-contained and can be presented independently. Section 11.5 should be presented before Section 11.6, "Introduction to Two-Way ANOVA."

How to Set Up the Test

STEP 1: Get Two Independent Random Samples, One from Each Population

We use the following notation:

Population I (larger s^2)	Population II (smaller s^2)
n_1 = sample size	n_2 = sample size
s_1^2 = sample variance	s_2^2 = sample variance
σ_1^2 = population variance	σ_2^2 = population variance

To simplify later discussion, we make the notational choice that

$$s_1^2 \geq s_2^2$$

This means that we *define* Population I as the population with the *larger* (or equal, as the case may be) sample variance. This is only a notational convention and does not affect the general nature of the test.

STEP 2: Set Up the Hypotheses

The null hypothesis will be that we have equal population variances.

$$H_0: \sigma_1^2 = \sigma_2^2$$

Reflecting on our notation setup, it makes sense to use an alternate hypothesis, either

$$H_1: \sigma_1^2 \neq \sigma_2^2 \qquad \text{or} \qquad H_1: \sigma_1^2 > \sigma_2^2$$

Notice that the test makes claims about variances. However, we can also use it for corresponding claims about standard deviations.

Hypotheses about Variances	Equivalent Hypotheses about Standard Deviations
$H_0: \sigma_1^2 = \sigma_2^2$	$H_0: \sigma_1 = \sigma_2$
$H_1: \sigma_1^2 \neq \sigma_2^2$	$H_1: \sigma_1 \neq \sigma_2$
$H_1: \sigma_1^2 > \sigma_2^2$	$H_1: \sigma_1 > \sigma_2$

STEP 3: Compute the Sample Test Statistic

$$F = \frac{s_1^2}{s_2^2}$$

For two normally distributed populations with equal variances ($H_0: \sigma_1^2 = \sigma_2^2$), the sampling distribution we will use is the *F distribution* (see Table 8 of Appendix II).

The *F* distribution depends on *two* degrees of freedom, but a typical *F* distribution graph is shown in Figure 11-12.

Properties of the F Distribution

- The *F* distribution is not symmetrical. It is skewed to the right.

- Values of the *F* distribution are always greater than or equal to zero.

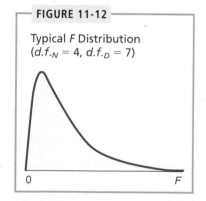

FIGURE 11-12

Typical *F* Distribution
($d.f._N = 4$, $d.f._D = 7$)

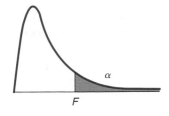

TABLE 11-13   Excerpt from Table 8 (Appendix II): The F Distribution

d.f._D	α	Degrees of Freedom for the Numerator			
		1	2	3	4 ···
⋮	⋮	⋮	⋮	⋮	⋮
	0.050	18.51	19.00	19.16	19.25
√2	0.025	38.51	39.00	39.17	39.25
	0.010	98.50	99.00	99.17	99.25

- A specific F distribution (see Table 8 in Appendix II) is determined from *two* degrees of freedom. These are called *degrees of freedom for the numerator $d.f._N$* and *degrees of freedom for the denominator $d.f._D$.*

For our tests of two variances, it can be shown that

$$d.f._N = n_1 - 1 \quad \text{and} \quad d.f._D = n_2 - 1$$

STEP 4: Use Table 8 in Appendix II

Use Table 8 with degrees of freedom for the numerator $d.f._N = n_1 - 1$, degrees of freedom for the denominator $d.f._D = n_2 - 1$, and the given level of significance α to find a critical F value. In Table 8, critical values for a *right-tailed* test are given, where α is the area of the right tail. For instance, using Table 11-13, which is an excerpt from Table 8, for $d.f._N = 3$, $d.f._D = 2$, and $\alpha = 0.05$, we see that the critical F value is 19.16. Once we find the critical F value from the table, we sketch the critical region. This will always be a right tail because of our notational setup $s_1^2 \geq s_2^2$. Finally, compare your sample F statistic with the critical value, and draw your conclusion.

Now that we have steps 1 to 4 as an outline, let's look at a specific example.

EXAMPLE 5

Testing two variances

Prehistoric Native Americans smoked pipes for ceremonial purposes. Most pipes were either carved-stone pipes or ceramic pipes made from clay. Clay pipes were easier to make, whereas stone pipes required careful drilling using hollow-core-bone drills and special stone reamers. An anthropologist claims that because clay pipes were easier to make, they show a greater variance in their construction. We want to test this claim using a 5% level of significance. Data for this example are taken from the Wind Mountain Archaeological Region. (Source: *Mimbres Mogollon Archaeology,* by A. I. Woosley and A. J. McIntyre, University of New Mexico Press.)

Ceramic Pipe Bowl Diameters (cm)

1.7	5.1	1.4	0.7	2.5	4.0
3.8	2.0	3.1	5.0	1.5	

Stone Pipe Bowl Diameters (cm)

1.6	2.1	3.1	1.4	2.2	2.1
2.6	3.2	3.4			

(a) Assume that the pipe bowl diameters follow normal distributions and that the given data make up independent random samples of pipe measurements taken from archaeological excavations at Wind Mountain. Use a calculator to verify the following:

Population I: Ceramic Pipes	Population II: Stone Pipes
$n_1 = 11$	$n_2 = 9$
$s_1^2 = 2.266$	$s_2^2 = 0.504$
$\sigma_1^2 =$ population variance	$\sigma_2^2 =$ population variance

Note: Because the sample variance for ceramic pipes (2.266) is larger than the sample variance for stone pipes (0.504), we designate Population I as ceramic pipes.

(b) Set up the null and alternate hypotheses.

$$H_0: \sigma_1^2 = \sigma_2^2 \quad \text{(or the equivalent, } \sigma_1 = \sigma_2\text{)}$$
$$H_1: \sigma_1^2 > \sigma_2^2 \quad \text{(or the equivalent, } \sigma_1 > \sigma_2\text{)}$$

The null hypothesis says that there is no difference. The alternate hypothesis supports the anthropologist's claim that clay pipes have a larger variance.

(c) The sample test statistic is

$$F = \frac{s_1^2}{s_2^2} = \frac{2.266}{0.504} = 4.496$$

Now if $\sigma_1^2 = \sigma_2^2$, then s_1^2 and s_2^2 also should be close in value. If this were the case, $F = s_1^2/s_2^2 \approx 1$. However, if $\sigma_1^2 > \sigma_2^2$, then we see that the sample statistic $F = s_1^2/s_2^2$ should be larger than 1. How much larger should F be to reject $H_0: \sigma_1^2 = \sigma_2^2$ at the 5% level of significance?

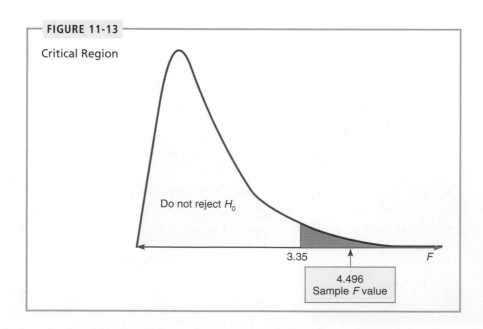

FIGURE 11-13

Critical Region

Do not reject H_0

3.35

4.496 Sample *F* value

F

(d) Now we use Table 8 of Appendix II with degrees of freedom for the numerator $d.f._N = n_1 - 1 = 11 - 1 = 10$ and degrees of freedom for the denominator $d.f._D = n_2 - 1 = 9 - 1 = 8$. Using $\alpha = 0.05$, we find that the critical value is 3.35. The critical region and sample test statistic are shown in Figure 11-13.

We use a right-tailed test ($H_1: \sigma_1^2 > \sigma_2^2$) and see that the sample F statistic is in the critical region (reject H_0). Therefore, we reject the claim $\sigma_1^2 = \sigma_2^2$ and at the 5% level of significance accept the anthropologist's claim that the variance for the ceramic pipes is larger.   ◇

GUIDED EXERCISE 10

Testing two variances

A large variance in blood chemistry components can result in health problems as the body attempts to return to equilibrium. J. B. O'Sullivan and C. M. Mahan conducted a study reported in the *American Journal of Clinical Nutrition* (vol. 19, pp. 345–351) that concerned the glucose (blood sugar) levels of pregnant and nonpregnant women at Boston City Hospital. For both groups, a fasting (12-hour fast) blood glucose test was done. The following data are in units of milligrams of glucose per 100 milliliters of blood.

Glucose Test: Nonpregnant Women

83	81	104	75	85	65	62	92	88	106

Glucose Test: Pregnant Women

62	84	90	100	66	68	75	93

Medical researchers question if the variance of the glucose test results for nonpregnant women is *different* (either way) compared with the variance for pregnant women. Let's conduct a test using a 5% level of significance.

(a) What assumptions must be made about the two populations and the samples?

⟹ The population measurements follow independent normal distributions. The samples are random samples from each population.

(b) Use a calculator to compute the sample variance for each data group. Then complete the following:

⟹ Recall that we choose our notation so that Population I has the *larger* sample variance.

Population I	Population II
$n_1 = $ _____	$n_2 = $ _____
$s_1^2 = $ _____	$s_2^2 = $ _____

Population I	Population II
$n_1 = 10$	$n_2 = 8$
$s_1^2 = 211.21$	$s_2^2 = 196.21$

(c) What is the null hypothesis?

⟹ The null hypothesis says that the population variances are equal.

$$H_0: \sigma_1^2 = \sigma_2^2$$

Continued

GUIDED EXERCISE 10 continued

(d) What is the alternate hypothesis?

⟹ We want to test the claim that the variances are different.

$H_1: \sigma_1^2 \neq \sigma_2^2$

(e) Compute the sample F statistic, $d.f._N$, and $d.f._D$.

⟹ $F = \dfrac{s_1^2}{s_2^2} = \dfrac{211.21}{196.21} = 1.08$

$d.f._N = n_1 - 1 = 10 - 1 = 9$

$d.f._D = n_2 - 1 = 8 - 1 = 7$

(f) Use Table 8 from Appendix II with level of significance $\alpha = 0.05$ to find a critical value for the two-tailed test.

⟹ We have a two-tailed test, so the probability in the right tail of the F distribution should be half the level of significance, or

$\dfrac{\alpha}{2} = \dfrac{0.05}{2} = 0.025$

Using $d.f._N = 9$ and $d.f._D = 7$ and Table 8, we see that the critical F value is 4.82.

(g) What about the left tail? Do we need a critical value for the left tail?

⟹ Our choice of notation puts the larger sample variance in Population I. Therefore, the sample test statistic $F = s_1^2/s_2^2$ will always be greater than or equal to 1. This implies that for a two-tailed test, we need to find only the right-tail critical value that puts an area of $\alpha/2$ in the right tail. We do not need to concern ourselves with a critical value for the left tail.

(h) Sketch the critical region on the F distribution and show the location of the sample F statistic on the diagram. Use the sketch to conclude the test.

⟹ **FIGURE 11-14** Critical Region

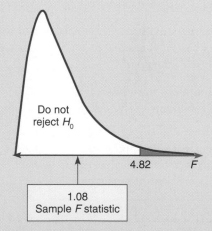

Do not reject H_0

4.82 F

1.08
Sample F statistic

The sample statistic $F = 1.08$ is not in the critical region, so we cannot reject H_0.

(i) What does the conclusion mean to a person doing medical research?

⟹ At the 5% level of significance, the claim that the glucose variance is the same for both pregnant and nonpregnant women cannot be rejected.

P values

◊ **COMMENT** We use Table 8 of Appendix II to estimate *P* values for tests of two variances. For instance, in Guided Exercise 10 the value of the sample test statistic was $F = 1.08$. Looking in the block with row header $d.f._D = 7$ and in the column with header $d.f._N = 9$, we see entries 3.68 for $\alpha = 0.050$, 4.82 for $\alpha = 0.025$, and 6.72 for $\alpha = 0.010$. Because the sample statistic $F = 1.08$ is less than the table entry 3.68, the area in the right tail beyond $F = 1.08$ is greater than $\alpha = 0.050$. Since the test is a two-tailed test, we double the area in the right tail to estimate the *P* value. Therefore, the *P* value is greater than 0.100.   ◊

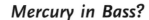

TECH NOTE   The TI-83Plus supports tests of two variances. Use **STAT**, **TESTS**, option **D:2-SampFTest**. For results consistent with the notational convention that the larger variance goes in the numerator of the sample *F* statistic, put the data with the larger variance in **List1**.

Variety is said to be the spice of life. However, in statistics, when we want to compare two populations, we will often need the assumption that the population variances are the same. As long as the two populations follow normal distributions, we can use methods of this section and random samples from the populations to determine if the assumption of equal variances is reasonable at a given level of significance.

VIEWP⊙INT

Mercury in Bass?

Largemouth bass were studied in 53 different lakes to examine factors that influence the level of mercury contamination. In many cases, the contamination was fairly low, except for older (trophy) fish, in which the mercury levels were much higher. Using information you have learned in this section, you can test variances of mercury contamination for different lakes and/or different regions. For more information, see an article by Lange, Royals, and Connor, *Transactions of the American Fisheries Society*. To find this article, visit the Brase/Brase statistics site at http://math.college.hmco.com/students and find a link to DASL, the Carnegie Mellon University Data and Story Library. From the DASL site, link to Nature under Data Subjects, and select Mercury in Bass Data file.

SECTION 11.4   PROBLEMS

For each of the following problems, please provide the requested information.
(a) State the null and alternate hypotheses. Is the test right-tailed or two-tailed? What is the level of significance?
(b) What sampling distribution will we use? What is the value of the sample test statistic? What are the appropriate degrees of freedom? What is the critical value?
(c) Sketch the critical region. Show the critical value and the sample test statistic on your sketch.

(d) Based on your answers to parts (a) to (c), shall we reject or fail to reject (i.e., accept) the null hypothesis at the specified level of significance? Explain your conclusion in the context of the problem.

Assume that data values in each problem come from independent populations that each follow a normal distribution.

1. *Agriculture: Wheat* Rothamsted Experimental Station (England) has studied wheat production since 1852. Each year many small plots of equal size but different soil/fertilizer conditions are planted with wheat. At the end of the growing season, the yield (in pounds) of the wheat on the plot is measured. The following data are based on information taken from an article by G. A. Wiebe in the *Journal of Agricultural Research* (vol. 50, pp. 331–357). For a random sample of years, one plot gave the following annual wheat production (in pounds):

4.15	4.21	4.27	3.55	3.50	3.79	4.09	4.42
3.89	3.87	4.12	3.09	4.86	2.90	5.01	3.39

Use a calculator to verify that, for this plot, the sample variance is $s^2 \approx 0.332$.

Another random sample of years for a second plot gave the following annual wheat production (in pounds):

4.03	3.77	3.49	3.76	3.61	3.72	4.13	4.01
3.59	4.29	3.78	3.19	3.84	3.91	3.66	4.35

Use a calculator to verify that the sample variance for this plot is $s^2 \approx 0.089$.

Test the claim that the population variance of annual wheat production for the first plot is larger than that for the second plot. Use a 1% level of significance.

2. *Agriculture: Wheat* Two plots at Rothamsted Experimental Station (see reference in Problem 1) were studied for production of wheat straw. For a random sample of years, the annual wheat straw production (in pounds) from one plot was as follows:

6.17	6.05	5.89	5.94	7.31	7.18
7.06	5.79	6.24	5.91	6.14	

Use a calculator to verify that, for the preceding data, $s^2 \approx 0.318$.

Another random sample of years for a second plot gave the following annual wheat straw production (in pounds):

6.85	7.71	8.23	6.01	7.22	5.58	5.47	5.86

Use a calculator to verify that, for these data, $s^2 \approx 1.078$.

Test the claim that there is a difference (either way) in the population variance of straw production for these two plots. Use a 5% level of significance.

3. *Economics: Productivity* An economist wonders if corporate productivity in some countries is more *volatile* than in other countries. One measure of a company's productivity is annual percentage yield based on total company assets. Data for this problem are based on information taken from *Forbes Top Companies,* edited by J. T. Davis. A random sample of leading companies in France gave the following percentage yield based on assets:

4.4	5.2	3.7	3.1	2.5	3.5	2.8	4.4	5.7	3.4	4.1
6.2	3.5	3.2	7.2	6.5	4.5	3.8	2.8	2.5	4.5	

Use a calculator to verify that $s^2 \approx 1.786$ for this sample of French companies.

Another random sample of leading companies in Germany gave the following percentage yield based on assets:

| 3.0 | 3.6 | 3.7 | 4.7 | 5.1 | 6.1 | 5.0 | 5.4 | 3.2 |
| 3.3 | 3.7 | 2.6 | 2.8 | 3.0 | 2.4 | 2.2 | 4.7 | 3.2 |

Use a calculator to verify that $s^2 \approx 1.285$ for this sample of German companies.

Test the claim that there is a difference (either way) in the population variance for leading companies in France and Germany. Use a 5% level of significance. How could your test conclusion relate to the economist's question regarding *volatility* (data spread) of corporate productivity of large companies in France compared with Germany?

4. *Economics: Productivity* A random sample of leading companies in South Korea gave the following percentage yield based on assets (see reference in Problem 3):

| 2.5 | 2.0 | 4.5 | 1.8 | 0.5 | 3.6 | 2.4 |
| 0.2 | 1.7 | 1.8 | 1.4 | 5.4 | 1.1 |

Use a calculator to verify that $s^2 = 2.247$ for these South Korean companies.

Another random sample of leading companies in Sweden gave the following percentage yield on assets:

| 2.3 | 3.2 | 3.6 | 1.2 | 3.6 | 2.8 | 2.3 | 3.5 | 2.8 |

Use a calculator to verify that $s^2 = 0.624$ for these Swedish companies.

Test the claim that the population variance of percentage yield on assets for South Korean companies is higher than for companies in Sweden. Use a 5% level of significance. How could your test conclusion relate to an economist's question regarding *volatility* of corporate productivity of large companies in South Korea compared with Sweden?

5. *Investing: Mutual Funds* You don't need to be rich to buy a few shares in a mutual fund. The question is how *reliable* are mutual funds as investments? This depends on the type of fund you buy. The following data are based on information taken from *Morningstar*, a mutual fund guide available in most libraries. A random sample of percentage annual returns for mutual funds holding stocks in aggressive-growth small companies is shown below.

| −1.8 | 14.3 | 41.5 | 17.2 | −16.8 | 4.4 | 32.6 | −7.3 | 16.2 | 2.8 | 34.3 |
| −10.6 | 8.4 | −7.0 | −2.3 | −18.5 | 25.0 | −9.8 | −7.8 | −24.6 | 22.8 |

Use a calculator to verify that $s^2 \approx 348.43$ for the sample of aggressive-growth small company funds.

Another random sample of percentage annual returns for mutual funds holding value (i.e., market underpriced) stocks in large companies is shown below.

| 16.2 | 0.3 | 7.8 | −1.6 | −3.8 | 19.4 | −2.5 | 15.9 | 32.6 | 22.1 | 3.4 |
| −0.5 | −8.3 | 25.8 | −4.1 | 14.6 | 6.5 | 18.0 | 21.0 | 0.2 | −1.6 |

Use a calculator to verify that $s^2 \approx 137.31$ for value stocks in large companies.

Test the claim that the population variance for mutual funds holding aggressive-growth small stocks is larger than the population variance for mutual funds holding value stocks in large companies. Use a 5% level of significance. How could your test conclusion relate to the question of *reliability* of returns for each type of mutual fund?

6. *Investing: Mutual Funds* How *reliable* are mutual funds that invest in bonds? Again, this depends on the bond fund you buy (see reference in Problem 5). A random sample of annual percentage returns for mutual funds holding short-term U.S. government bonds is shown below.

4.6	4.7	1.9	9.3	−0.8	4.1	10.5
4.2	3.5	3.9	9.8	−1.2	7.3	

Use a calculator to verify that $s^2 \approx 13.59$ for the preceding data.

A random sample of annual percentage returns for mutual funds holding intermediate-term corporate bonds is shown below.

−0.8	3.6	20.2	7.8	−0.4	18.8	−3.4	10.5
8.0	−0.9	2.6	−6.5	14.9	8.2	18.8	14.2

Use a calculator to verify that $s^2 \approx 72.06$ for returns from mutual funds holding intermediate-term corporate bonds.

Use $\alpha = 0.05$ to test the claim that the population variance for annual percentage return of mutual funds holding short-term government bonds is different from the population variance for mutual funds holding intermediate-term corporate bonds. How could your test conclusion relate to the question of *reliability* of returns for each type of mutual fund?

7. *Engineering: Fuel Injection* A new fuel injection system has been engineered for pickup trucks. The new system and the old system both produce about the same average miles per gallon. However, engineers question which system (old or new) will give better *consistency* in fuel consumption (miles per gallon) under a variety of driving conditions. A random sample of 31 trucks was fitted with the new fuel injection system and driven under different conditions. For these trucks, the sample variance of gasoline consumption was 51.4. Another random sample of 25 trucks was fitted with the old fuel injection system and driven under a variety of different conditions. For these trucks, the sample variance of gasoline consumption was 38.6. Test the claim that there is a difference in population variance of gasoline consumption for the two injection systems. Use a 5% level of significance. How could your test conclusion relate to the question regarding the *consistency* of fuel consumption for the two fuel injection systems?

8. *Engineering: Thermostats* A new thermostat has been engineered for the frozen food cases in large supermarkets. Both the old and new thermostats hold temperatures at an average of 25°F. However, it is hoped that the new thermostat might be more *dependable* in the sense that it will hold temperatures closer to 25°F. One frozen food case was equipped with the new thermostat, and a random sample of 21 temperature readings gave a sample variance of 5.1. Another similar frozen food case was equipped with the old thermostat, and a random sample of 16 temperature readings gave a sample variance of 12.8. Test the claim that the population variance of the old thermostat temperature readings is larger than that for the new thermostat. Use a 5% level of significance. How could your test conclusion relate to the question regarding the *dependability* of the temperature readings?

11.5
One-Way ANOVA: Comparing Several Sample Means

Introduction

In our past work, to determine the existence (or nonexistence) of a significant difference between population means, we restricted our attention to only two data groups representing the means in question. Many statistical applications in psychology, social science, business administration, and natural science involve many means and many data groups. Questions commonly asked are: Which of *several* alternative methods yields the best results in a particular setting? Which of *several* treatments leads to highest incidence of patient recovery? Which of *several* teaching methods leads to greatest student retention? Which of *several* investment schemes leads to greatest economic gain?

Using our previous methods (Sections 9.6 and 9.7) of comparing only *two* means would require many tests of significance to answer the preceding questions. For example, if we had only 5 variables, we would be required to perform 10 tests of significance in order to compare each variable to each of the other variables. If we had the time and patience, we could perform all 10 tests, but what about the risk of accepting a difference where there really is no difference (a type I error)? If the risk of a type I error on each test is $\alpha = 0.05$, then on 10 tests we expect the number of tests with a type I error to be 10(0.05) or 0.5 (see expected value, Section 5.3). This situation may not seem too serious to you, but remember that in a "real-world" problem and with the aid of a high-speed computer, a researcher may want to study the effect of 50 variables on the outcome of an experiment. Using a little mathematics, we could show that the study would require 1225 separate tests to check *each pair* of variables for a significant difference of means. At the $\alpha = 0.05$ level of significance of each test, we could expect (1225)(0.05) or 61.25 of the tests to have a type I error. In other words, these 61.25 tests would say there are differences between means when there really are no differences.

To avoid such problems, statisticians have developed a method called *analysis of variance* (abbreviated *ANOVA*). We will study single-factor analysis of variance (also called *one-way ANOVA*). With appropriate modification, methods of single-factor ANOVA generalize to *n*-dimensional ANOVA, but we leave that topic to more advanced studies.

EXAMPLE 6

One-way ANOVA test

Explanation of notation and formulas

A psychologist is studying the effect of dream deprivation on a person's anxiety level during waking hours. Brain waves, heart rate, and eye movements can be used to determine if a sleeping person is about to enter into a dream period. Three groups of subjects were randomly chosen from a large group of college students who volunteered to participate in the study. Group I subjects had their sleep interrupted four times each night but never during or immediately before a dream. Group II subjects had their sleep interrupted four times also, but on two occasions they were at the onset of a dream. Group III subjects were wakened four times, each time at the onset of a dream. This procedure was repeated 10 nights, and each day all subjects were given a test to determine level of anxiety. The data in Table 11-14 on the following page record the total of the test scores for each person over the entire project. Higher totals mean higher anxiety levels.

From Table 11-14, we see that group I had $n_1 = 6$ subjects, group II had $n_2 = 7$ subjects, and group III had $n_3 = 5$ subjects. For each subject, the anxiety score (x value) and the square of the test score (x^2 value) are also shown. In addition, special sums are shown.

TABLE 11-14   Dream Deprivation Study

Group I $n_1 = 6$ Subjects		Group II $n_2 = 7$ Subjects		Group III $n_3 = 5$ Subjects	
x_1	x_1^2	x_2	x_2^2	x_3	x_3^2
9	81	10	100	15	225
7	49	9	81	11	121
3	9	11	121	12	144
6	36	10	100	9	81
5	25	7	49	10	100
8	64	6	36		
		8	64		
$\Sigma x_1 = 38$	$\Sigma x_1^2 = 264$	$\Sigma x_2 = 61$	$\Sigma x_2^2 = 551$	$\Sigma x_3 = 57$	$\Sigma x_3^2 = 671$

$N = n_1 + n_2 + n_3 = 18$

$\Sigma x_{TOT} = \Sigma x_1 + \Sigma x_2 + \Sigma x_3 = 156$

$\Sigma x_{TOT}^2 = \Sigma x_1^2 + \Sigma x_2^2 + \Sigma x_3^2 = 1486$

We will outline the procedure for single-factor ANOVA in six steps. Each step will contain general methods and rationale appropriate to all single-factor ANOVA tests. As we proceed, we will use the data of Table 11-14 for a specific reference example.

Our application of ANOVA requires three basic assumptions. In a general problem with k groups:

Basic assumptions of ANOVA

1. We assume that each of our k groups of measurements is obtained from a population with a *normal* distribution.

2. Each group is randomly selected and is *independent* of all other groups. In particular, this means that we will not use the same subjects in more than one group and that the scores of one subject will not have an effect on the scores of another subject.

3. We assume that the variables from each group come from distributions with approximately the *same standard deviation*.

STEP 1: Determine the Null and Alternate Hypotheses

The purpose of an ANOVA test is to determine the existence (or nonexistence) of a statistically significant difference *among* the group means. In a general problem with k groups, we call the (population) mean of the first group μ_1, the population mean of the second group μ_2, and so forth. The null hypothesis is simply that *all* the group population means are the same. Since our basic assumptions say that each of the k groups of measurements comes from normal, independent distributions with common standard deviation, the null hypothesis says that all the sample groups come from *one and the same* population. The alternate hypothesis is that *not all* the group population means are equal. Therefore, in a problem with k groups we have

$$H_0: \mu_1 = \mu_2 = \cdots = \mu_k$$

H_1: At least two of the means μ_1, μ_2, . . . , μ_k are not equal.

Notice that the alternate hypothesis claims that *at least* two of the means are not equal. If more than two of the means are unequal, the alternate hypothesis is, of course, satisfied.

In our dream problem, we have $k = 3$; μ_1 is the population mean of group I, μ_2 is the population mean of group II, and μ_3 is the population mean of group III. Therefore,

$$H_0: \mu_1 = \mu_2 = \mu_3$$

H_1: At least two of the means μ_1, μ_2, μ_3 are not equal.

We will test the null and alternate hypotheses using an $\alpha = 0.05$ or $\alpha = 0.01$ level of significance. Notice that only one test is being performed even though we have $k = 3$ groups and three corresponding means. Using ANOVA avoids the problem of using multiple tests mentioned earlier.

STEP 2: Find SS_{TOT}

The concept of *sum of squares* is very important in statistics. We used a sum of squares in Chapter 3 to compute the sample standard deviation and sample variance.

$$s = \sqrt{\frac{\Sigma(x - \overline{x})^2}{n - 1}} \qquad \text{sample standard deviation}$$

$$s^2 = \frac{\Sigma(x - \overline{x})^2}{n - 1} \qquad \text{sample variance}$$

The numerator of the sample variance is a special sum of squares that plays a central role in ANOVA. Since this numerator is so important, we give it a special name SS (for sum of squares).

$$SS = \Sigma(x - \overline{x})^2 \qquad\qquad\qquad (2)$$

Using some college algebra, it can be shown that the following simpler formula is equivalent to Equation (2) and involves fewer calculations:

$$SS = \Sigma x^2 - \frac{(\Sigma x)^2}{n} \qquad\qquad\qquad (3)$$

where n is the sample size.

In future reference to SS, we will use Equation (3) because it is easier to use than Equation (2).

The *total sum of squares* SS_{TOT} can be found by using the entire collection of all data values in all groups:

$$SS_{TOT} = \Sigma x_{TOT}^2 - \frac{(\Sigma x_{TOT})^2}{N} \qquad\qquad\qquad (4)$$

where $N = n_1 + n_2 + \cdots + n_k$ is the total sample size from all groups.

Σx_{TOT} = sum of all data = $\Sigma x_1 + \Sigma x_2 + \cdots + \Sigma x_k$

Σx_{TOT}^2 = sum of all data squares = $\Sigma x_1^2 + \cdots + \Sigma x_k^2$

Using the specific data given in Table 11-14 for the dream example, we have

$k = 3$ total number of groups

$N = n_1 + n_2 + n_3 = 6 + 7 + 5 = 18$ total number of subjects

Σx_{TOT} = total sum of x values = $\Sigma x_1 + \Sigma x_2 + \Sigma x_3 = 38 + 61 + 57 = 156$

Σx_{TOT}^2 = total sum of x^2 values = $\Sigma x_1^2 + \Sigma x_2^2 + \Sigma x_3^2 = 264 + 551 + 671 = 1486$

Therefore, using Equation (4), we have

$$SS_{TOT} = \Sigma x_{TOT}^2 - \frac{(\Sigma x_{TOT})^2}{N} = 1486 - \frac{(156)^2}{18} = 134$$

The numerator for the total variation for all groups in our dream example is $SS_{TOT} = 134$. What interpretation can we give to SS_{TOT}? If we let $\bar{x}_{TOT}$ be the mean of all x values for all groups, then

$$\text{Mean of all } x \text{ values} = \bar{x}_{TOT} = \frac{\Sigma x_{TOT}}{N}$$

Under the null hypothesis (all groups come from the same normal distribution), $SS_{TOT} = \Sigma(x_{TOT} - \bar{x}_{TOT})^2$ represents the numerator of the sample variance for all groups. Therefore, SS_{TOT} represents total variability of the data. Total variability can occur in two ways:

1. Scores may differ from one another because they belong to *different groups* with different means (recall that the alternate hypothesis says that the means are not all equal). This difference is called *between-group variability* and is denoted SS_{BET}.

2. Inherent differences unique to each subject and differences due to chance may cause a particular score to be different from the mean of its *own group*. This difference is called *within-group variability* and is denoted SS_W.

Because total variability SS_{TOT} is a sum of between-group variability SS_{BET} and within-group variability SS_W, we may write

$$SS_{TOT} = SS_{BET} + SS_W$$

As we will see, SS_{BET} and SS_W are going to help us decide whether or not to reject the null hypothesis. Therefore, our next two steps are to compute these two quantities.

STEP 3: Find SS_{BET}

Recall that $\bar{x}_{TOT}$ is the mean of all x values from all groups. Between-group variability (SS_{BET}) measures the variability of group means. Because different groups may have different numbers of subjects, we must "weight" the variability contribution from each group by the group size n_i.

$$SS_{BET} = \sum_{\text{all groups}} n_i(\overline{x}_i - \overline{x}_{TOT})^2$$

where n_i = sample size of group i

$\overline{x}_i$ = sample mean of group i

$\overline{x}_{TOT}$ = mean for values from all groups

If we use algebraic manipulations, we can write the formula for SS_{BET} in the following computationally easier form:

$$SS_{BET} = \sum_{\text{all groups}} \left(\frac{(\Sigma x_i)^2}{n_i} \right) - \frac{(\Sigma x_{TOT})^2}{N} \tag{5}$$

where, as before, $N = n_1 + n_2 + \cdots + n_k$

Σx_i = sum of data in group i

Σx_{TOT} = sum of data from all groups

Using data from Table 11-14 for the dream example, we have

$$\begin{aligned} SS_{BET} &= \sum_{\text{all groups}} \left(\frac{(\Sigma x_i)^2}{n_i} \right) - \frac{(\Sigma x_{TOT})^2}{N} \\ &= \frac{(\Sigma x_1)^2}{n_1} + \frac{(\Sigma x_2)^2}{n_2} + \frac{(\Sigma x_3)^2}{n_3} - \frac{(\Sigma x_{TOT})^2}{N} \\ &= \frac{(38)^2}{6} + \frac{(61)^2}{7} + \frac{(57)^2}{5} - \frac{(156)^2}{18} \\ &= 70.038 \end{aligned}$$

Therefore, the numerator of between-group variation is

$$SS_{BET} = 70.038$$

STEP 4: Find SS_W

We could find the value of SS_W by using the formula relating SS_{TOT} with SS_{BET} and SS_W and solving for SS_W:

$$SS_W = SS_{TOT} - SS_{BET}$$

However, we prefer to compute SS_W a different way and use the preceding formula as a check on our calculations.

SS_W is the numerator of the variation within groups. Inherent differences unique to each subject and differences due to chance create the variability assigned to SS_W. In a general problem with k groups, the variability within the ith group could be represented by

$$SS_i = \Sigma(x_i - \overline{x}_i)^2$$

or by the mathematically equivalent formula

$$SS_i = \Sigma x_i^2 - \frac{(\Sigma x_i)^2}{n_i} \tag{6}$$

Because SS_i represents the variation within the ith group and we are seeking SS_W, the variability within *all* groups, we simply add SS_i for all groups:

$$SS_W = SS_1 + SS_2 + \cdots + SS_k \tag{7}$$

Using Equations (6) and (7) and the data of Table 11-14 with $k = 3$, we have

$$SS_1 = \Sigma x_1^2 - \frac{(\Sigma x_1)^2}{n_1} = 264 - \frac{(38)^2}{6} = 23.333$$

$$SS_2 = \Sigma x_2^2 - \frac{(\Sigma x_2)^2}{n_2} = 551 - \frac{(61)^2}{7} = 19.429$$

$$SS_3 = \Sigma x_3^2 - \frac{(\Sigma x_3)^2}{n_3} = 671 - \frac{(57)^2}{5} = 21.200$$

$$SS_W = SS_1 + SS_2 + SS_3 = 23.333 + 19.429 + 21.200 = 63.962$$

Let us check our calculation by using SS_{TOT} and SS_{BET}.

$$SS_{TOT} = SS_{BET} + SS_W$$
$$134 = 70.038 + 63.962 \qquad \text{(from Steps 2 and 3)}$$

We see that our calculation checks.

STEP 5: Find Variance Estimates (Mean Squares)

In Steps 3 and 4, we found SS_{BET} and SS_W. Although these quantities represent variability between groups and within groups, they are not yet the variance estimates we need for our ANOVA test. You may recall our study of Student's t distribution in which we introduced the concept of degrees of freedom. Degrees of freedom represent the number of values that are free to vary once we have placed certain restrictions on our data. In ANOVA, there are two types of degrees of freedom: $d.f._{BET}$, representing the degrees of freedom between groups, and $d.f._W$, representing degrees of freedom within groups. A theoretical discussion beyond the scope of this text would show

$d.f._{BET} = k - 1$ where k is the number of groups
$d.f._W = N - k$ where N is the total sample size

(*Note*: $d.f._{BET} + d.f._W = N - 1$.)

The variance estimates we are looking for are designated as follows:

MS_{BET}, the variance between groups (read *mean square between*)
MS_W, the variance within groups (read *mean square within*)

In the literature of ANOVA, the variances between and within groups are usually referred to as *mean squares* between and within groups, respectively. We will use the mean-square notation because it is used so commonly. However, remember that the notation MS_{BET} and MS_W both refer to *variances,* and you might occasionally see the variance notation S^2_{BET} and S^2_W used for these quantities. The formulas for the variances between and within samples follow the pattern of the basic formula for sample variance.

$$\text{Sample variance} = s^2 = \frac{\Sigma(x - \bar{x})^2}{n - 1} = \frac{SS}{n - 1}$$

Instead of using $n - 1$ in the denominator for MS_{BET} and MS_W variances, we use their respective degrees of freedom.

$$MS_{BET} = \frac{SS_{BET}}{d.f._{BET}} = \frac{SS_{BET}}{k - 1}$$

$$MS_W = \frac{SS_W}{d.f._{W}} = \frac{SS_W}{N - k}$$

Using these two formulas and the data of Table 11-14, we find the mean squares within and between variances for the dream deprivation example:

$$MS_{BET} = \frac{SS_{BET}}{k - 1} = \frac{70.038}{3 - 1} = 35.019$$

$$MS_W = \frac{SS_W}{N - k} = \frac{63.962}{18 - 3} = 4.264$$

STEP 6: Find the *F* Ratio and Complete the ANOVA Test

The logic of our ANOVA test rests on the fact that one of the variances, $MS_{BET,}$ *can* be influenced by population differences among means of the several groups, whereas the other variance, MS_W, *cannot* be so influenced. For instance, in the dream deprivation and anxiety study, the variance between groups MS_{BET} will be affected if any of the treatment groups has a population mean anxiety score *different* from any other group. On the other hand, the variance within groups MS_W compares anxiety scores of each treatment group to its own group anxiety mean, and the fact that group means might differ *does not* affect the MS_W value.

Recall that the null hypothesis claims that all the groups are samples from populations having the *same* (normal) distributions. The alternate hypothesis says that at least two of the sample groups come from populations with *different* (normal) distributions.

If the *null* hypothesis is *true,* MS_{BET} and MS_W should both estimate the *same* quantity. Therefore, if H_0 is true, the *F ratio*

$$F = \frac{MS_{BET}}{MS_W}$$

should be approximately 1, and variations away from 1 should occur only because of sampling errors. The variance within groups MS_W is a good estimate of the overall population variance, but the variance between groups MS_{BET} consists of the

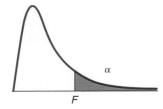

TABLE 11-15   *F* Distribution (Excerpt from Table 8 of Appendix II)

d.f.$_D$	α	Degrees of Freedom for Numerator			
		1	2	3	$\cdots$
1	0.050	161.45	199.50	215.71	$\cdots$
	0.025	647.79	699.50	864.16	$\cdots$
	0.010	4052.2	4999.5	5403.4	$\cdots$
2	0.050	18.51	19.00	19.16	$\cdots$
	0.025	38.51	39.00	39.17	$\cdots$
	0.010	98.50	99.00	99.17	$\cdots$
$\vdots$	$\vdots$	$\vdots$	$\vdots$	$\vdots$	
✓15	0.050	4.54	3.68	3.29	$\cdots$
	0.025	6.20	4.77	4.15	$\cdots$
	0.010	8.68	6.36	5.42	$\cdots$

Source: From *Biometrika Tables for Statisticians,* Vol. 1, by permission of the Biometrika Trustees.

population variance *plus* an additional variance stemming from the differences between samples. Therefore, if the *null* hypothesis is *false*, MS_{BET} will be larger than MS_W, and the *F* ratio will tend to be *larger* than 1.

The decision of whether or not to reject the null hypothesis is determined by the relative size of the *F* ratio. The *F* ratio and its corresponding probability distribution were invented by the English statistician Sir Ronald Fisher (1890–1962). Table 8 of Appendix II gives critical *F* values for levels of significance $\alpha = 0.050$, $\alpha = 0.025$, and $\alpha = 0.01$.

For our example about dreams, the computed *F* ratio is

$$F = \frac{MS_{BET}}{MS_W} = \frac{35.019}{4.264} = 8.213$$

Because larger *F* values tend to discredit the null hypothesis, our rejection (critical) region will be a *right tail* of the *F distribution*. To find a critical value, we use Table 8 in Appendix II. The table requires us to know *degrees of freedom for the numerator* and *degrees of freedom for the denominator* as well as the level of significance. The degrees of freedom are simply

$d.f._{BET}$ = degrees of freedom for the numerator

$d.f._W$ = degrees of freedom for the denominator

From Step 5, we know that

$d.f._N = d.f._{BET} = k - 1 = 3 - 1 = 2$

$d.f._D = d.f._W = N - k = 18 - 3 = 15$

Using $\alpha = 0.05$ as our level of significance, Table 8 gives $F_{0.05} = 3.68$ as our critical value (see Table 11-15). To find the critical value for $\alpha = 0.05$ and $d.f._N = 2$ and $d.f._D = 15$, we look in the column headed by 2 and the row block headed by 15. Since $\alpha = 0.05$, we use the row with that header for α.

As we can see from Figure 11–15, our observed *F* value of 8.213 lies in the critical or rejection region. Therefore, at the $\alpha = 0.05$ level of significance, we reject

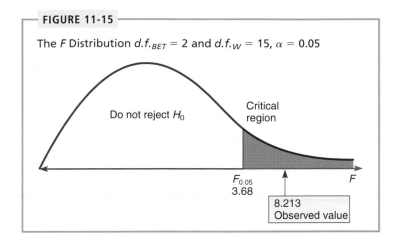

FIGURE 11-15

The F Distribution $d.f._{BET} = 2$ and $d.f._W = 15$, $\alpha = 0.05$

H_0 and conclude that not all the means are equal. The amount of dream deprivation *does* make a difference in mean anxiety level.   ◊

This completes our single-factor ANOVA test. Before we consider another example, let's summarize the main points.

Summary

If we have k independent data groups, each group belonging to a normal distribution and all groups having (approximately) the same standard deviation, then

1. H_0: $\mu_1 = \mu_2 = \cdots = \mu_k$

    H_1: not all of $\mu_1, \mu_2, \ldots, \mu_k$ are equal

    where μ_i is the population mean of group i

2. $SS_{TOT} = \Sigma x_{TOT}^2 - \dfrac{(\Sigma x_{TOT})^2}{N}$

    where Σx_{TOT} is the sum of all data elements from all groups

    Σx_{TOT}^2 is the sum of all data elements squared from all groups

    N is the total sample size

3. $SS_{TOT} = SS_{BET} + SS_W$

    where $SS_{BET} = \displaystyle\sum_{\text{all groups}} \left(\dfrac{(\Sigma x_i)^2}{n_i} \right) - \dfrac{(\Sigma x_{TOT})^2}{N}$

    n_i is the number of data elements in group i

    Σx_i is the sum of the data elements in group i

    $SS_W = \displaystyle\sum_{\text{all groups}} \left(\Sigma x_i^2 - \dfrac{(\Sigma x_i)^2}{n_i} \right)$

4. $MS_{BET} = \dfrac{SS_{BET}}{d.f._{BET}}$ where $d.f._{BET} = k - 1$

    $MS_W = \dfrac{SS_W}{d.f._W}$ where $d.f._W = N - k$

5. $F = \dfrac{MS_{BET}}{MS_W}$

TABLE 11-16   **Summary of ANOVA Results**

Source of Variation	Sum of Squares	Degrees of Freedom	Mean Square (Variance)	F Ratio	F Critical Value	Test Decision
			Basic Model			
Between groups	SS_{BET}	$d.f._{BET}$	MS_{BET}	$\dfrac{MS_{BET}}{MS_W}$	From table	Reject H_0 or fail to reject H_0
Within groups	SS_W	$d.f._W$	MS_W			
Total	SS_{TOT}	$N-1$				
			Summary of ANOVA Results from Dream Experiment			
Between groups	70.038	2	35.019	8.213	3.68	Reject H_0
Within groups	63.962	15	4.264			
Total	134	17				

6. We use a right-tailed test with critical region and critical values from Table 8 of Appendix II, the *F* distribution. To find the critical value, we use

(a) the given level of significance

(b) *degrees of freedom for numerator* $= d.f._{BET} = k - 1$

(c) *degrees of freedom for denominator* $= d.f._W = N - k$

Because an ANOVA test requires a number of calculations, we recommend that you summarize your results in a table such as Table 11-16.

GUIDED EXERCISE 11

One-way ANOVA test

A psychologist is studying pattern-recognition skills under four laboratory settings. In each setting, a fourth-grade child is given a pattern-recognition test with 10 patterns to identify. In setting I, the child is given *praise* for each correct answer and no comment about wrong answers. In setting II, the child is given *criticism* for each wrong answer and no comment about correct answers. In setting III, the child is given no praise or criticism, but the observer expresses *interest* in what the child is doing. In setting IV, the observer remains *silent* in an adjacent room watching the child through a one-way mirror. A random sample of fourth-grade children was used, and each child participated in the test only once. The test scores (number correct) for each group follow. (See Table 11-17.)

(a) Fill in the missing entries of Table 11-17.

$\Sigma x_{TOT} = $ _____

$\Sigma x^2{}_{TOT} = $ _____

$N = $ _____

$k = $ _____

$\Rightarrow$ $\Sigma x_{TOT} = \Sigma x_1 + \Sigma x_2 + \Sigma x_3 + \Sigma x_4$

      $= 41 + 14 + 38 + 28 = 121$

$\Sigma x^2{}_{TOT} = \Sigma x_1^2 + \Sigma x_2^2 + \Sigma x_3^2 + \Sigma x_4^2$

      $= 339 + 54 + 264 + 168 = 825$

$N = n_1 + n_2 + n_3 + n_4 = 5 + 4 + 6 + 5 = 20$

$k = 4$ groups

Continued

GUIDED EXERCISE 11 continued

TABLE 11-17   Pattern-Recognition Experiment

Group I (Praise) $n_1 = 5$		Group II (Criticism) $n_2 = 4$		Group III (Interest) $n_3 = 6$		Group IV (Silence) $n_4 = 5$	
x_1	x_1^2	x_2	x_2^2	x_3	x_3^2	x_4	x_4^2
9	81	2	4	9	81	5	25
8	64	5	25	3	9	7	49
8	64	4	16	7	49	3	9
9	81	3	9	8	64	6	36
7	49			5	25	7	49
				6	36		
$\Sigma x_1 = 41$		$\Sigma x_2 = 14$		$\Sigma x_3 = 38$		$\Sigma x_4 = 28$	
$\Sigma x_1^2 = 339$		$\Sigma x_2^2 = 54$		$\Sigma x_3^2 = 264$		$\Sigma x_4^2 = 168$	
$\Sigma x_{TOT} = $ _____		$\Sigma x^2_{TOT} = $ _____		$N = $ _____		$k = $ _____	

(b) What assumptions are we making about the data to apply a single-factor ANOVA test?

⇨ Because each of the groups comes from independent random samples (no child was tested twice), we need assume only that each group of data came from a normal distribution, and all the groups came from distributions with about the same standard deviation.

(c) What are the null and alternate hypotheses?

⇨ $H_0: \mu_1 = \mu_2 = \mu_3 = \mu_4$

In words, all the groups have the same population means, and this hypothesis, together with the basic assumptions of part (b), says that all the groups come from the same population.

H_1: not all the means μ_1, μ_2, μ_3, μ_4 are equal.

In words, not all the groups have the same population means, so at least one group did not come from the same population as the others.

(d) Find the value of SS_{TOT}.

⇨ $$SS_{TOT} = \Sigma x_{TOT}^2 - \frac{(\Sigma x_{TOT})^2}{N} = 825 - \frac{(121)^2}{20}$$
$$= 92.950$$

(e) Find SS_{BET}.

⇨ $$SS_{BET} = \sum_{\text{all groups}} \left(\frac{(\Sigma x_i)^2}{n_i} \right) - \frac{(\Sigma x_{TOT})^2}{N}$$
$$= \frac{(41)^2}{5} + \frac{(14)^2}{4} + \frac{(38)^2}{6}$$
$$+ \frac{(28)^2}{5} - \frac{(121)^2}{20} = 50.617$$

Continued

GUIDED EXERCISE 11 continued

(f) Find SS_W and check your calculations using the formula

$$SS_{TOT} = SS_{BET} + SS_W$$

➥ $$SS_W = \sum_{\text{all groups}} \left(\sum x_i^2 - \frac{(\sum x_i)^2}{n_i} \right)$$

$$SS_W = SS_1 + SS_2 + SS_3 + SS_4$$

$$SS_1 = \sum x_1^2 - \frac{(\sum x_1)^2}{n_1} = 339 - \frac{(41)^2}{5} = 2.800$$

$$SS_2 = \sum x_2^2 - \frac{(\sum x_2)^2}{n_2} = 54 - \frac{(14)^2}{4} = 5.000$$

$$SS_3 = \sum x_3^2 - \frac{(\sum x_3)^2}{n_3} = 264 - \frac{(38)^2}{6} = 23.333$$

$$SS_4 = \sum x_4^2 - \frac{(\sum x_4)^2}{n_4} = 168 - \frac{(28)^2}{5} = 11.200$$

$$SS_W = 42.333$$

Check: $SS_{TOT} = SS_{BET} + SS_W$

$$92.950 = 50.617 + 42.333 \text{ checks}$$

(g) Find $d.f._{BET}$ and $d.f._W$.

➥ $$d.f._{BET} = k - 1 = 4 - 1 = 3$$
$$d.f._W = N - k = 20 - 4 = 16$$

Check: $N - 1 = d.f._{BET} + d.f._W$

$$20 - 1 = 3 + 16 \text{ checks}$$

(h) Find the mean squares MS_{BET} and MS_W.

➥ $$MS_{BET} = \frac{SS_{BET}}{d.f._{BET}} = \frac{50.617}{3} = 16.872$$

$$MS_W = \frac{SS_W}{d.f._W} = \frac{42.333}{16} = 2.646$$

(i) Find the F ratio.

➥ $$F = \frac{MS_{BET}}{MS_W} = \frac{16.872}{2.646} = 6.376$$

(j) Find $F_{0.01}$, the critical value for an $\alpha = 0.01$ level of significance. Sketch the critical region and your observed F ratio. Does the test indicate we should reject the null hypothesis or not? Explain.

➥ $$d.f._N = d.f._{BET} = 3$$
$$d.f._D = d.f._W = 16$$

By Table 8 of Appendix II,

$$F_{0.01} = 5.29$$

The critical region is shown in Figure 11-16. Since the observed F ratio is in the critical region, we reject H_0 and conclude that there is a significant difference in population means of the four groups. The laboratory setting *does* affect the mean scores.

FIGURE 11-16 Critical Region

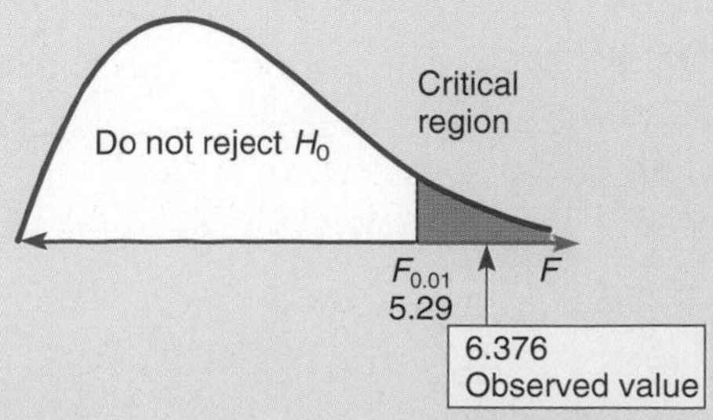

Continued

GUIDED EXERCISE 11 continued

(k) Make a summary table of this ANOVA test.   See Table 11-18.

TABLE 11-18   Summary of ANOVA Results for Pattern Recognition Experiment

Source of Variation	Sum of Squares	Degrees of Freedom	Mean Square (Variance)	F Ratio	F Critical Value	Test Decision
Between groups	50.617	3	16.872	6.376	5.29	Reject H_0
Within groups	42.333	16	2.646			
Total	92.950	19				

P values

◇ **COMMENT** We use Table 8 of Appendix II to estimate P values for ANOVA tests. For example, in Guided Exercise 11, the F ratio computed from the sample data is $F = 6.376$. To find an interval for the P value corresponding to $F = 6.376$, we look in Table 8 of Appendix II in the row headed by degrees of freedom in denominator $d.f. = 16$ and in the column headed by degrees of freedom in the numerator $d.f. = 3$. The table entry 3.24 corresponds to a P value of 0.05, while the entry 4.08 corresponds to a P value of 0.025 and the entry 5.29 corresponds to a P value of 0.01. Because the sample F ratio $F = 6.376$ is greater than the table entry 5.29, the P value for the sample statistic is less than 0.01:

$$P \text{ value} < 0.01$$

This means that we reject H_0 for all $\alpha \geq 0.01$. In particular, we reject H_0 when $\alpha = 0.01$. This conclusion is consistent with the results shown in Guided Exercise 11.   ◇

    TECH NOTE After you understand the process of ANOVA, technology tools offer valuable assistance in performing one-way ANOVA. The TI-83Plus, Excel, and Minitab all support one-way ANOVA. Both the TI-83Plus and Minitab use the terminology

Factor for Between Groups

Error for Within Groups

In all the technologies, enter data for each group in separate columns. The displays show results for Guided Exercise 11.

TI-83Plus Use **STAT**, **TESTS**, option **F:ANOVA** and enter the lists containing the data

```
One-way ANOVA
 F=6.376902887
 p=.0047646422
 Factor
  df=3
  SS=50.6166667
↓ MS=16.8722222
```

```
One-way ANOVA
↑ MS=16.8722222
 Error
  df=16
  SS=42.3333333
  MS=2.64583333
 Sxp=1.62660177
```

Excel Use Tools ➤ Data Analysis ➤ Anova: Single Factor

Anova: Single Factor						
SUMMARY						
Groups	*Count*	*Sum*	*Average*	*Variance*		
x1	5	41	8.2	0.7		
x2	4	14	3.5	1.666666667		
x3	6	38	6.333333333	4.666666667		
x4	5	28	5.6	2.8		
ANOVA						
Source of Variation	*SS*	*df*	*MS*	*F*	*P-value*	*F crit*
Between Groups	50.61666667	3	16.87222222	6.376902887	0.004764642	5.292235983
Within Groups	42.33333333	16	2.645833333			
Total	92.95	19				

Minitab Use Stat ➤ ANOVA ➤ Oneway(unstacked)

```
One-way Analysis of Variance

Analysis of Variance
Source    DF      SS      MS      F      P
Factor     3   50.62   16.87   6.38  0.005
Error     16   42.33    2.65
Total     19   92.95

                          Individual 95% CIs For Mean
                        Based on Pooled StDev
Level     N    Mean   StDev   ---+---------+---------+---------+---
x1        5   8.200   0.837                       (-----*-----)
x2        4   3.500   1.291   (------*------)
x3        6   6.333   2.160               (-----*-----)
x4        5   5.600   1.673           (------*------)
                              ---+---------+---------+---------+---
Pooled StDev = 1.627            2.5       5.0       7.5      10.0
```

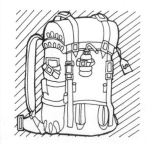

VIEWP◉INT *Gear Guide!*

How much does that backpack cost? Do brand names really make any difference in cost? One way to answer such a question is to construct a one-way ANOVA test to study list prices for various backpacks with similar features but different brand names. For more information, see *Backpacker Magazine*, vol. 25, issue 157.

SECTION 11.5   PROBLEMS

In each problem, assume that the distributions are normal and have approximately the same population standard deviations. In each problem, do the following:

(a) State the null and alternate hypotheses.

(b) Find SS_{TOT}, SS_{BET}, and SS_W and check that $SS_{TOT} = SS_{BET} + SS_W$.

(c) Find $d.f._{BET}$ and $d.f._W$.

(d) Find MS_{BET} and MS_W.

(e) Find the F ratio.

(f) Find the critical value $F_{0.01}$ or $F_{0.05}$ as the problem requires.

(g) Decide if the null hypothesis is to be rejected or not rejected for the given level of significance α.

(h) Make a summary table for your ANOVA test.

1. *Archaeology: Ceramics* Wind Mountain is an archaeological study area located in southwest New Mexico. Potsherds are broken pieces of prehistoric Native American clay vessels. One type of painted ceramic vessel is called *Mimbres classic black on white*. At three different sites the number of such sherds was counted in local dwelling excavations. (Source: Based on information from *Mimbres Mogollon Archaeology*, by A. I. Woosley and A. J. McIntyre, University of New Mexico Press.)

Site I	Site II	Site III
61	25	12
34	18	36
25	54	69
12	67	27
79		18
55		14
20		

Shall we reject or not reject the claim that there is no difference in population mean Mimbres classic black on white sherd counts for the three sites? Use a 1% level of significance.

2. *Archaeology: Ceramics* Another type of painted ceramic vessel is called *three circle red on white* (see reference in Problem 1). At four different sites in the Wind Mountain archaeological region, the number of such sherds was counted in local dwelling excavations.

Site I	Site II	Site III	Site IV
17	18	32	13
23	4	19	19
6	33	18	14
19	8	43	34
11	25		12
	16		15

Shall we reject or not reject the claim that there is no difference in the population mean three circle red on white sherd counts for the four sites? Use a 5% level of significance.

3. *Economics: Profits per Employee* How productive are U.S. workers? One way to answer this question is to study annual profits per employee. A random sample of companies in computers (I), aerospace (II), heavy equipment (III), and broadcasting

(IV) gave the following data regarding annual profits per employee (units in thousands of dollars). (Source: *Forbes Top Companies*, edited by J. T. Davis, John Wiley and Sons.)

I	II	III	IV
27.8	13.3	22.3	17.1
23.8	9.9	20.9	16.9
14.1	11.7	7.2	14.3
8.8	8.6	12.8	15.2
11.9	6.6	7.0	10.1
	19.3		9.0

Shall we reject or not reject the claim that there is no difference in population mean annual profits per employee in each of the four types of companies? Use a 5% level of significance.

4. *Economics: Profits per Employee* A random sample of companies in electric utilities (I), financial services (II), and food processing (III) gave the following information regarding annual profits per employee (units in thousands of dollars). (See reference in Problem 3.)

I	II	III
49.1	55.6	39.0
43.4	25.0	37.3
32.9	41.3	10.8
27.8	29.9	32.5
38.3	39.5	15.8
36.1		42.6
20.2		

Shall we reject or not reject the claim that there is no difference in population mean annual profits per employee in each of the three types of companies? Use a 1% level of significance.

5. *Ecology: Deer* Where are the deer? Random samples of square kilometer plots were taken in different ecological locations of Mesa Verde National Park. The deer count per square kilometer was recorded and is shown in the following table. (Source: *The Mule Deer of Mesa Verde National Park*, edited by G. W. Mierau and J. L. Schmidt, Mesa Verde Museum Association.)

Mountain Brush	Sagebrush Grassland	Pinon Juniper
30	20	5
29	58	7
20	18	4
29	22	9

Shall we reject or accept the claim that there is no difference among the mean number of deer per square kilometer in these different ecological locations? Use a 5% level of significance.

6. *Ecology: Vegetation* Wild irises are beautiful flowers found throughout the United States, Canada, and northern Europe. This problem concerns the length of the sepal (leaf-like part covering the flower) of different species of wild iris. Data are based on information taken from an article by R. A. Fisher in *Annals of*

Eugenics (vol. 7, part 2, pp. 179–188). Measurements of sepal length in centimeters from random samples of *Iris setosa* (I), *Iris versicolor* (II), and *Iris virginica* (III) are as follows:

I	II	III
5.4	5.5	6.3
4.9	6.5	5.8
5.0	6.3	4.9
5.4	4.9	7.2
4.4	5.2	5.7
5.8	6.7	6.4
5.7	5.5	
	6.1	

Shall we reject or not reject the claim that there are no differences among the population means of sepal length for the different species of iris? Use a 5% level of significance.

7. *Insurance: Sales* An executive at the home office of Big Rock Life Insurance is considering three branch managers as candidates for promotion to vice president. The branch reports include records showing sales volume for each salesperson in the branch (in hundreds of thousands of dollars). A random sample of these records was selected for salespersons in each branch. All three branches are located in cities where per capita income is the same. The executive wishes to compare these samples to see if there is a significant difference in performance of salespersons in the three different branches. If so, the information will be used to determine which of the managers to promote.

Branch Managed by Adams	Branch Managed by McDale	Branch Managed by Vasquez
7.2	8.8	6.9
6.4	10.7	8.7
10.1	11.1	10.5
11.0	9.8	11.4
9.9		
10.6		

Use an $\alpha = 0.01$ level of significance. Shall we reject the claim that there is no difference among the salespersons in the different branches or not?

8. *Ecology: Pollution* The quantity of dissolved oxygen is a measure of water pollution in lakes, rivers, and streams. Water samples were taken at four different locations in a river in an effort to determine if water pollution varied from location to location. Location I was 500 m above an industrial plant water discharge and near the shore. Location II was 200 m above the discharge point and in midstream. Location III was 50 m downstream from the discharge point and near the shore. Location IV was 200 m downstream from the discharge point and in midstream. The following table shows the results. Lower dissolved oxygen readings mean more pollution. Because of the difficulty in getting midstream samples, ecology students collecting the data had fewer of these samples. Use an $\alpha = 0.05$ level of significance. Do we reject the claim that the quantity of dissolved oxygen does not vary from one location to another or not?

Location I	Location II	Location III	Location IV
7.3	6.6	4.2	4.4
6.9	7.1	5.9	5.1
7.5	7.7	4.9	6.2
6.8	8.0	5.1	
6.2		4.5	

9. *Sociology: Ethnic Groups* A sociologist studying New York City ethnic groups wants to determine if there is a difference in income for immigrants from four different countries during their first year in the city. She obtained the data in the following table from a random sample of immigrants from these countries (incomes in thousands of dollars). Use a 0.05 level of significance to test the claim that there is no difference in the earnings of immigrants from the four different countries.

Country I	Country II	Country III	Country IV
12.7	8.3	20.3	17.2
9.2	17.2	16.6	8.8
10.9	19.1	22.7	14.7
8.9	10.3	25.2	21.3
16.4		19.9	19.8

11.6
Introduction to Two-Way ANOVA

FOCUS POINTS

✓ Learn the notation and setup for two-way ANOVA tests.

✓ Learn about the three main types of deviations and how they break into additional effects.

✓ Use mean-square values to compute different sample F statistics.

✓ Use the F distribution to conclude the test and estimate P values.

✓ Summarize experimental design features using a completely randomized design flowchart.

Suppose that Friendly Bank is interested in average customer satisfaction regarding the issue of obtaining bank balances and a summary of recent account transactions. Friendly Bank uses two systems, the first being a completely automated voice mail information system requiring customers to enter account numbers and passwords using the telephone keypad and the second being the use of bank tellers or bank representatives to give the account information personally to customers. In addition, Friendly Bank wants to learn if average customer satisfaction is the same regardless of the time of day of contact. Three times of day are under study: morning, afternoon, and evening.

Friendly Bank could do two studies: one regarding average customer satisfaction with regard to type of contact (automated or bank representative) and one regarding average customer satisfaction with regard to time of day. The first study could be done using a difference-of-means test because there are only two types of contact being studied. The second study could be accomplished using one-way ANOVA.

However, Friendly Bank could use just *one* study and the technique of *two-way analysis of variance* (known as *two-way ANOVA*) to simultaneously study average customer satisfaction not only with regard to the variable type of contact and to the variable time of day but also with regard to the *interaction* between the two variables. An interaction is present if, for instance, the difference in average customer satisfaction regarding type of contact is much more pronounced during the evening hours than, say, during the afternoon hours or, say, the morning hours.

Let's begin our study of two-way ANOVA by considering the organization of data appropriate to two-way ANOVA. Two-way ANOVA involves *two* variables. These variables are called *factors*. The *levels* of a factor are the different values the factor can assume. Example 7 demonstrates the use of this terminology for the information Friendly Bank is seeking.

TABLE 11-19   Table for Recording Average Customer Response

Factor 1: Time of Day	Factor 2: Type of Contact	
	Automated	Bank Representative
Morning	Morning Automated	Morning Bank Representative
Afternoon	Afternoon Automated	Afternoon Bank Representative
Evening	Evening Automated	Evening Bank Representative

EXAMPLE 7
Factors and levels

For the Friendly Bank study discussed earlier, identify the factors and their levels, and create a table displaying the information.

SOLUTION: There are two factors. Call factor 1 *time of day*. This factor has three levels: morning, afternoon, and evening. Factor 2 is *type of contact*. This factor has two levels: automated contact and personal contact through a bank representative. Table 11-19 shows how the information regarding customer satisfaction can be organized with respect to the two factors.

When we look at Table 11-19, we see six contact–time-of-day combinations. Each such combination is called a *cell* in the table. The number of cells in any two-way ANOVA data table equals the product of the number of levels of the row factor times the number of levels of the column factor. In the case illustrated by Table 11-19, we see that the number of cells is 3×2, or 6.   ◊

Basic assumptions of two-way ANOVA

Just as for one-way ANOVA, our application of two-way ANOVA requires some basic assumptions:

1. The measurements in each cell of a two-way ANOVA model are assumed to be drawn from a population with a normal distribution.

2. The measurements in each cell of a two-way ANOVA model are assumed to come from distributions with approximately the same variance.

3. The measurements in each cell come from independent random samples.

4. There are the *same number of measurements* in each cell.

Procedure to Conduct a Two-Way ANOVA Test (More Than One Measurement per Cell)

We will outline the procedure for two-way ANOVA in five steps. Each step will contain general methods and rationale appropriate to all two-way ANOVA tests with more than one data value in each cell. As we proceed, we will see how the method is applied to the Friendly Bank study.

Let's assume that Friendly Bank has taken random samples of customers fitting the criteria of each of the six cells described in Table 11-19. This means that a random sample of four customers fitting the morning-automated cell was surveyed. Another random sample of four customers fitting the afternoon-automated cell was surveyed, and so on. The bank measured customer satisfaction on a scale of 0 to 10 (10 representing highest customer satisfaction). The data appear in Table 11-20.

TABLE 11-20   Customer Satisfaction at Friendly Bank

Factor 1: Time of Day	Factor 2: Type of Contact		
	Automated	Bank Representative	Row Means
Morning	6, 5, 8, 4 $\bar{x} = 5.75$	8, 7, 9, 9 $\bar{x} = 8.25$	Row 1 $\bar{x} = 7.00$
Afternoon	3, 5, 6, 5 $\bar{x} = 4.75$	9, 10, 6, 8 $\bar{x} = 8.25$	Row 2 $\bar{x} = 6.50$
Evening	5, 5, 7, 5 $\bar{x} = 5.50$	9, 10, 10, 9 $\bar{x} = 9.50$	Row 3 $\bar{x} = 7.50$
Column means	Column 1 $\bar{x} = 5.33$	Column 2 $\bar{x} = 8.67$	Total $\bar{\bar{x}} = 7.00$

Table 11-20 also shows cell means, row means, column means, and the total mean $\bar{\bar{x}}$ computed for all 24 data values. We will use these means as we conduct the two-way ANOVA test.

As in any statistical test, the first task is to establish the hypotheses for the test. Then, as in one-way ANOVA, the *F* distribution is used to determine the test conclusion. To compute the sample *F* value for a given null hypothesis, many of the same kinds of computations are done as were done in one-way ANOVA. In particular, we will use degrees of freedom $d.f. = N - 1$ (where *N* is the total sample size) allocated among the row factor, the column factor, the interaction, and the error (corresponding to "within groups" of one-way ANOVA). We look at sum of squares *SS* (which measures variation) for the row factor, the column factor, the interaction, and the error. Then we compute the mean square *MS* for each category by taking the *SS* value and dividing by the corresponding degrees of freedom. Finally, we compute the sample *F* statistic for each factor and for the interaction by dividing the appropriate *MS* value by the *MS* value of the error.

STEP 1: Establish the Hypotheses

Because we have two factors, we have hypotheses regarding each of the factors separately (called *main effects*) and then hypotheses regarding the interaction between the factors.

These three sets of hypotheses are

1. H_0: There is no difference in population means among the levels of the row factor.
   H_1: At least two population means are different among the levels of the row factor.

2. H_0: There is no difference in population means among the levels of the column factor.
   H_1: At least two population means are different among the levels of the column factor.

3. H_0: There is no interaction between the factors.
   H_1: There is an interaction between the factors.

In the case of Friendly Bank, the hypotheses regarding the main effects are

H_0: There is no difference in population mean satisfaction depending on time of contact.

H_1: At least two population mean satisfaction measures are different depending on time of contact.

H_0: There is no difference in population mean satisfaction between the two types of customer contact.

H_1: There is a difference in population mean satisfaction between the two types of customer contact.

The hypotheses regarding interaction between factors are

H_0: There is no interaction between type of contact and time of contact.

H_1: There is interaction between type of contact and time of contact.

STEP 2: Compute *SS* Values

The calculations for the SS values are usually done on a computer. The main questions are whether population means differ according to the factors or the interaction of the factors. As we look at the Friendly Bank data in Table 11-20, we see that sample averages for customer satisfaction differ not only in each cell but also across the rows and across the columns. In addition, the total sample mean (designated $\bar{\bar{x}}$) differs from almost all the means. We know that different samples of the same size from the same population certainly can have different sample means. We need to decide if the differences are simply due to chance (sampling error) or are occurring because the samples have been taken from different populations with means that are not the same.

The tools we use to analyze the differences among the data values, the cell means, the row means, the column means, and the total mean are similar to those we used in Section 11.5 for one-way ANOVA. In particular, we first examine deviations of various measurements from the total mean $\bar{\bar{x}}$, and then we compute the sum of the squares *SS*.

There are basically three types of deviations:

Total deviation	= Treatment deviation	+ Error deviation
Compare each data value with the total mean $\bar{\bar{x}}$, $(x - \bar{\bar{x}})$.	For each data value, compare the mean of each cell with the total mean $\bar{\bar{x}}$, (cell $\bar{x} - \bar{\bar{x}}$).	Compare each data value with the mean of its cell, $(x - \text{cell } \bar{x})$.

The treatment deviation breaks down further as

Treatment deviation	= Deviation for main effect of factor 1	+ Deviation for main effect of factor 2	+ Deviation for interaction
For each data value, (cell $\bar{x} - \bar{\bar{x}}$).	For each data value, compare the row mean with the total mean $\bar{\bar{x}}$, (row $\bar{x} - \bar{\bar{x}}$).	For each data value, compare the column mean with the total mean $\bar{\bar{x}}$, (column $\bar{x} - \bar{\bar{x}}$).	For each data value, (cell $\bar{x}$ − corresponding row $\bar{x}$ − corresponding column $\bar{x}$ + $\bar{\bar{x}}$).

The deviations for each data value, row mean, column mean, or cell mean are then *squared and totaled over all the data*. This results in sum of squares or variations. The *treatment variations* correspond to *between-group variations* of one-way ANOVA. The *error variation* corresponds to the *within-group variation* of one-way ANOVA.

$$\text{Total variation} = \text{Treatment variation} + \text{Error variation}$$

$$\sum_{\text{all data}} (x - \overline{\overline{x}})^2 = \sum_{\text{all data}} (\text{cell } \overline{x} - \overline{\overline{x}})^2 + \sum_{\text{all data}} (x - \text{cell } \overline{x})^2$$

$$SS_{TOT} = SS_{TR} + SS_E$$

where

$$\text{Treatment variation} = \text{Factor 1 variation} + \text{Factor 2 variation} + \text{Interaction variation}$$

$$SS_{TR} = SS_{F1} + SS_{F2} + SS_{F1 \times F2}$$

$$\sum_{\text{all data}} (\text{cell } \overline{x} - \overline{\overline{x}})^2 = \sum_{\text{all data}} (\text{row } \overline{x} - \overline{\overline{x}})^2 + \sum_{\text{all data}} (\text{col } \overline{x} - \overline{\overline{x}})^2 + \sum_{\text{all data}} (\text{cell } \overline{x} - \text{row } \overline{x} - \text{col } \overline{x} + \overline{\overline{x}})^2$$

The actual calculation of all the required SS values is quite time-consuming. In most cases, computer packages are used to obtain the results. For the Friendly Bank data, the following table is a Minitab printout giving the sum of squares *SS* for the type-of-contact factor, the time-of-day factor, the interaction between time and type of contact, and the error.

```
Minitab Printout for Two-Way Analysis of Variance
Analysis of Variance for Response
Source            DF          SS          MS
Type              1           66.67       66.67
Time              2           4.00        2.00
Interaction       2           2.33        1.17
Error             18          29.00       1.61
Total             23          102.00
```

We see that $SS_{\text{type}} = 66.67$, $SS_{\text{time}} = 4.00$, $SS_{\text{interaction}} = 2.33$, $SS_{\text{error}} = 29.00$, and $SS_{TOT} = 102$ (the total of the other four sum of squares).

STEP 3: Compute the Mean Square (*MS*) Values

The calculations for the MS values are usually done on a computer. Although the sum of squares computed in Step 2 represents variation, we need to compute mean-square (*MS*) values for two-way ANOVA. As in one-way ANOVA, we compute *MS* values by dividing the *SS* values by respective degrees of freedom:

$$\text{Mean square } MS = \frac{\text{Corresponding sum of squares } SS}{\text{Respective degrees of freedom}}$$

For two-way ANOVA with more than one data value per cell, the degrees of freedom are

> **Degrees of freedom**
>
> d.f. of row factor $= r - 1$ d.f. of interaction $= (r - 1)(c - 1)$
>
> d.f. of column factor $= c - 1$ d.f. of error $= rc(n - 1)$
>
> d.f. of total $= nrc - 1$
>
> where $r =$ number of rows, $c =$ number of columns, $n =$ number of data in one cell.

The Minitab table shows the degrees of freedom and the MS values for the main effect factors, the interaction, and the error for the Friendly Bank study.

STEP 4: Compute the Sample F Statistic for Each Factor and for the Interaction

Under the assumption of the respective null hypothesis, we have

$$\text{Sample } F \text{ for row factor} = \frac{MS \text{ for row factor}}{MS \text{ for error}}$$

with degrees of freedom numerator, $d.f._N = d.f.$ of row factor
    degrees of freedom denominator, $d.f._D = d.f.$ of error

$$\text{Sample } F \text{ for column factor} = \frac{MS \text{ for column factor}}{MS \text{ for error}}$$

with degrees of freedom numerator, $d.f._N = d.f.$ of column factor
    degrees of freedom denominator, $d.f._D = d.f.$ of error

$$\text{Sample } F \text{ for interaction} = \frac{MS \text{ for interaction}}{MS \text{ for error}}$$

with degrees of freedom numerator, $d.f._N = d.f.$ for the interaction
    degrees of freedom denominator, $d.f._D = d.f.$ of error

For the Friendly Bank study, the sample F values are

$$\text{Sample } F \text{ for time:} \quad F = \frac{MS_{\text{time}}}{MS_{\text{error}}} = \frac{2.00}{1.61} = 1.24$$

$$d.f._N = 2 \quad \text{and} \quad d.f._D = 18$$

$$\text{Sample } F \text{ for type of contact:} \quad F = \frac{MS_{\text{type}}}{MS_{\text{error}}} = \frac{66.67}{1.61} = 41.41$$

$$d.f._N = 1 \quad \text{and} \quad d.f._D = 18$$

$$\text{Sample } F \text{ for interaction:} \quad F = \frac{MS_{\text{interaction}}}{MS_{\text{error}}} = \frac{1.17}{1.61} = 0.73$$

$$d.f._N = 2 \quad \text{and} \quad d.f._D = 18$$

STEP 5: Conclude the Test

Compare each sample test statistic with the critical value found in the F distribution table for a specified level of significance using the indicated degrees of freedom. If the sample test statistic is *larger* than the critical value, reject the null hypothesis at the specified level of significance. (See Figure 11-17 on the following page.)

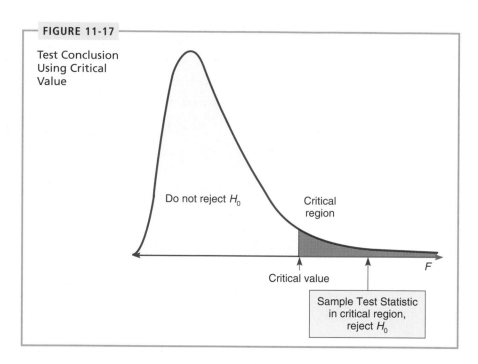

FIGURE 11-17

Test Conclusion Using Critical Value

Do not reject H_0

Critical region

Critical value

F

Sample Test Statistic in critical region, reject H_0

Some, but not all, statistical computer software packages provide *P* values for the sample test statistic. If the *P* value is available, it can be compared with the level of significance, and the test can be concluded in the usual way by comparing the *P* value with the level of significance.

Be sure to test for interaction between the factors *first*. If you *reject* the null hypothesis of no interaction, then you should *not* test for a difference of means in the levels of the row factors or a difference of means in the levels of the column factors because the interaction of the factors makes interpretation of the results of the main effects more complicated. A more extensive study of two-way ANOVA beyond the scope of this book shows how to interpret the results of the test of the main factors when there is interaction. For our purposes, we will simply stop the analysis rather than draw misleading conclusions.

If the test for interaction between the factors indicates that there is no evidence of interaction, then proceed to test the hypotheses regarding the levels of the row factor and the hypotheses regarding the levels of the column factor.

For the Friendly Bank study, we proceed as follows:

1. First, we determine if there is any evidence of interaction between the factors. The critical *F* value for interaction is found in Table 8 of Appendix II using $\alpha = 0.05$ and $d.f._N = 2$ and $d.f._D = 18$. The critical value is 3.55. Because the sample test statistic for interaction is $F = 0.73$ and this is smaller than the critical value, we do not reject H_0 for interaction. There is no evidence of interaction. Because there is no evidence of interaction between the main effects of type of contact and time of day, we proceed to test each factor for a difference of population mean satisfaction among the respective levels of the factors.

2. Next, we determine if there is a difference in mean satisfaction according to type of contact. The critical *F* value for type of contact is found in Table 8 of Appendix II using $\alpha = 0.05$ and $d.f._N = 1$ and $d.f._D = 18$. The critical value

is 4.41. Because the sample test statistic for type of contact is $F = 41.41$ and this is larger than the critical value, we see that there is evidence to reject H_0 that there is no difference in population mean satisfaction for the two types of contact. The test shows that there is a difference in average customer satisfaction between using an automated system and a bank representative.

3. Finally, we determine if there is a difference in mean satisfaction according to time of day. The critical F value for time of day is found in Table 8 of Appendix II using $\alpha = 0.05$ and $d.f._N = 2$ and $d.f._D = 18$. The critical value is 3.55. Because the sample test statistic for time of contact is $F = 1.24$ and this is smaller than the critical value, we see that there is no evidence to reject H_0 that there is no difference in population mean satisfaction for the time of day.

Special Case: One Observation in Each Cell with No Interaction

In the case where our data consist of only one value in each cell, there are no measures for sum of squares SS interaction or mean square MS interaction, and we cannot test for interaction of factors using two-way ANOVA. If it *seems reasonable* (based on other information) to assume that there is *no* interaction between the factors, then we can use two-way ANOVA techniques to test for average response differences due to the main effects. In Guided Exercise 12, we look at two-way ANOVA applied to the special case of only one measurement per cell and no interactions.

GUIDED EXERCISE 12

Special case two-way ANOVA

Let's use two-way ANOVA to test if the average fat content (grams of fat per 3-oz serving) of potato chips is different according to the brand or according to which laboratory made the measurement. Use $\alpha = 0.05$. (See the following Minitab tables.)

Average Grams of Fat in a 3-oz Serving of Potato Chips

Brand	Lab I	Lab II	Lab III
Texas Chips	32.4	33.1	32.9
Great Chips	37.9	37.7	37.8
Chip Ooh	29.1	29.4	29.5

(Laboratory)

Minitab Printout for *SS*, *MS*, and Degrees of Freedom

Analysis of Variance for Fat

Source	DF	SS	MS
Brand	2	108.7022	54.3511
Lab	2	0.1422	0.0711
Error	4	0.2244	0.0561
Total	8	109.0689	

(a) List the factors and the number of levels for each.

⟹ The factors are brand and laboratory. Each factor has three levels.

(b) Assuming that there is no interaction, list the hypotheses for each factor.

⟹ For brand,

H_0: There is no difference in population mean fat by brands.

H_1: At least two brands have different population mean fat content.

Continued

GUIDED EXERCISE 12 continued

For laboratory,

H_0: There is no difference in mean fat content as measured by the labs.

H_1: At least two of the labs give different mean fat measurements.

(c) Calculate the sample *F* statistic for brands, and compare it with the critical value found in Table 8 of Appendix II. What is your conclusion regarding average fat content among brands?

 For brand,

$$\text{Sample } F = \frac{MS_{\text{brand}}}{MS_{\text{error}}} \approx \frac{54.35}{0.056} \approx 970$$

The critical value for $d.f._N = 2$ and $d.f._D = 4$, and $\alpha = 0.05$ is 6.94. Because the sample *F* is greater than the critical value, we reject H_0 and conclude that at least two of the brands have different fat content.

(d) Calculate the sample *F* statistic for laboratories, and compare it with the critical value found in Table 8 of Appendix II. What is your conclusion regarding average fat content as measured by the different laboratories?

For laboratories,

$$\text{Sample } F = \frac{MS_{\text{lab}}}{MS_{\text{error}}} \approx \frac{0.0711}{0.056} \approx 1.27$$

The critical value for $d.f._N = 2$ and $d.f._D = 4$, and $\alpha = 0.05$ is 6.94. Because the sample *F* is less than the critical value, we fail to reject H_0 and conclude that there is no evidence of differences in average measurements of fat content as determined by the different laboratories.

TECH NOTE   The calculations involved with two-way ANOVA are usually done using statistical or spreadsheet software. Basic printouts from various software packages are similar. Specific instructions for using Excel and Minitab are in the Using Technology section at the end of the chapter.

Experimental Design

In the preceding section and in this section, we have seen aspects of one-way and two-way ANOVA, respectively. Now let's take a brief look at some experimental design features that are appropriate for the use of these techniques.

For one-way ANOVA, we have one factor. Different levels for the factor form the treatment groups under study. In a *completely randomized design*, independent random samples of experimental subjects or objects are selected for each treatment group. For example, suppose a researcher wants to study the effects of different treatments for the condition of slightly high blood pressure. Three treatments are under study: diet, exercise, and medication. In a completely randomized design, the people participating in the experiment are *randomly* assigned to each treatment group. Table 11-21 shows the process.

TABLE 11-21   Completely Randomized Design Flowchart

Subjects with slightly high blood pressure → Random assignment →
- Treatment 1: Diet
- Treatment 2: Exercise
- Treatment 3: Medication

TABLE 11-22   **Randomized Block Design Flowchart**

	Blocks		Treatments
Subjects with slightly high blood pressure	→ Under 30 →	Random assignment	→ Treatment 1: Diet → Treatment 2: Exercise → Treatment 3: Medication
	→ Age 31–50 →	Random assignment	→ Treatment 1: Diet → Treatment 2: Exercise → Treatment 3: Medication
	→ Over 50 →	Random assignment	→ Treatment 1: Diet → Treatment 2: Exercise → Treatment 3: Medication

For two-way ANOVA, there are *two* factors. When we *block* experimental subjects or objects together based on a similar characteristic that might affect responses to treatments, we have a *block design*. For example, suppose that the researcher studying treatments for slightly high blood pressure believes that the age of subjects might affect the response to the three treatments. In such a case, blocks of subjects in specified age groups are used. The factor age is used to form blocks. Suppose age has three levels: under age 30, ages 31–50, and over age 50. The same number of subjects are assigned to each block. Then the subjects in each block are randomly assigned to the different treatments of diet, exercise, or medication. Table 11-22 shows the *randomized block design*.

Experimental design is an essential component of good statistical research. The design of experiments can be quite complicated, and if the experiment is complex, the services of a professional statistician may be required. The use of blocks helps the researcher account for some of the most important sources of variability among the experimental subjects or objects. Then randomized assignments to different treatment groups help average out the effects of other remaining variability. In this way, differences among the treatment groups are more likely to be caused by the treatments themselves rather than by other sources of variability.

VIEWP◉INT   *Who Watches Cable TV?*

Consider the following claim: Average cable TV viewers are, generally speaking, as affluent as newspaper readers and better off than radio or magazine audiences. How do we know that this claim is true? One way to answer such a question is to construct a two-way ANOVA test where the rows represent household income levels and the columns represent cable TV viewers, newspaper readers, magazine audiences, and radio listeners. The response variable is an index that represents the ratio of the specific medium compared with U.S. averages. For more information and data, see *American Demographics* (vol. 17, no. 6).

SECTION 11.6 PROBLEMS

1. *Physical Therapy: Dual Tasks* Does talking while walking slow you down? A study reported in the journal *Physical Therapy* (vol. 72, no. 4) considered mean cadence (steps per minute) for subjects using no walking device, a standard walker, and a rolling walker. In addition, the cadence was measured when the subjects had to perform dual tasks. The second task was to respond vocally to a signal while walking. Cadence was measured for subjects who were just walking (using no device, a standard walker, or a rolling walker) and for subjects required to respond to a signal while walking. List the factors and the number of levels of each factor. How many cells are there in the data table?

2. *Professors: Salary Survey Academe, Bulletin of the American Association of University Professors* (vol. 83, no. 2) presents results of salary surveys (average salary) by rank of the faculty member (professor, associate, assistant, instructor) and by type of institution (public, private). List the factors and the number of levels of each factor. How many cells are there in the data table?

3. *Physical Therapy: Dual Tasks* For the study regarding mean cadence (see Problem 1), two-way ANOVA was used. Recall that the two factors were walking device (none, standard walker, rolling walker) and dual task (being required to respond vocally to a signal or no dual task required). Results of two-way ANOVA showed that there was no evidence of interaction between the factors. However, according to the article, "the ANOVA conducted on the cadence data revealed a main effect of walking device." When the hypothesis regarding no difference in mean cadence according to which, if any, walking device is used, the sample F was 30.94, with $d.f._N = 2$ and $d.f._D = 18$. Further, the P value for the result was reported to be less than 0.01. From this information, what is the conclusion regarding any difference in mean cadence according to the factor walking device used? Verify your conclusion by looking up the critical value for $\alpha = 0.01$.

4. *Education: Media Usage* In a study of media usage versus education level (*American Demographics*, vol. 17, no. 6), an index was used to measure media usage where a measurement of 100 represents the U.S. average. Values above 100 represent above average media usage.

Education Level	Media				
	Cable Network	Prime-Time TV	Radio	Newspaper	Magazine
Less than high school	80	112	87	76	85
High school graduate	103	105	100	99	101
Some college	107	94	106	105	107
College graduate	108	90	106	116	108

Source: From American Demographics, Vol. 17, No.6. Reprinted with permission. © 1995 American Demographics, Ithaca, NY.

(a) List the factors and the number of levels of each factor.
(b) Assume that there is no interaction between the factors. Use two-way ANOVA and the following Minitab printout to determine if there is a difference in population mean index based on education. Use $\alpha = 0.05$.
(c) Determine if there is a difference in population mean index based on media. Use $\alpha = 0.05$.

```
Minitab Printout for Media/Education Data
Analysis of Variance for Index
Source    DF      SS      MS
Edu        3     961     320
Media      4       5       1
Error     12    1299     108
Total     19    2264
```

5. *Income: Media Usage* In the same study described in Problem 4, media usage versus household income also was considered. The media usage indexes for the various media and income levels follow.

Income Level	Media				
	Cable Network	Prime-Time TV	Radio	Newspaper	Magazine
Less than $20,000	78	112	89	80	91
$20,000–$39,999	97	105	100	97	100
$40,000–$74,999	113	92	106	111	105
$75,000 or more	121	94	107	121	105

Source: From *American Demographics*, Vol. 17, No.6. Reprinted with permission. © 1995 *American Demographics*, Ithaca, NY.

(a) List the factors and the number of levels of each factor.
(b) Assume that there is no interaction between the factors. Use two-way ANOVA and the following Minitab printout to determine if there is a difference in population mean index based on income. Use $\alpha = 0.05$.
(c) Determine if there is a difference in population mean index based on media. Use $\alpha = 0.05$.

```
Minitab Printout for Media/Income Data
Analysis of Variance for Index
Source    DF      SS      MS
Income     3     925     308
Media      4      44      11
Error     12    1921     160
Total     19    2890
```

6. *Gender: Grade Point Average* Does college grade point average (GPA) depend on gender? Does it depend on class (freshman, sophomore, junior, senior)? In a study, the following GPA data were obtained for random samples of college students in each of the cells.

Gender	Class							
	Freshman		Sophomore		Junior		Senior	
Male	2.8	2.1	2.5	2.3	3.1	2.9	3.8	3.6
	2.7	3.0	2.9	3.5	3.2	3.8	3.5	3.1
Female	2.3	2.9	2.6	2.4	2.6	3.6	3.2	3.5
	3.5	3.9	3.3	3.6	3.3	3.7	3.8	3.6

(a) List the factors and the number of levels of each factor.

(b) Use two-way ANOVA and the following Minitab printout to determine if there is any evidence of interaction between the two factors at a level of significance of 0.05.

```
Minitab Printout of GPA Based on Gender and Class

Analysis of Variance for GPA
Source          DF      SS       MS
Gender           1    0.300    0.300
Class            3    2.153    0.718
Interaction      3    0.271    0.090
Error           24    5.312    0.221
Total           31    8.037
```

(c) If there is no evidence of interaction, use two-way ANOVA and the Minitab printout to determine if there is a difference in mean GPA based on class. Use $\alpha = 0.05$.

(d) If there is no evidence of interaction, use two-way ANOVA and the Minitab printout to determine if there is a difference in mean GPA based on gender. Use $\alpha = 0.05$.

7. *Education: Teaching Style* A researcher forms three blocks of students interested in taking a history course. The groups are based on grade point average (GPA). The first group consists of students with a GPA of less than 2.5, the second group consists of students with a GPA between 2.5 and 3.1, and the last group consists of students with a GPA greater than 3.1. History courses are taught three ways: traditional lecture, small-group collaborative method, and independent study. The researcher randomly assigns 10 students from each block to sections of history taught each of the three ways. Sections for each teaching style then have 10 students from each block. The researcher records scores on a common course final examination administered to each student. Draw a flowchart showing the design of this experiment. Does the design fit the model for randomized block design?

SUMMARY

In this chapter, we introduced applications of two new probability distributions: the chi-square and *F* distributions. We used the chi-square distribution to test for independence and goodness of fit, to estimate the variance σ^2, and to test hypotheses involving the variance σ^2.

We used the *F* distribution to test two variances. Another application of the *F* distribution was for a method called analysis of variance. Analysis of variance is used to determine if there are differences among means for several groups. In the case that the groups are based on values of only one variable, we have one-way ANOVA. In the case of groups formed using two variables, we use two-way ANOVA to test for differences of means based on either variable or an interaction between the variables.

IMPORTANT WORDS & SYMBOLS

Section 11.1

Independence test

Chi-square distribution, χ^2

Degrees of freedom, $d.f. = (R - 1)(C - 1)$ for χ^2 distribution and tests of independence

Contingency table with cells

Row total

Column total

Expected frequency of a cell, E

Observed frequency of a cell, O

Critical chi-square values, χ^2_α

Section 11.2

Goodness-of-fit test

Degrees of freedom $d.f.$ for χ^2 distribution and goodness-of-fit tests

Section 11.3

Hypotheses tests about σ^2

Confidence interval c for σ^2

χ^2_U, χ^2_L

Section 11.4

F distribution

F ratio

Hypothesis tests about two variances

Section 11.5

ANOVA

Sum of squares, SS

SS_{BET}, SS_W, SS_{TOT}

MS_{BET}, MS_W

F ratio

F distribution

Summary table for ANOVA

Degrees of freedom $d.f.$ for numerator for F distribution

Degrees of freedom $d.f.$ for denominator for F distribution

Section 11.6

Two-way ANOVA

Interaction

Factor

Level

Completely randomized design

Randomized block design

VIEWPOINT

Movies and Money!

Young adults are the movie industry's best customers. However, going to the movies is expensive, which may explain why attendance rates increase with household income. Using what you have learned in this chapter, you can create appropriate chi-square tests to determine how good a fit exists between national percentage rates of attendance by household income and attendance rates in your demographic area. For more information and national data, see *American Demographics* (vol. 18, no. 12).

CHAPTER REVIEW PROBLEMS

Before you solve a problem, first classify the problem as one of the following:
(a) Chi-square test of independence
(b) Chi-square goodness of fit

(c) Chi-square for testing or estimating σ^2 or σ
(d) *F* test for two variances
(e) One-way ANOVA
(f) Two-way ANOVA

Then, in each of the problems when a test is to be performed, do the following:

(i) State the null and alternate hypotheses.
(ii) Find the critical value(s).
(iii) Sketch the critical region and show the location of the critical value and sample statistic.
(iv) Decide whether you should reject or fail to reject the null hypothesis.
(v) If the problem is an ANOVA problem, make a summary table.
(vi) Interpret the results in the context of the problem.

1. *Sales: Packaging* The makers of Country Boy Corn Flakes are thinking about changing the packaging of the cereal with the hope of improving sales. In an experiment, five stores of similar size in the same region sold Country Boy Corn Flakes in different-shaped containers for 2 weeks. Total packages sold are given in the following table. Using a 0.05 significance, shall we reject or fail to reject the hypothesis that the mean sales are the same, no matter which shape box is used?

Cube	Cylinder	Pyramid	Rectangle
120	110	74	165
88	115	62	98
65	180	110	125
95	96	66	87
71	85	83	118

2. *Education: Exams* Professor Fair believes that extra time does not improve grades on exams. He randomly divided a group of 300 students into two groups and gave them all the same test. One group had exactly 1 hour in which to finish the test, and the other group could stay as long as desired. The results are shown in the following table. Test at the 0.01 level of significance that time to do a test and test results are independent.

Time	A	B	C	F	Row Total
1 h	23	42	65	12	142
Unlimited	17	48	85	8	158
Column total	40	90	150	20	300

3. *Tires: Blowouts* A consumer agency is investigating the blowout pressures of Soap Stone tires. A Soap Stone tire is said to blow out when it separates from the wheel rim due to impact forces usually caused by hitting a rock or a pothole in the road. A random sample of 30 Soap Stone tires was inflated to the recommended pressure, and then forces measured in foot-pounds were applied to each tire (one foot-pound is the force of one pound dropped from a height of one foot). The customer complaint is that some Soap Stone tires blow out under small-impact forces, while other tires seem to be well made and don't have this fault. For the 30 test tires, the sample standard deviation of blowout forces was 1353 foot-pounds.

(a) Soap Stone claims their tires will blow out at an average pressure of 20,000 foot-pounds with a standard deviation of 900 foot-pounds. The average blowout force is not in question, but the variability of blowout forces is in question. Using a 0.01 level of significance, test the claim that the variance of blowout pressures is more than Soap Stone claims it is.

(b) Find a 95% confidence interval for the variance of blowout pressures, using the information from the random sample.

4. *Computer Science: Data Processing* Anela is a computer scientist who is formulating a large and complicated program for a type of data processing. She has three ways of storing and retrieving data: CD, tape, and disks. As an experiment, she sets up her program three different ways: one using CDs, one using tape, and the other using disks. Then four test runs of this type of data processing are made on each program. The time required to execute each program is shown in the following table (in minutes). Use a 0.01 level of significance to test the hypothesis that the mean processing time is the same for each method.

CD	Tape	Disks
8.7	7.2	7.0
9.3	9.1	6.4
7.9	7.5	9.8
8.0	7.7	8.2

5. *Teacher Ratings: Grades* Professor Stone complains that student teacher ratings depend on the grade the student received. In other words, according to Professor Stone, a teacher who gives good grades gets good ratings, and a teacher who gives bad grades gets bad ratings. To test this claim, the Student Assembly took a random sample of 300 teacher ratings on which the student's grade for the course also was indicated. The results are given in the following table. Test the hypothesis that teacher ratings and student grades are independent at the 0.01 level of significance.

Rating	A	B	C	F (or withdrawal)	Row Total
Excellent	14	18	15	3	50
Average	25	35	75	15	150
Poor	21	27	40	12	100
Column total	60	80	130	30	300

6. *Packaging: Corn Flakes* A machine that puts corn flakes into boxes is adjusted to put an average of 15 oz into each box with standard deviation of 0.25 oz. If a random sample of 12 boxes gave a sample standard deviation of 0.38 oz, do these data support the claim that the variance has increased and the machine needs to be brought back into adjustment? (Use a 0.05 level of significance.)

7. *Sociology: Age Distribution* A sociologist is studying the age of the population in Blue Valley. Ten years ago the population was such that 20% were under 20 years old, 15% were in the 20 to 35-year-old bracket, 30% were between 36 and 50, 25% were between 51 and 65, and 10% were over 65. A study done this year used a random sample of 210 residents. This sample showed

Under 20	20–35	36–50	51–65	Over 65
15	25	70	80	20

At the 0.01 level of significance, has the age distribution of the population of Blue Valley changed?

8. *Engineering: Roller Bearings* Two processes for manufacturing large roller bearings are under study. In both cases, the diameters (in centimeters) are being examined. A random sample of 21 roller bearings from the old manufacturing process showed the sample variance of diameters to be $s^2 = 0.235$. Then another random sample of 26 roller bearings from the new manufacturing process showed the sample variance of their diameters to be $s^2 = 0.128$. Use a 5% level of significance to test the claim that there is a difference (either way) for the population variances for the old or new manufacturing processes.

9. *Engineering: Light Bulbs* Two processes for manufacturing 60-watt light bulbs are under study. In both cases, the life (in hours) of the bulb before it burns out is being examined. A random sample of 18 light bulbs manufactured using the old process showed the sample variance of life times to be $s^2 = 51.87$. Then another random sample of 16 light bulbs manufactured using the new process showed the sample variance of the life times to be $s^2 = 135.24$. Use a 5% level of significance to test the claim that the population variance of life times for the new manufacturing process is larger than that of the old process.

10. *Advertising: Newspapers* Does the section in which a newspaper ad is placed make a difference in the average daily number of people responding to the ad? Is there a difference in average daily number of people responding to the ad if it is placed in the Sunday newspaper as compared with the Wednesday newspaper? Video Entertainment is a video club that sells videotapes featuring movies of all kinds—instructional videos, videos of TV specials, etc. To attract new customers, Video Entertainment ran ads in the Wednesday newspaper and in the Sunday newspaper offering 6 free videotapes to new members. In addition, the ads were placed in the Sports Section, the Entertainment Section, and the Business Section of the local newspaper. Ads running in the different sections carried different promotion codes. Different codes also were used for Wednesday ads compared with Sunday ads. The number of people responding to the ads for days selected at random were recorded according to the promotion codes. The results follow:

Day	Sports			Entertainment			Business		
Wed	12	15	11	22	14	17	2	4	3
	12	15	20	22	18	12	5	0	1
Sun	20	23	25	32	26	28	13	16	13
	33	15	17	31	25	41	15	14	10

(a) List the factors and the number of levels for each factor.
(b) Use the following Minitab printout and the *F* distribution table (Table 8 of Appendix II) to test for interaction between the variables. Use a 1% level of significance.

```
Minitab Printout for Mean Number of
    Responses per Day

Analysis of Variance for Response
Source        DF       SS        MS
Day            1     1024.0    1024.0
Section        2     1573.6     786.8
Interaction    2       38.0      19.0
Error         30      555.7      18.5
Total         35     3191.2
```

(c) If there is no evidence of interaction between the factors, test for a difference in mean number of daily responses for the levels of the day factor. Use $\alpha = 0.01$.

(d) If there is no evidence of interaction between the factors, test for a difference in mean number of daily responses for the levels of the section factor. Use $\alpha = 0.01$.

DATA HIGHLIGHTS: GROUP PROJECTS

Break into small groups and discuss the following topics. Organize a brief outline in which you summarize the main points of your group discussion.

The *Statistical Abstract of the United States* reported information about the percentage of arrests of all drunk drivers according to age group. In the following table, the entry 3.7 in the first row means that in the entire United States about 3.7% of all people arrested for drunk driving were in the age group 16–17 years. The Freemont County Sheriff's Office obtained data about the number of drunk drivers arrested in each age group over the past several years. In the following table, the entry 8 in the first row means that eight people in the age group 16–17 years were arrested for drunk driving in Freemont County.

Distribution of Drunk Driver Arrests by Age

Age	National Percentage	Number in Freemont County
16–17	3.7	8
18–24	18.9	35
25–29	12.9	23
30–34	10.3	19
35–39	8.5	12
40–44	7.9	14
45–49	8.0	16
50–54	7.9	13
55–59	6.8	10
60–64	5.7	9
65 and over	9.4	15
	100%	174

Use a chi-square test with 5% level of significance to test the claim that the age distribution of drunk drivers arrested in Freemont County is the same as the national age distribution of drunk drivers arrested.

(a) State the null and alternate hypotheses.
(b) Find the value of the chi-square statistic from the sample.
(c) Find the degrees of freedom and the appropriate chi-square critical value.
(d) Sketch the critical region and locate your sample chi-square value and critical chi-square value on the sketch.
(e) Decide whether you should reject or not reject the null hypothesis. State your conclusion in the context of the problem.
(f) How could you gather data and conduct a similar test for the city or county in which you live? Explain.

LINKING CONCEPTS: WRITING PROJECTS

Discuss each of the following topics in class or review the topics on your own. Then write a brief but complete essay in which you summarize the main points. Please include formulas and graphs as appropriate.

1. In this chapter, you studied the chi-square distribution and three principal applications for the distribution.
   (a) Outline the basic ideas behind the chi-square test of independence. What is a contingency table? What are the null and alternate hypotheses? How are the test statistic and critical region constructed? What basic assumptions underlie this application of the chi-square distribution?
   (b) Outline the basic ideas behind the chi-square test of goodness of fit. What are the null and alternate hypotheses? How are the test statistic and critical region constructed? There are a number of direct similarities between tests of independence and tests for goodness of fit. Discuss and summarize these similarities.
   (c) Outline the basic ideas behind the chi-square method of testing and estimating a standard deviation. What basic assumptions underlie this process?

2. The Fisher *F* distribution is used to construct a one-way ANOVA test for comparing several sample means.
   (a) Outline the basic purpose of ANOVA. How does ANOVA avoid high risk due to multiple type I errors?
   (b) Outline the basic assumption for ANOVA.
   (c) What are the null and alternate hypotheses in an ANOVA test? If the test conclusion is to reject the null hypothesis, do we know which of the population means are different from each other?
   (d) What is the Fisher *F* distribution? How are the degrees of freedom for numerator and denominator determined?
   (e) What do we mean by a summary table of ANOVA results? What are the main components of such a table? How is the final decision made?

Using TECHNOLOGY

TI-83PLUS • EXCEL • MINITAB • COMPUTERSTAT

APPLICATION

Analysis of Variance (One-Way ANOVA)

The following data are a winter mildness/severity index for three European locations near 50° north latitude. For each decade, the number of unmistakably mild months minus the number of unmistakably severe months for December, January, and February is given.

Decade	Britain	Germany	Russia
1800	−2	−1	+1
1810	−2	−3	−1
1820	0	0	0
1830	−3	−2	−1
1840	−3	−2	+1
1850	−1	−2	+3
1860	+8	+6	+1
1870	0	0	−3
1880	−2	0	+1
1890	−3	−1	+1
1900	+2	0	+2
1910	+5	+6	+1
1920	+8	+6	+2
1930	+4	+4	+5
1940	+1	−1	−1
1950	0	+1	+2

Table is based on data from *Exchanging Climate* by H. H. Lamb; copyright © 1966. Reprinted by permission of Routledge, UK.

1. We wish to test the null hypothesis that the mean winter index for Britain, Germany, and Russia are all equal against the alternate hypothesis that they are not all equal. Use a 5% level of significance.

2. What is the sum of squares between groups? Within groups? What is the sample *F* ratio? What is the *P* value? Shall we reject or fail to reject the statement that the mean winter indexes for these locations in Britain, Germany, and Russia are the same?

3. What is the smallest level of significance at which we could conclude that the mean winter indexes for these locations are not all equal?

Technology Hints (One-Way ANOVA)

Computer Stat

Under the main menu item **Hypothesis Testing**, select **ANOVA**.

Technology Hints (Two-Way ANOVA)

Excel

Excel has two commands for two-way ANOVA, depending upon how many data values are in each cell. In both cases, use the menu selections **Tools ➤ Data Analysis**. Then use

ANOVA: Two-Factor with Replication if there are two or more sample measurements for each factor combination cell. Again, there must be the same number of data in each cell.

ANOVA: Two-Factor without Replication if there is only one data value in each factor combination cell.

Data entry is fairly straightforward. For example, look at the spreadsheet for the data of Guided Exercise 12 regarding the fat content of different brands of potato chips as measured by different labs. Using ANOVA without replication, we have

713

	A	B	C	D
1		**Lab 1**	**Lab 2**	**Lab 3**
2	Texas	32.4	33.1	32.9
3	Great	37.9	37.7	37.8
4	Chip Ooh	29.1	29.4	29.5

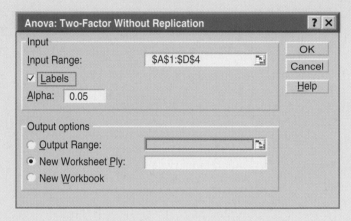

Anova: Two-Factor Without Replication   [?] [X]

Input

Input Range: $A$1:$D$4

☑ Labels

Alpha: 0.05

Output options

○ Output Range: _____

● New Worksheet Ply: _____

○ New Workbook

[OK] [Cancel] [Help]

Anova: Two-Factor Without Replication						
SUMMARY	*Count*	*Sum*	*Average*	*Variance*		
Texas	3	98.4	32.8	0.13		
Great	3	113.4	37.8	0.01		
Chip Ooh	3	88	29.33333333	0.043333333		
Lab 1	3	99.4	33.13333333	19.76333333		
Lab 2	3	100.2	33.4	17.29		
Lab 3	3	100.2	33.4	17.41		
ANOVA						
Source of Variation	*SS*	*df*	*MS*	*F*	*P-value*	*F crit*
Rows	108.7022222	2	54.35111111	968.6336634	4.2457E-06	6.944276265
Columns	0.142222222	2	0.071111111	1.267326733	0.374692378	6.944276265
Error	0.224444444	4	0.056111111			
Total	109.0688889	8				

Minitab

For Minitab, all the data for the response variable (in this case, fat content) are entered into a single column. Create two more columns, one for the row number of the cell containing the data value and one for the column number of the cell containing the data value. For the potato chip example, the rows correspond to the brand and the columns to the lab doing the analysis. Use menu choice **Stat ➤ ANOVA ➤ Two-Way.**

	C1	C2	C3
	Brand	Lab	Fat
1	1	1	32.4
2	1	2	33.1
3	1	3	32.9
4	2	1	37.9
5	2	2	37.7
6	2	3	37.8
7	3	1	29.1
8	3	2	29.4
9	3	3	29.5

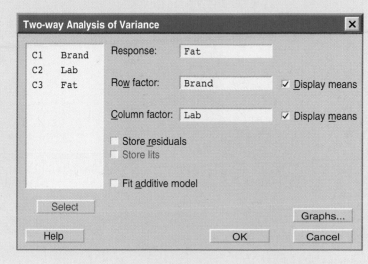

```
Two-Way Analysis of Variance

Analysis of Variance for Fat
Source        DF          SS          MS         F         P
Brand          2    108.7022     54.3511    968.63     0.000
Lab            2      0.1422      0.0711      1.27     0.375
Error          4      0.2244      0.0561
Total          8    109.0689

                            Individual 95% CI
Brand        Mean    -----+---------+---------+---------+------
1           32.80                        (*-)
2           37.80                                         (*-)
3           29.33      (*-)
                     -----+---------+---------+---------+------
                      30.00     32.50     35.00     37.50

                            Individual 95% CI
Lab          Mean    ---------+---------+---------+---------+-
1           33.13    (--------------*---------------)
2           33.40             (---------------*---------------)
3           33.40             (---------------*---------------)
                     ---------+---------+---------+---------+-
                      33.00     33.25     33.50     33.75
```

12 Nonparametric Statistics

Make everything as simple as possible, but no simpler.

—Albert Einstein

Albert Einstein (1879–1955)
This brilliant German-born American physicist
formulated the theory of relativity.

For on-line student resources, visit
math.college.hmco.com/students and
follow the Statistics links to the Brase/Brase,
Understandable Statistics, 7th edition web site.

◇ What if we cannot make assumptions about a population distribution? Can we still use statistical methods? What are the advantages and disadvantages? (SECTION 12.1)

◇ What are nonparametric tests? How do we handle a "before-and-after" situation? (SECTION 12.1)

◇ If we can't make assumptions about the population, and we have independent samples, how do we set up a nonparametric test? (SECTION 12.2)

◇ Suppose we are interested only in rank data (ordinal type data). If we have ordered pairs (*x*, *y*) of ranked data, is there a way to measure and test correlation? (SECTION 12.3)

FOCUS PROBLEM

How Cold? Compared to What?

Juneau is the capital of Alaska. The terrain surrounding Juneau is very rugged, and storms that sweep across the Gulf of Alaska usually hit Juneau. However, Juneau is located in southern Alaska, near the ocean, and temperatures are often comparable with those found in the lower 48 states. Madison is the capital of Wisconsin. The city is located between two large lakes. The climate of Madison is described as typical continental climate of interior North America. Consider the long-term average temperatures (in degrees Fahrenheit) paired by month for the two cities. (Source: National Weather Bureau.) Use a sign test with a 5% level of significance to test the claim that the overall temperature distribution of Madison is different (either way) from that of Juneau.

Month	Madison	Juneau
January	17.5	22.2
February	21.1	27.3
March	31.5	31.9
April	46.1	38.4
May	57.0	46.4
June	67.0	52.8
July	71.3	55.5
August	69.8	54.1
September	60.7	49.0
October	51.0	41.5
November	35.7	32.0
December	22.8	26.9

12.1
The Sign Test for Matched Pairs

FOCUS POINTS
✓ State the criteria for setting up a matched-pair sign test.
✓ Complete a matched-pair sign test.
✓ Interpret the results in the context of the application.

There are many situations where very little is known about the population from which samples are drawn. Therefore, we cannot make assumptions about the population distribution, such as the distribution is normal or binomial. In this chapter, we will study methods that come under the heading of *nonparametric statistics*. These methods are called *nonparametric* because they require no assumptions about the population distributions from which samples are drawn. The obvious advantages of these tests are that they are quite general and (as we shall see) not difficult to apply. The disadvantages are that they tend to waste information and tend to result in acceptance of the null hypothesis more often than they should; nonparametric tests are sometimes less sensitive than other tests.

The easiest of all the nonparametric tests is probably the *sign test*. The sign test is used when we compare sample distributions from two populations that are *not independent*. This occurs when we measure the sample twice, as in "before-and-after" studies. The following example shows how the sign test is constructed and used.

As part of their training, police cadets took a special course on identification awareness. To determine how the course affects a cadet's ability to identify a suspect, the 15 cadets were first given an identification awareness exam and then, after the course, were tested again. The police school would like to use the results of the two tests to see if the identification awareness course *improves* a cadet's score. Table 12-1 gives the scores for each exam.

Criteria for sign test

The sign of the difference is obtained by subtracting the precourse score from the postcourse score. If the difference is positive, we say the sign of the difference is +, and if the difference is negative, we indicate it with −. No sign is indicated if the scores are identical; in essence, such scores are ignored when using the sign test. To use the sign test, we need to compute the *proportion x of plus signs* to all signs. We ignore the pairs with no difference of signs. This is done in Guided Exercise 1.

TABLE 12-1   Scores for 15 Police Cadets

Cadet	Postcourse Score	Precourse Score	Sign of Difference
1	93	76	+
2	70	72	−
3	81	75	+
4	65	68	−
5	79	65	+
6	54	54	No difference
7	94	88	+
8	91	81	+
9	77	65	+
10	65	57	+
11	95	86	+
12	89	87	+
13	78	78	No difference
14	80	77	+
15	76	76	No difference

GUIDED EXERCISE 1

Proportion of plus signs

Look at Table 12-1 under the "Sign of Difference" column.

(a) How many plus signs do you see? ⟹ 10

(b) How many plus and minus signs do you see? ⟹ 12

(c) The *proportion of plus signs* is ⟹ $x = \dfrac{10}{12} = \dfrac{5}{6} = 0.833$

$$x = \frac{\text{Number of plus signs}}{\text{Total number of plus and minus signs}}$$

Use parts (a) and (b) to find x.

We observe that x is the sample proportion of plus signs, and we use p to represent the population proportion of plus signs (if *all* possible police cadets were used).

Null hypothesis

The null hypothesis is

H_0: $p = 0.5$ (the distributions of scores before and after the course are the same)

The null hypothesis says the identification awareness course does *not* affect the distribution of scores. Under the null hypothesis we expect the number of plus signs and minus signs to be about equal. This means the proportion of plus signs should be approximately 0.5.

Alternate hypothesis

The police department wants to see if the course *improves* a cadet's score. Therefore, the alternate hypothesis will be

H_1: $p > 0.5$ (the distribution of scores after the course is shifted higher than the distribution before the course)

The alternate hypothesis says the identification awareness course tends to improve scores. This means the proportion of plus signs should be greater than 0.5.

Critical value(s)

To test the null hypothesis H_0: $p = 0.5$ against the alternate hypothesis H_1: $p > 0.5$, we use methods of Section 9.5 for tests of proportions. As in Section 9.5, we will assume that all our samples are sufficiently large to permit a good normal approximation to the binomial distribution. For most practical work, this will be the case if the total number of plus and minus signs is 12 or more ($n \geq 12$).

When the total number of plus and minus signs is 12 or more, the sample statistic x (proportion of plus signs) has a distribution that is approximately normal with mean p and standard deviation $\sqrt{pq/n}$. Therefore, the critical values for the sign test are z values as shown in Table 12-2 on the following page.

In our example, the police department wants to test the hypothesis H_1: $p > 0.5$ at the $\alpha = 0.05$ level of significance. Table 12-2 indicates that the critical value for the right-tailed test is $z_0 = 1.645$.

TABLE 12-2   Critical Values for the Sign Test

H_0: $p = 0.5$ (the distributions of responses are the same)		
Alternate Hypothesis	$\alpha = 0.05$	$\alpha = 0.01$
H_1: $p > 0.5$ (the first distribution of responses is higher)	$z_0 = 1.645$	$z_0 = 2.33$
H_1: $p < 0.5$ (the first distribution of responses is lower)	$z_0 = -1.645$	$z_0 = -2.33$
H_1: $p \neq 0.5$ (the distributions of responses are different)	$z_0 = \pm 1.96$	$z_0 = \pm 2.58$
n = total number of plus and minus signs ($n \geq 12$)		

Sample test statistic

Under the null hypothesis H_0: $p = 0.5$, we assume the population proportion p of plus signs is 0.5. Therefore, the z value corresponding to the sample test statistic x is

$$z = \frac{x - p}{\sqrt{\dfrac{pq}{n}}} = \frac{x - 0.5}{\sqrt{\dfrac{(0.5)(0.5)}{n}}} = \frac{x - 0.5}{\sqrt{\dfrac{0.25}{n}}}$$

where n is the total number of plus and minus signs, and x is the total number of plus signs divided by n.

In the police cadet example, we found $x = 0.833$ in Guided Exercise 1. The value of n is 12. (Note that of the 15 cadets in the sample, 3 had no difference in precourse and postcourse test scores, so there are no signs for these 3.) The z value corresponding to $x = 0.833$ is then

$$z = \frac{0.833 - 0.5}{\sqrt{\dfrac{0.25}{12}}} = 2.31$$

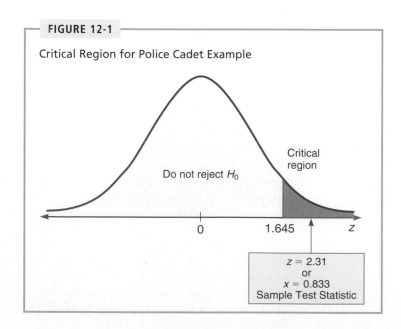

FIGURE 12-1

Critical Region for Police Cadet Example

Test conclusion

To conclude the test, we show the critical region on a sketch and locate the z value of the sample test statistic on the sketch. If the z value of the sample test statistic falls in the critical region, we reject H_0. The critical region for the police cadet example is shown in Figure 12-1. We see that the z value corresponding to $x = 0.833$ falls in the critical region. Therefore, we reject H_0 and conclude that the identification awareness course improves the cadets' scores.

GUIDED EXERCISE 2

Sign test

Dr. Kick-a-poo's Traveling Circus made a stop at Middlebury, Vermont, where the doctor opened a booth and sold bottles of Dr. Kick-a-poo's Magic Gasoline Additive. The additive is supposed to increase gas mileage when used according to instructions. Twenty local people purchased bottles of the additive and used it according to instructions. These people carefully recorded their mileage with and without the additive. The results are shown in Table 12-3.

TABLE 12-3 Mileage Before and After Kick-a-poo's Additive

Car	With Additive	Without Additive	Sign of Difference
1	17.1	16.8	+
2	21.2	20.1	+
3	12.3	12.3	No difference (N.D.)
4	19.6	21.0	−
5	22.5	20.9	+
6	17.0	17.9	——
7	24.2	25.4	——
8	22.2	20.1	——
9	18.3	19.1	——
10	11.0	12.3	——
11	17.6	14.2	——
12	22.1	23.7	——
13	29.9	30.2	——
14	27.6	27.6	——
15	28.4	27.7	——
16	16.1	16.1	——
17	19.0	19.5	——
18	38.7	37.9	——
19	17.6	19.7	——
20	21.6	22.2	——

TABLE 12-4 Completion of Table 12-3

Car	Sign of Difference
6	−
7	−
8	+
9	−
10	−
11	+
12	−
13	−
14	N.D.
15	+
16	N.D.
17	−
18	+
19	−
20	−

Continued

(a) In Table 12-3, complete the column headed "Sign of Difference." How many plus signs are there? How many total plus and minus signs are there? What is x, the proportion of plus signs?

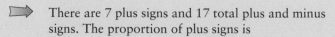 There are 7 plus signs and 17 total plus and minus signs. The proportion of plus signs is

$$x = \frac{7}{17} = 0.412$$

(b) Most people claim that the additive has no effect. Let's use a 0.05 level of significance to test this claim against the alternate hypothesis that the additive did have an effect (one way or the other). State the null and alternate hypotheses. Use Table 12-2 to find the critical values and shade the critical region. Convert the sample x value, $x = 0.412$, to a z value. Does the z value of the sample test statistic fall in the critical region or not? Do we reject or fail to reject H_0? What conclusion do we draw regarding the effect of the additive?

We use

H_0: $p = 0.5$ (mileage distributions are the same)

H_1: $p \neq 0.5$ (mileage distributions are different)

For $\alpha = 0.05$, the critical values for a two-tailed test are $z_0 = \pm 1.96$. To find the z value corresponding to $x = 0.412$, we use $n = 17$ (total number of signs):

$$z = \frac{x - 0.5}{\sqrt{0.25/n}} = \frac{0.412 - 0.5}{\sqrt{0.25/17}} = -0.73$$

Because the z value of the sample test statistic does not fall in the critical region, we do not reject H_0. We conclude that the additive seems to have no effect (at the 5% level of significance). See Figure 12-2.

FIGURE 12-2 Critical Region

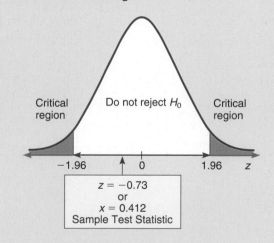

◊ **COMMENT** Because the sampling distribution for x, the proportion of plus signs, follows a normal distribution, you compute P values for the sample test statistic using the methods of Section 9.3. Recall that if the P value is less than or equal to the level of significance α, we reject H_0. ◊

Yukon News

The *Yukon News* featured an article entitled "Resurgence of the Dreaded White Plague," about the resurgence of tuberculosis (TB) in the far north. TB, also known as the white plague, has been present in Canada since it was brought in by European immigrants in the 17th century. Although antibiotics are widely used today, the disease has never been eradicated. Canadian National Health data suggest that TB is spreading faster in the Yukon than elsewhere in Canada. Because of this, the Canadian government has established many new TB clinics in remote Yukon villages. Using what you have learned in this section and Canadian National Health data, can you think of a way to use a sign test to study the claim that in these villages the rate of TB in the population dropped after the clinics were activated?

SECTION 12.1   PROBLEMS

For each of the problems, please provide the requested information.

(a) What is the null hypothesis? What is the alternate hypothesis? Will we use a left-tailed, right-tailed, or two-tailed test? What is the level of significance?

(b) What sampling distribution will we use? What is the critical value z_0 (or critical values $\pm z_0$)?

(c) Sketch the critical region, and show the critical value (or critical values).

(d) Calculate the z value corresponding to the sample test statistic x, and show its location on the sketch of part (c).

(e) Based on your answers for parts (a) to (d), shall we reject or fail to reject the null hypothesis? State the conclusion in the context of the problem.

1. *Product Design: Pens* The Daisy Pen Company has developed a new tip for its felt-tip pens. Twelve pens are filled with ink and fitted with the old tip. Each pen is attached to one of 12 motor-driven paper-covered drums, and the writing life of each pen is determined (in hours). Then each pen is refilled with ink and fitted with a new type tip. Again, the writing life of each pen is determined. The results follow (in hours):

Pen	New Tip	Old Tip	Pen	New Tip	Old Tip
1	52	50	7	47	46
2	47	55	8	57	53
3	56	51	9	56	52
4	48	45	10	46	40
5	51	57	11	56	49
6	59	54	12	47	51

Use a 0.05 level of significance to test the hypothesis that the writing life of the new tip is longer than that of the old tip.

2. *Psychology: Exams* A psychologist claims that students' pulse rates tend to increase just before an exam. To test this claim, she used a random sample of 14 psychology students and took their pulses before an ordinary class meeting and then again before a class meeting that consisted of an examination. The results follow.

Student	Pulse Rate Before Exam	Pulse Rate Before Ordinary Class	Student	Pulse Rate Before Exam	Pulse Rate Before Ordinary Class
1	88	81	8	80	73
2	77	77	9	68	71
3	72	75	10	75	73
4	74	79	11	82	76
5	81	79	12	61	66
6	70	68	13	77	68
7	75	77	14	64	60

Use a 0.05 level of significance to test the psychologist's claim.

3. *Education: Exams* A high-school science teacher decided to give a series of lectures on current events. To determine if the lectures had any effect on student awareness of current events, an exam was given to the class before the lectures, and a similar exam was given after the lectures. The scores follow. Use a 0.05 level of significance to test the claim that the lectures made no difference against the claim that the lectures did make some difference (either up or down).

Student	After Lectures	Before Lectures	Student	After Lectures	Before Lectures
1	107	111	10	44	40
2	115	110	11	119	115
3	120	93	12	130	101
4	78	75	13	91	110
5	83	88	14	99	90
6	56	56	15	96	98
7	71	75	16	83	76
8	89	73	17	100	100
9	77	83	18	118	109

4. *Production: Defects* A manufacturer of lens filters is using two production lines to make lens filters for cameras. The same production process is used on each line. The only difference is in the employees working the lines. Employees on line A are experienced at their work, whereas those on line B are new on the job. The number of defective lens filters produced by each line for a period of 15 days follows:

Day	Line A	Line B	Day	Line A	Line B
1	389	517	9	300	222
2	412	610	10	444	357
3	509	430	11	392	412
4	420	420	12	306	580
5	471	415	13	319	289
6	171	310	14	510	505
7	460	370	15	240	350
8	650	618			

Use a 0.01 level of significance to test the claim that there is no difference in defective production against the claim that there is a difference (one way or the other).

5. *Identical Twins: Reading Skills* To compare two elementary schools in teaching of reading skills, 12 sets of identical twins were used. In each case, one child was selected at random and sent to school A and his or her twin was sent to school B. Near the end of fifth grade, an achievement test was given to each child. The results follow:

Twin Pair	School A	School B	Twin Pair	School A	School B
1	177	86	7	86	93
2	150	135	8	111	77
3	112	115	9	110	96
4	95	110	10	142	130
5	120	116	11	125	147
6	117	84	12	89	101

Use a 0.05 level of significance to test the hypothesis that the two schools have the same effectiveness in teaching reading skills against the alternate hypothesis that the schools are not equally effective.

6. *Food Preservatives: Bread* A chemical company is testing two types of food preservatives to be used in bread. Twenty bakeries each bake two similar batches of bread, one with preservative A and the other with preservative B. The shelf life for each batch was determined as follows (shelf life in days):

Bakery	Preservative A	Preservative B	Bakery	Preservative A	Preservative B
1	5	6	11	7	6
2	7	3	12	5	8
3	3	3	13	6	4
4	5	7	14	9	7
5	5	6	15	3	4
6	3	5	16	5	7
7	5	4	17	8	6
8	6	8	18	4	6
9	5	7	19	5	8
10	6	4	20	7	3

Use a 0.05 level of significance to test the claim that bread with preservative B has a longer shelf life.

7. *Quitting Smoking: Hypnosis* One program to help people stop smoking cigarettes uses the method of posthypnotic suggestion to remind subjects to avoid smoking. A random sample of 18 subjects agreed to test the program. All subjects counted the number of cigarettes they usually smoke a day; then they counted the number of cigarettes smoked the day after hypnosis. (*Note:* It usually takes several weeks for the subject to stop smoking completely, and the method does not work for everyone.) The results follow.

Subject	Cigarettes Smoked per Day		Subject	Cigarettes Smoked per Day	
	After Hypnosis	Before Hypnosis		After Hypnosis	Before Hypnosis
1	28	28	10	5	19
2	15	35	11	12	32
3	2	14	12	20	42
4	20	20	13	30	26
5	31	25	14	19	37
6	19	40	15	0	19
7	6	18	16	16	38
8	17	15	17	4	23
9	1	21	18	19	24

Using a 1% level of significance, test the claim that the number of cigarettes smoked per day was less after hypnosis.

8. *Mosquito Repellant: Garlic* Some fishermen claim that eating a clove of garlic will help keep mosquitoes away from your skin. To test this claim, Gary collected a large, clear plastic tube of hungry mosquitoes. He took a random sample of 16 students, each of whom agreed to put a bare arm in the tube and count the number of mosquitoes that landed on the skin during a 2-minute time interval. Then each student ate a clove of garlic. After 3 hours (when the effect of the garlic was in the bloodstream), each student again put his or her arm in the tube and counted mosquitoes landing on the skin during a 2-minute time interval. The results follow. Using a 5% level of significance, test the claim that garlic tends to reduce the number of mosquitoes on the skin.

Student	Mosquitoes Landing on Skin		Student	Mosquitoes Landing on Skin	
	After Garlic	Before Garlic		After Garlic	Before Garlic
1	15	75	9	0	94
2	8	63	10	40	37
3	30	42	11	19	48
4	16	91	12	16	66
5	56	56	13	0	58
6	44	39	14	12	77
7	12	88	15	42	110
8	42	72	16	51	49

9. *Pediatrics: Pulse Rate* A pediatrician is studying the pulse rate of babies before and after birth. A random sample of 17 babies gave the following information (pulse rate = heart beats per minute). Using a 1% level of significance, test the claim that the pulse rates are different (either up or down).

Baby	Pulse Rate		Baby	Pulse Rate	
	24 Hours After Birth	24 Hours Before Labor Starts		24 Hours After Birth	24 Hours Before Labor Starts
1	70	61	10	60	52
2	73	72	11	63	59
3	82	80	12	58	64
4	80	83	13	79	67
5	58	58	14	71	71
6	88	77	15	73	60
7	80	80	16	72	75
8	60	65	17	85	81
9	71	73			

10. *Marketing: Ads* Enterprise Sales sent a team of 15 salespeople to Dodge City, Kansas, for door-to-door magazine sales. After one week of sales efforts, Enterprise Sales decided to buy local spot TV ads for its magazines and then continue the door-to-door sales effort another week. The results follow:

Salesperson	Number of Sales		Salesperson	Number of Sales	
	After TV Ads	Before TV Ads		After TV Ads	Before TV Ads
1	4	3	9	3	2
2	1	0	10	0	1
3	0	1	11	1	0
4	3	4	12	4	3
5	3	2	13	4	4
6	0	0	14	5	6
7	6	5	15	3	2
8	4	3			

Using a 5% level of significance, test the claim that sales before and after the TV ads are different (either up or down).

11. *Focus Problem: Meteorology* Solve the chapter Focus Problem found at the beginning of the chapter.

12.2
The Rank-Sum Test

The sign test is used when we have paired data values coming from dependent samples as in "before-and-after" studies. However, if the data values are *not* paired, the sign test should *not* be used.

For the situation in which we draw independent random samples from two populations, there is another nonparametric method for testing the difference between sample means; it is called the *rank-sum test* (also called the *Mann-Whitney test*). The rank-sum test can be used when assumptions about *normal* populations are not

TABLE 12-5   Decompression Times for 19 Navy Divers (in min)

Group A (had pill)	41	56	64	42	50	70	44	57	63	
					Mean time = 54.1 min					
Group B (no pill)	66	43	72	62	55	80	74	75	77	78
					Mean time = 68.2 min					

satisfied or when assumptions about *equal population variances* are not satisfied. To fix our thoughts on a definite problem, let's consider the following example.

When a scuba diver makes a deep dive, nitrogen builds up in the diver's blood. After returning to the surface, the diver must wait in a decompression chamber until the nitrogen level of the blood returns to normal. A physiologist working with the Navy has invented a pill that a diver takes 1 hour before diving. The pill is supposed to have the effect of reducing the waiting time spent in the decompression chamber. Nineteen Navy divers volunteered to help the physiologist determine if the pill has any effect. The divers were randomly divided into two groups: group A had 9 divers who each took the pill, and group B had 10 divers who did not take the pill. All the divers worked the same length of time on a deep salvage operation and returned to the decompression chamber. A monitoring device in the decompression chamber measured the waiting time for each diver's nitrogen level to return to normal. These times are recorded in Table 12-5.

Rank the data

The means of our two samples are 54.1 and 68.2 minutes. We will use the rank-sum test to decide whether the difference between the means is significant. First, we arrange the two samples jointly in order of increasing time. To do this, we use the data of groups A and B as if they were one sample. The times, groups, and ranks are shown in Table 12-6.

Group A occupies the ranks 1, 2, 4, 5, 7, 8, 10, 11, and 13, and group B occupies the ranks 3, 6, 9, 12, 14, 15, 16, 17, 18, and 19. We add up the ranks of the group with the *smaller* sample size, in this case group A.

The sum of the ranks is denoted by R:

$$R = 1 + 2 + 4 + 5 + 7 + 8 + 10 + 11 + 13 = 61$$

TABLE 12-6   Ranks for Decompression Time

Time	Group	Rank	Time	Group	Rank
41	A	1	64	A	11
42	A	2	66	B	12
43	B	3	70	A	13
44	A	4	72	B	14
50	A	5	74	B	15
55	B	6	75	B	16
56	A	7	77	B	17
57	A	8	78	B	18
62	B	9	80	B	19
63	A	10			

Let n_1 be the size of the *smaller sample* and n_2 be the size of the *larger sample*. In the case of the divers, $n_1 = 9$ and $n_2 = 10$. So R is the sum of the ranks from the smaller sample. If both samples are of the same size, then $n_1 = n_2$ and R is the sum of the ranks of either group (but not both groups).

Distribution of ranks

When both n_1 and n_2 are sufficiently large (each of size eight or more), advanced mathematical statistics can be used to show that R is approximately normally distributed with mean

$$\mu_R = \frac{n_1(n_1 + n_2 + 1)}{2}$$

and standard deviation

$$\sigma_R = \sqrt{\frac{n_1 n_2(n_1 + n_2 + 1)}{12}}$$

GUIDED EXERCISE 3

Mean and standard deviation of ranks

For the Navy divers, compute μ_R and σ_R. (Recall that $n_1 = 9$ and $n_2 = 10$.)

$$\mu_R = \frac{n_1(n_1 + n_2 + 1)}{2} = \frac{9(9 + 10 + 1)}{2} = 90$$

$$\sigma_R = \sqrt{\frac{n_1 n_2(n_1 + n_2 + 1)}{12}} = \sqrt{\frac{9 \cdot 10(9 + 10 + 1)}{12}} = \sqrt{150} = 12.25$$

Testing R

To determine if the difference between sample means is significant, we use a (two-sided) rank-sum test. Table 12-7 on the following page indicates the critical values to be used.

We convert the sample test statistic R to a z value using the formula

$$z = \frac{R - \mu_R}{\sigma_R}$$

Let's use the rank-sum test with $\alpha = 0.05$ level of significance for the problem of the Navy divers.

H_0: Decompression time distributions are the same.

H_1: Decompression time distributions are different.

Since $n_1 = 9$ and $n_2 = 10$, the samples are large enough to use Table 12-7 to find the critical values. For $\alpha = 0.05$, the critical values are $z_0 = \pm 1.96$. The critical region is shown in Figure 12-3 on the following page.

To find the z value corresponding to the sample rank $R = 61$, we use the values $\mu_R = 90$ and $\sigma_R = 12.25$ found in Guided Exercise 3. Then

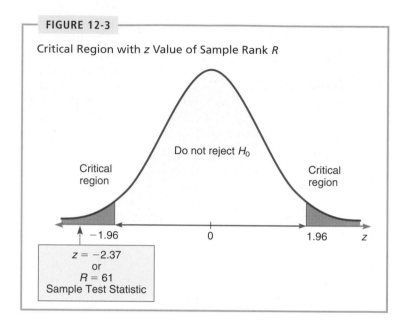

FIGURE 12-3

Critical Region with z Value of Sample Rank R

TABLE 12-7 Critical Values for the Rank-Sum Test (Each Sample Size ≥ 8)

When α Is	We Use the Critical Values
0.05	$z_0 = \pm 1.96$
0.01	$z_0 = \pm 2.58$

$$z = \frac{R - \mu_R}{\sigma_R} = \frac{61 - 90}{12.25} = -2.37$$

We note that the z value of the sample statistic R falls in the critical region shown in Figure 12-3. Therefore, we reject H_0 at the 5% level of significance and conclude that the pill does make a difference in decompression time.

◇ **COMMENT** Because the sampling distribution for the sum of ranks R follows a normal distribution, you compute P values for the sample test statistic using the methods of Section 9.3. Recall that if the P value is less than or equal to the level of significance α, we reject H_0. ◇

◇ **NOTE** For the decompression time data, there were no ties for any rank. If a tie does occur, then each of the tied observations is given the *mean* of the ranks that they occupy. For example, if we rank the numbers

41   42   44   44   44   44

we see that 44 occupies ranks three, four, five, and six. Therefore, we give each of the 44s a rank that is the mean of 3, 4, 5, and 6:

$$\text{Mean of ranks} = \frac{3 + 4 + 5 + 6}{4} = 4.5$$

TABLE 12-8

Observation	Rank
41	1
42	2
44	4.5
44	4.5
44	4.5
44	4.5

The final ranking would then be that shown in Table 12-8. ◇

For samples wherein n_1 or n_2 is less than 8, there are statistical tables that give appropriate critical values for the rank-sum test. Most libraries contain such tables, and the interested reader can find such information by looking under the *Mann-Whitney U Test*.

GUIDED EXERCISE 4

Rank-sum test

A biologist is doing research on elk in their natural Colorado habitat. Two regions are under study, both with about the same amount of forage and natural cover. However, region A seems to have more predators than region B. To determine if elk tend to live longer in either region, a sample of 10 elk from each region are tranquilized and have a tooth removed. A laboratory examination of the teeth reveals the ages of the elk. Results for each sample are given in Table 12-9.

TABLE 12-9   Ages of Elk

Group A	Group B
11	12
6	7
21	5
23	24
16	4
1	14
3	19
10	22
13	18
8	2

(a) Make a table showing the ages, groups, and ranks for the combined data.

TABLE 12-10   Ranks of Elk

Age	Group	Rank	Age	Group	Rank
1	A	1	12	B	11
2	B	2	13	A	12
3	A	3	14	B	13
4	B	4	16	A	14
5	B	5	18	B	15
6	A	6	19	B	16
7	B	7	21	A	17
8	A	8	22	B	18
10	A	9	23	A	19
11	A	10	24	B	20

(b) Find μ_R, σ_R, and R.

Since $n_1 = 10$ and $n_2 = 10$,

$$\mu_R = \frac{(10)(10 + 10 + 1)}{2} = 105$$

and $\sigma_R = \sqrt{\dfrac{10 \cdot 10(10 + 10 + 1)}{12}} = 13.23$

Since $n_1 = n_2 = 10$, we can use the sum of the ranks of either the A group or the B group. Let's use the sum of the ranks of the A group. The A group ranks are 1, 3, 6, 8, 9, 10, 12, 14, 17, and 19. Therefore,

$$R = 1 + 3 + 6 + 8 + 9 + 10 + 12 + 14 + 17 + 19 = 99$$

Continued

GUIDED EXERCISE 4 continued

(c) Using an $\alpha = 0.05$ level of significance, what critical values should we use for a (two-sided) rank-sum test? Compute the z value corresponding to the sample rank $R = 99$ found in part (b). Sketch the critical region, and show the location of the z value of sample rank on the sketch.

  The samples are large enough to use the critical values in Table 12-7. They are $z_0 = \pm 1.96$. Using the values for μ_R and σ_R found in part (b), we find the z value for $R = 99$ to be

$$z = \frac{R - \mu_R}{\sigma_R} = \frac{99 - 105}{13.23} = -0.45$$

FIGURE 12-4 Critical Region

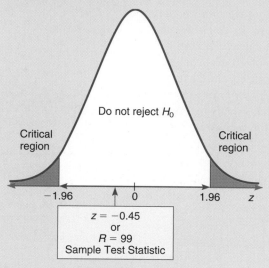

(d) Do the data support the null hypothesis (mean lifetimes are the same for the two regions)?

  Because the z value corresponding to the sample rank $R = 99$ does not fall in the critical region, we cannot reject the null hypothesis. The data do not indicate that elk live longer in one region.

 VIEWP⊙INT   *Point Barrow, Alaska*

Point Barrow is located very near the northernmost point of land in the United States. In 1935, Will Rogers (an American humorist, social critic, and philosopher) was killed with Wiley Post (a pioneer aviator) at a landing strip near Point Barrow. Since 1920, a weather station at the (now named) Wiley Post–Will Rogers Memorial Landing Strip has recorded daily high and low temperatures. From these readings, annual mean maximum and minimum temperatures have been computed. Is Point Barrow warming up, cooling down, or neither? Can you think of a way to gather data and construct a nonparametric test to investigate long-term temperature highs and lows at Point Barrow? For weather-related data, visit the Brase/Brase statistics site at http://math.college.hmco.com/students and find a link to the Geophysical Institute at the University of Alaska in Fairbanks. Then follow links to Point Barrow.

SECTION 12.2   PROBLEMS

In Problems 1–5, use a 0.05 level of significance to test the null hypothesis that there is no difference between group distributions against the alternate hypothesis that there is a difference either way.

1. *Education: Reading* Two groups of ninth-grade students are given a reading comprehension exam. Group A students are from Windy Heights Public School, and group B students are from Califf, a neighboring private school. The following table shows the results:

Group A	71	65	70	44	81	73	50	60	88	
Group B	69	45	66	85	75	90	63	84	77	55

2. *Psychology: Testing* A psychologist tested two adult groups for boredom tolerance. Group A consisted of 12 females, and group B consisted of 10 males. Their scores are in the following table.

Group A	73	68	41	103	92	88	50	111	120	66	75	115
Group B	150	99	85	77	35	69	100	135	54	72		

3. *Horse Trainer: Jumps* A horse trainer teaches horses to jump by using two methods of instruction. Horses being taught by method A have a lead horse that accompanies each jump. Horses being taught by method B have no lead horse. The table shows the number of training sessions required before each horse would do the jumps properly.

Method A	28	35	19	41	37	31	38	40	25	
Method B	42	33	26	24	44	46	34	20	48	39

4. *Education: French Verbs* A French teacher teaches verbs using two different methods. Two groups of 10 students were taught a list of verbs using the two different methods, one method for each group. The time required to learn the list using each method is shown in the following table (in minutes):

Method A	18	25	41	22	56	20	15	30	33	44
Method B	15	28	19	46	55	30	63	58	40	29

5. *Psychology: Testing* A cognitive aptitude test consists of putting together a puzzle. Nine people in group A took the test in a competitive setting (first and second to finish received a prize). Twelve people in group B took the test in a noncompetitive setting. The results follow (in minutes required to complete the puzzle):

Group A	7	12	10	15	22	17	18	13	8			
Group B	9	16	30	11	33	28	19	14	24	27	31	29

In Problems 6–10, use a 0.01 level of significance to test the null hypothesis that there is no difference between average performance of the groups against the alternate hypothesis that there is a difference either way.

6. *Psychology: Testing* A psychologist has developed a mental alertness test. She wishes to study the effects (if any) of type of food consumed on mental alertness. Twenty-one volunteers were randomly divided into two groups. Both groups were told to eat the amount they usually eat for lunch at noon. At 2:00 P.M., all subjects were given the alertness test. Group A had a low-fat lunch with no red meat, lots of vegetables, carbohydrates, and fiber. Group B had a high-fat lunch with red meat, vegetable oils, and low fiber. The only drink for both groups was water. The test scores are shown below:

Group A	76	93	52	81	68	79	88	90	67	85	60
Group B	44	57	60	91	62	86	82	65	96	42	

7. *Medical: Colds* Dr. Winchester is studying the effect of vitamin B_{52} on the common cold. A group of 19 Army personnel with common colds was randomly divided into two groups. Group A subjects were given 500-milligram doses of B_{52} three times a day. Group B subjects were given the same doses of placebos (sugar pills). For each group, the duration of the subject's cold is given (in days).

Group A	14	19	12	21	25	16	20	28	10
Group B	9	15	24	26	18	17	31	8	22	11

8. *Sports: Ski Racers* A group of 18 cross-country ski racers was randomly divided into two groups. Group A used modern no-wax "fish-scale" type Teflon ski bottoms. Group B used traditional wax. The times for these skiers to complete a 10-km run (with hills) are shown below in minutes:

Group A	45	41	52	47	58	55	40	38	33
Group B	44	39	30	61	37	42	36	50	31

9. *Education: Spelling* Sixteen fourth-grade children were randomly divided into two groups. Group A was taught spelling by a phonetic method. Group B was taught spelling by a memorization method. At the end of the fourth grade, all children were given a standard spelling exam. The scores are shown as follows:

Group A	77	95	83	69	85	92	61	79
Group B	62	90	70	81	63	75	80	72

10. *Industrial Chemistry: Cement* Dr. Hansen, an industrial chemist, has discovered a new catalyst that may affect the setting time of wet cement. Two groups of test slabs of cement were studied. Group A had no catalyst, and group B used the catalyst. The setting time for each slab was measured by Dr. Hansen. The results follow (in hours):

Group A	2.7	2.4	1.9	2.9	3.4	1.6	3.6	4.1
Group B	2.5	1.8	1.6	2.2	4.0	3.8	1.4	2.8

12.3
Spearman Rank Correlation

FOCUS POINTS
✓ Learn about monotone relations and the Spearman rank correlation coefficient.
✓ Compute the Spearman correlation coefficient and conduct statistical tests for significance.
✓ Interpret the results in the context of the application.

Data given in ranked form (ordinal type) are different from data given in measurement form (interval or ratio type). For instance, if we compare the test performance of three students and, say, Elizabeth did the best, Joel did next best, and Sally did the worst, we are giving the information in ranked form. We cannot say how much better Elizabeth did than Sally or Joel, but we do know how the three scores compare. If the actual test scores for the three tests were given, we would have data in measurement form and could tell exactly how much better Elizabeth did than Joel or Sally. In Chapter 10, we studied linear correlation of data in measurement form. In this section, we will study correlation of data in ranked form.

As a specific example of a situation in which we might want to compare ranked data from two sources, consider the following. Hendricks College has a new faculty position in its political science department. A national search to fill this position has resulted in a large number of qualified candidates. The political science faculty reserves the right to make the final hiring decision. However, the faculty is interested in comparing its opinion with student opinion about the teaching ability of the candidates. A random sample of nine equally qualified candidates was asked to give a classroom presentation to a large class of students. Both faculty and students attended the lectures. At the end of each lecture, both faculty and students filled out a questionnaire about the teaching performance of the candidate. Based on these questionnaires, each candidate was given an overall rank from the faculty and an overall rank from the students. The results are shown in Table 12-11. Higher ranks mean better teaching performance.

Using data in ranked form, how can we answer the following questions:

1. Do candidates getting higher ranks from the faculty tend to get higher ranks from students?

2. Is there any relation between faculty rankings and student rankings?

3. Do candidates getting higher ranks from faculty tend to get lower ranks from students?

We will use the Spearman rank correlation to answer such questions. In the early 1900s, Charles Spearman of the University of London developed the techniques that

TABLE 12-11   Faculty and Student Ranks of Candidates

Candidate	Faculty Rank	Student Rank
1	3	5
2	7	7
3	5	6
4	9	8
5	2	3
6	8	9
7	1	1
8	6	4
9	4	2

FIGURE 12-5

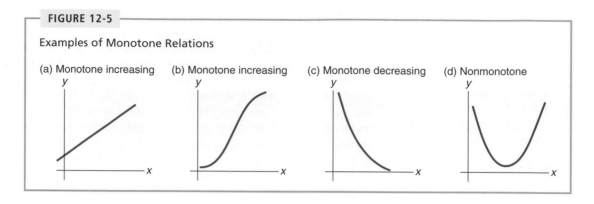

Examples of Monotone Relations

(a) Monotone increasing   (b) Monotone increasing   (c) Monotone decreasing   (d) Nonmonotone

now bear his name. The Spearman test of rank correlation requires us to use *ranked variables*. Because we are using only ranks, we cannot use the Spearman test to check on the existence of a linear relationship between the variables, as we did with the Pearson correlation coefficient (Section 10.3). The Spearman test checks only on the existence of a *monotone* relationship between the variables. (See Figure 12-5.) By a *monotone relationship** between variables x and y, we mean a relationship in which

1. as x increases, y also increases, or

2. as x increases, y decreases.

The relationship shown in Figure 12-5(d) is a nonmonotone relationship because as x increases, y at first decreases, but later starts to increase. Remember, for a relation to be monotone, as x increases, y must *always* increase or *always* decrease. In a nonmonotone relation, as x increases, y sometimes increases and sometimes decreases or stays unchanged.

GUIDED EXERCISE 5

Monotonic behavior

Identify each of the relations in Figure 12-6 as monotone increasing, monotone decreasing, or nonmonotone.

FIGURE 12-6

(a)   (b)   (c)   (d)

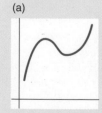   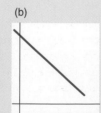   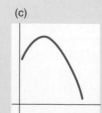   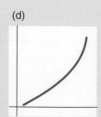

Answers: (a) nonmonotone, (b) monotone decreasing, (c) nonmonotone, (d) monotone increasing

*Some advanced texts call the monotone relationship we describe *strictly monotone*.

Spearman rank correlation coefficient

Before we can complete the solution of our problem about the political science department at Hendricks College, we need the following information.

If we have a sample of size n of randomly obtained ordered pairs (x, y), where both x and y values are from *ranked variables*, and if there are no ties in the ranks, then the Pearson product moment correlation coefficient (Section 10.3) can be reduced to the simpler equation.

$$r_s = 1 - \frac{6\Sigma d^2}{n(n^2 - 1)} \qquad \text{where } d = x - y$$

We call r_s the *Spearman rank correlation coefficient*. It has the following important properties:

1. $-1 \leq r_s \leq 1$. If $r_s = -1$, the relation between x and y is perfectly monotone decreasing. If $r_s = 0$, there is no monotone relation between x and y. If $r_s = 1$, the relation between x and y is perfectly monotone increasing. Values of r_s close to 1 or -1 indicate a strong tendency for x and y to have a monotone relationship (increasing or decreasing as the case may be), and values of r_s close to 0 indicate a very weak (or perhaps nonexistent) monotone relationship.

2. The probability distribution of r_s depends on the sample size n. Table 9 in Appendix II gives critical values for certain left- and right-tailed tests of r_s. It is important to note that we make no assumptions that x and y are normally distributed variables, and we make no assumption about the x and y relationship being linear.

3. The Spearman rank correlation coefficient r_s is our *sample* estimate for ρ_s, the *population* Spearman rank correlation coefficient. We will construct a test of significance for the Spearman rank correlation coefficient in much the same way that we tested the Pearson correlation coefficient (Section 10.4). The null hypothesis is

Hypothesis

$$H_0: \rho_s = 0$$

In effect, the null hypothesis says that there is no monotone relation (either increasing or decreasing). The alternate hypothesis depends on the type of test we want to use.

$H_1: \rho_s < 0$ (left-tailed test): The alternate hypothesis claims that there is a monotone-decreasing relation between x and y. The critical region is shown in Figure 12-7(a) on the following page.

$H_1: \rho_s > 0$ (right-tailed test): The alternate hypothesis claims that there is a monotone-increasing relation between x and y. The critical region is shown in Figure 12-7(b) on the following page.

$H_1: \rho_s \neq 0$ (two-tailed test): The alternate hypothesis claims that there is a monotone relation (either increasing or decreasing) between x and y. Figure 12-7(c) on the following page shows the corresponding critical regions.

Critical values

To find critical values and critical regions of the r_s distribution, we use Table 9 of Appendix II with the level of significance, sample size, and type of test (left-, right-, or two-tailed) employed.

EXAMPLE 1

Testing Spearman rank correlation coefficient

Using the information about the Spearman rank correlation coefficient, let's finish our problem about the search for a new member of the political science department at Hendricks College. Our work is organized in Table 12-12 on the next page, where the rankings given by students and faculty are listed for each of the nine candidates.

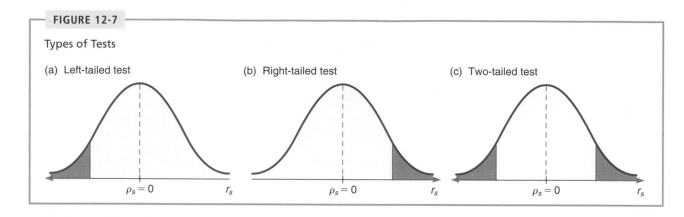

Types of Tests

(a) Left-tailed test   (b) Right-tailed test   (c) Two-tailed test

Since the sample size is $n = 9$, and $\Sigma d^2 = 16$, the Spearman rank correlation coefficient is

$$r_s = 1 - \frac{6\Sigma d^2}{n(n^2 - 1)} = 1 - \frac{6(16)}{9(81 - 1)} = 0.867$$

Let's test the claim that faculty and students tend to agree about the candidate's teaching ability. This means the x and y variables should be monotone increasing (as x increases, y increases). Since ρ_s is the population Spearman rank correlation coefficient, we have

$H_0: \rho_s = 0$ \qquad (There is no monotone relation.)

$H_1: \rho_s > 0$ \qquad (There is a monotone-increasing relation.)

If we use $\alpha = 0.005$ as our level of significance for a right-tailed test, then Table 9 of Appendix II with $n = 9$ gives a critical value of 0.834. Figure 12-8 shows the critical region.

Because the observed r_s value is in the critical region, we reject $H_0: \rho_s = 0$ and conclude that the relation between faculty and student ranks is monotonic increasing $\rho_s > 0$. This means that faculty and students tend to rank the teaching performance of candidates in the same way.

TABLE 12-12 Student and Faculty Ranks of Candidates and Calculations for the Spearman Rank Correlation Test

Candidate	Faculty Rank x	Student Rank y	$d = x - y$	d^2
1	3	5	−2	4
2	7	7	0	0
3	5	6	−1	1
4	9	8	1	1
5	2	3	−1	1
6	8	9	−1	1
7	1	1	0	0
8	6	4	2	4
9	4	2	2	4
				$\Sigma d^2 = 16$

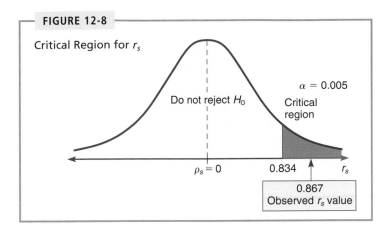

FIGURE 12-8

Critical Region for r_s

Do not reject H_0

$\alpha = 0.005$

Critical region

$\rho_s = 0$ 0.834 r_s

0.867
Observed r_s value

GUIDED EXERCISE 6

Testing Spearman rank correlation coefficient

Fishermen in the Adirondack Mountains are complaining that acid rain caused by air pollution is killing fish in their region. To study this claim, a biology research team studied a random sample of 12 lakes in the region. For each lake, they measured the level of acidity of rain in the drainage leading into the lake and the density of fish in the lake (number of fish per acre foot of water). Then they did a ranking of x = acidity and y = density of fish. The results are shown in Table 12-13. Higher x ranks mean more acidity, and higher y ranks mean higher density of fish.

TABLE 12-13   Acid Rain and Density of Fish

Lake	Acidity x	Fish Density y	$d = x - y$	d^2
1	5	8	−3	9
2	8	6	2	4
3	3	9	−6	36
4	2	12	−10	100
5	6	7	−1	1
6	1	10	−9	81
7	10	2	8	64
8	12	1	——	——
9	7	5	——	——
10	4	11	——	——
11	9	4	——	——
12	11	3	——	——
				$\Sigma d^2 = $ ——

(a) Complete the entries in the d and d^2 columns of Table 12-13, and find Σd^2.

Lake	x	y	d	d^2
8	12	1	11	121
9	7	5	2	4
10	4	11	−7	49
11	9	4	5	25
12	11	3	8	64
				$\Sigma d^2 = 558$

Continued

GUIDED EXERCISE 6 continued

(b) Compute r_s.

➡ $r_s = 1 - \dfrac{6\Sigma d^2}{n(n^2 - 1)} = 1 - \dfrac{6(558)}{12(144 - 1)} = -0.951$

(c) The fishermen claim that more acidity means lower density of fish. Would this claim say that x and y have a monotone-increasing or monotone-decreasing relation or no monotone relation?

➡ The claim says that as x increases, y decreases, so the claim is that the relation of x and y is monotone decreasing.

(d) To test the fishermen's claim, what should we use for the null hypothesis and for the alternate hypothesis?

➡ H_0: $\rho_s = 0$ (no monotone relation)
H_1: $\rho_s < 0$ (monotone-decreasing relation)

(e) Using a 0.001 level of significance, find a critical value and sketch the critical region for the test of part (d).

➡ $n = 12$; $\alpha = 0.001$

By Table 9 of Appendix II, the critical value for a left-tailed test is -0.826. See Figure 12-9.

FIGURE 12-9 Critical Region

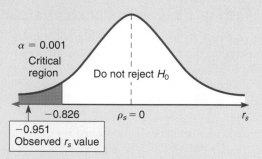

(f) Do the data indicate that we should reject the claim that more acidity means fewer fish? Explain.

➡ Because the observed value $r_s = -0.951$ is in the critical region, we reject H_0 and conclude that the level of acidity has a very significant ($\alpha = 0.001$) monotone-decreasing relation with fish density.

Ties of ranks

If ties occur in the assignment of ranks, we follow the usual method of averaging tied ranks. This method was discussed in Section 12.2 (The Rank-Sum Test). The next example illustrates the method.

◇ **COMMENT** Technically, the use of the given formula for r_s requires that there be no ties in rank. However, if the number of ties in rank is small relative to the number of ranks, the formula can be used with quite a bit of reliability. ◇

EXAMPLE 2

Tied ranks

Do people who smoke more tend to drink more cups of coffee? The following data were obtained from a random sample of $n = 10$ cigarette smokers who also drink coffee:

Person	Cigarettes per Day	Cups of Coffee per Day
1	8	4
2	15	7
3	20	10
4	5	3
5	22	9
6	15	5
7	15	8
8	25	11
9	30	18
10	35	18

To use the Spearman rank correlation test, we need to rank the data. It does not matter if we rank from smallest to largest or largest to smallest. The only requirement is that we be consistent in our rankings. Let us rank from smallest to largest.

First, we rank data for each variable as though there were no ties; then we average the ties as shown in Tables 12-14 and 12-15.

TABLE 12-14   Rankings of Cigarettes Smoked per Day

Person	Cigarettes per Day	Rank		Average Rank x	
4	5	1		1	
1	8	2		2	
2	15 ⎫	3 ⎫		4 ⎫	
6	15 ⎬ Ties	4 ⎬ Average rank is 4.		4 ⎬ Use the average	
7	15 ⎭	5 ⎭		4 ⎭ rank for tied data	
3	20	6		6	
5	22	7		7	
8	25	8		8	
9	30	9		9	
10	35	10		10	

TABLE 12-15   Rankings of Cups of Coffee per Day

Person	Cups of Coffee per Day	Rank		Average Rank y	
4	3	1		1	
1	4	2		2	
6	5	3		3	
2	7	4		4	
7	8	5		5	
5	9	6		6	
3	10	7		7	
8	11	8		8	
9	18 ⎫ Ties	9 ⎫ Average rank		9.5 ⎫ Use the average	
10	18 ⎭	10 ⎭ is 9.5.		9.5 ⎭ rank for tied data.	

TABLE 12-16   Ranks to Be Used for a Spearman Rank Correlation Test

Person	Cigarette Rank x	Coffee Rank y	$d = x - y$	d^2
1	2	2	0	0
2	4	4	0	0
3	6	7	−1	1
4	1	1	0	0
5	7	6	1	1
6	4	3	1	1
7	4	5	−1	1
8	8	8	0	0
9	9	9.5	−0.5	0.25
10	10	9.5	0.5	0.25
				$\Sigma d^2 = 4.5$

Next, we compute the observed value using results shown in Table 12-16:

$$r_s = 1 - \frac{6\Sigma d^2}{n(n^2 - 1)} = 1 - \frac{6(4.5)}{10(100 - 1)} = 0.973$$

Using 0.001 as a level of significance, test the claim that x and y have a monotone-increasing relationship. In other words, test the claim that people who tend to smoke more tend to drink more cups of coffee (Table 12-16).

$H_0: \rho_s = 0$ (There is no monotone relation.)

$H_1: \rho_s > 0$ (Right-tailed test)

From Table 9 of Appendix II, we find that when $n = 10$ and $\alpha = 0.001$, the critical value is $r_s = 0.879$. The resulting critical region is shown in Figure 12-10.

Because the observed value $r_s = 0.973$ is in the rejection region, we reject H_0 and conclude that x and y have a monotone-increasing relation. People who smoke more cigarettes tend to drink more coffee. ◇

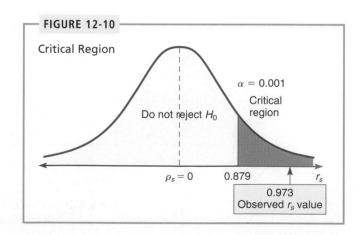

FIGURE 12-10

Critical Region

$\alpha = 0.001$

Critical region

Do not reject H_0

$\rho_s = 0$ 0.879 r_s

0.973
Observed r_s value

VIEWP⊙INT   *Rug Rats!*

When do babies start to crawl? Janette Benson, in her article "Infant Behavior and Development," claims crawling age is related to temperature during the month in which babies first try to crawl. To find a data file for this subject, visit the Brase/Brase statistics site at http://math.college.hmco.com/students and find a link to DASL, the Carnegie Mellon University Data and Story Library. Then look under Psychology in the Data Subjects and select the Crawling Datafile. Can you think of a way to gather data and construct a nonparametric test to study this claim?

SECTION 12.3   PROBLEMS

1. *Training Program: Sales* A data processing company has a training program for new salespeople. After completing the training program, each trainee is ranked by his or her instructor. After a year of sales, the same class of trainees is again ranked by a company supervisor according to net value of the contracts they have acquired for the company. The results for a random sample of 11 salespeople trained in the last year follow, where x is rank in training class and y is rank in sales after 1 year. Lower ranks mean higher standing in class and higher net sales.

Person	1	2	3	4	5	6	7	8	9	10	11
x rank	6	8	11	2	5	7	3	9	1	10	4
y rank	4	9	10	1	6	7	8	11	3	5	2

Using a 0.05 level of significance, test the claim that the relation between x and y is monotone (either increasing or decreasing).

2. *Economics: Stocks* As an economics class project, Debbie studied a random sample of 14 stocks. For each of these stocks, she found the cost per share (in dollars) and ranked each of the stocks according to cost. After 3 months, she found the earnings per share on each stock (in dollars). Again, Debbie ranked each of the stocks according to earnings. The way Debbie ranked, higher ranks mean higher cost and higher earnings. The results follow, where x is the rank in cost and y is the rank in earnings.

Stock	1	2	3	4	5	6	7	8	9	10	11	12	13	14
x	5	2	4	7	11	8	12	3	13	14	10	1	9	6
y	5	13	1	10	7	3	14	6	4	12	8	2	11	9

Using a 0.01 level of significance, test the claim that there is a monotone relation, either way, between ranks of cost and earnings.

3. *Psychology: Rat Colonies* A psychology professor is studying the relation between overcrowding and violent behavior in a rat colony. Eight colonies with different degrees of

overcrowding are being used. By using a television monitor, lab assistants record incidents of violence. Each colony has been ranked for crowdedness and violence. A rank of 1 means most crowded or most violent. The results for the eight colonies are in the following table, with x being the population density rank and y the violence rank.

Colony	1	2	3	4	5	6	7	8
x rank	3	5	6	1	8	7	4	2
y rank	1	3	5	2	8	6	4	7

Using a 0.05 level of significance, test the claim that lower crowding ranks mean lower violence ranks (i.e., the variables have a monotone-increasing relation).

4. *History: Exams* A history professor claims that students who finish exams quicker tend to get higher scores. The following data show the order of finish and score for 10 students selected at random:

Student	1	2	3	4	5	6	7	8	9	10
Order of finish	5	7	3	1	6	2	8	4	10	9
Score	73	90	82	95	65	82	78	75	80	55

(a) Ranking order of finish with 1 as first to finish, and ranking score with 1 as highest score, construct a table of ranks to be used for a Spearman rank correlation test.
(b) Using a 0.05 level of significance, test the claim that there is a monotone-increasing relation between rank of finish and rank of score.

5. *Psychology: Testing* An army psychologist gave a random sample of seven soldiers a test to measure sense of humor and another test to measure aggressiveness. High scores mean greater sense of humor or more aggressiveness.

Soldier	1	2	3	4	5	6	7
Score on humor test	60	85	78	90	93	45	51
Score on aggressiveness test	78	42	68	53	62	50	76

(a) Ranking the data with rank 1 for highest score on a test, make a table of ranks to be used in a Spearman rank correlation test.
(b) Using a 0.05 level of significance, test the claim that rank in humor has a monotone-decreasing relation to rank in aggressiveness.

6. *Consumers: Quality and Price* A consumer research group examined a random sample of eight stereo speaker systems selected from among major brands. They ranked each of the systems for overall quality of equipment and sound produced. They obtained the manufacturer's recommended price for each brand. A rank of 1 means highest quality.

System	1	2	3	4	5	6	7	8
Quality rank	4	8	5	2	7	6	1	3
Price, $	690	175	1200	970	225	785	470	850

(a) Using a rank of 1 for the highest price, make a table of ranks to be used in a Spearman rank correlation test.

(b) Using a 0.05 level of significance, test the claim that there is a monotone relation (either way) between rank of quality and rank of price.

7. *Cadets: Flying Aptitude* A group of 11 cadets selected at random were given a flying aptitude test before they went to flight training school. After graduation from training school, their commanding officer ranked each cadet according to his or her flying ability (higher ranks mean greater ability). The results were

Cadet	1	2	3	4	5	6	7	8	9	10	11
Aptitude score	720	390	710	480	970	480	517	830	690	850	480
Performance rank	7	1	8	4	10	2	5	11	6	9	3

(a) Using a rank of 1 for the lowest aptitude score, make a table of ranks to be used in a Spearman rank correlation test.

(b) Using a 0.005 level of significance, test the claim that there is a monotone-increasing relation between aptitude rank and performance rank.

8. *Management: Secretaries* At the Big Rock Insurance Company branch office, a pool of six secretaries work under two different managers. At the end of the year, both managers are asked to rank the secretaries. A rank of 1 means best secretary.

Secretary	1	2	3	4	5	6
Manager A rank	3	5	2	1	6	4
Manager B rank	1	3	6	2	5	4

Using a 0.05 level of significance, test the claim that x and y have a monotone-increasing relationship (i.e., the managers agree).

9. *Insurance: Sales* Big Rock Insurance Company did a study of per capita income and volume of insurance sales in eight Midwest cities. The volume of sales in the cities was ranked, with 1 being the largest volume. The per capita income was rounded to the nearest thousand dollars.

City	1	2	3	4	5	6	7	8
Volume of insurance sales rank	6	7	1	8	3	2	5	4
Per capita income in $1000	17	18	19	11	16	20	15	19

(a) Using a rank of 1 for the highest per capita income, make a table of ranks to be used for a Spearman rank correlation test.

(b) Using a 0.01 level of significance, test the claim that there is a monotone relation (either way) between rank in volume of sales and rank of per capita income.

SUMMARY

When we cannot assume that data come from a normal, binomial, or Student's *t* distribution, we can employ tests that make no assumptions about data distribution. Such tests are called nonparametric tests. We studied three widely used tests: the sign test, the rank-sum test, and the Spearman rank correlation coefficient test. Nonparametric tests have the advantage of being easy to use; however, they do tend to waste information and to be less sensitive than standard tests. It is usually good advice to use standard tests when possible, keeping nonparametric tests for situations wherein assumptions about the data distribution cannot be made.

IMPORTANT WORDS & SYMBOLS

Section 12.1
Nonparametric statistics
Sign test

Section 12.3
Spearman rank correlation coefficient r_s

Section 12.2
Rank-sum test

VIEWPOINT   *Lending a Hand*

Whom would you ask for help if you were sick? in need of money? upset with your spouse? depressed? Consider the claim: People look to sisters for emotional help and brothers for physical help. After that, people look to parents, clergy, or friends. Can you think of nonparametric tests to study such claims? For more information, see *American Demographics*, vol. 18, no. 8.

CHAPTER REVIEW PROBLEMS

For each problem, do the following:
(a) Decide whether you should use a sign test, rank-sum test, or Spearman test.
(b) State the null and alternate hypotheses.
(c) Find all critical values.
(d) Sketch the critical region, the critical values, and the sample statistic value.
(e) Decide whether you should reject or fail to reject the null hypothesis.
(f) Interpret the results in the context of the problem.

1. *Chemistry: Lubricant* In the production of synthetic motor lubricant from coal, a new catalyst has been discovered that seems to affect the viscosity index of the lubricant. In an experiment consisting of 21 production runs, 10 used the new catalyst and 11 did not. After each production run, the viscosity index of the lubricant was determined to be as follows:

With catalyst	1.6	3.2	2.9	4.4	3.7	2.5	1.1	1.8	3.8	4.2
Without catalyst	3.9	4.6	1.5	2.2	2.8	3.6	2.4	3.3	1.9	4.0	3.5

Use a 0.05 level of significance to test the null hypothesis that the viscosity index is unchanged by the catalyst against the alternate hypothesis that the viscosity index has changed.

2. *Self-Improvement: Memory* Professor Adams wrote a book called *Improving Your Memory*. The professor claims that if you follow the program outlined in the book, your memory will definitely improve. Fifteen people took the professor's course, in which the book and its program were used. On the first day of class, everyone took a memory exam; and on the last day, everyone took a similar exam. Their scores were as follows:

Last exam	225	120	115	275	85	76	114	200	99	135	170	110	216	280	78
First exam	175	110	115	200	60	85	160	190	70	110	140	10	190	200	92

Use a 0.05 level of significance to test the null hypothesis that the scores are the same whether or not people have taken the course against the alternate hypothesis that the scores of people who have taken the course are higher.

3. *Sales: Paint* A chain of hardware stores is trying to sell more paint by mailing pamphlets describing the paint. In 15 communities containing one of these hardware stores, the paint sales (in dollars) were recorded for the month before and the month after the ads were sent out. The results follow.

Sales After	Sales Before	Sales After	Sales Before
610	460	500	370
150	216	118	118
790	640	265	117
288	250	365	360
715	685	93	93
465	430	217	291
280	220	280	430
640	470		

Use a 0.01 level of significance to test the null hypothesis that the advertising had no effect on sales against the alternate hypothesis that it improved sales.

4. *Dogs: Obedience School* An obedience school for dogs experimented with two methods of training. One method involved rewards (food, praise); the other involved no rewards. The number of sessions required for training each of 19 dogs follows:

With rewards	12	17	15	10	16	20	9	23	8	14
No rewards	19	22	11	18	13	25	24	28	21	

Use a 0.05 level of significance to test the hypothesis that the number of sessions was the same for the two groups against the alternate hypothesis that they were not the same.

5. *Training Program: Fast Food* At McDouglas Hamburger stands, each employee must undergo a training program before he or she is hired. A group of nine people went through the training program and were hired to work in the Teton Park McDouglas Hamburger stand. Rankings in performance for the training program and after one month on the job are shown (a rank of 1 is for best performance).

Employee	1	2	3	4	5	6	7	8	9
Rank, training program	8	9	7	3	6	4	1	2	5
Rank on job	9	8	6	7	5	1	3	4	2

Using a 0.05 level of significance, test the claim that there is a monotone-increasing relation between rank in the training program and rank in performance on the job.

6. *Cooking School: Chocolate Mousse* Two expert French chefs judged chocolate mousse made by students in a Paris cooking school. Each chef ranked the best chocolate mousse as 1.

Student	1	2	3	4	5
Rank by Chef Pierre	4	2	3	1	5
Rank by Chef André	4	1	2	3	5

Use a 0.10 level of significance to test the claim that there is a monotone relation (either way) for ranks given by Chef Pierre and Chef André.

DATA HIGHLIGHTS: GROUP PROJECTS

Break into small groups and discuss the following topics. Organize a brief outline in which you summarize the main points of your group discussion.

In the world of business and economics, to what extent do assets determine profits? Do the big companies with large assets always make more profits? Is there a rank correlation between assets and profits? The following table is based on information taken from *Fortune* (vol. 135, no. 8). A rank of 1 means highest profits or highest assets. The companies are food service companies.

Company	Asset Rank	Profit Rank
Pepsico	4	2
McDonald's	1	1
Aramark	6	4
Darden Restaurants	7	5
Flagstar	11	11
VIAD	10	8
Wendy's International	2	3
Host Marriott Services	9	10
Brinker International	5	7
Shoney's	3	6
Food Maker	8	9

(a) Compute the Spearman rank correlation coefficient for these data.

(b) Using a 5% level of significance, test the claim that there is a monotone-increasing relation between the ranks of earnings and growth.

(c) Decide whether you should reject or not reject the null hypothesis. State your conclusion in the context of the problem.

(d) As an investor, what are some other features about food companies that you might be interested in ranking? Identify any such features that you think might have a monotone relation.

LINKING CONCEPTS: WRITING PROJECTS

Discuss each of the following topics in class or review the topics on your own. Then write a brief but complete essay in which you summarize the main points. Please include formulas and graphs as appropriate.

1. (a) What do we mean by the term *nonparametric statistics*? What do we mean by the term *parametric statistics*? How do nonparametric methods differ from the methods we have studied earlier?

   (b) What are the advantages of nonparametric statistical methods? How can they be used in problems where other methods we have learned would not apply?

   (c) Are there disadvantages to nonparametric statistical methods? What do we mean when we say nonparametric methods tend to be wasteful of information? Why do we say nonparametric methods are not as *sensitive* as parametric methods?

   (d) List three random variables from ordinary experience for which you think nonparametric methods would definitely apply and parametric methods would be questionable.

2. Outline the basic logic and ideas behind the sign test. Describe how the binomial probability distribution was used in the construction of the sign test. What assumptions must be made about the sign test? Why is the sign test so extremely general in its possible applications? Why is it a special test for "before-and-after" studies?

3. Outline the basic logic and ideas behind the rank-sum test. Under what conditions would you use the rank-sum test and *not* the sign test? What assumptions must be made about the rank-sum test? List two advantages the rank-sum test has that the methods of Section 9.7 do not have. List some advantages the methods of Section 9.7 have that the rank-sum test does not have.

4. What do we mean by a monotone relationship between two variables x and y? What do we mean by ranked variables? Give a graphic example of two variables x and y that have a monotonic relationship but do *not* have a linear relationship. Does the Spearman test check on a monotonic relationship or a linear relationship? Under what conditions does the Pearson product moment correlation coefficient reduce to the Spearman rank correlation coefficient? Summarize the basic logic and ideas behind the test for Spearman rank correlation. List variables x and y from daily experience for which you think a strong Spearman rank correlation coefficient exists but the variables are *not* linearly related.

Photo Credits

Chapter 1 p. 2: © Yuen Lee/Getty Images; p. 3: *(top)* © Bruce Coleman, Inc./PictureQuest, *(bottom)* Courtesy of Corrinne and Charles Brase; p. 6: © Geoff Bryant/Photo Researchers Inc.; p. 14: © Catherine Ursillo/Photo Researchers Inc.; p. 21: © Tim Davis/Photo Researchers Inc.; p. 24: © R. Lord/The Image Works

Chapter 2 p. 34: Brown Brothers; p. 35: *(top)* © Judy Gelles/Stock Boston, *(bottom)* © Jean-Marc Loubat/Photo Researchers Inc.; p. 37: © Michael Newman/PhotoEdit; p. 41: © Myrleen Cate/PhotoEdit; p. 48: © Russell D. Curtis/Photo Researchers Inc.; p. 57: © Ph. Royer/Explorer/Photo Researchers Inc.; p. 68: © Picture Press/Corbis; p. 69: © Michael J. Doolittle/The Image Works

Chapter 3 p. 88: © Wood River Gallery/PictureQuest; p. 89: *(top)* © Nancy Sheehan/PhotoEdit, *(bottom)* © Myrleen Ferguson Cate/PhotoEdit; p. 91: © Steve Skjold/PhotoEdit; p. 93: © Jose Carrillo/PhotoEdit; p. 100: © David Young-Wolff/PhotoEdit; p. 101: © Tony Freeman/PhotoEdit; p. 106: © Ron Watts/Corbis; p. 108: © Daemmrich/The Image Works; p. 115: © Rhoda Sidney/PhotoEdit; p. 127: © Gary Conner/PhotoEdit; p. 142: © Peter Southwick/Stock Boston

Chapter 4 p. 148: NorthWind Picture Archives; p. 149: *(top)* © David Young-Wolff/PhotoEdit, *(bottom)* © John Neubauer/PhotoEdit; p. 154: © Bonne Kamin/PhotoEdit; p. 157: © Randy Wells/Corbis; p. 172: © Mark Richards/PhotoEdit

Chapter 5 p. 204: National Portrait Gallery, Smithsonian Institution/Art Resource, NY; p. 205: *(top)* © Peter Vandermark/Stock Boston, *(bottom)* © Wojnarowicz/The Image Works; p. 211: © Kelly-Mooney Photography/Corbis; p. 227: © Ted Curtin/Stock Boston; p. 236: © Mark Richards/PhotoEdit Inc.; p. 249: © Richard Rowan/Photo Researchers Inc.

Chapter 6 p. 270: Historical Pictures/Stock Montage; p. 271: *(2 photos)* © Jeff Greenberg/PhotoEdit; p. 279: © Scott T. Smith/Corbis; p. 291: © Lois Ellen Frank/Corbis; p. 306: © David Young-Wolff/PhotoEdit; p. 320: © William H. Mullins/Photo Researchers Inc.

Chapter 7 p. 334: Ralph Steiner; p. 335: *(top)* © Syracuse Newspapers/David Lassman/The Image Works, *(bottom)* © Dana White/PhotoEdit; p. 343: © Picture Press/Corbis; p. 366: © Mike Mazzaschi/Stock Boston

Chapter 8 p. 372: Culver Pictures; p. 373: *(top)* © Paul Lally/Stock Boston, *(bottom)* © R. J. Erwin/Photo Researchers Inc.; p. 380: © Myrleen Ferguson Cate/PhotoEdit; p. 405: © Bob Daemmrich/Stock Boston; p. 413: © Francois Gohier/Photo Researchers Inc.; p. 423: © Jeff Foott/Discovery Images/PictureQuest; p. 445: © Judy Griesedieck/Corbis

Chapter 9 p. 452: Brown Brothers; p. 453: *(top)* © Will and Deni McIntyre/Photo Researchers Inc., *(bottom)* © Don Smetzer/PhotoEdit; p. 465: © J. H. Robinson/Photo Researchers Inc.; p. 483: © Daemmrich/The Image Works; p. 490: NASA; p. 502: © John Elk/Stock Boston; p. 508: © Fritz Hoffman/The Image Works; p. 531: © Daemmrich/The Image Works

Chapter 10 p. 550: © Archivo Iconografico, S.A./Corbis; p. 551: *(top)* © David Young-Wolff/PhotoEdit, *(bottom)* © Peter Menzel/Stock Boston; p. 553: © Sondra Davis/The Image Works; p. 570: © Jeff Greenberg/Photo Researchers Inc.; p. 584: © Jeff Greenberg/Photo Researchers Inc.; p. 601: © Douglas Faulkner/Photo Researchers Inc.

Chapter 11 p. 634: The Norman Rockwell Family Trust; p. 635: *(top)* © Nancy Richmond/The Image Works, *(bottom)* © Robert Brenner/PhotoEdit; p. 660: © Mark Richards/PhotoEdit; p. 669: © Richard A. Cooke/Corbis; p. 678: © Bob Daemmrich/Stock Boston

Chapter 12 p. 716: Keystone Press Agency; p. 717: *(top)* © James Shaffer/PhotoEdit, *(bottom)* © Peter Vandermark/Stock Boston; p. 718: © Richard Hutching/Photo Researchers Inc.; p. 728: © Andrew Wood/Photo Researchers Inc.; p. 738: © Getty Images/Kaluzny/Thatcher

Appendix I Additional Topics

PART I: BAYES'S THEOREM

The Reverend Thomas Bayes (1702–1761) was an English mathematician who discovered an important relation for conditional probabilities. This relation is referred to as *Bayes's rule* or *Bayes's theorem*. It uses conditional probabilities to adjust calculations so that we can accommodate new relevant information. We will restrict our attention to a special case of Bayes's theorem in which an event B is partitioned into only *two* mutually exclusive events (see Figure AI-1). The general formula is a bit complicated but is a straightforward extension of the basic ideas we will present here. Most advanced texts contain such an extension.

Note: We use the following compact notation in the statement of Bayes's theorem:

Notation	Meaning
A^c	complement of A; *not* A
$P(B\|A)$	probability of event B, *given* event A; P(B, *given* A)
$P(B\|A^c)$	probability of event B, *given* the complement of A; P(B, *given not* A)

We will use Figure AI-1 to motivate Bayes's theorem. Let A and B be events in a sample space that have probabilities not equal to zero or one. Let A^c be the complement of A.

Here is Bayes's theorem:
$$P(A|B) = \frac{P(B|A)P(A)}{P(B|A)P(A) + P(B|A^c)P(A^c)} \qquad (1)$$

Overview of Bayes's Theorem

Suppose we have an event A, and we calculate $P(A)$, the unconditional probability of A standing by itself. Now suppose we have a "new" event B and we know the probability of B, given that A occurs $P(B|A)$, as well as the probability of B, given

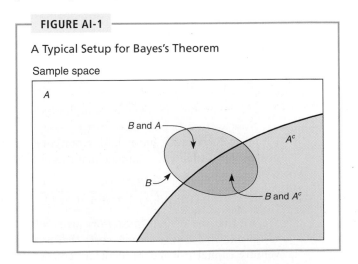

FIGURE AI-1

A Typical Setup for Bayes's Theorem

Sample space

that A does not occur $P(B|A^c)$. Where does such an event B come from? The event B can be constructed in many possible ways. For example, B can be constructed as the result of a consulting service or a testing procedure or a sorting activity. In the examples and problems, you will find more ways to construct such an event B.

How can we use this "new" information concerning the event B to adjust our calculation of the probability of event A, given B? That is, how can we make our calculation of the probability of A more realistic by including information about the event B? The answer is that we will use Equation (1) of Bayes's theorem.

Let's look at some examples using Equation (1) of Bayes's theorem. We are grateful to personal friends in the oil and natural gas business in Colorado who provided the basic information in the following example.

EXAMPLE 1

Bayes's theorem

A geologist has examined seismic data and other geologic formations in the vicinity of a proposed site for an oil well. Based on this information, the geologist reports a 65% chance of finding oil. The oil company decides to go ahead and start drilling. As the drilling progresses, sample cores are taken from the well and studied by the geologist. These sample cores have a history of predicting oil *when there is oil* about 85% of the time. However, about 6% of the time the sample cores will predict oil *when there is no oil*. (Note that these probabilities need not add up to one.) Our geologist is delighted because the sample cores predict oil for this well.

Use the "new" information from the sample cores to revise the geologist's original probability the well will hit oil. What is the new probability?

SOLUTION: To use Bayes's theorem, we need to identify the events A and B. Then we need to find $P(A)$, $P(A^c)$, $P(B|A)$, and $P(B|A^c)$. From the description of the problem, we have

A is the event the well strikes oil.

A^c is the event the well is dry (no oil).

B is the event the core samples indicate oil.

Again, from the description, we have

$$P(A) = 0.65, \quad \text{so} \quad P(A^c) = 1 - 0.65 = 0.35$$

These are our *prior* (before new information) probabilities. New information comes from the sample cores. Probabilities associated with the new information are

$$P(B|A) = 0.85$$

This is the probability that core samples indicate oil when there actually is oil.

$$P(B|A^c) = 0.06$$

This is the probability the core samples indicate oil when there is no oil (dry well).

Now we use Bayes's theorem to revise the probability that the well will hit oil based on the "new" information from core samples. The revised probability is the *posterior* probability we compute that uses the new information from the sample cores:

$$P(A|B) = \frac{P(B|A)P(A)}{P(B|A)P(A) + P(B|A^c)P(A^c)} = \frac{(0.85)(0.65)}{(0.85)(0.65) + (0.06)(0.35)} = 0.9634$$

We see that the revised (*posterior*) probability is about a 96% chance for the well to hit oil. This is why sample cores that are good can attract money in the form of venture capital (for independent drillers) on a big, expensive well. ◇

GUIDED EXERCISE 1

Bayes's theorem

Anasazi were prehistoric pueblo people in what is now the southwestern United States. Mesa Verde, Pecos Pueblo, and Chaco Canyon are beautiful national parks and monuments, but long ago they were home to many Anasazi. In prehistoric times, there were several Anasazi migrations, until finally their pueblo homes were completely abandoned. The delightful book *Proceedings of the Anasazi Symposium, 1981*, published by Mesa Verde Museum Association, contains a very interesting discussion about methods anthropologists use to (approximately) date Anasazi objects. There are two popular ways. One is to use environmental data in relation to other objects of known dates. The other is radioactive carbon dating.

Carbon dating has some variability in its accuracy, depending on how far back in time the age estimate goes and also depending on the condition of the specimen itself. Suppose experience has shown that the carbon method is correct 75% of the time it is used on an object of a known (given) time period. However, there is a 10% chance that the carbon method will predict that an object is from a certain period when we know the object is not from that period.

Using environmental data, an anthropologist reported the probability to be 40% that a fossilized deer bone bracelet was from a certain Anasazi migration period. Then, as a follow-up study, the carbon method also indicated that the bracelet was from this migration period. How can the anthropologist adjust her estimated probability to include the "new" information from the carbon dating?

(a) To use Bayes's theorem, we must identify the events A and B. From the description of the problem, what are A and B?

$\Rightarrow$ A is the event the bracelet is from the given migration period. B is the event that carbon dating indicates that the bracelet is from the given migration period.

(b) Find $P(A)$, $P(A^c)$, $P(B|A)$, and $P(B|A^c)$.

$\Rightarrow$ From the description,

$$P(A) = 0.40$$
$$P(A^c) = 0.60$$
$$P(B|A) = 0.75$$
$$P(B|A^c) = 0.10$$

(c) Compute $P(A|B)$, and explain the meaning of this number.

$\Rightarrow$ Using Bayes's theorem and the results of part (b), we have

$$P(A|B) = \frac{P(B|A)P(A)}{P(B|A)P(A) + P(B|A^c)P(A^c)}$$

$$= \frac{(0.75)(0.40)}{(0.75)(0.40) + (0.10)(0.60)} = 0.8333$$

The prior (before carbon dating) probabilities were only 40%. However, the carbon dating allowed us to revise this up to 83%. Thus, we are about 83% sure that the bracelet came from the given migration period. Perhaps additional research at the site may uncover more information for which Bayes's theorem could be used again.

PROBLEM: Bayes's theorem applied to quality control

A company that makes steel bolts knows from long experience that about 12% of its bolts are defective. If the company simply ships all bolts that it produces, then 12% of the shipment the customer receives will be defective. To decrease the percentage of defective bolts shipped to customers, an electronic scanner was installed. The scanner is positioned over the production line and is supposed to pick out the good bolts. However, the scanner itself is not perfect. To test the scanner, a large number of (pretested) "good" bolts were run under the scanner, and it accepted 90% of the bolts as good. Then a large number of (pretested) defective bolts were run under the scanner, and it accepted 3% of these as good bolts.

(a) If the company does not use the scanner, what percentage of a shipment is expected to be good? What percentage is expected to be defective?

(b) The scanner itself makes mistakes, and the company is questioning the value of using it. Suppose the company does use the scanner and ships only what the scanner passes as "good" bolts. In this case, what percentage of the shipment can be expected to be good? What percentage is expected to be defective?

Partial Answer

To solve this problem, we use Bayes's theorem. The result of using the scanner is a dramatic improvement of the shipped product. If the scanner is not used, only 88% of the shipped bolts will be good. However, if the scanner is used and only the bolts it passes as good are shipped, then 99.6% of the shipment is expected to be good. Even though the scanner itself makes a considerable number of mistakes, it is definitely worth using. Not only does it increase the quality of a shipment, the bolts it rejects can be recycled into new bolts.

PART II: THE HYPERGEOMETRIC PROBABILITY DISTRIBUTION

In Chapter 5, we examined the binomial distribution. The binomial probability distribution assumes *independent trials*. If the trials are constructed by drawing samples from a population, then we have two possibilities: We sample either *with replacement* or *without replacement*. If we draw random samples with replacement, the trials can be taken to be independent. If we draw random samples without replacement and the population is very large, then it is reasonable to say that the trials are approximately independent. In this case, we go ahead and use the binomial distribution. However, if the population is relatively small and we draw samples without replacement, the assumption of independent trials is not valid, and we should not use the binomial distribution.

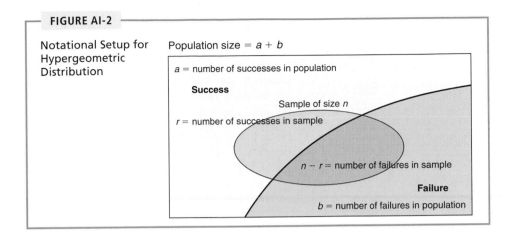

FIGURE AI-2

Notational Setup for Hypergeometric Distribution

Population size = $a + b$

a = number of successes in population

Success

Sample of size n

r = number of successes in sample

$n - r$ = number of failures in sample

Failure

b = number of failures in population

The *hypergeometric distribution* is a probability distribution of a random variable that has two outcomes when sampling is done *without replacement.*

Consider the following notational setup (see Figure AI-2). Suppose we have a population with only *two* distinct types of objects. Such a population might be made up of females and males, students and faculty, residents and nonresidents, defective and nondefective items, and so on. For simplicity of reference, let us call one type of object (your choice) "success" and the other "failure." Let's use the letter a to designate the number of successes in the population and the letter b to designate the number of failures in the population. Thus, the total population size is $a + b$. Next, we draw a random sample (without replacement) of size n from this population. Let r be the number of successes in this sample. Then $n - r$ is the number of failures in the sample. The hypergeometric distribution gives us the probability of r successes in the sample of size n.

Recall from Section 4.3 that the number of combinations of k objects taken j at a time can be computed as

$$C_{k,j} = \frac{k!}{j!(k - j)!}$$

Using the notation of Figure AI-2 and the formula for combinations, the hypergeometric distribution can be calculated.

Hypergeometric distribution

Given that the population has two distinct types of objects, success and failure,

   a counts the number of successes in the population.

   b counts the number of failures in the population.

For a random sample of size n taken *without replacement* from this population, the probability $P(r)$ of getting r successes in the *sample* is

$$P(r) = \frac{C_{a,r}C_{b,n-r}}{C_{(a+b),n}} \tag{2}$$

EXAMPLE 2

Hypergeometric distribution

A section of an Interstate 95 bridge across the Mianus River in Connecticut collapsed suddenly on the morning of June 28, 1983. (See *To Engineer Is Human: The Role of Failure in Successful Designs,* by Henry Petroski.) Three people were killed when their vehicles fell off the bridge. It was determined that the collapse was caused by the failure of a metal hanger design that left a section of the bridge with no support when something went wrong with the pins. Subsequent inspection revealed many cracked pins and hangers in bridges across the United States.

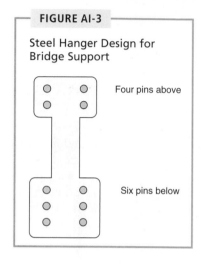

FIGURE AI-3

Steel Hanger Design for Bridge Support

Four pins above

Six pins below

(a) Suppose a hanger design uses 4 pins in the upper part and 6 pins in the lower part, as shown in Figure AI-3. The hangers come in a kit consisting of the hanger and 10 pins. When a work crew installs a hanger, they start with the top part and randomly select a pin, which is put into place. This is repeated until all 4 pins are in the top. Then they finish the lower part.

Assume that 3 pins in the kit are faulty. The other 7 are all right. What is the probability that all 3 faulty pins get put into the top part of the hanger? This means that the support is held up, in effect, by only one good pin.

SOLUTION: The population consists of 10 pins identical in appearance. However, 3 are faulty and 7 are good. The sampling of 4 pins for the top part of the hanger is done *without replacement*. Since we are interested in the faulty pins, let us label them "success" (only a convenient label). Using the notation of Figure AI-2 and the hypergeometric distribution, we have

a = number of successes in the population (bad pins) = 3

b = number of failures in the population (good pins) = 7

n = sample size (number of pins put in top) = 4

r = number of successes in sample (number of bad pins in top) = 3

The hypergeometric distribution applies because the population is relatively small (10 pins) and sampling is done without replacement. By Equation (2), we compute $P(r)$:

$$P(r) = \frac{C_{a,r}C_{b,n-r}}{C_{(a+b),n}}$$

Using the preceding information about a, b, n, and r, we get

$$P(r = 3) = \frac{C_{3,3}C_{7,1}}{C_{10,4}}$$

Using the formula for $C_{k,j}$, or Table 2 in Appendix II, or the combinations key on a calculator, we get

$$P(r = 3) = \frac{1 \cdot 7}{210} = 0.0333$$

We see that there is a better than 3.3% chance of getting 3 out of 4 bad pins in the top part of the hanger.

(b) Suppose that all the hanger kits are like the one described in part (a). On a long bridge that used 200 such hangers, how many do you expect are held up by only 1 good pin? How might this affect the safety of the bridge?

SOLUTION: We would expect

$$200(0.0333) = 6.66$$

That is, between 6 and 7 hangers are expected to be held up by only 1 good pin. As time goes on, this pin will corrode and show signs of wear as the bridge vibrates. With only one good pin, there is much less margin of safety.

Professor Petroski discusses the bridge on I-95 across the Mianus River in his book mentioned earlier. He points out that this dramatic accidental collapse resulted in better quality control (for hangers and pins) as well as better overall design of bridges. In addition to this, the government has greatly increased programs for maintenance and inspection of bridges.   ◊

GUIDED EXERCISE 2

Hypergeometric distribution

The biology club weekend outing has two groups. One group with 7 people will camp at Diamond Lake. The other group with 10 people will camp at Arapahoe Pass. Seventeen duffels were prepacked by the outing committee, but 6 of these had the tents accidentally left out of the duffel. The group going to Diamond Lake picked up their duffels at random from the collection and started off on the trail. The group going to Arapahoe Pass used the remaining duffels. What is the probability that all 6 duffels without tents were picked up by the group going to Diamond Lake?

(a) What is success? Are the duffels selected with or without replacement? Which probability distribution applies?

⟹ Success is taking a duffel without a tent. The duffels are selected without replacement. The hypergeometric distribution applies.

(b) Use the hypergeometric distribution to compute the probability of $r = 6$ successes in the sample of 7 people going to Diamond Lake.

⟹ To use the hypergeometric distribution, we need to know the values of

a = number of successes in population = 6

b = number of failures in population = 11

n = sample size = 7, since 7 people are going to Diamond Lake

r = number of successes in sample = 6

Then, $P(r = 6) = \dfrac{C_{6,6}\, C_{11,1}}{C_{17,7}} = \dfrac{1 \cdot 11}{19488} = 0.0006$

The probability that all 6 duffels without tents are taken by the 7 hikers to Diamond Lake is 0.0006.

PART III: ALTERNATE STANDARD NORMAL DISTRIBUTION TABLE: AREA FROM 0 TO z

Table 5—Areas of a Standard Normal Distribution (Appendix II)—provides the areas under the standard normal distribution that are to the left of a specified z value. Such areas are equivalent to the cumulative probability $P(z < z_0)$, where z is a standard normal variable and z_0 is a fixed value. Section 6.2 shows how to use Table 5.

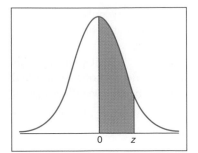

TABLE A   Areas of a Standard Normal Distribution (Alternate Version of Appendix II Table 5)

The table entries represent the area under the standard normal curve from 0 to the specified value of z.

z	.00	.01	.02	.03	.04	.05	.06	.07	.08	.09
0.0	.0000	.0040	.0080	.0120	.0160	.0199	.0239	.0279	.0319	.0359
0.1	.0398	.0438	.0478	.0517	.0557	.0596	.0636	.0675	.0714	.0753
0.2	.0793	.0832	.0871	.0910	.0948	.0987	.1026	.1064	.1103	.1141
0.3	.1179	.1217	.1255	.1293	.1331	.1368	.1406	.1443	.1480	.1517
0.4	.1554	.1591	.1628	.1664	.1700	.1736	.1772	.1808	.1844	.1879
0.5	.1915	.1950	.1985	.2019	.2054	.2088	.2123	.2157	.2190	.2224
0.6	.2257	.2291	.2324	.2357	.2389	.2422	.2454	.2486	.2517	.2549
0.7	.2580	.2611	.2642	.2673	.2704	.2734	.2764	.2794	.2823	.2852
0.8	.2881	.2910	.2939	.2967	.2995	.3023	.3051	.3078	.3106	.3133
0.9	.3159	.3186	.3212	.3238	.3264	.3289	.3315	.3340	.3365	.3389
1.0	.3413	.3438	.3461	.3485	.3508	.3531	.3554	.3577	.3599	.3621
1.1	.3643	.3665	.3686	.3708	.3729	.3749	.3770	.3790	.3810	.3830
1.2	.3849	.3869	.3888	.3907	.3925	.3944	.3962	.3980	.3997	.4015
1.3	.4032	.4049	.4066	.4082	.4099	.4115	.4131	.4147	.4162	.4177
1.4	.4192	.4207	.4222	.4236	.4251	.4265	.4279	.4292	.4306	.4319
1.5	.4332	.4345	.4357	.4370	.4382	.4394	.4406	.4418	.4429	.4441
1.6	.4452	.4463	.4474	.4484	.4495	.4505	.4515	.4525	.4535	.4545
1.7	.4554	.4564	.4573	.4582	.4591	.4599	.4608	.4616	.4625	.4633
1.8	.4641	.4649	.4656	.4664	.4671	.4678	.4686	.4693	.4699	.4706
1.9	.4713	.4719	.4726	.4732	.4738	.4744	.4750	.4756	.4761	.4767
2.0	.4772	.4778	.4783	.4788	.4793	.4798	.4803	.4808	.4812	.4817
2.1	.4821	.4826	.4830	.4834	.4838	.4842	.4846	.4850	.4854	.4857
2.2	.4861	.4864	.4868	.4871	.4875	.4878	.4881	.4884	.4887	.4890
2.3	.4893	.4896	.4898	.4901	.4904	.4906	.4909	.4911	.4913	.4916
2.4	.4918	.4920	.4922	.4925	.4927	.4929	.4931	.4932	.4934	.4936
2.5	.4938	.4940	.4941	.4943	.4945	.4946	.4948	.4949	.4951	.4952
2.6	.4953	.4955	.4956	.4957	.4959	.4960	.4961	.4962	.4963	.4964
2.7	.4965	.4966	.4967	.4968	.4969	.4970	.4971	.4972	.4973	.4974
2.8	.4974	.4975	.4976	.4977	.4977	.4978	.4979	.4979	.4980	.4981
2.9	.4981	.4982	.4982	.4983	.4984	.4984	.4985	.4985	.4986	.4986
3.0	.4987	.4987	.4987	.4988	.4988	.4989	.4989	.4989	.4990	.4990
3.1	.4990	.4991	.4991	.4991	.4992	.4992	.4992	.4992	.4993	.4993
3.2	.4993	.4993	.4994	.4994	.4994	.4994	.4994	.4995	.4995	.4995
3.3	.4995	.4995	.4995	.4996	.4996	.4996	.4996	.4996	.4996	.4997
3.4	.4997	.4997	.4997	.4997	.4997	.4997	.4997	.4997	.4997	.4998
3.5	.4998	.4998	.4998	.4998	.4998	.4998	.4998	.4998	.4998	.4998
3.6	.4998	.4998	.4998	.4999	.4999	.4999	.4999	.4999	.4999	.4999

For values of z greater than or equal to 3.70, use 0.4999 to approximate the shaded area under the standard normal curve.

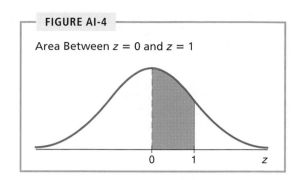

FIGURE AI-4

Area Between $z = 0$ and $z = 1$

An alternate standard normal distribution table provides areas under the standard normal distribution that are between 0 and a specified positive z value. Table A on the preceding page is such a table.

EXAMPLE 3

Area between 0 and z

Find the area under the standard normal curve between $z = 0$ and $z = 1$. This area is shown in Figure AI-4.

SOLUTION: In the upper-left corner of the table we see the letter z. The column under z gives us the units value and tenths for z. The other column headings indicate the hundredths value of z. The table entries give the areas under the normal curve from the mean $z = 0$ to a specified value of z. To find the area from $z = 0$ to $z = 1$, we observe that if $z = 1$, then the units value of z is 1 and the tenths value is 0. So we look in the column labeled z for 1.0. The area from $z = 0$ to $z = 1$ is given in the corresponding row of the column with heading 0.00 because $z = 1$ is the same as $z = 1.00$. The area we read from the table for $z = 1.00$ is 0.3413.   ◇

Table A gives areas under the normal curve for regions *beginning* at $z = 0$ and extending to a specified positive z value. However, because the normal curve is symmetrical, we also can use the table directly to find areas beginning with a negative z value and extending to $z = 0$. Example 4 shows this process.

EXAMPLE 4

Area between 0 and negative z value

Find the area under the standard normal curve from $z = -2.34$ to 0.

SOLUTION: The area from $z = -2.34$ to 0 is the same as the area from $z = 0$ to 2.34. (See Figure AI-5.) By Table A, the area from 0 to 2.34 is 0.4904. Therefore, the area from $z = -2.34$ to 0 is also 0.4904.   ◇

To find areas other than those between a given z value and $z = 0$, we use Table A together with addition or subtraction of areas we find in Table A. Figure AI-6 on the next page shows how to combine areas. As you study the figure, notice that

FIGURE AI-5

Area from $z = -2.34$ to 0 Equals Area from $z = 0$ to 2.34

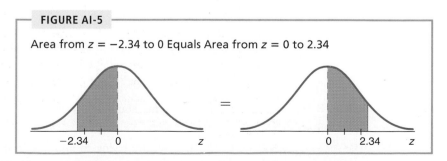

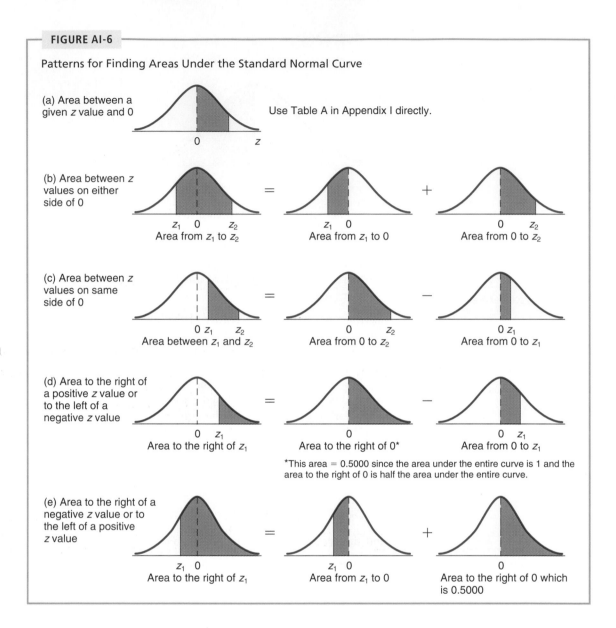

FIGURE AI-6

Patterns for Finding Areas Under the Standard Normal Curve

(a) Area between a given z value and 0

Use Table A in Appendix I directly.

(b) Area between z values on either side of 0

Area from z_1 to z_2 = Area from z_1 to 0 + Area from 0 to z_2

(c) Area between z values on same side of 0

Area between z_1 and z_2 = Area from 0 to z_2 − Area from 0 to z_1

(d) Area to the right of a positive z value or to the left of a negative z value

Area to the right of z_1 = Area to the right of 0* − Area from 0 to z_1

*This area = 0.5000 since the area under the entire curve is 1 and the area to the right of 0 is half the area under the entire curve.

(e) Area to the right of a negative z value or to the left of a positive z value

Area to the right of z_1 = Area from z_1 to 0 + Area to the right of 0 which is 0.5000

1. For areas extending from one side of the mean $z = 0$ to the other side, we *add* areas found in Table A.

2. For areas completely on one side of the mean $z = 0$ (but not bordering $z = 0$), we *subtract* areas found in Table A.

3. The area extending from $z = 0$ and including the entire right half of the graph is 0.5000. Likewise, the area extending from $z = 0$ and including the entire left half of the graph is 0.5000.

EXAMPLE 5

Area between two positive z values

Find the area under the standard normal curve in Figure AI-7 from $z = 1.00$ to $z = 2.70$.

SOLUTION: The area we are trying to find lies entirely to the right of $z = 0$ and does not border $z = 0$. Therefore, we need to subtract component areas.

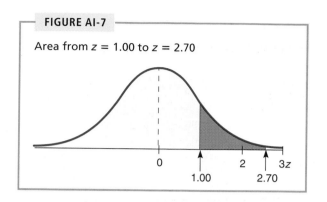

FIGURE AI-7

Area from $z = 1.00$ to $z = 2.70$

$$\begin{pmatrix} \text{Area from} \\ 1.00 \text{ to } 2.70 \end{pmatrix} = \begin{pmatrix} \text{Area from} \\ 0 \text{ to } 2.70 \end{pmatrix} - \begin{pmatrix} \text{Area from} \\ 0 \text{ to } 1.00 \end{pmatrix}$$

$$= \quad 0.4965 \quad - \quad 0.3413 = 0.1552 \qquad \diamond$$

EXAMPLE 6

Area to the left of a negative z value

Find the area under the standard normal curve to the left of $z = -0.94$.

SOLUTION: We sketch the area and notice that the area to the left of -0.94 is the same as the area to the right of 0.94 (see Figure AI-8).

To find the area to the right of 0.94, we observe

$$\begin{pmatrix} \text{Area to the} \\ \text{right of } 0.94 \end{pmatrix} = \begin{pmatrix} \text{Area to the} \\ \text{right of } 0 \end{pmatrix} - \begin{pmatrix} \text{Area from} \\ 0 \text{ to } 0.94 \end{pmatrix}$$

$$= \quad 0.5000 \quad - \quad 0.3264 = 0.1736 \qquad \diamond$$

We have practiced the skill of finding areas under the standard normal curve for various intervals along the z axis. This skill is important, since *the probability that z lies in an interval is given by the area* under the standard normal curve above that interval.

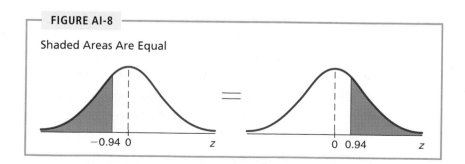

FIGURE AI-8

Shaded Areas Are Equal

GUIDED EXERCISE 3

Area between a negative z value and a positive z value

Find the area from $z = -3.00$ to $z = 2.65$. First, we *draw the picture* (see Figure AI-9) and observe the location of the requested area. Next, we find component areas from Table A and combine them appropriately.

FIGURE AI-9 Area Between $z = -3.00$ and $z = 2.65$

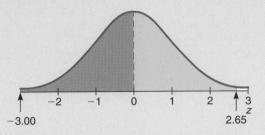

(a) Look at Figure AI-9. Should we add or subtract component areas?

⟹ Since the area extends from the left side of $z = 0$ to the right side, we add the component areas.

(b) Find the area under the standard normal curve between $z = 0$ and $z = 2.65$.

⟹ We look under the z column of Table A until we find 2.6; then we stay in this row and move to the right until we are in the column headed by 0.05. The area from $z = 0$ to $z = 2.65$ is given by the entry 0.4960.

(c) Find the area under the standard normal curve between $z = -3.00$ and $z = 0$.

⟹ Since the area from $z = -3.00$ to $z = 0$ is the same as that from $z = 0$ to $z = 3.00$, we look down the z column until we find 3.0. Then we move to the right in this row until we are in the column headed by 0.00. This entry is 0.4987, which is the area from $z = -3.00$ to $z = 0$.

(d) Use parts (b) and (c) to find the area under the standard normal curve between $z = -3.00$ and $z = 2.65$.

⟹ $$\begin{pmatrix} \text{Area from} \\ -3.00 \text{ to } 2.65 \end{pmatrix} = \begin{pmatrix} \text{Area from} \\ -3.00 \text{ to } 0 \end{pmatrix} + \begin{pmatrix} \text{Area from} \\ 0 \text{ to } 2.65 \end{pmatrix}$$
$$\qquad\qquad\qquad\downarrow \qquad\qquad\quad \downarrow$$
$$= \quad 0.4987 \quad + \quad 0.4960$$

The desired area is 0.9947.

Appendix II Tables

1. Random Numbers
2. Binomial Coefficients $C_{n,r}$
3. Binomial Probability Distribution $C_{n,r}p^r q^{n-r}$
4. Poisson Probability Distribution
5. Areas of a Standard Normal Distribution
6. Student's t Distribution
7. The χ^2 Distribution
8. The F Distribution
9. Critical Values for Spearman Rank Correlation, r_s

TABLE 1 Random Numbers

92630	78240	19267	95457	53497	23894	37708	79862	76471	66418
79445	78735	71549	44843	26104	67318	00701	34986	66751	99723
59654	71966	27386	50004	05358	94031	29281	18544	52429	06080
31524	49587	76612	39789	13537	48086	59483	60680	84675	53014
06348	76938	90379	51392	55887	71015	09209	79157	24440	30244
28703	51709	94456	48396	73780	06436	86641	69239	57662	80181
68108	89266	94730	95761	75023	48464	65544	96583	18911	16391
99938	90704	93621	66330	33393	95261	95349	51769	91616	33238
91543	73196	34449	63513	83834	99411	58826	40456	69268	48562
42103	02781	73920	56297	72678	12249	25270	36678	21313	75767
17138	27584	25296	28387	51350	61664	37893	05363	44143	42677
28297	14280	54524	21618	95320	38174	60579	08089	94999	78460
09331	56712	51333	06289	75345	08811	82711	57392	25252	30333
31295	04204	93712	51287	05754	79396	87399	51773	33075	97061
36146	15560	27592	42089	99281	59640	15221	96079	09961	05371
29553	18432	13630	05529	02791	81017	49027	79031	50912	09399
23501	22642	63081	08191	89420	67800	55137	54707	32945	64522
57888	85846	67967	07835	11314	01545	48535	17142	08552	67457
55336	71264	88472	04334	63919	36394	11196	92470	70543	29776
10087	10072	55980	64688	68239	20461	89381	93809	00796	95945
34101	81277	66090	88872	37818	72142	67140	50785	21380	16703
53362	44940	60430	22834	14130	96593	23298	56203	92671	15925
82975	66158	84731	19436	55790	69229	28661	13675	99318	76873
54827	84673	22898	08094	14326	87038	42892	21127	30712	48489
25464	59098	27436	89421	80754	89924	19097	67737	80368	08795
67609	60214	41475	84950	40133	02546	09570	45682	50165	15609
44921	70924	61295	51137	47596	86735	35561	76649	18217	63446
33170	30972	98130	95828	49786	13301	36081	80761	33985	68621
84687	85445	06208	17654	51333	02878	35010	67578	61574	20749
71886	56450	36567	09395	96951	35507	17555	35212	69106	01679
00475	02224	74722	14721	40215	21351	08596	45625	83981	63748
25993	38881	68361	59560	41274	69742	40703	37993	03435	18873

TABLE 1 *continued*

92882	53178	99195	93803	56985	53089	15305	50522	55900	43026
25138	26810	07093	15677	60688	04410	24505	37890	67186	62829
84631	71882	12991	83028	82484	90339	91950	74579	03539	90122
34003	92326	12793	61453	48121	74271	28363	66561	75220	35908
53775	45749	05734	86169	42762	70175	97310	73894	88606	19994
59316	97885	72807	54966	60859	11932	35265	71601	55577	67715
20479	66557	50705	26999	09854	52591	14063	30214	19890	19292
86180	84931	25455	26044	02227	52015	21820	50599	51671	65411
21451	68001	72710	40261	61281	13172	63819	48970	51732	54113
98062	68375	80089	24135	72355	95428	11808	29740	81644	86610
01788	64429	14430	94575	75153	94576	61393	96192	03227	32258
62465	04841	43272	68702	01274	05437	22953	18946	99053	41690
94324	31089	84159	92933	99989	89500	91586	02802	69471	68274
05797	43984	21575	09908	70221	19791	51578	36432	33494	79888
10395	14289	52185	09721	25789	38562	54794	04897	59012	89251
35177	56986	25549	59730	64718	52630	31100	62384	49483	11409
25633	89619	75882	98256	02126	72099	57183	55887	09320	73463
16464	48280	94254	45777	45150	68865	11382	11782	22695	41988

Source: Reprinted from *A Million Random Digits with 100,000 Normal Deviates* by the Rand Corporation (New York: The Free Press, 1955). Copyright 1955 and 1983 by the Rand Corporation. Used by permission.

TABLE 2 Binomial Coefficients $C_{n,r}$

r \\ n	0	1	2	3	4	5	6	7	8	9	10
1	1	1									
2	1	2	1								
3	1	3	3	1							
4	1	4	6	4	1						
5	1	5	10	10	5	1					
6	1	6	15	20	15	6	1				
7	1	7	21	35	35	21	7	1			
8	1	8	28	56	70	56	28	8	1		
9	1	9	36	84	126	126	84	36	9	1	
10	1	10	45	120	210	252	210	120	45	10	1
11	1	11	55	165	330	462	462	330	165	55	11
12	1	12	66	220	495	792	924	792	495	220	66
13	1	13	78	286	715	1,287	1,716	1,716	1,287	715	286
14	1	14	91	364	1,001	2,002	3,003	3,432	3,003	2,002	1,001
15	1	15	105	455	1,365	3,003	5,005	6,435	6,435	5,005	3,003
16	1	16	120	560	1,820	4,368	8,008	11,440	12,870	11,440	8,008
17	1	17	136	680	2,380	6,188	12,376	19,448	24,310	24,310	19,448
18	1	18	153	816	3,060	8,568	18,564	31,824	43,758	48,620	43,758
19	1	19	171	969	3,876	11,628	27,132	50,388	75,582	92,378	92,378
20	1	20	190	1,140	4,845	15,504	38,760	77,520	125,970	167,960	184,756

TABLE 3 Binomial Probability Distribution $C_{n,r}\, p^r q^{n-r}$

This table shows the probability of r successes in n independent trials, each with probability of success p.

n	r	.01	.05	.10	.15	.20	.25	.30	.35	.40	.45	.50	.55	.60	.65	.70	.75	.80	.85	.90	.95
2	0	.980	.902	.810	.723	.640	.563	.490	.423	.360	.303	.250	.203	.160	.123	.090	.063	.040	.023	.010	.002
	1	.020	.095	.180	.255	.320	.375	.420	.455	.480	.495	.500	.495	.480	.455	.420	.375	.320	.255	.180	.095
	2	.000	.002	.010	.023	.040	.063	.090	.123	.160	.203	.250	.303	.360	.423	.490	.563	.640	.723	.810	.902
3	0	.970	.857	.729	.614	.512	.422	.343	.275	.216	.166	.125	.091	.064	.043	.027	.016	.008	.003	.001	.000
	1	.029	.135	.243	.325	.384	.422	.441	.444	.432	.408	.375	.334	.288	.239	.189	.141	.096	.057	.027	.007
	2	.000	.007	.027	.057	.096	.141	.189	.239	.288	.334	.375	.408	.432	.444	.441	.422	.384	.325	.243	.135
	3	.000	.000	.001	.003	.008	.016	.027	.043	.064	.091	.125	.166	.216	.275	.343	.422	.512	.614	.729	.857
4	0	.961	.815	.656	.522	.410	.316	.240	.179	.130	.092	.062	.041	.026	.015	.008	.004	.002	.001	.000	.000
	1	.039	.171	.292	.368	.410	.422	.412	.384	.346	.300	.250	.200	.154	.112	.076	.047	.026	.011	.004	.000
	2	.001	.014	.049	.098	.154	.211	.265	.311	.346	.368	.375	.368	.346	.311	.265	.211	.154	.098	.049	.014
	3	.000	.000	.004	.011	.026	.047	.076	.112	.154	.200	.250	.300	.346	.384	.412	.422	.410	.368	.292	.171
	4	.000	.000	.000	.001	.002	.004	.008	.015	.026	.041	.062	.092	.130	.179	.240	.316	.410	.522	.656	.815
5	0	.951	.774	.590	.444	.328	.237	.168	.116	.078	.050	.031	.019	.010	.005	.002	.001	.000	.000	.000	.000
	1	.048	.204	.328	.392	.410	.396	.360	.312	.259	.206	.156	.113	.077	.049	.028	.015	.006	.002	.000	.000
	2	.001	.021	.073	.138	.205	.264	.309	.336	.346	.337	.312	.276	.230	.181	.132	.088	.051	.024	.008	.001
	3	.000	.001	.008	.024	.051	.088	.132	.181	.230	.276	.312	.337	.346	.336	.309	.264	.205	.138	.073	.021
	4	.000	.000	.000	.002	.006	.015	.028	.049	.077	.113	.156	.206	.259	.312	.360	.396	.410	.392	.328	.204
	5	.000	.000	.000	.000	.000	.001	.002	.005	.010	.019	.031	.050	.078	.116	.168	.237	.328	.444	.590	.774
6	0	.941	.735	.531	.377	.262	.178	.118	.075	.047	.028	.016	.008	.004	.002	.001	.000	.000	.000	.000	.000
	1	.057	.232	.354	.399	.393	.356	.303	.244	.187	.136	.094	.061	.037	.020	.010	.004	.002	.000	.000	.000
	2	.001	.031	.098	.176	.246	.297	.324	.328	.311	.278	.234	.186	.138	.095	.060	.033	.015	.006	.001	.000
	3	.000	.002	.015	.042	.082	.132	.185	.236	.276	.303	.312	.303	.276	.236	.185	.132	.082	.042	.015	.002
	4	.000	.000	.001	.006	.015	.033	.060	.095	.138	.186	.234	.278	.311	.328	.324	.297	.246	.176	.098	.031
	5	.000	.000	.000	.000	.002	.004	.010	.020	.037	.061	.094	.136	.187	.244	.303	.356	.393	.399	.354	.232
	6	.000	.000	.000	.000	.000	.000	.001	.002	.004	.008	.016	.028	.047	.075	.118	.178	.262	.377	.531	.735
7	0	.932	.698	.478	.321	.210	.133	.082	.049	.028	.015	.008	.004	.002	.001	.000	.000	.000	.000	.000	.000
	1	.066	.257	.372	.396	.367	.311	.247	.185	.131	.087	.055	.032	.017	.008	.004	.001	.000	.000	.000	.000
	2	.002	.041	.124	.210	.275	.311	.318	.299	.261	.214	.164	.117	.077	.047	.025	.012	.004	.001	.000	.000
	3	.000	.004	.023	.062	.115	.173	.227	.268	.290	.292	.273	.239	.194	.144	.097	.058	.029	.011	.003	.000
	4	.000	.000	.003	.011	.029	.058	.097	.144	.194	.239	.273	.292	.290	.268	.227	.173	.115	.062	.023	.004
	5	.000	.000	.000	.001	.004	.012	.025	.047	.077	.117	.164	.214	.261	.299	.318	.311	.275	.210	.124	.041
	6	.000	.000	.000	.000	.000	.001	.004	.008	.017	.032	.055	.087	.131	.185	.247	.311	.367	.396	.372	.257
	7	.000	.000	.000	.000	.000	.000	.000	.001	.002	.004	.008	.015	.028	.049	.082	.133	.210	.321	.478	.698

p

TABLE 3 *continued*

n	r	.01	.05	.10	.15	.20	.25	.30	.35	.40	.45	.50	.55	.60	.65	.70	.75	.80	.85	.90	.95
8	0	.923	.663	.430	.272	.168	.100	.058	.032	.017	.008	.004	.002	.001	.000	.000	.000	.000	.000	.000	.000
	1	.075	.279	.383	.385	.336	.267	.198	.137	.090	.055	.031	.016	.008	.003	.001	.000	.000	.000	.000	.000
	2	.003	.051	.149	.238	.294	.311	.296	.259	.209	.157	.109	.070	.041	.022	.010	.004	.001	.000	.000	.000
	3	.000	.005	.033	.084	.147	.208	.254	.279	.279	.257	.219	.172	.124	.081	.047	.023	.009	.003	.000	.000
	4	.000	.000	.005	.018	.046	.087	.136	.188	.232	.263	.273	.263	.232	.188	.136	.087	.046	.018	.005	.000
	5	.000	.000	.000	.003	.009	.023	.047	.081	.124	.172	.219	.257	.279	.279	.254	.208	.147	.084	.033	.005
	6	.000	.000	.000	.000	.001	.004	.010	.022	.041	.070	.109	.157	.209	.259	.296	.311	.294	.238	.149	.051
	7	.000	.000	.000	.000	.000	.000	.001	.003	.008	.016	.031	.055	.090	.137	.198	.267	.336	.385	.383	.279
	8	.000	.000	.000	.000	.000	.000	.000	.000	.001	.002	.004	.008	.017	.032	.058	.100	.168	.272	.430	.663
9	0	.914	.630	.387	.232	.134	.075	.040	.021	.010	.005	.002	.001	.000	.000	.000	.000	.000	.000	.000	.000
	1	.083	.299	.387	.368	.302	.225	.156	.100	.060	.034	.018	.008	.004	.001	.000	.000	.000	.000	.000	.000
	2	.003	.063	.172	.260	.302	.300	.267	.216	.161	.111	.070	.041	.021	.010	.004	.001	.000	.000	.000	.000
	3	.000	.008	.045	.107	.176	.234	.267	.272	.251	.212	.164	.116	.074	.042	.021	.009	.003	.001	.000	.000
	4	.000	.001	.007	.028	.066	.117	.172	.219	.251	.260	.246	.213	.167	.118	.074	.039	.017	.005	.001	.000
	5	.000	.000	.001	.005	.017	.039	.074	.118	.167	.213	.246	.260	.251	.219	.172	.117	.066	.028	.007	.001
	6	.000	.000	.000	.001	.003	.009	.021	.042	.074	.116	.164	.212	.251	.272	.267	.234	.176	.107	.045	.008
	7	.000	.000	.000	.000	.000	.001	.004	.010	.021	.041	.070	.111	.161	.216	.267	.300	.302	.260	.172	.063
	8	.000	.000	.000	.000	.000	.000	.000	.001	.004	.008	.018	.034	.060	.100	.156	.225	.302	.368	.387	.299
	9	.000	.000	.000	.000	.000	.000	.000	.000	.000	.001	.002	.005	.010	.021	.040	.075	.134	.232	.387	.630
10	0	.904	.599	.349	.197	.107	.056	.028	.014	.006	.003	.001	.000	.000	.000	.000	.000	.000	.000	.000	.000
	1	.091	.315	.387	.347	.268	.188	.121	.072	.040	.021	.010	.004	.002	.000	.000	.000	.000	.000	.000	.000
	2	.004	.075	.194	.276	.302	.282	.233	.176	.121	.076	.044	.023	.011	.004	.001	.000	.000	.000	.000	.000
	3	.000	.010	.057	.130	.201	.250	.267	.252	.215	.166	.117	.075	.042	.021	.009	.003	.001	.000	.000	.000
	4	.000	.001	.011	.040	.088	.146	.200	.238	.251	.238	.205	.160	.111	.069	.037	.016	.006	.001	.000	.000
	5	.000	.000	.001	.008	.026	.058	.103	.154	.201	.234	.246	.234	.201	.154	.103	.058	.026	.008	.001	.000
	6	.000	.000	.000	.001	.006	.016	.037	.069	.111	.160	.205	.238	.251	.238	.200	.146	.088	.040	.011	.001
	7	.000	.000	.000	.000	.001	.003	.009	.021	.042	.075	.117	.166	.215	.252	.267	.250	.201	.130	.057	.010
	8	.000	.000	.000	.000	.000	.000	.001	.004	.011	.023	.044	.076	.121	.176	.233	.282	.302	.276	.194	.075
	9	.000	.000	.000	.000	.000	.000	.000	.000	.002	.004	.010	.021	.040	.072	.121	.188	.268	.347	.387	.315
	10	.000	.000	.000	.000	.000	.000	.000	.000	.000	.000	.001	.003	.006	.014	.028	.056	.107	.197	.349	.599
11	0	.895	.569	.314	.167	.086	.042	.020	.009	.004	.001	.000	.000	.000	.000	.000	.000	.000	.000	.000	.000
	1	.099	.329	.384	.325	.236	.155	.093	.052	.027	.013	.005	.002	.001	.000	.000	.000	.000	.000	.000	.000
	2	.005	.087	.213	.287	.295	.258	.200	.140	.089	.051	.027	.013	.005	.002	.001	.000	.000	.000	.000	.000

p

TABLE 3 *continued*

											p										
n	r	.01	.05	.10	.15	.20	.25	.30	.35	.40	.45	.50	.55	.60	.65	.70	.75	.80	.85	.90	.95
11	3	.000	.014	.071	.152	.221	.258	.257	.225	.177	.126	.081	.046	.023	.010	.004	.001	.000	.000	.000	.000
	4	.000	.001	.016	.054	.111	.172	.220	.243	.236	.206	.161	.113	.070	.038	.017	.006	.002	.000	.000	.000
	5	.000	.000	.002	.013	.039	.080	.132	.183	.221	.236	.226	.193	.147	.099	.057	.027	.010	.002	.000	.000
	6	.000	.000	.000	.002	.010	.027	.057	.099	.147	.193	.226	.236	.221	.183	.132	.080	.039	.013	.002	.000
	7	.000	.000	.000	.000	.002	.006	.017	.038	.070	.113	.161	.206	.236	.243	.220	.172	.111	.054	.016	.001
	8	.000	.000	.000	.000	.000	.001	.004	.010	.023	.046	.081	.126	.177	.225	.257	.258	.221	.152	.071	.014
	9	.000	.000	.000	.000	.000	.000	.001	.002	.005	.013	.027	.051	.089	.140	.200	.258	.295	.287	.213	.087
	10	.000	.000	.000	.000	.000	.000	.000	.000	.001	.002	.005	.013	.027	.052	.093	.155	.236	.325	.384	.329
	11	.000	.000	.000	.000	.000	.000	.000	.000	.000	.000	.000	.001	.004	.009	.020	.042	.086	.167	.314	.569
12	0	.886	.540	.282	.142	.069	.032	.014	.006	.002	.001	.000	.000	.000	.000	.000	.000	.000	.000	.000	.000
	1	.107	.341	.377	.301	.206	.127	.071	.037	.017	.008	.003	.001	.000	.000	.000	.000	.000	.000	.000	.000
	2	.006	.099	.230	.292	.283	.232	.168	.109	.064	.034	.016	.007	.002	.001	.000	.000	.000	.000	.000	.000
	3	.000	.017	.085	.172	.236	.258	.240	.195	.142	.092	.054	.028	.012	.005	.001	.000	.000	.000	.000	.000
	4	.000	.002	.021	.068	.133	.194	.231	.237	.213	.170	.121	.076	.042	.020	.008	.002	.001	.000	.000	.000
	5	.000	.000	.004	.019	.053	.103	.158	.204	.227	.223	.193	.149	.101	.059	.029	.011	.003	.001	.000	.000
	6	.000	.000	.000	.004	.016	.040	.079	.128	.177	.212	.226	.212	.177	.128	.079	.040	.016	.004	.000	.000
	7	.000	.000	.000	.001	.003	.011	.029	.059	.101	.149	.193	.223	.227	.204	.158	.103	.053	.019	.004	.000
	8	.000	.000	.000	.000	.001	.002	.008	.020	.042	.076	.121	.170	.213	.237	.231	.194	.133	.068	.021	.002
	9	.000	.000	.000	.000	.000	.000	.001	.005	.012	.028	.054	.092	.142	.195	.240	.258	.236	.172	.085	.017
	10	.000	.000	.000	.000	.000	.000	.000	.001	.002	.007	.016	.034	.064	.109	.168	.232	.283	.292	.230	.099
	11	.000	.000	.000	.000	.000	.000	.000	.000	.000	.001	.003	.008	.017	.037	.071	.127	.206	.301	.377	.341
	12	.000	.000	.000	.000	.000	.000	.000	.000	.000	.000	.000	.001	.002	.006	.014	.032	.069	.142	.282	.540
15	0	.860	.463	.206	.087	.035	.013	.005	.002	.000	.000	.000	.000	.000	.000	.000	.000	.000	.000	.000	.000
	1	.130	.366	.343	.231	.132	.067	.031	.013	.005	.002	.000	.000	.000	.000	.000	.000	.000	.000	.000	.000
	2	.009	.135	.267	.286	.231	.156	.092	.048	.022	.009	.003	.001	.000	.000	.000	.000	.000	.000	.000	.000
	3	.000	.031	.129	.218	.250	.225	.170	.111	.063	.032	.014	.005	.002	.000	.000	.000	.000	.000	.000	.000
	4	.000	.005	.043	.116	.188	.225	.219	.179	.127	.078	.042	.019	.007	.002	.001	.000	.000	.000	.000	.000
	5	.000	.001	.010	.045	.103	.165	.206	.212	.186	.140	.092	.051	.024	.010	.003	.001	.000	.000	.000	.000
	6	.000	.000	.002	.013	.043	.092	.147	.191	.207	.191	.153	.105	.061	.030	.012	.003	.001	.001	.000	.000
	7	.000	.000	.000	.003	.014	.039	.081	.132	.177	.201	.196	.165	.118	.071	.035	.013	.003	.003	.000	.000
	8	.000	.000	.000	.001	.003	.013	.035	.071	.118	.165	.196	.201	.177	.132	.081	.039	.014	.013	.002	.000
	9	.000	.000	.000	.000	.001	.003	.012	.030	.061	.105	.153	.191	.207	.191	.147	.092	.043	.045	.010	.001
	10	.000	.000	.000	.000	.000	.001	.003	.010	.024	.051	.092	.140	.186	.212	.206	.165	.103	.045	.010	.001

TABLE 3 *continued*

n	r	.01	.05	.10	.15	.20	.25	.30	.35	.40	.45	.50	.55	.60	.65	.70	.75	.80	.85	.90	.95
15	11	.000	.000	.000	.000	.000	.000	.001	.002	.007	.019	.042	.078	.127	.179	.219	.225	.188	.116	.043	.005
	12	.000	.000	.000	.000	.000	.000	.000	.000	.002	.005	.014	.032	.063	.111	.170	.225	.250	.218	.129	.031
	13	.000	.000	.000	.000	.000	.000	.000	.000	.000	.001	.003	.009	.022	.048	.092	.156	.231	.286	.267	.135
	14	.000	.000	.000	.000	.000	.000	.000	.000	.000	.000	.000	.002	.005	.013	.031	.067	.132	.231	.343	.366
	15	.000	.000	.000	.000	.000	.000	.000	.000	.000	.000	.000	.000	.000	.002	.005	.013	.035	.087	.206	.463
16	0	.851	.440	.185	.074	.028	.010	.003	.001	.000	.000	.000	.000	.000	.000	.000	.000	.000	.000	.000	.000
	1	.138	.371	.329	.210	.113	.053	.023	.009	.003	.001	.000	.000	.000	.000	.000	.000	.000	.000	.000	.000
	2	.010	.146	.275	.277	.211	.134	.073	.035	.015	.006	.002	.001	.000	.000	.000	.000	.000	.000	.000	.000
	3	.000	.036	.142	.229	.246	.208	.146	.089	.047	.022	.009	.003	.001	.000	.000	.000	.000	.000	.000	.000
	4	.000	.006	.051	.131	.200	.225	.204	.155	.101	.057	.028	.011	.004	.001	.000	.000	.000	.000	.000	.000
	5	.000	.001	.014	.056	.120	.180	.210	.201	.162	.112	.067	.034	.014	.005	.001	.000	.000	.000	.000	.000
	6	.000	.000	.003	.018	.055	.110	.165	.198	.198	.168	.122	.075	.039	.017	.006	.001	.000	.000	.000	.000
	7	.000	.000	.000	.005	.020	.052	.101	.152	.189	.197	.175	.132	.084	.044	.019	.006	.001	.000	.000	.000
	8	.000	.000	.000	.001	.006	.020	.049	.092	.142	.181	.196	.181	.142	.092	.049	.020	.006	.001	.000	.000
	9	.000	.000	.000	.000	.001	.006	.019	.044	.084	.132	.175	.197	.189	.152	.101	.052	.020	.005	.000	.000
	10	.000	.000	.000	.000	.000	.001	.006	.017	.039	.075	.122	.168	.198	.198	.165	.110	.055	.018	.003	.000
	11	.000	.000	.000	.000	.000	.000	.001	.005	.014	.034	.067	.112	.162	.201	.210	.180	.120	.056	.014	.001
	12	.000	.000	.000	.000	.000	.000	.000	.001	.004	.011	.028	.057	.101	.155	.204	.225	.200	.131	.051	.006
	13	.000	.000	.000	.000	.000	.000	.000	.000	.001	.003	.009	.022	.047	.089	.146	.208	.246	.229	.142	.036
	14	.000	.000	.000	.000	.000	.000	.000	.000	.000	.001	.002	.006	.015	.035	.073	.134	.211	.277	.275	.146
	15	.000	.000	.000	.000	.000	.000	.000	.000	.000	.000	.000	.001	.003	.009	.023	.053	.113	.210	.329	.371
	16	.000	.000	.000	.000	.000	.000	.000	.000	.000	.000	.000	.000	.000	.001	.003	.010	.028	.074	.185	.440
20	0	.818	.358	.122	.039	.012	.003	.001	.000	.000	.000	.000	.000	.000	.000	.000	.000	.000	.000	.000	.000
	1	.165	.377	.270	.137	.058	.021	.007	.002	.000	.000	.000	.000	.000	.000	.000	.000	.000	.000	.000	.000
	2	.016	.189	.285	.229	.137	.067	.028	.010	.003	.001	.000	.000	.000	.000	.000	.000	.000	.000	.000	.000
	3	.001	.060	.190	.243	.205	.134	.072	.032	.012	.004	.001	.000	.000	.000	.000	.000	.000	.000	.000	.000
	4	.000	.013	.090	.182	.218	.190	.130	.074	.035	.014	.005	.001	.000	.000	.000	.000	.000	.000	.000	.000
	5	.000	.002	.032	.103	.175	.202	.179	.127	.075	.036	.015	.005	.001	.000	.000	.000	.000	.000	.000	.000
	6	.000	.000	.009	.045	.109	.169	.192	.171	.124	.075	.036	.015	.005	.001	.000	.000	.000	.000	.000	.000
	7	.000	.000	.002	.016	.055	.112	.164	.184	.166	.122	.074	.037	.015	.005	.001	.000	.000	.000	.000	.000
	8	.000	.000	.000	.005	.022	.061	.114	.161	.180	.162	.120	.073	.035	.014	.004	.001	.000	.000	.000	.000
	9	.000	.000	.000	.001	.007	.027	.065	.116	.160	.177	.160	.119	.071	.034	.012	.003	.000	.000	.000	.000

TABLE 3 *continued*

n	r											p										
		.01	.05	.10	.15	.20	.25	.30	.35	.40	.45	.50	.55	.60	.65	.70	.75	.80	.85	.90	.95
20	10	.000	.000	.000	.000	.002	.010	.031	.069	.117	.159	.176	.159	.117	.069	.031	.010	.002	.000	.000	.000
	11	.000	.000	.000	.000	.000	.003	.012	.034	.071	.119	.160	.177	.160	.116	.065	.027	.007	.001	.000	.000
	12	.000	.000	.000	.000	.000	.001	.004	.014	.035	.073	.120	.162	.180	.161	.114	.061	.022	.005	.000	.000
	13	.000	.000	.000	.000	.000	.000	.001	.005	.015	.037	.074	.122	.166	.184	.164	.112	.055	.016	.002	.000
	14	.000	.000	.000	.000	.000	.000	.000	.001	.005	.015	.037	.075	.124	.171	.192	.169	.109	.045	.009	.000
	15	.000	.000	.000	.000	.000	.000	.000	.000	.001	.005	.015	.036	.075	.127	.179	.202	.175	.103	.032	.002
	16	.000	.000	.000	.000	.000	.000	.000	.000	.000	.001	.005	.014	.035	.074	.130	.190	.218	.182	.090	.013
	17	.000	.000	.000	.000	.000	.000	.000	.000	.000	.000	.001	.004	.012	.032	.072	.134	.205	.243	.190	.060
	18	.000	.000	.000	.000	.000	.000	.000	.000	.000	.000	.000	.001	.003	.010	.028	.067	.137	.229	.285	.189
	19	.000	.000	.000	.000	.000	.000	.000	.000	.000	.000	.000	.000	.000	.002	.007	.021	.058	.137	.270	.377
	20	.000	.000	.000	.000	.000	.000	.000	.000	.000	.000	.000	.000	.000	.000	.001	.003	.012	.039	.122	.358

TABLE 4 Poisson Probability Distribution

For a given value of λ, entry indicates the probability
of obtaining a specified value of r.

r	λ .1	.2	.3	.4	.5	.6	.7	.8	.9	1.0
0	.9048	.8187	.7408	.6703	.6065	.5488	.4966	.4493	.4066	.3679
1	.0905	.1637	.2222	.2681	.3033	.3293	.3476	.3595	.3659	.3679
2	.0045	.0164	.0333	.0536	.0758	.0988	.1217	.1438	.1647	.1839
3	.0002	.0011	.0033	.0072	.0126	.0198	.0284	.0383	.0494	.0613
4	.0000	.0001	.0003	.0007	.0016	.0030	.0050	.0077	.0111	.0153
5	.0000	.0000	.0000	.0001	.0002	.0004	.0007	.0012	.0020	.0031
6	.0000	.0000	.0000	.0000	.0000	.0000	.0001	.0002	.0003	.0005
7	.0000	.0000	.0000	.0000	.0000	.0000	.0000	.0000	.0000	.0001

r	λ 1.1	1.2	1.3	1.4	1.5	1.6	1.7	1.8	1.9	2.0
0	.3329	.3012	.2725	.2466	.2231	.2019	.1827	.1653	.1496	.1353
1	.3662	.3614	.3543	.3452	.3347	.3230	.3106	.2975	.2842	.2707
2	.2014	.2169	.2303	.2417	.2510	.2584	.2640	.2678	.2700	.2707
3	.0738	.0867	.0998	.1128	.1255	.1378	.1496	.1607	.1710	.1804
4	.0203	.0260	.0324	.0395	.0471	.0551	.0636	.0723	.0812	.0902
5	.0045	.0062	.0084	.0111	.0141	.0176	.0216	.0260	.0309	.0361
6	.0008	.0012	.0018	.0026	.0035	.0047	.0061	.0078	.0098	.0120
7	.0001	.0002	.0003	.0005	.0008	.0011	.0015	.0020	.0027	.0034
8	.0000	.0000	.0001	.0001	.0001	.0002	.0003	.0005	.0006	.0009
9	.0000	.0000	.0000	.0000	.0000	.0000	.0001	.0001	.0001	.0002

r	λ 2.1	2.2	2.3	2.4	2.5	2.6	2.7	2.8	2.9	3.0
0	.1225	.1108	.1003	.0907	.0821	.0743	.0672	.0608	.0550	.0498
1	.2572	.2438	.2306	.2177	.2052	.1931	.1815	.1703	.1596	.1494
2	.2700	.2681	.2652	.2613	.2565	.2510	.2450	.2384	.2314	.2240
3	.1890	.1966	.2033	.2090	.2138	.2176	.2205	.2225	.2237	.2240
4	.0992	.1082	.1169	.1254	.1336	.1414	.1488	.1557	.1622	.1680
5	.0417	.0476	.0538	.0602	.0668	.0735	.0804	.0872	.0940	.1008
6	.0146	.0174	.0206	.0241	.0278	.0319	.0362	.0407	.0455	.0504
7	.0044	.0055	.0068	.0083	.0099	.0118	.0139	.0163	.0188	.0216
8	.0011	.0015	.0019	.0025	.0031	.0038	.0047	.0057	.0068	.0081
9	.0003	.0004	.0005	.0007	.0009	.0011	.0014	.0018	.0022	.0027
10	.0001	.0001	.0001	.0002	.0002	.0003	.0004	.0005	.0006	.0008
11	.0000	.0000	.0000	.0000	.0000	.0001	.0001	.0001	.0002	.0002
12	.0000	.0000	.0000	.0000	.0000	.0000	.0000	.0000	.0000	.0001

TABLE 4 *continued*

					λ					
r	3.1	3.2	3.3	3.4	3.5	3.6	3.7	3.8	3.9	4.0
0	.0450	.0408	.0369	.0334	.0302	.0273	.0247	.0224	.0202	.0183
1	.1397	.1304	.1217	.1135	.1057	.0984	.0915	.0850	.0789	.0733
2	.2165	.2087	.2008	.1929	.1850	.1771	.1692	.1615	.1539	.1465
3	.2237	.2226	.2209	.2186	.2158	.2125	.2087	.2046	.2001	.1954
4	.1734	.1781	.1823	.1858	.1888	.1912	.1931	.1944	.1951	.1954
5	.1075	.1140	.1203	.1264	.1322	.1377	.1429	.1477	.1522	.1563
6	.0555	.0608	.0662	.0716	.0771	.0826	.0881	.0936	.0989	.1042
7	.0246	.2078	.0312	.0348	.0385	.0425	.0466	.0508	.0551	.0595
8	.0095	.0111	.0129	.0148	.0169	.0191	.0215	.0241	.0269	.0298
9	.0033	.0040	.0047	.0056	.0066	.0076	.0089	.0102	.0116	.0132
10	.0010	.0013	.0016	.0019	.0023	.0028	.0033	.0039	.0045	.0053
11	.0003	.0004	.0005	.0006	.0007	.0009	.0011	.0013	.0016	.0019
12	.0001	.0001	.0001	.0002	.0002	.0003	.0003	.0004	.0005	.0006
13	.0000	.0000	.0000	.0000	.0001	.0001	.0001	.0001	.0002	.0002
14	.0000	.0000	.0000	.0000	.0000	.0000	.0000	.0000	.0000	.0001

					λ					
r	4.1	4.2	4.3	4.4	4.5	4.6	4.7	4.8	4.9	5.0
0	.0166	.0150	.0136	.0123	.0111	.0101	.0091	.0082	.0074	.0067
1	.0679	.0630	.0583	.0540	.0500	.0462	.0427	.0395	.0365	.0337
2	.1393	.1323	.1254	.1188	.1125	.1063	.1005	.0948	.0894	.0842
3	.1904	.1852	.1798	.1743	.1687	.1631	.1574	.1517	.1460	.1404
4	.1951	.1944	.1933	.1917	.1898	.1875	.1849	.1820	.1789	.1755
5	.1600	.1633	.1662	.1687	.1708	.1725	.1738	.1747	.1753	.1755
6	.1093	.1143	.1191	.1237	.1281	.1323	.1362	.1398	.1432	.1462
7	.0640	.0686	.0732	.0778	.0824	.0869	.0914	.0959	.1002	.1044
8	.0328	.0360	.0393	.0428	.0463	.0500	.0537	.0575	.0614	.0653
9	.0150	.0168	.0188	.0209	.0232	.0255	.0280	.0307	.0334	.0363
10	.0061	.0071	.0081	.0092	.0104	.0118	.0132	.0147	.0164	.0181
11	.0023	.0027	.0032	.0037	.0043	.0049	.0056	.0064	.0073	.0082
12	.0008	.0009	.0011	.0014	.0016	.0019	.0022	.0026	.0030	.0034
13	.0002	.0003	.0004	.0005	.0006	.0007	.0008	.0009	.0011	.0013
14	.0001	.0001	.0001	.0001	.0002	.0002	.0003	.0003	.0004	.0005
15	.0000	.0000	.0000	.0000	.0001	.0001	.0001	.0001	.0001	.0002

TABLE 4 *continued*

	λ									
r	5.1	5.2	5.3	5.4	5.5	5.6	5.7	5.8	5.9	6.0
0	.0061	.0055	.0050	.0045	.0041	.0037	.0033	.0030	.0027	.0025
1	.0311	.0287	.0265	.0244	.0225	.0207	.0191	.0176	.0162	.0149
2	.0793	.0746	.0701	.0659	.0618	.0580	.0544	.0509	.0477	.0446
3	.1348	.1293	.1239	.1185	.1133	.1082	.1033	.0985	.0938	.0892
4	.1719	.1681	.1641	.1600	.1558	.1515	.1472	.1428	.1383	.1339
5	.1753	.1748	.1740	.1728	.1714	.1697	.1678	.1656	.1632	.1606
6	.1490	.1515	.1537	.1555	.1571	.1584	.1594	.1601	.1605	.1606
7	.1086	.1125	.1163	.1200	.1234	.1267	.1298	.1326	.1353	.1377
8	.0692	.0731	.0771	.0810	.0849	.0887	.0925	.0962	.0998	.1033
9	.0392	.0423	.0454	.0486	.0519	.0552	.0586	.0620	.0654	.0688
10	.0200	.0220	.0241	.0262	.0285	.0309	.0334	.0359	.0386	.0413
11	.0093	.0104	.0116	.0129	.0143	.0157	.0173	.0190	.0207	.0225
12	.0039	.0045	.0051	.0058	.0065	.0073	.0082	.0092	.0102	.0113
13	.0015	.0018	.0021	.0024	.0028	.0032	.0036	.0041	.0046	.0052
14	.0006	.0007	.0008	.0009	.0011	.0013	.0015	.0017	.0019	.0022
15	.0002	.0002	.0003	.0003	.0004	.0005	.0006	.0007	.0008	.0009
16	.0001	.0001	.0001	.0001	.0001	.0002	.0002	.0002	.0003	.0003
17	.0000	.0000	.0000	.0000	.0000	.0000	.0001	.0001	.0001	.0001

	λ									
r	6.1	6.2	6.3	6.4	6.5	6.6	6.7	6.8	6.9	7.0
0	.0022	.0020	.0018	.0017	.0015	.0014	.0012	.0011	.0010	.0009
1	.0137	.0126	.0116	.0106	.0098	.0090	.0082	.0076	.0070	.0064
2	.0417	.0390	.0364	.0340	.0318	.0296	.0276	.0258	.0240	.0223
3	.0848	.0806	.0765	.0726	.0688	.0652	.0617	.0584	.0552	.0521
4	.1294	.1249	.1205	.1162	.1118	.1076	.1034	.0992	.0952	.0912
5	.1579	.1549	.1519	.1487	.1454	.1420	.1385	.1349	.1314	.1277
6	.1605	.1601	.1595	.1586	.1575	.1562	.1546	.1529	.1511	.1490
7	.1399	.1418	.1435	.1450	.1462	.1472	.1480	.1486	.1489	.1490
8	.1066	.1099	.1130	.1160	.1188	.1215	.1240	.1263	.1284	.1304
9	.0723	.0757	.0791	.0825	.0858	.0891	.0923	.0954	.0985	.1014
10	.0441	.0469	.0498	.0528	.0558	.0588	.0618	.0649	.0679	.0710
11	.0245	.0265	.0285	.0307	.0330	.0353	.0377	.0401	.0426	.0452
12	.0124	.0137	.0150	.0164	.0179	.0194	.0210	.0227	.0245	.0264
13	.0058	.0065	.0073	.0081	.0089	.0098	.0108	.0119	.0130	.0142
14	.0025	.0029	.0033	.0037	.0041	.0046	.0052	.0058	.0064	.0071
15	.0010	.0012	.0014	.0016	.0018	.0020	.0023	.0026	.0029	.0033
16	.0004	.0005	.0005	.0006	.0007	.0008	.0010	.0011	.0013	.0014
17	.0001	.0002	.0002	.0002	.0003	.0003	.0004	.0004	.0005	.0006
18	.0000	.0001	.0001	.0001	.0001	.0001	.0001	.0002	.0002	.0002
19	.0000	.0000	.0000	.0000	.0000	.0000	.0000	.0001	.0001	.0001

TABLE 4 *continued*

	λ									
r	7.1	7.2	7.3	7.4	7.5	7.6	7.7	7.8	7.9	8.0
0	.0008	.0007	.0007	.0006	.0006	.0005	.0005	.0004	.0004	.0003
1	.0059	.0054	.0049	.0045	.0041	.0038	.0035	.0032	.0029	.0027
2	.0208	.0194	.0180	.0167	.0156	.0145	.0134	.0125	.0116	.0107
3	.0492	.0464	.0438	.0413	.0389	.0366	.0345	.0324	.0305	.0286
4	.0874	.0836	.0799	.0764	.0729	.0696	.0663	.0632	.0602	.0573
5	.1241	.1204	.1167	.1130	.1094	.1057	.1021	.0986	.0951	.0916
6	.1468	.1445	.1420	.1394	.1367	.1339	.1311	.1282	.1252	.1221
7	.1489	.1486	.1481	.1474	.1465	.1454	.1442	.1428	.1413	.1396
8	.1321	.1337	.1351	.1363	.1373	.1382	.1388	.1392	.1395	.1396
9	.1042	.1070	.1096	.1121	.1144	.1167	.1187	.1207	.1224	.1241
10	.0740	.0770	.0800	.0829	.0858	.0887	.0914	.0941	.0967	.0993
11	.0478	.0504	.0531	.0558	.0585	.0613	.0640	.0667	.0695	.0722
12	.0283	.0303	.0323	.0344	.0366	.0388	.0411	.0434	.0457	.0481
13	.0154	.0168	.0181	.0196	.0211	.0227	.0243	.0260	.0278	.0296
14	.0078	.0086	.0095	.0104	.0113	.0123	.0134	.0145	.0157	.0169
15	.0037	.0041	.0046	.0051	.0057	.0062	.0069	.0075	.0083	.0090
16	.0016	.0019	.0021	.0024	.0026	.0030	.0033	.0037	.0041	.0045
17	.0007	.0008	.0009	.0010	.0012	.0013	.0015	.0017	.0019	.0021
18	.0003	.0003	.0004	.0004	.0005	.0006	.0006	.0007	.0008	.0009
19	.0001	.0001	.0001	.0002	.0002	.0002	.0003	.0003	.0003	.0004
20	.0000	.0000	.0001	.0001	.0001	.0001	.0001	.0001	.0001	.0002
21	.0000	.0000	.0000	.0000	.0000	.0000	.0000	.0000	.0001	.0001

	λ									
r	8.1	8.2	8.3	8.4	8.5	8.6	8.7	8.8	8.9	9.0
0	.0003	.0003	.0002	.0002	.0002	.0002	.0002	.0002	.0001	.0001
1	.0025	.0023	.0021	.0019	.0017	.0016	.0014	.0013	.0012	.0011
2	.0100	.0092	.0086	.0079	.0074	.0068	.0063	.0058	.0054	.0050
3	.0269	.0252	.0237	.0222	.0208	.0195	.0183	.0171	.0160	.0150
4	.0544	.0517	.0491	.0466	.0443	.0420	.0398	.0377	.0357	.0337
5	.0882	.0849	.0816	.0784	.0752	.0722	.0692	.0663	.0635	.0607
6	.1191	.1160	.1128	.1097	.1066	.1034	.1003	.0972	.0941	.0911
7	.1378	.1358	.1338	.1317	.1294	.1271	.1247	.1222	.1197	.1171
8	.1395	.1392	.1388	.1382	.1375	.1366	.1356	.1344	.1332	.1318
9	.1256	.1269	.1280	.1290	.1299	.1306	.1311	.1315	.1317	.1318
10	.1017	.1040	.1063	.1084	.1104	.1123	.1140	.1157	.1172	.1186
11	.0749	.0776	.0802	.0828	.0853	.0878	.0902	.0925	.0948	.0970
12	.0505	.0530	.0555	.0579	.0604	.0629	.0654	.0679	.0703	.0728
13	.0315	.0334	.0354	.0374	.0395	.0416	.0438	.0459	.0481	.0504
14	.0182	.0196	.0210	.0225	.0240	.0256	.0272	.0289	.0306	.0324
15	.0098	.0107	.0116	.0126	.0136	.0147	.0158	.0169	.0182	.0194

TABLE 4 *continued*

					λ					
r	8.1	8.2	8.3	8.4	8.5	8.6	8.7	8.8	8.9	9.0
16	.0050	.0055	.0060	.0066	.0072	.0079	.0086	.0093	.0101	.0109
17	.0024	.0026	.0029	.0033	.0036	.0040	.0044	.0048	.0053	.0058
18	.0011	.0012	.0014	.0015	.0017	.0019	.0021	.0024	.0026	.0029
19	.0005	.0005	.0006	.0007	.0008	.0009	.0010	.0011	.0012	.0014
20	.0002	.0002	.0002	.0003	.0003	.0004	.0004	.0005	.0005	.0006
21	.0001	.0001	.0001	.0001	.0001	.0002	.0002	.0002	.0002	.0003
22	.0000	.0000	.0000	.0000	.0001	.0001	.0001	.0001	.0001	.0001

					λ					
r	9.1	9.2	9.3	9.4	9.5	9.6	9.7	9.8	9.9	10
0	.0001	.0001	.0001	.0001	.0001	.0001	.0001	.0001	.0001	.0000
1	.0010	.0009	.0009	.0008	.0007	.0007	.0006	.0005	.0005	.0005
2	.0046	.0043	.0040	.0037	.0034	.0031	.0029	.0027	.0025	.0023
3	.0140	.0131	.0123	.0115	.0107	.0100	.0093	.0087	.0081	.0076
4	.0319	.0302	.0285	.0269	.0254	.0240	.0226	.0213	.0201	.0189
5	.0581	.0555	.0530	.0506	.0483	.0460	.0439	.0418	.0398	.0378
6	.0881	.0851	.0822	.0793	.0764	.0736	.0709	.0682	.0656	.0631
7	.1145	.1118	.1091	.1064	.1037	.1010	.0982	.0955	.0928	.0901
8	.1302	.1286	.1269	.1251	.1232	.1212	.1191	.1170	.1148	.1126
9	.1317	.1315	.1311	.1306	.1300	.1293	.1284	.1274	.1263	.1251
10	.1198	.1210	.1219	.1228	.1235	.1241	.1245	.1249	.1250	.1251
11	.0991	.1012	.1031	.1049	.1067	.1083	.1098	.1112	.1125	.1137
12	.0752	.0776	.0799	.0822	.0844	.0866	.0888	.0908	.0928	.0948
13	.0526	.0549	.0572	.0594	.0617	.0640	.0662	.0685	.0707	.0729
14	.0342	.0361	.0380	.0399	.0419	.0439	.0459	.0479	.0500	.0521
15	.0208	.0221	.0235	.0250	.0265	.0281	.0297	.0313	.0330	.0347
16	.0118	.0127	.0137	.0147	.0157	.0168	.0180	.0192	.0204	.0217
17	.0063	.0069	.0075	.0081	.0088	.0095	.0103	.0111	.0119	.0128
18	.0032	.0035	.0039	.0042	.0046	.0051	.0055	.0060	.0065	.0071
19	.0015	.0017	.0019	.0021	.0023	.0026	.0028	.0031	.0034	.0037
20	.0007	.0008	.0009	.0010	.0011	.0012	.0014	.0015	.0017	.0019
21	.0003	.0003	.0004	.0004	.0005	.0006	.0006	.0007	.0008	.0009
22	.0001	.0001	.0002	.0002	.0002	.0002	.0003	.0003	.0004	.0004
23	.0000	.0001	.0001	.0001	.0001	.0001	.0001	.0001	.0002	.0002
24	.0000	.0000	.0000	.0000	.0000	.0000	.0000	.0001	.0001	.0001

TABLE 4 *continued*

					λ					
r	11	12	13	14	15	16	17	18	19	20
0	.0000	.0000	.0000	.0000	.0000	.0000	.0000	.0000	.0000	.0000
1	.0002	.0001	.0000	.0000	.0000	.0000	.0000	.0000	.0000	.0000
2	.0010	.0004	.0002	.0001	.0000	.0000	.0000	.0000	.0000	.0000
3	.0037	.0018	.0008	.0004	.0002	.0001	.0000	.0000	.0000	.0000
4	.0102	.0053	.0027	.0013	.0006	.0003	.0001	.0001	.0000	.0000
5	.0224	.0127	.0070	.0037	.0019	.0010	.0005	.0002	.0001	.0001
6	.0411	.0255	.0152	.0087	.0048	.0026	.0014	.0007	.0004	.0002
7	.0646	.0437	.0281	.0174	.0104	.0060	.0034	.0018	.0010	.0005
8	.0888	.0655	.0457	.0304	.0194	.0120	.0072	.0042	.0024	.0013
9	.1085	.0874	.0661	.0473	.0324	.0213	.0135	.0083	.0050	.0029
10	.1194	.1048	.0859	.0663	.0486	.0341	.0230	.0150	.0095	.0058
11	.1194	.1144	.1015	.0844	.0663	.0496	.0355	.0245	.0164	.0106
12	.1094	.1144	.1099	.0984	.0829	.0661	.0504	.0368	.0259	.0176
13	.0926	.1056	.1099	.1060	.0956	.0814	.0658	.0509	.0378	.0271
14	.0728	.0905	.1021	.1060	.1024	.0930	.0800	.0655	.0514	.0387
15	.0534	.0724	.0885	.0989	.1024	.0992	.0906	.0786	.0650	.0516
16	.0367	.0543	.0719	.0866	.0960	.0992	.0963	.0884	.0772	.0646
17	.0237	.0383	.0550	.0713	.0847	.0934	.0963	.0936	.0863	.0760
18	.0145	.0256	.0397	.0554	.0706	.0830	.0909	.0936	.0911	.0844
19	.0084	.0161	.0272	.0409	.0557	.0699	.0814	.0887	.0911	.0888
20	.0046	.0097	.0177	.0286	.0418	.0559	.0692	.0798	.0866	.0888
21	.0024	.0055	.0109	.0191	.0299	.0426	.0560	.0684	.0783	.0846
22	.0012	.0030	.0065	.0121	.0204	.0310	.0433	.0560	.0676	.0769
23	.0006	.0016	.0037	.0074	.0133	.0216	.0320	.0438	.0559	.0669
24	.0003	.0008	.0020	.0043	.0083	.0144	.0226	.0328	.0442	.0557
25	.0001	.0004	.0010	.0024	.0050	.0092	.0154	.0237	.0336	.0446
26	.0000	.0002	.0005	.0013	.0029	.0057	.0101	.0164	.0246	.0343
27	.0000	.0001	.0002	.0007	.0016	.0034	.0063	.0109	.0173	.0254
28	.0000	.0000	.0001	.0003	.0009	.0019	.0038	.0070	.0117	.0181
29	.0000	.0000	.0001	.0002	.0004	.0011	.0023	.0044	.0077	.0125
30	.0000	.0000	.0000	.0001	.0002	.0006	.0013	.0026	.0049	.0083
31	.0000	.0000	.0000	.0000	.0001	.0003	.0007	.0015	.0030	.0054
32	.0000	.0000	.0000	.0000	.0001	.0001	.0004	.0009	.0018	.0034
33	.0000	.0000	.0000	.0000	.0000	.0001	.0002	.0005	.0010	.0020
34	.0000	.0000	.0000	.0000	.0000	.0000	.0001	.0002	.0006	.0012
35	.0000	.0000	.0000	.0000	.0000	.0000	.0000	.0001	.0003	.0007
36	.0000	.0000	.0000	.0000	.0000	.0000	.0000	.0001	.0002	.0004
37	.0000	.0000	.0000	.0000	.0000	.0000	.0000	.0000	.0001	.0002
38	.0000	.0000	.0000	.0000	.0000	.0000	.0000	.0000	.0000	.0001
39	.0000	.0000	.0000	.0000	.0000	.0000	.0000	.0000	.0000	.0001

Source: Extracted from William H. Beyer (ed.), *CRC Basic Statistical Tables* (Cleveland, Ohio: The Chemical Rubber Co., 1971).

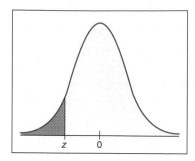

Table entry for z is the area to the left of z.

TABLE 5   Areas of a Standard Normal Distribution

(a) Table of Areas to the Left of z

z	.00	.01	.02	.03	.04	.05	.06	.07	.08	.09
−3.4	.0003	.0003	.0003	.0003	.0003	.0003	.0003	.0003	.0003	.0002
−3.3	.0005	.0005	.0005	.0004	.0004	.0004	.0004	.0004	.0004	.0003
−3.2	.0007	.0007	.0006	.0006	.0006	.0006	.0006	.0005	.0005	.0005
−3.1	.0010	.0009	.0009	.0009	.0008	.0008	.0008	.0008	.0007	.0007
−3.0	.0013	.0013	.0013	.0012	.0012	.0011	.0011	.0011	.0010	.0010
−2.9	.0019	.0018	.0018	.0017	.0016	.0016	.0015	.0015	.0014	.0014
−2.8	.0026	.0025	.0024	.0023	.0023	.0022	.0021	.0021	.0020	.0019
−2.7	.0035	.0034	.0033	.0032	.0031	.0030	.0029	.0028	.0027	.0026
−2.6	.0047	.0045	.0044	.0043	.0041	.0040	.0039	.0038	.0037	.0036
−2.5	.0062	.0060	.0059	.0057	.0055	.0054	.0052	.0051	.0049	.0048
−2.4	.0082	.0080	.0078	.0075	.0073	.0071	.0069	.0068	.0066	.0064
−2.3	.0107	.0104	.0102	.0099	.0096	.0094	.0091	.0089	.0087	.0084
−2.2	.0139	.0136	.0132	.0129	.0125	.0122	.0119	.0116	.0113	.0110
−2.1	.0179	.0174	.0170	.0166	.0162	.0158	.0154	.0150	.0146	.0143
−2.0	.0228	.0222	.0217	.0212	.0207	.0202	.0197	.0192	.0188	.0183
−1.9	.0287	.0281	.0274	.0268	.0262	.0256	.0250	.0244	.0239	.0233
−1.8	.0359	.0351	.0344	.0336	.0329	.0322	.0314	.0307	.0301	.0294
−1.7	.0446	.0436	.0427	.0418	.0409	.0401	.0392	.0384	.0375	.0367
−1.6	.0548	.0537	.0526	.0516	.0505	.0495	.0485	.0475	.0465	.0455
−1.5	.0668	.0655	.0643	.0630	.0618	.0606	.0594	.0582	.0571	.0559
−1.4	.0808	.0793	.0778	.0764	.0749	.0735	.0721	.0708	.0694	.0681
−1.3	.0968	.0951	.0934	.0918	.0901	.0885	.0869	.0853	.0838	.0823
−1.2	.1151	.1131	.1112	.1093	.1075	.1056	.1038	.1020	.1003	.0985
−1.1	.1357	.1335	.1314	.1292	.1271	.1251	.1230	.1210	.1190	.1170
−1.0	.1587	.1562	.1539	.1515	.1492	.1469	.1446	.1423	.1401	.1379
−0.9	.1841	.1814	.1788	.1762	.1736	.1711	.1685	.1660	.1635	.1611
−0.8	.2119	.2090	.2061	.2033	.2005	.1977	.1949	.1922	.1894	.1867
−0.7	.2420	.2389	.2358	.2327	.2296	.2266	.2236	.2206	.2177	.2148
−0.6	.2743	.2709	.2676	.2643	.2611	.2578	.2546	.2514	.2483	.2451
−0.5	.3085	.3050	.3015	.2981	.2946	.2912	.2877	.2843	.2810	.2776
−0.4	.3446	.3409	.3372	.3336	.3300	.3264	.3228	.3192	.3156	.3121
−0.3	.3821	.3783	.3745	.3707	.3669	.3632	.3594	.3557	.3520	.3483
−0.2	.4207	.4168	.4129	.4090	.4052	.4013	.3974	.3936	.3897	.3859
−0.1	.4602	.4562	.4522	.4483	.4443	.4404	.4364	.4325	.4286	.4247
−0.0	.5000	.4960	.4920	.4880	.4840	.4801	.4761	.4721	.4681	.4641

For values of z less than −3.49, use 0.000 to approximate the area.

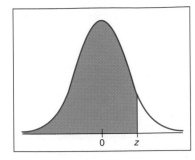

Table entry for z is the area to the left of z.

TABLE 5(a) *continued*

z	.00	.01	.02	.03	.04	.05	.06	.07	.08	.09
0.0	.5000	.5040	.5080	.5120	.5160	.5199	.5239	.5279	.5319	.5359
0.1	.5398	.5438	.5478	.5517	.5557	.5596	.5636	.5675	.5714	.5753
0.2	.5793	.5832	.5871	.5910	.5948	.5987	.6026	.6064	.6103	.6141
0.3	.6179	.6217	.6255	.6293	.6331	.6368	.6406	.6443	.6480	.6517
0.4	.6554	.6591	.6628	.6664	.6700	.6736	.6772	.6808	.6844	.6879
0.5	.6915	.6950	.6985	.7019	.7054	.7088	.7123	.7157	.7190	.7224
0.6	.7257	.7291	.7324	.7357	.7389	.7422	.7454	.7486	.7517	.7549
0.7	.7580	.7611	.7642	.7673	.7704	.7734	.7764	.7794	.7823	.7852
0.8	.7881	.7910	.7939	.7967	.7995	.8023	.8051	.8078	.8106	.8133
0.9	.8159	.8186	.8212	.8238	.8264	.8289	.8315	.8340	.8365	.8389
1.0	.8413	.8438	.8461	.8485	.8508	.8531	.8554	.8577	.8599	.8621
1.1	.8643	.8665	.8686	.8708	.8729	.8749	.8770	.8790	.8810	.8830
1.2	.8849	.8869	.8888	.8907	.8925	.8944	.8962	.8980	.8997	.9015
1.3	.9032	.9049	.9066	.9082	.9099	.9115	.9131	.9147	.9162	.9177
1.4	.9192	.9207	.9222	.9236	.9251	.9265	.9279	.9292	.9306	.9319
1.5	.9332	.9345	.9357	.9370	.9382	.9394	.9406	.9418	.9429	.9441
1.6	.9452	.9463	.9474	.9484	.9495	.9505	.9515	.9525	.9535	.9545
1.7	.9554	.9564	.9573	.9582	.9591	.9599	.9608	.9616	.9625	.9633
1.8	.9641	.9649	.9656	.9664	.9671	.9678	.9686	.9693	.9699	.9706
1.9	.9713	.9719	.9726	.9732	.9738	.9744	.9750	.9756	.9761	.9767
2.0	.9772	.9778	.9783	.9788	.9793	.9798	.9803	.9808	.9812	.9817
2.1	.9821	.9826	.9830	.9834	.9838	.9842	.9846	.9850	.9854	.9857
2.2	.9861	.9864	.9868	.9871	.9875	.9878	.9881	.9884	.9887	.9890
2.3	.9893	.9896	.9898	.9901	.9904	.9906	.9909	.9911	.9913	.9916
2.4	.9918	.9920	.9922	.9925	.9927	.9929	.9931	.9932	.9934	.9936
2.5	.9938	.9940	.9941	.9943	.9945	.9946	.9948	.9949	.9951	.9952
2.6	.9953	.9955	.9956	.9957	.9959	.9960	.9961	.9962	.9963	.9964
2.7	.9965	.9966	.9967	.9968	.9969	.9970	.9971	.9972	.9973	.9974
2.8	.9974	.9975	.9976	.9977	.9977	.9978	.9979	.9979	.9980	.9981
2.9	.9981	.9982	.9982	.9983	.9984	.9984	.9985	.9985	.9986	.9986
3.0	.9987	.9987	.9987	.9988	.9988	.9989	.9989	.9989	.9990	.9990
3.1	.9990	.9991	.9991	.9991	.9992	.9992	.9992	.9992	.9993	.9993
3.2	.9993	.9993	.9994	.9994	.9994	.9994	.9994	.9995	.9995	.9995
3.3	.9995	.9995	.9995	.9996	.9996	.9996	.9996	.9996	.9996	.9997
3.4	.9997	.9997	.9997	.9997	.9997	.9997	.9997	.9997	.9997	.9998

For z values greater than 3.49, use 1.000 to approximate the area.

TABLE 5 *continued*

(b) Confidence Interval Critical Values z_c

Level of Confidence c	Critical Value z_c
0.75, or 75%	1.15
0.80, or 80%	1.28
0.85, or 85%	1.44
0.90, or 90%	1.645
0.95, or 95%	1.96
0.98, or 98%	2.33
0.99, or 99%	2.58

TABLE 5 *continued*

(c) Hypothesis Testing, Critical Values z_0

Level of Significance	$\alpha = 0.05$	$\alpha = 0.01$
Critical value z_0 for a left-tailed test	−1.645	−2.33
Critical value z_0 for a right-tailed test	1.645	2.33
Critical values $\pm z_0$ for a two-tailed test	±1.96	±2.58

TABLE 6 Student's *t* Distribution

	c	0.750	0.800	0.850	0.900	0.950	0.980	0.990
	α'	0.125	0.100	0.075	0.050	0.025	0.010	0.005
d.f.	α"	0.250	0.200	0.150	0.100	0.050	0.020	0.010
1		2.414	3.078	4.165	6.314	12.706	31.821	63.657
2		1.604	1.886	2.282	2.920	4.303	6.965	9.925
3		1.423	1.638	1.924	2.353	3.182	4.541	5.841
4		1.344	1.533	1.778	2.132	2.776	3.747	4.604
5		1.301	1.476	1.699	2.015	2.571	3.365	4.032
6		1.273	1.440	1.650	1.943	2.447	3.143	3.707
7		1.254	1.415	1.617	1.895	2.365	2.998	3.499
8		1.240	1.397	1.592	1.860	2.306	2.896	3.355
9		1.230	1.383	1.574	1.833	2.262	2.821	3.250
10		1.221	1.372	1.559	1.812	2.228	2.764	3.169
11		1.214	1.363	1.548	1.796	2.201	2.718	3.106
12		1.209	1.356	1.538	1.782	2.179	2.681	3.055
13		1.204	1.350	1.530	1.771	2.160	2.650	3.012
14		1.200	1.345	1.523	1.761	2.145	2.624	2.977
15		1.197	1.341	1.517	1.753	2.131	2.602	2.947
16		1.194	1.337	1.512	1.746	2.120	2.583	2.921
17		1.191	1.333	1.508	1.740	2.110	2.567	2.898
18		1.189	1.330	1.504	1.734	2.101	2.552	2.878
19		1.187	1.328	1.500	1.729	2.093	2.539	2.861
20		1.185	1.325	1.497	1.725	2.086	2.528	2.845
21		1.183	1.323	1.494	1.721	2.080	2.518	2.831
22		1.182	1.321	1.492	1.717	2.074	2.508	2.819
23		1.180	1.319	1.489	1.714	2.069	2.500	2.807
24		1.179	1.318	1.487	1.711	2.064	2.492	2.797
25		1.178	1.316	1.485	1.708	2.060	2.485	2.787
26		1.177	1.315	1.483	1.706	2.056	2.479	2.779
27		1.176	1.314	1.482	1.703	2.052	2.473	2.771
28		1.175	1.313	1.480	1.701	2.048	2.467	2.763
29		1.174	1.311	1.479	1.699	2.045	2.462	2.756
30		1.173	1.310	1.477	1.697	2.042	2.457	2.750
35		1.170	1.306	1.472	1.690	2.030	2.438	2.724
40		1.167	1.303	1.468	1.684	2.021	2.423	2.704
45		1.165	1.301	1.465	1.679	2.014	2.412	2.690
50		1.164	1.299	1.462	1.676	2.009	2.403	2.678
55		1.163	1.297	1.460	1.673	2.004	2.396	2.668
60		1.162	1.296	1.458	1.671	2.000	2.390	2.660
90		1.158	1.291	1.452	1.662	1.987	2.369	2.632
120		1.156	1.289	1.449	1.658	1.980	2.358	2.617
∞		1.150	1.282	1.440	1.645	1.960	2.326	2.58

Student's *t* values generated by Minitab Version 9.2

c is a confidence level:

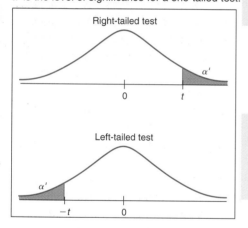

α' is the level of significance for a one-tailed test:

Right-tailed test

Left-tailed test

α" is the level of significance for a two-tailed test:

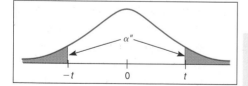

TABLE 7   The χ^2 Distribution

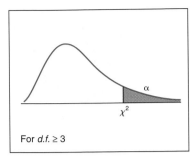

For d.f. ≥ 3

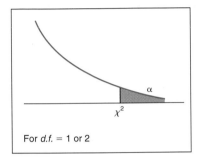

For d.f. = 1 or 2

d.f.	.995	.990	.975	.950	.900	.100	.050	.025	.010	.005
1	0.0^4393	0.0^3157	0.0^3982	0.0^2393	0.0158	2.71	3.84	5.02	6.63	7.88
2	0.0100	0.0201	0.0506	0.103	0.211	4.61	5.99	7.38	9.21	10.60
3	0.072	0.115	0.216	0.352	0.584	6.25	7.81	9.35	11.34	12.84
4	0.207	0.297	0.484	0.711	1.064	7.78	9.49	11.14	13.28	14.86
5	0.412	0.554	0.831	1.145	1.61	9.24	11.07	12.83	15.09	16.75
6	0.676	0.872	1.24	1.64	2.20	10.64	12.59	14.45	16.81	18.55
7	0.989	1.24	1.69	2.17	2.83	12.02	14.07	16.01	18.48	20.28
8	1.34	1.65	2.18	2.73	3.49	13.36	15.51	17.53	20.09	21.96
9	1.73	2.09	2.70	3.33	4.17	14.68	16.92	19.02	21.67	23.59
10	2.16	2.56	3.25	3.94	4.87	15.99	18.31	20.48	23.21	25.19
11	2.60	3.05	3.82	4.57	5.58	17.28	19.68	21.92	24.72	26.76
12	3.07	3.57	4.40	5.23	6.30	18.55	21.03	23.34	26.22	28.30
13	3.57	4.11	5.01	5.89	7.04	19.81	22.36	24.74	27.69	29.82
14	4.07	4.66	5.63	6.57	7.79	21.06	23.68	26.12	29.14	31.32
15	4.60	5.23	6.26	7.26	8.55	22.31	25.00	27.49	30.58	32.80
16	5.14	5.81	6.91	7.96	9.31	23.54	26.30	28.85	32.00	34.27
17	5.70	6.41	7.56	8.67	10.09	24.77	27.59	30.19	33.41	35.72
18	6.26	7.01	8.23	9.39	10.86	25.99	28.87	31.53	34.81	37.16
19	6.84	7.63	8.91	10.12	11.65	27.20	30.14	32.85	36.19	38.58
20	7.43	8.26	8.59	10.85	12.44	28.41	31.41	34.17	37.57	40.00
21	8.03	8.90	10.28	11.59	13.24	29.62	32.67	35.48	38.93	41.40
22	8.64	9.54	10.98	12.34	14.04	30.81	33.92	36.78	40.29	42.80
23	9.26	10.20	11.69	13.09	14.85	32.01	35.17	38.08	41.64	44.18
24	9.89	10.86	12.40	13.85	15.66	33.20	36.42	39.36	42.98	45.56
25	10.52	11.52	13.12	14.61	16.47	34.38	37.65	40.65	44.31	46.93
26	11.16	12.20	13.84	15.38	17.29	35.56	38.89	41.92	45.64	48.29
27	11.81	12.88	14.57	16.15	18.11	36.74	40.11	43.19	46.96	49.64
28	12.46	13.56	15.31	16.93	18.94	37.92	41.34	44.46	48.28	50.99
29	13.21	14.26	16.05	17.71	19.77	39.09	42.56	45.72	49.59	52.34
30	13.79	14.95	16.79	18.49	20.60	40.26	43.77	46.98	50.89	53.67
40	20.71	22.16	24.43	26.51	29.05	51.80	55.76	59.34	63.69	66.77
50	27.99	29.71	32.36	34.76	37.69	63.17	67.50	71.42	76.15	79.49
60	35.53	37.48	40.48	43.19	46.46	74.40	79.08	83.30	88.38	91.95
70	43.28	45.44	48.76	51.74	55.33	85.53	90.53	95.02	100.4	104.2
80	51.17	53.54	57.15	60.39	64.28	96.58	101.9	106.6	112.3	116.3
90	59.20	61.75	65.65	69.13	73.29	107.6	113.1	118.1	124.1	128.3
100	67.33	70.06	74.22	77.93	82.36	118.5	124.3	129.6	135.8	140.2

Source: From H. L. Herter, *Biometrika*, June 1964. Printed by permission of Biometrika Trustees.

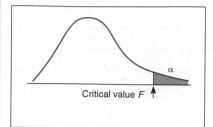

Critical value F ↑

TABLE 8   The F Distribution

Degrees of Freedom for Denominator	α	Degrees of Freedom for Numerator								
		1	2	3	4	5	6	7	8	9
1	.050	161.45	199.50	215.71	224.58	230.16	233.99	236.77	238.88	240.54
1	.025	647.79	799.50	864.16	899.58	921.85	937.11	948.22	956.66	963.28
1	.010	4052.2	4999.5	5403.4	5624.6	5763.6	5859.0	5928.4	5981.1	6022.5
2	.050	18.51	19.00	19.16	19.25	19.30	19.33	19.35	19.37	19.38
2	.025	38.51	39.00	39.17	39.25	39.30	39.33	39.36	39.37	39.39
2	.010	98.50	99.00	99.17	99.25	99.30	99.33	99.36	99.37	99.39
3	.050	10.13	9.55	9.28	9.12	9.01	8.94	8.89	8.85	8.81
3	.025	17.44	16.04	15.44	15.10	14.88	14.73	14.62	14.54	14.47
3	.010	34.12	30.82	29.46	28.71	28.24	27.91	27.67	27.49	27.35
4	.050	7.71	6.94	6.59	6.39	6.26	6.16	6.09	6.04	6.00
4	.025	12.22	10.65	9.98	9.60	9.36	9.20	9.07	8.98	8.90
4	.010	21.20	18.00	16.69	15.98	15.52	15.21	14.98	14.80	14.66
5	.050	6.61	5.79	5.41	5.19	5.05	4.95	4.88	4.82	4.77
5	.025	10.01	8.43	7.76	7.39	7.15	6.98	6.85	6.76	6.68
5	.010	16.26	13.27	12.06	11.39	10.97	10.67	10.46	10.29	10.16
6	.050	5.99	5.14	4.76	4.53	4.39	4.28	4.21	4.15	4.10
6	.025	8.81	7.26	6.60	6.23	5.99	5.82	5.70	5.60	5.52
6	.010	13.75	10.92	9.78	9.15	8.75	8.47	8.26	8.10	7.98
7	.050	5.59	4.74	4.35	4.12	3.97	3.87	3.79	3.73	3.68
7	.025	8.07	6.54	5.89	5.52	5.29	5.12	4.99	4.90	4.82
7	.010	12.25	9.55	8.45	7.85	7.46	7.19	6.99	6.84	6.72
8	.050	5.32	4.46	4.07	3.84	3.69	3.58	3.50	3.44	3.39
8	.025	7.57	6.06	5.42	5.05	4.82	4.65	4.53	4.43	4.36
8	.010	11.26	8.65	7.59	7.01	6.63	6.37	6.18	6.03	5.91
9	.050	5.12	4.26	3.86	3.63	3.48	3.37	3.29	3.23	3.18
9	.025	7.21	5.71	5.08	4.72	4.48	4.32	4.20	4.10	4.03
9	.010	10.56	8.02	6.99	6.42	6.06	5.80	5.61	5.47	5.35
10	.050	4.96	4.10	3.71	3.48	3.33	3.22	3.14	3.07	3.02
10	.025	6.94	5.46	4.83	4.47	4.24	4.07	3.95	3.85	3.78
10	.010	10.04	7.56	6.55	5.99	5.64	5.39	5.20	5.06	4.94
11	.050	4.84	3.98	3.59	3.36	3.20	3.09	3.01	2.95	2.90
11	.025	6.72	5.26	4.63	4.28	4.04	3.88	3.76	3.66	3.59
11	.010	9.65	7.21	6.22	5.67	5.32	5.07	4.89	4.74	4.63
12	.050	4.75	3.89	3.49	3.26	3.11	3.00	2.91	2.85	2.80
12	.025	6.55	5.10	4.47	4.12	3.89	3.73	3.61	3.51	3.44
12	.010	9.33	6.93	5.95	5.41	5.06	4.82	4.64	4.50	4.39

TABLE 8 *continued*

Degrees of Freedom for Denominator α		Degrees of Freedom for Numerator										
		10	12	15	20	25	30	40	50	60	120	1000
1	.050	241.88	243.91	245.95	248.01	249.26	250.10	251.14	251.77	252.20	253.25	254.19
1	.025	968.63	976.71	984.87	993.10	998.08	1001.4	1005.6	1008.1	1009.8	1014.0	1017.7
1	.010	6055.8	6106.3	6157.3	6208.7	6239.8	6260.6	6286.8	6302.5	6313.0	6339.4	6362.7
2	.050	19.40	19.41	19.43	19.45	19.46	19.46	19.47	19.48	19.48	19.49	19.49
2	.025	39.40	39.41	39.43	39.45	39.46	39.46	39.47	39.48	39.48	39.49	39.50
2	.010	99.40	99.42	99.43	99.45	99.46	99.47	99.47	99.48	99.48	99.49	99.50
3	.050	8.79	8.74	8.70	8.66	8.63	8.62	8.59	8.58	8.57	8.55	8.53
3	.025	14.42	14.34	14.25	14.17	14.12	14.08	14.04	14.01	13.99	13.95	13.91
3	.010	27.23	27.05	26.87	26.69	26.58	26.50	26.41	26.35	26.32	26.22	26.14
4	.050	5.96	5.91	5.86	5.80	5.77	5.75	5.72	5.70	5.69	5.66	5.63
4	.025	8.84	8.75	8.66	8.56	8.50	8.46	8.41	8.38	8.36	8.31	8.26
4	.010	14.55	14.37	14.20	14.02	13.91	13.84	13.75	13.69	13.65	13.56	13.47
5	.050	4.74	4.68	4.62	4.56	4.52	4.50	4.46	4.44	4.43	4.40	4.37
5	.025	6.62	6.52	6.43	6.33	6.27	6.23	6.18	6.14	6.12	6.07	6.02
5	.010	10.05	9.89	9.72	9.55	9.45	9.38	9.29	9.24	9.20	9.11	9.03
6	.050	4.06	4.00	3.94	3.87	3.83	3.81	3.77	3.75	3.74	3.70	3.67
6	.025	5.46	5.37	5.27	5.17	5.11	5.07	5.01	4.98	4.96	4.90	4.86
6	.010	7.87	7.72	7.56	7.40	7.30	7.23	7.14	7.09	7.06	6.97	6.89
7	.050	3.64	3.57	3.51	3.44	3.40	3.38	3.34	3.32	3.30	3.27	3.23
7	.025	4.76	4.67	4.57	4.47	4.40	4.36	4.31	4.28	4.25	4.20	4.15
7	.010	6.62	6.47	6.31	6.16	6.06	5.99	5.91	5.86	5.82	5.74	5.66
8	.050	3.35	3.28	3.22	3.15	3.11	3.08	3.04	3.02	3.01	2.97	2.93
8	.025	4.30	4.20	4.10	4.00	3.94	3.89	3.84	3.81	3.78	3.73	3.68
8	.010	5.81	5.67	5.52	5.36	5.26	5.20	5.12	5.07	5.03	4.95	4.87
9	.050	3.14	3.07	3.01	2.94	2.89	2.86	2.83	2.80	2.79	2.75	2.71
9	.025	3.96	3.87	3.77	3.67	3.60	3.56	3.51	3.47	3.45	3.39	3.34
9	.010	5.26	5.11	4.96	4.81	4.71	4.65	4.57	4.52	4.48	4.40	4.32
10	.050	2.98	2.91	2.85	2.77	2.73	2.70	2.66	2.64	2.62	2.58	2.54
10	.025	3.72	3.62	3.52	3.42	3.35	3.31	3.26	3.22	3.20	3.14	3.09
10	.010	4.85	4.71	4.56	4.41	4.31	4.25	4.17	4.12	4.08	4.00	3.92
11	.050	2.85	2.79	2.72	2.65	2.60	2.57	2.53	2.51	2.49	2.45	2.41
11	.025	3.53	3.43	3.33	3.23	3.16	3.12	3.06	3.03	3.00	2.94	2.89
11	.010	4.54	4.40	4.25	4.10	4.01	3.94	3.86	3.81	3.78	3.69	3.61
12	.050	2.75	2.69	2.62	2.54	2.50	2.47	2.43	2.40	2.38	2.34	2.30
12	.025	3.37	3.28	3.18	3.07	3.01	2.96	2.91	2.87	2.85	2.79	2.73
12	.010	4.30	4.16	4.01	3.86	3.76	3.70	3.62	3.57	3.54	3.45	3.37

TABLE 8 *continued*

Degrees of Freedom for Denominator	α	Degrees of Freedom for Numerator								
		1	2	3	4	5	6	7	8	9
13	0.050	4.67	3.81	3.41	3.18	3.03	2.92	2.83	2.77	2.71
13	0.025	6.41	4.97	4.35	4.00	3.77	3.60	3.48	3.39	3.31
13	0.010	9.07	6.70	5.74	5.21	4.86	4.62	4.44	4.30	4.19
14	0.050	4.60	3.74	3.34	3.11	2.96	2.85	2.76	2.70	2.65
14	0.025	6.30	4.86	4.24	3.89	3.66	3.50	3.38	3.29	3.21
14	0.010	8.86	6.51	5.56	5.04	4.69	4.46	4.28	4.14	4.03
15	0.050	4.54	3.68	3.29	3.06	2.90	2.79	2.71	2.64	2.59
15	0.025	6.20	4.77	4.15	3.80	3.58	3.41	3.29	3.20	3.12
15	0.010	8.68	6.36	5.42	4.89	4.56	4.32	4.14	4.00	3.89
16	0.050	4.49	3.63	3.24	3.01	2.85	2.74	2.66	2.59	2.54
16	0.025	6.12	4.69	4.08	3.73	3.50	3.34	3.22	3.12	3.05
16	0.010	8.53	6.23	5.29	4.77	4.44	4.20	4.03	3.89	3.78
17	0.050	4.45	3.59	3.20	2.96	2.81	2.70	2.61	2.55	2.49
17	0.025	6.04	4.62	4.01	3.66	3.44	3.28	3.16	3.06	2.98
17	0.010	8.40	6.11	5.19	4.67	4.34	4.10	3.93	3.79	3.68
18	0.050	4.41	3.55	3.16	2.93	2.77	2.66	2.58	2.51	2.46
18	0.025	5.98	4.56	3.95	3.61	3.38	3.22	3.10	3.01	2.93
18	0.010	8.29	6.01	5.09	4.58	4.25	4.01	3.84	3.71	3.60
19	0.050	4.38	3.52	3.13	2.90	2.74	2.63	2.54	2.48	2.42
19	0.025	5.92	4.51	3.90	3.56	3.33	3.17	3.05	2.96	2.88
19	0.010	8.18	5.93	5.01	4.50	4.17	3.94	3.77	3.63	3.52
20	0.050	4.35	3.49	3.10	2.87	2.71	2.60	2.51	2.45	2.39
20	0.025	5.87	4.46	3.86	3.51	3.29	3.13	3.01	2.91	2.84
20	0.010	8.10	5.85	4.94	4.43	4.10	3.87	3.70	3.56	3.46
21	0.050	4.32	3.47	3.07	2.84	2.68	2.57	2.49	2.42	2.37
21	0.025	5.83	4.42	3.82	3.48	3.25	3.09	2.97	2.87	2.80
21	0.010	8.02	5.78	4.87	4.37	4.04	3.81	3.64	3.51	3.40
22	0.050	4.30	3.44	3.05	2.82	2.66	2.55	2.46	2.40	2.34
22	0.025	5.79	4.38	3.78	3.44	3.22	3.05	2.93	2.84	2.76
22	0.010	7.95	5.72	4.82	4.31	3.99	3.76	3.59	3.45	3.35
23	0.050	4.28	3.42	3.03	2.80	2.64	2.53	2.44	2.37	2.32
23	0.025	5.75	4.35	3.75	3.41	3.18	3.02	2.90	2.81	2.73
23	0.010	7.88	5.66	4.76	4.26	3.94	3.71	3.54	3.41	3.30
24	0.050	4.26	3.40	3.01	2.78	2.62	2.51	2.42	2.36	2.30
24	0.025	5.72	4.32	3.72	3.38	3.15	2.99	2.87	2.78	2.70
24	0.010	7.82	5.61	4.72	4.22	3.90	3.67	3.50	3.36	3.26

TABLE 8 *continued*

Degrees of Freedom for Denominator	α	Degrees of Freedom for Numerator										
		10	12	15	20	25	30	40	50	60	120	1000
13	0.050	2.67	2.60	2.53	2.46	2.41	2.38	2.34	2.31	2.30	2.25	2.21
13	0.025	3.25	3.15	3.05	2.95	2.88	2.84	2.78	2.74	2.72	2.66	2.60
13	0.010	4.10	3.96	3.82	3.66	3.57	3.51	3.43	3.38	3.34	3.25	3.18
14	0.050	2.60	2.53	2.46	2.39	2.34	2.31	2.27	2.24	2.22	2.18	2.14
14	0.025	3.15	3.05	2.95	2.84	2.78	2.73	2.67	2.64	2.61	2.55	2.50
14	0.010	3.94	3.80	3.66	3.51	3.41	3.35	3.27	3.22	3.18	3.09	3.02
15	0.050	2.54	2.48	2.40	2.33	2.28	2.25	2.20	2.18	2.16	2.11	2.07
15	0.025	3.06	2.96	2.86	2.76	2.69	2.64	2.59	2.55	2.52	2.46	2.40
15	0.010	3.80	3.67	3.52	3.37	3.28	3.21	3.13	3.08	3.05	2.96	2.88
16	0.050	2.49	2.42	2.35	2.28	2.23	2.19	2.15	2.12	2.11	2.06	2.02
16	0.025	2.99	2.89	2.79	2.68	2.61	2.57	2.51	2.47	2.45	2.38	2.32
16	0.010	3.69	3.55	3.41	3.26	3.16	3.10	3.02	2.97	2.93	2.84	2.76
17	0.050	2.45	2.38	2.31	2.23	2.18	2.15	2.10	2.08	2.06	2.01	1.97
17	0.025	2.92	2.82	2.72	2.62	2.55	2.50	2.44	2.41	2.38	2.32	2.26
17	0.010	3.59	3.46	3.31	3.16	3.07	3.00	2.92	2.87	2.83	2.75	2.66
18	0.050	2.41	2.34	2.27	2.19	2.14	2.11	2.06	2.04	2.02	1.97	1.92
18	0.025	2.87	2.77	2.67	2.56	2.49	2.44	2.38	2.35	2.32	2.26	2.20
18	0.010	3.51	3.37	3.23	3.08	2.98	2.92	2.84	2.78	2.75	2.66	2.58
19	0.050	2.38	2.31	2.23	2.16	2.11	2.07	2.03	2.00	1.98	1.93	1.88
19	0.025	2.82	2.72	2.62	2.51	2.44	2.39	2.33	2.30	2.27	2.20	2.14
19	0.010	3.43	3.30	3.15	3.00	2.91	2.84	2.76	2.71	2.67	2.58	2.50
20	0.050	2.35	2.28	2.20	2.12	2.07	2.04	1.99	1.97	1.95	1.90	1.85
20	0.025	2.77	2.68	2.57	2.46	2.40	2.35	2.29	2.25	2.22	2.16	2.09
20	0.010	3.37	3.23	3.09	2.94	2.84	2.78	2.69	2.64	2.61	2.52	2.43
21	0.050	2.32	2.25	2.18	2.10	2.05	2.01	1.96	1.94	1.92	1.87	1.82
21	0.025	2.73	2.64	2.53	2.42	2.36	2.31	2.25	2.21	2.18	2.11	2.05
21	0.010	3.31	3.17	3.03	2.88	2.79	2.72	2.64	2.58	2.55	2.46	2.37
22	0.050	2.30	2.23	2.15	2.07	2.02	1.98	1.94	1.91	1.89	1.84	1.79
22	0.025	2.70	2.60	2.50	2.39	2.32	2.27	2.21	2.17	2.14	2.08	2.01
22	0.010	3.26	3.12	2.98	2.83	2.73	2.67	2.58	2.53	2.50	2.40	2.32
23	0.050	2.27	2.20	2.13	2.05	2.00	1.96	1.91	1.88	1.86	1.81	1.76
23	0.025	2.67	2.57	2.47	2.36	2.29	2.24	2.18	2.14	2.11	2.04	1.98
23	0.010	3.21	3.07	2.93	2.78	2.69	2.62	2.54	2.48	2.45	2.35	2.27
24	0.050	2.25	2.18	2.11	2.03	1.97	1.94	1.89	1.86	1.84	1.79	1.74
24	0.025	2.64	2.54	2.44	2.33	2.26	2.21	2.15	2.11	2.08	2.01	1.94
24	0.010	3.17	3.03	2.89	2.74	2.64	2.58	2.49	2.44	2.40	2.31	2.22

TABLE 8   *continued*

Degrees of Freedom for Denominator	α	Degrees of Freedom for Numerator								
		1	2	3	4	5	6	7	8	9
25	0.05	4.24	3.39	2.99	2.76	2.60	2.49	2.40	2.34	2.28
25	0.025	5.69	4.29	3.69	3.35	3.13	2.97	2.85	2.75	2.68
25	0.01	7.77	5.57	4.68	4.18	3.85	3.63	3.46	3.32	3.22
26	0.05	4.23	3.37	2.98	2.74	2.59	2.47	2.39	2.32	2.27
26	0.025	5.66	4.27	3.67	3.33	3.10	2.94	2.82	2.73	2.65
26	0.01	7.72	5.53	4.64	4.14	3.82	3.59	3.42	3.29	3.18
27	0.05	4.21	3.35	2.96	2.73	2.57	2.46	2.37	2.31	2.25
27	0.025	5.63	4.24	3.65	3.31	3.08	2.92	2.80	2.71	2.63
27	0.01	7.68	5.49	4.60	4.11	3.78	3.56	3.39	3.26	3.15
28	0.05	4.20	3.34	2.95	2.71	2.56	2.45	2.36	2.29	2.24
28	0.025	5.61	4.22	3.63	3.29	3.06	2.90	2.78	2.69	2.61
28	0.01	7.64	5.45	4.57	4.07	3.75	3.53	3.36	3.23	3.12
29	0.05	4.18	3.33	2.93	2.70	2.55	2.43	2.35	2.28	2.22
29	0.025	5.59	4.20	3.61	3.27	3.04	2.88	2.76	2.67	2.59
29	0.01	7.60	5.42	4.54	4.04	3.73	3.50	3.33	3.20	3.09
30	0.05	4.17	3.32	2.92	2.69	2.53	2.42	2.33	2.27	2.21
30	0.025	5.57	4.18	3.59	3.25	3.03	2.87	2.75	2.65	2.57
30	0.01	7.56	5.39	4.51	4.02	3.70	3.47	3.30	3.17	3.07
40	0.05	4.08	3.23	2.84	2.61	2.45	2.34	2.25	2.18	2.12
40	0.025	5.42	4.05	3.46	3.13	2.90	2.74	2.62	2.53	2.45
40	0.01	7.31	5.18	4.31	3.83	3.51	3.29	3.12	2.99	2.89
50	0.05	4.03	3.18	2.79	2.56	2.40	2.29	2.20	2.13	2.07
50	0.025	5.34	3.97	3.39	3.05	2.83	2.67	2.55	2.46	2.38
50	0.01	7.17	5.06	4.20	3.72	3.41	3.19	3.02	2.89	2.78
60	0.05	4.00	3.15	2.76	2.53	2.37	2.25	2.17	2.10	2.04
60	0.025	5.29	3.93	3.34	3.01	2.79	2.63	2.51	2.41	2.33
60	0.01	7.08	4.98	4.13	3.65	3.34	3.12	2.95	2.82	2.72
100	0.05	3.94	3.09	2.70	2.46	2.31	2.19	2.10	2.03	1.97
100	0.025	5.18	3.83	3.25	2.92	2.70	2.54	2.42	2.32	2.24
100	0.01	6.90	4.82	3.98	3.51	3.21	2.99	2.82	2.69	2.59
200	0.05	3.89	3.04	2.65	2.42	2.26	2.14	2.06	1.98	1.93
200	0.025	5.10	3.76	3.18	2.85	2.63	2.47	2.35	2.26	2.18
200	0.01	6.76	4.71	3.88	3.41	3.11	2.89	2.73	2.60	2.50
1000	0.05	3.85	3.00	2.61	2.38	2.22	2.11	2.02	1.95	1.89
1000	0.025	5.04	3.70	3.13	2.80	2.58	2.42	2.30	2.20	2.13
1000	0.01	6.66	4.63	3.80	3.34	3.04	2.82	2.66	2.53	2.43

Source: From *Biometrika Tables for Statisticians,* Vol. 1, by permission of the Biometrika Trustees.

TABLE 8   *continued*

Degrees of Freedom for Denominator	α	Degrees of Freedom for Numerator										
		10	12	15	20	25	30	40	50	60	120	1000
25	0.05	2.24	2.16	2.09	2.01	1.96	1.92	1.87	1.84	1.82	1.77	1.72
25	0.025	2.61	2.51	2.41	2.30	2.23	2.18	2.12	2.08	2.05	1.98	1.91
25	0.01	3.13	2.99	2.85	2.70	2.60	2.54	2.45	2.40	2.36	2.27	2.18
26	0.05	2.22	2.15	2.07	1.99	1.94	1.90	1.85	1.82	1.80	1.75	1.70
26	0.025	2.59	2.49	2.39	2.28	2.21	2.16	2.09	2.05	2.03	1.95	1.89
26	0.01	3.09	2.96	2.81	2.66	2.57	2.50	2.42	2.36	2.33	2.23	2.14
27	0.05	2.20	2.13	2.06	1.97	1.92	1.88	1.84	1.81	1.79	1.73	1.68
27	0.025	2.57	2.47	2.36	2.25	2.18	2.13	2.07	2.03	2.00	1.93	1.86
27	0.01	3.06	2.93	2.78	2.63	2.54	2.47	2.38	2.33	2.29	2.20	2.11
28	0.05	2.19	2.12	2.04	1.96	1.91	1.87	1.82	1.79	1.77	1.71	1.66
28	0.025	2.55	2.45	2.34	2.23	2.16	2.11	2.05	2.01	1.98	1.91	1.84
28	0.01	3.03	2.90	2.75	2.60	2.51	2.44	2.35	2.30	2.26	2.17	2.08
29	0.05	2.18	2.10	2.03	1.94	1.89	1.85	1.81	1.77	1.75	1.70	1.65
29	0.025	2.53	2.43	2.32	2.21	2.14	2.09	2.03	1.99	1.96	1.89	1.82
29	0.01	3.00	2.87	2.73	2.57	2.48	2.41	2.33	2.27	2.23	2.14	2.05
30	0.05	2.16	2.09	2.01	1.93	1.88	1.84	1.79	1.76	1.74	1.68	1.63
30	0.025	2.51	2.41	2.31	2.20	2.12	2.07	2.01	1.97	1.94	1.87	1.80
30	0.01	2.98	2.84	2.70	2.55	2.45	2.39	2.30	2.25	2.21	2.11	2.02
40	0.05	2.08	2.00	1.92	1.84	1.78	1.74	1.69	1.66	1.64	1.58	1.52
40	0.025	2.39	2.29	2.18	2.07	1.99	1.94	1.88	1.83	1.80	1.72	1.65
40	0.01	2.80	2.66	2.52	2.37	2.27	2.20	2.11	2.06	2.02	1.92	1.82
50	0.05	2.03	1.95	1.87	1.78	1.73	1.69	1.63	1.60	1.58	1.51	1.45
50	0.025	2.32	2.22	2.11	1.99	1.92	1.87	1.80	1.75	1.72	1.64	1.56
50	0.01	2.70	2.56	2.42	2.27	2.17	2.10	2.01	1.95	1.91	1.80	1.70
60	0.05	1.99	1.92	1.84	1.75	1.69	1.65	1.59	1.56	1.53	1.47	1.40
60	0.025	2.27	2.17	2.06	1.94	1.87	1.82	1.74	1.70	1.67	1.58	1.49
60	0.01	2.63	2.50	2.35	2.20	2.10	2.03	1.94	1.88	1.84	1.73	1.62
100	0.05	1.93	1.85	1.77	1.68	1.62	1.57	1.52	1.48	1.45	1.38	1.30
100	0.025	2.18	2.08	1.97	1.85	1.77	1.71	1.64	1.59	1.56	1.46	1.36
100	0.01	2.50	2.37	2.22	2.07	1.97	1.89	1.80	1.74	1.69	1.57	1.45
200	0.05	1.88	1.80	1.72	1.62	1.56	1.52	1.46	1.41	1.39	1.30	1.21
200	0.025	2.11	2.01	1.90	1.78	1.70	1.64	1.56	1.51	1.47	1.37	1.25
200	0.01	2.41	2.27	2.13	1.97	1.87	1.79	1.69	1.63	1.58	1.45	1.30
1000	0.05	1.84	1.76	1.68	1.58	1.52	1.47	1.41	1.36	1.33	1.24	1.11
1000	0.025	2.06	1.96	1.85	1.72	1.64	1.58	1.50	1.45	1.41	1.29	1.13
1000	0.01	2.34	2.20	2.06	1.90	1.79	1.72	1.61	1.54	1.50	1.35	1.16

TABLE 9   Critical Values for Spearman Rank Correlation, r_s

For a right- (left-) tailed test, use the positive (negative) critical value found in the table under significance level for a one-tailed test. For a two-tailed test, use both the positive and negative of the critical value found in the table under significance level for a two-tailed test, n = number of pairs.

	Significance level for a one-tailed test at			
	0.05	0.025	0.005	0.001
	Significance level for a two-tailed test at			
n	0.10	0.05	0.01	0.002
5	0.900	1.000		
6	0.829	0.886	1.000	
7	0.715	0.786	0.929	1.000
8	0.620	0.715	0.881	0.953
9	0.600	0.700	0.834	0.917
10	0.564	0.649	0.794	0.879
11	0.537	0.619	0.764	0.855
12	0.504	0.588	0.735	0.826
13	0.484	0.561	0.704	0.797
14	0.464	0.539	0.680	0.772
15	0.447	0.522	0.658	0.750
16	0.430	0.503	0.636	0.730
17	0.415	0.488	0.618	0.711
18	0.402	0.474	0.600	0.693
19	0.392	0.460	0.585	0.676
20	0.381	0.447	0.570	0.661
21	0.371	0.437	0.556	0.647
22	0.361	0.426	0.544	0.633
23	0.353	0.417	0.532	0.620
24	0.345	0.407	0.521	0.608
25	0.337	0.399	0.511	0.597
26	0.331	0.391	0.501	0.587
27	0.325	0.383	0.493	0.577
28	0.319	0.376	0.484	0.567
29	0.312	0.369	0.475	0.558
30	0.307	0.363	0.467	0.549

Source: From G. J. Glasser and R. F. Winter, "Critical Values of the Coefficient of Rank Correlation for Testing the Hypothesis of Independence," *Biometrika, 48,* 444 (1961). Reprinted by permission of Biometrika Trustees.

Answers and Key Steps to Odd-Numbered Problems

CHAPTER 1

Section 1.1

1. (a) Response regarding frequency of eating at fast-food restaurants. (b) Qualitative. (c) Responses for *all* adults in the U.S.
3. (a) Student fees. (b) Quantitative. (c) Student fees at *all* colleges and universities in the U.S.
5. (a) Time interval between check arrival and clearance. (b) Quantitative. (c) Time interval between check arrival and clearance for *all* checks from offices in the five-state region.
7. (a) Ratio. (b) Interval. (c) Nominal. (d) Ordinal. (e) Ratio. (f) Ratio.
9. (a) Nominal. (b) Ratio. (c) Interval. (d) Ordinal. (e) Ratio. (f) Interval.

Section 1.2

1. See text.
3. Select a starting place in the table and group the digits in groups of 4. Scan the table by rows and include the first 6 groups with numbers between 0001 and 8615.
5. (a) One way is to number the subjects 01 through 40. Make groups of two digits in the random-number table. The first valid number is included in Group 1 while the second valid number is included in Group 2. Continue alternating the group assignments until both groups are filled. (b) Use the process of part a, limited to numbers 01 through 22. (c) Label the two groups of healthy subjects A and B and the two groups of sick subjects C and D. Select a starting place in the random-number table. If the first digit is even, use healthy Group A in the first combined group. If the second digit is even, use Group C of the sick subjects in the first combined group. If either digit is odd, use the other group of healthy or sick subjects. The second combined group contains the two groups not selected for the first combined group.
7. (a) Yes, when a die is rolled several times, the same number may appear more than once. Outcome on the 4th roll is 2. (b) No, for a fair die, the outcomes are random.

9. Use a random-number table to select four distinct numbers corresponding to people in your class.
   (a) Reasons may vary. For instance, the first four students may make a special effort to get to class on time. (b) Reasons may vary. For instance, four students who come in late might all be nursing students enrolled in an anatomy and physiology class that meets the hour before in a far-away building. They may be more motivated than other students to complete a degree requirement. (c) Reasons may vary. For instance, four students sitting in the back row might be less inclined to participate in class discussions. (d) Reasons may vary. For instance, the tallest students might all be male.
11. (a) Assign each sheep a distinct number from 1 to 250. Because the largest number has three digits, read groups of three digits from a random-number table. Select a starting place, and proceed until you have 15 distinct numbers from 001 to 250. Numbers included in the sample will vary according to the starting place and pattern for reading the table. (b) Because the largest number has four digits, read groups of four digits from a random-number table. Select a starting place, and proceed until you have 10 distinct numbers from 1024 to 8342. Numbers included in the sample will vary according to the starting place and pattern for reading the table. (c) Number the pieces of luggage as they come off the conveyor belt in the next 25 minutes. Use a random-number table to choose five distinct numbers that correspond to pieces of luggage. Numbers will vary according to the starting place and pattern for reading the table. (d) Give the survey to as many people as will take the time to complete the survey between 6 and 7 P.M. Number all the surveys obtained. Use a random-number table to choose 12 distinct numbers corresponding to surveys. Note that the population you are sampling consists of survey data from people who are willing to complete surveys, not from everyone who comes to the shopping center during the designated time period.
13. Since there are five possible outcomes for each question, read single digits from a random-number table. Select a starting place, and proceed until you have 10 digits from 1 to 5. Repetition is required. The correct answer for

each question will be the letter choice corresponding to the digit chosen for that question.

15. (a) Simple random sample.   (b) Cluster sample.
    (c) Convenience sample.   (d) Systematic sample.
    (e) Stratified sample.

Section 1.3

1. (a) Observational study.   (b) Experiment.
   (c) Experiment.   (d) Observational study.
3. (a) Sampling.   (b) Simulation.   (c) Census.
   (d) Experiment.
5. (a) Use random selection to pick 10 calves to inoculate; test all calves; no placebo.   (b) Use random selection to pick 9 schools to visit; survey all schools; no placebo.
   (c) Use random selection to pick 40 volunteers for skin patch with drug; survey all volunteers; placebo used.

CHAPTER REVIEW PROBLEMS

1. Depends on article.
3. See text.
5. In the random-number table use groups of 2 digits. Select the first six distinct groups of 2 digits that fall in the range from 01 to 42. Choices vary according to the starting place in the random-number table.
7. (a) Observational study.   (b) Experiment.
9. Possible directions on survey questions: Give height in inches, give age as of last birthday, give GPA to one decimal place, and so forth. Think about the types of responses you wish to have on each question.

CHAPTER 2

Section 2.1

1. Highest Level of Education and Average Annual Household Income (in thousands of dollars)

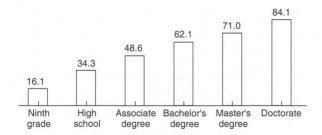

3. Number of People Who Died in a Calendar Year from Listed Causes—Pareto Chart

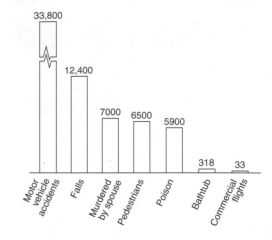

5. Where We Hide the Mess

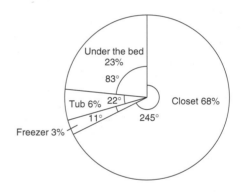

7. How College Professors Spend Time

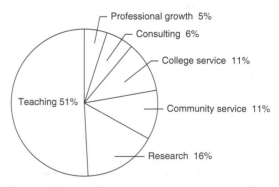

9. Percentage of Households with Telephone Gadgets

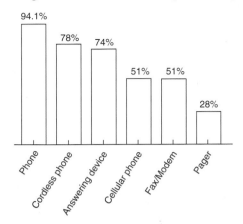

No, percents total more than 100.

11. Elevation of Pyramid Lake Surface—Time Plot

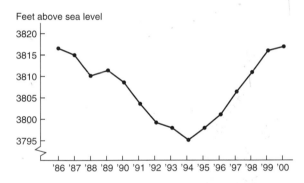

13. Both stocks ended down for the one-year period. Coca-Cola ranged from a high of about $64 to a low of about $42.50. McDonald's ranged from a high of about $37 to a low of about $25. McDonald's attained its high during the first week and was never as high again. From the high of the first week to the high of the last week shown, Coca-Cola declined about 15% while McDonald's declined about 19%.

Section 2.2

1. (a) Class width = 25.
   (b)

Class Limits	Boundaries	Midpoint	Frequency	Relative Frequency	Cumulative Frequency
236–260	235.5–260.5	248	4	0.07	4
261–285	260.5–285.5	273	9	0.16	13
286–310	285.5–310.5	298	25	0.44	38
311–335	310.5–335.5	323	16	0.28	54
336–360	335.5–360.5	348	3	0.05	57

(c–e) Hours to Complete the Iditarod—Histogram, Frequency Polygon, Relative-Frequency Histogram

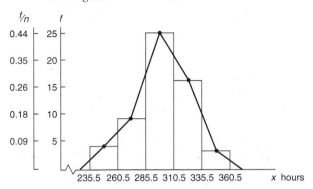

(f) Hours to Complete the Iditarod—Ogive

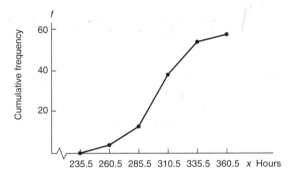

3. (a) Class width = 7.
   (b)

Class Limits	Boundaries	Midpoint	Frequency	Relative Frequency	Cumulative Frequency
5–11	4.5–11.5	8	4	0.08	4
12–18	11.5–18.5	15	7	0.14	11
19–25	18.5–25.5	22	12	0.24	23
26–32	25.5–32.5	29	12	0.24	35
33–39	32.5–39.5	36	12	0.24	47
40–46	39.5–46.5	43	2	0.04	49
47–53	46.5–53.5	50	1	0.02	50

(c–e) Percentage of Children in Neighborhood—
Histogram, Frequency Polygon, Relative-
Frequency Histogram

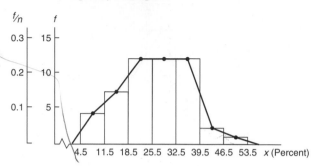

(f) Percentage of Children in Neighborhood—Ogive

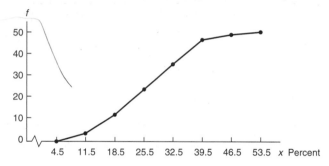

5. (a) Class width = 17.
   (b) Nurses' Night Workload

Class Limits	Boundaries	Midpoint	Frequency	Relative Frequency	Cumulative Frequency
18–34	17.5–34.5	26	1	0.03	1
35–51	34.5–51.5	43	2	0.06	3
52–68	51.5–68.5	60	5	0.14	8
69–85	68.5–85.5	77	15	0.43	23
86–102	85.5–102.5	94	12	0.34	35

(c–e) Nurses' Night Workload—Histogram, Frequency
Polygon, and Relative-Frequency Histogram

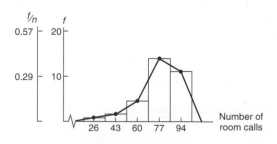

(f) Nurses' Night Workload—Ogive

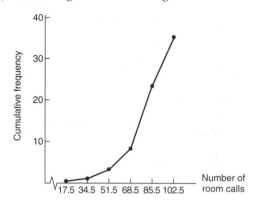

7. (a) Class Midpoints: 34.5; 44.5; 54.5; 64.5; 74.5; 84.5.
   (b) Age of Senators—Frequency Polygon

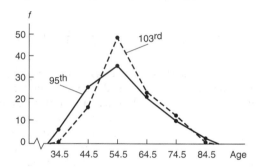

   (c) In general, the ages in the 103rd Congress are older.
9. (a)

**Profit as a Percent of Sales—Food
Companies** (class width = 3)

Class	Frequency	Midpoint
−3–−1	2	−2
0–2	16	1
3–5	10	4
6–8	9	7
9–11	2	10

**Profit as a Percent of Sales—Electronic
Companies** (class width = 5)

Class	Frequency	Midpoint
−6–−2	3	−4
−1–3	13	1
4–8	20	6
9–13	7	11
14–18	1	16

Profit as a Percent of Sales

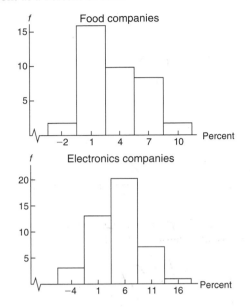

(b) Because the classes and class widths are different for the two company types, it is difficult to compare profits. We can notice that for the electronic companies the percentage of profits does extend as high as 18, while for the food companies the highest percent of profits is 11. On the other hand, some of the electronic companies also have greater losses than the food companies. Had we made the class limits the same for both company types, it would be easier to compare the data.

11. (a) 85; 84.2%.   (b) 25; 24.8%.

13. (a) Version 1 is skewed left; version 2 is uniform; version 3 is symmetrical; version 4 is bimodal; version 5 is skewed right.   (b) Answers will vary.

15. (a) Class width = 0.40.
   (b, c)

Class Limits	Boundaries	Midpoint	Frequency
0.46–0.85	0.455–0.855	0.655	4
0.86–1.25	0.855–1.255	1.055	5
1.26–1.65	1.255–1.655	1.455	10
1.66–2.05	1.655–2.055	1.855	5
2.06–2.45	2.055–2.455	2.255	5
2.46–2.85	2.455–2.855	2.655	3

(c) Tonnes of Wheat—Histogram

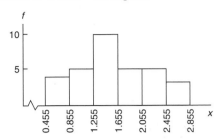

17. (a) One.   (b) 5/51 or 9.8%.   (c) Interval from 650 to 750.

19. Dotplot for Percentage of Children in Neighborhood

Section 2.3

1. (a) Longevity of Cowboys

4		7 = 47 years
4		7
5		2 7 8 8
6		1 6 6 8 8
7		0 2 2 3 3 5 6 7
8		4 4 4 5 6 6 7 9
9		0 1 1 2 3 7

(b) Yes, certainly these cowboys lived long lives.

3. Average Length of Hospital Stay

5		2 = 5.2 days
5		2 3 5 5 6 7
6		0 2 4 6 6 7 7 8 8 8 8 9 9
7		0 0 0 0 0 1 1 1 2 2 2 3 3 3 3 4 4 5 5 6 6 8
8		4 5 7
9		4 6 9
10		0 3
11		1

The distribution is skewed right.

5. (a) Minutes Beyond 2 Hours (1961–1980)

0		9 = 9 minutes past 2 hours
0˙		9 9
1*		0 0 2 3 3 4
1˙		5 5 6 6 7 8 8 9
2*		0 2 3 3

(b) Minutes Beyond 2 Hours (1981–2000)

0		7 = 7 minutes past 2 hours
0˙		7 7 7 8 8 8 8 9 9 9 9 9 9 9 9
1*		0 0 1 1 4

(c) In more recent times, the winning times have been closer to 2 hours, with all the times between 7 and 14 minutes over two hours. In the earlier period, more than half the times were more than 2 hours and 14 minutes.

7. Angular Momentum of Stars

00		14 = 0.014 arc sec/century		
00		08 14 38 42 50 57	11	69 69
01		73	12	60 60
02		16 19 51	13	
03		51 69	14	38
04		30	15	
05			16	16 60
06		23 67	17	
07		59 88	18	08
08		88		
09				
10		24 57		

No large gaps. There are a greater number with angular momentum below 0.888 than above.

9. Milligrams of Tar per Cigarette

1		0 = 1.0 mg tar		
1		0	11	4
2			12	0 4 8
3			13	7
4		1 5	14	1 5 9
5			15	0 1 2 8
6			16	0 6
7		3 8	17	0
8		0 6 8		
9		0		
10			29	8

11. Milligrams of Nicotine per Cigarette

0		1 = 0.1 milligram
0*		1 4 4
0˙		5 6 6 6 7 7 7 8 8 9 9 9
1*		0 0 0 0 0 0 0 1 2
1˙		
2*		0

13. (a) $49,000 to $126,000 for California; $45,000 to $120,000 for New York.
    (b) New York; California.
    (c) California has slightly higher average salaries.

Chapter 2 Review

1. Figure 2-1(a) is essentially a bar graph, but the bar lengths do not have the same scales. Figure 2-1(b) is a time plot. This figure gives clear, undistorted representation of fuel economy standards by year.

3. Problems with Tax Returns

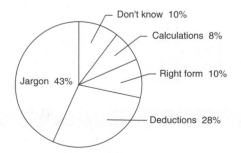

5. (a) Class width = 6.

Class Limits	Boundaries	Midpoint	Frequency	Relative Frequency	Cumulative Frequency
59–64	58.5–64.5	61.5	1	0.02	1
65–70	64.5–70.5	67.5	7	0.13	8
71–76	70.5–76.5	73.5	6	0.11	14
77–82	76.5–82.5	79.5	13	0.25	27
83–88	82.5–88.5	85.5	18	0.34	45
89–94	88.5–94.5	91.5	7	0.13	52
95–100	94.5–100.5	97.5	1	0.02	53

(b–d) Glucose Level—Histogram, Frequency Polygon, and Relative-Frequency Histogram

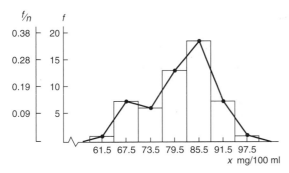

(e) Glucose Level—Ogive

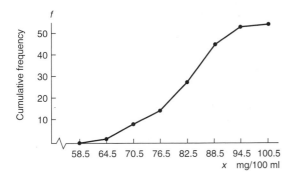

7. (a) 1240s had 40 data. (b) 75. (c) From 1203 to 1212. Little if any repairs or new construction.
9. (a) 65 to 84. (b) 56%. (c) 44%.

CHAPTER 3

Section 3.1

1. Mean = 156.33; median = 157; mode = 157. A gardener in Colorado should look at seed and plant descriptions to determine if the plant can thrive and mature in the designated number of frost-free days. The mean, median, and mode are all close. About half the locations have 157 or fewer frost-free days.
3. $\bar{x} \approx 167.3$ °F; median = 171 °F; mode = 178 °F.
5. (a) $\bar{x} \approx 3.27$; median = 3; mode = 3. (b) $\bar{x} \approx 4.21$; median = 2; mode = 1. (c) Lower canyon mean is greater; median and mode are less. (d) Trimmed mean = 3.75 and is closer to upper canyon mean.
7. (a) Mean = 28.83 thousand dollars. (b) Median = 18.5 thousand dollars. The median best describes the salary of the majority of employees, since the mean is influenced by the high salaries of the president and vice president. (c) Mean = 17.3 thousand dollars; median = 17 thousand dollars. (d) Without the salaries for the two executives, the mean and the median are closer, and both reflect the salary of most of the other workers more accurately. The mean changed quite a bit, while the median did not, a difference that indicates that the mean is more sensitive to the absence or presence of extreme values.
9. (a) For the first week, mean = 14.57 meteoroids; median = 15 meteoroids; mode = 15 meteoroids. (b) For all nine nights, mean = 24.56 meteoroids; median = 15 meteoroids; mode = 15 meteoroids. (c) Extreme values affect the mean far more than they do the median or mode.
11. (a) $\bar{x} \approx \$136.15$; median = \$66.50; mode = \$60. (b) 5% trimmed mean = \$121.28; yes, but still higher than the median. (c) Median. The low and high prices would be useful.
13. (a) Mode. (b) Mean, median, mode if it exists. (c) Mode if it exists; median, mean if 24-hour clock is used.
15. (a) If the largest data value is *replaced* by a larger value, the mean will increase because the sum of the data values will increase, but the number of them will remain the same. The median will not change. The same value will still be in the eighth position when the data are ordered. (b) If the largest value is replaced by a value that is smaller (but still higher than the median), the mean will decrease because the sum of the data values will decrease. The median will not change. The same value will be in the eighth position in increasing order. (c) If the largest value is replaced by a value that is smaller than the median, the mean will decrease because the sum of the data values will decrease. The median also will decrease because the former value in the eighth position will move to the ninth position in increasing order. The median will be the new value in the eighth position.
17. Answers will vary according to data collected.

Section 3.2

1. (a) Range = 54 deer/km²; sample mean $\bar{x} = 20.9$ deer/km²; sample variance $s^2 = 225.0$; sample standard deviation $s = 15.0$ deer/km². (b) CV = 71.8%. Since

the standard deviation is about 72% of the mean, there is considerable variation in the distribution of deer from one part of the park to another.

3. (a) Range = 77.6; $\bar{x}$ = 55.7. (b) $s^2 \approx 537$; $s \approx 23.17$. (c) $CV \approx 41.6\%$. Larger than for geese and so greater relative standard deviation.

5. (a) Pax $CV \approx 98.9\%$; Vanguard $CV \approx 222.8\%$; Pax seems less risky. (b) Pax: -11.43% to 34.81%; Vanguard -19.39% to 30.61%; The performance range for Pax seems better than that for Vanguard (based on this historical data).

7. (a) 75% of the cycles should fall within 2 standard deviations of the mean: 6.67 to 15.35 years. (b) 93.8% of the cycles should fall within 4 standard deviations of the mean: 2.33 to 19.69 years.

9. (a) Range = 737; $\bar{x} \approx 566.9$. (b) $s^2 \approx 71{,}202$; $s \approx 266.8$. (c) $CV \approx 47.1\%$. (d) 33 to 1100.

11. 4.1% to 14.5%.

13. Construction artifacts have highest average and lowest relative standard deviation.

15. (a) Use calculator. (b) Wal-Mart $CV \approx 2\%$; Disney $CV \approx 3\%$; yes. (c) Wal-Mart 48.85 to 55.21; Disney 29.29 to 35.17.

17. Answers vary.

Section 3.3

1. Midpoints: 25.5, 35.5, 45.5; $\bar{x} \approx 35.80$; $s^2 \approx 61.1$; $s \approx 7.82$.

3. Midpoints: 10.55, 14.55, 18.55, 22.55, 26.55; $\bar{x} \approx 15.6$; $s^2 \approx 23.4$; $s \approx 4.8$.

5. Midpoints: 21, 29.5, 39.5, 49.5, 59.5, 72.5; $\bar{x} \approx 39.12$; $s \approx 17.01$; $CV \approx 43.5\%$.

7. Midpoints and frequencies are shown on the figure. $\bar{x} = 7.9$ hours; $s = 1.05$ hours; $CV = 13.29$.

9. (a) $\bar{x} = 3.97$; $s = 2.415$. (b) The results of entering the 31 individual values into a calculator are the same. The grouped-data method should give the same results, since each group consists of only one value. When you have many repeated values in a data set, you might consider using the method of grouped data. It is generally faster.

11. $\bar{x} \approx 4.11$; $s^2 \approx 3.02$; $s \approx 1.74$.

13. $\Sigma wx = 87.65$; $\Sigma w = 1$; weighted average = 87.65.

15. $\Sigma wx = 85$; $\Sigma w = 10$; weighted average = 8.5.

17. $\Sigma wx = 12.26$; $\Sigma w = 1$; weighted average = 12.26 lb.

Section 3.4

1. 82% or more of the scores were at or below Angela's score; 18% or fewer of the scores were above Angela's score.

3. No, the score 82 might have a percentile rank less than 70.

5. Low = 0.52; Q_1 = 0.735; median = 0.875; Q_3 = 1.325; high = 1.92; IQR = 0.59.

Cost of Serving of Pizza

7. Low = 2; Q_1 = 9.5; median = 23; Q_3 = 28.5; high = 42; IQR = 19.

Nurses' Length of Employment (months)

9. Suburban: low = 808; Q_1 = 972; median = 1081; Q_3 = 1216; high = 1292; IQR = 244.
Urban: low = 1768; Q_1 = 1968; median = 2231.5; Q_3 = 2674; high = 2910; IQR = 706.

Auto Insurance Premiums for Suburban and Urban Customers (dollars)

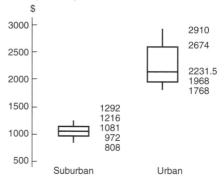

11. (a) Low = 17; Q_1 = 22; median = 24; Q_3 = 27; high = 38; IQR = 5.
(b) 3rd quartile, since it is between the median and Q_3.

Bachelor's Degree Percentage by State

13. (a) California has the lowest premium. (b) Pennsylvania has the highest median premium. (c) California has the smallest range. Texas has the smallest interquartile range. (d) Part a is the five-number summary for Texas. It has the smallest IQR. Part b is the five-number summary for Pennsylvania. It has the largest minimum. Part c is the five-number summary for California. It has the lowest minimum.

15. (a) Smallest median, assistant; highest increase, associate.
  (b) Instructor.
  (c) Assistant.
  (d) Professor; yes.
  (e) Associate, 9.16; instructor, 10.23; yes, the points indicated by asterisks are above the upper limit for each rank.

Chapter 3 Review

1. (a) $\bar{x} = 109.5$; $s = 31.7$; $CV = 28.9\%$; range $= 69$.
  (b) $\bar{x} = 110.125$; $s = 7.2$; $CV = 6.5\%$; range $= 20$.
  (c) The first distribution is more spread than the second.

3. (a) Low $= 31$; Percentage of Democratic Vote by
  $Q_1 = 40$; County
  median $= 45$;
  $Q_3 = 52.5$;
  high $= 68$;
  $IQR = 12.5$.

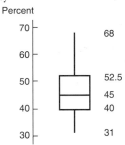

  (b) Class width $= 8$.

Class	Midpoint	f
31–38	34.5	11
39–46	42.5	24
47–54	50.5	15
55–62	58.5	7
63–70	66.5	3

  $\bar{x} \approx 46.1$; $s \approx 8.64$; 28.82 to 63.38.
  (c) $\bar{x} \approx 46.15$; $s \approx 8.63$.

5. Mean weight $= 156.25$ lb.

7. (a) $\bar{x} = 7.83$ lb; $s = 2.32$ lb; $CV = 29.6\%$; range $= 4.8$ lb. (b) $\bar{x} = 9.95$ lb; $s = 0.29$ lb; $CV = 2.9\%$; range $= 0.7$ lb. (c) Second line has more consistent performance as reflected by the smaller standard deviation, CV, and range.

9. (a) Midpoints: 50, 61, 72, 83, 94, 105; $\bar{x} \approx 78.3$; $s^2 \approx 149$; $s \approx 12.2$; $CV \approx 15.6\%$. (b) 53.9 to 102.7.

11. $\Sigma w = 16$, $\Sigma wx = 121$, average $= 7.56$.

13. (a) It is possible for the range and the standard deviation to be the same. For instance, for data values that are all the same, such as 1, 1, 1, 1, 1, the range and standard deviation are both 0. (b) It is possible for the mean, median, and mode to be all the same. For instance, the data set 1, 2, 3, 3, 3, 4, 5 has mean, median, and mode all equal to 3. The averages can all be different, as in the data set 1, 2, 3, 3. In this case, the mean is 2.25, the median is 2.5, and the mode is 3.

CHAPTER 4

Section 4.1

1. See text.
3. b, since 4.1 is greater than 1; d, since -0.5 is less than 0; h, since 150% is greater than 100% or 1.
5. Answers vary. Probability as a relative frequency. One concern is whether the students in the class are more or less adept at wiggling their ears than people in the general population.
7. (a) $P(0) = 15/375$; $P(1) = 71/375$; $P(2) = 124/375$; $P(3) = 131/375$; $P(4) = 34/375$. (b) Yes, the listed number of similar preferences form the sample space.
9. (a) $P(\text{best idea 6 A.M.–12 noon}) = 290/966 = 0.30$; $P(\text{best idea 12 noon–6 P.M.}) = 135/966 = 0.14$; $P(\text{best idea 6 P.M.–12 midnight}) = 319/966 = 0.33$; $P(\text{best idea from 12 midnight–6 A.M.}) = 222/966 = 0.23$. (b) The probabilities add up to 1. They should add up to 1 provided that the intervals do not overlap and each inventor chose only one interval. The sample space is the set of four time intervals.
11. (a) These events form a sample space. Everyone in the survey gave a response that fell into one of the three categories: $P(\text{left alone}) = 770/1000 = 0.77$; $P(\text{waited on}) = 160/1000 = 0.16$; $P(\text{different treatment}) = 70/1000 = 0.07$. (b) $P(\text{not left alone}) = 1 - P(\text{left alone}) = 1 - 0.77 = 0.23$; $P(\text{not waited on}) = 1 - P(\text{waited on}) = 1 - 0.16 = 0.84$.
13. (b) $P(\text{success}) = 2/17$ or 0.118. (c) $P(\text{make shot}) = 3/8$ or 0.375.
15. (a) $P(\text{enter if walks by}) = 58/127 \approx 0.46$. (b) $P(\text{buy if entered}) = 25/58 \approx 0.43$. (c) $P(\text{walk in and buy}) = 25/127 \approx 0.20$. (d) $P(\text{not buy}) = 1 - P(\text{buy}) = 1 - 0.43 = 0.57$.

Section 4.2

1. (a) 20%; yes (b) 40%; yes (c) $100\% - 30\% = 70\%$
3. (a) 20% (no orange) (b) 40% (c) $100\% - 20\% = 80\%$
5. (a) Yes (b) $P(5 \text{ on green } and \text{ 3 on red}) = P(5) \cdot P(3) = (1/6)(1/6) = 1/36 \approx 0.028$. (c) $P(3 \text{ on green } and \text{ 5 on}$

red) $= P(3) \cdot P(5) = (1/6)(1/6) = 1/36 \approx 0.028.$ (d)
$P((5$ on green *and* 3 on red) *or* (3 on green *and* 5 on
red)) $= (1/36) + (1/36) = 1/18 \approx 0.056.$

7. (a) $P(\text{sum of } 6) = P(1 \text{ and } 5) + P(2 \text{ and } 4) +$
$P(3 \text{ and } 3) + P(4 \text{ and } 2) + P(5 \text{ and } 1) = (1/36) + (1/36)$
$+ (1/36) + (1/36) + (1/36) = 5/36.$ (b) $P(\text{sum of } 4) =$
$P(1 \text{ and } 3) + P(2 \text{ and } 2) + P(3 \text{ and } 1) = (1/36) +$
$(1/36) + (1/36) = 3/36$ or $1/12.$ (c) $P(\text{sum of } 6 \text{ or sum}$
of 4) $= P(\text{sum of } 6) + P(\text{sum of } 4) = (5/36) + (3/36) =$
$8/36$ or $2/9$; yes.

9. (a) No, after the first draw the sample space becomes
smaller and probabilities for events on the second draw
change.   (b) $P(\text{ace on 1st } and \text{ king on 2nd}) = P(\text{ace}) \cdot$
$P(\text{king, } given \text{ ace}) = (4/52)(4/51) = 4/663.$ (c) $P(\text{king}$
on 1st *and* ace on 2nd) $= P(\text{king}) \cdot P(\text{ace, } given \text{ king}) =$
$(4/52)(4/51) = 4/663.$ (d) $P(\text{ace and king in either or-}$
der) $= P(\text{ace on 1st } and \text{ king on 2nd}) + P(\text{king on 1st}$
and ace on 2nd) $= (4/663) + (4/663) = 8/663.$

11. (a) Yes, replacement of the card restores the sample space
and all probabilities for the second draw remain unchanged regardless of the outcome of the first card.
(b) $P(\text{ace on 1st } and \text{ king on 2nd}) = P(\text{ace}) \cdot P(\text{king}) =$
$(4/52)(4/52) = 1/169.$ (c) $P(\text{king on 1st } and \text{ ace on}$
2nd) $= P(\text{king}) \cdot P(\text{ace}) = (4/52)(4/52) = 1/169.$
(d) $P(\text{ace and king in either order}) = P(\text{ace on 1st } and$
king on 2nd) $+ P(\text{king on 1st } and \text{ ace on 2nd}) = (1/169)$
$+ (1/169) = 2/169.$

13. (a) $P(6 \text{ years } or \text{ older}) = P(6{-}9) + P(10{-}12) +$
$P(13 \text{ and over}) = 0.27 + 0.14 + 0.22 = 0.63.$ (b)
$P(12 \text{ years } or \text{ younger}) = P(2 \text{ and under}) + P(3{-}5) +$
$P(6{-}9) + P(10{-}12) = 0.15 + 0.22 + 0.27 + 0.14 =$
$0.78.$ (c) $P(\text{between 6 and 12}) = P(6{-}9) +$
$P(10{-}12) = 0.27 + 0.14 = 0.41.$ (d) $P(\text{between 3 and}$
9) $= P(3{-}5) + P(6{-}9) = 0.22 + 0.27 = 0.49.$ The category 13 and over contains far more ages than the group
10–12. It is not surprising that more toys are purchased
for this group, since there are more children in this
group.

15. The information from James Burke can be viewed as
conditional probabilities. $P(\text{report lie, } given \text{ person is ly-}$
ing) $= 0.72$ and $P(\text{report lie, } given \text{ person is not lying}) =$
$0.07.$ (a) $P(\text{person is not lying}) = 0.90; P(\text{person is not}$
lying *and* polygraph reports lie) $= P(\text{person is not lying})$
$\times P(\text{reports lie, } given \text{ person not lying}) = (0.90)(0.07) =$
0.063 or $6.3\%.$ (b) $P(\text{person is lying}) = 0.10; P(\text{person}$
is lying *and* polygraph reports lie) $= P(\text{person is lying}) \times$
$P(\text{reports lie, } given \text{ person is lying}) = (0.10)(0.72) =$
0.072 or $7.2\%.$ (c) $P(\text{person is not lying}) = 0.5; P(\text{per-}$
son is lying) $= 0.5; P(\text{person is not lying } and \text{ polygraph}$
reports lie) $= P(\text{person is not lying}) \times P(\text{reports lie, } given$
person not lying) $= (0.50)(0.07) = 0.035$ or $3.5\%.$
$P(\text{person is lying } and \text{ polygraph reports lie}) = P(\text{person}$
is lying) $\times P(\text{reports lie, } given \text{ person is lying}) =$
$(0.50)(0.72) = 0.36$ or $36\%.$ (d) $P(\text{person is not lying})$
$= 0.15; P(\text{person is lying}) = 0.85; P(\text{person is not lying}$
and polygraph reports lie) $= P(\text{person is not lying}) \times$

$P(\text{reports lie, } given \text{ person is not lying}) = (0.15)(0.07) =$
0.0105 or $1.05\%.$ $P(\text{person is lying } and \text{ polygraph re-}$
ports lie) $= P(\text{person is lying}) \times P(\text{reports lie, } given \text{ per-}$
son is lying) $= (0.85)(0.72) = 0.612$ or $61.2\%.$

17. (a) $P(\text{glasses } and \text{ woman}) = P(\text{glasses}) \times P(\text{woman,}$
given glasses) $= (0.56)(0.554) = 0.310.$
(b) $P(\text{glasses } and \text{ man}) = P(\text{glasses}) \times P(\text{man, } given$
glasses) $= (0.56)(0.446) = 0.250.$ (c) $P(\text{contacts } and$
woman) $= P(\text{contacts}) \times P(\text{woman, } given \text{ contacts}) =$
$(0.036)(0.631) = 0.023.$ (d) $P(\text{contacts } and \text{ man}) =$
$P(\text{contacts}) \times P(\text{man, } given \text{ contacts}) = (0.036)(0.369) =$
$0.013.$ (e) Since the preceding events are all disjoint,
$P(\text{any one of the above}) = 0.310 + 0.250 + 0.023 +$
$0.013 = 0.596; P(\text{none}) = 1 - 0.596 = 0.404.$

19. (a) 686/1160; 270/580; 416/580.   (b) No.
(c) 270/1160; 416/1160.   (d) 474/1160; 310/580.
(e) No.   (f) $686/1160 + 580/1160 - 270/1160 =$
996/1160.

21. (a) 72/154.   (b) 82/154.   (c) 79/116.   (d) 37/116.
(e) 72/270.   (f) 82/270.

23. (a) 932/1894.   (b) 353/739.   (c) 142/1894.
(d) 22/224.   (e) 1007/1894.   (f) 39/66.   (g) No; Probabilities computed in parts (a) and (b) are not equal. (Note:
more than once is two or more.)

25. (a) $P(A) = 0.65.$ (b) $P(B) = 0.71.$ (c) $P(B, given$
$A) = 0.87.$ (d) $P(A \text{ and } B) = P(A) \cdot P(B, given \text{ } A) =$
$(0.65)(0.87) \approx 0.57.$ (e) $P(A \text{ or } B) = P(A) + P(B) -$
$P(A \text{ and } B) = 0.65 + 0.71 - 0.57 = 0.79.$ (f) $P(\text{not}$
close) $= P(\text{profit 1st year } or \text{ profit 2nd year}) = P(A \text{ or}$
$B) = 0.79; P(\text{close}) = 1 - P(\text{not close}) = 1 - 0.79 =$
$0.21.$

27. (a) $P(\text{TB } and \text{ positive}) = P(\text{TB})(\text{positive, } given \text{ TB}) =$
$(0.04)(0.82) = 0.033.$ (b) $P(\text{does not have TB}) = 1 -$
$P(\text{TB}) = 1 - 0.04 = 0.96.$ (c) $P(\text{no TB } and \text{ positive}) =$
$P(\text{no TB})P(\text{positive, } given \text{ no TB}) = (0.96)(0.09) =$
$0.086.$

Section 4.3

1. (a) Outcomes for Tossing a Coin Three Times

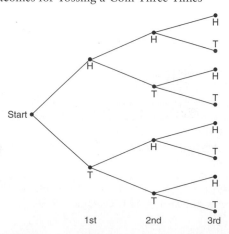

(b) 3   (c) 3/8

3. (a) Outcomes for Drawing Two Balls (without replacement)

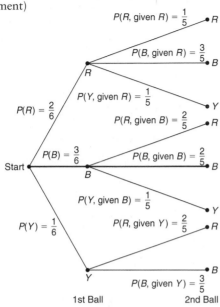

1st Ball   2nd Ball

(b) $P(R$ and $R) = 2/6 \cdot 1/5 = 1/15$.
$P(R$ 1st and B 2nd$) = 2/6 \cdot 3/5 = 1/5$.
$P(R$ 1st and Y 2nd$) = 2/6 \cdot 1/5 = 1/15$.
$P(B$ 1st and R 2nd$) = 3/6 \cdot 2/5 = 1/5$.
$P(B$ 1st and B 2nd$) = 3/6 \cdot 2/5 = 1/5$.
$P(B$ 1st and Y 2nd$) = 3/6 \cdot 1/5 = 1/10$.
$P(Y$ 1st and R 2nd$) = 1/6 \cdot 2/5 = 1/15$.
$P(Y$ 1st and B 2nd$) = 1/6 \cdot 3/5 = 1/10$.

5. (a) Choices for Three True/False Questions

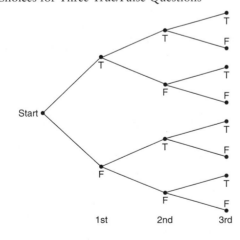

1st   2nd   3rd

(b) 1/8.

7. $4 \cdot 3 \cdot 2 \cdot 1 = 24$ ways.

9. (a) $52 \cdot 52 = 2704$ (b) $4 \cdot 4 = 16$.
   (c) $16/2704 = 0.006$.

11. $4 \cdot 3 \cdot 3 = 36$.

13. $P_{5,2} = (5!/3!) = 5 \cdot 4 = 20$.

15. $P_{7,7} = (7!/0!) = 7! = 5040$.

17. $C_{5,2} = (5!/(2!3!)) = 10$.

19. $C_{7,7} = (7!/(7!0!)) = 1$.

21. $P_{15,3} = 2730$.

23. (a) $8! = 40,320$.   (b) $8 \cdot 7 \cdot 6 \cdot 5 \cdot 4 = 6720$.

25. $5 \cdot 4 \cdot 3 = 60$.

27. $C_{15,5} = (15!/(5!10!)) = 3003$.

29. (a) $C_{12,6} = (12!/(6!6!)) = 924$.
   (b) $C_{7,6} = (7!/(6!1!)) = 7$.   (c) $7/924 = 0.008$.

Chapter 4 Review

1. $P(\text{asked}) = 24\%$; $P(\text{received, } given \text{ asked}) = 45\%$;
$P(\text{ask } and \text{ receive}) = (0.24)(0.45) = 10.8\%$.

3. (a) If the first card is replaced, the outcomes are independent. Replacing the first card restores the original sample space. If the first card is not replaced, the outcomes are not independent, because removing the first card changes the sample space.
(b) $P(\text{heart } and \text{ heart}) = (13/52)(13/52) = 0.063$.
(c) $P(\text{heart } and \text{ heart}) = (13/52)(12/51) = 0.059$.

5. (a) Drop a fixed number of tacks and count how many land flat side down. Then form the ratio of the number landing flat side down to the total number dropped.
(b) Up, down.   (c) $P(\text{up}) = 160/500 = 0.32$; $P(\text{down}) = 340/500 = 0.68$.

7. (a)

Outcomes x	2	3	4	5	6
$P(x)$	0.028	0.056	0.083	0.111	0.139

(b)

x	7	8	9	10	11	12
$P(x)$	0.167	0.139	0.111	0.083	0.056	0.028

9. $C_{8,2} = (8!/(2!6!)) = (8 \cdot 7/2) = 28$.

11. $3 \cdot 2 \cdot 1 = 6$.

13. $4 \cdot 4 \cdot 4 \cdot 4 \cdot 4 = 1024$ choices; $P(\text{all correct}) = 1/1024 = 0.00098$.

15. $10 \cdot 10 \cdot 10 = 1000$.

CHAPTER 5

Section 5.1

1. (a) Discrete.   (b) Continuous.   (c) Continuous.
   (d) Discrete.   (e) Continuous.

3. (a) Yes.   (b) No; probabilities total to more than 1.

5. (a) Yes, events are distinct and probabilities total 1.
   (b) Income Distribution ($1000).

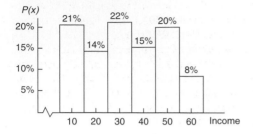

   (c) 32.3 thousand dollars.   (d) 16.12 thousand dollars.
7. (a)

x	36	37	38	39	40	41
$P(x)$	0.029	0.048	0.053	0.096	0.125	0.154

x	42	43	44	45
$P(x)$	0.163	0.135	0.120	0.077

   (b)

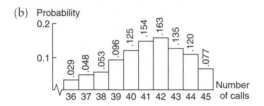

   (c) 0.673.   (d) 0.351.   (e) 41.288.   (f) 2.326.
9. (a) Number of Fish Caught in a 6-Hour Period at
   Pyramid Lake, Nevada

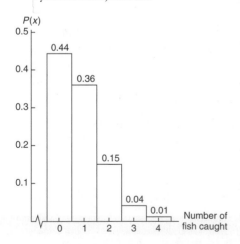

   (b) 0.56.   (c) 0.20.   (d) 0.82.   (e) 0.899.

11. (a) 15/719; 704/719.   (b) $0.73; $14.27.
13. (a) 0.01191; $595.50.   (b) $646; $698; $751.50;
    $806.50; $3497.50 total.   (c) $4197.50.
    (d) $1502.50.
15. (a) $\mu_W = 15$; $\sigma_W^2 = 208$; $\sigma_W \approx 14.4$.
    (b) $\mu_W = 107.5$; $\sigma_W^2 = 52$; $\sigma_W \approx 7.2$.
    (c) $\mu_L = 90$; $\sigma_L^2 = 92.16$; $\sigma_L = 9.6$.
    (d) $\mu_L = 90$; $\sigma_L^2 = 57.76$; $\sigma_L = 7.6$.
17. (a) $\mu_W = 50.2$; $\sigma_W^2 = 66.125$; $\sigma_W \approx 8.13$.
    (b) The means are the same.   (c) The standard devia-
    tion for two policies is smaller.   (d) As we include more
    policies, the coefficients in W decrease, resulting in
    smaller σ_W^2 and σ_W. For instance, for 3 policies, $W = (\mu_1$
    $+ \mu_2 + \mu_3)/3 \approx 0.33\mu_1 + 0.33\mu_2 + 0.33\mu_3$ and $\sigma_W^2 \approx$
    $(0.33)^2\sigma_1^2 + (0.33)^2\sigma_2^2 + (0.33)^2\sigma_3^2$. Yes, the risk appears
    to decrease by a factor of $1/\sqrt{n}$.

Section 5.2

1. A trial is one flip of a fair quarter. Success = coin shows
   heads. Failure = coin shows tails. $n = 3$; $p = 0.5$; $q =$
   0.5.   (a) $P(r = 3 \text{ heads}) = C_{3,3}p^3q^0 = 1(0.5)^3(0.5)^0 =$
   0.125. To find this value in Table 3 of Appendix II, use
   the group in which $n = 3$, the column headed by $p =$
   0.5, and the row headed by $r = 3$.   (b) $P(r = 2$
   heads$) = C_{3,2}p^2q^1 = 3(0.5)^2(0.5)^1 = 0.375$. To find this
   value in Table 3 of Appendix II, use the group in which
   $n = 3$, the column headed by $p = 0.5$, and the row
   headed by $r = 2$.   (c) $P(r \text{ is } 2 \text{ or more}) = P(r = 2$
   heads$) + P(r = 3 \text{ heads}) = 0.375 + 0.125 = 0.500$.
   (d) The probability of getting three tails when you toss a
   coin three times is the same as getting no or zero heads.
   Therefore, $P(3 \text{ tails}) = P(r = 0 \text{ heads}) = C_{3,0}p^0q^3 =$
   $1(0.5)^0(0.5)^3 = 0.125$. To find this value in Table 3 of
   Appendix II, use the group in which $n = 3$, the column
   headed by $p = 0.5$, and the row headed by $r = 0$.
3. (a) A trial is a man's response to the question,
   "Would you marry the same woman again?" Success = a
   positive response. Failure = a negative response. $n = 10$;
   $p = 0.80$; $q = 0.20$. Using values in Table 3 of Appendix
   II, $P(r \text{ is at least } 7) = P(r = 7) + P(r = 8) + P(r = 9) +$
   $P(r = 10) = 0.201 + 0.302 + 0.268 + 0.107 = 0.878$.
   $P(r \text{ is less than half of } 10) = P(r < 5) = P(r = 0) +$
   $P(r = 1) + P(r = 2) + P(r = 3) + P(r = 4) = 0.000 +$
   $0.000 + 0.000 + 0.001 + 0.006 = 0.007$.   (b) A trial
   is a woman's response to the question, "Would you
   marry the same man again?" Success = a positive response. Failure = a negative response. $n = 10$; $p = 0.5$;
   $q = 0.5$. Using values in Table 3 of Appendix II, $P(r \text{ is at}$
   least 7$) = P(r = 7) + P(r = 8) + P(r = 9) + P(r = 10) =$
   $0.117 + 0.044 + 0.010 + 0.001 = 0.172$. $P(r \text{ is less}$
   than half of 10$) = P(r < 5) = P(r = 0) + P(r = 1) +$
   $P(r = 2) + P(r = 3) + P(r = 4) = 0.001 + 0.010 +$
   $0.044 + 0.117 + 0.205 = 0.377$.

5. A trial consists of a woman's response regarding her mother-in-law. Success = dislike. Failure = like. $n = 6$; $p = 0.90$; $q = 0.10$. (a) $P(r = 6) = 0.531$. (b) $P(r = 0) = 0.000$ (to 3 digits). (c) $P(r \geq 4) = P(r = 4) + P(r = 5) + P(r = 6) = 0.098 + 0.354 + 0.531 = 0.983$. (d) $P(r \leq 3) = 1 - P(r \geq 4) \approx 1 - 0.983 = 0.017$ or 0.016 directly from table.

7. A trial is taking a polygraph exam. Success = pass. Failure = fail. $n = 9$; $p = 0.85$; $q = 0.15$. (a) $P(r = 9) = 0.232$. (b) $P(r \geq 5) = P(r = 5) + P(r = 6) + P(r = 7) + P(r = 8) + P(r = 9) = 0.028 + 0.107 + 0.260 + 0.368 + 0.232 = 0.995$. (c) $P(r \leq 4) = 1 - P(r \geq 5) \approx 1 - 0.995 = 0.005$ or 0.006 directly from table. (d) $P(r = 0) = 0.000$ (to 3 digits).

9. A trial consists of checking the gross receipts of the Green Parrot Italian Restaurant for one business day. Success = gross is over \$2200. Failure is gross is at or below \$2200. $p = 0.85$; $q = 0.15$. (a) $n = 7$; $P(r \geq 5) = P(r = 5) + P(r = 6) + P(r = 7) = 0.210 + 0.396 + 0.321 = 0.927$. (b) $n = 10$; $P(r \geq 5) = P(r = 5) + P(r = 6) + P(r = 7) + P(r = 8) + P(r = 9) + P(r = 10) = 0.008 + 0.040 + 0.130 + 0.276 + 0.347 + 0.197 = 0.998$. (c) $n = 5$; $P(r < 3) = P(r = 0) + P(r = 1) + P(r = 2) = 0.000 + 0.002 + 0.024 = 0.026$. (d) $n = 10$; $P(r < 7) = P(r = 0) + P(r = 1) + P(r = 2) + P(r = 3) + P(r = 4) + P(r = 5) + P(r = 6) = 0.000 + 0.000 + 0.000 + 0.000 + 0.001 + 0.008 + 0.040 = 0.049$. (e) $n = 7$; $P(r < 3) = P(r = 0) + P(r = 1) + P(r = 2) = 0.000 + 0.000 + 0.001 = 0.001$. Yes. If p were really 0.85, then the event of a 7-day period with gross income exceeding \$2200 fewer than 3 days would be very rare. If it happened again, we would suspect that $p = 0.85$ is too high.

11. A trial is catching and releasing a pike. Success = pike dies. Failure = pike lives. $n = 16$; $p = 0.05$; $q = 0.95$. (a) $P(r = 0) = 0.440$. (b) $P(r < 3) = 0.957$. (c) $P(r = 0) = 0.440$ (all live is equivalent to none die). (d) Change success to live; $p = 0.95$; $P(r > 14) = 0.811$.

13. (a) A trial consists of using the Myers-Briggs instrument to determine if a person in marketing is an extrovert. Success = extrovert. Failure = not extrovert. $n = 15$; $p = 0.75$; $q = 0.25$. $P(r \geq 10) = 0.851$; $P(r \geq 5) = 0.999$; $P(r = 15) = 0.013$. (b) A trial consists of using the Myers-Briggs instrument to determine if a computer programmer is an introvert. Success = introvert. Failure = not introvert. $n = 5$; $p = 0.60$; $q = 0.40$. $P(r = 0) = 0.010$; $P(r \geq 3) = 0.683$; $P(r = 5) = 0.078$.

15. (a) $n = 10$; $p = 0.40$; $P(r = 0) = 0.006$. (b) $n = 10$; $p = 0.40$; $P(r < 5) = 0.633$. (c) $n = 10$; $p = 0.30$; $P(r \leq 2) = 0.382$. (d) $n = 10$; $p = 0.70$; $P(r \geq 6) = 0.849$.

17. $n = 8$; $p = 0.53$; $q = 0.47$. (a) 0.812515; yes, truncated at 5 digits. (b) 0.187486; 0.18749; yes, rounded to 5 digits.

19. A trial consists of determining the kind of stone in a chipped stone tool. (a) $n = 11$. Success = obsidian. Failure = not obsidian. $p = 0.15$; $q = 0.85$; $P(r \geq 3) = 0.221$. (b) $n = 5$. Success = basalt. Failure = not basalt. $p = 0.55$; $q = 0.45$; $P(r \geq 2) = 0.869$. (c) $n = 10$. Success = neither obsidian nor basalt. Failure = either obsidian or basalt. The two outcomes, tool is obsidian or tool is basalt, are mutually exclusive. Therefore, $P(\text{obsidian } or \text{ basalt}) = 0.55 + 0.15 = 0.70$. $P(\text{neither obsidian nor basalt}) = 1 - 0.70 = 0.30$. Therefore, $p = 0.30$; $P(r \geq 4) = 0.350$.

21. (a) They are the same. (b) They are the same. (c) $r = 1$. (d) The one headed by $p = 0.80$.

Section 5.3

1. (a) Binomial Distribution
The distribution is symmetrical.

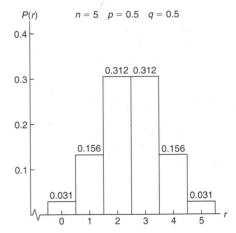

(b) Binomial Distribution
The distribution is skewed right.

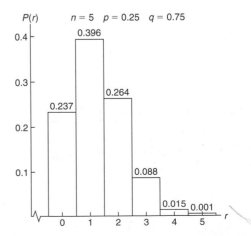

(c) Binomial Distribution
The distribution is skewed left.

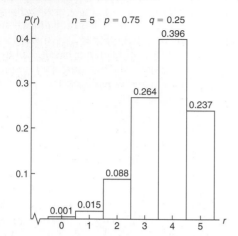

(d) The distributions are mirror images of one another.
(e) The distribution would be skewed left for $p = 0.73$ because the more likely number of successes are to the right of the middle.

3. (a) Households with Children Under 2 That Buy Film

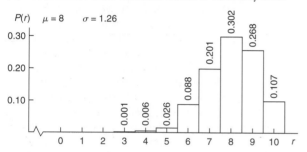

(b) Households with No Children Under 21 That Buy Film

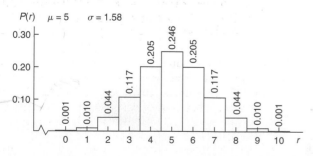

(c) Yes. Adults with children seem to buy more film.

5. (a) Binomial Distribution for Number of Addresses Found

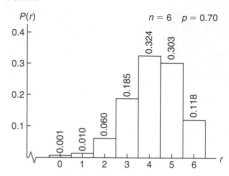

(b) $\mu = 4.2$; $\sigma = 1.122$.
(c) $n = 5$. Note that $n = 5$ gives $P(r \geq 2) = 0.97$.

7. (a) Binomial Distribution for Number of Illiterate People

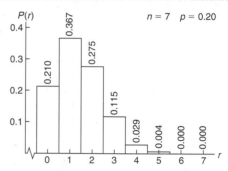

(b) $\mu = 1.4$; $\sigma = 1.058$.
(c) $n = 12$. Note that $n = 12$ gives $P(r \geq 7) = 0.98$, where success = literate and $p = 0.80$.

9. (a) Binomial Distribution for Number of Gullible Consumers

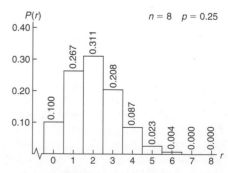

(b) $\mu = 2$; $\sigma = 1.225$.
(c) $n = 16$. Note that $n = 16$ gives $P(r \geq 1) = 0.99$.

11. $n = 20$ gives $P(r \geq 5) = 0.95$.

13. $P(r \geq 1) = 1 - P(r = 0)$. From the formula for the binomial distribution with $p = 0.1$ and $q = 0.9$, $P(r = 0) = C_{n,0}p^0 q^n = 1(0.1)^0(0.9)^n$. Therefore, $P(r \geq 1) = 1 - 0.9^n$. Computing this probability for various values of n shows that $n = 22$ is the smallest value for which $P(r \geq 1)$ is at least 0.90.

15. (a) $P(r = 0) = 0.004$; $P(r = 1) = 0.047$; $P(r = 2) = 0.211$; $P(r = 3) = 0.422$; $P(r = 4) = 0.316$.
    (b) Binomial Distribution for Number of Parolees Who Do Not Become Repeat Offenders

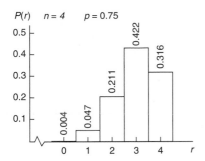

(c) $\mu = 3$; $\sigma = 0.866$.
    (d) $n = 7$. Note that $n = 7$ gives $P(r \geq 3) = 0.987$.
17. $n = 12$; $p = 0.25$ not serve; $q = 0.75$ serve   (a) $P(r = 12 \text{ serve}) = 0.032$.   (b) $P(r \geq 6 \text{ not serve}) = 0.053$.
    (c) For serving, $\mu = 9$; $\sigma \approx 1.50$.   (d) To be at least 95.9% that 12 are available to serve, call 20.
19. $n = 6$; $p = 0.80$ not solve; $q = 0.20$ solve   (a) $P(r = 6 \text{ not solved}) = 0.262$.   (b) $P(r \geq 1 \text{ solved}) = 0.738$.
    (c) For solving crime, $\mu = 1.2$; $\sigma \approx 0.98$.   (d) To be 90% sure of solving 1 or more crimes, investigate $n = 11$ crimes.
21. (a) $P(r = 7 \text{ guilty in U.S.}) = 0.028$; $P(r = 7 \text{ guilty in Japan}) = 0.698$.   (b) For guilty in Japan, $\mu = 6.65$, $\sigma \approx 0.58$; For guilty in U.S., $\mu = 4.2$; $\sigma \approx 1.30$.   (c) To be 99% sure of at least 2 guilty convictions in the U.S., look at $n = 8$ trials. To be 99% sure of at least 2 guilty convictions in Japan, look at $n = 3$ trials.
23. (a) 9.   (b) 10.

Section 5.4

1. (a) $p = 0.77$; $P(n) = (0.77)(0.23)^{n-1}$.   (b) $P(1) = 0.77$.   (c) $P(2) = 0.1771$.   (d) $P(3 \text{ or more tries}) = 1 - P(1) - P(2) = 0.0529$.   (e) 1.29 or 1.
3. (a) $P(n) = (0.05)(0.95)^{n-1}$.   (b) $P(5) = 0.0407$.
    (c) $P(10) = 0.0315$.   (d) $P(n > 3) = 1 - P(1) - P(2) - P(3) = 1 - 0.05 - 0.0475 - 0.0451 = 0.8574$.   (e) 20.
5. (a) $P(n) = (0.71)(0.29)^{n-1}$.   (b) $P(1) = 0.71$; $P(2) = 0.2059$; $P(n \geq 3) = 1 - P(1) - P(2) = 0.0841$.   (c) $P(n) = (0.83)(0.17)^{n-1}$; $P(1) = 0.83$; $P(2) = 0.1411$; $P(n \geq 3) = 1 - P(1) - P(2) = 0.0289$.
7. (a) $P(n) = (0.30)(0.70)^{n-1}$.   (b) $P(3) = 0.147$.
    (c) $P(n > 3) = 1 - P(1) - P(2) - P(3) = 1 - 0.300 - 0.210 - 0.147 = 0.343$.   (d) 3.33 or 3.
9. (a) $\lambda = (1.7/10) \times (3/3) = 5.1$ per 30-minute interval; $P(r) = e^{-5.1}(5.1)^r/r!$.   (b) Using Table 4 in Appendix II with $\lambda = 5.1$, we find $P(4) = 0.1719$; $P(5) = 0.1753$;

$P(6) = 0.1490$.   (c) $P(r \geq 4) = 1 - P(0) - P(1) - P(2) - P(3) = 1 - 0.0061 - 0.0311 - 0.0793 - 0.1348 = 0.7487$.   (d) $P(r < 4) = 1 - P(r \geq 4) = 1 - 0.7487 = 0.2513$.
11. (a) Births and deaths occur somewhat rarely in a group of 1000 people in a given year. For 1000 people, $\lambda = 16$ births; $\lambda = 8$ deaths.   (b) By Table 4 in Appendix II, $P(10 \text{ births}) = 0.0341$; $P(10 \text{ deaths}) = 0.0993$; $P(16 \text{ births}) = 0.0992$; $P(16 \text{ deaths}) = 0.0045$.
    (c) $\lambda(\text{births}) = (16/1000) \times (1500/1500) = 24$ per 1500 people. $\lambda(\text{deaths}) = (8/1000) \times (1500/1500) = 12$ per 1500 people. By the table, $P(10 \text{ deaths}) = 0.1048$; $P(16 \text{ deaths}) = 0.0543$. Since $\lambda = 24$ is not in the table, use the formula for $P(r)$ to find $P(10 \text{ births}) = 0.00066$; $P(16 \text{ births}) = 0.02186$.   (d) $\lambda(\text{births}) = (16/1000) \times (750/750) = 12$ per 750 people. $\lambda(\text{deaths}) = (8/1000) \times (750/750) = 6$ per 750 people. By Table 4 of Appendix II, $P(10 \text{ births}) = 0.1048$; $P(10 \text{ deaths}) = 0.0413$; $P(16 \text{ births}) = 0.0543$; $P(16 \text{ deaths}) = 0.0003$.
13. (a) The Poisson distribution is a good choice for r because gale-force winds occur rather rarely. The occurrences are usually independent.   (b) Interval of 108 hours $\lambda = (1/60) \times (108/108) = 1.8$ per 108 hours. Using Table 4 of Appendix II, we find that $P(2) = 0.2678$; $P(3) = 0.1607$; $P(4) = 0.0723$; $P(r < 2) = P(0) + P(1) = 0.1653 + 0.2975 = 0.4628$.   (c) Interval of 180 hours $\lambda = (1/60) \times (180/180) = 3$ per 180 hours. Table 4 of Appendix II gives $P(3) = 0.2240$; $P(4) = 0.1680$; $P(5) = 0.1008$; $P(r < 3) = P(0) + P(1) + P(2) = 0.0498 + 0.1494 + 0.2240 = 0.4232$.
15. (a) The sales of large buildings are rare events. It is reasonable to assume that they are independent. The variable r = number of sales in a fixed time interval.
    (b) For a 60-day period, $\lambda = (8/275) \times (60/60) = 1.7$ per 60 days. By Table 4 of Appendix II, $P(0) = 0.1827$; $P(1) = 0.3106$; $P(r \geq 2) = 1 - P(0) - P(1) = 0.5067$.
    (c) For a 90-day period, $\lambda = (8/275) \times (90/90) = 2.6$ per 90 days. By Table 4 of Appendix II, $P(0) = 0.0743$; $P(2) = 0.2510$; $P(r \geq 3) = 1 - P(0) - P(1) - P(2) = 1 - 0.0743 - 0.1931 - 0.2510 = 0.4816$.
17. (a) The problem satisfies the conditions for a binomial experiment with small $p = 0.0018$ and large $n = 1000$. $np = 1.8$, which is less than 10, so the Poisson approximation to the binomial distribution would be a good choice. $\lambda = np = 1.8$.   (b) By Table 4, Appendix II, $P(0) = 0.1653$.   (c) $P(r > 1) = 1 - P(0) - P(1) = 1 - 0.1653 - 0.2975 = 0.5372$.   (d) $P(r > 2) = 1 - P(0) - P(1) - P(2) = 1 - 0.1653 - 0.2975 - 0.2678 = 0.2694$.   (e) $P(r > 3) = 1 - P(0) - P(1) - P(2) - P(3) = 1 - 0.1653 - 0.2975 - 0.2678 - 0.1607 = 0.1087$.
19. (a) The problem satisfies the conditions for a binomial experiment with n large, $n = 175$, and p small. $np = (175)(0.005) = 0.875 < 10$. The Poisson distribution

would be a good approximation to the binomial. $n = 175$; $p = 0.005$; $\lambda = np = 0.9$.   (b) By Table 4 of Appendix II, $P(0) = 0.4066$.   (c) $P(r \geq 1) = 1 - P(0) = 0.5934$.   (d) $P(r \geq 2) = 1 - P(0) - P(1) = 0.2275$.

21. (a) $n = 100$; $p = 0.02$, $r = 2$; $P(2) = C_{100,2}(0.02)^2$ $(0.98)^{98} = 0.2734$.   (b) $\lambda = np = 2$; $P(2) = [e^{-2}(2)^2]/2! = 0.2707$.   (c) The approximation is correct to two decimal places.   (d) $n = 100$; $p = 0.02$; $r = 3$. By the formula for the binomial distribution, $P(3) = 0.1823$. By the Poisson approximation, $P(3) = 0.1804$. The approximation is correct to two decimal places.

Chapter 5 Review

1. (a) 38; 11.6
   (b) Leases in Months

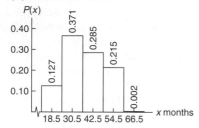

3. (a) Claimants Under 25

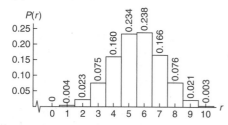

   (b) $P(r \geq 6) = 0.504$.   (c) $\mu = 5.5$; $\sigma = 1.57$.
5. (a) 0.039.   (b) 0.403.   (c) 8.
7. (a) Number of Good Grapefruit

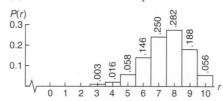

   (b) 0.244, 0.999.   (c) 7.5.   (d) 1.37.
9. $P(r \leq 2) = 0.000$ (to 3 digits). The data seem to indicate that the percent favoring the increase in fees is less than 85%.
11. (a) Coughs are a relatively rare occurrence. It is reasonable to assume that they are independent events, and the variable is the number of coughs in a fixed time interval. (b) $\lambda = 11$ coughs per minute; $P(r \leq 3) = P(0) + P(1) + P(2) + P(3) = 0.000 + 0.002 + 0.0010 + 0.0037 =$

0.0049.   (c) $\lambda = (11/1) \times (0.5/0.5) = 5.5$ coughs per 30-second period. $P(r \geq 3) = 1 - P(0) - P(1) - P(2) = 1 - 0.0041 - 0.0225 - 0.0618 = 0.9116$.

13. The loan-default problem satisfies the conditions for a binomial experiment. Moreover, p is small, n is large, and $np < 10$. Use of the Poisson approximation to the binomial distribution is appropriate. $n = 300$; $p = 1/350 = 0.0029$, and $\lambda = np = 300(0.0029) = 0.86 \approx 0.9$; $P(r \geq 2) = 1 - P(0) - P(1) = 1 - 0.4066 - 0.3659 = 0.2275$.

15. (a) Use the geometric distribution with $p = 0.5$. $P(n = 2) = (0.5)(0.5) = 0.25$. As long as you toss the coin at least twice, it does not matter how many more times you toss it. To get the first head on the second toss, you must get a tail on the first and a head on the second.
   (b) $P(n = 4) = (0.5)(0.5)^3 = 0.0625$; $P(n > 4) = 1 - P(1) - P(2) - P(3) - P(4) = 1 - 0.5 - 0.5^2 - 0.5^3 - 0.5^4 = 0.0625$.

CHAPTER 6

Section 6.1

1. (a) No, it's skewed.   (b) No, it crosses the horizontal axis.   (c) No, it has three peaks.   (d) No, the curve is not smooth.
3. Figure 6-16 has the larger standard deviation. The mean of Figure 6-16 is $\mu = 10$. The mean of Figure 6-17 is $\mu = 4$.
5. (a) 50%.   (b) 68%.   (c) 99.7%.
7. (a) 50%.   (b) 50%.   (c) 68%.   (d) 95%.
9. (a) From 1207 to 1279.   (b) From 1171 to 1315.
   (c) From 1135 to 1351.
11. (a) From 1.70 mA to 4.60 mA.   (b) From 0.25 mA to 6.05 mA.
13. (a) Tri-County Bank Monthly Loan Request—First Year (thousands of dollars)

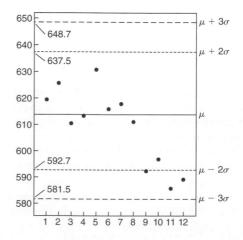

The process is out of control with type III warning signal, since 2 of 3 consecutive points are more than 2 standard deviations below the mean. The trend is down.
(b) Tri-County Bank Monthly Loan Requests—Second Year (thousands of dollars).

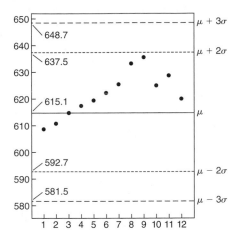

The process shows warning signal II, a run of nine consecutive points above the mean. It would say the economy is heating up.
15. Visibility Standard Index

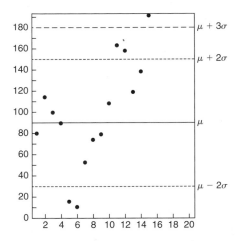

There is one point above 3σ. Thus control signal I indicates "out of control." Control signal III is present. There are two consecutive points below $\mu - 2\sigma$ and two consecutive points above $\mu + 2\sigma$. The out-of-control signals that cause the most concern are those above the mean. Special pollution regulations may be appropriate for those periods.

Section 6.2

1. (a) Robert, Jan, and Linda each scored above the mean.
(b) Joel scored on the mean. (c) Susan and John scored below the mean. (d) Robert, 172; Jan, 184; Susan, 110; Joel, 150; John, 134; Linda, 182.

3. (a) $-4.00 < z < 4.00$. (b) $z < -1.6$. (c) $1.00 < z$.
(d) $81.75°F < x$. (e) $x < 63.5°F$.
(f) $64°F < x < 81.25°F$.
5. (a) $-1.77 < z$. (b) $z < 1.61$. (c) $-1.45 < z < 1.45$.
(d) $3706 < x < 5907$. (e) $x < 5615$. (f) $6000 < x$.
(g) A population of 2800 deer corresponds to a z value of -2.58. Data values this far below the mean occur less than 2.5% of the time. This would be an unusually low number. The population 6300 corresponds to a z value of 3.06. Fall deer populations are practically never that large. Such a population would be considered an unusually high population.
7. (a) $-1.00 < z$. (b) $z < -2.00$.
(c) $-2.67 < z < 2.33$. (d) $x < 4.4$. (e) $5.2 < x$.
(f) $4.1 < x < 4.5$.
(g) A red blood cell count of 5.9 or higher corresponds to a standard z score of 3.67. Practically no data values occur this far above the mean. Such a count would be considered unusually high for a healthy female.
9. 0.5000. 11. 0.0934. 13. 0.6736. 15. 0.0643.
17. 0.8888. 19. 0.4993. 21. 0.4778. 23. 0.8953.
25. 0.3471. 27. 0.0306. 29. 0.5000. 31. 0.4483.
33. 0.8849. 35. 0.0885. 37. 0.8849. 39. 0.8808.
41. 0.3226. 43. 0.4474. 45. 0.2939. 47. 0.6704.

Section 6.3

1. $P(3 \leq x \leq 6) = P(-0.50 \leq z \leq 1.00) = 0.5328$.
3. $P(50 \leq x \leq 70) = P(0.67 \leq z \leq 2.00) = 0.2286$.
5. $P(8 \leq x \leq 12) = P(-2.19 \leq z \leq -0.94) = 0.1593$.
7. $P(x \geq 30) = P(z \geq 2.94) = 0.0016$.
9. $P(x \geq 90) = P(z \geq -0.67) = 0.7486$.
11. -1.555. 13. 0.13. 15. 1.41. 17. -0.92.
19. ± 2.33.
21. (a) $P(x > 60) = P(z > -1) = 0.8413$. (b) $P(x < 110)$
$= P(z < 1) = 0.8413$. (c) $P(60 \leq x \leq 110) =$
$P(-1.00 \leq z \leq 1.00) = 0.8413 - 0.1587 = 0.6826$.
(d) $P(x > 140) = P(z > 2.20) = 0.0139$.
23. (a) $P(x > 675) = P(z > 1.75) = 0.0401$.
(b) $P(x < 450) = P(z < -0.50) = 0.3085$.
(c) $P(450 \leq x \leq 675) = P(-0.50 \leq z \leq 1.75) =$
0.6514. (d) $P(x > 28) = P(z > 1.67) = 0.0475$.
(e) $P(x > 12) = P(z > -1.00) = 0.8413$.
(f) $P(12 \leq x \leq 28) = P(-1.00 \leq z \leq 1.67) = 0.7938$.
25. (a) $P(x < 3.0 \text{ mm}) = P(z < -2.33) = 0.0099$.
(b) $P(x > 7.0 \text{ mm}) = P(z > 2.11) = 0.0174$.
(c) $P(3.0 \text{ mm} < x < 7.0 \text{ mm}) = P(-2.33 < z < 2.11) =$
0.9727.
27. (a) $P(3000 < x < 3500) = P(-1.18 < z < 1.59) =$
0.8251. (b) $P(x < 3000) = P(z < -1.18) = 0.1190$.
(c) $P(x > 3500) = P(z > 1.59) = 0.0559$.
29. (a) $P(x < 36 \text{ mo}) = P(z < -1.13) = 0.1292$. They will replace 13% of their batteries. (b) $P(z < z_0) = 10\%$ for $z_0 = -1.28$; $x = -1.28(8) + 45 = 34.76$. Guarantee the batteries for 35 months.

31. (a) According to the empirical rule, about 95% of the data lies between $\mu - 2\sigma$ and $\mu + 2\sigma$. Since this interval is 4σ wide, we have $4\sigma \approx 6$ years, so $\sigma = 1.5$ years.   (b) $P(x > 5) = P(z > -2.00) = 0.9772$.   (c) $P(x < 10) = P(z < 1.33) = 0.9082$.   (d) $P(z < z_0) = 0.10$ for $z_0 = -1.28$; $x = -1.28(1.5) + 8 = 6.08$ years. Guarantee the TVs for about 6.1 years.

33. (a) $\sigma \approx 12$ beats/min.   (b) $P(x < 25) = P(z < -1.75) = 0.0401$.   (c) $P(x > 60) = P(z > 1.17) = 0.1210$.   (d) $P(25 \leq x \leq 60) = P(-1.75 \leq z \leq 1.17) = 0.8389$.   (e) $P(z \leq z_0) = 0.90$ for $z = 1.28$; $x = 1.28(12) + 46 = 61.36$ beats/min. A heart rate of 61 beats/min corresponds to the 90% cutoff point of the distribution.

35. (a) $P(z \geq z_0) = 0.99$ for $z_0 = -2.33$; $x = -2.33(3.7) + 90 \approx 81.38$ mo. Guarantee the microchips for 81 months.   (b) $P(x \leq 84) = P(z \leq -1.62) = 0.0526$.   (c) Expected loss $= (50{,}000{,}000)(0.0526) = \$2{,}630{,}000$.   (d) Profit $= \$370{,}000$.

37. (a) In general, $P(A,\ given\ B) = P(A\ and\ B)/P(B)$; $P(x > 20) = P(z > 0.50) = 0.3085$; $P(x > 15) = P(z > -0.75) = 0.7734$; $P(x > 20,\ given\ x > 15) = 0.3989$.   (b) $P(x > 25) = P(z > 1.75) = 0.0401$; $P(x > 18) = P(z > 0.00) = 0.5000$; $P(x > 25,\ given\ x > 18) = 0.0802$.

39. (a) $P(x > 646) = P(z > 0.74) = 0.2296$.   (b) $z = 1.28$; $x \approx \$669$.

Section 6.4

Note: Answers may differ slightly depending on how many digits are carried in the computation of the standard deviation and the computation of z.

1. (a) $P(r \geq 50) = P(x \geq 49.5) = P(z \geq -27.53) \approx 1$ or almost certain.   (b) $P(r \geq 50) = P(x \geq 49.5) = P(z \geq 7.78) \approx 0$ or almost impossible for a random sample.

3. (a) $P(r \geq 15) = P(x \geq 14.5) = P(z \geq -1.61) = 0.9463$.   (b) $P(r \geq 28) = P(x \geq 27.5) = P(z \geq 1.49) = 0.0681$.   (c) $P(15 \leq r \leq 28) = P(14.5 \leq x \leq 28.5) = P(-1.61 \leq z \leq 1.73) = 0.9045$.   (d) Since both np and nq are larger than 5, the normal approximation is appropriate.

5. (a) $P(r \geq 15) = P(x \geq 14.5) = P(z \geq -2.35) = 0.9906$.   (b) $P(r \geq 30) = P(x \geq 29.5) = P(z \geq 0.62) = 0.2676$.   (c) $P(25 \leq r \leq 35) + P(24.5 \leq x \leq 35.5) = P(-0.37 \leq z \leq 1.81) = 0.6092$.   (d) $P(r > 40) = P(r \geq 41) = P(x \geq 40.5) = P(z \geq 2.80) = 0.0026$.

7. (a) $P(r \geq 47) = P(x \geq 46.5) = P(z \geq -1.94) = 0.9738$.   (b) $P(r \leq 58) = P(x \leq 58.5) = P(z \leq 1.75) = 0.9599$. In parts (c) and (d), let r be the number of products that succeed, and use $p = 1 - 0.80 = 0.20$.   (c) $P(r \geq 15) = P(x \geq 14.5) = P(z \geq 0.40) = 0.3446$.   (d) $P(r < 10) = P(r \leq 9) = P(x \leq 9.5) = P(z \leq -1.14) = 0.1271$.

9. (a) $P(r > 280) = P(r \geq 281) = P(x > 280.5) = P(z \geq -2.16) = 0.9846$.   (b) $P(r \geq 320) =$

$P(x \geq 319.5) = P(z \geq 1.95) = 0.0256$.   (c) $P(280 \leq r \leq 320) = P(279.5 \leq x \leq 320.5) = P(-2.26 \leq z \leq 2.05) = 0.9679$.   (d) $n = 430$; $p = 0.70$; $q = 0.30$; np and nq are both greater than 5. These conditions mean that the normal approximation to the binomial is appropriate.

11. (a) $P(r \geq 540) = P(x \geq 539.5) = P(z \geq 3.81) \approx 0.000$.   (b) $P(r \leq 500) = P(x \leq 500.5) = P(z \leq 1.11) = 0.8665$.   (c) $P(485 \leq r \leq 525) = P(484.5 \leq x \leq 525.5) = P(0 \leq z \leq 2.84) = 0.4977$.

13. (a) $P(r > 180) = P(x \geq 180.5) = P(z > -1.11) = 0.8665$.   (b) $P(r < 200) = P(x \leq 199.5) = P(z \leq 1.07) = 0.8577$).   (c) $P(\text{take sample and buy product}) = P(\text{take sample}) \cdot P(\text{buy, given take sample}) = 0.222$.   (d) $P(60 \leq r \leq 80) = P(59.5 \leq x \leq 80.5) = P(-1.47 \leq z \leq 1.37) = 0.8439$.

15. (a) 0.94.   (b) $P(r \leq 255)$.   (c) $P(r \leq 255) = P(x \leq 255.5) = P(z \leq 1.16) = 0.8770$.

Chapter 6 Review

1. (a) 0.4599.   (b) 0.4015.   (c) 0.0384.   (d) 0.0104.   (e) 0.0250.   (f) 0.8413.
3. (a) 0.9821.   (b) 0.3156.   (c) 0.2977.
5. 1.645.
7. $z = \pm 1.96$.
9. (a) 0.89.   (b) 0.   (c) 0.2514.
11. (a) 0.0166.   (b) 0.975.
13. (a) 0.9772.   (b) 17.3 hr.
15. (a) From \$1.81 to \$3.51 is a 68% range of errors.   (b) From \$0.96 to \$4.36 is a 95% range of errors.   (c) Almost all errors are from \$0.11 to \$5.21.
17. (a) 0.5812.   (b) 0.0668.   (c) 0.0122.
19. (a) 0.8665.   (b) 0.7330.

CHAPTER 7

Section 7.1

1. A set of measurements or counts either existing or conceptual. For example, the population of all ages of all people in Colorado; the population of weights of all students in your school; the population count of all antelope in Wyoming.

3. A numerical descriptive measure of a population, such as μ, the population mean; σ, the population standard deviation; σ^2, the population variance.

5. A statistical inference is a conclusion about the value of a population parameter. We will do both estimation and testing.

7. They help us visualize the sampling distribution by using tables and graphs that approximately represent the sampling distribution.

9. We studied the sampling distribution of mean trout lengths based on samples of size 5. Other such sampling distributions abound.

Section 7.2

Note: Answers may differ slightly depending on the number of digits carried in the standard deviation.

1. (a) $\mu_{\bar{x}} = 15$; $\sigma_{\bar{x}} = 2.0$; $P(15 \leq \bar{x} \leq 17) = P(0 \leq z \leq 1.00) = 0.3413$. (b) $\mu_{\bar{x}} = 15$; $\sigma_{\bar{x}} = 1.75$; $P(15 \leq \bar{x} \leq 17) = P(0 \leq z \leq 1.14) = 0.3729$. (c) The standard deviation is smaller in part b because of the larger sample size. Therefore, the distribution about $\mu_{\bar{x}}$ is narrower in part b.

3. (a) No; the sample size is only 9 and so is too small. (b) Yes; the $\bar{x}$ distribution also will be normal with $\mu_{\bar{x}} = 25$; $\sigma_{\bar{x}} = 3.5/3$; $P(23 \leq \bar{x} \leq 26) = P(-1.71 \leq z \leq 0.86) = 0.7615$.

5. (a) $P(x < 74.5) = P(z < -0.63) = 0.2643$. (b) $P(\bar{x} < 74.5) = P(z < -2.79) = 0.0026$. (c) No. If the weight of only one car were less than 74.5 tons, we cannot conclude that the loader is out of adjustment. If the mean weight for a sample of 20 cars were less than 74.5 tons, we would suspect that the loader is malfunctioning. As we see in part b, the probability of this happening is very low if the loader is correctly adjusted.

7. (a) $P(x < 40) = P(z < -1.80) = 0.0359$. (b) Since the x distribution is approximately normal, the $\bar{x}$ distribution is approximately normal with mean 85 and standard deviation 17.678. $P(\bar{x} < 40) = P(z < -2.55) = 0.0054$. (c) $P(\bar{x} < 40) = P(z < -3.12) = 0.0009$. (d) $P(\bar{x} < 40) = P(z < -4.02) < 0.0002$. (e) Yes; if the average value based on five tests were less than 40, the patient is almost certain to have excess insulin.

9. (a) $P(x < 54) = P(z < -1.27) = 0.1020$. (b) The expected number undernourished is 2200(0.1020), or about 224. (c) $P(\bar{x} \leq 60) = P(z \leq -2.99) = 0.0014$. (d) $P(\bar{x} < 64.2) = P(z < 1.20) = 0.8849$. Since the sample average is above the mean, it is quite unlikely that the doe population is undernourished.

11. (a) Since x itself represents a sample mean return based on a large (random) sample of stocks, x has a distribution that is approximately normal (central limit theorem). (b) $P(1\% \leq \bar{x} \leq 2\%) = P(-1.63 \leq z \leq 1.09) = 0.8105$. (c) $P(1\% \leq \bar{x} \leq 2\%) = P(-3.27 \leq z \leq 2.18) = 0.9849$. (d) Yes. The standard deviation decreases as the sample size increases. (e) $P(\bar{x} < 1\%) = P(z < -3.27) = 0.0005$. This is very unlikely if $\mu = 1.6\%$. One would suspect that μ has slipped below 1.6%.

13. (a) Since x itself represents a sample mean from a large (random) sample of bonds, x is approximately normally

distributed according to the central limit theorem. (b) $P(\bar{x} < 6\%) = P(z < -2.19) = 0.0143$. Yes, it is very unlikely that $\bar{x}$ would be less than 6% if $\mu = 10.8\%$. The junk bond market appears to be weaker. (c) $P(\bar{x} > 16\%) = P(z > 2.37) = 0.0089$. Yes, it is very unlikely that $\bar{x}$ would be greater than 16% if $\mu = 10.8\%$. The junk bond market may be heating up.

15. (a) 30 or more. (b) No.

17. (a) The total checkout time for 30 customers is the sum of the checkout times for each individual customer. Thus, $w = x_1 + x_2 + \cdots + x_{30}$, and the probability that the total checkout time for the next 30 customers is less than 90 is $P(w < 90)$. (b) $w < 90$ is equivalent to $x_1 + x_2 + \cdots + x_{30} < 90$. Divide both sides by 30 to get $\bar{x} < 3$ for samples of size 30. Therefore, $P(w < 90) = P(\bar{x} < 3)$. (c) By the central limit theorem, $\bar{x}$ is approximately normal with $\mu_{\bar{x}} = 2.7$ min and $\sigma_{\bar{x}} = 0.1095$. (d) $P(\bar{x} < 3) = P(z < 2.74) = 0.9969$.

19. (a) $P(w < 9,500 \text{ g}) = P(\bar{x} < 211.11) = P(z < -2.31) = 0.0104$. (b) $P(w > 12,000 \text{ g}) = P(\bar{x} > 266.67) = P(z > 2.13) = 0.0166$. (c) $P(9,500 \leq w \leq 12,000) = P(211.11 \leq \bar{x} \leq 266.67) = P(-2.31 \leq z \leq 2.13) = 0.9730$.

Section 7.3

1. (a) Answers vary. (b) When np and nq both exceed 5; $\mu_{\hat{p}} = p$; $\sigma_{\hat{p}} = \sqrt{pq/n}$. (c) Yes; both np and nq exceed 5; $\mu_{\hat{p}} = 0.21$; $\sigma_{\hat{p}} \approx 0.071$; continuity correction ≈ 0.015. $P(0.15 \leq \hat{p} \leq 0.25) \approx P(0.135 \leq x \leq 0.265) \approx P(-1.06 \leq z \leq 0.77) \approx 0.6348$. (d) No; $np < 5$. (e) Yes; both np and nq exceed 5; $\mu_{\hat{p}} = 0.15$; $\sigma_{\hat{p}} \approx 0.052$; continuity correction ≈ 0.010. $P(\hat{p} \geq 0.22) \approx P(x \geq 0.21) \approx P(z \geq 1.15) \approx 0.1251$.

3. $\mu_{\hat{p}} = 0.60$; $\sigma_{\hat{p}} \approx 0.089$; continuity correction ≈ 0.017. (a) $P(\hat{p} \geq 0.5) = P(x \geq 0.483) \approx P(z \geq -1.31) \approx 0.9049$. (b) $P(\hat{p} \geq 0.667) \approx P(x \geq 0.65) \approx P(z \geq 0.56) \approx 0.2877$. (c) $P(\hat{p} \leq 0.33) \approx (x \leq 0.35) \approx P(z \leq -2.81) = 0.0025$. (d) Yes; both np and nq exceed 5.

5. $\mu_{\hat{p}} = 0.11$; $\sigma_{\hat{p}} \approx 0.042$; continuity correction ≈ 0.009. (a) $P(\hat{p} \leq 0.15) \approx (x \leq 0.159) \approx P(z \leq 1.17) \approx 0.8790$. (b) $P(0.10 \leq \hat{p} \leq 0.15) \approx P(0.091 \leq x \leq 0.159) = P(-0.45 \leq z \leq 1.17) = 0.5526$. (c) Yes, both np and nq exceed 5.

7. (a) Both np and nq exceed 5; $\mu_{\hat{p}} = 0.06$; $\sigma_{\hat{p}} = 0.024$; continuity correction = 0.005. (b) $P(\hat{p} \geq 0.07) \approx P(x \geq 0.065) \approx P(z \geq 0.21) = 0.4168$. (c) $P(\hat{p} \geq 0.11) \approx P(x \geq 0.105) \approx P(z \geq 1.88) \approx 0.0301$. Yes, the probability of producing this proportion of defective toys is only about 3%.

9.

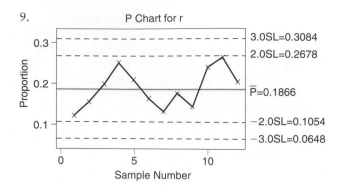

P Chart for r

No out-of-control signals.

11.

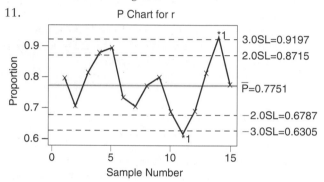

P Chart for r

Out-of-control signal III occurs on days 4 and 5; out-of-control signal I occurs on day 11 on the low side and day 14 on the high side. Out-of-control signals on the low side are of most concern for the homeless seeking work. The foundation should look to see what happened on that day. The foundation might take a look at the out-of-control periods on the high side to see if there is a possibility of cultivating more jobs.

Chapter 7 Review

1. (a) A normal distribution. (b) The mean μ of the x distribution. (c) $\sigma/\sqrt{n}$, where σ is the standard deviation of the x distribution. (d) They will both be approximately normal with the same mean, but the standard deviations will be $\sigma/\sqrt{50}$ and $\sigma/\sqrt{100}$, respectively.
3. (a) $P(x \geq 40) = P(z \geq 0.71) = 0.2389$. (b) $P(\bar{x} \geq 40)$ $= P(z \geq 2.14) = 0.0162$.
5. $P(98 \leq \bar{x} \leq 102) = P(-1.33 \leq z \leq 1.33) = 0.8164$.
7. (a) $P(\bar{x} \geq 750) = P(z \geq 0) = 0.5000$.
(b) $P(745 \leq \bar{x} \leq 755) = P(-2.00 \leq z \leq 2.00) =$ 0.9544.
9. (a) $\mu_{\hat{p}} = 0.22$; $\sigma_{\hat{p}} \approx 0.059$; continuity correction = 0.01; $P(0.20 \leq \hat{p} \leq 0.25) \approx P(0.19 \leq \hat{p} \leq 0.26) \approx$ $P(-0.51 \leq z \leq 0.68) \approx 0.4467$. (b) $\mu_{\hat{p}} = 0.27$; $\sigma_{\hat{p}} \approx 0.072$; continuity correction ≈ 0.013; $P(\hat{p} \geq 0.35) \approx$ $P(\hat{p} \geq 0.337) \approx P(z \geq 0.93) \approx 0.1762$. (c) No, $np < 5$.

Section 8.1

1. (a) The mean and standard deviation round to the values given. (b) Using the rounded values for the mean and standard deviation given in part a, the interval is from 143.8 to 149.2. (c) Using the rounded values for the mean and standard deviation given in part a, the interval is from 143.0 to 150.0. (d) Using the rounded values for the mean and standard deviation given in part a, the interval is from 141.0 to 152.0. (e) The lengths increase as c increases because the values of z_c increase as c increases. If we want to be more certain that μ is in the interval based on the given sample, we have to make the interval wider.
3. (a) The mean and standard deviation round to the values given. (b) Using the rounded values of part a, the 75% interval is from 34.19 thousand to 37.81 thousand. (c) Yes. 30 thousand dollars is below the mean annual profit per employee. We can say with 75% confidence that the mean lies between 34.19 thousand and 37.81 thousand. (d) Yes. 40 thousand is above the mean annual profit per employee. (e) 33.41 thousand to 38.59 thousand. We can say with 90% confidence that the mean lies between 33.4 thousand and 38.6 thousand dollars. 30 thousand is below the mean and 40 thousand is above the mean.
5. (a) The values for the mean and standard deviation round to those given. (b) Using the rounded values for the mean and standard deviation given in part a, the interval is from 50.37 to 51.95. (c) Using the rounded values for the mean and standard deviation given in part a, the interval is from 49.92 to 52.40. (d) We can be 99% sure that the average January temperature is between 49.92 and 52.4°F. It is possible that the average is 53°F, but not very likely. It is possible that a few more years of observation might be needed before such a claim could be made.
7. (a) 1.10 sec to 1.30 sec. (b) 560.87 Hz to 657.13 Hz.
9. (a) 2.3 to 2.7 min. (b) 13.9 to 16.5 min. (c) 23.5 to 27.9 min. (d) The intervals got longer as s increased. This is to be expected, since the endpoints are $\pm z_c s/\sqrt{n}$ from the sample mean and z_c and n were fixed. The length of each interval is $2z_c s/\sqrt{n}$.
11. 3.17 to 3.63 yr.
13. 15,342 to 16,658.
15. 11.3 to 12.5 mg/l.
17. (a) The mean and standard deviation round to the values given. (b) 283.9 lb to 290.9 lb. (c) Yes. 200 lb is below the mean. We can say with 85% confidence that the mean lies between 283.9 lb and 290.9 lb. (d) Yes. 291 lb is just above the upper limit of the confidence interval. This player is heavier than average. (e) 282.6 lb to 292.2 lb; 200 lb is still well below the lower confidence limit. 291 is near the upper bound of the confidence interval.

Section 8.2

1. 2.110.
3. 1.721.
5. (a) The mean and standard deviation round to the values given.   (b) Using the rounded values for the mean and standard deviation given in part a, the interval is from 1249 to 1295.
7. (a) Use calculator.   (b) $120.84 to $175.82.
9. (a) Use calculator.   (b) 74.7 lb to 107.3 lb.
11. 76.04 to 82.46 cm.
13. (a) Use calculator.   (b) $10.84 to $13.86.
15. (a) Use calculator; 8.5 to 11.1 in.   (b) Use calculator; 15.7 to 18.5 in.
17. (a) Use calculator; 29.4 thousand to 36.8 thousand.   (b) Use calculator; 18.0 thousand to 23.6 thousand.
19. (a) Boxplots differ in length of interquartile box, location of median, and length of whiskers. The boxplots come from different samples.   (b) Yes; No; For 95% confidence intervals, we expect about 95% of the samples to generate intervals that contain the mean of the population.

Section 8.3

1. (a) $\hat{p} = 39/62 = 0.6290$.   (b) 0.51 to 0.75. If this experiment were repeated many times, about 95% of the intervals would contain p.   (c) Both np and nq are greater than 5. If either is less than 5, the normal curve will not necessarily give a good approximation to the binomial.
3. (a) $\hat{p} = 1619/5222 = 0.3100$.   (b) 0.29 to 0.33. If we repeat the survey with many different samples of 5222 dwellings, about 99% of the intervals will contain p.   (c) Both np and nq are greater than 5. If either is less than 5, the normal curve will not necessarily give a good approximation to the binomial.
5. (a) $\hat{p} = 0.5420$.   (b) 0.53 to 0.56.   (c) Yes. np and nq are greater than 5.
7. (a) $\hat{p} = 17/99 = 0.1717$.   (b) 0.12 to 0.23.   (c) Yes. Both np and nq are greater than 5.
9. (a) $\hat{p} = 0.0304$.   (b) 0.02 to 0.05.   (c) Yes. np and nq are greater than 5.
11. (a) $\hat{p} = 0.0600$.   (b) 0.04 to 0.08.
13. (a) $\hat{p} = 0.1910$.   (b) 0.18 to 0.20.
15. (a) $\hat{p} = 0.8603$.   (b) 0.84 to 0.89.   (c) A recent study shows that 86% of women shoppers remained loyal to their favorite supermarket last year. The margin of error was 2.5 percentage points.
17. (a) $\hat{p} = 0.25$.   (b) 0.22 to 0.28.   (c) A survey of 1000 large corporations has shown that 25% will choose a nonsmoking job candidate over an equally qualified smoker. The margin of error was 2.7%.

Section 8.4

1. Estimate a mean. Use 75 plots.
3. (a) Estimate a proportion; 208.   (b) 40.
5. Estimate a mean; 120 total or 64 more.
7. (a) Estimate a proportion; 666.   (b) 662.
9. Estimate a mean; 117 or 34 more.
11. (a) Estimate a proportion; 144.   (b) 127 total or 69 more.
13. (a) Estimate a proportion; 16,641.   (b) 14,718.
15. Estimate a mean; 385 or 218 more.
17. (a) Estimate a proportion; 68.   (b) 50.
19. (a) $1/4 - (p - 1/2)^2 = 1/4 - (p^2 - p + 1/4) = -p^2 + p = p(1 - p)$.   (b) Since $(p - 1/2)^2 \geq 0$, then $1/4 - (p - 1/2)^2 \leq 1/4$ because we are subtracting $(p - 1/2)^2$ from 1/4.
21. (a) 1,664,100.   (b) No. When $p = 0.5$, the formula for sample size without a preliminary estimate is the same as the formula with a preliminary estimate.

Section 8.5

1. (a) $E = 0.1446$; interval from -0.41 to -0.13.   (b) Interval contains values that are all negative. The football players appear to be shorter.   (c) $E = 0.1904$; interval from -0.46 to -0.08; no. The interval contains values that are all negative. We can say with 99% confidence that football players are shorter than basketball players.
3. (a) Use calculator to check that the means and standard deviations round to the values given.   (b) $s = 0.9573$; interval from 2.07 to 4.13.   (c) The interval consists of all positive numbers. At the 99% confidence level it appears that electronics companies have a higher profit as a percentage of revenue than food and drug companies.
5. (a) $\hat{\sigma} = 0.0232$; $E = 0.0599$; the interval is from 0.67 to 0.79.   (b) The confidence interval contains values that are all positive, so we can be 99% sure that $p_1 > p_2$.
7. (a) $E = 0.3201$; the interval is from -9.12 to -8.48.   (b) The interval consists of negative values only. At the 99% confidence level we can conclude that $\mu_1 < \mu_2$.
9. (a) Use calculator to show that the means and standard deviations round to the values given.   (b) $s = 2.6435$; interval from -1.28 to 2.68.   (c) The interval consists of numbers that are of different signs. No. There does not seem to be a difference in the average number of children for high- or low-income families.
11. (a) $\hat{p}_1 = 0.3095$; $\hat{p}_2 = 0.1184$; $\hat{\sigma} = 0.0413$; interval from 0.085 to 0.297.   (b) The interval contains numbers that are all positive. A greater proportion of hogans occurs in Fort Defiance.
13. (a) $E = 7.9690$; interval from -28.46 to -12.52.   (b) The interval contains numbers that are all negative. We conclude with 95% confidence that the average

weight of bucks in the Cache la Poudre Region is less than the average weight of bucks in Mesa Verde.

15. (a) $\bar{x}_1 - \bar{x}_2 = 0.52$; $s = 3.3563$; interval from -1.29 to 2.33. (b) $\bar{x}_1 - \bar{x}_3 = 1.96$; $s = 3.4214$; interval from 0.11 to 3.81. (c) $\bar{x}_2 - \bar{x}_3 = 1.44$; $s = 3.6805$; interval from -0.55 to 3.43. (d) At the 85% confidence level we can say that the mean index of self-esteem based on competence is greater than the mean index of self-esteem based on physical attractiveness. We cannot conclude that there is a difference in mean index of self-esteem based on competence and that based on social acceptance. We cannot conclude that there is a difference in the mean indices based on social acceptance and physical attractiveness.

17. (a) Based on the same data, a 99% confidence interval is longer than a 95% confidence interval. Therefore, if the 95% confidence interval has both positive and negative values, so will the 99% confidence interval. However, for the same data, a 90% confidence interval is shorter than a 95% confidence interval. The 90% confidence interval might contain only positive or only negative values even if the 95% interval contains both. (b) Based on the same data, a 99% confidence interval is longer than a 95% confidence interval. Even if the 95% confidence interval contains values that are all positive, the longer 99% interval could contain both positive and negative values. Since, for the same data, a 90% confidence interval is shorter than a 95% confidence interval, if the 95% confidence interval contains only positive values, so will the 90% confidence interval.

19. (a) $n = 896.1$ or 897 couples in each sample. (b) $n = 768.3$ or 769 couples in each sample.

Chapter 8 Review

1. See text.
3. Interval for mean, large sample; 176.91 to 180.49.
5. Interval for mean, small sample. (a) Use calculator. (b) 64.1 to 84.3.
7. Interval for proportion; 0.50 to 0.54.
9. Interval for proportion. (a) $\hat{p} = 0.4072$. (b) 0.333 to 0.482.
11. Difference of means, large samples. (a) -0.332 to 0.932. (b) No. Interval contains both positive and negative values. At the 90% level of confidence, we cannot say that the average percentage salary increase for professors in the East is greater.
13. Difference of means, small samples. (a) $s = 6.9712$; 5.46 to 10.54. (b) Yes. The interval contains values that are all positive. At the 75% level of confidence it appears that the average weight of adult male wolves from the Northwest Territories is greater.

15. Difference of proportions. (a) $\hat{p}_1 = 0.8495$; $\hat{p}_2 = 0.8916$; -0.1409 to 0.0567. (b) No. The interval contains both negative and positive numbers. We do not detect a difference in the proportions at the 95% confidence level.

17. (a) $P(A_1 < \mu_1 < B_1 \text{ and } A_2 < \mu_2 < B_2) = (0.80)(0.80) = 0.64$. The complement of the event $A_1 < \mu_1 < B_1$ and $A_2 < \mu_2 < B_2$ is that either μ_1 is not in the first interval or μ_2 is not in the second interval or both. Thus, $P(\text{at least one interval fails}) = 1 - P(A_1 < \mu_1 < B_1 \text{ and } A_2 < \mu_2 < B_2) = 1 - 0.64 = 0.36$. (b) Suppose $P(A_1 < \mu_1 < B_1) = c$ and $P(A_2 < \mu_2 < B_2) = c$. If we want the probability that both hold to be 90%, and if x_1 and x_2 are independent, then $P(A_1 < \mu_1 < B_1 \text{ and } A_2 < \mu_2 < B_2) = 0.90$ means $P(A_1 < \mu_1 < B_1) \cdot P(A_2 < \mu_2 < B_2) = 0.90$ so $c^2 = 0.90$ or $c = 0.9487$. (c) In order to have a high probability of success for the whole project, the probability that each component will perform as specified must be significantly higher.

CHAPTER 9

Section 9.1

1. See text.
3. No, if we fail to reject the null hypothesis, we have not proven it beyond all doubt. We have failed only to find sufficient evidence to reject it.
5. (a) H_0: $\mu = 60$ kg. (b) H_1: $\mu < 60$ kg. (c) H_1: $\mu > 60$ kg. (d) H_1: $\mu \neq 60$ kg. (e) For part b, the critical region is on the left. For part c, the critical region is on the right. For part d, the critical region is on both sides of the mean.
7. (a) H_0: $\mu = 16.4$ feet. (b) H_1: $\mu > 16.4$ feet. (c) H_1: $\mu < 16.4$ feet. (d) H_1: $\mu \neq 16.4$ feet. (e) For part b, the critical region is on the right. For part c, the critical region is on the left. For part d, the critical region is on both sides of the mean.
9. (a) H_0: $\mu = 288$ lb. (b) Higher: H_1: $\mu > 288$ lb; lower: H_1: $\mu < 288$ lb; different: H_1: $\mu \neq 288$ lb. For higher, the critical region is the right tail; for lower, the critical region is the left tail; for different, the critical region is both tails.

Section 9.2

1. H_0: $\mu = 16.4$ feet; H_1: $\mu < 16.4$ feet; left-tailed; normal distribution; $z_0 = -2.33$; sample $z = -2.44$; Reject H_0; The storm is lessening; Results statistically significant.
3. H_0: $\mu = 31.8$ calls/day; H_1: $\mu \neq 31.8$ calls/day; two-tailed; normal distribution; $z_0 = \pm 2.58$; sample $z = -2.45$; Do not reject H_0; There is not enough evidence

that the mean number of messages has decreased; Results not statistically significant.

5. H_0: $\mu = \$4.75$; H_1: $\mu > \$4.75$; right-tailed; normal distribution; $z_0 = 2.33$; sample $z = 3.14$; Reject H_0; There is evidence that her average tip is more than $4.75; Results statistically significant.

7. H_0: $\mu = 10.2$ seconds; H_1: $\mu < 10.2$ seconds; left-tailed; normal distribution; $z_0 = -1.645$; sample $z = -1.52$; Do not reject H_0. There is not enough evidence to conclude that the mean acceleration time is less; Results not statistically significant.

9. H_0: $\mu = 19.0$; H_1: $\mu > 19.0$; right-tailed; normal distribution; $z_0 = 2.33$; sample $z = 1.19$; Do not reject H_0. There is not sufficient evidence to conclude that the average oxygen capacity has increased; Results not statistically significant.

11. H_0: $\mu = 7.4$; H_1: $\mu \neq 7.4$; two-tailed; normal distribution; $z_0 = \pm1.96$; sample $z = 2.12$; Reject H_0; The drug has changed the pH of the blood; Results are statistically significant.

13. The mean and standard deviation round to the value given; H_0: $\mu = \$13.9$ thousand; H_1: $\mu > \$13.9$ thousand; right-tailed; normal distribution; $z_0 = 1.645$; sample $z = 2.34$; Reject H_0; The mean franchise costs for pizza businesses are significantly higher; Results are statistically significant.

15. Essay or class discussion.

17. H_0: $\mu = 21$; H_1: $\mu \neq 21$; $\alpha = 0.01$; two-tailed test. (a) $c = 1 - 0.01 = 0.99$ or 99% confidence interval; interval is from 20.28 to 23.72 based on the sample statistics. Since the hypothesized value $\mu = 21$ falls in the confidence interval, we do not reject H_0. (b) $z_0 = \pm2.58$; sample $z = 1.50$; Do not reject H_0; The conclusions are the same.

Section 9.3

1. Sample $z = 2.78$; P value $= 0.0027$; The data are significant at the 1% level.

3. Sample $z = -1.18$; P value $= 2(0.1190) = 0.238$; No, not significant at the 5% level.

5. H_0: $\mu = 1.75$; H_1: $\mu > 1.75$; sample $z = 3.02$; P value $= 0.0013$; Yes, the data are significant at the 1% level.

7. H_0: $\mu = 19$ inches; H_1: $\mu < 19$ inches; sample $z = -0.80$; P value $= 0.2119$; No, the data are not significant at the 5% level. We cannot conclude that the average length of trout caught in Pyramid Lake is different from 19 inches.

9. H_0: $\mu = \$15.35$; H_1: $\mu < 15.35$; sample $z = -3.29$; P value $= 0.0005$; Yes, the data are significant at the 5% level. There is evidence that college students' average daily ownership expenses are less than the national average.

Section 9.4

1. $t_0 = -1.860$.

3. $t_0 = \pm2.807$.

5. $t_0 = \pm2.201$.

7. (a) Rounded answers are used in part b.   (b) H_0: $\mu = 4.8$; H_1: $\mu < 4.8$; left-tailed; $t_0 = -2.015$; sample $t = -4.06$; P value < 0.005; Reject H_0; At the 5% significance level, this patient's average red blood count is less than 4.8.

9. (a) Rounded answers are used in part b.   (b) H_0: $\mu = 16.5$ days; H_1: $\mu \neq 16.5$ days; two-tailed; $t_0 = \pm2.110$; sample $t = 2.565$; $0.020 < P$ value < 0.050; Reject H_0; At the 5% level the mean incubation time is different from 16.5 days.

11. (a) Rounded answers are used in part b.   (b) H_0: $\mu = 1300$; H_1: $\mu \neq 1300$; two-tailed; $t_0 = \pm3.250$; sample $t = -2.71$; $0.020 < P$ value < 0.050; Do not reject H_0 at the 1% level of significance. There is not enough evidence to conclude that the population mean of tree-ring dates is different from 1300.

13. (a) Rounded answers are used in part b.   (b) H_0: $\mu = 77$ years; H_1: $\mu < 77$ years; left-tailed; $t_0 = -1.729$; sample $t = -0.6304$; P value > 0.125; Do not reject H_0. There is not enough evidence to conclude that the population mean life span is less than 77 years.

15. (a) Rounded answers used in part b.   (b) H_0: $\mu = 7.3$; H_1: $\mu > 7.3$; right-tailed; $t_0 = 1.729$; sample $t = 2.555$; $0.005 < P$ value < 0.010; Reject at 5% level. The evidence supports the claim that the average time women with children spend shopping in houseware stores in the Cherry Creek Mall is higher than the national average.

Section 9.5

1. H_0: $p = 0.70$; H_1: $p \neq 0.70$; two-tailed; normal; $z_0 = \pm2.58$; sample $\hat{p} = 0.75$ with $z = 0.62$; P value $= 2(0.2676) = 0.5352$; Do not reject H_0; The population proportion of such arrests is not significantly different from 0.70.

3. H_0: $p = 0.77$; H_1: $p < 0.77$; left-tailed; normal; $z_0 = -2.33$; sample $\hat{p} = 0.5556$ with $z = -2.65$; P value $= 0.004$; Reject H_0; The population proportion of driver fatalities related to alcohol is less than 77% at the 1% significance level.

5. H_0: $p = 0.50$; H_1: $p < 0.50$; left-tailed; normal; $z_0 = -2.33$; sample $\hat{p} = 0.2941$; $z = -2.40$; P value $= 0.0082$; Reject H_0. The population proportion of female wolves is less than 50% at the 1% significance level.

7. H_0: $p = 0.261$; H_1: $p \neq 0.261$; two-tailed; normal; $z_0 = \pm2.58$; sample $\hat{p} = 0.1924$ with $z = -2.78$; P value $= 2(0.0027) = 0.0054$; Reject H_0. The population proportion of this type of five-syllable sequence is significantly different from that of Plato's *Republic*.

9. $H_0: p = 0.47$; $H_1: p > 0.47$; right-tailed; normal; $z_0 = 2.33$; sample $\hat{p} = 0.4871$ with $z = 1.09$; P value $= 0.1379$; Do not reject H_0. The population loyalty of Chevrolet owners is not significantly greater than 47%.

11. $H_0: p = 0.092$; $H_1: p > 0.092$; right-tailed; normal; sample $\hat{p} = 0.1480$ with $z = 2.71$; P value $= 0.0034$; Reject H_0. The population proportion of students with hypertension during final exams week is higher than 9.2% at the 5% significance level.

13. $H_0: p = 0.82$; $H_1: p \neq 0.82$; two-tailed; normal; $z_0 = \pm 2.58$; sample $\hat{p} = 0.7671$ with $z = -1.18$; P value $= 2(0.1190) = 0.2380$; Do not reject H_0. The population proportion of extroverts among college leaders is not different from 82% at the 1% significance level.

15. $H_0: p = 0.76$; $H_1: p \neq 0.76$; two-tailed; normal; $z_0 = \pm 2.58$; $\hat{p} = 0.7966$ with $z = 0.66$; P value $= 0.5092$; Do not reject H_0. There is insufficient evidence to conclude that the proportion of Colorado college professors who would choose the career again is different from the national rate.

Section 9.6

1. $H_0: \mu_d = 0$; $H_1: \mu_d \neq 0$; two-tailed; t distribution with $d.f. = 7$; $t_0 = \pm 2.365$; $\bar{d} = 2.25$; $s_d = 7.78$; sample $t = 0.818$; P value > 0.25; Do not reject H_0. There is not a significant difference between the population mean percentage increase in corporate revenue and the population mean percentage increase of CEO salary.

3. $H_0: \mu_d = 0$; $H_1: \mu_d > 0$; right-tailed; t distribution with $d.f. = 4$; $t_0 = 3.747$; $\bar{d} = 12.6$; $s_d = 22.66$; sample $t = 1.243$; P value > 0.125; Fail to reject H_0; At the 1% level of significance, we do not conclude that average peak wind gusts are higher in January than they are in April.

5. $H_0: \mu_d = 0$; $H_1: \mu_d > 0$; right-tailed; $t_0 = 1.895$; $\bar{d} = 6.125$; $s_d = 9.83$; sample $t = 1.76$; $0.050 < P$ value < 0.075; Fail to reject H_0. The evidence does not support the claim that the winter proportion of males in the packs is higher than the summer.

7. $H_0: \mu_d = 0$; $H_1: \mu_d > 0$; right-tailed; t distribution with $d.f. = 7$; $t_0 = 1.895$; $\bar{d} = 6$; $s_d = 21.5$; sample $t = 0.789$; P value > 0.125; Do not reject H_0; We do not have enough evidence to conclude that the average number of inhabited houses is greater than the average number of hogans on the Navajo Reservation.

9. $H_0: \mu_d = 0$; $H_1: \mu_d \neq 0$; two-tailed; t distribution with $d.f. = 4$; $t_0 = \pm 2.776$; $\bar{d} = 1.0$; $s_d = 5.24$; sample $t = 0.427$; P value > 0.250. Do not reject H_0. There is not enough evidence to conclude that there is a difference in the average number of service ware shards in subarea 1 compared to subarea 2.

11. $H_0: \mu_d = 0$; $H_1: \mu_d < 0$; left-tailed; t distribution with $d.f. = 5$; $t_0 = -2.015$; $\bar{d} = -10.5$; $s_d = 5.17$; sample $t = -4.97$; P value < 0.005; Reject H_0; The population

mean heart rate after the test is higher than that before the test.

13. $H_0: \mu_d = 0$; $H_1: \mu_d > 0$; right-tailed; t distribution with $d.f. = 8$; $t_0 = 1.860$; $\bar{d} = 2.0$; $s_d = 4.5$; sample $t = 1.33$; $0.10 < P$ value < 0.125; Do not reject H_0; The population mean score on the last round is not significantly higher than the population mean score on the first round.

15. $H_0: \mu_d = 0$; $H_1: \mu_d > 0$; right-tailed; t distribution with $d.f. = 7$; $t_0 = 1.895$; $\bar{d} = 0.775$; $s_d = 1.0539$; sample $t = 2.080$; $0.025 < P$ value < 0.050; Reject H_0; At the 5% level of significance it seems that rats receiving larger rewards perform better on the ladder climb.

Section 9.7

1. $H_0: \mu_1 = \mu_2$; $H_1: \mu_1 > \mu_2$; right-tailed test; $z_0 = 2.33$. Sample test statistic $\bar{x}_1 - \bar{x}_2 = 0.7$ with $z = 4.22$. P value ≈ 0.0001. Since the sample test statistic falls in the critical region, we reject H_0 and conclude that, on average, 10-year-old children have more REM sleep than do 35-year-old adults.

3. $H_0: \mu_1 = \mu_2$; $H_1: \mu_1 \neq \mu_2$; two-tailed; $z_0 = \pm 2.58$; Sample test statistic $\bar{x}_1 - \bar{x}_2 = 0.5$ with $z = 3.49$; P value $= 0.0004$; Reject H_0; The data indicate a difference in preferences for camping or fishing at the 1% level of significance.

5. $H_0: \mu_1 = \mu_2$; $H_1: \mu_1 < \mu_2$; left-tailed test; $z_0 = -1.645$. Sample test statistic $\bar{x}_1 - \bar{x}_2 = -2.3$ with $z = -2.60$. P value $= 0.0047$. Since the sample test statistic falls in the critical region, we reject H_0 and conclude that the average sick leave for night workers is more than that for day workers.

7. $H_0: \mu_1 = \mu_2$; $H_1: \mu_1 \neq \mu_2$; two-tailed test; $z_0 = \pm 1.96$. Sample test statistic $\bar{x}_1 - \bar{x}_2 = -1.4$ with $z = -0.11$. P value $= 0.9124$. Since the sample test statistic does not fall in the critical region, we do not reject H_0. There is virtually no evidence of any difference in the average scores of the two groups.

9. (a) The means and standard deviations round to the results given.   (b) $H_0: \mu_1 = \mu_2$; $H_1: \mu_1 \neq \mu_2$; two-tailed test; $d.f. = 29$; $t_0 = \pm 2.045$; $s = 2.6389$. Sample test statistic $\bar{x}_1 - \bar{x}_2 = 0.82$ with $t = 0.865$; $0.25 < P$ value. Since the sample test statistic does not fall in the critical region, we do not reject H_0. There is not enough evidence to conclude that the average number of cases for fox rabies is different in the two regions.

11. (a) The means and standard deviations round to the results given.   (b) $H_0: \mu_1 = \mu_2$; $H_1: \mu_1 \neq \mu_2$; two-tailed; $d.f. = 12$; $t_0 = \pm 2.179$; $s = 3.6745$; sample test statistic $\bar{x}_1 - \bar{x}_2 = 0.15$ with $t = 0.0764$; P value > 0.25; Do not reject H_0. The data do not indicate that the population mean time lost for hot tempers is different from that lost due to disputes.

13. (a) The means and standard deviations round to the results given.   (b) H_0: $\mu_1 = \mu_2$; H_1: $\mu_1 < \mu_2$; left-tailed tests; $d.f. = 20$; $t_0 = -1.725$; $s = 2.592$. Sample test statistic $\bar{x}_1 - \bar{x}_2 = -3.6$ with $t = -3.244$. P value < 0.005. Since the sample test statistic falls in the critical region, we reject H_0 and conclude that the average water temperature has increased.

15. (a, b) The means and standard deviations round to the results given.   (c) H_0: $\mu_1 = \mu_2$; H_1: $\mu_1 \neq \mu_2$; two-tailed test; $d.f. = 12$; $t_0 = \pm 3.055$; $s = 10.626$. Sample test statistic $\bar{x}_1 - \bar{x}_2 = -5.7$ with $t = -1.004$; $0.25 < P$ value. Since the sample test statistic does not fall in the critical region, we do not reject H_0. There is not sufficient evidence to conclude that the average number of emergency calls during the day differs from the average number at night.

17. H_0: $p_1 = p_2$; H_1: $p_1 \neq p_2$; two-tailed; $z_0 = \pm 2.58$; $\hat{p} = 0.0676$; sample test statistic $\hat{p}_1 - \hat{p}_2 = 0.0237$ with $z = 0.79$; P value $= 0.4296$; Do not reject H_0; The data do not indicate that the population proportions are different.

19. Let p_1 = proportion who did not attend college and who believe in extraterrestrials and p_2 = proportion who did attend college and who believe in extraterrestrials; H_0: $p_1 = p_2$; H_1: $p_1 < p_2$; left-tailed test; $z_0 = -2.33$; $\hat{p} = 0.42$. Sample test statistic $\hat{p}_1 - \hat{p}_2 = -0.10$ with $z = -1.43$. P value $= 0.0764$. Since the sample test statistic does not fall in the critical region, we do not reject H_0. There is not enough evidence to conclude that the first proportion is less than the second.

21. H_0: $p_1 = p_2$; H_1: $p_1 < p_2$; left-tailed; $z_0 = -1.645$; $\hat{p} = 0.5570$; sample test statistic $\hat{p}_1 - \hat{p}_2 = -0.1493$ with $z = -4.44$; P value < 0.0001; Reject H_0; The data indicate that the population proportion of hotel guests requesting nonsmoking rooms has increased.

23. H_0: $p_1 = p_2$; H_1: $p_1 > p_2$; right-tailed; $\hat{p} = 0.0405$; $z_0 = 1.645$; sample test statistic $\hat{p}_1 - \hat{p}_2 = 0.0259$ with $z = 1.70$; P value $= 0.0446$; Reject H_0; The data indicate that the population proportion of domestic food items and staple items is higher near site 1.

Chapter 9 Review

1. Single mean; large sample; H_0: $\mu = 11.1$; H_1: $\mu \neq 11.1$; two-tailed; $z_0 = \pm 1.96$; sample $z = -3.00$; P value $= 0.0026$; Reject H_0; The average number of miles driven per vehicle is different from the national average at the 5% level of significance.

3. Single mean, small sample; H_0: $\mu = 0.8$; H_1: $\mu > 0.8$; right-tailed test; $d.f. = 8$; $t_0 = 2.896$; t value corresponding to $\bar{x}$ is $t = 5.854$; P value < 0.005; Reject H_0. The Toylot claim is too low.

5. Single proportion; H_0: $p = 0.60$; H_1: $p < 0.60$; left-tailed; $z_0 = -2.33$; z value corresponding to $\hat{p} = 0.4444$ is -3.01; P value $= 0.0013$; Reject H_0; The data indicate that the population mortality rate has dropped.

7. Single proportion; H_0: $p = 0.20$; H_1: $p > 0.20$; right-tailed test; $z_0 = 1.645$; z value corresponding to $\hat{p} = 0.30$ is $z = 4.00$; P value < 0.0002; Reject H_0. It seems that the proportion of students who read the poetry magazine is greater than 20%.

9. Single mean, large sample; H_0: $\mu = 40$; H_1: $\mu > 40$; right-tailed test; $z_0 = 2.33$; z value corresponding to $\bar{x}$ is $z = 3.34$; P value $= 0.0004$; Reject H_0. It seems that the mean number of matches per box is over 40.

11. Difference of means, small independent samples; H_0: $\mu_1 = \mu_2$; H_1: $\mu_1 \neq \mu_2$; two-tailed test; $d.f. = 28$; $t_0 = \pm 2.048$; $s = 1.9097$; t value corresponding to sample test statistic $\bar{x}_1 - \bar{x}_2 = -0.4$ is $t = -0.5723$; $0.25 < P$ value; Fail to reject H_0. There does not seem to be any difference.

13. Single mean, small sample; H_0: $\mu = 7$; H_1: $\mu \neq 7$; two-tailed test; $d.f. = 7$; $t_0 = \pm 2.365$; t value corresponding to $\bar{x}$ is $t = 1.697$; $0.1 < P$ value < 0.15; Fail to reject H_0. There is not enough evidence to conclude that the machine has slipped out of adjustment.

15. Paired difference test; H_0: $\mu_d = 0$; H_1: $\mu_d < 0$; left-tailed; $d.f. = 4$; $t_0 = -2.132$; t value corresponding to $\bar{d} = -4.94$ is -2.832; $0.010 < P$ value < 0.025; Reject H_0; The data indicate that the average net sales have improved.

17. Difference of means; large independent samples; H_0: $\mu_1 = \mu_2$; H_1: $\mu_1 \neq \mu_2$; two-tailed; $z_0 = \pm 1.96$; $z = 1.82$ for $\bar{x}_1 - \bar{x}_2 = 0.3$; P value $= 0.0688$; Fail to reject H_0. The data do not indicate a difference in the population mean lengths of the two types of projectile points.

19. (a) Reject H_0; P value < 0.01.   (b) Reject H_0; P value < 0.05.

CHAPTER 10

Section 10.1

1. Moderate or low linear correlation.

3. High linear correlation.

5. High linear correlation.

7. (a) List and Dealer Price Pontiac Grand Am (thousands of dollars)

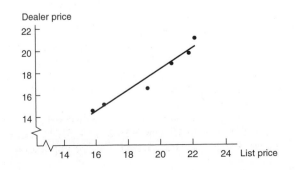

(b) Draw the line you think fits best. (Method to find equation is in Section 10.2.)   (c) High.

9. (a) Ages and Average Weights of Shetland Ponies

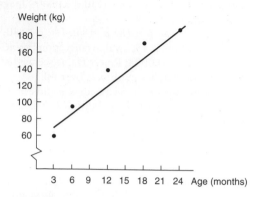

(b) Draw line you think best.   (c) High.

11. (a) Change in Wages and in Consumer Prices in Various Countries (%)

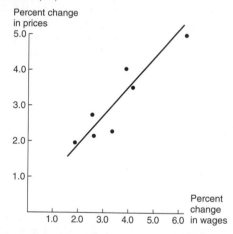

(b) Just draw the line that you think best. In Section 10.2, you will learn how to get the equation for the line.
(c) The linear correlation appears to be moderate.

13. (a) Body Diameter and Weight of Prehistoric Pottery

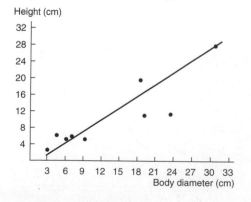

(b) Draw line you think best.   (c) Moderate.

15. (a) Unit Length on y Same as That on x

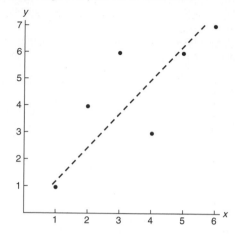

(b) Unit Length on y Twice That on x

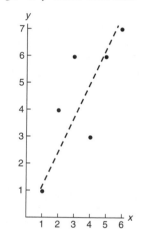

(c) Unit Length on y Half That on x

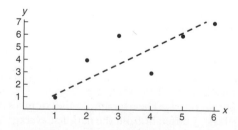

(d) The line in part b appears steeper than in part a, while the line in part c appears flatter than in part a. The slopes actually are all the same, but the lines look different because of the change in unit lengths on the y and x axes.

Section 10.2

Note: In this section and the next two, answers may vary slightly, depending on how many significant digits are used throughout the calculations.

1. (a) Absenteeism and Number of Assembly Line Defects

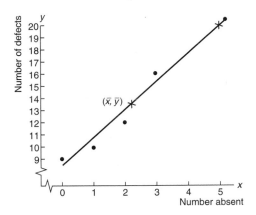

(b) $\bar{x} = 2.2$; $\bar{y} = 13.4$; $b \approx 2.337838$; $y = 2.338x + 8.26$.   (c) See figure of part a.   (d) $S_e = 0.877649$.
(e) 17.6.   (f) 14.3 to 20.9.

3. (a) Weight of Cars and Gasoline Mileage

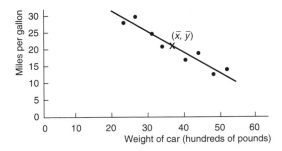

(b) $\bar{x} = 37.375$; $\bar{y} = 20.875$; $b = -0.6007$; $y = 43.3263 - 0.6007x$.   (c) See figure of part a.   (d) $S_e = 2.2361$.
(e) 20.5.   (f) 17.1 to 23.9.

5. (a) Education and Income in Small Cities

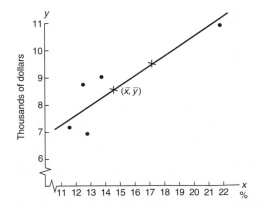

(b) $\bar{x} = 14.48$; $\bar{y} = 8.54$; $b \approx 0.31969$; $y = 0.320x + 3.91$.   (c) See figure of part a.   (d) $S_e = 0.9248$.
(e) 10.3 thousand.   (f) 8.4 to 12.2 thousand.

7. (a) Per Capita Income and Per Capita Retail Sales in Small Cities (thousands of dollars)

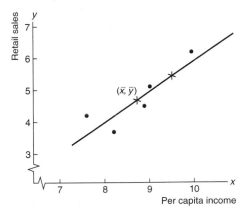

(b) $\bar{x} = 8.72$; $\bar{y} = 4.74$; $b \approx 0.966314$; $y = 0.966x - 3.69$.   (c) See figure of part a.   (d) $S_e = 0.536832$.
(e) 5.49 thousand.   (f) 4.45 to 6.53 thousand.

9. (a, c) List and Dealer Price, Chevrolet Silverado (thousands of dollars)

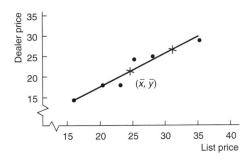

(b) $\bar{x} = 24.73$; $\bar{y} = 21.35$; $b = 0.8062$; $y = 1.410 + 0.806x$.   (d) $S_e = 1.526$.   (e) 19.872 thousand dollars.
(f) 17.326 to 22.417 thousand dollars.

11. (a) Cultural Affiliation and Elevation of Archaeological Sites

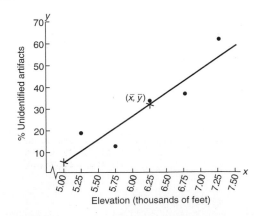

(b) $\bar{x} = 6.25$; $\bar{y} = 32.8$; $b \approx 22.0$; $y = 22.0x - 104.7$.
(c) See figure of part a.   (d) $S_e = 8.996296$.   (e) 38.3.
(f) 24 to 52 (to nearest whole number). (Depending on how you round, the decimal parts of the interval endpoints will vary.)

13. (a) Elevation and the Number of Frost-Free Days

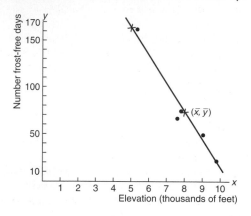

(b) $\bar{x} = 7.92$; $\bar{y} = 73.6$; $b \approx -30.878$; $y = -30.878x + 318.16$.   (c) See figure of part a.   (d) $S_e = 11.860286$.
(e) 132.89.   (f) 104.7 to 161.1.

15. (a) Solubility of Carbon Dioxide in Water

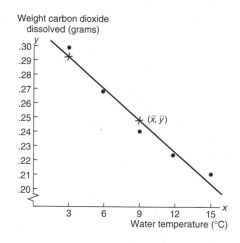

(b) $\bar{x} = 9.0$; $\bar{y} = 0.248$; $b \approx -0.00733$; $y = -0.00733x + 0.314$.   (c) See figure in part a.   (d) $S_e = 0.006928$.
(e) 0.241.   (f) 0.223 to 0.259.

17. (a) Yes. The pattern of residuals appears randomly scattered around the horizontal line at 0.   (b) No. There do not appear to be any outliers.

Section 10.3

1. (a) No.   (b) Increase in population.
3. (a) No.   (b) Better medical treatment.

5. (a) Number of Jobs (in hundreds)

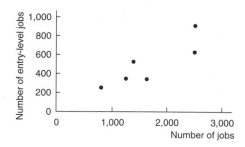

(b) Closer to 1.   (c) $r = 0.860$; $r^2 = 0.740$; 74% explained; 26% unexplained.
7. (a) Percentage of 16- to 19-Year-Olds Not in School and Number of Violent Crimes per 1000

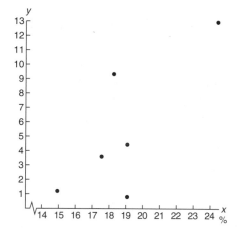

(b) Closer to 1.   (c) $r = 0.764$; $r^2 = 0.584$; 58.4% explained; 41.6% unexplained.
9. (a) Per Capita Income and Number of Medical Doctors per 10,000 Residents

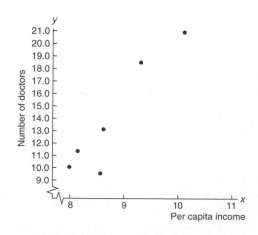

(b) Closer to 1.   (c) $r = 0.934$; $r^2 = 0.872$; 87.2% explained; 12.8% unexplained.

11. (a) Drivers' Ages and Fatal Accident Rate Due to Speeding

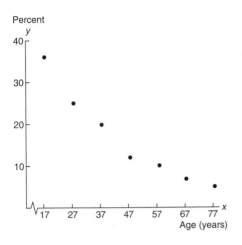

(b) Closer to -1.   (c) $r = -0.959$; $r^2 = 0.920$; 92% explained; 8% unexplained.

13. (a) Body Height and Bone Size

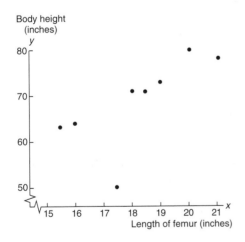

(b) Closer to 1.   (c) $r = 0.7061$; $r^2 = 0.499$; 49.9% explained; 50.1% unexplained.

15. (a) $SS_{xy} = SS_{yx}$.   (b) Same.   (c) Same.
    (d) $r = 0.6184$ in both cases; least-squares equations are not necessarily the same.

Section 10.4

1. (a) Use calculator.   (b) H_0: $\rho = 0$; H_1: $\rho < 0$; $d.f. = 6$; $t_0 = -1.943$; $r = -0.377$ with $t = -0.998$; P value > 0.125; Fail to reject H_0. The sample evidence does not

support a negative correlation.   (c) H_0: $\beta = 0$; H_1: $\beta \neq 0$; $d.f. = 6$; $t_0 = \pm 1.943$; $b = -0.468$ with $t = -0.998$; P value > 0.250; Fail to reject H_0. The sample evidence does not support a nonzero slope.   (d) $S_e = 7.443$; interval from -1.064 to 0.128. Since the confidence interval includes both positive and negative values, we conclude that the slope is zero and so faculty salaries are not tied to tuition.

3. (a) Use calculator.   (b) H_0: $\rho = 0$; H_1: $\rho < 0$; $d.f. = 5$; $t_0 = -3.365$; $r = -0.976$ with $t = -10.06$; P value < 0.005; Reject H_0. The sample evidence supports a negative correlation.   (c) H_0: $\beta = 0$; H_1: $\beta < 0$; $d.f. = 5$; $t_0 = -3.365$; $b = -0.054$ with $t = -10.06$; P value < 0.005; Reject H_0. The sample evidence supports a negative correlation.   (d) $S_e = 0.166$; -0.065 to -0.044; For every meter more of depth, the optimal time decreases from about 0.04 to 0.07 hour.

5. (a) Use calculator.   (b) H_0: $\rho = 0$; H_1: $\rho > 0$; $d.f. = 4$; $t_0 = 3.747$; $r = 0.956$ with $t = 6.534$; P value < 0.005; Reject H_0. The sample evidence supports a positive correlation.   (c) H_0: $\beta = 0$; H_1: $\beta > 0$; $d.f. = 4$; $t_0 = 3.747$; $b = 0.758$ with $t = 6.534$; P value < 0.005; Reject H_0. The sample evidence supports a positive correlation.   (d) $S_e = 0.1527$; from 0.510 to 1.005; Every $1,000 increase in list price shows an increase in dealer price of between $510 and $1005.

7. (a) H_0: $\rho = 0$; H_1: $\rho \neq 0$; $d.f. = 4$; $t_0 = \pm 4.604$; sample $t = 4.129$; $0.01 < P$ value < 0.02; Do not reject H_0; r is not significant at the 0.01 level of significance.   (b) H_0: $\rho = 0$; H_1: $\rho \neq 0$; $d.f. = 8$; $t_0 = \pm 3.355$; sample $t = 5.840$; P value < 0.01; Reject H_0; r is significant at the 0.01 level of significance.   (c) As n increases, the degrees of freedom increase, resulting in a critical value with smaller absolute value. In addition, as n increases, the t value corresponding to r also increases, resulting in a smaller P value.

Section 10.5

1. (a) Response variable is x_1. Explanatory variables are x_2, x_3, x_4.   (b) 1.6 is the constant term; 3.5 is the coefficient of x_2; -7.9 is the coefficient of x_3; and 2.0 is the coefficient of x_4.   (c) $x_1 = 10.7$.   (d) 3.5 units; 7 units; -14 units.   (e) $d.f. = 8$; $t = 1.860$; 2.72 to 4.28.
    (f) H_0: $\beta_2 = 0$; H_1: $\beta_2 \neq 0$; $d.f. = 8$; $t_0 = \pm 2.306$. The sample test statistic has $t = 8.35$. Since the sample test statistic falls into the critical region, we reject H_0. The explanatory variable x_2 should be included in the regression equation.

3. (a) $CVx_1 \approx 9.08$; $CVx_2 \approx 14.59$; $CVx_3 \approx 8.88$; x_2 has greatest spread; x_3 has smallest.   (b) $r^2x_1x_2 \approx 0.958$; $r^2x_1x_3 \approx 0.942$; $r^2x_2x_3 \approx 0.895$; x_2; Yes; 95.8%;

94.2%.   (c) 97.7%.   (d) $x_1 = 30.99 + 0.861x_2 + 0.335x_3$; 3.35; 8.61.   (e) H_0: coefficient = 0; H_1: coefficient ≠ 0; $d.f. = 8$; $t_0 = \pm 2.306$. For β_2 the sample test statistic is $t = 3.47$; For β_3, $t = 2.56$. Reject H_0 for each coefficient and conclude that the coefficients for x_2 and x_3 are not zero.   (f) $d.f. = 8$; $t = 1.86$; C.I. for β_2 is 0.40 to 1.32; C.I. for β_3 is 0.09 to 0.58.   (g) 153.9; 148.3 to 159.4.

5. (a) $CVx_1 \approx 39.64$; $CVx_2 \approx 44.45$; $CVx_3 \approx 50.62$; $CVx_4 \approx 52.15$; x_4, x_1 has a small CV because we divide by a large mean.   (b) $r^2x_1x_2 \approx 0.842$; $r^2x_1x_3 \approx 0.865$; $r^2x_1x_3 \approx 0.225$; $r^2x_2x_3 \approx 0.624$; $r^2x_2x_4 \approx 0.184$; $r^2x_3x_4 \approx 0.089$; x_4; 84.2%.   (c) 96.7%.   (d) $x_1 = 7.68 + 3.66x_2 + 7.62x_3 + 0.83x_4$; 7.62 million dollars.   (e) H_0: coefficient = 0; H_1: coefficient ≠ 0; $d.f. = 6$; $t_0 = \pm 2.447$. For β_2 test, the sample test statistic is $t = 3.28$; For β_3, $t = 4.60$; for β_4, $t = 1.54$. For β_2 and β_3, reject H_0 and conclude that the coefficients of x_2 and x_3 are not zero. For β_4, fail to reject H_0 and conclude that the coefficient of x_4 could be zero.   (f) $d.f. = 6$; $t = 1.943$; C.I. for β_2 is 1.49 to 5.83; C.I. for β_3 is 4.40 to 10.84; C.I. for β_4 is −0.22 to 1.88.   (g) 91.95; 77.6 to 106.3.   (h) 5.63; 4.21 to 7.04.

Chapter 10 Review

1. (a) Age and Mortality Rate for Bighorn Sheep

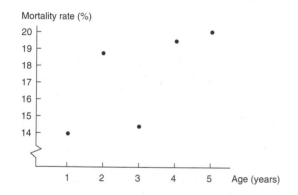

(b) $\bar{x} = 3$; $\bar{y} = 17.38$; $b = 1.27$; $y = 13.57 + 1.27x$.
(c) $r = 0.685$; $r^2 = 0.469$.   (d) H_0: $\rho = 0$; H_1: $\rho > 0$; $d.f. = 3$; $t_0 = 4.541$; sample $t = 1.629$; Do not reject H_0. There does not seem to be a positive correlation between age and mortality rate of bighorn sheep.   (e) H_0: $\beta = 0$; H_1: $\beta > 0$; $d.f. = 3$; $t_0 = 4.541$; sample $t = 1.629$; Do not reject H_0.

3. (a) Weight of One-Year-Old versus Weight of Adult

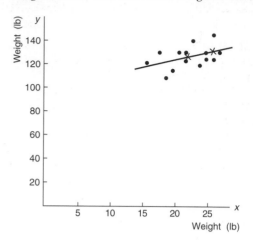

(b) $\bar{x} = 21.43$; $\bar{y} = 126.79$; $b = 1.285$; $y = 99.25 + 1.285x$.   (c) See figure of part a.   (d) 124.95.   (e) $S_e = 8.38$.   (f) 105.91 to 143.99.   (g) See figure of part a.   (h) $r = 0.47$; $r^2 = 0.221$.   (i) H_0: $\rho = 0$; H_1: $\rho > 0$; $d.f. = 12$; $t_0 = 2.681$; sample $t = 1.845$; Fail to reject H_0. There does not seem to be any significant correlation at the 1% level.   (j) H_0: $\beta = 0$; H_1: $\beta > 0$; $d.f. = 12$; $t_0 = 2.681$; sample $t = 1.845$; Fail to reject H_0.

5. (a) Weight of Mail versus Number of Employees Required

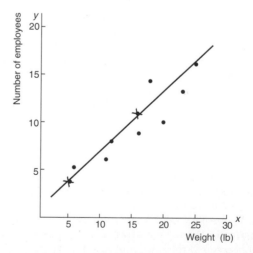

(b) $\bar{x} = 16.38$; $\bar{y} = 10.13$; $b = 0.554$; $y = 1.051 + 0.554x$.   (c) See line in figure of part a.   (d) 9.36.   (e) $S_e = 1.73$.   (f) 4.87 to 13.86.   (g) $r = 0.91$; $r^2 = 0.828$.   (h) H_0: $\rho = 0$; H_1: $\rho > 0$; $d.f. = 6$; $t_0 = 3.143$; sample $t = 5.376$; Reject H_0.   (i) H_0: $\beta = 0$; H_1: $\beta > 0$; $d.f. = 6$; $t_0 = 3.143$; sample $t = 5.376$; Reject H_0.
(j) 0.41 to 0.70; For each additional pound of mail, assign 1 employee to work from 41% to 70% of a work day on mail.

CHAPTER 11

Section 11.1

1. H_0: Myers-Briggs preference and profession are independent; H_1: Myers-Briggs preference and profession are not independent; $\chi^2 = 43.5562$; $d.f. = 2$; $\chi^2_{0.01} = 9.21$; The sample statistic falls in the critical region; Reject H_0. Myers-Briggs preference and profession are not independent.

3. H_0: Site type and pottery type are independent; H_1: Site type and pottery type are not independent; $\chi^2 = 0.5552$; $d.f. = 4$; $\chi^2_{0.05} = 9.49$; The sample statistic falls outside the critical region; Do not reject H_0; There is not sufficient evidence to conclude that site type and pottery type are not independent.

5. H_0: Age distribution and location are independent; H_1: Age and location are not independent; $\chi^2 = 0.6704$; $d.f. = 4$; $\chi^2_{0.05} = 9.49$; The sample statistic falls outside the critical region; Do not reject H_0; Age distribution and location appear to be independent.

7. H_0: Ages of young adults and movie preferences are independent; H_1: Ages of young adults and movie preferences are not independent; $\chi^2 = 3.6230$; $d.f. = 4$; $\chi^2_{0.05} = 9.49$; The sample statistic falls outside the critical region; Do not reject H_0; Age of young adults and movie preference appear to be independent.

9. H_0: Ticket sales and type of billings are independent; H_1: Ticket sales and type of billing are not independent; $\chi^2 = 1.8685$; $d.f. = 3$; $\chi^2_{0.05} = 7.81$; The sample statistic falls outside the critical region; Do not reject H_0; Ticket sales and type of billing appear to be independent.

11. H_0: Stone tool construction material and site are independent; H_1: Stone tool construction material and site are not independent; $\chi^2 = 609.845$; $d.f. = 3$; $\chi^2_{0.01} = 11.34$; The sample statistic falls in the critical region; Reject H_0; construction material and site are not independent.

Section 11.2

1. H_0: The distributions are the same; H_1: The distributions are different; $\chi^2 = 11.788$; $d.f. = 3$; $\chi^2_{0.05} = 7.81$; Reject H_0; The distributions are different.

3. H_0: The distributions are the same; H_1: The distributions are different; $\chi^2 = 0.1984$; $d.f. = 4$; $\chi^2_{0.01} = 13.28$; Do not reject H_0; There is no evidence that the distributions are different.

5. (i) Essay. (ii) H_0: The distribution of temperatures fits a normal distribution; H_1: The distribution is not normal; $\chi^2 = 1.4567$; $d.f. = 5$; $\chi^2_{0.01} = 15.09$; Do not reject H_0.

7. H_0: The distributions are the same; H_1: The distributions are different; $\chi^2 = 9.333$; $d.f. = 3$; $\chi^2_{0.05} = 7.81$; Reject H_0; The fish distribution has changed.

9. H_0: The distributions are the same; H_1: The distributions are different; $\chi^2 = 13.70$; $\chi^2_{0.01} = 15.09$; Do not reject H_0; The distributions are the same.

Section 11.3

1. H_0: $\sigma^2 = 42.3$; H_1: $\sigma^2 > 42.3$; right-tailed; $d.f. = 22$; $\chi^2_{0.05} = 33.92$; $\chi^2 = 23.98$; Do not reject H_0; $\chi^2_U = 36.78$; $\chi^2_L = 10.98$; Interval for σ^2 is from 27.57 to 92.37.

3. H_0: $\sigma^2 = 136.2$; H_1: $\sigma^2 < 136.2$; left-tailed; $d.f. = 7$; $\chi^2_{0.99} = 1.24$; $\chi^2 = 5.92$; Do not reject H_0. The variance has not decreased; $\chi^2_U = 14.07$; $\chi^2_L = 2.17$; Interval for σ^2 is from 57.26 to 371.29.

5. H_0: $\sigma^2 = 9$; H_1: $\sigma^2 < 9$; left-tailed; $d.f. = 23$; $\chi^2_{0.95} = 13.09$; $\chi^2 = 9.23$; Reject H_0. The new typhoid shot has a smaller variance of protection times. $\chi^2_U = 35.17$; $\chi^2_L = 13.09$; Interval for σ is from 1.54 to 2.52.

7. H_0: $\sigma^2 = 0.15$; H_1: $\sigma^2 > 0.15$; right-tailed; $d.f. = 60$; $\chi^2_{0.01} = 88.38$; $\chi^2 = 108$; Reject H_0; The variance is greater than 0.15 mm^2. All fan blades should be replaced; $\chi^2_U = 79.08$; $\chi^2_L = 43.19$; Interval for σ is from 0.45 mm to 0.61 mm.

9. (a) H_0: $\sigma^2 = 15$; H_1: $\sigma^2 \neq 15$; $\chi^2 = 20.02$; the critical values are 35.48 and 10.28. Because the observed value $\chi^2 = 20.02$ does not fall into the critical region, we do not reject H_0. There is no difference in variance.
(b) $\chi^2_U = 32.67$; $\chi^2_L = 11.59$; $9.19 < \sigma^2 < 25.91$.
(c) $3.03 < \sigma < 5.09$.

Section 11.4

1. H_0: $\sigma_1^2 = \sigma_2^2$; H_1: $\sigma_1^2 > \sigma_2^2$; sample $F = 3.73$; $F_0 = 3.52$; Reject H_0. The variance in annual wheat production from the first plot is greater than that from the second plot.

3. H_0: $\sigma_1^2 = \sigma_2^2$; H_1: $\sigma_1^2 \neq \sigma_2^2$; Population I has data from France; $d.f._N = 20$; $d.f._D = 17$; sample $F = 1.390$; $F_0 = 2.62$; Do not reject H_0. There is no significant difference in the volatility of corporate productivity of large companies in France and in Germany.

5. H_0: $\sigma_1^2 = \sigma_2^2$; H_1: $\sigma_1^2 > \sigma_2^2$; Population I has data from aggressive growth; $d.f._N = 20$; $d.f._D = 20$; sample $F = 2.54$; $F_0 = 2.12$; Reject H_0; The population variance for mutual funds holding aggressive-growth small stocks is larger than that for funds holding value stocks.

7. H_0: $\sigma_1^2 = \sigma_2^2$; H_1: $\sigma_1^2 \neq \sigma_2^2$; Population I has data from new system; $d.f._N = 30$; $d.f._D = 24$; sample $F = 1.33$; $F_0 = 2.21$; Do not reject H_0. There is no significant difference in the population variance of gasoline consumption for the two injection systems.

Section 11.5

1. (a) $H_0: \mu_1 = \mu_2 = \mu_3$; H_1: Not all the means are equal.
   (b–h)

Source of Variation	Sum of Squares	Degrees of Freedom	MS	F Ratio	F Critical Value	Test Decision
Between groups	520.280	2	260.14	0.48	6.51	Do not reject H_0
Within groups	7544.190	14	538.87			
Total	8064.470	16				

3. (a) $H_0: \mu_1 = \mu_2 = \mu_3 = \mu_4$; H_1: Not all the means are equal.
   (b–h)

Source of Variation	Sum of Squares	Degrees of Freedom	MS	F Ratio	F Critical Value	Test Decision
Between groups	89.637	3	29.879	0.846	3.16	Do not reject H_0
Within groups	635.827	18	35.324			
Total	725.464	21				

5. (a) $H_0: \mu_1 = \mu_2 = \mu_3$; H_1: Not all the means are equal.
   (b–h)

Source of Variation	Sum of Squares	Degrees of Freedom	MS	F Ratio	F Critical Value	Test Decision
Between groups	1303.167	2	651.58	5.005	4.26	Reject H_0
Within groups	1171.750	9	130.19			
Total	2474.917	11				

7. (a) $H_0: \mu_1 = \mu_2 = \mu_3$; H_1: Not all the means are equal.
   (b–h)

Source of Variation	Sum of Squares	Degrees of Freedom	MS	F Ratio	F Critical Value	Test Decision
Between groups	2.042	2	1.021	0.336	7.21	Do not reject H_0
Within groups	33.428	11	3.039			
Total	35.470	13				

9. (a) $H_0: \mu_1 = \mu_2 = \mu_3 = \mu_4$; H_1: Not all the means are equal.
   (b–h)

Source of Variation	Sum of Squares	Degrees of Freedom	MS	F Ratio	F Critical Value	Test Decision
Between groups	238.225	3	79.408	4.611	3.29	Reject H_0
Within groups	258.340	15	17.223			
Total	496.565	18				

Section 11.6

1. 2 factors: walking device with 3 levels and task with 2 levels; data table has 6 cells.
3. Since the P value is less than 0.01, there is a significant difference in mean cadence according to the factor walking device used. The critical value is $F_{0.01} = 6.01$. Since the sample $F = 30.94$ is greater than $F_{0.01}$, F lies in the critical region and we reject H_0.
5. (a) 2 factors: income with 4 levels and media type with 5 levels. (b) For income level, H_0: There is no difference in population mean index based on income level; H_1: At least two income levels have different population mean

indices; $F_{\text{income}} = 1.925$; $F_{0.05} = 3.49$; Do not reject H_0. The data do not indicate any differences in population mean index according to income level.   (c) For media, H_0: There is no difference in population mean index according to media type; H_1: At least two media types have different population mean indices; $F_{\text{media}} = 0.069$; $F_{0.05} = 3.26$; Do not reject H_0. The data do not indicate any differences in population mean index according to media type.

7. Randomized Block Design

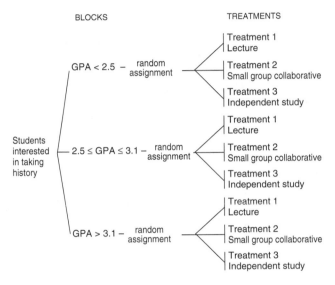

Yes, the design fits the model for randomized block design.

Chapter 11 Review

1. H_0: $\mu_1 = \mu_2 = \mu_3 = \mu_4$; H_1: Not all the means are equal.

Chapter 11 Review Problem 1

Source of Variation	Sum of Squares	Degrees of Freedom	MS
Between groups	6149.75	3	2049.917
Within groups	12454.80	16	778.425
Total	18604.55	19	

F Ratio	F Critical	Test Decision
2.633	3.24	Do not reject H_0

3. (a) H_0: $\sigma^2 > 810000$; H_1: $\sigma^2 > 810000$; $\chi^2 = 65.54$; critical value is $\chi^2_{0.01} = 49.59$. Because the observed value $\chi^2 = 65.54$ is in the critical region, we reject H_0. The variance is more than claimed.   (b) $\chi^2_U = 45.72$; $\chi^2_L = 16.05$; $1161147.4 < \sigma^2 < 3307642.4$.

5. H_0: Student grade and teacher rating are independent; H_1: Student grade and teacher rating are not independent; $\chi^2 = 9.80$; $\chi^2_{0.01} = 16.81$. Do not reject H_0. They are independent.

7. H_0: The distributions are the same; H_1: The distributions are different: $\chi^2 = 33.93$; $\chi^2_{0.01} = 13.28$. Reject H_0. The distribution has changed.

9. H_0: $\sigma_1^2 = \sigma_2^2$; H_1: $\sigma_1^2 > \sigma_2^2$; sample $F = 2.61$; $F_0 = 2.31$; Because the sample statistic is greater than the critical value, we reject H_0. The data indicate that the population variance of life times for bulbs made by the new process is greater than that for bulbs made by the old process.

Section 12.1

1. H_0: Writing time distributions are the same; H_1: Writing time distribution for new tip is higher; right-tailed test; $z_0 = 1.645$. The z value for sample statistic $x = 0.75$ is $z = 1.73$. Because the sample test statistic falls in the critical region, we reject H_0 and conclude that the writing life of the new tip is longer than that of the old.

3. H_0: Distributions are the same; H_1: Distributions are different; two-tailed tests; $z_0 = \pm 1.96$. The z value for sample statistic $x = 0.6250$ is $z = 1.00$. Because the sample test statistic does not fall in the critical region, we do not reject H_0. There is not enough evidence to conclude that the scores before and after the lectures are different.

5. H_0: Distributions are the same; H_1: Distributions are different; two-tailed test; $z_0 = \pm 1.96$. The z value for sample statistic $x = 0.5833$ is $z = 0.58$. Because the sample test statistic does not fall in the critical region, do not reject H_0. There is not enough evidence to conclude that there is any difference in effectiveness in teaching reading at the two schools.

7. H_0: Distributions are the same; H_1: Distribution after hypnosis is lower; $z_0 = -2.33$. The z value for the sample test statistic $x = 0.1875$ is $z = -2.50$. Because the sample test statistic falls in the critical region, reject H_0 and conclude the number of cigarettes smoked per day after hypnosis is less than the mean number before hypnosis.

9. H_0: Distributions are the same; H_1: Distributions are different; two-tailed test; $z_0 = \pm 2.58$. The z value for the sample test statistic $x = 0.6429$ is $z = 1.07$. Because the sample test statistic does not fall in the critical region, we

do not reject H_0. There is not enough evidence to conclude that there is a difference in the pulse rate of babies before and after birth.

11. H_0: Distributions are the same; H_1: Distributions are different; two-tailed test; $z_0 = \pm 1.96$; The z value for the sample test statistic $x = 0.667$ is $z = 1.16$. Because the sample test statistic does not fall in the critical region, we do not reject H_0. There is not enough evidence to conclude that the temperature in Madison is different from that in Juneau.

Section 12.2

1. H_0: Distributions are the same; H_1: Distributions are different; two-tailed test; $z_0 = \pm 1.96$; $R = 82$; $\mu_R = 90$; $\sigma_R = 12.25$. The z value of sample rank R is $z = -0.65$. Because the sample test statistic does not fall in the critical region, we do not reject H_0. There is not enough evidence to conclude that there is a difference in performance.

3. H_0: Distributions are the same; H_1: Distributions are different; two-tailed test; $z_0 = \pm 1.96$; $R = 80$; $\mu_R = 90$; $\sigma_R = 12.25$. The z value of sample rank R is $z = -0.82$. Because the sample test statistic does not fall in the critical region, we do not reject H_0. There is not enough evidence to conclude that the number of training sessions for the two methods are different.

5. H_0: Distributions are the same; H_1: Distributions are different; two-tailed test; $z_0 = \pm 1.96$; $R = 66$; $\mu_R = 99$; $\sigma_R = 14.07$. The z value of sample rank R is $z = -2.35$. The sample test statistic falls in the critical region, so we reject H_0. It seems that the time to take the test in a competitive setting is different from that in a noncompetitive setting.

7. H_0: Distributions are the same; H_1: Distributions are different; two-tailed test; $z_0 = \pm 2.58$; $R = 92$; $\mu_R = 90$; $\sigma_R = 12.25$. The z value of sample rank R is $z = 0.16$. Because the sample test statistic does not fall in the critical region, we do not reject H_0. There is not enough evidence to conclude that there is a difference.

9. H_0: Distributions are the same; H_1: Distributions are different; two-tailed test; $z_0 = \pm 2.58$; $R_A = 78$; $\mu_R = 68$; $\sigma_R = 9.52$. The z value of sample rank $R_A = 1.05$. Because the sample test statistic does not fall in the critical region, we do not reject H_0. There is not enough evidence to conclude that there is a difference.

Section 12.3

1. H_0: $\rho_s = 0$; H_1: $\rho_s \neq 0$; $r_s = 0.682$; the critical values are 0.619 and -0.619. Because the observed value $r_s = 0.682$ falls in the critical region, we reject H_0 and conclude that there is a monotone relation (either increasing or decreasing).

3. H_0: $\rho_s = 0$; H_1: $\rho_s > 0$; $r_s = 0.571$; the critical value is 0.620. Because the observed value $r_s = 0.571$ falls in the acceptance region, we accept H_0 and conclude that there is no monotone relation between crowding and violence.

5. H_0: $\rho_s = 0$; H_1: $\rho_s < 0$; $r_s = -0.214$; the critical value is -0.715. Because the observed value $r_s = -0.214$ falls in the acceptance region, we accept H_0 and conclude that there is no monotone relation.

7. H_0: $\rho_s = 0$; H_1: $\rho_s > 0$; $r_s = 0.955$; the critical value is 0.764. Because the observed value $r_s = 0.955$ falls in the critical region, we reject H_0 and conclude that there is a monotone-increasing relation between aptitude rank and performance rank.

9. H_0: $\rho_s = 0$; H_1: $\rho_s \neq 0$; $r_s = 0.661$; the critical values are ± 0.881. Because the observed value $r_s = 0.661$ falls in the acceptance region, we accept H_0 and conclude that there is no monotone relation.

Chapter 12 Review

1. H_0: Distributions are the same; H_1: Distributions are different; two-tailed test; $z_0 = \pm 1.96$; $R = 107$; $\mu_R = 110$; $\sigma_R = 14.20$. The z value of sample rank R is $z = -0.21$. Because the sample test statistic does not fall in the critical region, we do not reject H_0. There is not enough evidence to show a difference.

3. H_0: Distributions are the same; H_1: Distribution after ads is higher; right-tailed test; $z_0 = 2.33$. The z value of sample test statistic $x = 0.77$ is $z = 1.95$. Because the sample test statistic does not fall in the critical region, we do not reject H_0. There is not enough evidence to conclude that the sales improved after advertising.

5. H_0: $\rho_s = 0$; H_1: $\rho_s > 0$; critical value is 0.600. Because the observed value $r_s = 0.617$ falls in the critical region, we reject H_0 and conclude that there is a monotone-increasing relation in the ranks.

Index

FREQUENTLY USED FORMULAS

n = sample size N = population size f = frequency

Chapter 2

Class Width = $\dfrac{\text{high} - \text{low}}{\text{number classes}}$ (increase to next integer)

Class Midpoint = $\dfrac{\text{upper limit} + \text{lower limit}}{2}$

Lower boundary = lower boundary of previous class
   + class width

Chapter 3

Sample mean $\bar{x} = \dfrac{\Sigma x}{n}$

Population mean $\mu = \dfrac{\Sigma x}{N}$

Weighted average = $\dfrac{\Sigma xw}{\Sigma w}$

Range = largest data value − smallest data value

Sample standard deviation $s = \sqrt{\dfrac{\Sigma(x - \bar{x})^2}{n - 1}}$

Computation formula $s = \sqrt{\dfrac{SS_x}{n - 1}}$ where
   $SS_x = \Sigma x^2 - \dfrac{(\Sigma x)^2}{n}$

Population standard deviation $\sigma = \sqrt{\dfrac{\Sigma(x - \mu)^2}{N}}$

Sample variance s^2

Population variance σ^2

Sample Coefficient of Variation $CV = \dfrac{s}{\bar{x}} \cdot 100$

Sample mean for grouped data $\bar{x} = \dfrac{\Sigma xf}{n}$

Sample standard deviation for grouped data
   $s = \sqrt{\dfrac{\Sigma(x - \bar{x})^2 f}{n - 1}} = \sqrt{\dfrac{\Sigma x^2 f - (\Sigma xf)^2/n}{n - 1}}$

Chapter 4

Probability of the complement of event A
   $P(not\ A) = 1 - P(A)$

Multiplication rule for independent events
   $P(A\ and\ B) = P(A) \cdot P(B)$

General multiplication rules
   $P(A\ and\ B) = P(A) \cdot P(B,\ given\ A)$
   $P(A\ and\ B) = P(B) \cdot P(A,\ given\ B)$

Addition rule for mutually exclusive events
   $P(A\ or\ B) = P(A) + P(B)$

General addition rule
   $P(A\ or\ B) = P(A) + P(B) - P(A\ and\ B)$

Permutation rule $P_{n,r} = \dfrac{n!}{(n - r)!}$

Combination rule $C_{n,r} = \dfrac{n!}{r!(n - r)!}$

Chapter 5

Mean of a discrete probability distribution $\mu = \Sigma x P(x)$

Standard deviation of a discrete probability distribution
   $\sigma = \sqrt{\Sigma(x - \mu)^2 P(x)}$

Given $L - a + bx$
   $\mu_L = a + b\mu$ $\sigma_L = |b|\sigma$

Given $W = ax_1 + bx_2$ (x_1 and x_2 independent)
   $\mu_W = a\mu_1 + b\mu_2$
   $\sigma_W = \sqrt{a^2\sigma_1^2 + b^2\sigma_2^2}$

For Binomial Distributions

r = number of successes; p = probability of success;
   $q = 1 - p$

Binomial probability distribution $P(r) = C_{n,r}p^r q^{n-r}$

Mean $\mu = np$

Standard deviation $\sigma = \sqrt{npq}$

Geometric Probability Distribution

n = number of trial on which first success occurs
   $P(n) = p(1 - p)^{n-1}$

Poisson Probability Distribution

λ = mean number of successes over given interval
   $P(r) = \dfrac{e^{-\lambda}\lambda^r}{r!}$

Chapter 6

Raw score $x = z\sigma + \mu$ Standard score $z = \dfrac{x - \mu}{\sigma}$

Chapter 7

Mean of $\bar{x}$ distribution $\mu_{\bar{x}} = \mu$

Standard deviation of $\bar{x}$ distribution $\sigma_{\bar{x}} = \dfrac{\sigma}{\sqrt{n}}$

Standard score for $\bar{x}$ $z = \dfrac{\bar{x} - \mu}{\sigma/\sqrt{n}}$

Mean of $\hat{p}$ distribution $\mu_{\hat{p}} = p$

Standard deviation of $\hat{p}$ distribution $\sigma_{\hat{p}} = \sqrt{\dfrac{pq}{n}}$; $q = 1 - p$

Chapter 10

Regression and Correlation

In all these formulas $SS_x = \Sigma x^2 - \dfrac{(\Sigma x)^2}{n}$,

$SS_y = \Sigma y^2 - \dfrac{(\Sigma y)^2}{n}$, $SS_{xy} = \Sigma xy - \dfrac{(\Sigma x)(\Sigma y)}{n}$

Least squares line $y = a + bx$ where $b = \dfrac{SS_{xy}}{SS_x}$ and

$a = \bar{y} - b\bar{x}$

Standard error of estimate $S_e = \sqrt{\dfrac{SS_y - bSS_{xy}}{n-2}}$

where $b = \dfrac{SS_{xy}}{SS_x}$

Pearson product moment correlation coefficient

$r = \dfrac{SS_{xy}}{\sqrt{SS_x SS_y}}$

Coefficient of determination $= r^2$

Confidence interval for y

$y_p - E < y < y_p + E$ where y_p is the predicted y value for x

$E = t_c S_e \sqrt{1 + \dfrac{1}{n} + \dfrac{(x - \bar{x})^2}{SS_x}}$ with $d.f. = n - 2$

Sample test statistic for r

$t = \dfrac{r\sqrt{n-2}}{\sqrt{1 - r^2}}$ with $d.f. = n - 2$

Sample test statistic test for b

$t = \dfrac{b - \beta}{S_e / \sqrt{SS_x}}$ with $d.f. = n - 2$

Confidence interval for β

$b - t_c \dfrac{S_e}{\sqrt{SS_x}} < \beta < b + t_c \dfrac{S_e}{\sqrt{SS_x}}$ with $d.f. = n - 2$

Chapter 11

$\chi^2 = \Sigma \dfrac{(O - E)^2}{E}$ where $E = \dfrac{(\text{row total})(\text{column total})}{\text{sample size}}$

Tests of Independence $d.f. = (R - 1)(C - 1)$

Goodness of fit $d.f. = (\text{number of entries}) - 1$

Confidence Interval for σ^2; $d.f. = n - 1$

$\dfrac{(n-1)s^2}{\chi^2_U} < \sigma^2 < \dfrac{(n-1)s^2}{\chi^2_L}$

Sample test statistic for H_0: $\sigma^2 = k$; $d.f. = n - 1$

$\chi^2 = \dfrac{(n-1)s^2}{\sigma^2}$

Testing Two Variances

Sample test statistic $F = \dfrac{s_1^2}{s_2^2}$

where $s_1^2 \geq s_2^2$

$d.f._N = n_1 - 1$; $d.f._D = n_2 - 1$

ANOVA

$k = $ number of groups; $N = $ total sample size

$SS_{TOT} = \Sigma x^2_{TOT} - \dfrac{(\Sigma x_{TOT})^2}{N}$

$SS_{BET} = \displaystyle\sum_{all\ groups} \left(\dfrac{(\Sigma x_i)^2}{n_i} \right) - \dfrac{(\Sigma x_{TOT})^2}{N}$

$SS_W = \displaystyle\sum_{all\ groups} \left(\Sigma x_i^2 - \dfrac{(\Sigma x_i)^2}{n_i} \right)$

$SS_{TOT} = SS_{BET} + SS_W$

$MS_{BET} = \dfrac{SS_{BET}}{d.f._{BET}}$ where $d.f._{BET} = k - 1$

$MS_W = \dfrac{SS_W}{d.f._W}$ where $d.f._W = N - k$

$F = \dfrac{MS_{BET}}{MS_W}$ where $d.f.$ numerator $= d.f._{BET} = k - 1$;

$d.f.$ denominator $= d.f._W = N - k$

Two-Way ANOVA

$r = $ number of rows; $c = $ number of columns

Row factor F: $\dfrac{MS\ \text{row factor}}{MS\ \text{error}}$

Column factor F: $\dfrac{MS\ \text{column factor}}{MS\ \text{error}}$

Interaction F: $\dfrac{MS\ \text{interaction}}{MS\ \text{error}}$

with degrees of freedom for
row factor $= r - 1$ interaction $= (r - 1)(c - 1)$
column factor $= c - 1$ error $= rc(n - 1)$

Chapter 12

Sample test statistic for $x = $ proportion of plus signs to all signs ($n \geq 12$)

$z = \dfrac{x - 0.5}{\sqrt{0.25/n}}$

Sample test statistic for $R = $ sum of ranks

$z = \dfrac{R - \mu_R}{\sigma_R}$ where $\mu_R = \dfrac{n_1(n_1 + n_2 + 1)}{2}$ and

$\sigma_R = \sqrt{\dfrac{n_1 n_2 (n_1 + n_2 + 1)}{12}}$

Spearman rank correlation coefficient

$r_s = 1 - \dfrac{6\Sigma d^2}{n(n^2 - 1)}$ where $d = x - y$

Chapter 8

Confidence Interval

for $\mu(n \geq 30)$

$$\bar{x} - z_c\frac{\sigma}{\sqrt{n}} < \mu < \bar{x} + z_c\frac{\sigma}{\sqrt{n}}$$

for $\mu(n < 30)$

$$d.f. = n - 1$$

$$\bar{x} - t_c\frac{s}{\sqrt{n}} < \mu < \bar{x} + t_c\frac{s}{\sqrt{n}}$$

for $p(np > 5$ and $nq > 5)$

$$\hat{p} - z_c\sqrt{\frac{\hat{p}(1 - \hat{p})}{n}} < p < \hat{p} + z_c\sqrt{\frac{\hat{p}(1 - \hat{p})}{n}}$$

where $\hat{p} = r/n$

for difference of means ($n_1 \geq 30$ and $n_2 \geq 30$)

$$(\bar{x}_1 - \bar{x}_2) - z_c\sqrt{\frac{\sigma_1^2}{n_1} + \frac{\sigma_2^2}{n_2}} < \mu_1 - \mu_2 < (\bar{x}_1 - \bar{x}_2)$$
$$+ z_c\sqrt{\frac{\sigma_1^2}{n_1} + \frac{\sigma_2^2}{n_2}}$$

for difference of means ($n_1 < 30$ and/or $n_2 < 30$ and $\sigma_1 \approx \sigma_2$)

$$d.f. = n_1 + n_2 - 2$$

$$(\bar{x}_1 - \bar{x}_2) - t_c s\sqrt{\frac{1}{n_1} + \frac{1}{n_2}} < \mu_1 - \mu_2 < (\bar{x}_1 - \bar{x}_2)$$
$$+ t_c s\sqrt{\frac{1}{n_1} + \frac{1}{n_2}}$$

where $s = \sqrt{\dfrac{(n_1 - 1)s_1^2 + (n_2 - 1)s_2^2}{n_1 + n_2 - 2}}$

for difference of proportions

where $\hat{p}_1 = r_1/n_1$; $\hat{p}_2 = r_2/n_2$; $\hat{q}_1 = 1 - \hat{p}_1$; $\hat{q}_2 = 1 - \hat{p}_2$

$$(\hat{p}_1 - \hat{p}_2) - z_c\sqrt{\frac{\hat{p}_1\hat{q}_1}{n_1} + \frac{\hat{p}_2\hat{q}_2}{n_2}} < p_1 - p_2 < (\hat{p}_1 - \hat{p}_2)$$
$$+ z_c\sqrt{\frac{\hat{p}_1\hat{q}_1}{n_1} + \frac{\hat{p}_2\hat{q}_2}{n_2}}$$

Sample Size for Estimating

means $n = \left(\dfrac{z_c\sigma}{E}\right)^2$

proportions

$n = p(1 - p)\left(\dfrac{z_c}{E}\right)^2$ with preliminary estimate for p

$n = \dfrac{1}{4}\left(\dfrac{z_c}{E}\right)^2$ without preliminary estimate for p

Chapter 9

Sample Test Statistics for Tests of Hypotheses

for $\mu(n \geq 30)$ $\quad z = \dfrac{\bar{x} - \mu}{\sigma\sqrt{n}}$

for $\mu(n < 30)$; $d.f. = n - 1$ $\quad t = \dfrac{\bar{x} - \mu}{s/\sqrt{n}}$

for p $\quad z = \dfrac{\hat{p} - p}{\sqrt{pq/n}}$ where $q = 1 - p$

for paired differences d $\quad t = \dfrac{\bar{d} - \mu_d}{s_d/\sqrt{n}}$ with $d.f. = n - 1$

difference of means, large sample

$$z = \frac{(\bar{x}_1 - \bar{x}_2) - (\mu_1 - \mu_2)}{\sqrt{\dfrac{\sigma_1^2}{n_1} + \dfrac{\sigma_2^2}{n_2}}}$$

difference of means, small sample with $\sigma_1 \approx \sigma_2$;
$d.f. = n_1 + n_2 - 2$

$$t = \frac{(\bar{x}_1 - \bar{x}_2) - (\mu_1 - \mu_2)}{s\sqrt{\dfrac{1}{n_1} + \dfrac{1}{n_2}}}$$

where $s = \sqrt{\dfrac{(n_1 - 1)s_1^2 + (n_2 - 1)s_2^2}{n_1 + n_2 - 2}}$

difference of proportions

$$z = \frac{\hat{p}_1 - \hat{p}_2}{\sqrt{\dfrac{\bar{p}\bar{q}}{n_1} + \dfrac{\bar{p}\bar{q}}{n_2}}} \text{ where } \bar{p} = \frac{r_1 + r_2}{n_1 + n_2}; \bar{q} = 1 - \bar{p};$$

$$\hat{p}_1 = r_1/n_1; \hat{p}_2 = r_2/n_2$$